OCEANIC SOUND SCATTERING PREDICTION

MARINE SCIENCE

Coordinating Editor: Ronald J. Gibbs, *University of Delaware*

Volume 1 — *Physics of Sound in Marine Sediments*
Edited by Loyd Hampton

An Office of Naval Research symposium
Consulting Editors: Alexander Malahoff and Donald Heinrichs
Department of the Navy

Volume 2 — *Deep-Sea Sediments: Physical and Mechanical Properties*
Edited by Anton L. Inderbitzen

An Office of Naval Research symposium
Consulting Editors: Alexander Malahoff and Donald Heinrichs
Department of the Navy

Volume 3 — *Natural Gases in Marine Sediments*
Edited by Isaac R. Kaplan
An Office of Naval Research symposium
Consulting Editors: Alexander Malahoff and Donald Heinrichs
Department of the Navy

Volume 4 — *Suspended Solids in Water*
Edited by Ronald J. Gibbs

An Office of Naval Research symposium
Consulting Editors: Alexander Malahoff and Donald Heinrichs
Department of the Navy

Volume 5 — *Oceanic Sound Scattering Prediction*
Edited by Neil R. Andersen and Bernard J. Zahuranec

An Office of Naval Research symposium

A Continuation Order Plan is available for this series. A continuation order will bring delivery of each new volume immediately upon publication. Volumes are billed only upon actual shipment. For further information please contact the publisher.

OCEANIC SOUND SCATTERING PREDICTION

Edited by

Neil R. Andersen
National Science Foundation
Washington, D.C.

and

Bernard J. Zahuranec
Office of Naval Research
Arlington, Virginia

PLENUM PRESS • NEW YORK AND LONDON

Library of Congress Cataloging in Publication Data

Main entry under title:

Oceanic sound scattering prediction.

(Marine science; v. 5)
Proceedings of a symposium conducted by the Ocean Science and Technology Division of the Office of Naval Research held in Monterey, Calif. Nov. 10-14, 1975.
Includes indexes.
1. Underwater acoustics—Congresses. 2. Sound-waves—Scattering—Congresses. 3. Bioacoustics—Congresses. 4. Marine biology—Congresses. I. Andersen, Neil R. II. Zahuranec, Bernard J. III. United States. Office of Naval Research. Ocean Science and Technology Division.
QC242.025   551.4'601   77-3445
ISBN 0-306-35505-1

Proceedings of a symposium conducted by the Ocean Science and Technology
Division of the Office of Naval Research on Oceanic Sound Scattering
Prediction held in Monterey, California, November 10–14, 1975

© 1977 Plenum Press, New York
A Division of Plenum Publishing Corporation
227 West 17th Street, New York, N.Y. 10011

All rights reserved

No part of this book may be reproduced, stored in a retrieval system, or transmitted,
in any form or by any means, electronic, mechanical, photocopying, microfilming,
recording, or otherwise, without written permission from the Publisher

Printed in the United States of America

Preface

The history of the meeting from which this publication grew dates back a number of years. The idea of holding a conference occurred in the late sixties somewhere in the Equatorial Atlantic Ocean during discussions between G. Brooke Farquhar and Neil R. Andersen, who were at the time conducting research on the biology of scattering layers, and the marine chemical and biological factors affecting nuclear fallout distributions, respectively. In the course of discussions about the biological nature of sound scattering layers and the dependence of cell growth on available food, a question arose. Since the chemical and/or biochemical makeup of a specific ocean area may control the biological populations comprising the scattering layers, why not utilize that makeup as an indicator of future volume reverberation characteristics in a particular geographical location?

The basic thesis evolved as follows: Acoustic volume reverberation is primarily caused by organisms. All organisms are dependent on their food supply, and therefore, ultimately on the chemical precursors in the food chain. It therefore follows that 1) if there is a relationship between the volume reverberation properties of an ocean area and the biological makeup of sound scattering layers giving rise to such properties; and 2) if there is a relationship between the biological makeup of a given ocean area and the initial chemical and/or biochemical conditions of the water giving rise to the biological community, it then follows that there must be a relationship between the chemical and/or biochemical nature of an ocean area and the future volume reverberation characteristics of either the same area, or another whose properties are determined by the physical transport from the initial location of parameters giving rise to the sound scattering phenomenon. The only problem that was foreseen in clearly defining such relationships was that the data to develop such a prediction capability did not appear to exist. Moreover, there appeared to be so little interdisciplinary appreciation of the problem, it might not be possible to even assess the state of the art for sound scattering forecasting. It was concluded that it would be useful to acquire information in order to determine the possibility of developing such a capability. Thus, planning began for the meeting which resulted in this publication.

Understanding and predicting volume reverberation may have practical applications such as finding submerged objects, studying schools of fish, and using oceanic acoustics in other ways to investigate the marine environment. However, the primary problem is to gain a better appreciation of the interaction between the chemistry and biology of the ocean, how the physics of the medium maintains the boundary conditions, and in what manner the systems can be modeled, so that given the proper information, one can predict what will follow. This is an extremely difficult task, but because of its potential importance, it should not be shied away from.

In 1968 there was a conference on Biological Sound Scattering convened at the Airlie House in Warrenton, Virginia, which was attended by several of the contributors of papers in this volume. At that meeting volume reverberation was discussed in detail and the proceedings were published. However, those discussions primarily considered the biological and physical aspects of the acoustics phenomenon. It was not intended that this conference be a duplication or update of the material covered in 1968. Rather, the resulting publication is meant to look into the future, and better understand the basic interrelationships of this problem area. Such considerations cover the spectrum from the credibility of the chemical and physical measurements being employed to the numerical characteristics of the eventual model derived.

In 1973, specific plans for this conference were finally set in motion with an organizational meeting at the Naval Postgraduate School, Monterey, California. This meeting was attended by Drs. Andre Cobet, Naval Biomedical Research Laboratory; Richard Johnson, Oregon State University; Theodore Packard, then of the University of Washington and presently of the Bigelow Laboratory of Ocean Sciences; William Pearcy, Oregon State University; Richard Pieper, University of Southern California; Francis Richards, University of Washington; Paul Scully-Power, Royal Australian Naval Research Laboratory, New South Wales, Australia; Eugene Traganza, Naval Postgraduate School; and Ned Ostenso of the Office of Naval Research, in addition to the editors. The structure of the conference, and therefore this volume, was essentially guided by the nature of the problem being addressed: 1) the physical and chemical environment, 2) biological considerations (i.e. food chains), 3) bioacoustics, and 4) modeling, each with appropriate chairmen. The object of the meeting was not to be an exhaustive presentation of papers, but rather, a discussion based on the preparation of individual papers containing research results of the analysis of specific problem areas, which were to be available to the participants prior to the meeting. The information in these papers was then incorporated into the initial discussions in plenary sessions, followed by working group deliberations. The results of these deliberations, along with the individual papers, constitute this volume.

We wish to acknowledge and thank the above named individuals who played a central role in developing plans for the meeting, and to Mrs. Mary-Francis Thompson and Mrs. Linda Gale of the American Institute

PREFACE

of Biological Sciences who were responsible for all logistic details for the meeting at the Asilomar Conference Center, Pacific Grove, California, where this conference was convened, 10-14 November, 1975. We also wish to acknowledge the assistance and continued pleasant personality of Miss Sarah Teachout in numerous grim situations, without which this publication would not have been possible. We finally wish to gratefully recognize the able assistance of Rose Marie Baldwin in typing the manuscripts and organizing this volume.

Neil R. Andersen
Bernard J. Zahuranec
Washington, D.C.
1977

Contents

I. PHYSICAL AND CHEMICAL ENVIRONMENT

Recommendations of the Working Group on the Physical and Chemical Environment 3
   R. C. Dugdale

Recommendations of the Subgroup on Mapping 11
   R. H. Backus, J. A. McGowan, J. L. Reid

Some Thoughts on the Dependence of Sound Speed and the Scattering Layers Upon Ocean Circulation 15
   J. L. Reid

The Chemistry of Plankton Production and Decomposition in Seawater 65
   D. Dyrssen

The Cycling of Labile Organic Compounds: Sterols in the North Atlantic Ocean 85
   R. B. Gagosian

Deep-Sea Metabolism in the Eastern Tropical North Pacific Ocean 101
   T. T. Packard, H. J. Minas, T. Owens, A. Devol

Marine Areas of Anomalous Chemistry Resulting from Oxygen Deficiencies 117
   F. A. Richards

Turbulence as a Factor in Sound Scattering in the Upper Ocean 129
   J. D. Woods

Optical Parameters that may Affect Vertical Movement of Animals in the Ocean 147
   J. R. V. Zaneveld

II. FOOD CHAINS

Recommendations of the Working Group on Food Chains 161
    J. Mauchline

Estimating Production of Midwater Organisms 177
    J. Mauchline

Windows into a Sea of Confusion: Sampling Limitations to the Measurement of Ecological Parameters in Oceanic Mid-Water Environments 217
    M. V. Angel

On the Distribution of Midwater Fishes in the Eastern North Atlantic 249
    J. Badcock, N. R. Merrett

Studies on Pelagic Animal Biomasses 283
    M. Blackburn

Physiological Approaches to the Biology of Midwater Organisms 301
    J. J. Childress

Aspects of the Feeding Ecology of Oceanic Midwater Fishes 325
    T. L. Hopkins, R. C. Baird

The Lanternfish *Lobianchia dofleini*: An Example of the Importance of Life-History Information in Prediction of Oceanic Sound Scattering 361
    C. Karnella, R. H. Gibbs, Jr.

Observations on Feeding Habits of the Mesopelagic Fish *Benthosema glaciale* (Myctophidae) off NW Africa . . . 381
    J. Kinzer

Possible Factors Affecting Succession in Plankton Communities 393
    A. G. Lewis

Growth and Moulting of Crustacea, Especially Euphausiids 401
    J. Mauchline

What Regulates Pelagic Community Structure in the Pacific? 423
    J. A. McGowan

The Role of Microorganisms in the Marine Environment 445
    R. Y. Morita

Food and Feeding Structures of Deep-Sea *Thysanopoda* Euphausiids 457
    T. Nemoto

III. BIOACOUSTICS

Recommendations of the Working Group on Bioacoustics 483
    G. B. Farquhar, D. V. Holliday

Biological Sound Scattering in the Oceans: A Review 493
    G. B. Farquhar

Pelagic Faunal Provinces and Sound-Scattering Levels in the Atlantic Ocean 529
    R. H. Backus, J. E. Craddock

Sound Scattering and Oceanic Midwater Fishes 549
    R. C. Baird, D. F. Wilson

A Study of the Swimbladders of Selected Mesopelagic Fish Species 565
    A. L. Brooks

The Minox Program: An Example of a Multidisciplinary Oceanic Investigation 591
    W. A. Friedl, G. V. Pickwell, R. J. Vent

Extracting Bio-Physical Information from the Acoustic Signatures of Marine Organisms 619
    D. V. Holliday

Spectral Models for Biological Sound Scattering . . . 625
    R. K. Johnson

Progress in the Correlation of Volume Scattering Strengths to Biological Data 631
    R. H. Love

Variations in Abundance of Sound Scattering Animals Off Oregon 647
    W. G. Pearcy

Some Comparisons Between Oceanographic Measurements
and High-Frequency Scattering of Underwater Sound . . 667
    R. E. Pieper

On the Prediction of Sound Scattering in the Oceans
from Fish Capture Data 679
    P. Scully-Power

Acoustic Volume Scattering Measurements with Related
Biological and Chemical Observations in the Northeastern Tropical Pacific 697
    R. J. Vent, G. V. Pickwell

IV. MODELING

Recommendations of the Working Group on Modeling . . . 719
    J. J. O'Brien

On the Large Scale Circulation of the Ocean: A Discussion for the Unfamiliar 723
    J. C. McWilliams

On Persistence of Aquatic Ecosystems 749
    T. G. Hallam

Mesoscale Oceanic Phytoplankton Patchiness Caused by
Hurricane Effects on Nutrient Distribution in the Gulf
of Mexico 767
    R. L. Iverson

How Can the Methodologies of Pattern Recognition Aid
in Modeling Volume Reverberation Processes? 779
    P. T. McElroy

Biological Prediction in Pelagic Marine Ecosystems . . 803
    T. Platt, K. L. Denman

Daytime Depths of Sound Scattering Layers in the Major
Biogeographic Regions of the Pacific Ocean 811
    S. A. Tont

Vertically Migrating Herbivorous Plankton - Their
Possible Role in the Creation of Small Scale Phytoplankton Patchiness in the Ocean 817
    J. S. Wroblewski

LIST OF CONTRIBUTORS AND PARTICIPANTS 849

INDEX 853

I

PHYSICAL AND CHEMICAL ENVIRONMENT

Recommendations of the Working Group on the Physical and Chemical Environment

R. C. Dugdale, Chairman

Bigelow Laboratory for Ocean Sciences

INTRODUCTION

The working group addressed itself to two areas of concern to those needing to predict acoustic reverberation in the sea caused by organisms:
1. The extraction of generalities about the major physical and chemical domains of the oceans.
2. The processes leading to inhomogeneities in space and time within these domains, especially within the large central gyres.

Within these broad topics, several subgroups considered the following aspects:
   a. Large scale distributions and anomalous areas (Reid, Richards, Roden).
   b. Optics (Zaneveld, Holm-Hansen).
   c. Chemical control of biological processes (Gagosian, Dyrssen, Morita, Packard).
   d. Nutrient supply to near surface waters in central gyres (Woods, Barber, Holm-Hansen).
   e. Turbulence and variability of nutrient and biological factors (McWilliams).

LARGE SCALE DISTRIBUTIONS

The subgroup on large-scale distributions of physical, chemical, and biological properties of the oceans recognized the need to outline the major acoustic reverberation domains and their boundaries. Although such domains depend upon time, depth and frequency, it was concluded that it is possible to outline the major reverberation domains from intelligent use of presently existing information for a limited number of frequencies and depths, on a seasonal basis.

There appears to be a relation between the larger scale circulation systems of the ocean and the biomass. The subtropical anticyclonic gyres are poorer in both nutrients and productivity and hence lower in biomass than the cyclonic gyres of higher latitudes and the system of zonal flows near the equator. It appears also that each of these systems contains populations of organisms that are widespread throughout the system but markedly different from the populations of the other systems. There is, of course, both time and space variation within each system, or domain, and the boundaries between the domains are not sharp in all cases.

Because so many species appear to be confined to, yet widespread within, a subtropical gyre, it appears worthwhile to investigate carefully whether there are patterns of acoustic reverberation that also correspond to the gyres. While the volume of the biomass itself may not be directly related to the reverberation, if the biomass within a gyre consists of approximately the same species everywhere, then perhaps some generalizations about the nature of the reverberation patterns can be made that will apply throughout the gyre. The rich equatorial zone might also be examined in this way. However, other areas, such as the eastern boundary currents, are subject to large advective influx from more than one source, have more heterogeneous populations, and thus might not lend themselves to such generalizations. The eastern part of the subequatorial zone may also be less tractable, not only because of the advection of forms from higher latitudes, but because of the shallow and extreme oxygen minimum found there. It may be that the patterns of vertical migration are altered or broken down there or that different species, with different sonic characteristics, appear there in significant quantities.

The subgroup feels that future progress will come from studies of the nonseasonal variations of the large-scale distributions, particularly as manifested at the boundaries. Therefore, it recommends that increased effort be directed toward theoretical and observational endeavors leading to the understanding and eventual prediction of the reverberation domain fluctuations.

ANOMALOUS AREAS

Anomalous areas of the ocean can here be defined as regions where chemical or other features are persistently so different from the rest of the system that they can be expected to produce, attract, or eliminate characteristic communities. The principal such regions now defined are those of the eastern tropical Pacific where enormous lobes of oxygen poor water extend seaward from the coasts of Mexico,

Central and South America across a significant fraction of the oceanic width. They are characterized by nitrate reduction, a deep nitrite maximum, and denitrification. The response of various marine organisms to the conditions in these regions are not known in detail, but it seems safe to assume that the conditions exercise some control over the organisms that inhabit the regions.

After respiratory processes have stripped all the dissolved oxygen, nitrate and nitrite from a marine domain, the respiratory decomposition of additional organic matter will be at the expense of sulfate ion and will be accompanied by sulfide production. This toxic substance will eliminate almost all organisms more advanced than bacteria and result in acoustically different environments. Extremes of these conditions arise in many weakly circulating basins and fjords and in two large marine systems -- the Black Sea and the Cariaco Trench. The smaller systems may be sulfide bearing or they may be flushed annually, occasionally, and catastrophically at long and irregular time intervals. Hydrogen sulfide also has been observed in the coastal undercurrent off Peru. Much could be learned about the acoustic properties of marine environments by comparing oxygen-deficient and anoxic, sulfide bearing systems with otherwise comparable environments.

OPTICS

The sun is the energy source for the oceanic ecosystem. The light flux at a point as a function of wavelength (spectral scalar irradiance), or a related quantity, the photosynthetically active light must be known as a function of depth if models of the ecosystem are to be constructed or if phytoplankton growth is to be understood. Scalar irradiance of all solar energy at a given depth also governs the vertical migration of a large number of sound scattering species. The study of scalar irradiance in the ocean is thus of central significance in the eventual understanding of sound scattering patterns and should be encouraged.

Only very few measurements of spectral scalar or vector irradiance as a function of depth have been made in the open ocean. Seasonal variations are not known, nor are variations on intermediate and small space scales (10 m to 10 km). It is recommended that scalar irradiance be routinely included in studies of sound-scattering species and of phytoplankton. Furthermore, studies of spectral scalar irradiance should be carried out in the deep ocean to ascertain variability in both space and time.

Optical parameters are extremely useful in the study of suspended particulate matter (including phytoplankton). The beam attenuation coefficient in the red (650 nm) as well as volume scattering measurements at a constant angle ($4°$ appears best, but $45°$ and $90°$ have been used more widely) show high correlations with total suspended particulate volume in regions where the nature of the suspensoids does not fluctuate greatly. Fluorometers also should be used in studies of suspended matter. Such optical parameters are useful in the study of both horizontal and vertical phytoplankton distributions (patchiness). Instruments can either be towed, in which case vertical profiling underway is possible, or aircraft-mounted instrumentation that measures related parameters can be employed. The next decade should provide us with the ability to measure fluorescence from aircraft as well as light attenuation of the upper layers of the ocean. Spacecraft or airborne spectrometry of "ocean color" may eventually permit assessment of small scale temporal and spatial variations in chlorophyll, particle concentration and perhaps phytoplankton biomass. The problem of ocean color has by no means been resolved and additional studies should be encouraged.

The relationship between the measurable parameters fluorescence and suspended particulate volume on the one hand and phytoplankton biomass on the other will depend on the speciation in a given domain and should be studied for a region of interest, so that remote or underway measurements can give more reliable phytoplankton biomass information.

In a domain in which the relative abundance of species does not fluctuate greatly with time one may postulate that sound scattering is proportional to the number of sound scattering animals present, which in turn can be postulated to be related to the concentration in a domain of reasonably constant speciation is an indication of volume reverberation should be tested.

There are various biological parameters which should also be studied concomitantly with the above optical measurements in the water column. These include: 1) a description of the light-sensory organs of the animal species, together with other characteristics (e.g., color, photophores, etc.) related to ambient light fields, 2) biochemical studies of the visual pigments (spectral sensitivity) in the light organs, as compared to the ambient spectral irradiance, and 3) a study of the functioning of light organs in regard to ontogenetic sequences and migratory patterns.

CHEMICAL CONTROL OF BIOLOGICAL PROCESSES

The regulation of biological processes such as phytoplankton growth, grazing and predation and the control of the distribution of pelagic populations by organic compounds are poorly known. Work is needed to reveal mesoscale patterns in deep-sea metabolism and microbial patchiness as well as microscale patterns of the organic compounds that serve as indices of deep-sea populations. The dynamics of these populations are controlled not only by nutrient levels (phosphate, nitrate, ammonia and silica), but also by ectocrine compounds such as vitamins (B_{12}, biotin, thiamine, etc.), hormonal-like compounds, antibiotics, toxins, and allomones (chemical compounds used for both inter-and intraspecies communication).

Studies of the role of the compounds named above have been retarded in the past by insensitive methodology. However, new chemical techniques and instrumentation such as gas chromatography and mass spectrometry are proving useful in examining the kinetics of growth as well as the dynamics of grazing and predation. Utilizing enzyme analysis, nutrient regeneration and assimilation rates can be measured. Both the concentrations and the mechanisms by which controlling messenger compounds regulate trophic interactions can be determined by organic analysis and biological assays. In addition to this role in studying the dynamic properties of pelagic ecosystems, chemical techniques can be used to determine the static properties as well. The recent lipopoly-saccharide and epifluorescent techniques can facilitate the assessment of microbial populations which may indicate the magnitude and the spatial and temporal distribution of mid-water communities. These techniques can help with the understanding of pelagic ecosystems, but only a multi-disciplinary effort between investigators such as marine organic geochemists, biochemists, microbiologists, plankton ecologists and chemical oceanographers will be successful.

NUTRIENT SUPPLY TO NEAR SURFACE WATERS IN CENTRAL GYRES

In the central oceanic gyres remote from coastal upwelling and shallow water mixing and away from strong currents, a weak average upward flux of nutrients nourishes a correspondingly low rate of production in the surface layer. This upward nutrient flux supplied from deep sources passes through the thermocline, where turbulent mixing is weak and has a different character from the more vigorous mixing in the surface layer.

Frontal Upwelling

The subgroup, reviewing current ideas concerning turbulent transport in the thermocline, came to the conclusion that frontal upwelling as described by *Woods* (1975) is probably the most likely mechanism. However, the model is at present supported only by observation in the central Mediterranean Sea and while there is some evidence of similar fronts in the central gyres (e.g., *Katz*, 1972), further investigation is needed.

If the vertical supply of nutrients through the thermocline is carried by frontal upwelling, then there would be considerable intermittency (an areal concentration of 1% of upwelling tongues of water is probably a high estimate). This spatial inhomogeneity should be a central feature of models of primary production in central gyres, since it will affect not only the prediction of local patchiness in the production rate, but also the mean production estimated for areas containing a representative sample of upwelling events.

The subgroup made the following recommendations.
1. that field experiments be undertaken to test the hypothesis that primary production in central gyres is nourished by frontal upwelling, combining the techniques of investigation of the physical structures developed by *Woods* (1970) with continuous sampling of the chemical and biological constituents (as in *Kelley et al.*, 1975).
2. that the possibility of such spatial and temporal inhomogeneities in production rates be taken into account when estimating mean values for the central gyres from local measurements which may not be a representative sample.
3. that theoretical models of primary production in central gyres be developed to test the implications of nutrients being supplied by relatively rare events.

Measurements of Nutrient Distribution

It is important to combine measurements of the space-time variability of physical (e.g., light, temperature) and biological (e.g., chlorophyll, plankton) variables with simultaneous measurements of nutrient distribution in the same spectral window (i.e., the same ranges of length and time scales), in order to produce a comprehensive data base for stimulating and testing models of patchiness in productivity in the central gyres. Since the nutrient concentrations in central ocean gyres sometimes fall below the threshold of existing measurement techniques, encouragement should be given to those seeking to improve the resolution of nutrient measurements.

Transport of Nutrient by Grazers

The subgroup reviewed current ideas on the role of vertical migration in the general nutrient dynamics of the central gyres. Since the biological and chemical character of these regions is determined in a large degree by the magnitude of the nutrient supply to the surface waters, it is important to consider the impact that a biological process may have in complementing the physical supply processes described above. Recognizing that in the central gyres recycling of nutrients by grazer and microbial regeneration is the major process satisfying the nutrient demand of primary producers, additional research is needed to better understand the small daily inputs and losses. In addition to the transports resulting from vertical migration, research must include the topics of nitrogen fixation by blue-green algae and the inputs of combined inorganic nitrogen by rainfall.

There was no clear concensus on the net impact vertical migration will have on the nutrient dynamics. Two opposing ideas can be developed. There may be a net upward transport of new or nitrate nitrogen by zooplankton grazing on phytoplankton in the deep layer of maximum nitrate uptake and subsequent regeneration of ammonia-nitrogen by these same zooplankton as they move upward into the zone of maximum primary production. This coupled process would provide a small, daily supply of new nitrogen to the surface, depending on a behaviorally controlled biological process: vertical migration. This process provides a biological conveyor belt tying together two parts of the zone of active phytoplankton growth. On the other hand, when vertical migration is considered as a process tying together the zone of phytoplankton growth with the aphotic zone where no net production of new organic matter occurs, the impact of vertical migration may be a net downward transport of nutrient elements. In their vertical movements herviborous organisms are grazing mainly in the euphotic zone where phytoplankton are available but they are continuously excreting regenerated material, for example ammonia and urea, so that about half of the excretion by migration occurs well below the euphotic zone. In the central gyres, the nitrogen consumed as phytoplankton, carried downwards below the euphotic zone, through the pycnocline, and excreted at depth, could be a significant loss.

What is needed to answer these questions is a nitrogen budget in Z (depth) for a central gyre region that ignores X and Y processes and attempts to establish mass balance as a function of depth alone. Referring to the spatial events that we suspect may characterize gyres, such a study should be done in conjunction with resolution of the spatial inhomogeneities, but the first pass should not be a budget in X, Y, and Z until the attempt for Z alone has been carried out. Such a study will include monitoring of the migration so that if there is a chemical wake left by the vertical excursion, these inhomogeneities in Z can be resolved for the nitrogen budget. Nitrogen is recommended as the primary element to be studied since it provides the most information about the causative relations. However, operationally, it would be cost effective to attempt to determine simultaneously the silicon, carbon, oxygen and phosphorous budgets.

TURBULENCE AND THE VARIABILITY OF NUTRIENT AND BIOLOGICAL FACTORS

Currently, the variability of physical quantities is a subject of great interest to physical oceanographers, not only for their own sake but also for possible relationships with large scale, very low frequency ocean currents and heating distributions. This variability is energetically dominated in the mid-ocean by the class of motion called mesoscale eddies with time periods of 40-250 days, horizontal wavelengths around 400 km, and depth scales comparable to the ocean depths; however, less energetic variability ranges down to seconds and millimeters.

The distribution and density of biological populations and the nutrients upon which they feed ought to be strongly influenced by this physical turbulence, whether through systematic current advection or a more turbulent dispersion. Under the most extreme situations, the organisms and nutrients can simply be passively carried by the turbulence, and knowledge of the turbulence would permit a successful prediction of the variability of the biology and nutrients. However, additional observations of biological and nutrient variability on a variety of scales and in a variety of environments are needed to allow a fair comparison with physical variability.

REFERENCES

Katz, E. 1972. Further investigation of a front in the Sayano Sea. *Deep-Sea Research*.

Kelley, J. C., T. E. Whitledge, and R. C. Dugdale. 1975. Results of sea surface mapping in the Peru upwelling system. *Limnol. Oceanogr.* 20: 784-794.

Woods, J. D. 1970. Measuring thermocline fronts from the air. *Under water J.* 2: 90-99.

Woods, J. D. 1975. Diffusion due to fronts in the rotation subrange of turbulence in the seasonal thermocline. *La Houille Blanche* 7/8: 589-597.

Recommendations of the Subgroup on Mapping

SOUND-SCATTERING PATTERNS, THE OCEAN CIRCULATION, AND BIOLOGICAL PROVINCES

As a result of the ocean-wide study of the distribution of certain pelagic animals in the Pacific, Atlantic, and Indian Oceans, a number of patterns of principles have emerged. For example, most of the animals studied have more or less continuous distributions over hundreds of thousands of square kilometers, yet there are often repeated geographical distribution patterns -- i.e., while there are many species there are few patterns because many different species have the same distributional patterns. The distribution of animals corresponds to the distribution of physico-chemical properties of the ocean -- i.e., the patterns of animal distribution correspond to the broad-scale circulational features. The patterns detected in the Pacific Ocean based on the study of zooplankton species and those detected in the Atlantic Ocean based on the study of mesopelagic fishes are very similar -- i.e., in both oceans there are patterns that can be called "equatorial", "subtropical gyre" and so on. Included among those organisms that are so distributed are groups of animals known to be important sound-scatterers -- e.g., mesopelagic fishes with swimbladders including the abundant lanternfishes (Myctophidae).

The above has suggested that the faunal provinces of the world ocean can to a large extent be taken as sound-scattering provinces -- that is, as the ocean changes faunally, so will it change in its sound-scattering properties, more or less. This hypothesis has been

EDITORS'NOTE: During the deliberations of the four working groups, an interdisciplinary group comprised of Drs. Reid, McGowan and Backus formed another discussion group on Mapping. This report represents the results of their discussions.

tested in part by *Chapman, Bluy, Adlington, and Robison* (1974), who made broad-band sound-scattering observations over wide reaches of the Atlantic Ocean and in the eastern Pacific Ocean and compared their results with schemes for dividing the oceans based upon faunal data. The first paragraph of their summary says:

> "Spectra of column strength have been used as a tool for comparing reverberation conditions in a wide range of oceanographic areas. As the spectra often maintained consistent features over distances of hundreds of kilometers but changed dramatically in the neighborhood of known faunal or oceanographic boundaries, the present study supports the concept of dividing the world's oceans into reverberation provinces."

Thus, world ocean maps of reverberation provinces could be prepared for the interim guidance of planners of biological and acoustical surveys, or for the guidance of naval and other operators at sea (in letting them anticipate changed conditions, for instance). Regional acoustical information would be used to the extent that it is available; this would be supplemented principally by zoogeographic data, but, where such data are scanty or wanting, the atlases of physical oceanography could be resorted to because of the faunal-physical correspondence noted above. Such maps would constitute a first-order gross picture and, in most cases, would not attempt to provide details on seasonal and other variability that might be important in volume reverberation. Furthermore, as our knowledge regarding the kinds, sizes, and abundance of organisms and their relationship to the frequency dependence of sound-scattering increases, so will our ability to greatly refine such maps.

We note that a number of pertinent maps have been prepared already. They are the basis for our belief that a certain large-scale order exists within the ocean and is reflected in the distribution of its organisms. As such, they constitute a first guess as to the domains which might be examined for similarity in sound-scattering characteristcs. We cannot provide a complete list of such maps, but we call attention to those at hand and presented or reviewed at this meeting. These include maps prepared by Backus for the Atlantic, McGowan for the Pacific, Brinton for the world ocean, the various maps presented by Beklemeshev, and the primary productivity and zooplankton maps of the world ocean in the FAO Atlas. These largely reflect the major gyres and the effect of the zonal circulation near the equator. We call attention also to the many others that we cannot cite immediately that deal with organisms of smaller distribution, such as within the eastern boundary currents and the eastern tropical areas of very high productivity, where the biomass is larger and apparently more variable.

While these maps constitute a first-order gross picture, in most cases they do not attempt to provide details on the seasonal and non-seasonal variability or of the degree of local patchiness or short-term fluctuations that may be of importance in variations in acoustic reverberation.

REGIONAL, BROAD-SCALE OCEANOGRAPHIC MAPPING AND ACOUSTIC CONSIDERATIONS

In the last ten years of the acoustic scattering measurement effort in the United States and elsewhere, it has become clear that the partitioning of the oceans on relatively large scales, in which regions are defined upon similar properties -- that is, having some degree of "order" to them -- is an extremely useful concept for understanding large-scale variations in acoustic scattering. Most scattering measurements have been made in the frequency band between about 1 and 25 kHz. When the broad-scale patterns of variation in this frequency band are examined, it is clear that the observed changes in pattern are closely associated with the water mass region, gyre, or zonal system where the observations are made, and that marked changes are related to the crossing of boundaries between two identifiable regions (systems, water masses, gyres, etc.).

With the techniques and approaches now available for establishing or delineating large, identifiable oceanic provinces or systems, the utility of maps of such major regions as an indication of different acoustic scattering provinces is clear.

The largest difficulty in the mapping of such "acoustic provinces" is related to the frequency dependence of scattering. As our base of understanding improves regarding the kinds, sizes, and abundance of various organisms, and how they relate to frequency dependence, there will be a basis for refinement of such maps of provinces, and an improvement in their application to a broader band of acoustic frequencies.

The validity of such maps will depend upon the relative constancy of the pertinent characteristics within each area. Samples from two areas, one presumed to be relatively constant and one presumed to be relatively inconstant, were tested using several simple indices (McGowan, 1977). The results support the expectations. This kind of "testing" can be expanded to help resolve questions concerning the coherence (i.e., size, shape, and temporal persistence) of areas of relative constancy. Additional testing may be performed on certain other areas with selected samples and data already in hand. However, expanding the testing procedure to resolve questions of temporal and areal coherence may require additional sampling effort if the validity of the tests is not to be compromised.

The preparation of maps of such orderly regimes in the ocean will today provide an extremely useful tool for the acoustician, and the applied oceanographer as well, in designing acoustic measurement programs or as a predictive tool. This effort should be carried out as soon as possible.

REFERENCES

Chapman, R. P., O. Z. Bluy, R. H. Adlington, and A. E. Robison. 1974. Deep scattering layer spectra in the Atlantic and Pacific Oceans and adjacent seas. *J. Acoust. Soc. Am.*, 56 (6): 1722-1734.

McGowan, J. A. 1977. What regulates pelagic community structure in the Pacific? In: *"Oceanic Sound Scattering Prediction"*, N. R. Andersen and B. J. Zahuranec (Eds.), Plenum, N.Y., (this volume).

Some Thoughts on the Dependence of Sound Speed and the Scattering Layers Upon Ocean Circulation

Joseph L. Reid

Scripps Institution of Oceanography

ABSTRACT

The pattern of sound speed in the ocean is shown to be related to the circulation through the quasi-geostrophic balance to the major flow-field, which requires certain patterns in the large-scale density field and hence the temperature structure. The resulting temperature structure defines the depth and sound speed of the Sofar channel, which show correspondence to the major anticyclonic and cyclonic gyres of the ocean. This circulation, with the associated surface convergence and divergence within the anticyclonic and cyclonic gyres, also defines the field of plant nutrients, which are highly concentrated within the cyclonic gyres and very low within the anticyclonic gyres. These nutrient patterns largely define the primary productivity and are closely related to the distribution of zooplankton volume. This suggests that the major gyres may constitute separate biological domains, each inhabited by special groups of organisms. It is proposed that these physical-biological domains may define the major patterns of sound-scattering organisms which should be most numerous and lie at shallower depths within the cyclonic gyres than within the anticyclonic gyres. There are some data in support of this proposal but not nearly enough to be conclusive at this time.

EDITORS' NOTE: The term sound velocity was used in the underwater acoustic literature during the sixties. Since "velocity" is a vector quantity and the meaning is, more accurately, a scalar quantity, sound "speed" is now commonly used in the scientific literature and will be used in this paper.

INTRODUCTION

In the following discussion an attempt is made to indicate very briefly and in a very general way how the density structure and hence the temperature structure and sound speed are related to the general circulation of the ocean. Considerations are also given as to how this affects the distribution of nutrients and hence marine life in general, and what patterns might be expected in the distribution of organisms as a result of this system of circulation. A final consideration is devoted to what patterns might occur in the intensity and depth of the scattering layers.

FLOW, DENSITY, AND TEMPERATURE

Figure 1 is a map of the geopotential anomaly at the sea surface relative to the 1000-decibar surface for the Pacific Ocean in

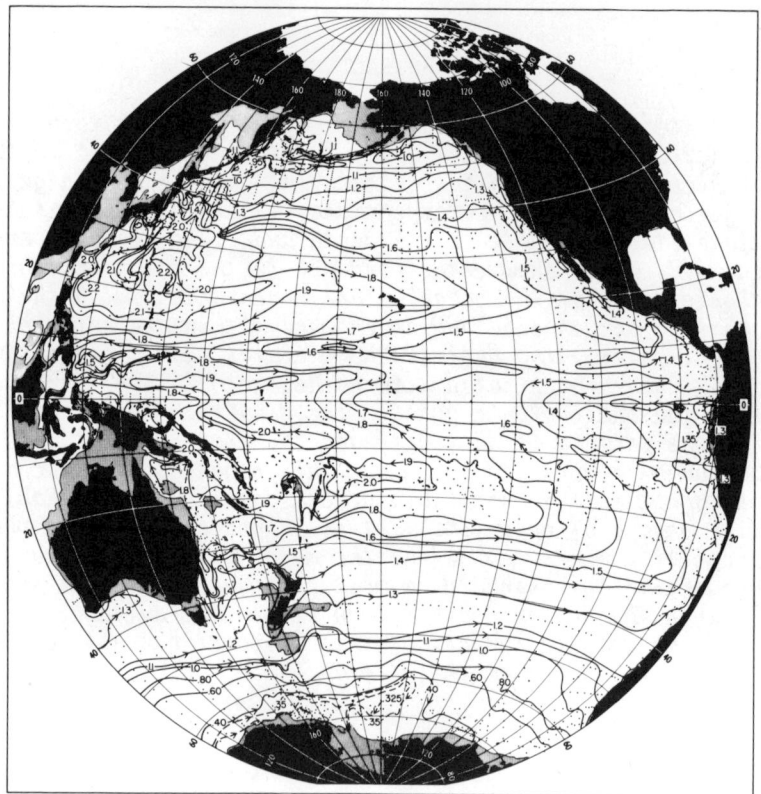

Fig. 1. The geopotential anomaly at the sea-surface relative to the 1000-decibar surface, in dynamic meters (10 m^2/sec^2 or 10 J/kg). In the shaded areas the ocean depth is less than 1000 m [*Reid and Arthur*, 1975, Fig. 1].

northern winter. If the ocean is quasi-geostrophically balanced, the horizontal gradients might be expected to reflect the sense and intensity of the flow relative to 1000 decibars. If the flow at 1000 decibars is substantially weaker, then it might be anticipated that this map would give a useful picture of the surface circulation. Indeed, it does compare well enough with maps made from other sorts of measurements.

The deeper flow can also be examined (Figure 2; see also the maps at still greater depths presented by *Reid and Arthur* [1975]) and it is observed that the circulation weakens, and particularly that the great subtropical anticyclonic gyres lie farther poleward at greater depths.

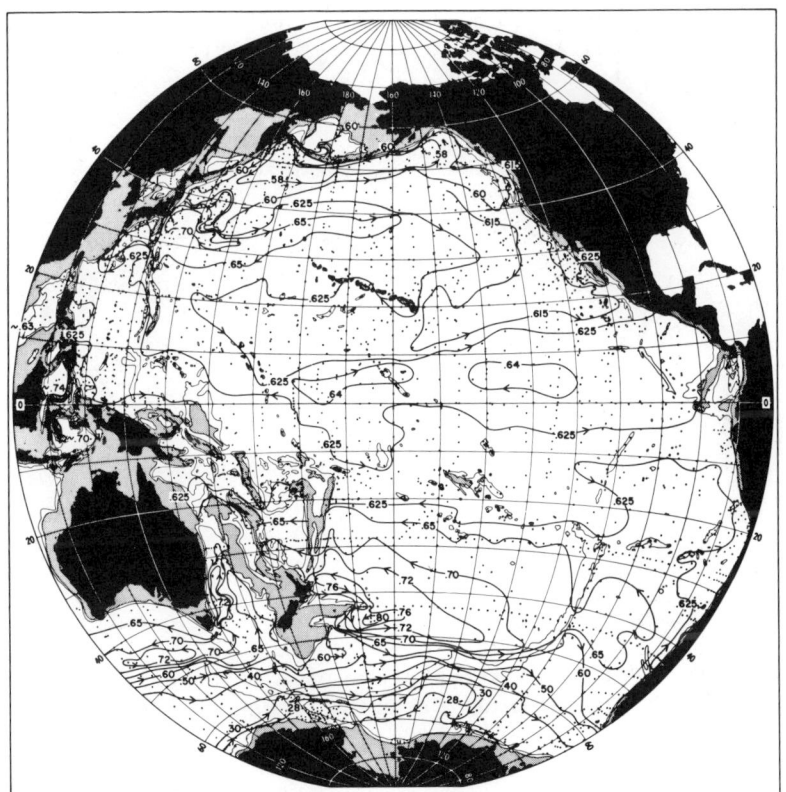

Fig. 2. The geopotential anomaly at the 1000-decibar surface relative to the 2000-decibar surface, in dynamic meters (10 m^2/sec^2 or 10 J/kg). In the shaded areas the ocean depth is less than 2000 m and the light line is the 3000-m depth contour [*Reid and Arthur*, 1975, Fig. 2].

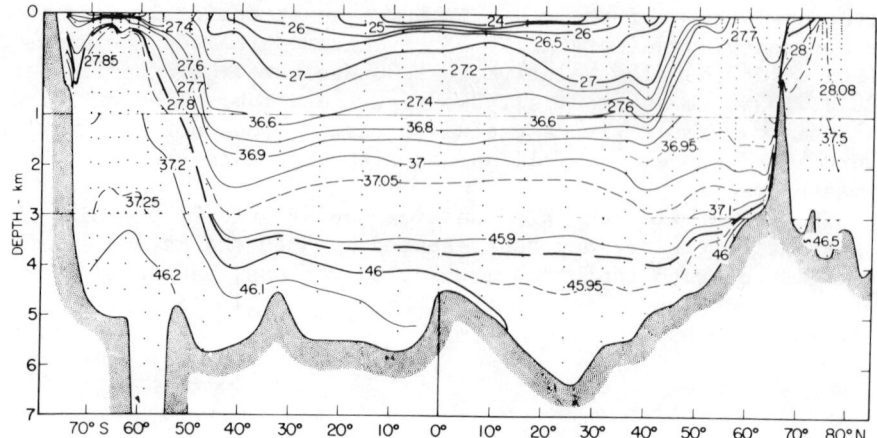

Fig. 3a. σ_0 (0-1000 m), σ_2 (1000-3000 m), σ_4 (>3000 m) on a meridional vertical section in the western Atlantic Ocean [*Reid and Lynn*, 1971, from Fig. 3a].

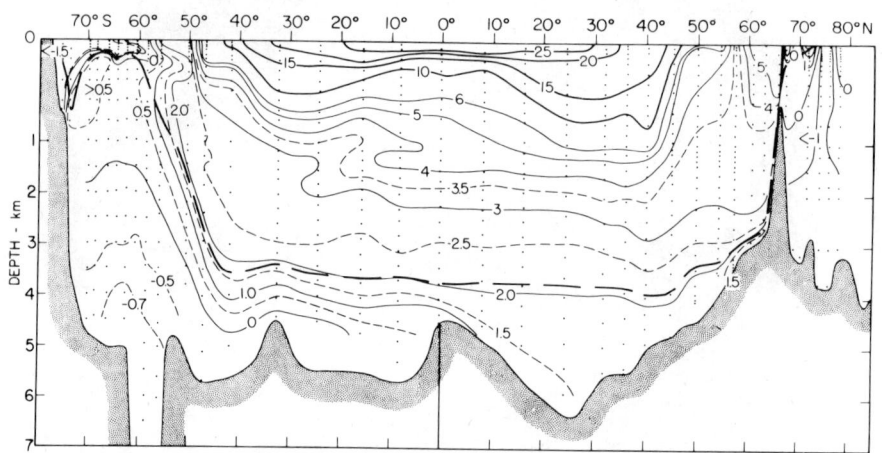

Fig. 3b. Potential temperature on a meridional vertical section in the western Atlantic Ocean [*Reid and Lynn*, 1971, from Fig. 3a].

The Atlantic and Indian Oceans also demonstrate this quasi-geostrophic flow, and corresponding patterns are seen in the density field there [*Defant*, 1941; *Wyrtki*, 1971]. This sort of flow requires a density distribution to provide the horizontal pressure gradients, and the shape can be seen in vertical sections in all

OCEAN CIRCULATION   19

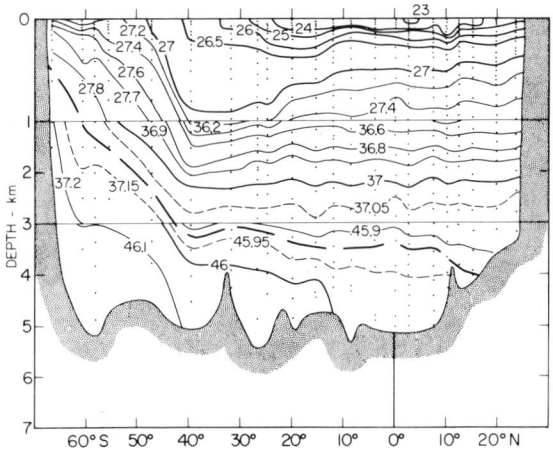

Fig. 4a. σ_0 (0-1000 m), σ_2 (1000-3000 m), σ_4 (>3000 m) on a meridional vertical section in the western Indian Ocean [*Reid and Lynn*, 1971, from Fig. 3b].

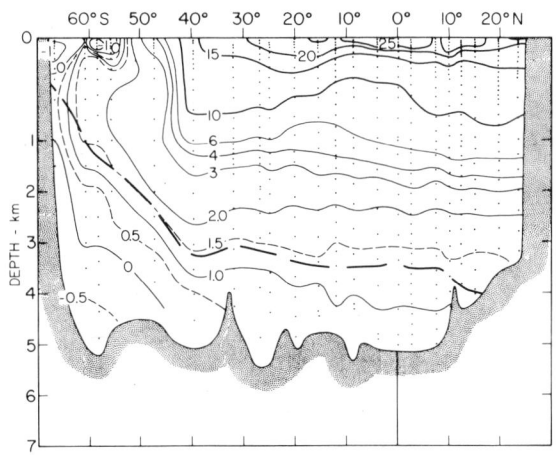

Fig. 4b. Potential temperature on a meridional vertical section in the western Indian Ocean [*Reid and Lynn*, 1971, from Fig. 3b].

oceans. Figures 3a, 4a and 5a are north-south sections for the Atlantic, Indian, and Pacific Oceans. The great subtropical anticyclonic gyres are seen in the deeper positions of the isopycnals in middle latitudes and their shallow positions beneath the high-latitude cyclonic gyres and at the equator. The temperature

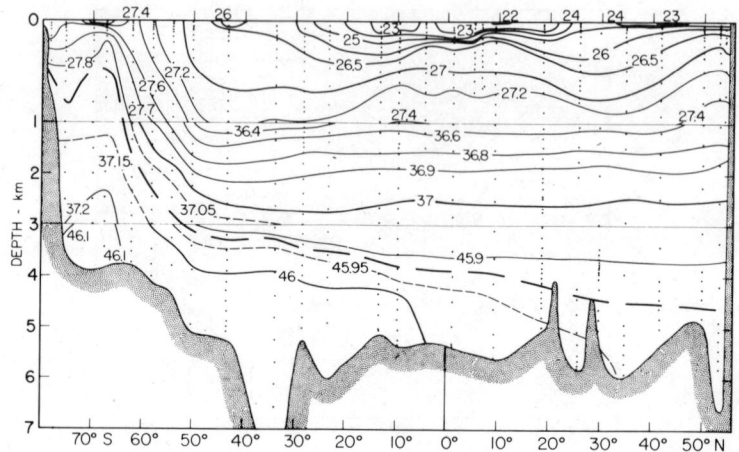

Fig. 5a. σ_0 (0-1000 m), σ_2 (1000-3000 m), σ_4 (>3000 m) on a meridional vertical section in the western South Pacific and central North Pacific [*Reid and Lynn*, 1971, from Fig. 3c].

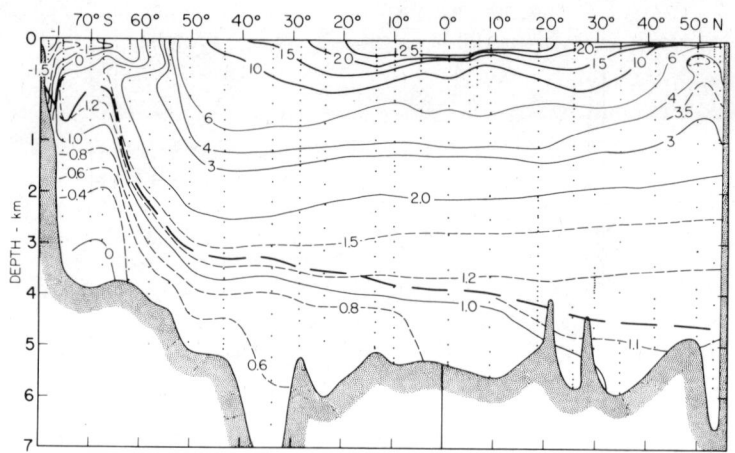

Fig. 5b. Potential temperature on a meridional vertical section in the western South Pacific and central North Pacific [*Reid and Lynn*, 1971, from Fig. 3c].

(Figures 3b, 4b, 5b), of course, shows a closely related pattern, with warmer waters extending deepest in mid-latitudes, and cold waters near the surface not only in high latitudes but also near the equator.

Figure 6a is a similar section for the Pacific Ocean, with the

OCEAN CIRCULATION

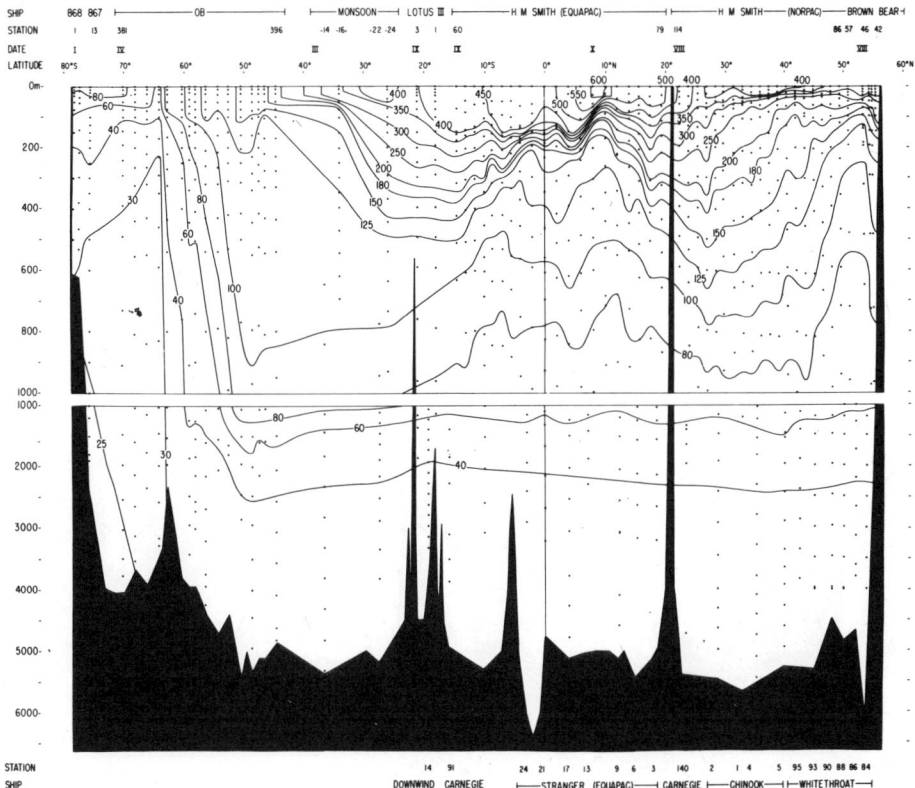

Fig. 6a. Thermosteric anomaly (in centiliters per ton) along approximately 160°W from Antarctica to Alaska [Reid, 1965, Fig. 6].

upper 1000 m expanded. Figure 6b is the corresponding temperature field. It can be seen that the temperature field is not just the consequence of equatorial heating and polar cooling. Below the surface the equatorial zone is colder than the subtropical regions. Beneath the subarctic gyre the deeper water is cold, but not from local surface cooling: no overturn to great depths takes place there, and the deeper, colder water is separated from the cold surface water by a warm layer. The cold deep water in the North Pacific Ocean has come from the Antarctic and has risen beneath the cyclonic gyre.

Likewise, at depths from 800 to 2000 m the warmest water is found near 45°S, certainly not a zone of maximum surface warming. The point is that the geostrophic balance to the major current systems is important in the distribution of characteristics, including heat.

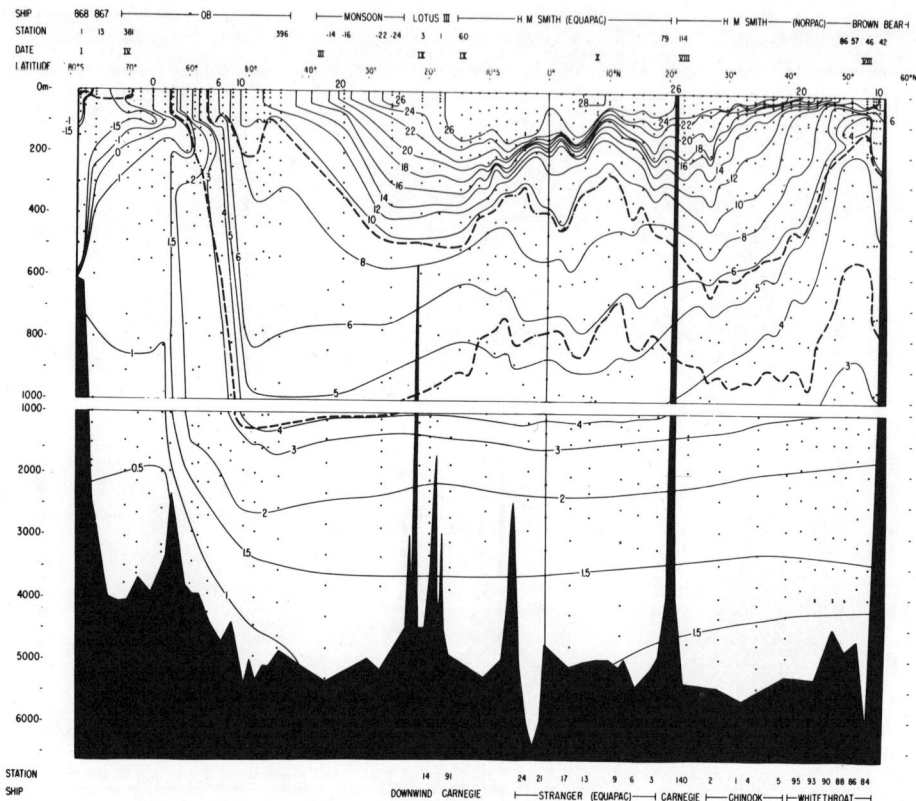

Fig. 6b. Temperature (Celsius) along approximately 160°W from Antarctica to Alaska [Reid, 1965, Fig. 2].

SOUND SPEED

A north-south section of sound speed in the Pacific Ocean (Figure 7) is not symmetric about the equator; this is clearly because the flow is not symmetric. The anticyclonic gyres are slightly different in size and the cyclonic gyres are very different in size (the southern cyclonic gyre surrounding Antarctica), and the density field must reflect these differences. So, therefore, does the temperature field as seen before, and thus the sound speed. In the upper 1500 m the sound speeds are lower at 50°N than at 50°S, and the speed at the sound channel is highest near 40°S-50°S. The section is for northern winter, and the channel lies at the surface in the far north, but just beneath the surface in the far south.

Likewise, in an east-west section along about 28°S the temperatures (Figure 8) are lower in the east, because of the equatorward flow and the associated pressure gradient, and thus the channel (Figure 9) is both shallower and of lower speed in that area.

OCEAN CIRCULATION

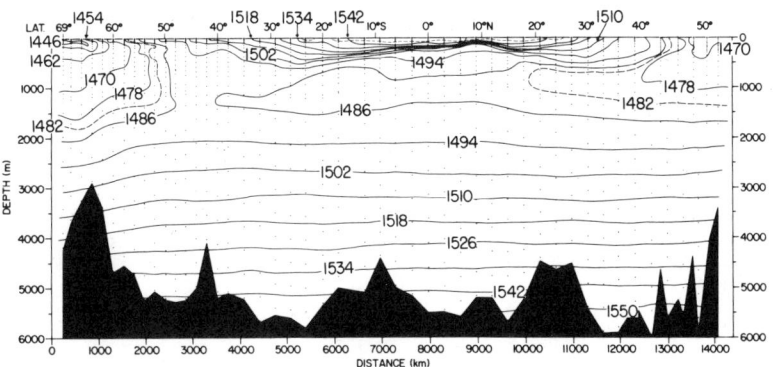

Fig. 7. North-south section of sound speed (m/sec) along 170° W in the Pacific Ocean.

Maps have been made of the depth of the Sofar channel (Figure 10) and of the speed there (Figure 11). The maps could be improved, but the general features, even in the depth map, are consonant with the circulation maps referred to earlier--the high-latitude cyclonic flow and the subtropical anticyclonic flow are reflected in these maps. The enormous anticyclonic gyre of the South Pacific Ocean dominates the pattern of sound speed.

It would appear that there should be more detail in the subequatorial zone. Maps of temperature and sound speed at various depths emphasize this point. At 200 m (Figure 12) and at 500 m (Figure 13), the detail is clear, and at 1000 m (Figure 14), which is about the depth of the Sofar channel near the equator, the systematic zonal flows are reflected in a corresponding series of zonal maxima and minima. These should be reflected in some zonal highs and lows in the depth of the channel.

At the greatest depths the system of gyres seen at the surface, and frequently referred to as wind-driven, yields to the thermohaline flow patterns, and the pattern of speed at 4000 m, which must resemble that of temperature (Figure 15), is quite different from the shallower patterns, though equally systematic. The range of sound speed at 4000 m in the open Pacific is from about 1516 m/sec in the Antarctic Ocean to about 1524 m/sec in the North Pacific, but in the basins separated from the great ocean at 4000 m the values reach 1525 m/sec in the Chilean Basin, 1533 in the Celebes Sea, and 1556 in the Sulu Sea. *Fenner and Bucca* [1971] present an excellent study of the relation of the sound speed in the North Atlantic Ocean to the circulation and water masses there.

Seasonal Variations of Sound Speed in the Northwest Pacific

The Northwest Pacific Ocean has been misrepresented in some earlier studies on sound speed. *Johnson and Norris* (1968) did not

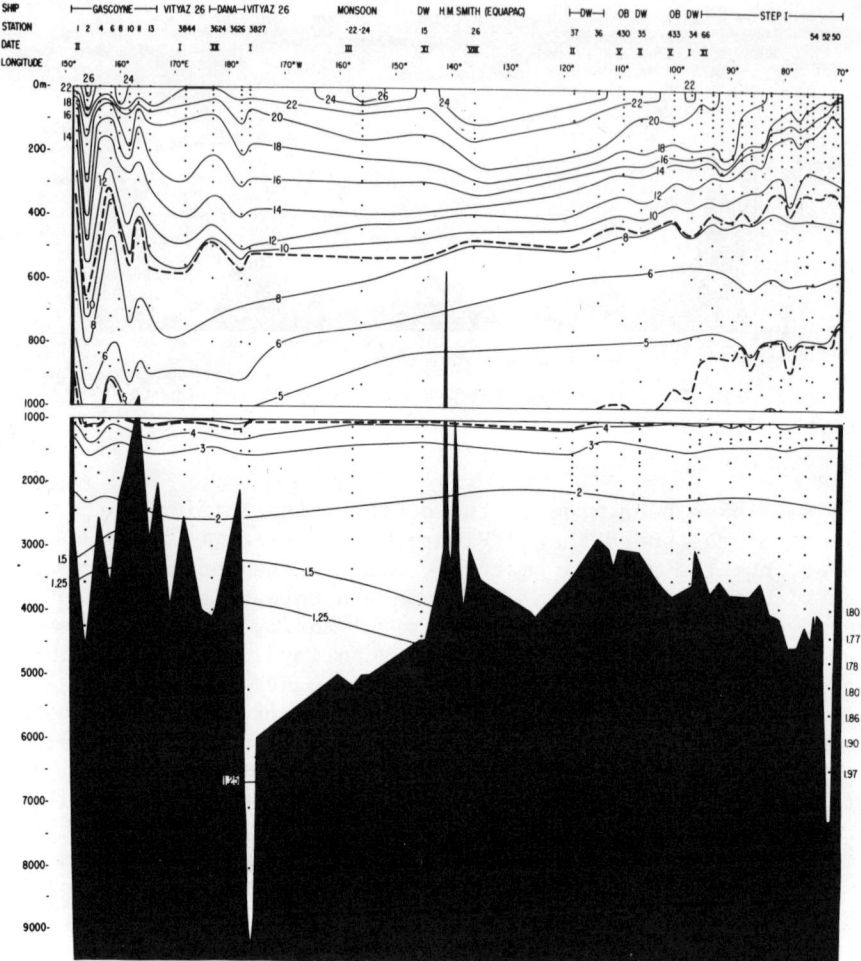

Fig. 8. Temperature (Celsius) along approximately 27°S from Australia to South America (Reid, 1965, Fig. 12).

consider mid-winter data in their work there and found no systematic seasonal variation in depth of the channel or the speed. Their map (Figure 16a) finds the channel to be subsurface, with minimum speeds of about 1453 m/sec. Actually, it intersects the surface there in winter (Figure 16b), rising abruptly from 600 m depth into the subarctic gyre, and the speed (Figure 16c) drops as much as 15 m/sec from their fall and spring values.

This seasonal change in the pattern of sound speed is not the result of local cooling alone but of the seasonal change in circulation as well. The subsurface temperature maximum is found in the

OCEAN CIRCULATION

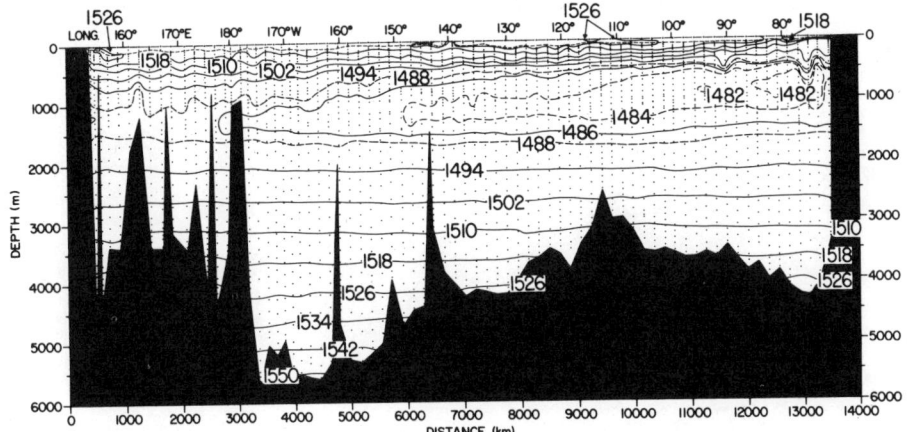

Fig. 9. East-west section of sound speed (m/sec) along 28°S in the Pacific Ocean.

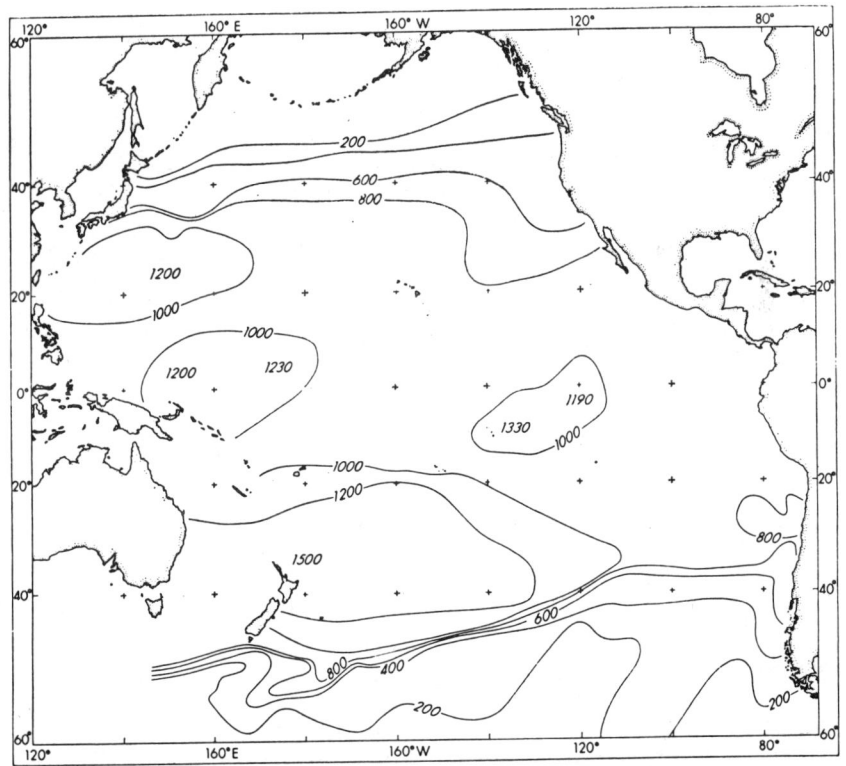

Fig. 10. Depth (m) of the Sofar channel in the Pacific Ocean [*Johnson and Norris*, 1968, Fig. 6].

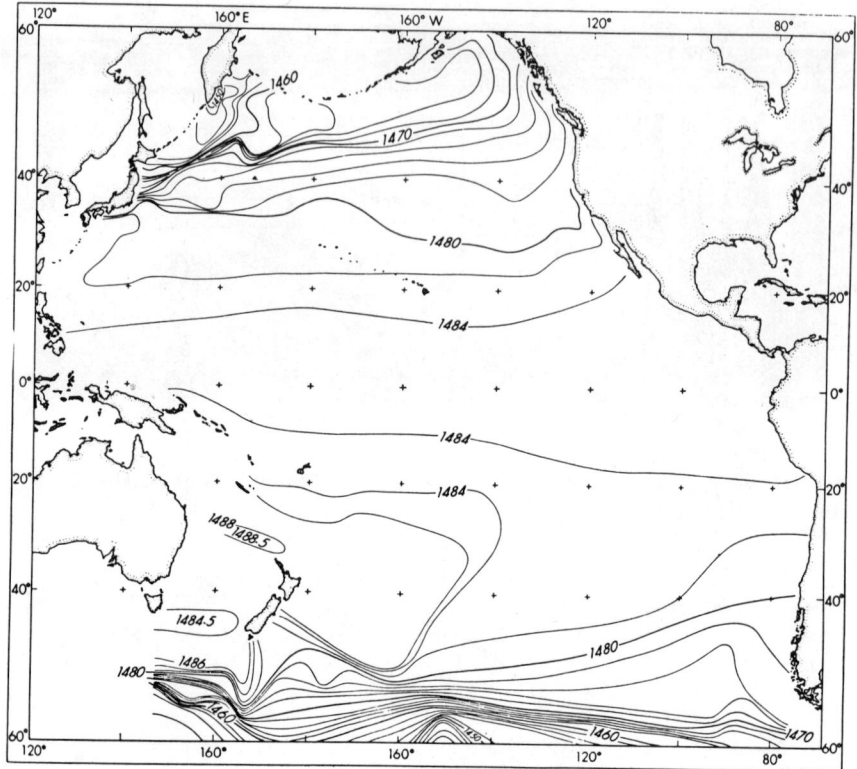

Fig. 11. Sound speed (m/sec) at the Sofar channel in the Pacific Ocean [Johnson and Norris, 1968, Fig. 5].

northwestern part of the North Pacific in both winter and summer (Figures 17 and 18). In the offshore areas this feature, lying well into the pycnocline, varies but little in its characteristics, though the surface temperatures change seasonally by several degrees. This sort of change is illustrated in Figure 19a. But the Kamchatka Current undergoes a major change in winter; it strengthens considerably, and the associated density structure is remarkably different from summer. Such changes in the temperature and salinity fields are illustrated in Figure 19b; the temperature and salinity at the temperature maximum layer are not changed, but the layer lies about 400 m deeper in winter. The change in sound speed (Figure 19c) is more than 15 m/sec at 400 m depth.

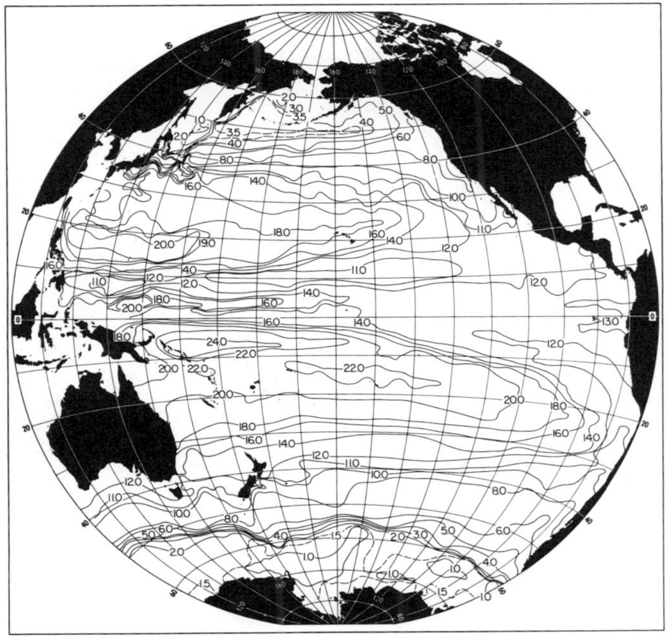

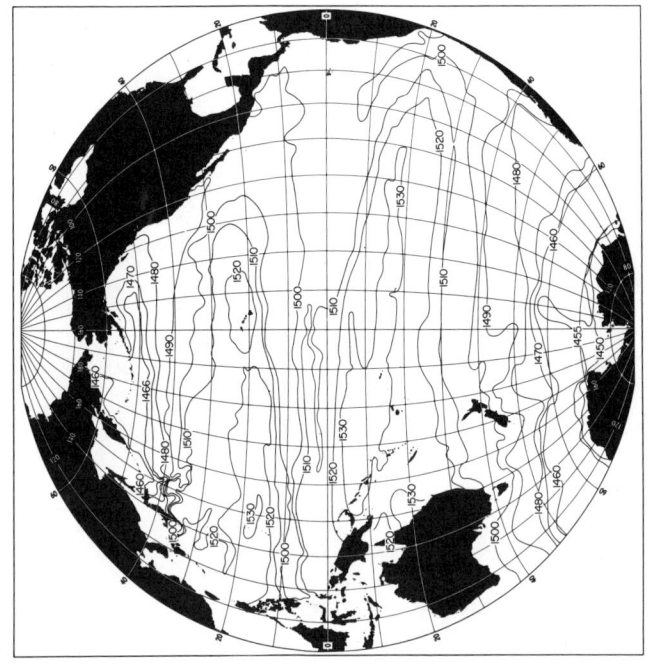

Fig. 12a. 200-m temperature (Celsius), northern winter, Pacific Ocean.

Fig. 12b. 200-m sound speed (m/sec), northern winter, Pacific Ocean.

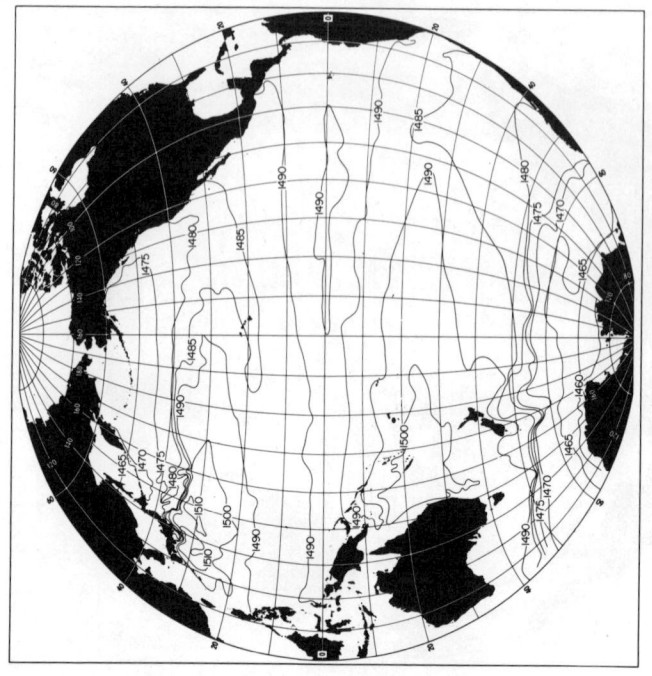

Fig. 13b. 500-m sound speed (m/sec), northern winter, Pacific Ocean.

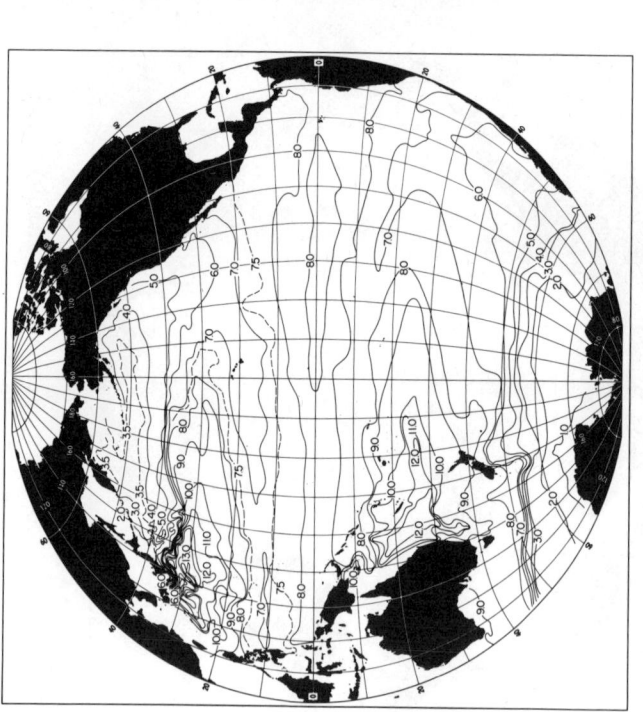

Fig. 13a. 500-m temperature (Celsius), northern winter, Pacific Ocean.

OCEAN CIRCULATION

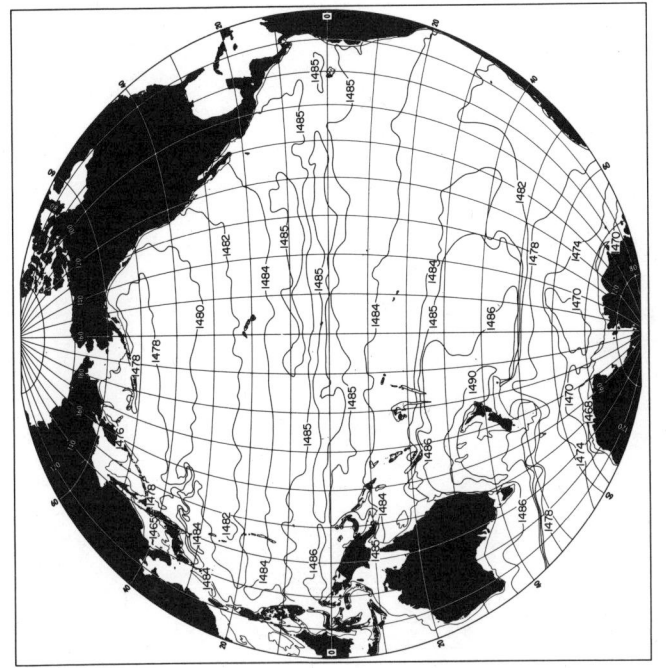

Fig. 14b. 1000-m sound speed (m/sec), Pacific Ocean.

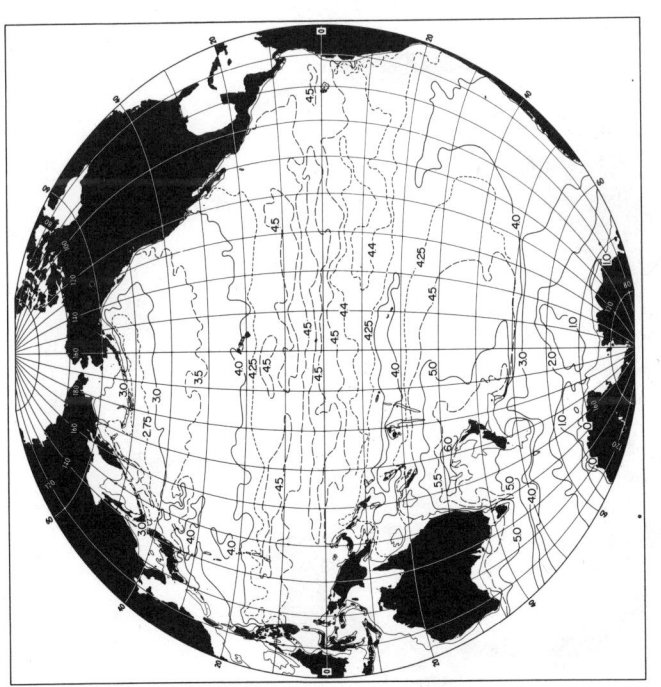

Fig. 14a. 1000-m temperature (Celsius), Pacific Ocean.

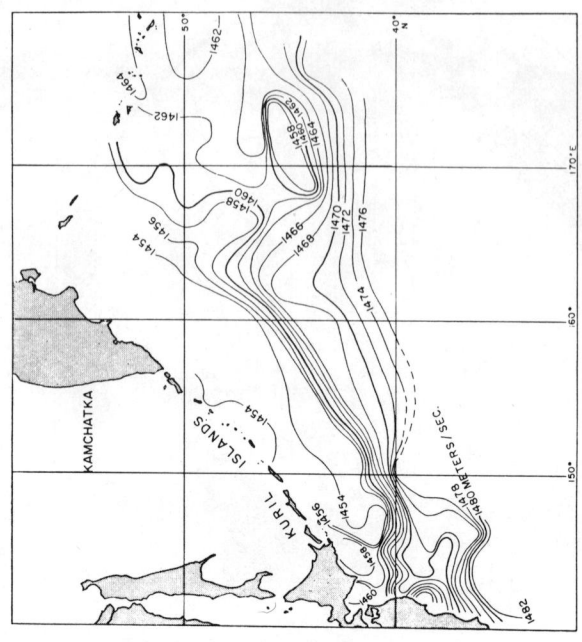

Fig. 16a. Sound speed (m/sec) at Sofar channel, northwest Pacific Ocean, 16 May to 15 July (*Johnson and Norris*, 1968, Fig. 1).

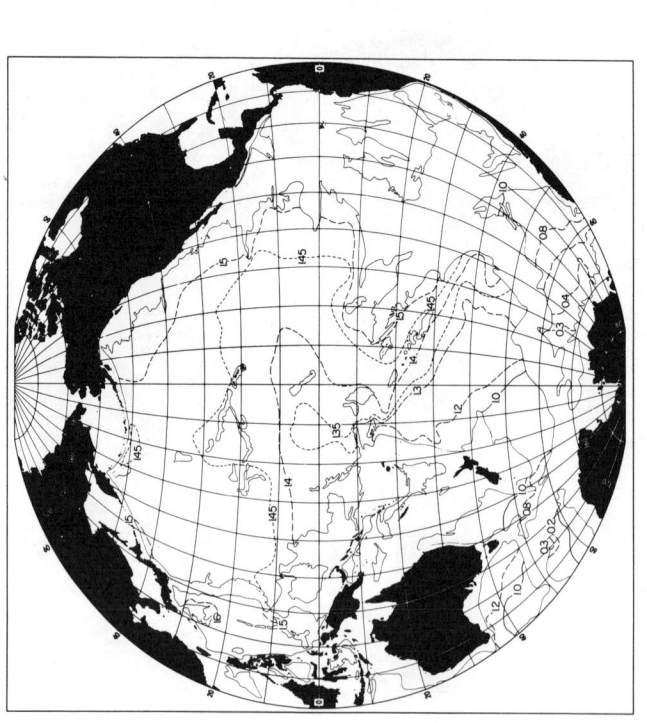

Fig. 15. 4000-m temperature (Celsius), Pacific Ocean.

OCEAN CIRCULATION

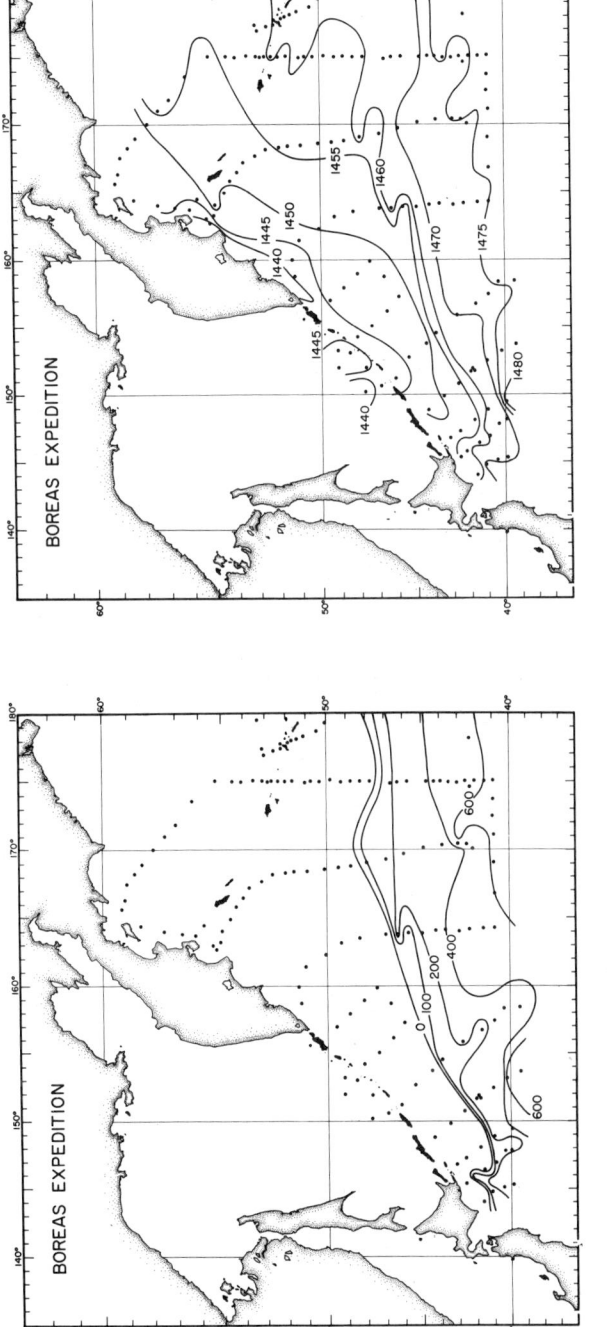

Fig. 16c. Sound speed (m/sec at Sofar channel, northwest Pacific Ocean, February-April 1966, from the Boreas data.

Fig. 16b. Depth (m) of Sofar channel, northwest Pacific Ocean, February-April 1966, from the Boreas data.

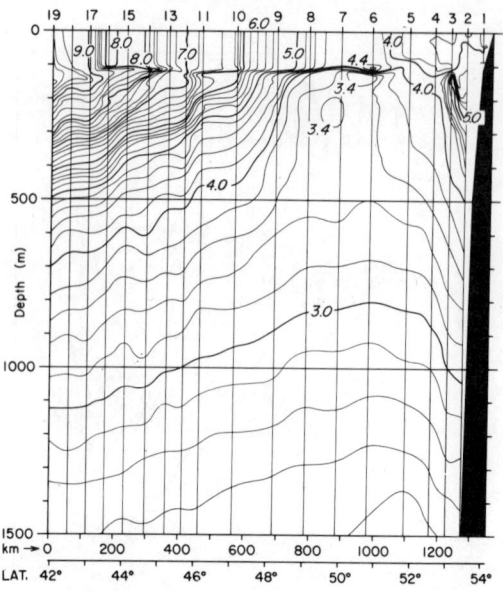

Fig. 17a. Winter temperature (Celsius) along 165°W [*Reid*, 1973, Fig. 32a].

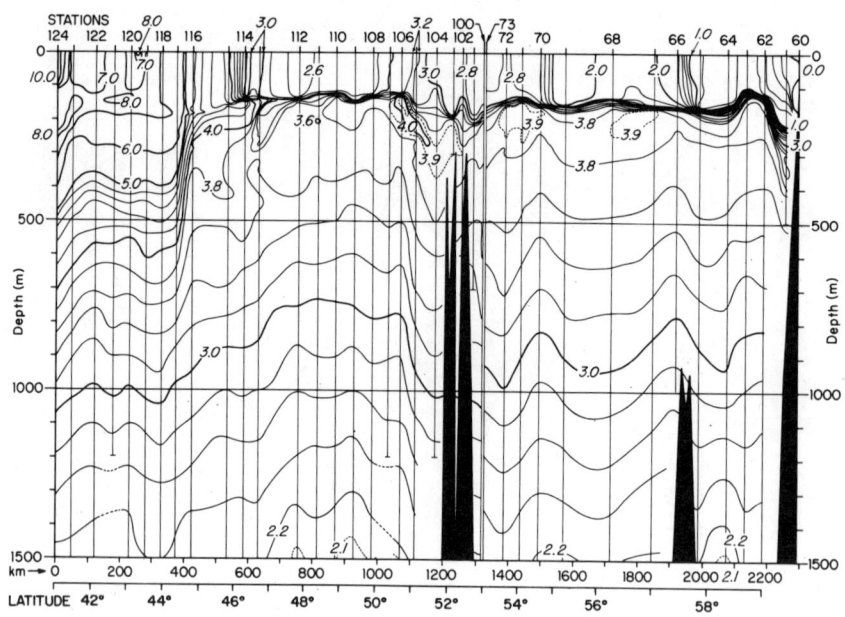

Fig. 17b. Winter temperature (Celsius) along 175°E [*Reid*, 1973, Fig. 35a].

OCEAN CIRCULATION

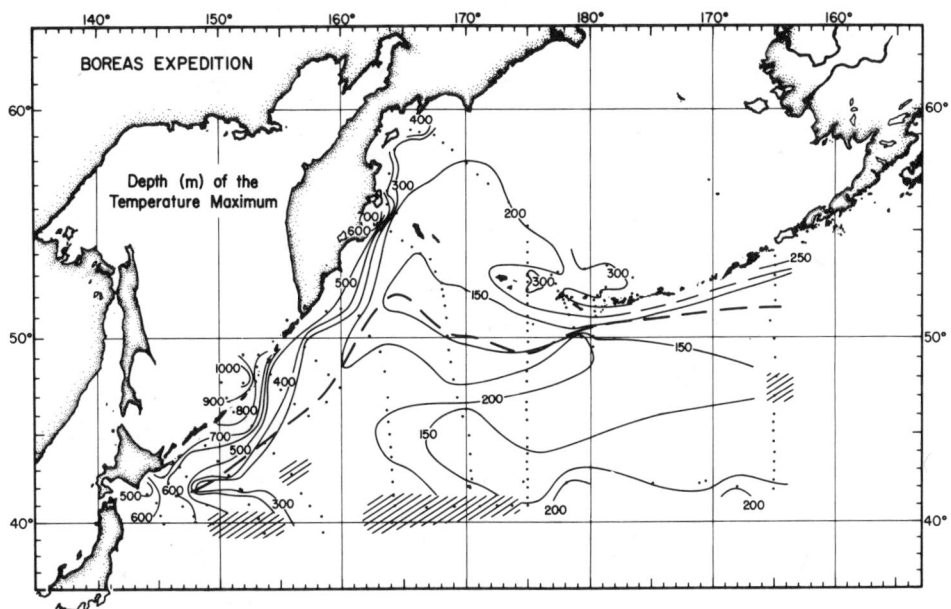

Fig. 18a. Depth (m) of the subsurface temperature maximum [Reid, 1973, Fig. 40b].

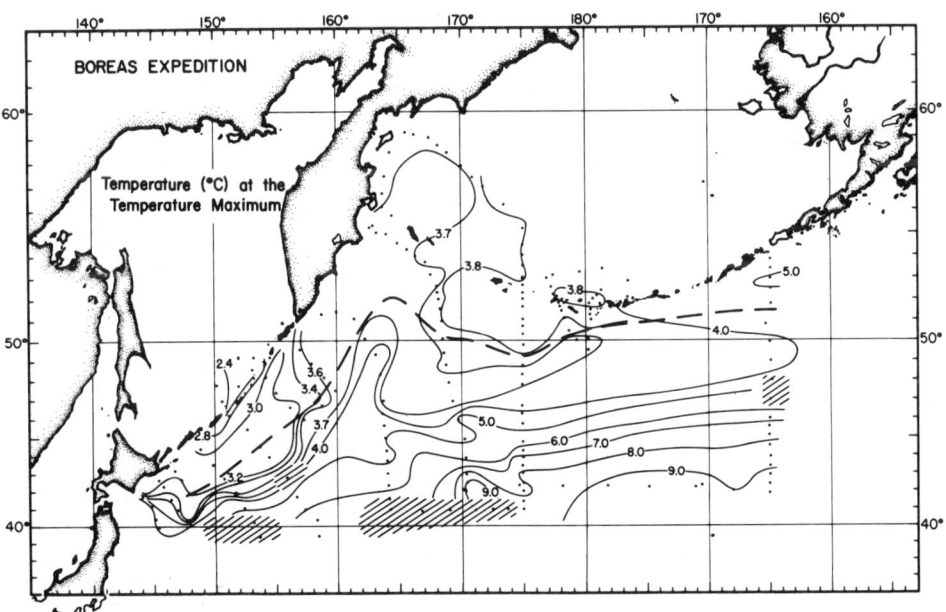

Fig. 18b. Temperature (Celsius) at the subsurface temperature maximum [Reid, 1973, Fig. 40a].

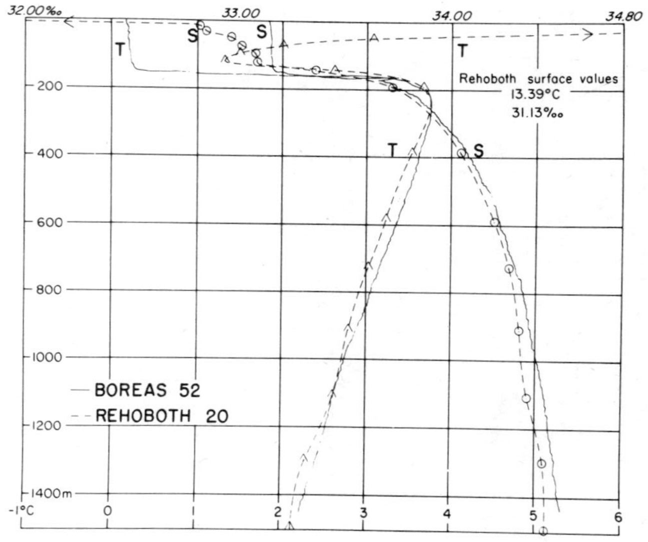

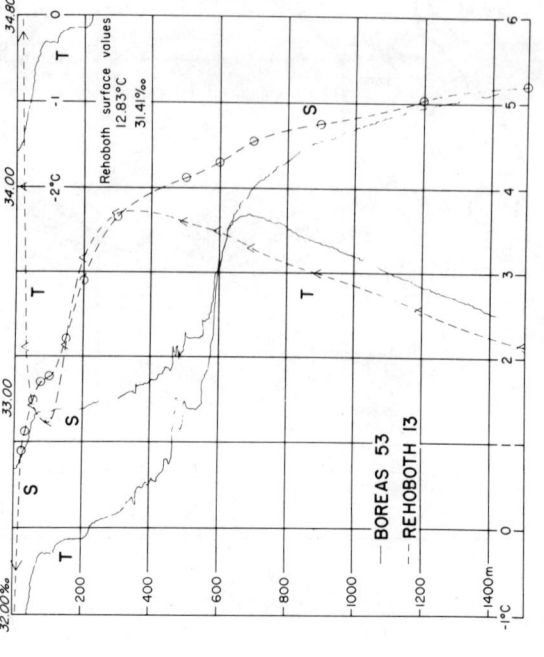

Fig. 19a. Temperature and salinity at a summer station, 54° 58.9'N 162°59.0'E, compared with those at a winter station, 55°08'N 163°17'E (Reid, 1973, Fig. 24).

Fig. 19b. Temperature and salinity at a summer station, 55°28.6'N 163°02.1'E, compared with those at a winter station, 55°18'N 162°45'E (Reid, 1973, Fig. 28).

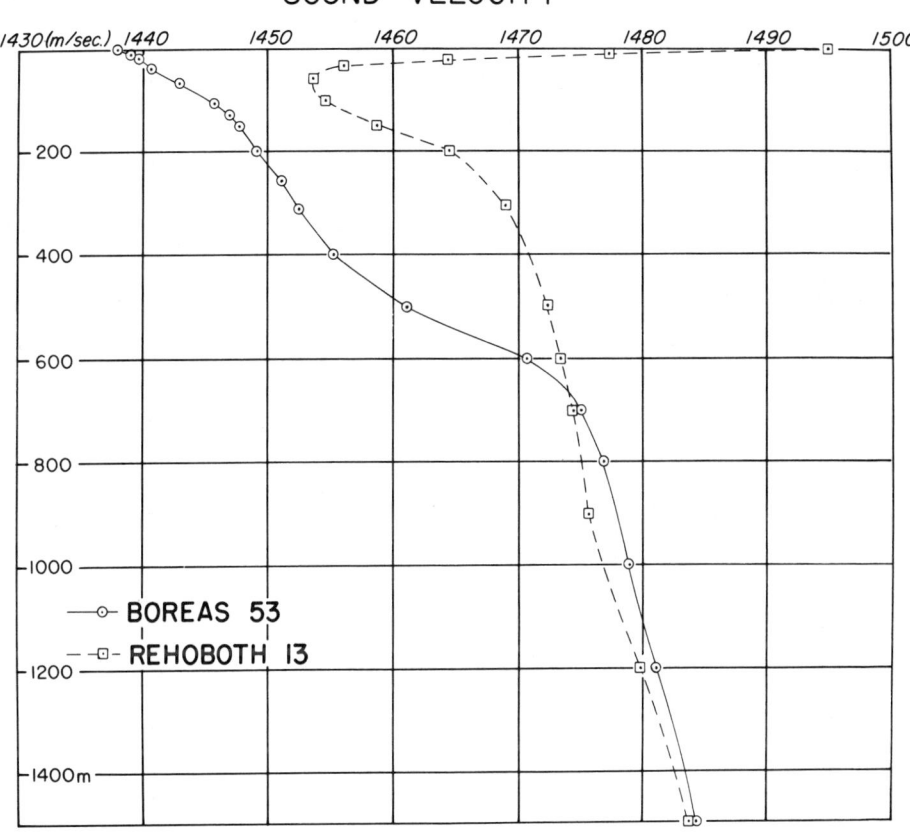

Fig. 19c. Sound speed at the two stations shown in Fig. 19b.

EFFECT OF THE CIRCULATION UPON PRODUCTIVITY

Now--to the effect of the circulation features upon marine life. The distribution of nutrients will be addressed first. Figure 20 is a north-south section of phosphate in the Pacific Ocean. A correspondence to the circulation field can be seen in the sense that near the sea surface the anticyclonic gyres are poorest in nutrients and the cyclonic gyres are richest. The asymmetry noted before in flow, density, and temperature is enhanced here--the North Pacific holds much more phosphate than does the South Pacific. This is related to that thermal cap--the water below that cap in the North Pacific has come from at least as far away as 50°S, along a subsurface route, and has accumulated nutrients from the sinking and decay of the overlying organisms all along the way. It has risen near the equator and in high latitudes because of the nature of the circulation that brings the higher-density water nearest the surface there.

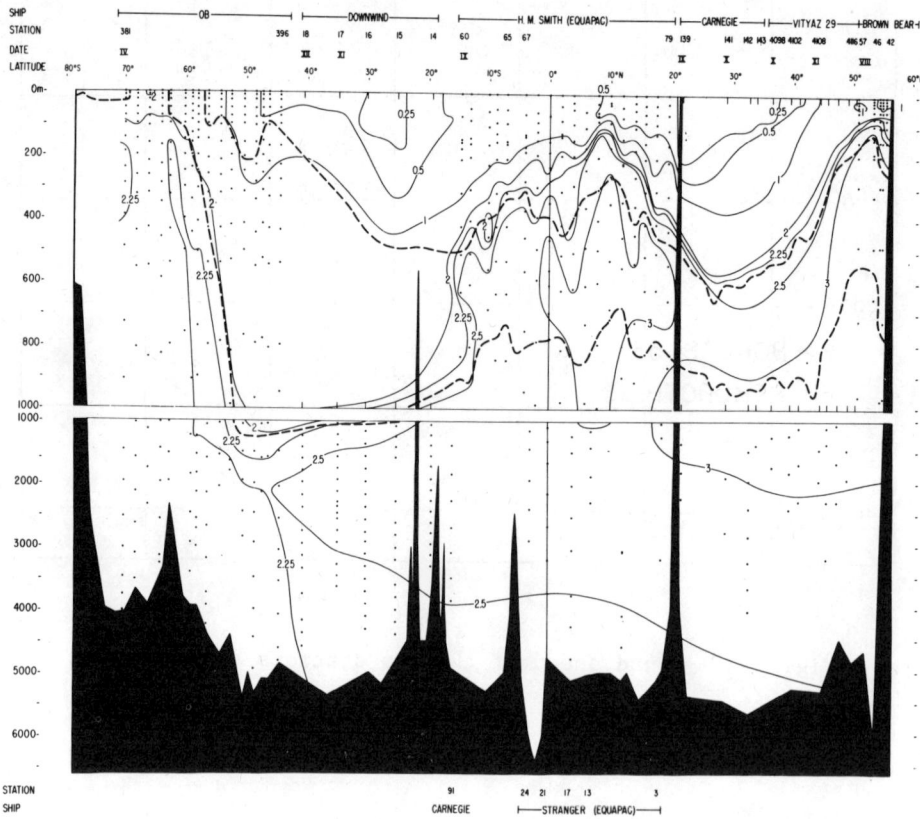

Fig. 20. Inorganic phosphate-phosphorus (µg-at/ℓ) along approximately 160°W from Antarctica to Alaska [Reid, 1965, Fig. 5].

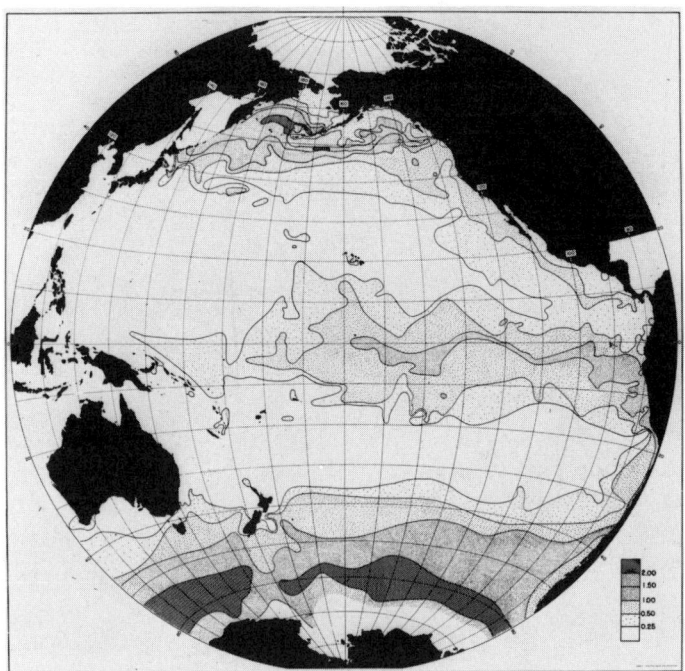

Fig. 21. Distribution of phosphate-phosphorus (µg-at/ℓ) at surface of the Pacific Ocean [Reid, 1962, Fig. 2b].

The surface features of the Atlantic are much the same--poorest in the anticyclonic gyres and richest in the cyclonic gyres, but the underlying waters are vastly different. This is because of the nature of the exchange of water between the Atlantic and Pacific Oceans [Berger, 1970]. The deep waters that have accumulated nutrients in the Atlantic pour into the Indian and Pacific, and it is the less dense, poorer waters of the Pacific that return to the Atlantic.

Imagine what phosphate would look like if it were a conservative characteristic--it would be distributed in a somewhat opposite fashion, like the salinity, with highest values in the convergent, evaporative anticyclonic gyres and low values within the high-latitude, divergent, cyclonic gyres. It is the phytoplankton that effect this vast rearrangement--they remove the nutrients that converge into the anticyclonic gyres and cause them to sink to greater depths, where they emerge into the lighted zone again within the cyclonic gyres. They really perform quite a task.

As a consequence of these circulations and usages, the surface phosphate (Figure 21) looks much like the gyral systems, with the highs and lows inverted from the salinities. Likewise, the 100-m

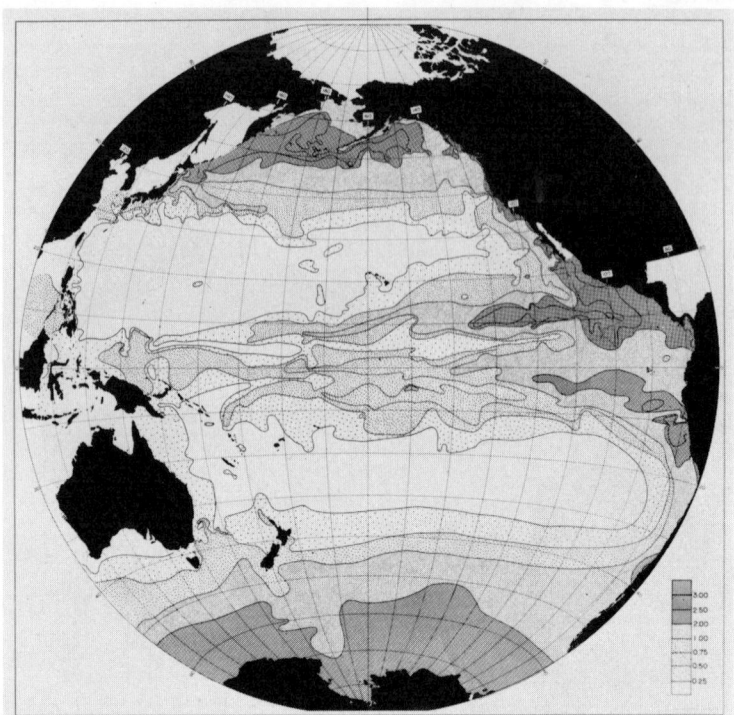

Fig. 22. Distribution of phosphate-phosphorus (µg-at/ℓ) at 100 m in the Pacific Ocean [Reid, 1962, Fig. 3b].

phosphate (Figure 22) describes the convergences and divergences, and hence the circulation, very well, in considerably greater detail than that at the surface. And as a consequence of the high nutrients and high productivity within these domains, the zooplankton volume is high (Figure 23).

This sort of comparison has, of course, been made by *Hentschel and Wattenberg* [1930] for part of the Atlantic. However, they were unable to do the whole Atlantic because of lack of nutrient data and lack of comparable plankton data. The *F.A.O. Department of Fisheries* [1972] has prepared maps of phytoplankton production and zooplankton for the world ocean from the heterogeneous assemblages of data available to them. They bear out, in general, what might be expected from the circulation.

OCEANIC HABITATS DEFINED BY THE CIRCULATION

In an earlier study [*Reid*, 1962] it was suggested that in the Pacific these gyral systems might easily represent different habitats, not only in the sense of being densely or sparsely populated,

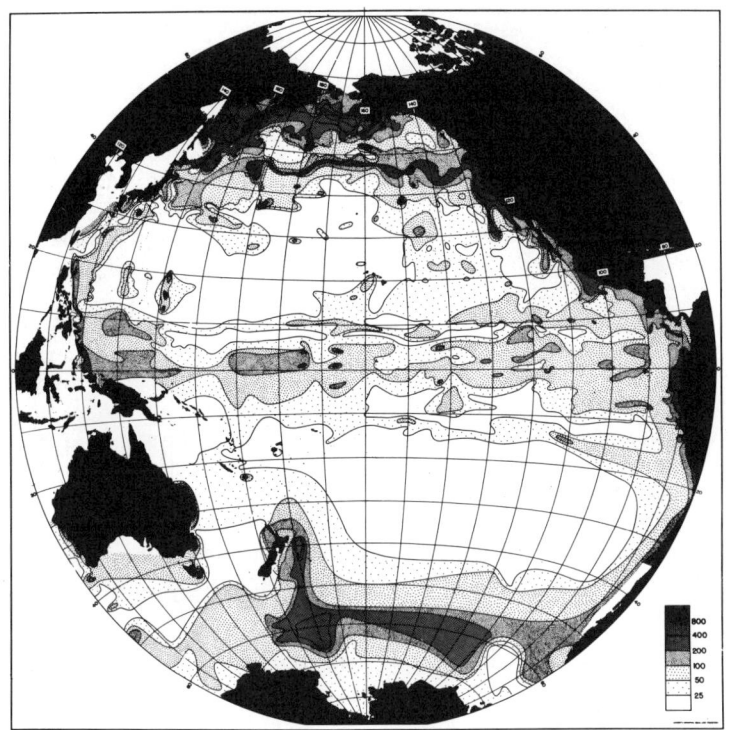

Fig. 23. Distribution of zooplankton volume (parts per 10^9 by volume) in approximately the upper 150 m of the Pacific Ocean [*Reid*, 1962, Fig. 4b].

but also in the kinds of organisms that they contain. One might think of the subarctic gyres as rich but cool and poorly lit in the winter months, the subtropical gyres as poor but warm and well lit, and the equatorial zone as both rich and well lit. In addition, the waters within the gyres are to some extent physically contained-- the high-latitude gyres by the continents, the subtropical by the Ekman convergence, and the equatorial by the system of zonal flows.

This concept has been discussed in greater detail by *McGowan* [1971 and 1974]. Space does not permit a review of his discussion, but it seems well worthwhile to show some actual distributions of various species.

Brinton's [1962] work on *Euphausia pacifica* (Figure 24) and other species (Figure 25) certainly fits the subarctic concept, and *E. brevis* (Figure 26) fits the subtropical anticyclonic gyres, not only in the Pacific but also in the Atlantic and Indian Oceans (Figure 27). *E. diomedeae* (Figure 28) and *Nematoscelis gracilis* (Figure 29) are beautifully equatorial.

Some species suggest a finer division. *E. distinguenda* (Figure 30) seems to inhabit only the richest part of the subequatorial zone, in the east. *E. tenera* (Figure 31) can spread considerably beyond the subequatorial zone. And *N. microps* (Figure 32), *Stylocheiron elongatum* (Figure 33), and *S. maximum* (Figure 34) can inhabit both the subequatorial and subtropical systems and a substantial part of the subarctic domains. From this distribution, one might suppose that they are limited only by low temperature, and thus can extend much farther northward in the North Atlantic, which is much warmer than the North Pacific (Figure 35).

There is a problem with a number of transition species such as *Thysanoessa gregaria* (Figure 36), which seem to be found entirely within the eastward-flowing zone. How these planktonic animals can maintain their distribution is not clear. One suggestion has been made by *Kling* [in press] for certain Radiolaria and may apply to *T. gregaria* and perhaps others: It is that they may lie deeper underneath the warmer water to the south (Figure 37), where the westward flow normally associated with the North Equatorial Current lies much farther north. At the surface the westward flow is south of 25°N, but at 600 m it extends to 30°N and near 1000 m up to 35°N (Figure 38).

There is, of course, a notable variation in these distributions, not only local, as can be seen in Figure 23, but also seasonally and year-to-year. It appears, however, that the gross patterns do not change very much. That is, the anticyclonic gyres appear to be low both in nutrients and zooplankton relative to the cyclonic gyres and the equatorial zone and eastern boundary currents in all seasons and years observed.

In the California Current the zooplankton volume is lower in winter than in summer (Figure 39) in all years observed, yet even at the lowest season it is higher than the waters of the anticyclonic gyres. Likewise, some years are substantially lower in zooplankton volume than other years (Figure 40), but they are still higher than those of the anticyclonic gyres.

The seasonal variation of phosphate may be out of phase with that of zooplankton. In the Northwest Pacific the phosphate values are higher in winter (Figure 41) than in summer (Figure 21), yet in both cases the values are higher than in the central Pacific. And the zooplankton volume is lower in winter (Figure 42) than in summer (Figure 23), but still higher than in the region to the south.

OCEAN CIRCULATION

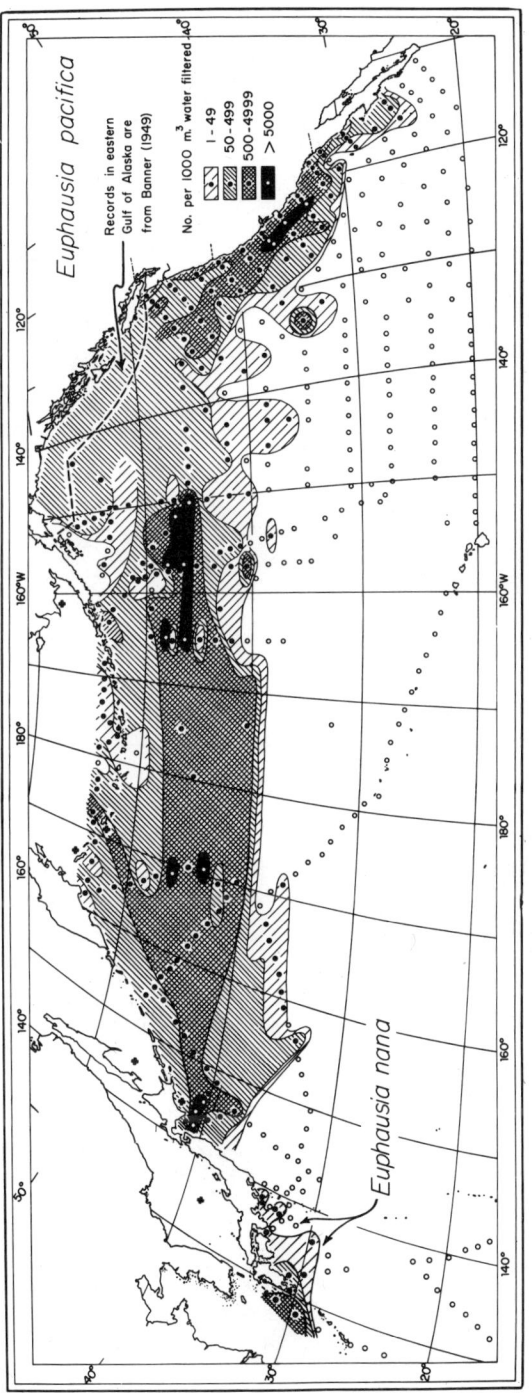

Fig. 24. Geographical distributions of *Euphausia pacifica* and *E. nana* [Brinton, 1962, Fig. 28].

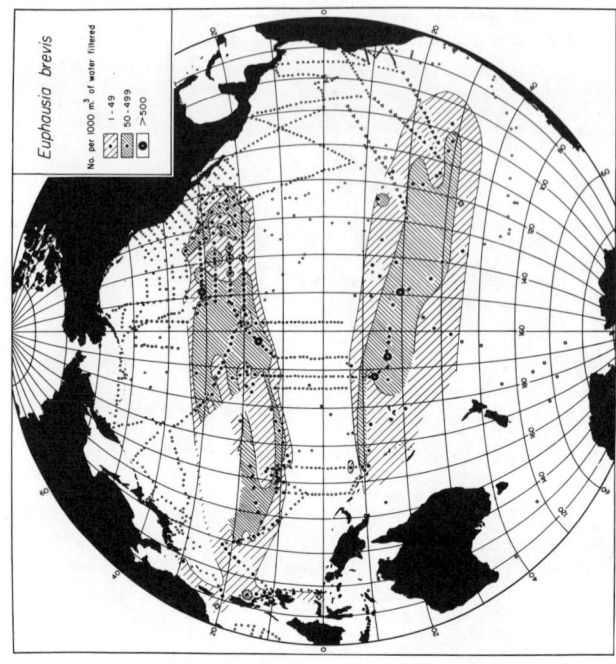

Fig. 26. Geographical distribution of *Euphausia brevis* (Brinton, 1962, Fig. 37).

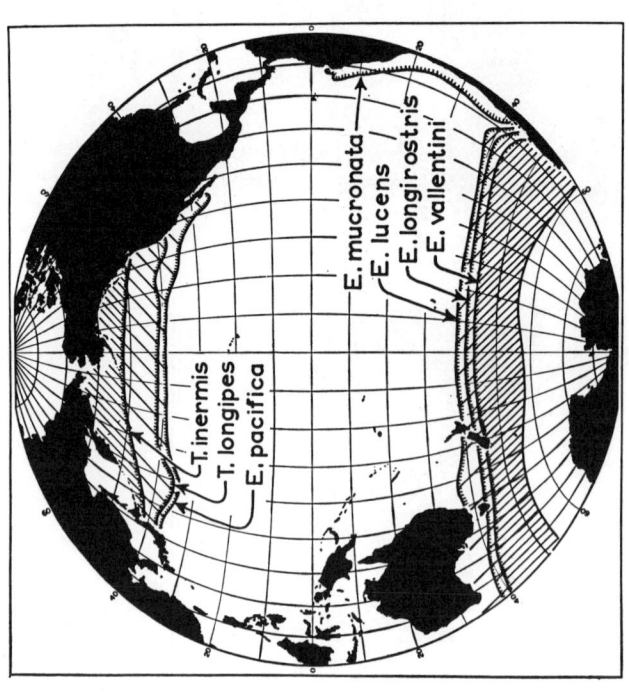

Fig. 25. Distributions of subarctic and subantarctic euphausiid species and *Euphausia mucronata*, a Peru Current species (Brinton, 1962, Fig. 103).

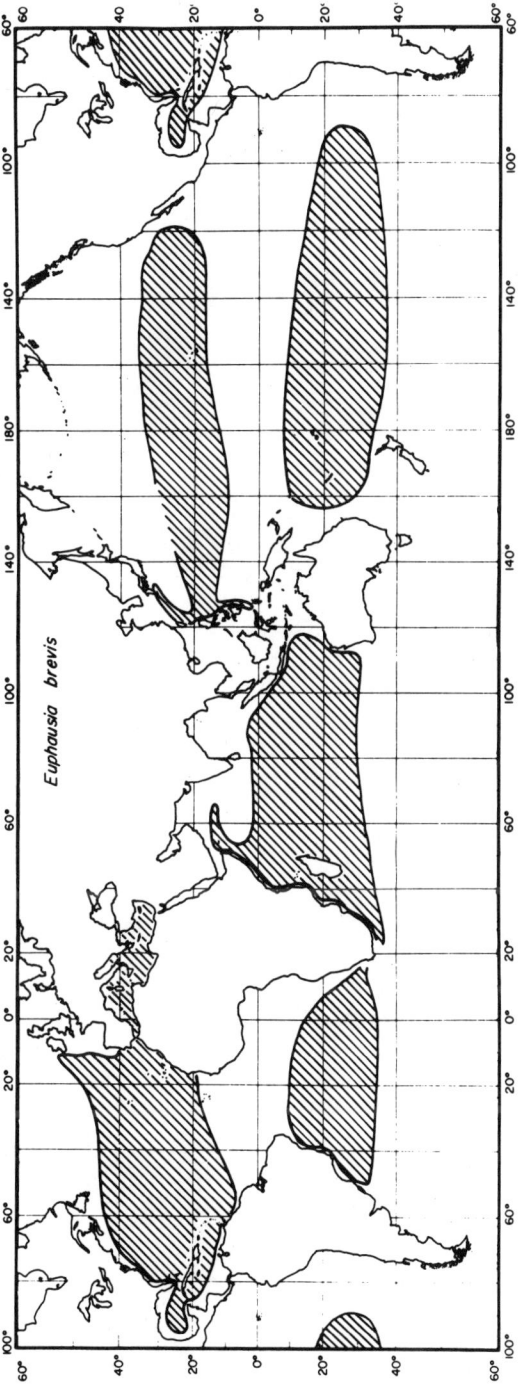

Fig. 27. Geographical distribution in world ocean of *Euphausia brevis* [Brinton, 1975, Fig. 53a].

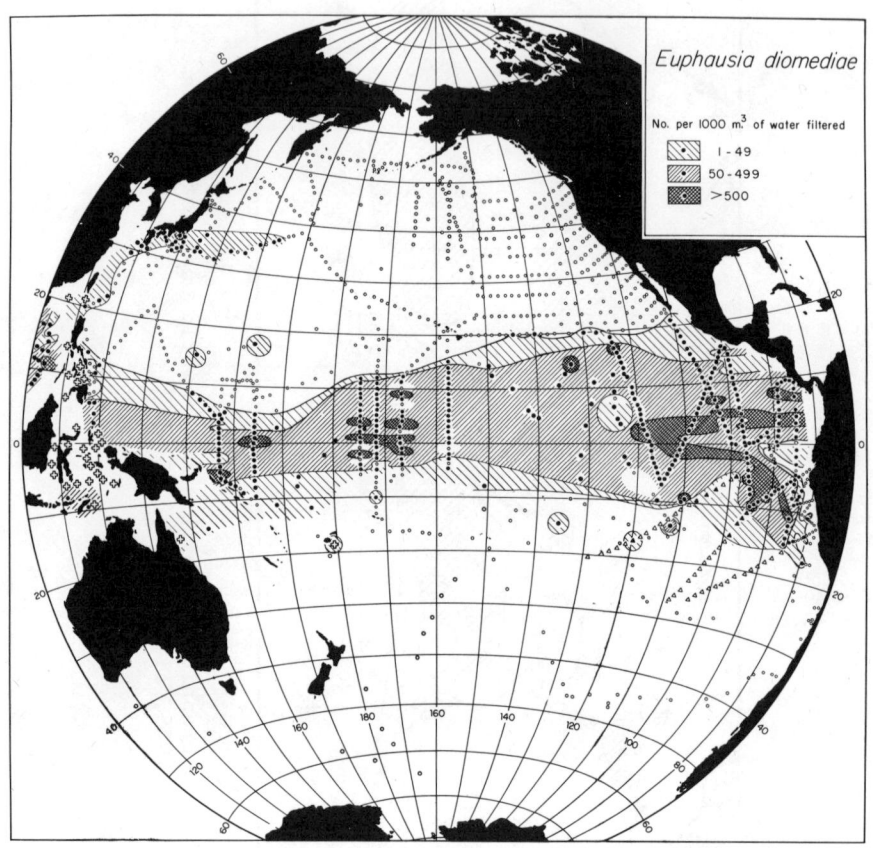

Fig. 28. Geographical distribution of *Euphausia diomedeae* (Brinton, 1962, Fig. 42).

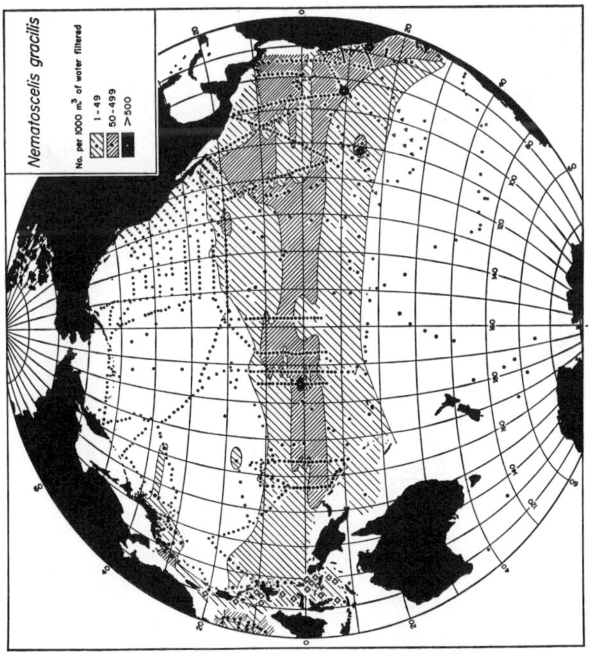

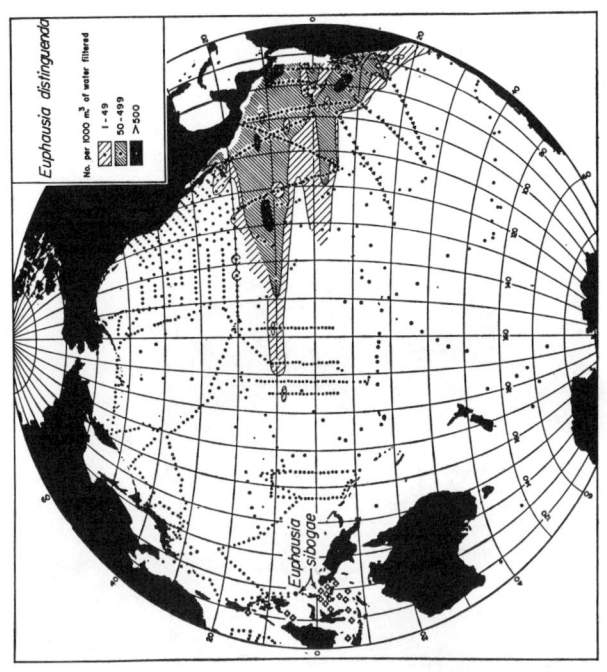

Fig. 29. Geographical distribution of *Nematoscelis gracilis* (Brinton, 1962, Fig. 70).

Fig. 30. Geographical distribution of *Euphausia distinguenda* (Brinton, 1962, Fig. 47).

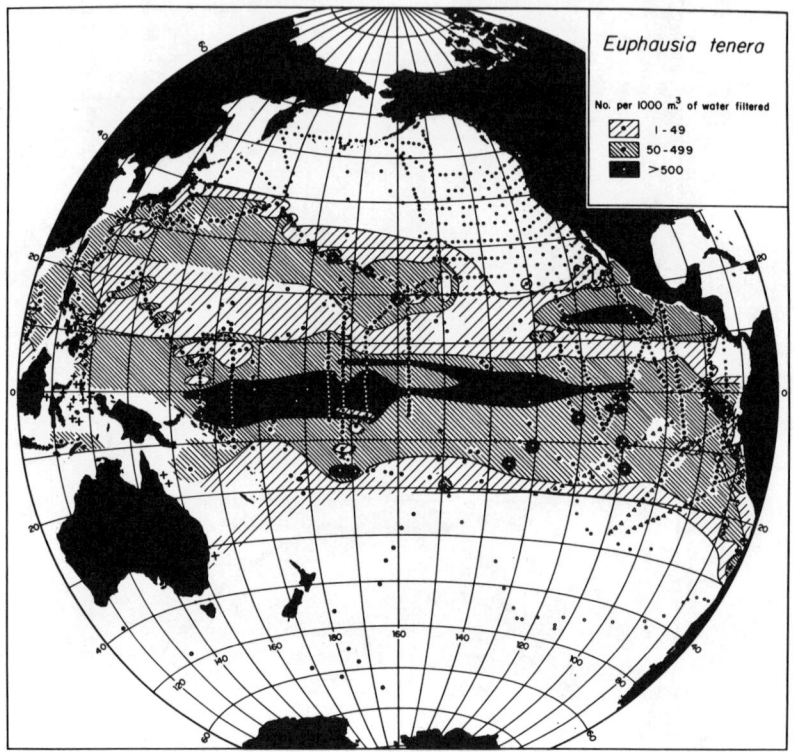

Fig. 31. Geographical distribution of *Euphausia tenera* (*Brinton*, 1962, Fig. 43).

OCEAN CIRCULATION

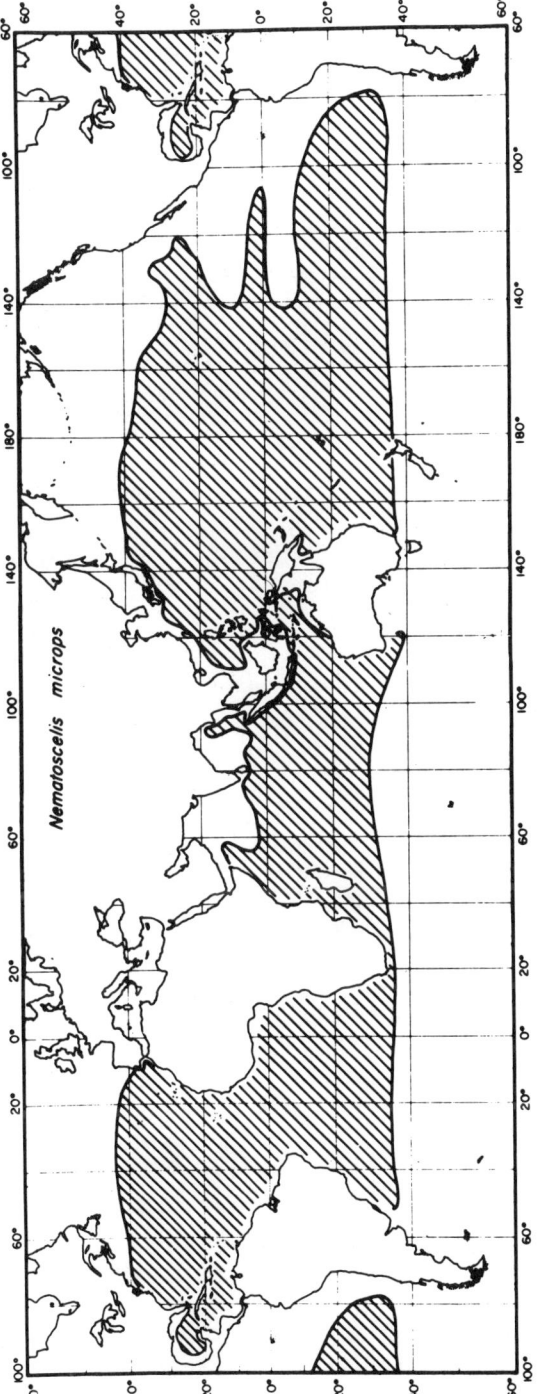

Fig. 32. Geographical distribution in world ocean of *Nematoscelis microps* [Brinton, 1975, Fig. 61a].

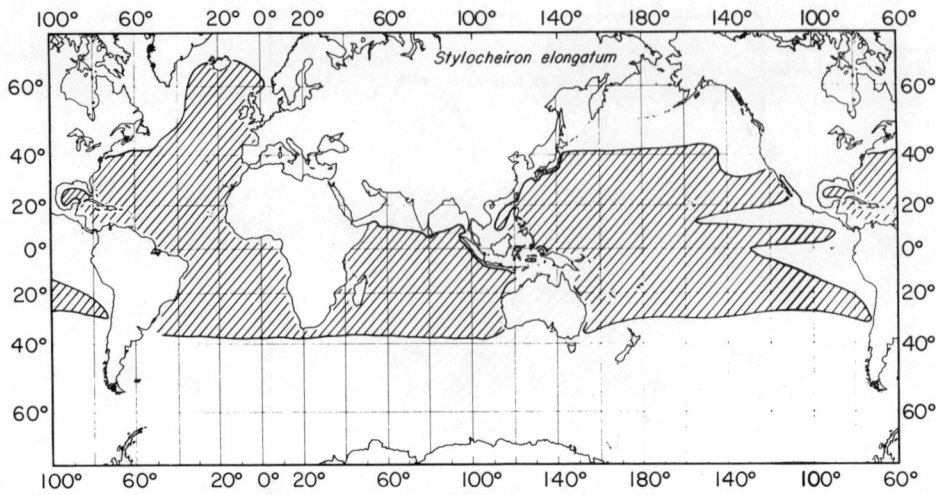

Fig. 33. Geographical distribution in world ocean of *Stylocheiron elongatum* (Brinton, 1975, Fig. 72a).

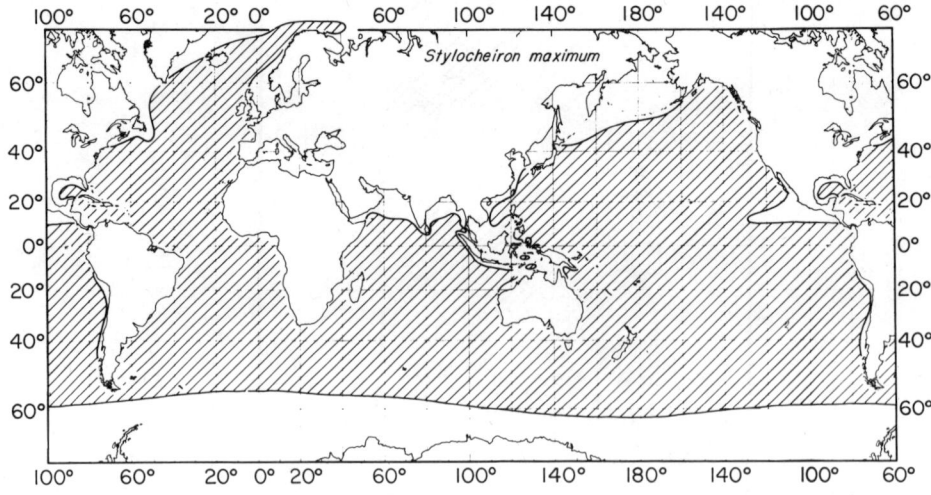

Fig. 34. Geographical distribution in world ocean of *Stylocheiron maximum* (Brinton, 1975, Fig. 74a).

OCEAN CIRCULATION 49

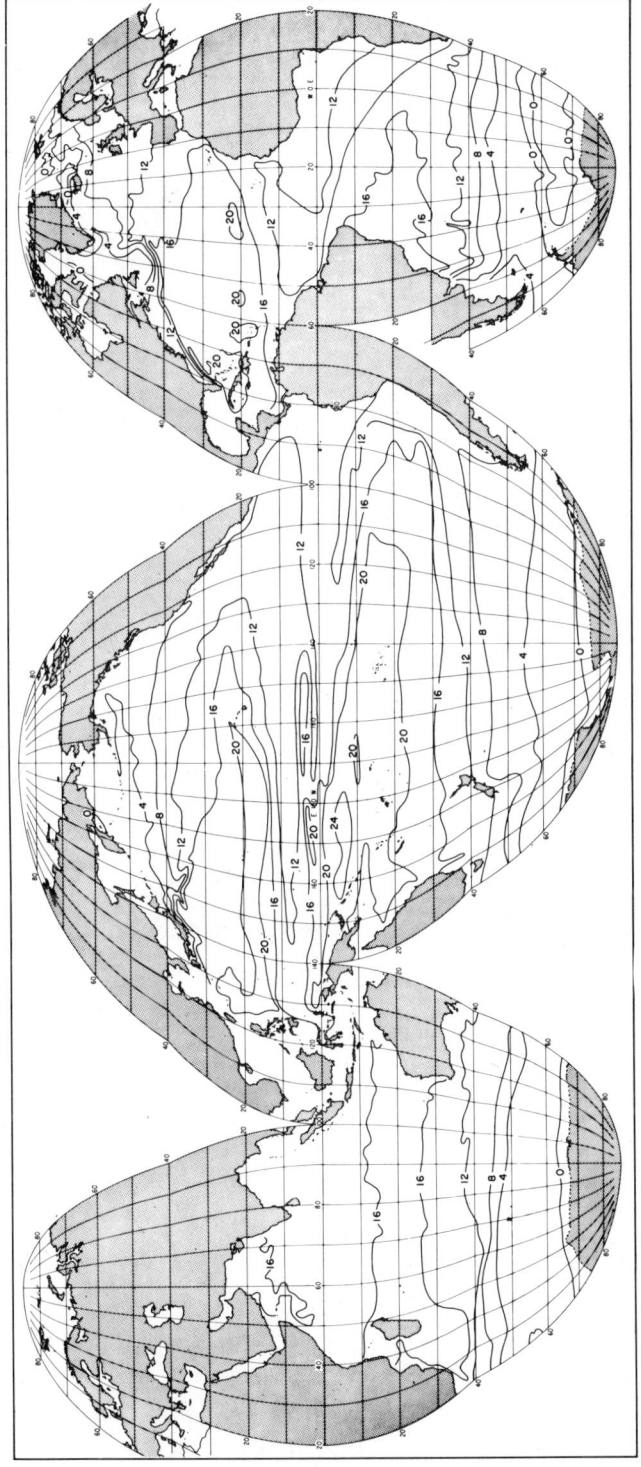

Fig. 35. 200-m temperature (Celsius) in Indian Ocean [*Wyrtki*, 1971], Pacific Ocean, northern winter (from Reid, unpublished), and Atlantic Ocean, northern winter [*Wüst and Defant*, 1936; *Dietrich*, 1969].

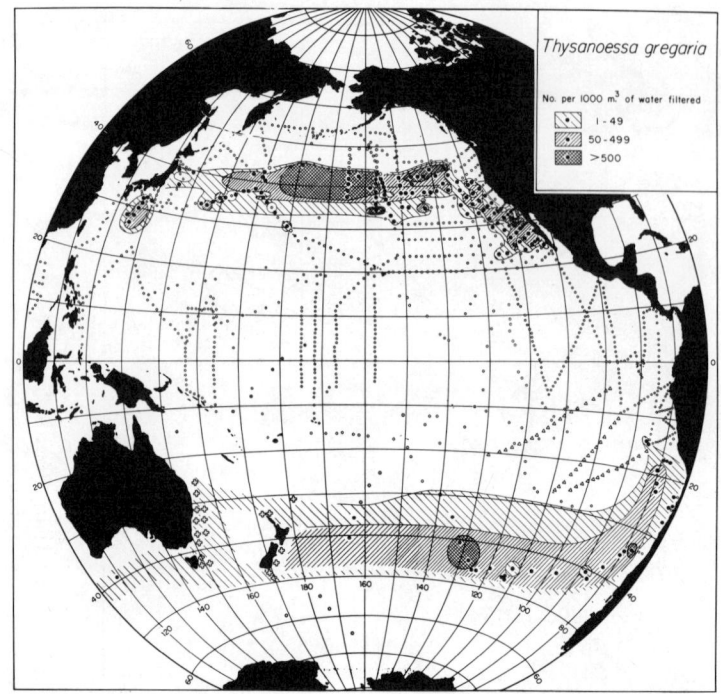

Fig. 36. Geographical distribution of *Thysanoessa gregaria* (*Brinton*, 1962, Fig. 57).

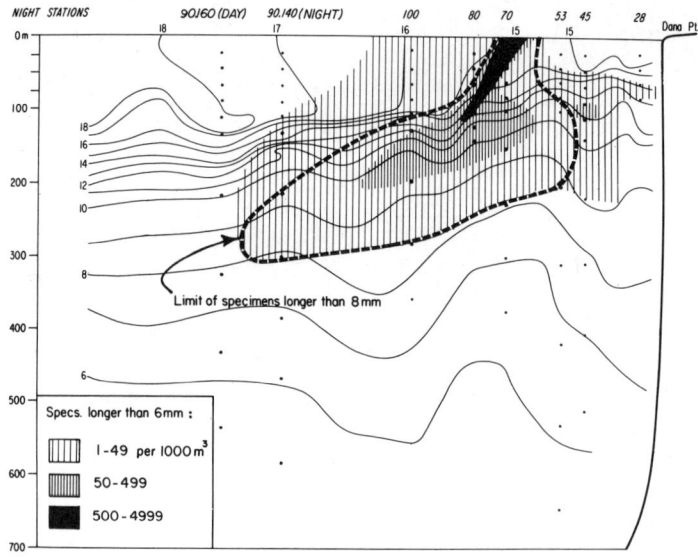

Fig. 37. Observations of *Thysanoessa gregaria* taken during January-February 1964 along a line, 1100 km long, extending 240°T from the coast of California at 33°N (*Brinton*, personal communication).

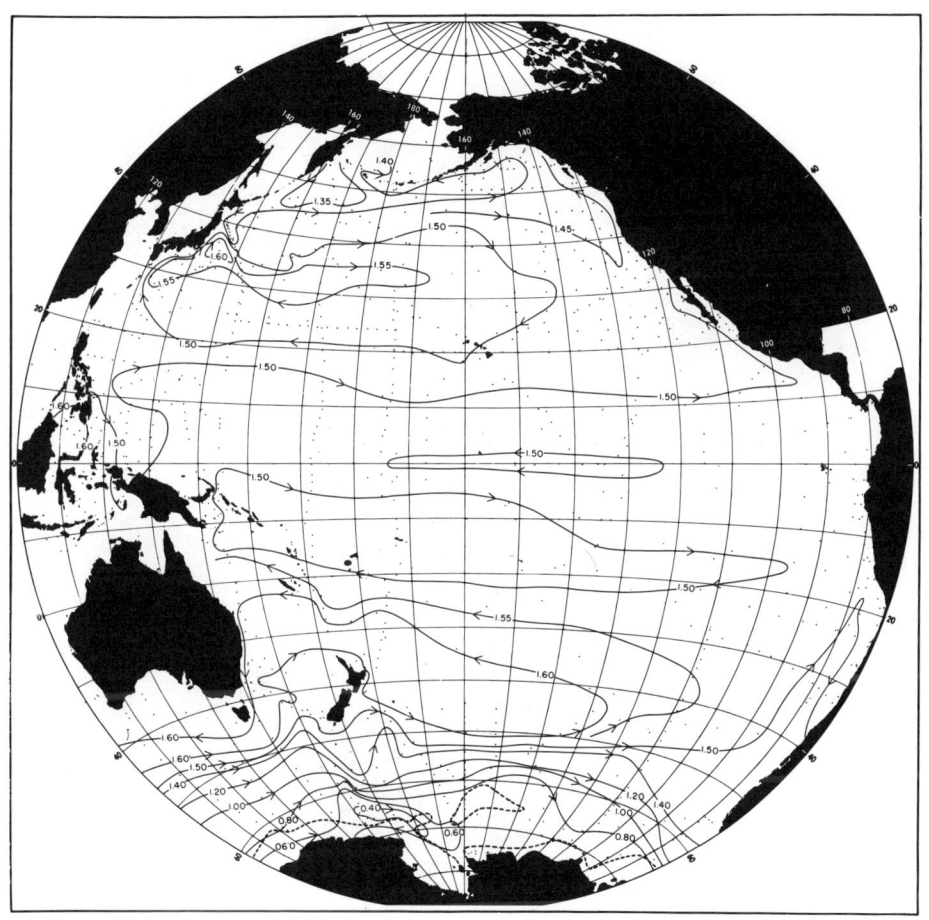

Fig. 38. Acceleration potential on the surface where $\delta_t = 80$ cl/t (*Reid*, 1965, Fig. 23).

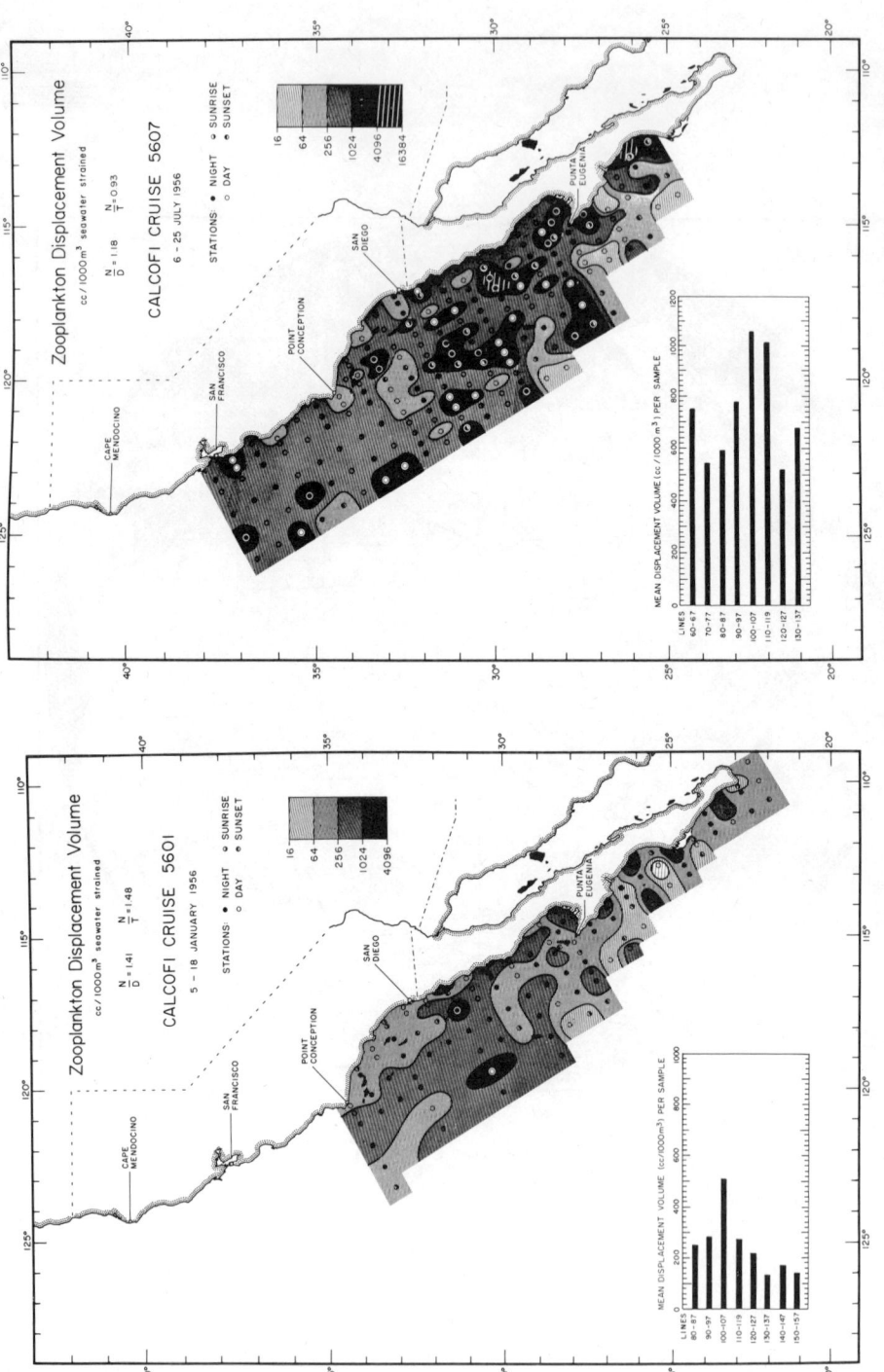

Fig. 39. Zooplankton displacement volume from CalCOFI Cruise 5601 (left) and from CalCOFI Cruise 5607 (right) [*Smith*, 1971, Charts 7 and 82].

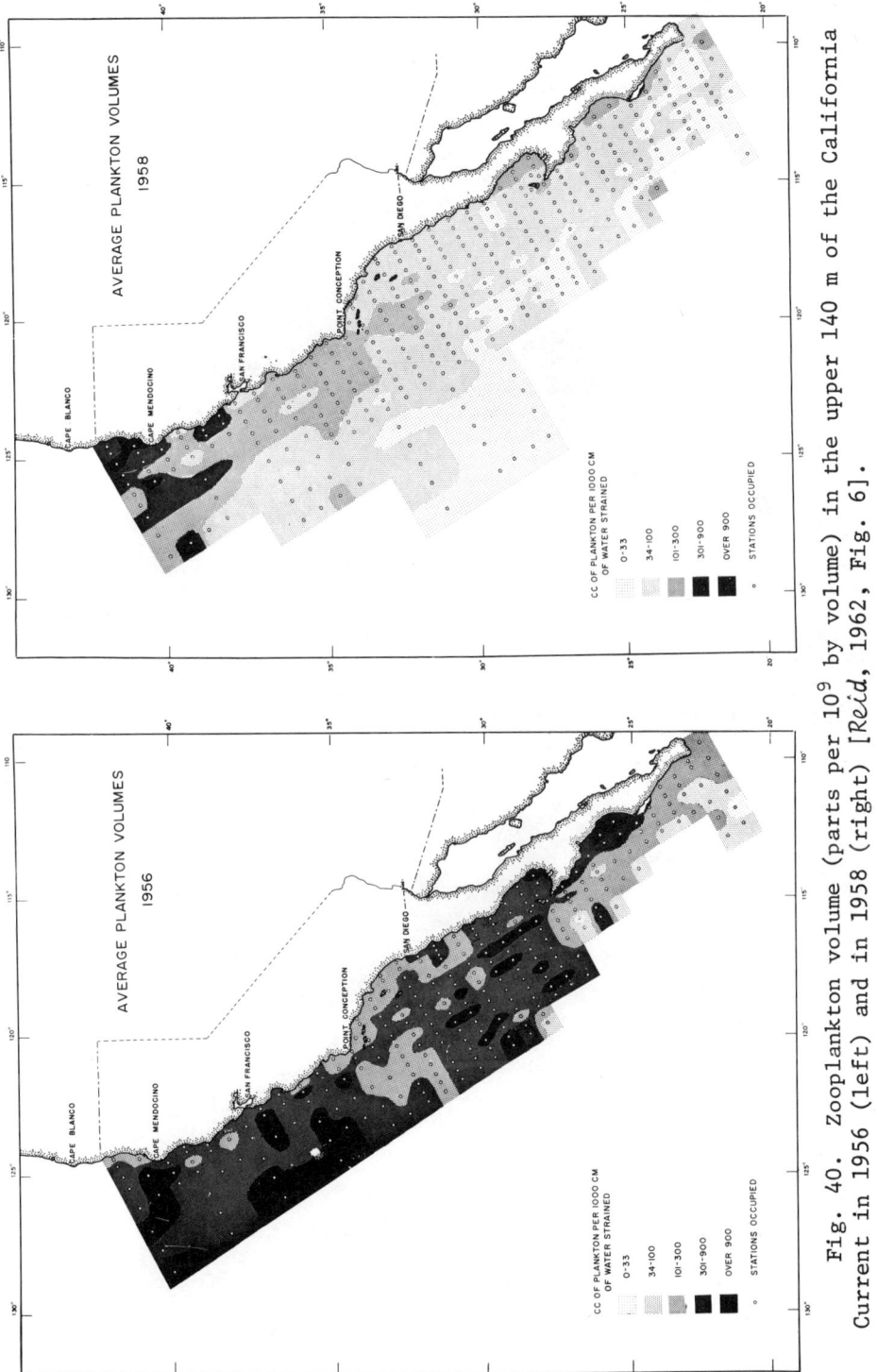

Fig. 40. Zooplankton volume (parts per 10^9 by volume) in the upper 140 m of the California Current in 1956 (left) and in 1958 (right) [*Reid*, 1962, Fig. 6].

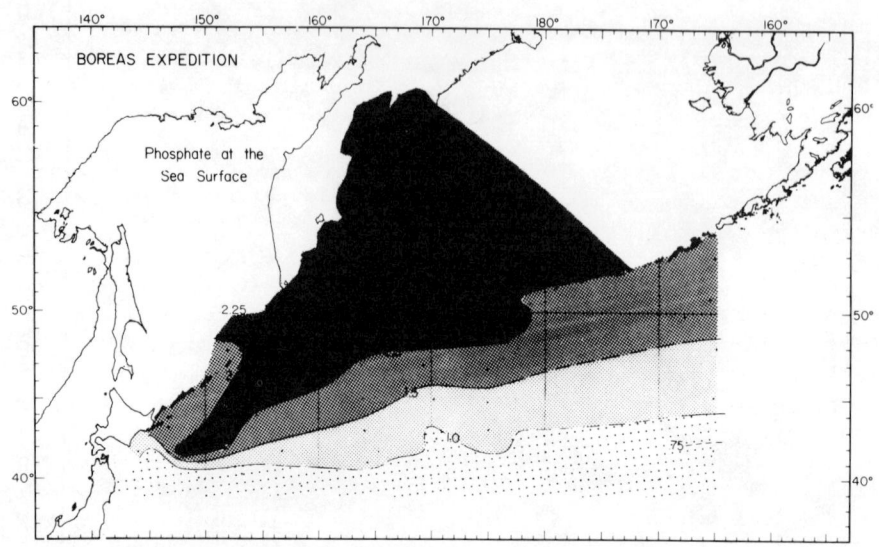

Fig. 41. Phosphate-phosphorus at the sea surface, northwest Pacific Ocean (Reid, 1973, adapted from Fig. 21).

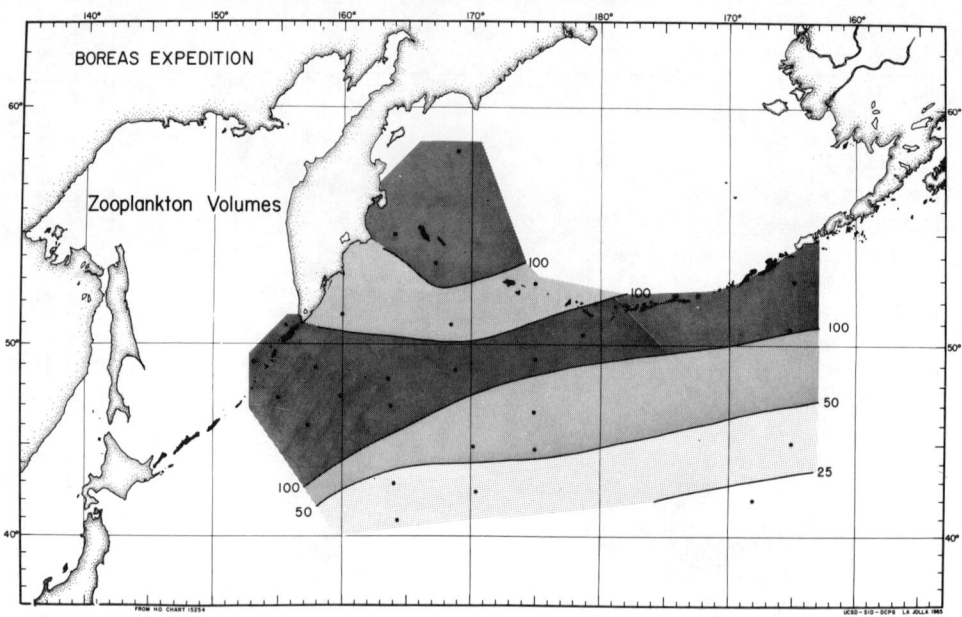

Fig. 42. Zooplankton volumes, northwest Pacific Ocean.

THE DISTRIBUTION OF SOUND-SCATTERERS

It was anticipated when this discussion was initiated that the final stage would be to show a comparable series of maps that defined the abundance and depth of the sound-scatterers. Although it was not expected that a large amount of materials would be found ready at hand in the literature, it was felt that there would be a sufficient amount present to demonstrate at least roughly what is obviously the conjecture being presented: That is, that the plankton domains created by the circulation systems are also the domains of fishes that may act as sound-scatterers; that regions high in nutrients, primary productivity, and zooplankton are also high in numbers of sound-scatterers; perhaps, that the more intense scattering layers may lie nearer the sea surface; that the scattering layers should lie deeper within the anticyclonic gyres, and the fishes there should be less abundant than in the subarctic and subantarctic cyclonic gyres and along the equator.

In the event, not as much has been found as is considered desirable. This may be because there isn't much, or because better sources were not found, or perhaps because it isn't so. It is very difficult to go beyond *Beklemishev's* [1964] study. He points out that the occurrence and number of layers vary with a certain pattern and this pattern is related to the food chain, and that where sound-scattering layers are well developed the productivity is high. While he has subdivided the northern area (Figure 43) into more domains than one might on the basis of the materials shown here, his equatorial domain and southern central domain can be seen clearly, and his subantarctic domains fit the original concept quite well.

His exposition of the data available at that time is meticulous, but the varying nature of the measurements and the complexities of the layers defy any quantitative result. He does find more and thicker scattering-layer material in the richer areas, but does not have the data to determine just how much more. He notes from earlier work that the central areas are known to be poorest and the subarctic and equatorial areas to be richest. In his data he finds scattering layers in the anticyclonic areas less than half the time and near the equator more than half the time. Along the equator itself in the eastern Pacific he finds them all the time. He finds also from his samples that the number of species in the sound-scattering layer in the tropical regions is larger than the number in the subarctic regions, though he does not remark on the differences in biomass.

Let us consider a few of the maps that have been found. Hopefully there are others. *Parin* [1970] has mapped six myctophidae that seem to fit this set of patterns (Figure 44). Number one may be considered one of the subtropicals in the Atlantic that may extend northward into the cyclonic gyre in those warmer waters. Number two is clearly a Pacific subarctic gyre type, and three a subantarctic. Four and five are subtropical gyre types and six a subequatorial type. This may fit the general scheme in terms of areas, but their abundance is not known.

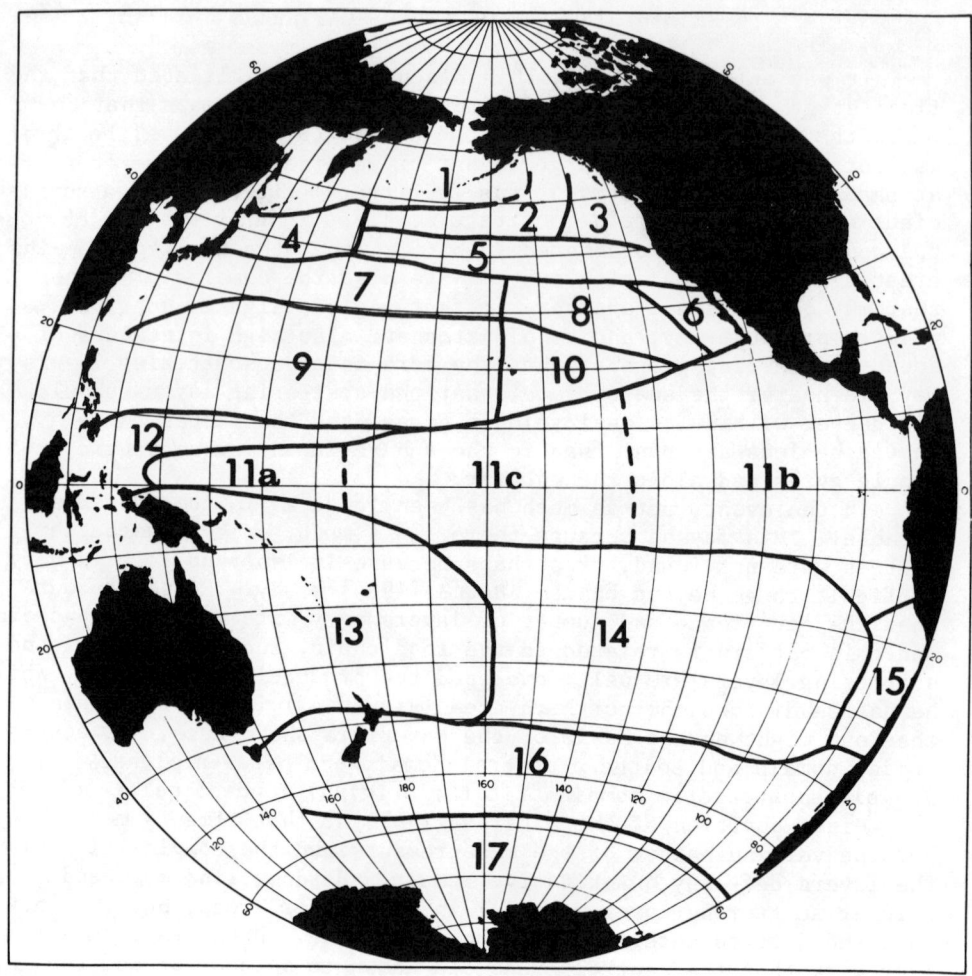

Fig. 43. Areas of the Pacific Ocean where averaging of data is based by considering the sound-scattering layers (from *Beklemishev*, 1964, Fig. 2; translation, 1967):

1. Subarctic area
2. Aleutian Current area
3. Alaskan Current area
4. Oyashio and Kuroshio areas
5. N. Pacific Current area
6. California area
7. NW subtropical area
8. NE subtropical area
9. NW tropical area
10. NE tropical area
11a. Equatorial area - western part
11b. Equatorial area - eastern part
11c. Equatorial area - central part
12. Transitional area between the Equatorial and Tropical areas
13. SW Central area
14. SE Central area
15. Peru Central area
16. Subantarctic area
17. Antarctic area

OCEAN CIRCULATION

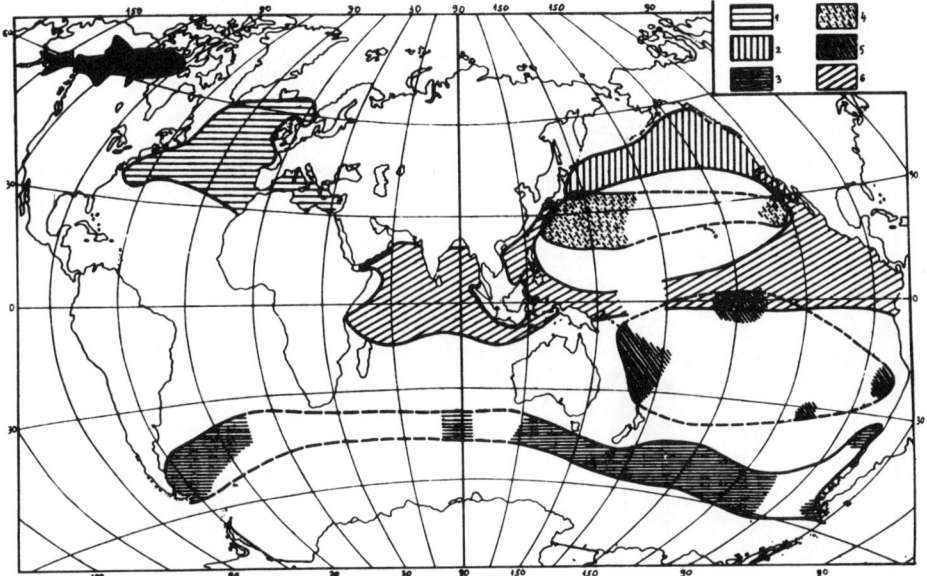

Fig. 44. Geographical distribution of six Myctophidae in the world ocean [Parin, 1968, Fig. 29].

Backus et al. [1970, 1971] show what may be a subarctic type in the Atlantic, and provide abundance also (Figure 45). They give some numbers for various species that they characterize as tropical, broadly tropical, Sargasso Sea, Amazonian and Caribbean, and "northern pattern". They have a "widespread" category which they note is more numerous in the highly productive waters. (Note also the distributions of various species presented in the contribution of Backus and Craddock.)

In the California Current we can see the subtropical *Vinciguerria lucetia* larvae (Figure 46) impinging into the coastal regions and the subarctic *Leuroglossus stilbius* larvae (Figure 47) extending southward. However, total abundance of sound-scatterers is not available to plot on a broad scale.

As for the depth of the scattering layer, the only large-area map that was found was that prepared by Haigh [1971] (Figure 48) for the 1970 symposium. The depth of his layer D map has been compared with the surface flow of the Atlantic (Figure 49), and some resemblance has been found. His layer lies deepest within the subtropical anticyclonic gyre, but this layer, at least as he defines it from his material, disappears in the Sargasso Sea. And it lies shallowest along the left-hand edge of the North Atlantic Current, or along the axis of the subarctic cyclonic gyre, and deeper again around the perimeter.

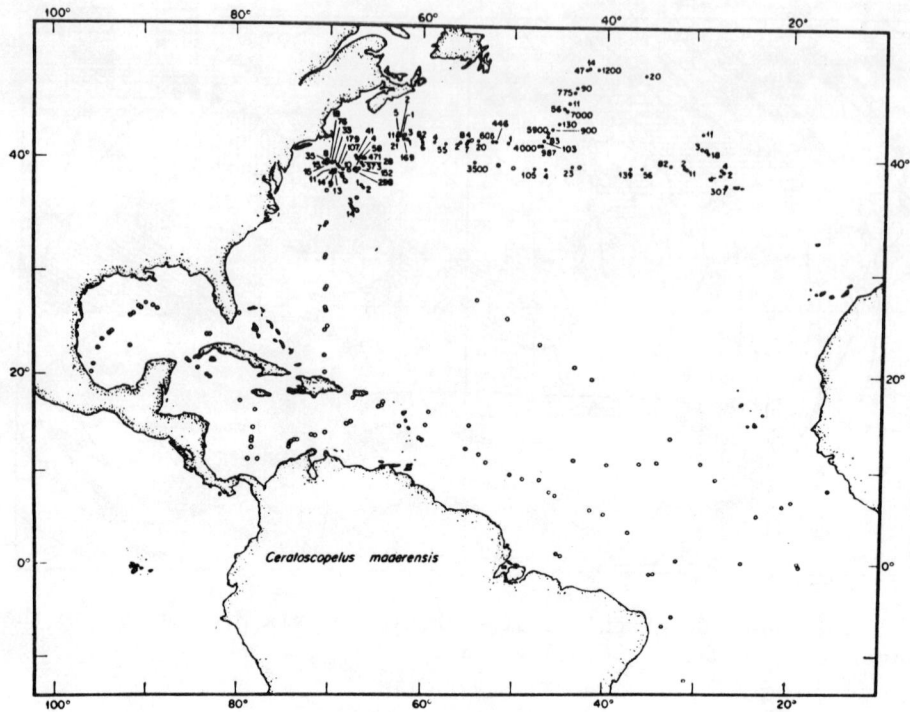

Fig. 45. Distribution of *Ceratoscopelus maderensis*, showing the northern pattern (northern edge of Gulf Stream forms southern limit in west) (Backus et al., 1971, Fig. 13).

CONCLUSION

An attempt has been made to show briefly the relation of the sound speed field to the geostrophically balanced nature of the major circulation systems of the ocean, and an attempt has also been made to show the effects of the flow, including the convergences and divergences, upon the nutrient field, and hence the distribution of biomass. The data available appears to support these general concepts. However, the next and most critical conjecture--that these concepts imply that the sound-scattering fishes are distributed according to this system of circulation--is only partly supported by the available data on fish distribution and scattering-layers. These data do not refute this concept, but those that have been located up to this point are not sufficient to make the point clear. Studies of the sort reported by Backus and Craddock in this publication lend major support, however. Perhaps in the near future enough materials will become available to provide a clear answer.

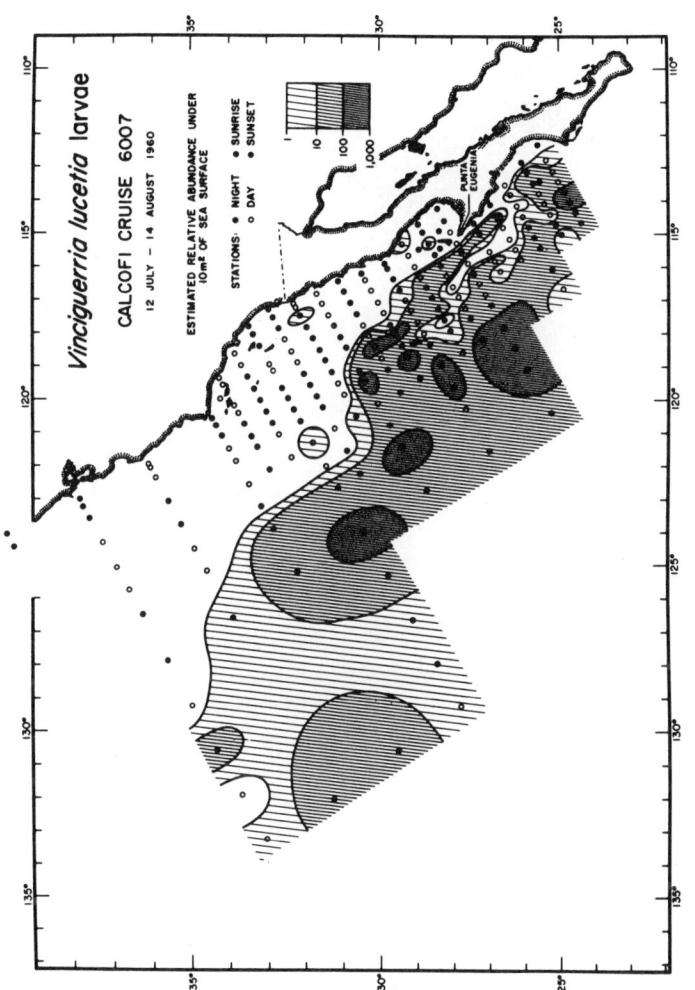

Fig. 46. Estimated relative abundance under 10 m² of sea surface of *Vinciguerria lucetia* larvae, CalCOFI Cruise 6007 [*Ahlstrom*, 1972, Chart 43].

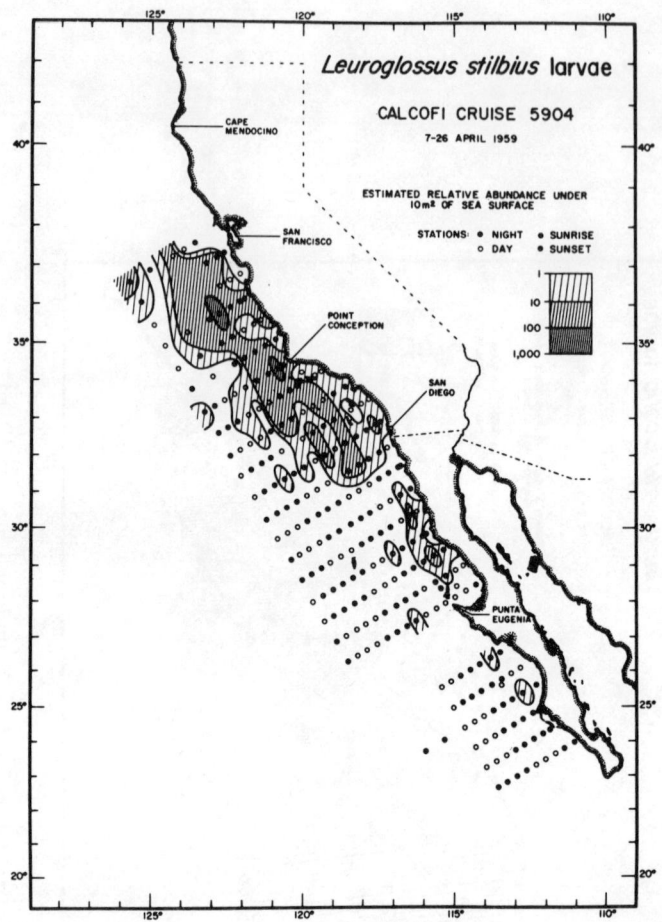

Fig. 47. Estimated relative abundance under 10 m^2 of sea surface of *Leuroglossus stilbius* larvae, CalCOFI Cruise 5904 [Ahlstrom, 1972, Chart 196].

Meanwhile, it appears that the concept provides a framework for examination not only of the fish distributions but also of the sound-scatterers, their areal and vertical distributions and their relative abundance within the various oceanic domains.

ACKNOWLEDGMENTS

In the preparation of this manuscript I have had many useful

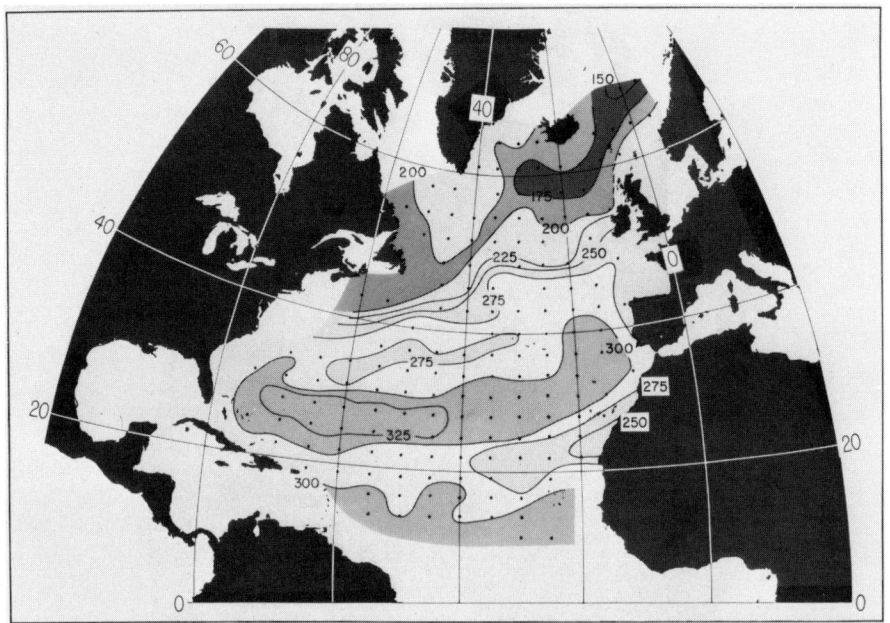

Fig. 48. Maximum midlayer depth of layer D (*Haigh*, 1971, adapted from Fig. 7).

discussions with John A. McGowan, Edward Brinton, and Abraham Fleminger. They have pointed out many features of the plankton distributions and many concepts that are unknown, unfortunately, to most physical oceanographers. In the illustrations, I have drawn very heavily upon the work of Edward Brinton, who has kindly made available many of his original maps. I am particularly grateful to John McGowan for discussions of the work he has done in the central gyre, which goes much farther toward justifying the concepts mentioned here than it would have been possible to do from the physical oceanographic data alone. The work reported here was supported by the Office of Naval Research. I thank the University of California Press for permission to reprint Figures 24-26, 28-34, and 35, and I thank the University of Chicago Press for permission to use the Goode's homolosine equal-area projection on which Figure 35 is prepared (copyright by the University of Chicago Department of Geography).

Note: After the Asilomar Symposium and during the final preparation of this manuscript, my attention was called to a paper by *Chapman et al.* (1974) which deals with the geographical variation of sound-scattering with much more recent and better data than most

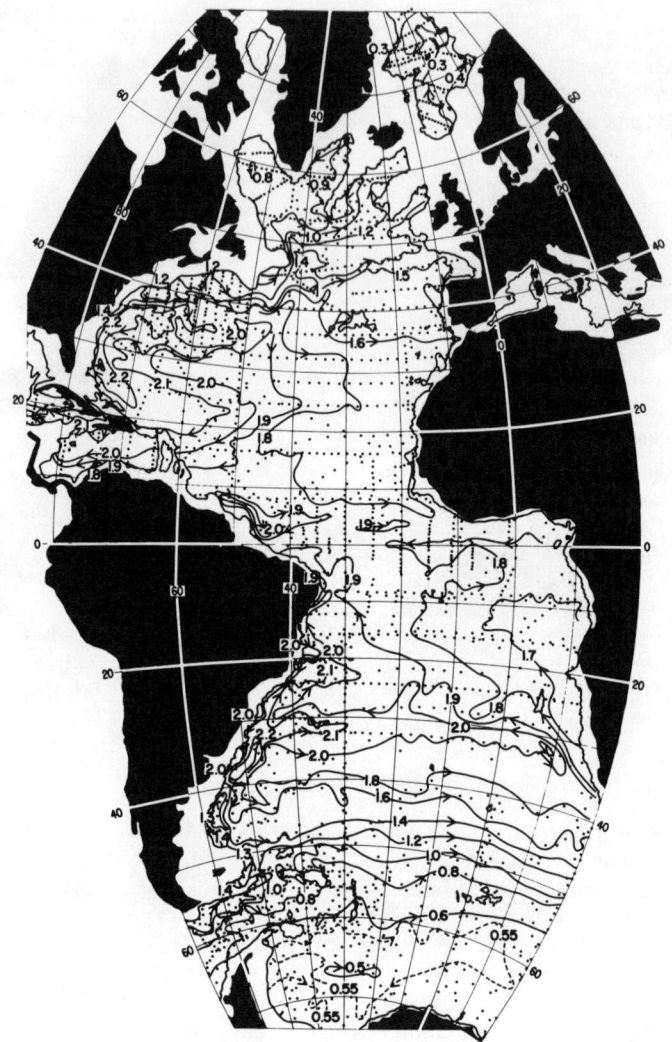

Fig. 49. Geopotential anomaly in dynamic meters (10 J/kg) at the sea surface relative to the 2000-decibar surface, Atlantic Ocean.

of the investigations I had been able to locate. They discuss reverberation conditions in a wide range of ocean areas and point out dramatic changes near known oceanic and faunal boundaries. I regret that I did not know of this before so that I could make use of it in my discussion.

REFERENCES

Ahlstrom, E. H. 1972. Distributional atlas of fish larvae in the California current region: Six common mesopelagic fishes--*Vinciguerria lucetia, Triphoturus mexicanus, Stenobrachius leucopsarus, Leuroglossus stilbius, Bathylagus wesethi,* and *Bathylagus ochotensis,* 1955 through 1960, *CalCOFI Atlas No. 17,* i-xv, plates 1-306.

Backus, R. H., J. E. Craddock, R. L. Haedrich, and D. L. Shores. 1970. The distribution of mesopelagic fishes in the equatorial and western North Atlantic Ocean. *J. Mar. Res.*, 28, 179-201.

Backus, R. H., J. E. Craddock, R. L. Haedrich, and D. L. Shores. 1971. The distribution of mesopelagic fishes in the equatorial and western North Atlantic Ocean. *Proceedings of an International Symposium on Biological Sound Scattering in the Ocean,* 20-40, Maury Center for Ocean Science, Washington, D. C.

Beklemishev, K. V. 1964. Echo-sounding records of macroplankton concentrations and their distribution in the Pacific Ocean. *Trud. Inst. Okeanol.*, 65, 197-229. Translated in 1967 by M. Slessers, U.S. Naval Oceanographic Office, Transl. No. 343.

Berger, W. H. 1970. Fractionation by deep-sea circulation. *Bull. Geol. Soc. Amer.*, 81, 1385-1402.

Brinton, E. 1962. The distribution of Pacific euphausiids. *Bull. Scripps Inst. Oceanogr.*, 8, 51-270.

Brinton, E. 1975. Euphausiids of southeast Asian waters. *Naga Rep.*, 4, Pt. 5, 287 pp.

Chapman, R. P., O. Z. Bluy, R. H. Adlington, and A. E. Robison. 1974. Deep scattering layer spectra in the Atlantic and Pacific Oceans and adjacent seas. *J. Acoust. Soc. Amer.*, 56, 1722-1734.

Defant, A. 1941. Die relative Topographie einzelner Druckflachen in Atlantischen Ozean. *Wiss. Ergebn. dtsch. atlant. Exped. "Meteor"*, 6, 183-190.

Dietrich, G. 1969. *Atlas of the Hydrography of the Northern North Atlantic Ocean.* Charlottenlund Slot, Conseil International pour l'Exploration de la Mer - Service Hydrographique, 140 pp.

FAO Department of Fisheries. 1972. *Atlas of the Living Resources of the Seas.* Food and Agriculture Organization of the United Nations, Rome, i-vii, 1-12, 62 plates, 13-19.

Fenner, D. F., and P. J. Bucca. 1971. The sound velocity structure of the North Atlantic Ocean. *Naval Oceanographic Office Informal Rep.*, No. 71-13, 86 pp.

Haigh, K. K. R. 1971. Geographic, seasonal, and annual patterns of midwater scatterers between latitudes 10° and 68° north in the Atlantic. *Proceedings of an International Symposium on Biological Sound Scattering in the Ocean,* pp. 268-280.

Hentschel, E., and H. Wattenberg. 1930. Plankton und Phosphat in der Oberflächenschicht des Südatlantischen Ozeans. *Ann. Hydrogr. Mar. Meteor.*, 58, 273-277.

Johnson, R. H., and R. A. Norris. 1968. Geographic variation of Sofar speed and axis depth in the Pacific Ocean. *J. Geophys. Res.*, 73, 4695-4700.

Kling, S. A. Relation of radiolarian distributions to subsurface hydrography in the North Pacific. *Deep-Sea Res.*, *in press*.

McGowan, J. A. 1971. Oceanic biogeography of the Pacific, in *The Micropaleontology of Oceans*, edited by B. M. Funnell and W. R. Riedel, 3-74, Cambridge University Press.

McGowan, J. A. 1974. The nature of oceanic ecosystems, in *The Biology of the Oceanic Pacific* (Proceedings of the 33rd Annual Biology Colloquium, 1972, Oregon State Univ.), 9-28, Oregon State Univ. Press, Corvallis.

Parin, N. V. 1968. *Ichthyofauna of the Epipelagic Zone*, Izdatel' stvo "Nauka", Moscow.

Reid, J. L., Jr. 1962. On the circulation, phosphate-phosphorus content and zooplankton volumes in the upper part of the Pacific Ocean. *Limnol. & Oceanogr.* 7, 287-306.

Reid, J. L., Jr. 1965. Intermediate waters of the Pacific Ocean. *Johns Hopkins Oceanogr. Stud.*, 2, 85 pp., 32 figs.

Reid, J. L. 1973. Northwest Pacific Ocean waters in winter. *Johns Hopkins Oceanogr. Stud.*, 5, 96 pp.

Reid, J. L., and R. J. Lynn. 1971. On the influence of the Norwegian-Greenland and Weddell Seas upon the bottom waters of the Indian and Pacific Oceans. *Deep-Sea Res.*, 18, 1063-1088.

Reid, J. L., and R. S. Arthur. 1975. Interpretation of maps of geopotential anomaly for the deep Pacific Ocean. *J. Mar. Res.*, 33 (Suppl.), 37-52.

Smith, P. E. 1971. Distributional atlas of zooplankton volume in the California current region, 1951 through 1966. *CalCOFI Atlas No. 13*, i-xvi, plates 1-144.

Wüst, G., and A. Defant. 1936. Atlas zur Schichtung und Zirkulation des Atlantischen Ozeans. Schnitte und Karten von Temperatur, Salzgehalt und Dichte. *Wiss. Ergebn. dtsch. atlant. Exped. "Meteor", 1925-1927*, 6 (Atlas), Beilagen I-CIII.

Wyrtki, K. 1971. *Oceanographic Atlas of the International Indian Ocean Expedition*, National Science Foundation, Washington, D. C., 531 pp.

The Chemistry of Plankton Production and Decomposition in Seawater

David Dyrssen

University of Gothenburg, Sweden

ABSTRACT

A model is presented for associating inorganic constituent concentrations to the production and decomposition of plankton. The variation in seawater pH caused by plankton growth and decay is also discussed. It is concluded that refertilization of the photic zone by upwelling may not be a sufficient chemical condition for a plankton bloom. A suggestion to increase studies on trace organics which may inhibit or promote production is made.

FACTORS AFFECTING PLANKTON PRODUCTION

The Supply of Inorganic Matter

In growing plankton, inorganic matter is transferred through cell membranes to form organic tissue and cell liquid. Inorganic carbonate is never limiting for the growth of phytoplankton in seawater since no more than 15% of that available is used. The limiting substances are nitrate and phosphate, and silicate for the plankton that form biogenic opal. Trace metals seem only to be limiting in nutrient-enriched coastal waters. On the contrary, the copper concentration in seawater may be high enough to inhibit plankton growth (*Steemann Nielsen*, 1975).

The biochemical role of some of the trace metals (such as titanium and aluminum) found in phytoplankton (*Brewer*, 1975, p. 483) is obscure and their presence may be due to metal hydroxides adsorbed on the surface of the planktonic matter. This also seems to be the case with cadmium if it is introduced in a plankton bloom, such as in the situation with the CEPEX bag experiment in Saanich Inlet, Victoria, Canada (*Kremling*, 1975).

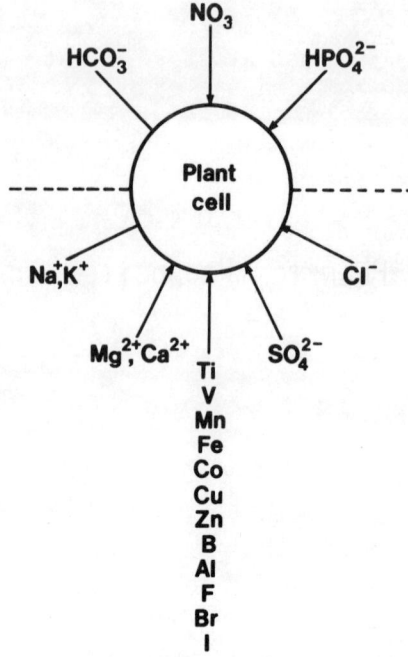

Fig. 1. Inorganic substances that have to be transported into a growing plankton cell for the formation of soft tissue and cell liquid.

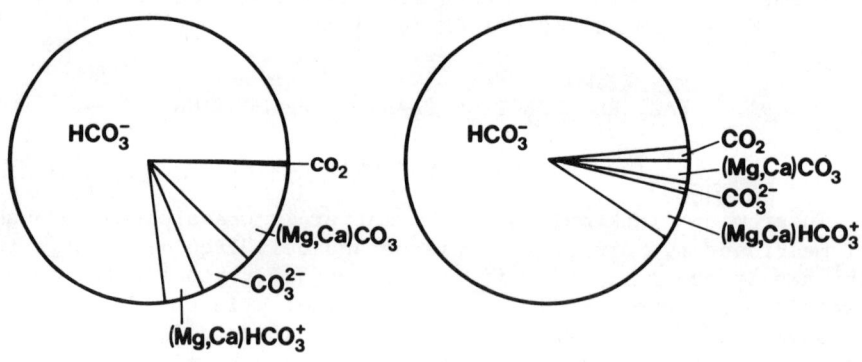

Fig. 2. Carbonate speciation in the upper production zone (left) of the sea calculated with a total carbonate concentration of 2.000 mM and carbonate alkalinity of 2.350 mM giving a pH of 8,280, and in the lower regeneration zone (right) with a total carbonate concentration of 2.350 mM, a carbonate alkalinity of 2.400 mM, and pH of 7.856.

Membrane Transport

The membrane transport of uncharged species may take place through diffusion, especially if the species has a lipophilic character like CO_2. It is therefore reasonable to believe that the plankton assimilate CO_2 instead of HCO_3^-, in spite of the fact that the latter is the dominating inorganic species in seawater.

The enzyme carbonic anhydrase accelerates the nucleophilic attach of oxygen on CO_2 to form carbonate ions as well as the reverse reaction (*Lindskög et al.*, 1971; *Campbell and Dwek*, 1975). It seems that most phytoplankton make use of this zinc enzyme. However, in coccolithophorida (the oldest form of plankton) it seems as if the carbohydrates are formed through membrane transport of HCO_3^- ions (*Paasche*, 1964). The alkali formed by the reaction

$$HCO_3^- + H_2O \rightarrow CH_2O + O_2 + OH^- \tag{1}$$

is accommodated by the formation of calcium carbonate, which can be regarded as a slag product.

$$Ca^{2+} + HCO_3^- + OH^- \rightarrow CaCO_3(s) + H_2O \tag{2}$$

In general, it appears that ions are transported through biological membranes with the aid of carriers. Thus, it is not biologically correct to visualize the membranes as filters with small holes. A dialysis cell made of tubing with a pore diameter of 4.8 nm (*Benes and Steinnes*, 1974) may be useful for differentiating ionic and particulate matter, but it cannot serve as a model for biological membranes. Such cells may even release enzymes to digest particulate matter.

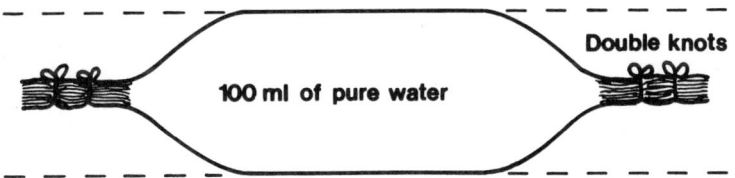

Fig. 3. Dialysis cell made of a 20-cm length of 1.5 in Visking tubing, the mean pore diameter of which is 4.8 nm.

The Supply of Light

Water of course is only a limiting factor for photosynthesis on land. Additionally, temperature effects are more serious on land.

Shadowing effects on sunlight are also quite different on land than in seawater, even if the latitude and cloud effects are the same (or similar). In nutrient-poor clear seawater, the light is attenuated to such a degree that it is not sufficient to maintain a photosynthetic crop below 150 m (the compensation depth). Thus nutrients released in the regeneration zone below this depth are of no use unless upwelling occurs. In nutrient-rich waters, the production itself will have a shadowing effect and the compensation depth drops to 30 m, or even 10 m in coastal areas, such as in the North Sea region. A similar shadowing effect should occur in upwelling areas.

The Regeneration Zone

When light is not sufficient to maintain the phytoplankton production, descending particles will disintegrate by oxidation of the soft tissues (mainly catalyzed by enzymes supplied by bacteria) and dissolution of the hard parts (calcium carbonate and silica). In

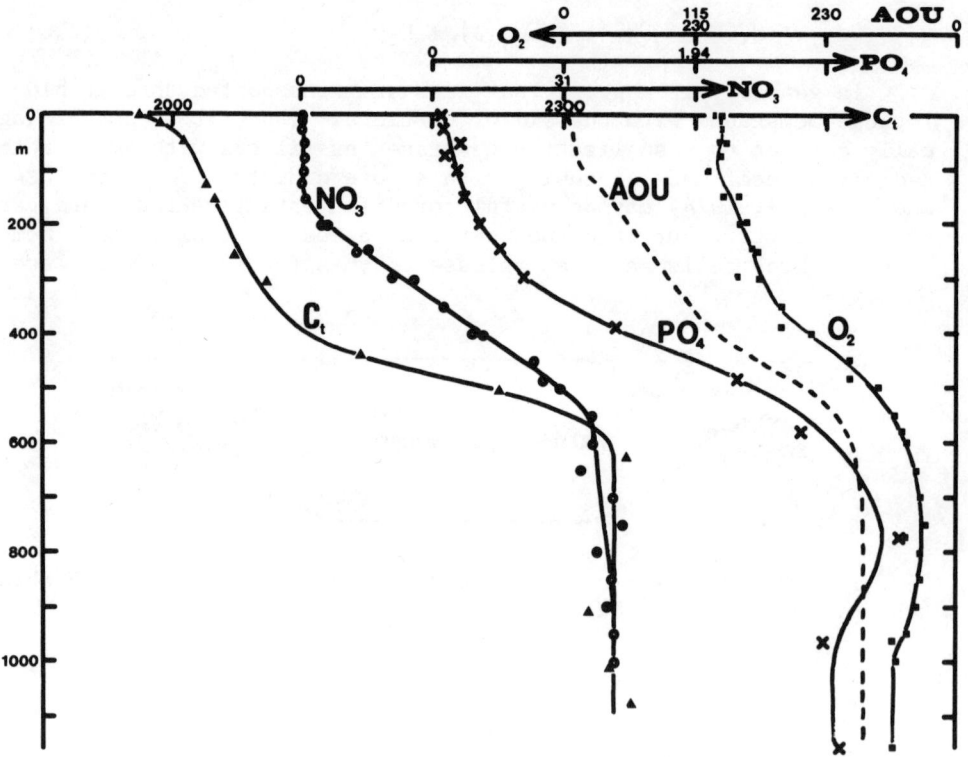

Fig. 4. Depth profiles for total carbonate (C_t), nitrate, phosphate and oxygen in Hawaiian waters. The concentration scale is micromoles per kilogram seawater (μM).

Hawaiian waters the reactions are almost completed at about 600 m for the soft tissue (organic matter) and calcium carbonate, while the dissolution of silicious material will continue down to 2000 m (cf. Broecker, 1974, p. 6).

Very little particulate matter of planktonic origin reaches the bottom of the deep sea. However, in regions where depths are less than 200 m this material will reach the ocean floor and will decay in the bottom sediments. The regenerated nutrients formed are spread into the overlying water by vertical turbulent diffusion.

MODELING CONSIDERATIONS

A General Model Substance for Decaying Planktonic Matter

To relate the depth profiles in Figure 4 with one another and with the profiles for silica, alkalinity, and calcium, a model substance is needed. This model should work in spite of the fact that the surface water around the Hawaiian Islands has lost its nutrients previously in the trade wind current, and that the deeper water may have been enriched in material that was produced in the Atlantic and Indian Oceans (cf. Broecker, 1974, p. 25). A model substance that includes soft tissue in the form of carbohydrates, lipids, peptides, phosphate esters and trace metals, as well as hard parts, may be formulated as follows when normalized to one phosphate ester:

$$(CH_2O)_x (CH_2)_y (NHCO)_z C(H_2PO_4) (TE)_{0.1} (CaCO_3)_s (SiO_2)_t \qquad (3)$$

where TE denotes trace elements. Upon oxidation and dissolution the following reactions will take place:

$$CH_2O + O_2 \rightarrow CO_2 + H_2O \qquad (4)$$

$$CH_2 + 1.5\, O_2 \rightarrow CO_2 + H_2O \qquad (5)$$

$$NHCO + 2O_2 \rightarrow CO_2 + NO_3^- + H^+ \qquad (6)$$

$$C(H_2PO_4)^- + O_2 \rightarrow CO_2 + H_2PO_4^- \qquad (7)$$

$$CaCO_3(s) \rightarrow Ca^{2+} + CO_3^{2-} \qquad (8)$$

$$SiO_2(s) + H_2O \rightarrow Si(OH)_4 \qquad (9)$$

The chemical shifts associated with the decomposition are:

$$\Delta C_t = x + y + z + 1 + s \qquad (10)$$

$$\Delta A_t = -z + 2s \qquad (11)$$

$$\Delta N = z \tag{12}$$

$$\Delta P = 1 \tag{13}$$

$$\Delta O_2 = -(x + 1.5y + 2z + 1) = -AOU \tag{14}$$

$$\Delta Ca_t = s \tag{15}$$

$$\Delta SiO_2 = t \tag{16}$$

In Hawaiian waters, the AOU (apparent oxygen utilization) is about 265 µM while ΔC_t is about 310 µM. The number of oxygens released per molecule of CO_2 assimilated is 1.20 to 1.33. Using the midpoint of this range for the oxidation of the soft tissue, AOU = 265 µM corresponds to $\Delta C_t \simeq 210$ µM. Thus, $\Delta Ca_t = 100$ µM (equals ΔC_t from dissolved $CaCO_3$).

According to Broecker (1974, p. 12) $\Delta C_t/\Delta Ca_t = 3$, which agrees with our data. Also according to Broecker (1974, p. 12), C:N:P = 120:15:1, which means $\Delta N = 39$ µM and $\Delta P = 2.6$ µM. We obtained 36 and 3 µM, respectively.

ΔSi_t is approximately 150 µM$_w$ (cf. Broecker, 1974, p. 6) and thus $t/s = 1.5$. The oxidation of the peptide and phosphate parts of the soft tissue will use 36 x 2 + 3 = 75 µM of O_2 leaving 190 µM for the oxidation of the carbohydrate and fatty parts. Thus

$$\frac{x + y}{x + 1.5y} = \frac{210 - 36 - 3}{190} = 0.9 \text{ or } x/y = 3.5 \tag{17}$$

Our model substance then has the composition:

$$(CH_2O)_{50}(CH_2)_{14}(NHCO)_{15}C(H_2PO_4)(TE)_{0.1}(CaCO_3)_{40}(SiO_2)_{60} \tag{18}$$

The shifts upon decomposition are:

$$\Delta C_t = 120 \tag{19}$$

$$\Delta A_t = 65 \tag{20}$$

$$\Delta N = 15 \tag{21}$$

$$\Delta P = 1 \tag{22}$$

$$\Delta O_2 = 102 \tag{23}$$

$$\Delta Ca_t = 40 \tag{24}$$

$$\Delta Si_t = 60 \tag{25}$$

This means that the uptake and release of nitrate will influence the alkalinity (cf. *Gundersen and Mountain*, 1973) and $\Delta A_t/\Delta Ca_t$ will not be two, but $65/40 = 1.625$. We found 1.62 ± 0.38 on the ninth cruise with R/V Dmitry Mendeleev (*Almgren, Dyrssen, and Strandberg*, 1975a). *Brewer et al.* (1975) found 1.45 ± 0.19 which agrees with our value within the rather large limits of error.

If the soft tissue decays somewhat more readily than the calcium carbonate dissolves, this should be observable in graphs where the concentrations of total carbonate, nitrate, phosphate, and AOU at each 50 m depth interval down to 1000 m are plotted against each other and compared with the stoichiometric model. This is done in Figures 5a - 5f. In Figures 5a - 5c, it is obvious that the data follow the soft tissue slopes (80:15, 80:102, and 80:1, respectively) down to 450 m. At this point C_t increases more rapidly, so that the 600 m values agree with the total particle decomposition including C_t from dissolved $CaCO_3$ (slopes 120:15, 120:102, and 120:1, respectively). Of course this effect does not show up in the graphs of soft tissue constituents only (Figures 5d - 5f). Below 600 m there are only small changes in the concentrations of the inorganic constituents.

The model substance explains the overall shifts in some inorganic constituents, but not all details. For example, C_t decreases between 0 and 150 m when the nitrate and phosphate concentrations are still very low (Figures 5a and 5c). The nitrate and phosphate data (Figure 5d) do not follow the slope 15:1 exactly, and the deviations may reflect selective microbial processes (cf. *Gundersen and Mountain*, 1973). Of course it is possible to refine the model by bringing in minor decay products such as NH_4^+, NO_2^- and dissolved organics. But, the purpose of the model is only to give an overview of the main chemical shifts in inorganic constituents upon plankton production and decomposition (respiration).

A General Model for Primary Production

The model substances for the decomposition of sinking plankton may differ from the model for the production of phytoplankton. In this case the ratio C:N:P of 106:15:1 is used and 1.20 to 1.33 oxygens are released per molecule of CO_2 assimilated. A model that fits these conditions as well as phytoplankton analysis by *Parsons and Takahashi* (1973) is

$$(CH_2O)_{40}(CH_2)_{20}(CH_2OCH_2NHCO)_{15}CMgHPO_4. \qquad (26)$$

The formation reactions are just opposite the decomposition reactions.

Alkalinity shifts generated by the growth of phytoplankton have been studied by *Brewer and Goldman* (1975) considering both NO_3^- and NH_4^+ as nitrogen sources. The shifts are somewhat different from

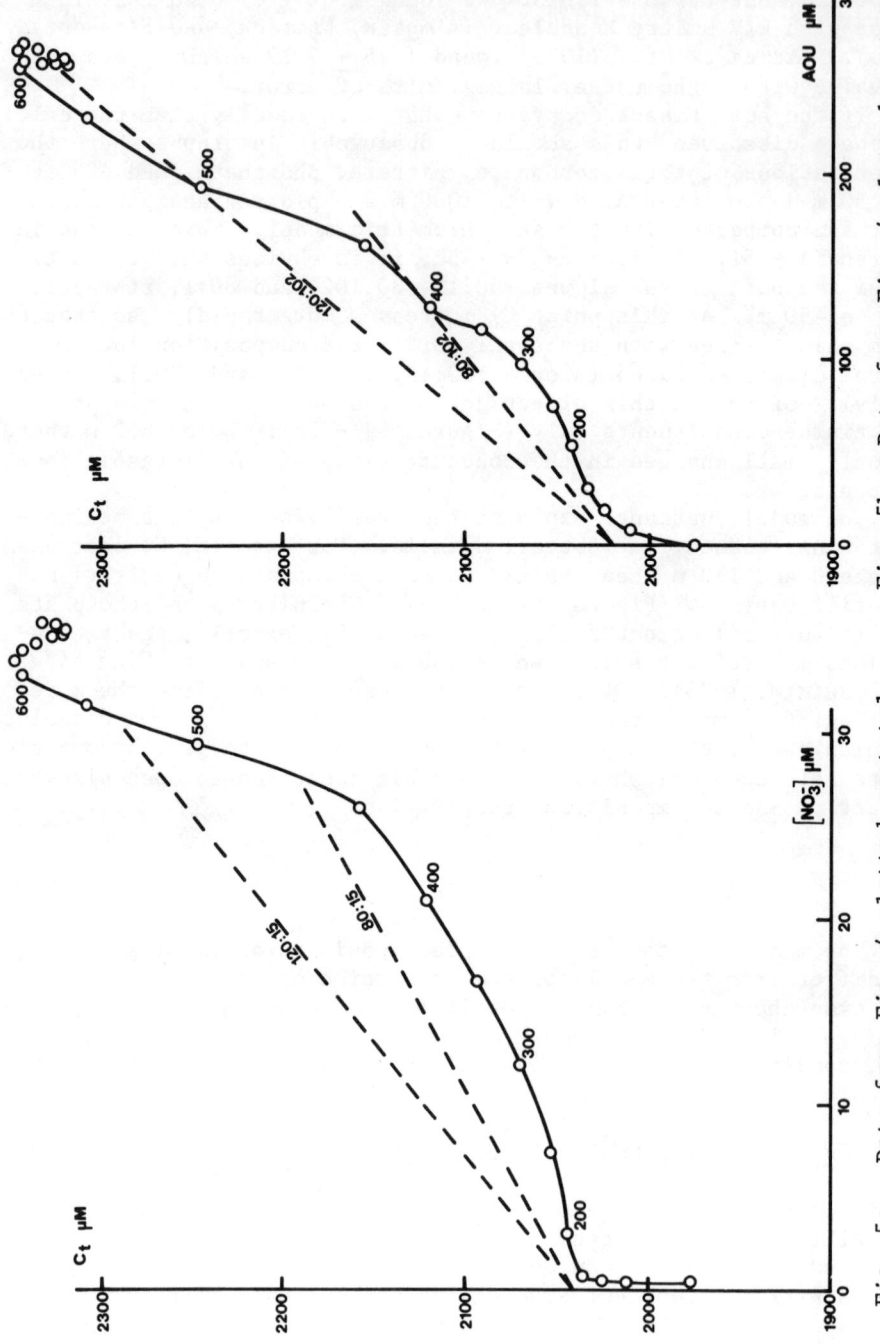

Fig. 5a. Data from Fig. 4 plotted as total carbonate against nitrate. The slope 80:15 includes only the soft tissue, while the slope 120:15 also includes the calcium carbonate.

Fig. 5b. Data from Fig. 4 plotted as total carbonate against the apparent oxygen utilization. The slopes are drawn according to the model substance.

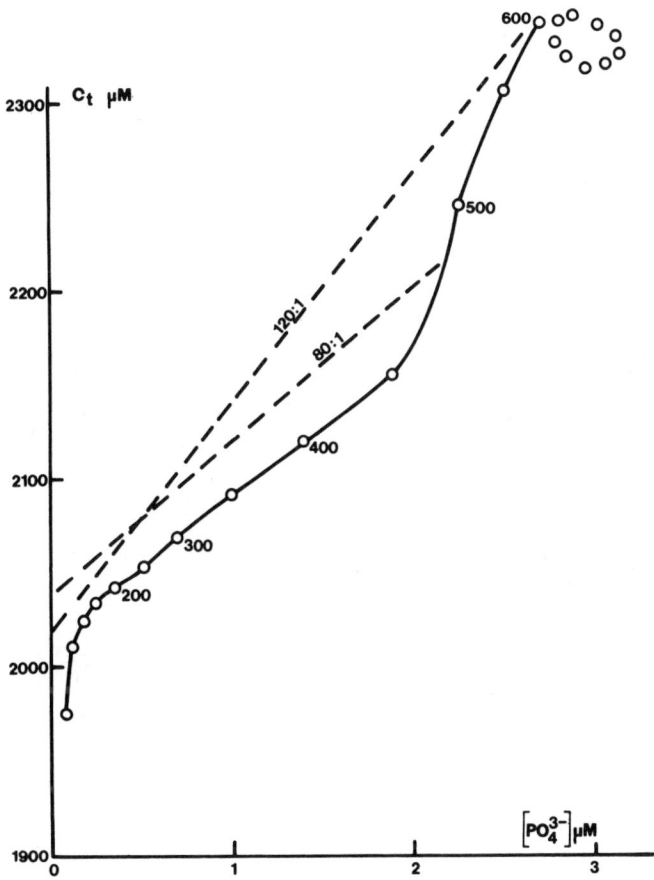

Fig. 5c. Data from Fig. 4 plotted as total carbonate against phosphate. The slope 80:1 only includes carbonate from the oxidation of soft tissue, while the slope 120:1 also includes carbonate released by dissolution of the calcium carbonate parts.

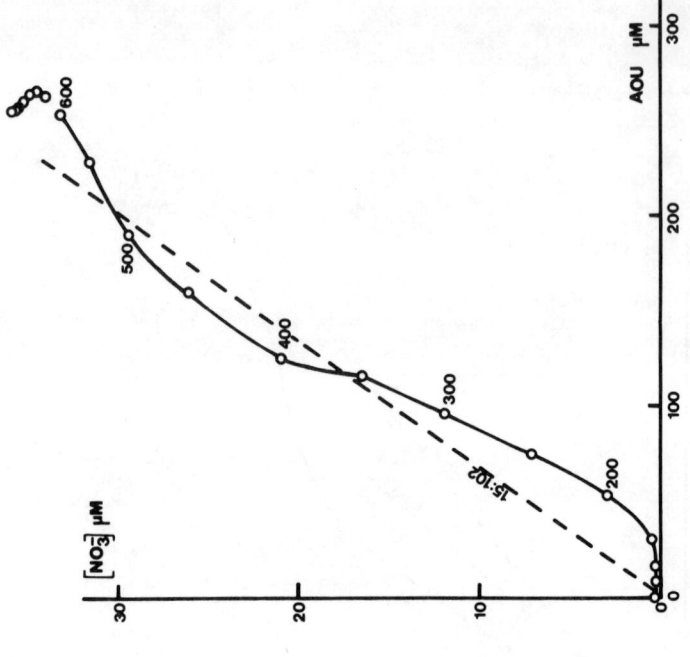
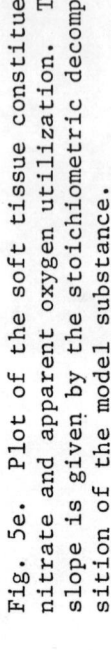

Fig. 5e. Plot of the soft tissue constituents nitrate and apparent oxygen utilization. The slope is given by the stoichiometric decomposition of the model substance.

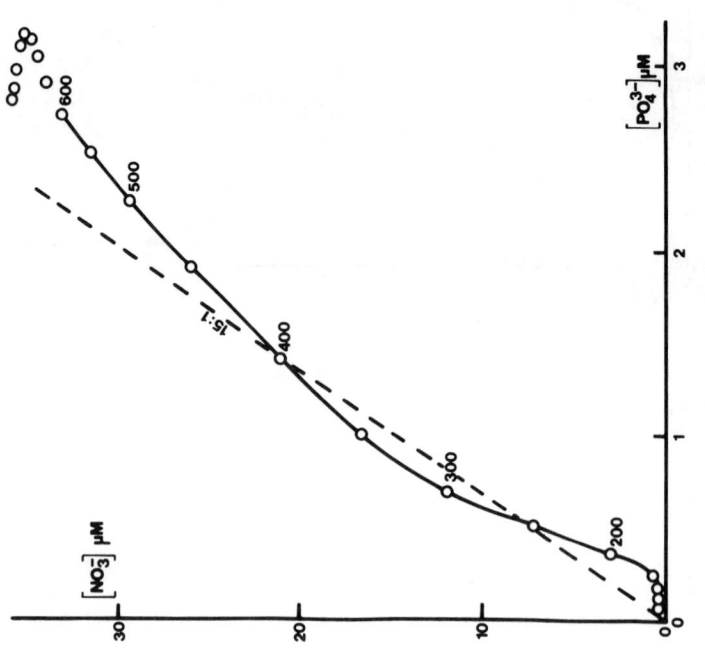

Fig. 5d. Plot of the soft tissue constituents nitrate and phosphate. The slope is given by the stoichiometric decomposition of the model substance.

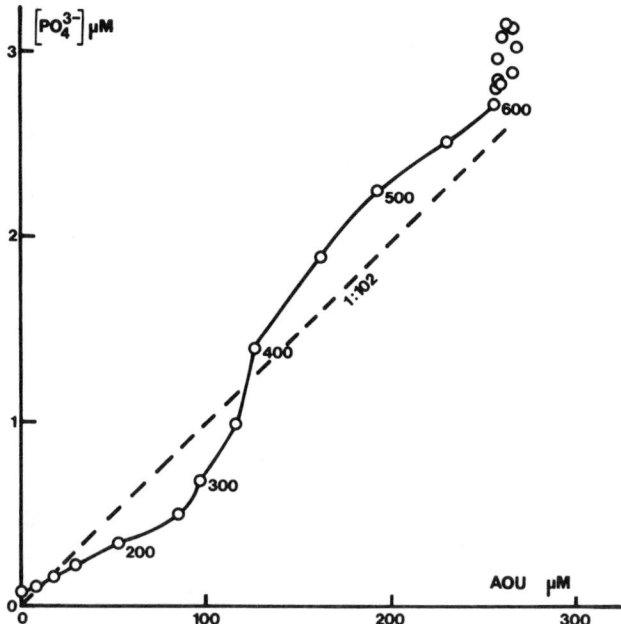

Fig. 5f. Plot of the soft tissue constituents phosphate and apparent oxygen utilization. The slope is given by the stoichiometric decomposition of the model substance.

those predicted by the stoichiometric modeling, which gives some support to protein binding of amino acids (Morita, 1975). Thus, part of the protein could be in the form of amino acids and this transformation could change the alkalinity depending on the charge, e.g.

$$CH_2NHCO + H_2O + Na^+ \rightarrow NH_2CH_2COO^-Na^+ + H^+. \tag{27}$$

The effect of such a reaction would be to decrease the alkalinity shift for nitrate uptake and increase the reverse shift for the uptake of ammonium ion.

SOME ON-BOARD TRACE METAL ANALYSES

The concentration of three different trace metals were determined during the ninth cruise of R/V Dmitry Mendeleev (Almgren, Dyrssen, and Strandberg, 1975). They represent divalent transition metals with different mean concentrations in seawater: 1 µg/l for Cu, 5 µg/l for Zn, and 0.03 µg/l for Hg. The proportions are approximately the same (cf. Brewer, 1975, p. 483) in phytoplankton (although their concentrations are 7000 times larger, i.e. µg per g day weight). Deep seawater contains about 5-20 µg particulate matter per liter (i.e. material that is retained on a 0.5 µm Nuclearpore filter) (Chesselet, 1974). The concentrations of different trace metals are in the order of 500 ppm for Cu, 1000 ppm for Zn and 100 ppm for Hg in such material. Therefore, their contributions to the total trace metal concentrations in seawater will be no more than 0.01, 0.02, and 0.002 µg/l, respectively.

Surface seawater may contain much more particulate matter in the form of living plankton and detritus. However, since the trace metal concentrations are reported to be much smaller (cf. Brewer, 1975, p. 483) the amount of trace metals bound to particulate (filterable) matter is still only a small fraction of the total concentration. There is, however, one indication that colloidal material may adsorb (hold) most of the mercury. We determined mercury on board by flameless atomic absorption using tin(II) chloride as a reducing agent (Almgren, Dyrssen, and Strandberg, 1975). Without acid pretreatment of the seawater samples for one week (2.5 ml conc. nitric acid per 500 ml) the samples showed less than 0.004 µg Hg/l (the detection limit). The CEPEX bag experiments also indicate that most of the trace metals are adsorbed on the plankton during the bloom (Kremling, 1975).

The analytical difficulties in trace metal determinations are well known (cf. Brewer, 1975, p. 438). If the coefficients of variation of the Trace Element Intercalibration Study are plotted against the concentration in seawater, as in Fig. 6, one finds that the coefficients of variation are much smaller for the ions (circles) that have well-defined ionic forms for which the main sea salts can act as hold-back carriers. Within this group the coefficient of variation decreases with increasing concentration of the trace element. The other ions (squares) are characterized by two factors: They form uncharged hydroxo species in seawater and they are biochemically active metals. Their tendency to be adsorbed onto and to be built into inert complexes is therefore much more pronounced.

Our analyses of copper and zinc were done on board directly after sampling. As a result, errors due to sample storage were avoided. The dithizone extraction of copper and zinc with nitrobenzene seemed to give more precise results (standard deviations less than 10%) than the hexone extraction with pyrrolidine dithiocarbamate. The variations in Table I may, therefore, reflect real variations in the Pacific Ocean. Certainly, there is no depth variation as in the case with nitrate and phosphate. However, according to the stoichiometric model above, the total trace element content was only 1/10 of the phosphorus content. As a result, plankton formation and decomposition does not appear to cause any marked variations in the concentrations of copper, zinc, and mercury. Deeper waters seem to

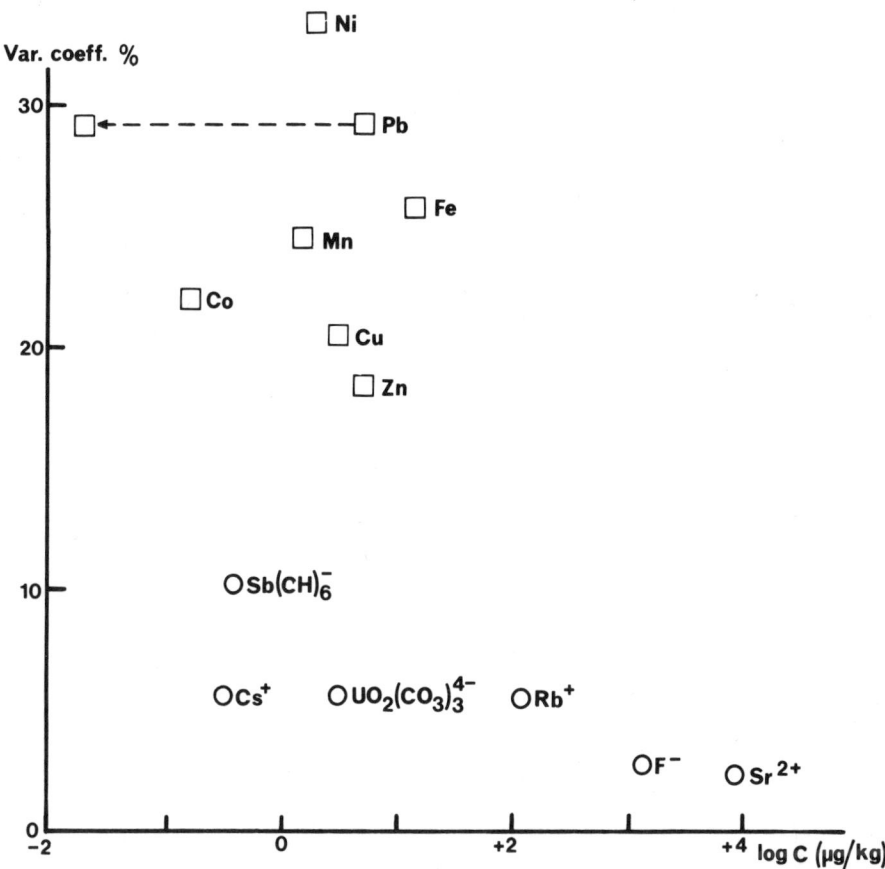

Fig. 6. Coefficients of variation of the analyses of trace elements in seawater (cf. Brewer, 1975, p. 438). The lead value of 5 µg/kg is much too high.

have low copper and mercury concentrations while the mean zinc concentration shows no depth variation.

Mercury was also determined at 4 m within the Pacific tropical zone during the whole cruise. The mean value north of the equator was 0.027 µg/l (width 0.012 to 0.090 µg/l; 12 determinations) and the mean value south of the equator was 0.027 µg/l (width 0.010 to 0.061 µg/l, 23 determinations). The highest value was 0.090 µg/l outside Hawaii. This supports the suggestion that mercury is carried into the sea mainly as wet fall-out (cf. Goldberg, 1972) from volcanic debris.

TABLE I

Trace Element Data of Pacific Ocean Samples

	0-100 m	125-301 m	435-535 m	865-1074 m	1740-5300 m
Cu (μg/l) width std. dev.	1.29 / 0.3-2.8 / 0.71	1.35 / 0.2-4.5 / 1.51	1.47 / 0.5-4.3 / 1.50	0.91 / 0.4-1.3 / 0.33	0.59 / 0.2-1 / 0.29
Zn (μg/l) width std. dev.	3.96 / 0.4-9.7 / 2.66	5.06 / 2-13.3 / 3.96	7.73 / 2.4-20.2 / 7.04	4.66 / 2.2-8.5 / 2.33	6.02 / 1.4-13 / 3.75
Hg (μg/l) width std. dev.	0.054 / 0.013-0.104 / 0.030	0.052 / 0.017-0.100 / 0.033	0.032 / 0.016-0.047 / 0.018	0.027 / 0.010-0.050 / 0.016	0.027 / 0.005-0.071 / 0.021
omitted value	0.200	0.165	0.177	none	none

CHEMISTRY OF PLANKTON PRODUCTION AND DECOMPOSITION

pH AS AN UPWELLING INDICATOR

Variations in the pH of seawater are mainly the result of shifts in alkalinity and total carbonate caused by plankton production and decomposition. On a longer time scale, the static value of pH is probably set by sediment ion exchange equilibria as suggested by Sillén (1963). In order to understand the variations of pH due to chemical shifts, it should be stressed that the main carbonate species in seawater is HCO_3^- and the concentrations of CO_2 and CO_3^{2-} are set by the disproportionation equilibrium (cf. Dyrssen and Hansson, 1974):

$$2HCO_3^- \rightleftharpoons CO_2 + CO_3^{2-} + H_2O \tag{28}$$

where the equilibrium constant is defined by

$$K = \frac{[CO_2][CO_3^{2-}]}{[HCO_3^-]^2} \tag{29}$$

since $[HCO_3^-]$ is rather constant, the deviations in $[CO_2]$ and $[CO_3^{2-}]$ will be proportional to the concentrations, i.e.

$$\frac{d[CO_2]}{[CO_2]} \approx \frac{-d[CO_3^{2-}]}{[CO_3^{2-}]}. \tag{30}$$

Furthermore, pH is regulated by the equilibrium constant

$$K_1 = \frac{[H^+][HCO_3^-]}{[CO_2]}. \tag{31}$$

Thus,

$$\frac{d[H^+]}{[H^+]} \approx \frac{d[CO_2]}{[CO_2]}. \tag{32}$$

The shift in pH is related to the shift in $[H^+]$ by

$$dpH = \frac{1}{2.3}\left(\frac{d[H^+]}{[H^+]}\right). \tag{33}$$

The carbonate alkalinity, which is close to A_t, is defined by

$$A_C = [HCO_3^-] + 2[CO_3^{2-}], \tag{34}$$

and the total carbonate by

$$C_t = [CO_2] + [HCO_3^-] + [CO_3^{2-}]. \tag{35}$$

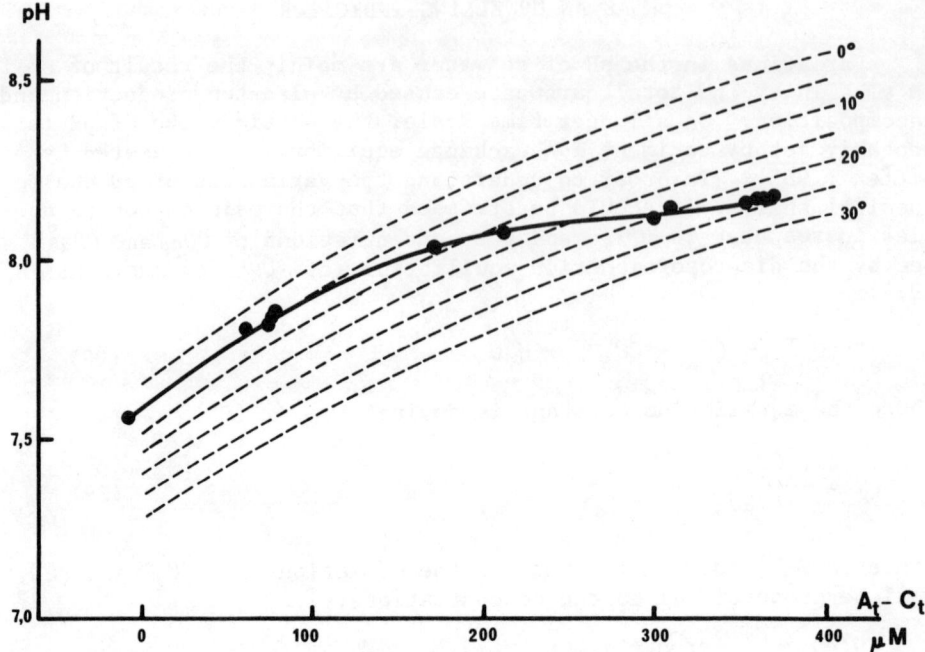

Fig. 7. A pH versus $A_t - C_t$ diagram based on our definition of $[H^+]_t$ and on our constants of K_{1C}, K_{2C}, and K_B (cf. *Dyrssen and Hansson*, 1974). The thick line represents *in situ* data for different depths close to Oahu and Hawaii in the Pacific. The point with $A_t - C_t = -7$ μM corresponding to pH = 7.559 at 5.83°C and 625 m is quite extreme.

Thus,

$$A_C - C_t = [CO_3^{2-}] - [CO_2] \tag{36}$$

and

$$d(A_C - C_t) = d[CO_3^{2-}] - d[CO_2]. \tag{37}$$

Using the approximate relations above,

$$d(A_C - C_t) = 2.3 dpH([CO_2] + [CO_3^{2-}]). \tag{38}$$

Figure 7 shows that pH is not a linear function of $A_C - C_t$, since $[CO_2] + [CO_3^{2-}]$ is not constant (the values are higher in the production zone). However, the purpose of the curves is to show the relation between pH and the chemical shifts due to plankton production and decomposition according to the stoichiometric model above. Calculated values of $[CO_2]$ and $[CO_3^{2-}]$ from our determinations of alkalinity and total carbonate (cf. Almgren, Dyrssen, and Strandberg, 1975b) show that the surface of the sea is practically at equilibrium with the partial atmospheric pressure of CO_2, and the deeper waters are at equilibrium with calcium carbonate. However, if water from the regeneration (decomposition) zone between 150 and 600 m with excess CO_2 (low pH and $A_t - C_t$) reaches the surface through upwelling, pH will decrease until $[CO_2]$ is adjusted by giving off CO_2 to the atmosphere, or by photosynthetic uptake. Thus, such areas may be localized by the determination of pH and alkalinity (cf. Anfält, Granéli, and Strandberg, 1975). Surface (4 m) pH determinations during the ninth cruise of R/V Dmitry Mendeleev show (Figure 8) that the surface water is slightly undersaturated with respect to CO_2, except in areas close to the equator. This could be due to the rapid transportation of coastal upwelling water by trade wind currents or by upwelling in the middle of the Pacific.

At a recent conference on "The Nature of Seawater" (Goldberg, 1975) it was stressed how important it is to measure the pH of seawater with special seawater buffers (also see Hansson, 1973, and Almgren, Dyrssen, and Strandberg, 1975b). The use of National Bureau of Standards buffers introduces junction potentials which depend on the type of reference electrode used. Furthermore, the standardization is slow since it takes time for the glass electrode to adjust from the NBS buffer media to seawater. The difference between Hansson's pH scale and the use of NBS buffers is shown in Figure 9, which is calculated from two sets of determinations of the acidity constant of hydrogen sulfide (Goldhaber and Kaplan, 1975, and Almgren, Dyrssen, Elgquist, and Johansson, 1975).

In order to calculate the concentration of CO_2, the set of dissociation constants used must be consistent with the pH scale. Using seawater buffers, it is possible to determine pH with the precision (and accuracy) of ±0.003 pH units. To match this precision, the alkalinity should be determined with a precision of 0.1%, which can be done with a titration technique. However, even with an approximate value of the alkalinity it is possible to trace upwelling areas by pH measurements. The in situ temperature is needed, but it is not necessary to correct pH for pressure since it is the photic zone which is of most interest.

The patchiness of plankton blooms and other examples of biocoenosis are due to many environmental factors. Refertilization of the photic zone by upwelling may not be a sufficient chemical condition for a plankton bloom. One set of chemical factors that need to be studied much more are trace organics which may inhibit and promote production. It will be necessary to analyze fresh, unstored seawater samples with modern analytical techniques such as gas chromatography employing capillary columns, high performance chromatography with sensitive detectors, and mass spectrometry.

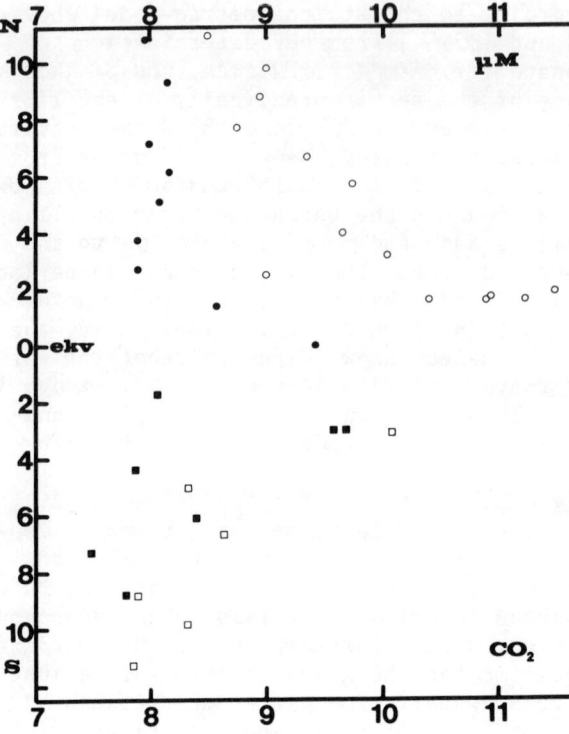

Fig. 8. The concentration of carbon dioxide (in μM) from pH determinations at a depth of 4 m. The top figure shows the latitude dependence; the bottom figure the temperature dependence. The straight lines in the bottom figure represent equilibrium values calculated for 325 ppm carbon dioxide in the atmosphere at three different salinities (34, 35 and 36 o/oo). Circles: Honolulu-Fanning-equator at 164° 20.5 W. Squares: Equator at 164° 20.5 W.-Apia. Triangles: Apia-Funafuti-Suva-Port Vila. Blocks: Port Vila-equator at 159° 33' E. Dots: North of equator at 159° 33' E.

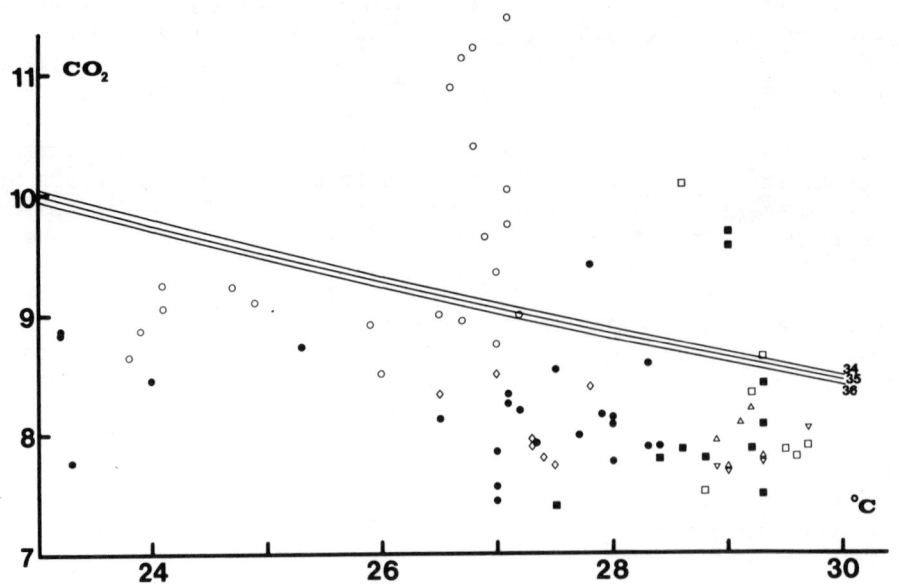

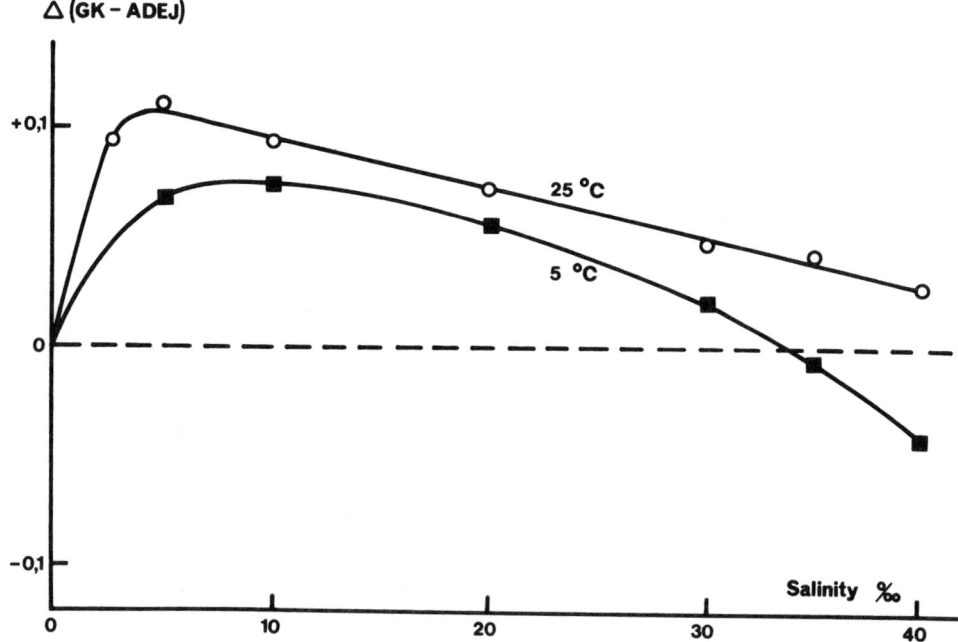

Fig. 9. The difference between pH determined by the NBS buffers (GK) and by the Gran extrapolation method as a function of temperature and salinity.

REFERENCES

Almgren, T., D. Dyrssen, and M. Strandberg; 1975a. Data from the Swedish participation of the ninth cruise of R/V Dmitry Mendeleev. *Anal. Chem. Department Report on the Chemistry of Sea Water XV*, Univ. of Gothenburg.

Almgren, T., D. Dyrssen, and M. Strandberg. 1975b. Determination of pH on the moles per kg seawater scale (M_w). *Deep-Sea Research* 22: 635-646.

Almgren, T., D. Dyrssen, B. Elgquist, and O. Johansson. 1975. Dissociation of Hydrogen Sulfide in Sea Water. *Mar. Chem.* (in press.)

Anfält, T., A. Granéli and M. Strandberg. 1975. A new probe photometer based on opto-electronic components, application to total alkalinity in seawater. *Anal. Chem.* (in print).

Benes, P. and E. Steinnes. 1974. In Situ Dialysis for the Determination of the state of trace elements in natural waters. *Water Research* 8: 947-953.

Brewer, P. G. 1975. Minor Elements in Sea Water. *Chemical Oceanography* (2d ed, Ch. 7) by J. P. Riley and G. Skirrow, Academic Press, New York.

Brewer, P. G., and J. C. Goldman. 1976. Alkalinity changes generated by phytoplankton growth. *Limnol & Ocean 21:* 108-117.

Brewer, P. G., G. T. F. Wong, M. P. Bacon, and D. W. Spencer. 1975. An Oceanic Calcium Problem? *Earth and Planetary Science Letters 26:* 81-87.

Broecker, W. S. 1974. *Chemical Oceanography.* Harcourt, Brace, Jovanovich; New York.

Campbell, I. D. and R. A. Dwek. 1975. Application of Physical Methods to the Determination of Structure in Solution. *The Nature of Seawater:* 165-189. Ed. by E. D. Goldberg. Dahlem Konferenzen. Abakon Verlagsgesellschaft, Berlin.

Chesselet, R. 1974. Deep Ocean Suspended Matter Chemistry. Lecture at III Int. Symp. on the Chemistry of the Mediterranean, Rovinj, Yugoslavia, May 5-8, 1974.

Dyrssen, D. and I. Hansson. 1974. Junctions and Activities before the Dahlem Conference. *Anal. Chem. Department Report on the Chemistry of Sea Water XIII.* Univ. of Gothenburg.

Goldberg, E. D. 1972. Man's Role in Major Sedimentary Cycle. *The Changing Chemistry of the Oceans:* 267-290. Ed. by D. Dyrssen and D. Jagner. Almqvist et Wiksell, Stockholm.

Goldberg, E. D. 1972. *The Nature of Seawater.* Dahlem Konferenzen. Abakon Verlagsgesellschaft, Berlin.

Goldhaber, M. B. and I. R. Kaplan. 1975. Apparent Dissociation Constants of Hydrogen Sulfied in Chloride Solutions. *Mar. Chem. 3:2:* 83-104.

Gundersen, K. and C. W. Mountain. 1973. Oxygen Utilization and pH Change in the Ocean Resulting from Biological Nitrate Formation. *Deep-Sea Research 20:* 1083-1091.

Hansson, I. 1973. A New Set of pH-Scales and Standard Buffers for Sea Water. *Deep-Sea Research 20:* 479-491.

Kremling, K. 1975. Personal Communication.

Lindskog, S., L. E. Henderson, K. K. Kannau, A. Liljas, P. O. Nyman, and B. Strandberg. 1971. Carbonic Anhydrase. *The Enzymes,* Vol. 5, 3d ed., Ch. 21. Academic Press.

Morita, R. Y. 1975. Personal Communication.

Paasche, E. 1964. A Tracer Study of the Inorganic Carbon Uptake During Coccolith Formation and Photosynthesis in the Coccolitophorid. *Coccolithus hyxleyi. Physiologia Pl.* (Suppl.) III, 1-82.

Parsons, T. and M. Takahashi. 1973. Biological Oceanographic Processes. Pergamon Press, p. 43.

Sillén, L. G. 1963. How has Sea Water Got its Present Composition? *Svensk Kem. Tidskr., 75:* 161-177.

Steemann Nielsen, E. 1975. Lecture series at the Goteborg Center for Marine Research and Techniques. Cf. also *Marine Photosynthesis. Elsevier Oceanogr. Ser. 13.*

The Cycling of Labile Organic Compounds: Sterols in the North Atlantic Ocean

Robert B. Gagosian

Woods Hole Oceanographic Institution

ABSTRACT

Seawater samples collected from the continental shelf and slope waters of the western North Atlantic Ocean and Sargasso Sea have been analyzed for a class of biogenic compounds, the sterols. The sterol concentrations found ranged from 0.1 to 1.3 µg/l seawater. Cholesterol is the major free and esterified sterol in both the surface and deep water. β-Sitosterol, fucosterol, brassicasterol, 22-dehydrocholesterol, campesterol, 24-methylenecholesterol, norcholestadienol and stigmasterol are found in lower concentrations at the surface and in the deep sea. Several sterols, e.g. brassicasterol, appear to be produced and consumed in the upper 1000 m of the water column, whereas few sterols, e.g. cholesterol, were found in the entire water column and may be examples of more resistant organic compounds.

Several mechanisms can be postulated for the injection of sterols into the deep sea. Vertical fluxes of organic particles from the surface appear to deliver sterols into the mid-depth waters of the Sargasso Sea. However, some other process(es), e.g. physical transport (viz. horizontal and vertical advection and diffusion), resuspension of sediments, or _in situ_ deep water biological production and consumption, is controlling deep water sterol distributions. Detailed profiles of ancillary data such as particulate organic carbon, hydrographic and total particulate matter collected at the same station are necessary to complement detailed profiles of specific organic compounds in order to discern the transport mechanisms of these biochemicals to the deep sea.

INTRODUCTION

The cycling of organic matter in the sea has received a great deal of attention in recent years. For the most part, this has involved the collection of extensive data concerned with the properties of mixtures of organic comounds such as particulate and dissolved organic carbon (POC and DOC) (*Sharp*, 1973; *Menzel*, 1974), delta C^{13} (*Williams and Gordon*, 1970), spectral parameters (*Mattson et al.*, 1974) and biological oxygen demand (*Zsolnay*, 1975). This "bulk" approach, however, describes the integrated result of several simultaneously occurring processes affecting the specific classes of organic compounds in the organic matter pool. It would, therefore, be fruitful to study the cycling of individual, specific classes of organic compounds in order to use them as tracers in separating the myriad of mechanisms and processes governing the cycling of bulk organic matter. In addition, it is the labile organic compounds, rather than the more resistant organic matter, which take part in the nutritional and hormonal processes of deep sea biota. Hence, knowledge of the origin, transport, and sinks of labile organic compounds is most important to our understanding of deep sea biological activity.

There is, however, very little information on the origin, distribution and fate of organic compounds in the sea. Even less is known about their geographical and temporal variations. Data concerning the rates and mechanisms of organic chemical processes in the oceans are practically non-existent and the mechanisms that govern the transport of organic compounds to the deep sea are not clear. Two of the major reasons for this paucity of information are the complexity of the analytical techniques required for the quantification of organic compounds from seawater and the contamination problems associated with these analyses.

The chemical structures of only 10 to 20% of the dissolved organic matter in seawater have been elucidated (*Hood*, 1971; *Riley and Chester*, 1971; *Wagner*, 1969; *Wangersky*, 1972; *Williams*, 1971). This fraction consists of amino acids, sugars, urea, hydrocarbons, organic pigments and acids, and fatty acids and alcohols. Recently, another class of marine biochemicals, the sterols, has been studied (*Gagosian*, 1975a; *Gagosian*, 1976). The sterols are among the most important hormone regulators of growth, respiration and reproduction in organisms. Cholesterol and related steroid alcohols are not only the immediate biosynthetic precursors for all steroidal hormones, but they also have hormonal activity themselves (*Kanazawa and Teshima*, 1971a). These steroidal alcohols, the sterols, are of special interest from a geochemical point of view. Their chemical stability and structural diversity, along with their inherent optical activity, allow them to be used as indicators of biological activity in the oceans.

In seawater, sterols act as good tracers of biogenic material in the complex mixture of dissolved and particulate matter. After deposition to the sediments, sterols should become useful indicators of geochemical processes in recent sediments. Sterols may provide infor-

THE CYCLING OF LABILE ORGANIC COMPOUNDS 87

mation on the marine or terrestrial, plant or animal origin of sedimentary organic matter.

It is known that most marine invertebrates are unable to biosynthesize sterols (*Zandee*, 1964, 1967). Therefore, they must obtain them from exogenous sources, such as by adsorption or filtration from seawater. Hence, the presence of dissolved sterols in seawater is of interest as they have a role in the food chain, particularly with regard to zooplankton. The sterols of particulate matter may be of equal importance and they may be utilized as a dietary supplement for the larvae of marine animals.

In this manuscript the results of three cruises concerned with the sterol concentrations and individual sterol distributions in the North Atlantic Ocean are reported. From the vertical profiles of these compounds, along with POC, hydrographic and total particulate matter data, an attempt to understand sterol distributions in the ocean and the transport mechanisms governing their movement into the deep sea was undertaken.

Sampling and Analyses

In 1973 and 1974 Cruise 33 of R/V Knorr and Cruises 82 and 85 of the R/V Atlantis II were conducted in the Sargasso Sea and slope waters of the North Atlantic Ocean. Five stations of large volume samples were occupied, the last being the most detailed. The large volume samples for sterols and particulate organic carbon (POC) were obtained with 60 liter aluminum Bodman bottles (*Bodman et al.*, 1961) as previously described (*Gagosian*, 1975a). At each location a complete hydrographic station was simultaneously made, using Teflon-lined Nansen bottles.

Sterols were extracted from seawater and analyzed by the methods previously described (*Gagosian*, 1975a). Briefly, 20 liter samples were transferred from the Bodman bottles to thoroughly precleaned five gallon glass carboys. The samples were then doubly extracted in 5 liter glass separatory funnels with hexane. The hexane extracts were placed in precleaned pint bottles with Teflon caps and stored in a freezer for shore-based analysis. Blanks were run for the entire process. After returning to shore the hexane extracts were concentrated on a rotary evaporator, lyophilized, and derivatized to make the trimethylsilyl ether of the alcohol functional group. Quantification and structural elucidation were then determined by gas chromatography, and gas chromatography-mass spectrometry. The lower limit of sterol detectability is 1 ng of each sterol/l seawater. The reproducibility of replicate samples is ± 5%; sample variability is ± 15% for samples taken at the same location and depth in the Sargasso Sea. The blank value for sterols is 30 ng/l seawater.

Samples for POC were analyzed as described by *Gagosian* (*Gagosian*, 1976). Briefly, the seawater samples (4 liter surface water and 10 liter deep water) were transferred to a 12 liter fiber glass container and pressurized to less than 5 lbs. N_2. The samples were then filtered through Gelman Type A (1-2 μ pore size) glass fiber filters which had been precombusted at 450°C for 24 hours. After filtration, the filters were carefully separated, dried, and stored in a freezer for shore-based analysis. After returning to shore, the filters were again dried, and total organic C was determined by dry combustion on a Perkin Elmer CHN analyzer. The reproducibility for the combined sampling and analysis, as determined by replicate analyses is ± 7% and the blank for entire procedure is 2.5 μg C/l for deep water and 5 μg C/l for surface water. Total particulate matter samples were taken from 30 liter PVC Niskin bottles and analyzed by Dr. D. W. Spencer of this institution. DOC analyses were not made due to controversy on the validity of DOC determination methods (Menzel and Vaccaro, 1964; Sharp, 1973).

RESULTS AND DISCUSSION

Sterol Sources in Seawater

There are several sources of sterols in the surface waters of the ocean. The majority of marine sterols contain 27-29 carbon atoms, with the C_{29} sterols being predominant. However, recent discoveries of a norcholesterol and of gorgosterol and its analogs have increased the range of marine sterols from C_{25} to C_{30}. It should be noted that the variation in number of carbon atoms occurs almost exclusively on the side chain, mainly at C-24 (see Figure 1, structure 2). Other structural diversity that exists is found almost entirely in the side chain, except for the saturation of ring B to form the reduced sterols, the stanols (Figure 1, structure 4), and the dehydration and the saturation of rings A and B to form the cyclo-alkane steranes.

General reviews have recently appeared on sterols in marine organisms (Scheuer, 1973) and sterols of marine invertebrates and plants (Austin, 1970). More specific reviews have been prepared on sterols in mollusca (Idler and Wiseman, 1971a), echinoderms (Goad et al., 1972), and crustacea (Idler and Wiseman, 1971b). The distribution and function of sterols in algae have been reviewed by Heftmann (1971) and Patterson (1971). Sterols in fungi have been reviewed by Weete (1973).

From these reviews, we find a large diversity of structures and a wide distribution of sterols in a concentration range of 0.01% to 2% for marine invertebrates, and 0.005% to 0.5% for marine plants. Cholesterol was found to be the most abundant sterol in the advanced invertebrates, whereas the more primitive invertebrates have much more diversified sterol compositions (Austin, 1970; Gagosian, 1975b; Idler and Wiseman, 1971b). The primary sterol found in brown algae (Phaephyta) is fucosterol (Safe et al., 1974; Smith et al., 1973). Red algae (Rhodophyta) contain mainly cholesterol and desmosterol with a few exceptions. Green algae (Chlorophyta) have a very complex

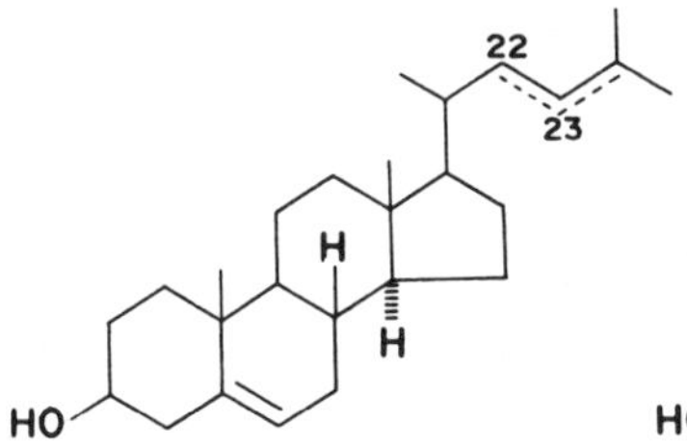

THE CYCLING OF LABILE ORGANIC COMPOUNDS

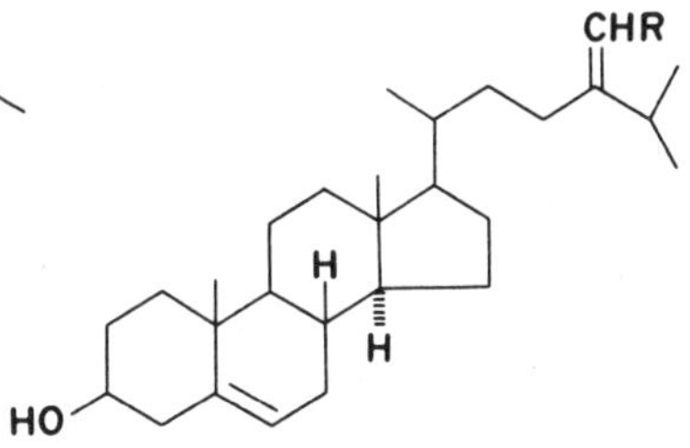

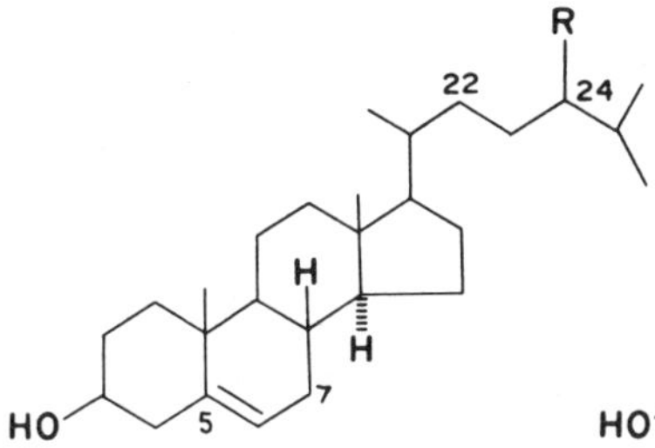

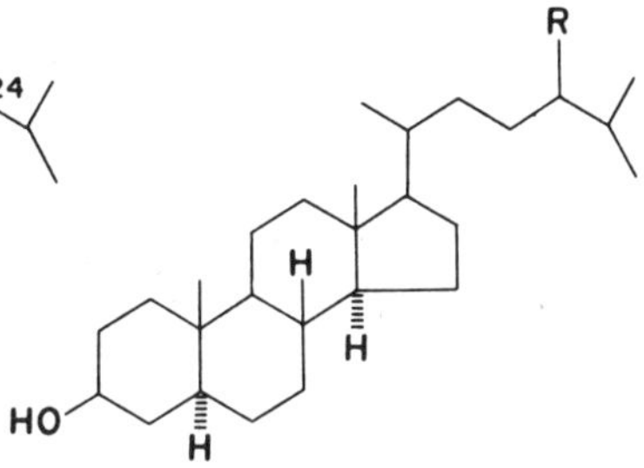

Fig. 1. Structures of sterols isolated from seawater of the North Atlantic Ocean.

sterol composition of some twenty to twenty-five compounds. Two species of marine diatoms, *Cyclotella nana*, and *Nitzschia closterium*, were found to contain only brassicasterol (Kanazawa et al., 1971).

A few species of bacteria (Schubert et al., 1968) and blue green algae (DeSouza and Nes, 1968; Reitz and Hamilton, 1968; Teshima and Kanazawa, 1972) contain sterols, but at very low concentrations. Several species of bacteria have no sterols at all (Schubert et al., 1968). The apparent lack of steroidal compounds in bacteria potentially makes this class of compounds particularly useful as a tracer of organic processes in the sea since other organic compounds such as amino acids, sugars and fatty acids are present in bacteria. Marine yeasts contain from .01% to 0.1% sterols, ergosterol and campesterol being the major components (Teshima and Kanazawa, 1971). Boutry (1967, 1970) has recently calculated a concentration of 0.4% sterols in phytoplankton collected from the Mediterranean Sea, with cholesterol and 24-methylenecholesterol being the major components, and campesterol, stigmasterol and β-sitosterol concentrations ranging from 1-5% of the total. Cholesterol was found to be the major component of the sterol mixture in

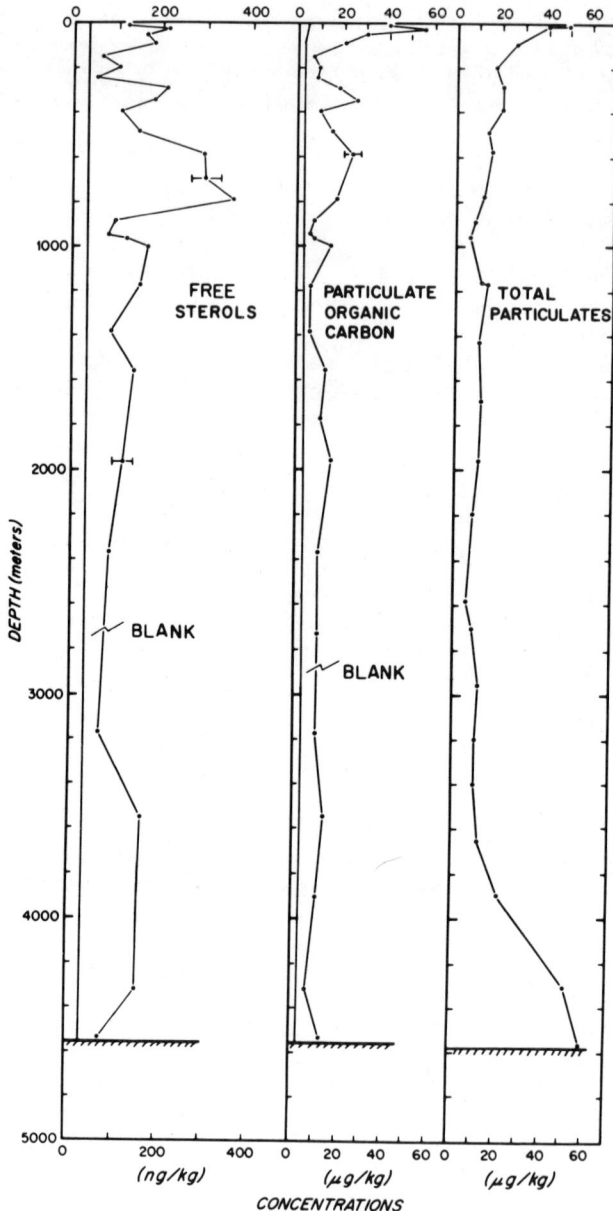

Fig. 2. Total free sterol, particulate organic carbon, and total particulate matter concentrations as a function of depth from 33°40.6'N; 57°36.8'W - September, 1974. POC and sterol samples were taken from the same sampler, whereas those for total particulate matter were taken from Niskin bottles on the same cast. Therefore, POC values found to be greater than total particulate matter at the same depth probably reflect sample variability.

zooplankton collected from the Mediterranean Sea. β-Sitosterol is a common terrestrial plant sterol isolated in high concentration, and may enter the marine environment through river runoff or aeolian transport on particulates. This sterol has also been found in coastal grasses in the Gulf of Mexico (Attaway et al., 1971).

Sterol Distributions in Seawater

During the past two years we have analyzed seawater samples collected on two cruises concerned with the sterol concentrations and individual sterol distributions in both the surface and deep waters of the continental shelf and slope waters of the North Atlantic Ocean and the Sargasso Sea (Gagosian, 1975a). In addition, we have analyzed samples and constructed detailed vertical profiles of free sterols (extractable into organic solvent), POC and total particulate matter at a station in the Sargasso Sea ($33°40.6'N$; $57°36.8'W$) (Figure 2). The sterol profiles constructed from this data are some of the most detailed profiles reported for any single class of biochemicals in seawater. Hydrographic data for this station is presented elsewhere (Gagosian, 1976).

Assigned structures of the sterols isolated are shown in Figure 1. Table 1 lists the molecular formulas and common names of these compounds. The peak numbers refer to the peaks in the gas chromatograms (Gagosian, 1975a).

Sterol Concentrations in the Water Column. The sterol concentrations of the deep water samples are listed in Table 2, with corresponding station numbers, potential temperatures, salinities, and depths. From the temperature-salinity data of Table 2, one concludes that sample K-33-II-1 contained a mixture of overlying water, Labrador Sea water, and possibly some Mediterranean Sea water (Spencer, 1972; Worthington and Wright, 1970). K-33-II-2, sampled at 3095 m, contained North Atlantic Deep water composed of Iceland-Scotland Overflow and Denmark Strait Overflow water. K-33-II-3 and K-33-I-8 samples collected from 4630 m and 4142 m respectively, contained Antarctic Bottom water mixed with North Atlantic Deep water (Spencer, 1972).

Samples collected in the Sargasso Sea in September, 1973, at Stations 6 and 7 exhibit bottom water sterol concentrations approximately equal to sterol concentrations of the surface samples (Table 2). However, deep water samples collected in February from the edge and rise of the continental shelf south of George's Banks at Stations C and D exhibit lower sterol concentrations than those found in the surface waters of these stations. Sample K-33-I-6 (Station 6), taken at 80 m in the summer thermocline of the Sargasso Sea, is almost a factor of two higher in both free and total sterol concentrations than the surface and deep water samples collected at this station. Also, at Station 6 particulate organic carbon (POC) values from just below the summer thermocline (80-90 m) are higher than their surface water values. Thus, a correlation of sterol concen-

TABLE 1. Sterols Isolated and Identified from North Atlantic Ocean Samples by Gas Chromatography-Mass Spectrometry

Peak No.	Formula	Assignment	Structure (Figure 1)
1	$C_{26}H_{42}O$	Norcholestadienol	$1\Delta^{22}$ or Δ^{23}
2	$C_{27}H_{44}O$	22-Dehydrocholesterol	2a Δ^{22}
3	$C_{27}H_{46}O$	Cholesterol	2a
4	$C_{28}H_{46}O$	Brassicasterol	2b Δ^{22}
5[1]	$C_{28}H_{44}O$	Ergosterol	2b Δ^{7}, Δ^{22}
6 (a)	$C_{28}H_{48}O$	Campesterol	2b
(b)	$C_{28}H_{46}O$	24-Methylenecholesterol	3a
7	$C_{29}H_{48}O$	Stigmasterol	2c Δ^{22}
8 (a)	$C_{29}H_{50}O$	β-Sitosterol	2c
(b)	$C_{29}H_{48}O$	Fucosterol	3b

[1] Quantities isolated were too small for absolute structural assignment.

tration with POC does exist for these samples from the euphotic zone. This correlation can be seen further from the total free sterol (extractable into organic solvents and not saponified) and POC concentrations plotted as a function of depth in Figure 2. The blank values for sterols and POC have not been subtracted from the data presented on the profiles, but are shown on the profiles. In the upper 1000 m, variations in sterol concentration correlate closely with POC and considerably less, if at all, with total particulate matter. The sterol water samples were unfiltered. As a

TABLE 2. Sterol Concentrations in Subsurface Seawater from the North Atlantic Ocean (in ng/l)

Sample No.	Depth (m)	Potential Temperature (°C)	Salinity (°/oo)	Free	Total Sterols Free and Esterified
Location - Station No. 7 (27°50'N, 67°26'W) 4938 m water column					
K-33-II-4	7	27.40	36.503	200	380
-1	993	6.47	35.214	120	330
-2	3095	2.50	34.944	170	320
-3	4630	1.80	34.891	230	450
Location - Station No. 6 (32°20'N, 63°00'W) 4380 m water column					
K-33-I-5	7	27.40	36.554	390	610
-6	80	20.38	36.495	700	1350
-7	702	12.74	35.870	260	340
-8	4142	1.86	34.887	360	710
Location - Station D (39°58'N, 67°00'W) 3050 m water column					
AII-81-5	10	10.15	34.116	660	710
-6	110	14.33	35.448	390	610
-7	1989	3.36	34.964	340	-
Location - Station C (40°00'N, 69°02'W) 1150 m water column					
AII-81-4	8	11.13	34.196	710	900
-3	1100	4.88	34.991	390	570

result, this good correlation suggests that most sterols in the upper 1000 m are associated with POC. However, this correlation of sterols with POC is not as pronounced in the deep water, suggesting some other process may be controlling the sterol distributions. At 3600 m an increase in total sterol concentration is observed. This increase is very slight for POC which is roughly constant from 1200 m to the bottom at about 6 µg/l.

The lack of comparable data from other stations makes it difficult to draw any conclusions from a single vertical profile. However, several features of the sterol profile (Figure 2) are worthy of mention. The maxima at 40-50 m for both sterols and POC may be due to either *in situ* biological productivity or a change in the density gradient at the bottom of the summer thermocline. Sargasso Sea 18°C water is between 250-350 m at these stations. Therefore, the maxima at 300 m in the vertical profile for both sterols and POC may be due to this advective feature. The oxygen minimum and phosphate and nitrate maxima occur at 750 m at this station where a large sterol maximum occurs. Remineralization of biogenic detritus releasing several bound organic compounds which would not generally be extractable from seawater by organic solvents (for example, due to entrapment in carbonate particles) may be occurring, thus yielding higher concentrations of soluble sterols. However, *in situ* production or advective processes cannot be ruled out.

A minimum in the profile occurs at 950 m. This is where the influence of Mediterranean Sea water is the the strongest. One would not expect to see this small input (approximately 2%) reflected in sterol concentrations even if Mediterranean Sea water sterol concentrations were zero because the sterol sample analytical uncertainty is $\pm$ 15%. The reason for this 950 m minimum is therefore unknown. However, it is interesting to note that the minimum is also present in the POC and total particulate matter profiles. The increase of sterols in the deep water at 3600 m and 4350 m does not appear to correlate linearly with POC. Since the appearance of the nepheloid layer in this region of the North Atlantic Ocean occurs at this depth (*Biscaye and Eittreim*, 1974), correlation of sterol concentrations with the resuspension of particles from bottom sediments is tempting. Advective and diffusive processes from North Atlantic Ocean deep water and Antarctic Ocean bottom water may also be responsible for the deep sea sterol concentrations observed. Data concerning sterol concentrations in the dissolved and particulate fractions of seawater are necessary to further elucidate the processes outlined above.

<u>Distribution of Individual Sterols</u>. The distributions for the individual sterols are given in Table 3 and Figure 3. The percent compositions of each sterol in both the free and total (free + esterified) sterol fractions are listed in Table 3. Cholesterol and β-sitosterol are the most abundant free sterols in the deep water. The norcholestadienol and 22-dehydrocholesterol are next in order of abundance while stigmasterol, campesterol, 22-methylenecholesterol, and ergosterol have lower percentages. Brassicasterol, not only has the lowest concentration of all the sterols, but decreases to a lower percent composition with depth in the water column at all four stations (Table 3) and in the detailed profile (Figure 3). As mentioned earlier, brassicasterol is an algal sterol. Hence, its low concentration in the deep water is not surprising. The large maximum in brassicasterol concentration at 600-800 m may be due to remineral-

TABLE 3. Sterol Distributions in Subsurface Seawater from the North Atlantic Ocean

Sample Number	Depth (m)	% Composition — Sterols							
		Norcholesta-dienol	22-Dehydro-cholesterol	Chole-sterol	Brassi-casterol	Ergo-sterol	Campe-sterol	Stigma-sterol	β-Sito-sterol
Location — Station No. 7 — 4938 m Water Column									
Free Sterols									
K-33-II-4	7	9	13	23	12	—	—	8	31
-1	993	11	—	43	—	—	—	—	47
-2	3095	9	—	48	—	—	—	4	39
-3	4630	9	7	24	—	5	—	4	37
Total Sterols[1]									
K-33-II-4	7	5	8	56	10	—	—	4	17
-1	993	4	—	62	—	—	—	6	28
-2	3095	2	—	75	—	—	—	2	21
-3	4630	10	4	46	—	3	—	3	25
Location — Station No. 6 — 4380 m Water Column									
Free Sterols									
K-33-I-5	7	3	8	39	15	—	1	6	28
-6	80[2]	10	9	30	6	5	3	3	22
-7	702	5	3	45	—	3	5	5	24
-8	4142[2]	6	—	54	—	2	2	4	21
Total Sterols[1]									
K-33-I-5	7	2	5	50	17	—	2	6	19
-6	80[3]	5	6	50	8	3	2	5	14
-7	702	2	2	54	3	3	3	4	19
-8	4142[3]	9	6	50	2	1	1	5	19
Location — Station D — 3050 m Water Column									
Free Sterols									
AII-81-5	10	7	12	14	17	—	8	4	29
-6	110	12	11	21	8	3	3	5	37
-7	1989	17	11	22	—	3	3	6	39
Total Sterols[1]									
AII-81-5	10	6	10	32	10	—	6	4	25
-6	110	7	7	49	6	2	2	7	20
Location — Station C — 1150 m Water Column									
Free Sterols									
AII-81-4	8[2]	10	11	26	13	—	7	4	22
-3	1100	14	9	23	1	3	2	5	35
Total Sterols[1]									
AII-81-4	8[3]	6	10	24	10	—	7	5	26
-3	1100	5	9	40	2	2	4	5	25

[1] Total Sterol Percentage = Free and Esterified Sterols.
[2] A Sterol Peak at 2.55 Retention Time Representing 3% was Present.
[3] A Sterol Peak at 2.55 Retention Time Representing 2% was Present.

ization of sterols bound to organic matter derived from algae, or to physical mixing processes, the same explanations given for the total free sterol maximum at 600-800 m. The fact that brassicasterol is not found deeper than 800 m leads one to believe that this is a fairly labile sterol and is recycled in the top 1000 m of the ocean (Figure 3).

From the gas chromatograms of samples analyzed from Station 6 at 7 m, 80 m, 702 m and 4142 m (*Gagosian*, 1975a), Table 3 and Figure 3, one concludes that below 800 m all the sterol concentrations decrease with increasing depth in the water column except for cholesterol. The reason for the sharp decrease for cholesterol 10 m above the sediment-water interface is now known. Bacteria are known to reduce cholesterol (mw 386) to cholestanol (mw 388) (Figure 1, structure 4a) (*Eyssen*, 1974). Surface sediments in this study area contain many of these reduced sterols (*Gagosian*, unpublished data). It is possible, therefore, that resuspended sediment just above the bottom provides the sites for this biochemical reduction of sterols, and this is reflected by the low cholesterol concentration 10 m above the bottom. Alternatively, these deep water distributions may be a reflection of the surface sterol production at the original site of deep water formation. On the other hand, the distributional differences of individual sterols may be due to the decomposition or structural rearrangement of sterols by organisms during transport to the deep sea. In the euphotic zone phytoplankton produce several C-27, C-28 and C-29 sterols. The organisms living in the deeper water are cholesterol producers. They have the capability of transforming C-28 and C-29 sterols from phytoplankton through dealkylation processes into the C-27 sterol, cholesterol (*Teshima*, 1972). One or more of

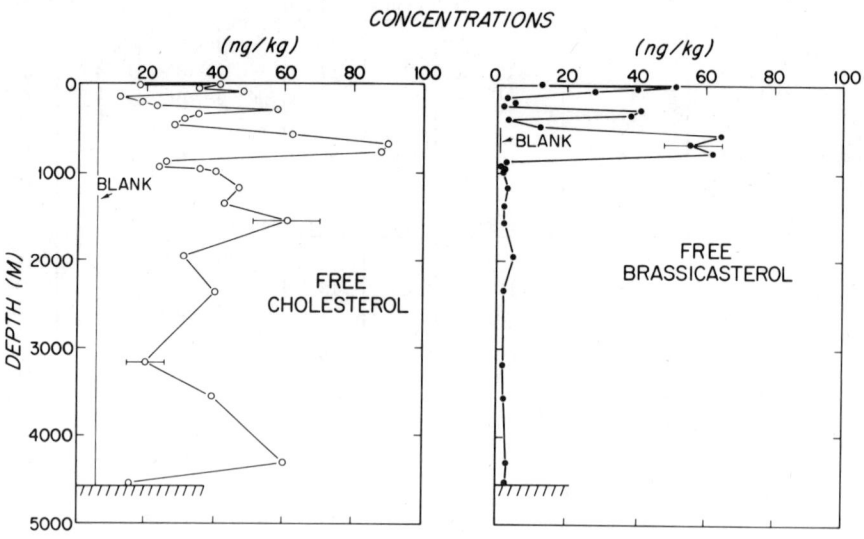

Fig. 3. The individual sterol concentrations of cholesterol and brassicasterol, as a function of depth from 33°40.6'N; 57°36.8'W.

the processes outlined above may be responsible for deep water sterol distributions.

Sterol Esters. By comparing the free and total (free and esterified) sterol percentages for each sterol peak in each sample in Table 3, it is evident that the only sterol percentages which are significantly higher for the total sterols are those of cholesterol (i.e., the ester fraction is made up almost exclusively of cholesterol esters). In the deep water samples cholesterol (free and esterified) is approximately half of the total sterol mixture. For the surface water samples, cholesterol (free and esterified) is from 30 to 40% of the total sterol fraction.

In both the surface and deep water samples, sterol esters comprise only about one-third of the total sterols. This observation is unexpected because plant and animal sterols are usually found in the esterified state with fatty acids or as sulfate conjugates. It is clear, then, that the residence times of sterol esters in the surface waters of the open ocean are quite low. Some esters, however, are not hydrolyzed, even by the time they are transported to the deep water. These compounds may be bound either on or in organic or inorganic detritus which protects the sterol esters from chemical or biochemical hydrolysis. The presence of sterol esters in the deep sea may be due not only to vertical transport on detritus, but also to *in situ* production. Sterol ester production in the deep sea should be considerably lower than in the highly productive euphotic zone. However, the hydrolysis rates of sterol esters in the deep sea may be considerably slower than those in the surface waters. If this is true, then the steady-state concentration of sterol esters in both the deep and surface waters could be approximately the same.

SUMMARY

We have observed from this work that:

(1) Detailed profiles of ancillary data such as POC, hydrographic, and total particulate matter collected at the same station are necessary to complement profiles of specific organic compounds in order to discern the transport mechanisms of these biochemicals to the deep sea.

(2) Free and esterified sterols are present in the Sargasso Sea and the continental shelf and slope water of the western North Atlantic Ocean in the 0.1 - 1.3 µg/l range.

(3) Cholesterol is the major free sterol in both the surface and deep water. β-Sitosterol, fucosterol, brassicasterol, 22-dehydrocholesterol, campesterol, 22-methylenecholesterol, norcholestadienol and stigmasterol are found in lower concentrations at the surface and in the deep sea.

(4) The major sterol esters found in both the surface and deep water are cholesterol esters with very low concentrations of other sterol esters.

(5) Several sterols (e.g., brassicasterol) appear to be produced and consumed in the upper 1000 m of the water column, whereas few sterols (e.g., cholesterol) were found in the entire water column and may be examples of more resistant organic compounds.

Several mechanisms can be postulated for the injection of sterols into the deep sea. Vertical fluxes of organic particles from the surface appear to deliver sterols into the mid-depth waters of the Sargasso Sea. However, some other process(es) is controlling deep water distributions. Physical transport processes (viz. horizontal and vertical advection and diffusion) may be responsible for control of these deep water concentrations. In addition, *in situ* deep water biological production and consumption cannot be ruled out and inputs from sediment resuspension must be considered.

Further work concerning more detailed profiles of sterols and comparisons with nutrients, primary productivity, hydrographic and POC data along with samples from varying oceanic environments (e.g., upwelling areas and anoxic basins) are needed in order to further evaluate the cycling of these most characteristic biogenic substances.

ACKNOWLEDGEMENTS

This research was supported by the Office of Sea Grant of the Department of Commerce (22-2000) and the National Science Foundation (GA-44224). Ms. Gale Nigrelli was responsible for a large amount of the extraction and analytical work described here. This paper is Woods Hole Oceanographic Institution Contribution No. 3667.

REFERENCES

Attaway, D. H., P. Haug, and P. L. Parker. 1971. Sterols in five coastal spermatophytes. *Lipids* 6: 687-691.

Austin, J. 1970. The sterols of marine invertebrates and plants. In: *Advances in Steroid Biochemistry and Pharmacology* (Ed. M. H. Briggs), Academic Press. 73-96.

Biscaye, P. E. and S. L. Eittreim. 1974. Variations in benthic boundary layer phenomena: Nepheloid layer in the North American basin. In: *Suspended Solids in Water* (Ed. R. J. Gibbs), Plenum Publ. Corp. 227-260.

Bodman, R. H., L. W. Slabaugh, and V. T. Bowen. 1961. A multipurpose large volume seawater sampler. *J. Mar. Res.* 19: 141-148.

Boutry, J. and C. Baron. 1967. Etude biochimique des planctons. II. Insaponifiables et sterols d'un plancton marin animal. *Bull. Soc. Chim. Biol.* 49: 1399-1401.

Boutry, J. and G. Jacques. 1970. Etude biochimique des planctons. III. Insaponifiables et sterols de plancton marine vegetal. *Bull. Soc. Chim. Biol.* 52: 349-352.

DeSouza, N. J. and W. R. Nes. 1968. Sterols: Isolation from a blue-green alga. *Science 162*: 363.

Eyssen, H. 1974. Biohydrogenation of sterols by Eubacterium 21,408. *Abstracts from AOCS meeting, October 1974, Philadelphia, Pa. Sterol Symposium.*

Gagosian, R. B. 1975a. Sterols in western North Atlantic Ocean. *Geochim. Cosmochim. Acta,* 39: 1443-1454.

Gagosian, R. B. 1975b. Sterols of the lobster (*Homarus americanus*) and shrimp (*Pandalus borealis*). *Experientia, 31*: 878-880.

Gagosian, R. B. 1976. A detailed vertical profile of sterols in the Sargasso Sea. *Limnol. and Oceanogr., 21*: 702-711.

Goad, L. J., I. Rubenstein, and A. G. Smith. 1972. Sterols of echinoderms. *Proc. Roy. Soc. Ser. B, 180*: 223-246.

Heftmann, E. 1971. Functions of sterols in plants. *Lipids 6*: 128-133.

Hood, D. W., ed. 1971. *Organic Matter in Natural Waters*, University of Alaska.

Idler, D. R. and P. Wiseman. 1971a. Sterols of molluscs. *Int. J. Biochem. 2*: 516-528.

Idler, D. R. and P. Wiseman. 1971b. Sterols of crustacea. *Int. J. Biochem. 2*: 91-98.

Kanazawa, A. and S. Teshima. 1971a. In vivo conversion of cholesterol to steroid hormones in the spiny lobster, *Panulirus japonica*. *Bull. Jap. Soc. Sci. Fish. 37*: 891-898.

Kanazawa, A., M. Yoshioka, and S. Teshima. 1971. The occurrence of brassicasterol in the diatoms *Cyclotella nana*, and *Nitzschia closterium*. *Bull. Jap. Soc. Sci. Fish. 37*: 899-903.

Mattson, J. S., C. A. Smith, T. T. Jones, S. M. Gerchakov, and B. D. Epstein. 1974. Continuous monitoring of dissolved organic matter by UV-visible photometry. *Limnol. Oceanogr. 19*: 530-535.

Menzel, D. W. 1974. Primary productivity, dissolved and particulate organic matter, and the sites of oxidation of organic matter. In: *The Sea, vol. 5, Marine Chemistry* (Ed. E. D. Goldberg), John Wiley and Sons. 659-678.

Menzel, D. W. and R. F. Vaccaro. 1964. Measurement of dissolved organic and particulate carbon in seawater. *Limnol. Oceanogr. 9*: 138-142.

Patterson, G. W. 1971. The distribution of sterols in algae. *Lipids 6*: 120-127.

Reitz, R. C. and J. G. Hamilton. 1968. The isolation and identification of two sterols from two species of blue-green algae. *Comp. Biochem. Physiol. 25*: 401-416.

Riley, J. P. and R. Chester. 1971. Dissolved and particulate organic carbon in the sea. In: *Introduction to Marine Chemistry*, Academic Press. 182-218.

Safe, L. M., C. J. Wong, and R. F. Chandler. 1974. Sterols of marine algae. *J. Pharm. Sci.* 63: 464-466.

Scheuer, P. J. 1973. *Chemistry of Marine Natural Products*, Academic Press. 60-87.

Schubert, K., G. Rose, H. Wachtel, C. Horhold, and N. Ikekawa. 1968. Zum Vorkommen von sterinen in bakterien. *European J. Biochem.* 5: 246-251.

Sharp, J. H. 1973. Size classes of organic carbon in seawater. *Limnol. and Oceanogr.* 441-447.

Smith, L. L., A. K. Dhar, J. L. Gilchrist, and Y. Y. Lin. 1973. Sterols of the brown alga *Sargassum fluitans*. *Phytochem.* 12: 2727-2732.

Spencer, D. W. 1972. GEOSECS II, the 1970 North Atlantic station: Hydrographic features, oxygen and nutrients. *Earth and Planet. Sci. Lett.* 16: 91-102.

Teshima, S. 1972. Studies on the sterol metabolism in marine crustaceans. *Mem. Fac. Fish.*, Kagoshima Univ. 21: 69-147.

Teshima, S. and A. Kanazawa. 1971. Sterol composition of marine occurring yeast. *Bull. Jap. Soc. Sci. Fish.*, 37: 68-72.

Teshima, S. and A. Kanazawa. 1972. Occurrence of sterols in the blue-green alga, *Anabaena cylindrica*. *Bull. Jap. Soc. Scien. Fish.* 38: 1197-1202.

Wagner, F. S. 1969. Composition of the dissolved organic compounds in seawater: A review. *Contrib. Mar. Science* 14: 115-153.

Wangersky, P. J. 1972. The cycle of organic carbon in seawater. *Chimia*, 26: 559-564.

Weete, J. D. 1973. Sterols of the fungi. *Phytochem.* 12: 1843-1864.

Williams, P. M. 1971. The distribution and cycling of organic matter in the ocean. In: *Organic Compounds in Aquatic Environments*. (Eds. S. D. Faust and J. V. Hunter), Marcel Dekker. 145-163.

Williams, P. M. and L. I. Gordon. 1970. Carbon-13:carbon-12 ratios in dissolved and particulate organic matter in the sea. *Deep-Sea Res.* 17: 19-27.

Worthington, L. V. and W. R. Wright. 1970. *North Atlantic Ocean Atlas*, Woods Hole Oceanographic Institution Atlas Series, Vol. 2.

Zandee, D. I. 1964. Absence of sterol synthesis in some arthropods. *Nature* 202: 1335-1336.

Zandee, D. I. 1967. Absence of cholesterol synthesis as contrasted with the presence of fatty acid synthesis in some arthropods. *Comp. Biochem. Physiol.*, 20: 811-822.

Zsolnay, A. 1975. Total labile carbon in the euphotic zone of the Baltic Sea as measured by BOD. *Marine Biology* 29: 125-128.

Deep-Sea Metabolism in the Eastern Tropical North Pacific Ocean

T. T. Packard, H. J. Minas, T. Owens, and A. Devol

Bigelow Laboratory for Ocean Sciences; Centre Universitaire de Luminy, Marseille, France; University of Washington; University of Washington

ABSTRACT

Electron Transport System (ETS) activity in the particulate matter was measured at depths to 3000 m under the Costa Rica Dome. The activity was found to decrease with depth in an exponential form described by the equation, $ETS = ETS_0 e^{-\alpha \ln z}$. A contour plot of the ETS activity at 3000 m reveals gradients that range from 0.3 to 20 pM O_2 hr^{-1} ℓ^{-1} km^{-1}. At 2000 m, the ETS activity was compared to rates of O_2 consumption that were calculated from a vertical advection-diffusion model of the O_2 distribution. The ratio of ETS activity to O_2 consumption was 18.3 ± 12.7. From this ratio and from the directly measured ETS activity, rates of O_2 consumption were calculated. A mechanism is proposed for coupling observed variations in deep metabolism with variations in surface productivity based on the relaxation of upwelling.

INTRODUCTION

The consumption of oxygen and production of CO_2 by the metabolism of marine organisms is a process that occurs throughout the ocean at all depths. In the wind mixed euphotic zone this catabolic process is balanced by the photosynthetic production of oxygen and fixation of CO_2, and by the gaseous exchange with the atmosphere across the sea-surface interface. Below this shallow epipelagic zone, catabolic processes occur unbalanced by *in-situ* production and are held in equilibrium only by the meager supply of oxygen from oceanic mixing and diffusion (*Richards*, 1957). These processes occur at extremely low rates; yet they have an observable effect on the deep-sea fields of O_2 and CO_2. *Craig* (1971) has convincingly demonstrated that without considering these processes the abyssal distribution of O_2 and CO_2 cannot be accurately modeled. From this finding and from the change in the distribution of O_2 caused by the metabolism term in his model, *Craig* (1971) has calculated rates of deep-sea O_2 consumption and CO_2 production. He finds a rate of 179 nM O_2 yr^{-1} ℓ^{-1} for the Pacific Ocean, which agrees with a calculation by *Munk* (1966). *Riley* (1951) finds a significantly lower value of 8.9 nM O_2 yr^{-1} ℓ^{-1} for the deep waters of the Atlantic Ocean.

Calculations of this type are useful in deep-sea studies and have been used by physicists and biologists to determine, among other things, patterns of deep-ocean circulation and the bio-dynamics of deep-sea communities. For example, *Kuo and Veronis* (1973) used an average rate of abyssal O_2 consumption, derived from the nutrient model of *Riley* (1951) to test the abyssal circulation model of *Stommel* (1958). *Carmack and Aagaard* (1973) used the O_2 consumption rate predicted by a box model of the Atlantic deep water (*Wright*, 1969) to calculate the renewal rate of Greenland Sea deep water. On the biological side, O_2 consumption rates were used by *Munk* (1966) to calculate the bioenergetics of deep-sea mixing by nekton and zooplankton, and by *Riley* (1951) to determine the relative importance of the metabolism of the abyssal communities *vis a vis* the metabolism of the epipelagic communities.

Common to all of these studies are the assumptions that the oxygen consumption rate is uniform in the horizontal plane of ocean space and is described in the vertical direction by an exponential equation of the form,

$$J = J_0 e^{-\alpha z} \tag{1}$$

in which J is the *in-situ* oxygen utilization; J_0 is the oxygen utilization in the surface waters, α is a constant and z is depth, positive downward from the sea surface. Both of these assumptions are largely untested, because, except for our earlier work (*Packard, Healy, and Richards*, 1971), no direct measurements of deep-sea metabolism have been made. Thus, many paradigms of deep-sea biology, circulation, or chemistry are simply heuristic exercises and will remain as such until direct measurements of the temporal and spatial field of deep sea metabolic activity are made. This paper describes such measurements, made on the microplankton in deep waters under the Costa Rica Dome in the eastern tropical Pacific Ocean. The measurements were made on the PINTA-I expedition of the R/V T. G. Thompson of the University of Washington. From them, horizontal and vertical gradients in deep-ocean metabolism are calculated, and by comparison with oxygen utilization calculations made with a vertical-advection diffusion model (*Craig*, 1971), oxygen consumption rate profiles are predicted.

METHODS

Seawater samples were taken with a 50 liter PVC bottle, filtered through a 216 μ net into a large polypropylene bottle and transferred to a cold room, maintained at 3°C. The transfer process normally required no more than 10 minutes. The sample suffered no more than a 2° rise in temperature during transfer. Thirty to forty ℓ of the sample were filtered at 3°C and 0.3 atm through a 47 mm Gelman (Type A) glass fiber filter. The filtration was accomplished within 45 min. The filter was ground in a teflon-glass homogenizer at 0 - 4°C and assayed at 20°C for respiratory electron transfer system (ETS) activity (Packard, 1971). The volumes of the homogenate and the extracted formazan solutions were minimized to gain sensitivity. The control consisted of the sum of the pigment blank (homogenate and tetrazolium dye incubated without substrates) and the reagent blank (reagents incubated without homogenate). The precision of the assay was within 1% of the mean value when replicates were measured on the same homogenate (Packard, 1971). When sampling natural populations, a filtration error is introduced and the precision is less. Nevertheless, the standard error of six subsamples of surface water taken from the same water bottle was ±13% at the 28 nM O_2 hr^{-1} ℓ^{-1} level. The standard error of triplicate deep-sea samples taken from the same depth was somewhat less. At the 0.54 nM O_2 hr^{-1} ℓ^{-1} level (500 m samples) it was ± 16% and at the 0.36 nM O_2 hr^{-1} ℓ^{-1} level (3000 m samples) it was ±10%.

The ETS activity at *in situ* temperature was calculated from the equation

$$A = \frac{64.3\,HSE}{V}\left[\exp\frac{Ea}{R}\left(\frac{1}{T_1} - \frac{1}{T_2}\right)\right] \quad (2)$$

in which A is the ETS activity in nM O_2 hr^{-1} ℓ^{-1}, H is the homogenate volume in $m\ell$, S is the volume of the extracted formazan solution in $m\ell$, E is the corrected absorbance of the formazan solution at 4900 Å in a 1 cm spectrophotometer cell, V is the volume of the seawater filtered, Ea = 15 Kcal/mole, R = 1.987 Kcal deg^{-1} $mole^{-1}$, T_1 is the incubation temperature (°K) and T_2 is the *in-situ* sea water temperaure (°K) at the depth of the sample.

RESULTS

ETS activity was measured at 100, 500, 1000, 2000, 3000 m and occasionally other depths in the vicinity of the Costa Rica Dome during the cruise. The location of each station is shown in Figure 1, and the ETS activities are presented in Table 1. Individual ETS profiles do not always decrease with depth in a simple predictable manner. Analysis of the upper portion (between 100 m and 1000 m) of 13 profiles reveals that 5 profiles exhibit no change in activity with depth, 5 exhibit a minimum between 300 and 800 m, 2 decrease logarithmically with depth, and one actually shows a maximum at 500 m (Station 10). At greater depths (between 1 km and 3 km) 8 out

Table 1. ETS activity (nM O_2 hr^{-1} ℓ^{-1}) in the particulate matter in the northeastern tropical Pacific Ocean.

Station	Depth (km)	ETS	Station	Depth (km)	ETS	Station	Depth (km)	ETS	Station	Depth (km)	ETS
2	0.5	1.027	10	0.1	0.134	15	0.1	1.027	18	0.1	1.607
	1.0	1.071		0.5	1.473		0.2	0.402		0.3	0.223
3	1.0	2.768		1.0	0.402		0.3	0.022		0.5	0.625
4	1.0	1.205		2.0	0.084		0.4	0.268		1.0	0.402
6	0.1	2.232		3.0	0.491		0.5	2.009		2.0	0.179
	0.5	2.455	11	0.1	3.169		1.0	0.759		3.0	0.089
	1.0	1.562		0.5	0.804		2.0	0.179	19	0.1	3.705
	2.0	0.223		1.0	0.848		3.0	0.009		0.2	0.536
	3.0	0.089		2.0	0.670	16	0.1	3.571		0.3	0.223
7	0.1	3.616		3.0	0.223		0.3	0.313		0.4	0.446
	0.5	1.294	13	0.1	1.920		0.5	0.223		0.5	0.313
	1.0	0.401		0.5	1.071		1.0	0.536		1.0	0.067
	2.0	0.223		1.0	0.580		2.0	0.027		2.0	0.067
	2.5	0.178		2.0	0.040		3.0	0.134		3.0	0.063
9	0.1	3.348		3.0	0.313	17	0.1	3.750	20	0.1	3.036
	0.5	2.232	14	0.1	5.313		0.3	0.759		0.3	0.313
	1.0	0.536		0.3	1.161		0.5	0.536		0.5	0.402
	2.0	0.267		0.5	0.179		1.0	0.446		1.0	0.313
	3.5	0.491		1.0	0.134		2.0	0.268		2.0	0.134
				2.0	0.223		3.0	0.058			
				3.0	0.402						

DEEP-SEA METABOLISM 105

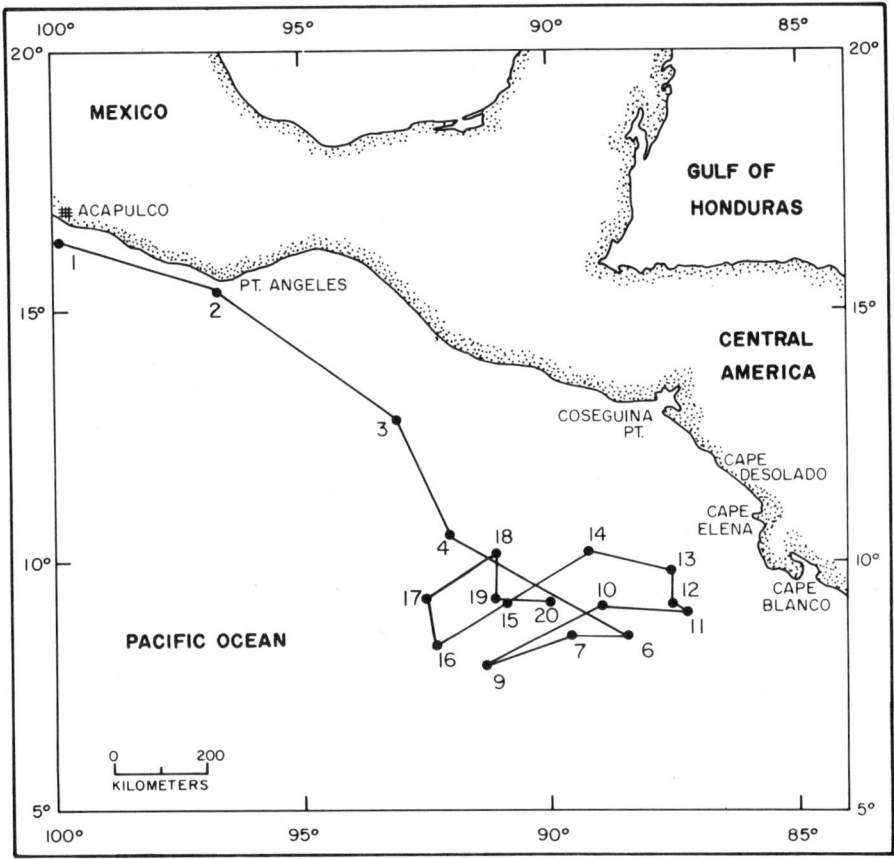

Fig. 1. Track of the R/V T. G. Thompson (Cruise 75) in the vicinity of the Costa Rica Dome. Station numbers are indicated.

of the 13 profiles exhibit a continual decrease with depth, 4 show a minimum at 2000 m and one shows a continual increase with depth (Station 14). In spite of these individual variations, a generalized ETS profile (Figure 2), characterized by a logarithmic decrease with depth, emerges when the data are averaged. Regression analysis (Figure 2) reveals that the equation $ETS = ETS_0 e^{-\alpha \ln z}$ fits the profile better than $ETS = ETS_0 e^{-\alpha z}$. In both equations ETS_0 is the ETS activity in the surface waters, α is a constant derived from the slope of the line in Figure 2, and z is the depth of the ETS measurement in km from the surface. *Wyrtki* (1962) and *Ben-Yaakov* (1972) discuss values of α; *Wyrtki* (1962) finds that a value between 3-4 km^{-1} describes the Atlantic Ocean data of *Riley* (1951), and *Ben-Yaakov* (1972) finds a lower value of 1.9 to 1.5 km^{-1} for the

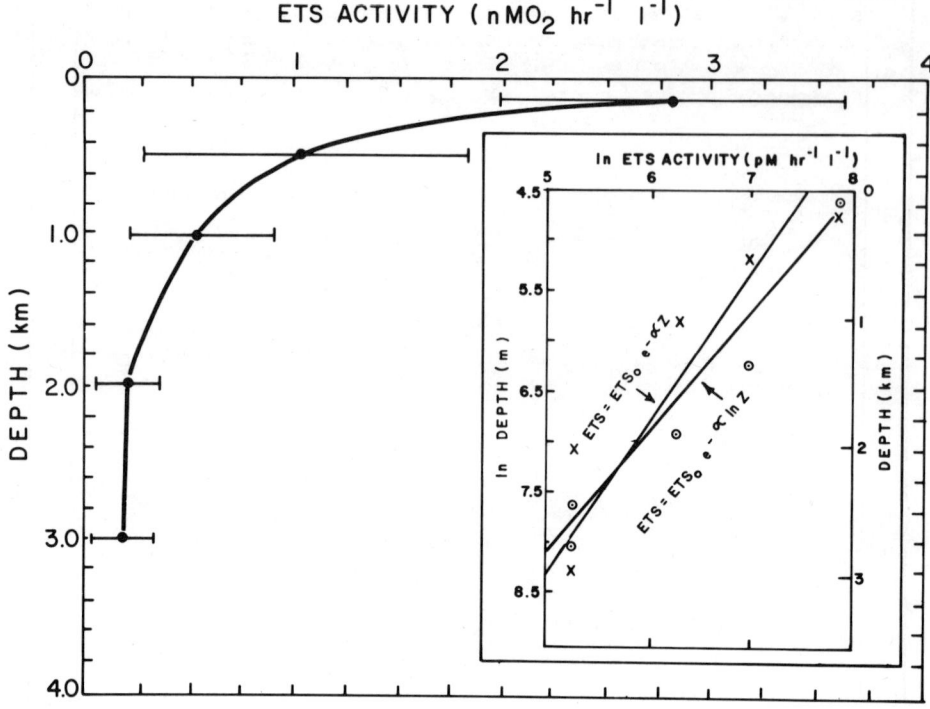

Fig. 2. Depth profile of ETS activity (Table 4) in the vicinity of the Costa Rica Dome. The inset figure compares a plot of $\ln$ ETS versus depth (x) with a plot of $\ln$ ETS versus log depth (o). The former plot is described by $\ln$ ETS $= -0.90\ z + 7.5$ ($r = 0.93$) while the latter is described by $\ln$ ETS $= -0.84\ \ln z + 11.9$ ($r = 0.97$). Taking the antilogarithms, these equations become: ETS $= 1.86 e^{-0.9z}$ and ETS $= 150 e^{-0.84\ \ln z}$, in nM units.

Northeastern Pacific. We find a value of 0.9 km^{-1} for the simple exponential equation (ETS $= ETS_0 e^{-\alpha z}$). For the log-log form (ETS $= ETS_0 e^{-\alpha \ln z}$) we find values of 0.84 km^{-1} and 0.15 $\mu M\ O_2\ hr^{-1}\ \ell^{-1}$ for α and ETS_0, respectively.

The horizontal variations of deep-sea ETS activity at 3000 m and in the waters between 100 and 3000 m are shown in Figures 3 and 4, respectively. At 3000 m (Figure 3) the ETS activity ranges from 58 to 491 $pM\ O_2\ hr^{-1}\ \ell^{-1}$ while the gradients of the ETS activity range from 0.3 to 20 $pM\ O_2\ hr^{-1}\ \ell^{-1}\ km^{-1}$. The ETS activity in the water column (Figure 4) ranges by a factor of 3 from 0.94 $mM\ O_2\ hr^{-1}\ m^{-2}$ in the northwest corner of the region to 2.99 $mM\ O_2\ hr^{-1}\ m^{-2}$ in the southeast corner. Along the same transect, but calculated on a volume basis, the mean consumption of the water column ranges from 0.18 to 1.03 $nM\ O_2\ hr^{-1}\ \ell^{-1}$. A metabolic frontal zone extends

DEEP-SEA METABOLISM

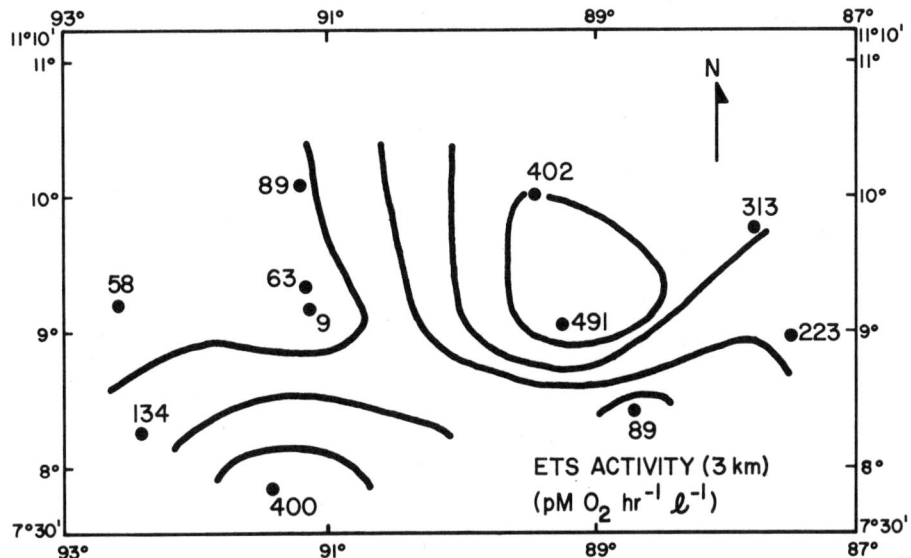

Fig. 3. Horizontal gradients of ETS activity at 3 km.

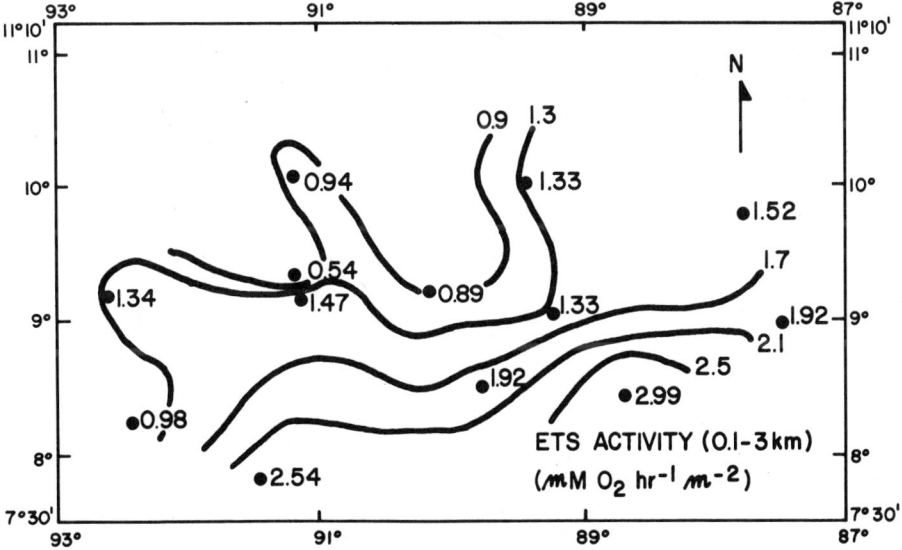

Fig. 4. Horizontal gradients of the total ETS activity in the water column between 0.1 and 3 km.

from the southwest to the northeast corner of the region. Its position is not readily explicable but since 9°N, 89°W is the reputed center of the Costa Rica Dome (*Wrytki*, 1964), large "deep-metabolism" gradients (*Craig*, 1971), associated with the enhanced surface production could be expected.

DISCUSSION

Respiratory ETS activity is an index of catabolic metabolism. The effect of this metabolic activity on oxygenated sea water is to decrease the O_2 and increase the CO_2 concentrations. But, calculating the decrease in O_2 from ETS activity involves a calibration problem that is similar to that associated with the conversion of ATP to living carbon biomass and chlorophyll to phytoplankton carbon biomass. In the euphotic zone this problem is relatively easy to solve because the microplankton is dominated by phytoplankton and furthermore, the relationship between ETS activity and respiratory oxygen consumption has been investigated (*Kenner and Ahmed*, 1975). In the deep sea both the population composition and the relationship between ETS activity and respiration in the components of this microplankton population are largely unknown.

Our original solution to this problem was to use a respiratory control ratio (RCR) empirically determined on terrestrial mammals or enteric bacteria. A preferable solution is the direct calibration with independently derived deep-sea oxygen utilization rates. Recently, such rate calculations have been made by *Craig* (1971) and *Kroopnick* (1974) and using their approach oxygen utilization rates have been calculated at 2 km from the O_2 and hydrographic data at each station (*Kuntz et al.*, 1975) and compared with corresponding ETS activities (Table 1). These calculations (Table 2) require the solution of the following equation:

$$J = w\{[(C - C_o) - (C_m - C_o)\delta(z)]/[z - z_m\delta(3)]\} \qquad (3)$$

in which J is the oxygen utilization rate in $\mu M\ O_2\ yr^{-1}\ \ell^{-1}$, C is the oxygen concentration at depth (z) and C_m and C_o are the oxygen concentrations at the top and the bottom of the linear portion of the T-S diagram for each station. Depth (z) is taken as positive upward with $z_o = 0$ and $z = z_m$ at the top and bottom of this same diagram. The calculation of $\delta(3)$ requires knowledge of the mixing parameter, K/w, in which K is the vertical eddy diffusion coefficient and w is the vertical advection velocity. Further discussion and description of equation (3), $\delta(3)$, etc., may be found in a series of papers by *Craig* (1969, 1971) and *Craig and Weiss* (1970). The 2 km level was chosen for the calibration depth because it was in the linear portion of the T-S diagram for this region and because the suite of ETS and oxygen data was most complete at this depth. Depth profiles of salinity in this depth range were convex upward indicating that w was posi-

Table 2. Calculation of the ETS/J ratio at the 2 km level in the vicinity of the Costa Rica Dome. K/w and J/w were calculated by the methods of Wyrtki (1962) and Craig (1969), respectively. A value of 4 m yr^{-1} was used for w (Kroopnick, 1974).

Station	K/w (Km)	J/w ($\mu M\ O_2\ \ell^{-1} km^{-1}$)	J (pM $O_2\ hr^{-1} \ell^{-1}$)	ETS (pM $O_2\ hr^{-1} \ell^{-1}$)	ETS/J
6	0.63	−14.96	6.70	223.2	33.3
7	0.63	−21.70	9.82	223.2	22.7
13	0.60	−11.61	5.36	40.2	7.5
14	0.63	−28.79	13.39	223.2	16.7
15	0.61	−12.81	5.80	223.2	38.5
16	0.55	−34.96	16.07	26.8	1.7
18	0.60	−23.84	11.16	178.6	16.0
19	0.58	−15.04	6.70	67.0	10.0

Mean value of ETS/J = 18.3

Standard deviation = ±12.7

Coefficicent of variation = ±69%

tive. K/w was calculated for each station from the slope of the depth profile of $\ln(\Theta - \Theta_0)$ where Θ and Θ_0 are potential temperatures at depth z and depth $z = z_0$ (Wyrtki, 1962). The horizontal oxygen gradients were smaller than the vertical gradients by a factor of 6×10^{-3}, fulfilling the final requirement of the vertical-advection diffusion model.

The ratio of ETS to J (Table 2) varies from 1.7 to 38.5 with a mean value of 18.3 ± 12.7. One would expect a ratio of 2 if the measured ETS activity is the maximum activity (v_{max}) of the respiratory electron transport system and if this system is constrained by substrate availability to operate at its K_m, i.e., at $1/2$ v_{max}. However, since the biochemistry and physiology of deep-sea plankton are largely unknown, these assumptions, as well as the ratio of 2, may be invalid; but, for heuristic purposes one can accept the empirical validity of 18.3 for the deep waters under the Costa Rica Dome and calculate rates of oxygen consumption from the ETS data in Table 1. The results of these calculations are shown in Table 3.

Both the oxygen consumption rates and the ETS activities are slightly higher than the ones reported on an earlier cruise (Packard et al., 1971). Here the ETS activity at 3 km averaged 0.10 nM O_2 hr^{-1} ℓ^{-1}; whereas recent measurements yield 0.19 nM O_2 hr^{-1} ℓ^{-1} (Table 4). However, the oxygen consumption calculations were essentially the same, 10.7 pM O_2 hr^{-1} ℓ^{-1} and 10 pM O_2 hr^{-1} ℓ^{-1} (Table 4). At shallower depths the discrepancy is greater but is probably not of serious consequence due to the paucity of the shallow-water measurements in the Costa Rica Dome during the earlier cruise. The aberrance, however, does increase the calculated rate of carbon consumption. A previous calculation (Packard et al., 1971) showed that only 3% of the surface productivity was consumed below 100 m. It is now calculated that between 100 and 3000 m, 13 gr yr^{-1} m^{-2} carbon is consumed. This value represents 18% and 9% of the productivity of the region as estimated by Broenkow's nutrient model of the Costa Rica Dome (Broenkow, 1965) and as measured by the Eastropac Program (Owen and Zeitschel, 1970).

Our original calculation of the deep water residence time was 900 years. By following the same argument, using 4.5 mℓ/ℓ as the initial oxygen concentration and using 2.63 mℓ/ℓ for the mean O_2 concentration at the 3000 m level under the Costa Rica Dome, and 2.5 $\mu\ell$ yr^{-1} ℓ^{-1} for the O_2 consumption rate one can calculate a residence time of 750 years. This calculation is sensitive to the initial value of the oxygen. If one uses Duedall and Coote's (1972) value of 4.8 mℓ/ℓ at 3000 m in the South Pacific, for the initial oxygen concentration, the residence time is 870 years.

Even though a logarithmic profile described the depth related decrease of the ETS activity below the Costa Rica Dome, individual profiles displayed great variety of shape. It is unlikely that this variation was caused by imprecision because the observed deviations in the profiles were greater than ±16%. Since ETS variations normally reflect biomass variations (Packard et al., 1974) and since the productivity in the surface waters is the only source of deep-sea biomass

Table 3. Oxygen consumption ($nM\ O_2\ hr^{-1}\ \ell^{-1}$) in the particulate matter in the northeastern tropical Pacific Ocean. These rates were calculated from the ETS data in Table 1 by dividing by the factor 18.3.

Station	Depth (km)	R	Station	Depth (km)	R	Station	Depth (km)	R	Station	Depth (km)	R
2	0.5	0.056	10	0.1	0.007	15	0.1	0.056	18	0.1	0.088
	1.0	0.059		0.5	0.080		0.2	0.022		0.3	0.012
3	1.0	0.151		1.0	0.022		0.3	0.001		0.5	0.034
4	1.0	0.066		2.0	0.005		0.4	0.015		1.0	0.022
	0.1	0.122		3.0	0.027		0.5	0.110		2.0	0.010
6	0.1	0.134	11	0.1	0.173		1.0	0.041		3.0	0.005
	0.5	0.085		0.5	0.044		2.0	0.010	19	0.1	0.202
	1.0	0.012		1.0	0.046		3.0	0.001		0.2	0.029
	2.0	0.005		2.0	0.037	16	0.1	0.195		0.3	0.012
	3.0	0.196		3.0	0.012		0.3	0.017		0.4	0.024
7	0.1	0.071	13	0.1	0.104		0.5	0.012		0.5	0.017
	0.5	0.022		0.5	0.058		1.0	0.029		1.0	0.004
	1.0	0.012		1.0	0.032		2.0	0.001		2.0	0.004
	2.0	0.010		2.0	0.002		3.0	0.007		3.0	0.003
	2.5	0.183		3.0	0.017	17	0.1	0.205	20	0.1	0.166
9	0.1	0.122	14	0.1	0.290		0.3	0.041		0.3	0.017
	0.5	0.029		0.3	0.063		0.5	0.029		0.5	0.022
	1.0	0.015		0.5	0.010		1.0	0.024		1.0	0.017
	2.0	0.027		1.0	0.007		2.0	0.015		2.0	0.007
	3.5			2.0	0.012		3.0	0.003			
				3.0	0.022						

Table 4. Mean values of the ETS activity, the oxygen consumption rate, and the carbon oxidation rate in the vicinity of the Costa Rica Dome. The oxygen consumption was calculated from the expression, $R = ETS/18.3$; the carbon oxidation rate (J_{CO_2}) was calculated from the expression $J_{CO_2} = 10.2\ ETS/18.3$ (R.Q. = 0.85 was assumed). ΣJ_{CO_2} between 0.1 and 3 km was 8.5 C $yr^{-1} m^{-2}$.

Depth (km)	ETS Activity ($nM\ O_2\ hr^{-1}\ \ell^{-1}$)	Oxygen Consumption ($pM\ O_2\ hr^{-1}\ \ell^{-1}$)	Oxygen Consumption ($\mu M\ O_2\ yr^{-1}\ \ell^{-1}$)	Carbon Oxidation ($\mu g\ C\ yr^{-1}\ \ell^{-1}$)
0.1	2.80 ± 0.14	153	1.34	13.67
0.5	1.05 ± 0.79	57	0.50	5.13
1.0	0.54 ± 0.38	29	0.26	2.64
2.0	0.20 ± 0.16	11	0.10	0.98
3.0	0.19 ± 0.16	10	0.09	0.93

in the oceanic realm, it is possible that these abyssal variations in the ETS profiles are related to changes in the surface productivity of the Costa Rica Dome. Seasonal variations of the carbon productivity in the vicinity of the Costa Rica Dome have been noted by *Owen and Zeitzschel* (1970) and yearly variations by successive cruises into the area by the R/V. T. G. Thompson in January-February, 1968, 1969, and 1973 (*Dugdale and Healy*, 1970; *MacIsaac and Dugdale*, 1970; and *Kuntz et al.*, 1975). These productivity variations are caused by alternate relaxations in the upwelling which couple to provide both a source for deep-sea particulate organic matter and an explanation of the variety of shapes of deep-sea ETS profiles. This coupling between upwelling, productivity, and the delivery of particulate organic matter to the deep-sea is illustrated in Figure 5. The mechanism is based on the assumption that a relaxation of upwelling will entrain plankton in the subsiding waters and that once these organisms are isolated from light they will begin to die and sink. The sinking of this submerged bloom would then serve as food for the deep-sea pelagic and benthic organisms. It would be characterized by maxima in the profiles of the organic matter indices, such as the maxima observed in the ETS activity (Table 1). Now, if the upwelling in the Costa Rica Dome occurs as sporadic patches on top of the subsurface permanent thermal dome, as *Wyrtki* (1964) suggests, then the mechanism shown in Figure 5 helps to explain the variety of shapes that we observed in the ETS profiles (Table 1). Furthermore, it helps to explain the deviations in the ETS-J ratios. Since the time scale of the life of the ETS maxima would be on the order of months and since the time scale inherent in the calculations of J/w, K/w and w is on the order of years, if not decades, it would be only fortuitous to find systematic changes in the ETS-R ratio from station to station. Only in regions of unvarying productivity, such as the central gyres, would the ETS-J ratio show little variation.

SUMMARY AND CONCLUSIONS

Deep metabolism as indicated by ETS activity was demonstrated to vary both vertically and horizontally in the bathyal waters under the Costa Rica Dome. The average depth variation was described by an exponential equation of the form: $ETS = ETS_o e^{-\alpha \ln z}$. The deviation of individual profiles from the average may have been caused by variations in the surface productivity that in turn, were related to intermittant upwelling. Horizontal variations in the total ETS activity between 100 m and 3000 m were characterized by a frontal zone extending under the Dome from the southwest to the northeast with elevated activity in the southwest sector. Horizontal gradients in the ETS activity at the 3000 m level ranged from 0.3 to 20 $pM\ O_2\ hr^{-1}\ \ell^{-1}\ km^{-1}$.

A calibration of the ETS method was made at the 2000 m level with the vertical advection-diffusion model of *Craig* (1969). An ETS-R ratio of 18.3 was obtained. From this ratio and the ETS profiles,

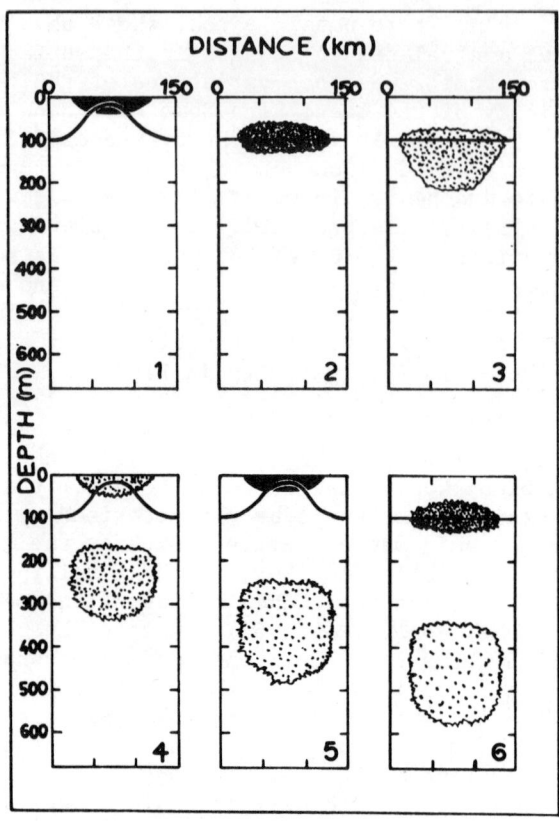

Fig. 5. A mechanism for pumping particulate organic matter into the deep-sea. Panel 1 shows the ascendance of subsurface water into the euphotic zone and the plankton bloom zone that these rising waters stimulate. The black line represents a hypothetical isopycnal that would normally be found at 100 m in the surrounding oligotrophic waters. Panel 2 shows a relaxation of the upwelling with a return of the isopycnal to its original depth and the isolation of the partially diluted bloom in the dimly lit subsurface waters. The warmer oligotrophic waters of the surrounding seas have responded to the relaxation by spreading over the previously upwelled, denser, and now plankton-rich waters. Panel 3 shows the dispersion, through advection of this subsurface plankton patch and the onset of sinking in the moribund plankton population. Panel 4 shows a return to upwelling conditions with doming of the isopycnals and entrainment of the upper part of the patch into the euphotic zone. The elevated portion of the patch serves as a seed for a new plankton bloom while the lower portion continues to sink into the deep sea. Panel 5 shows the growth response of the seed population in the euphotic zone and the continued dispersion and sinking of the deep patch. Panel 6 shows a second relaxation and the birth of another subsurface plankton patch that will serve as another pulse in the food supply for deep-sea organisms.

rates of oxygen utilization were calculated between the 100 m and 3000 m level and were shown to be 90% and 7% higher than previous calculations at the 100 and 3000 m levels, respectively. These differences have a large effect on the rate of carbon consumption in the deep water column, but little on age calculations for the water at 3000 m.

ACKNOWLEDGEMENTS

We thank Dr. D. Blasco for discussing the effects of intermittent upwelling on phytoplankton populations, Ms. M. McCarty for her original contour maps, Ms. C. Johnson for typing the manuscript and Dr. A. Zirino for reading it. This work was funded by ONR contract N-00014-67-A-0103-0014 to T. Packard. Contribution No. 75017 from the Bigelow Laboratory for Ocean Sciences, W. Boothbay Harbor, Maine 04575.

REFERENCES

Ben-Yaakov, S. 1972. On the CO_2-O_2 System in the Northeastern Pacific. *Mar. Chem.*, 1(1): 3-26.

Broenkow, W. W. 1965. The distribution of nutrients in the Costa Rica Dome in the eastern tropical Pacific Ocean. *Limnol. Oceanogr.*, 10(1): 40-52.

Carmack, E. and K. Aagaard. 1973. On the deep water of the Greenland Sea. *Deep-Sea Res.*, 20: 687-715.

Craig. H. 1969. Abyssal carbon and radiocarbon in the Pacific. *J. Geophys. Res.*, 74: 5491-5506.

Craig, H. and R. F. Weiss. 1970. The GEOSECS 1969 intercalibration stations; introduction, hydrographic features, the total CO_2-O_2 relationships. *G. Geophys. Res.*, 75: 7641-7647.

Craig, H. 1971. The deep metabolism: oxygen consumption in abyssal ocean water. *G. Geophys. Res.*, 76(21): 5078-5086

Duedall, I. W. and A. R. Coote. 1972. Oxygen distribution in the Pacific Ocean. *J. Geophys. Res.*, 77(12): 2201-2203.

Dugdale, R. C. and M. L. Healy. 1970. *Tech. Rep.*, Dept. Oceanogr., University of Wash., 250: 39-162.

Kenner, R. A. and S. I. Ahmed. The correlation between oxygen utilization and electron transport activity in marine phytoplankton. *Mar. Biol.*, 33: 129-133.

Kroopnick, P. 1974. The dissolved O_2-CO_2-^{13}C system in the eastern equatorial Pacific. *Deep-Sea Res.*, 21: 211-227.

Kuntz, D., T. Packard, A. Devol, and J. Anderson. 1975. Chemical, physical, and biological observations in the vicinity of the Costa Rica Dome (January-February, 1973). *Tech. Rep. No. 321*, Dept. Oceanogr., University of Wash., 187 pp.

Kuo, H. H. and G. Veronis. 1973. The use of oxygen as a test for an abyssal circulation model. *Deep-Sea Res.*, 20: 871-888.

MacIsaac, J. and R. C. Dugdale. 1970. University of Washington. *Spec. Rep. Dept. Oceanogr.*, 42: 61-97.

Munk, W. H. 1966. Abyssal recipes. *Deep-Sea Res.*, 13: 707-730.

Owen, R. W. and B. Zeitschel. 1970. Phytoplankton production: seasonal change in the oceanic eastern tropical Pacific. *Mar. Biol.* 7: 32-36.

Packard, T. T. 1971. The measurement of respiratory electron transport activity in marine phytoplankton. *J. Mar. Res.*, 29(3): 235-244.

Packard, T. T., M. L. Healy, and F. A. Richards. 1971. Vertical distribution of the activity of the respiratory electron transport system in marine plankton. *Limnol. Oceanogr.* 16: 60-70.

Packard, T. T., D. Harmon, and J. Boucher. 1974. Respiratory electron transport activity in plankton. *Tethys*, 6: 213-222.

Richards, F. A. 1957. Oxygen in the ocean. *Geol. Soc. Amer. Mem. 1*: 185-238.

Riley, G. A. 1951. Oxygen, phosphate, and nitrate in the Atlantic Ocean. *Bingham Oceanogr. Coll. Bull.* 13(1): 1-128.

Stommel, H. 1958. The abyssal circulation. *Deep-Sea Res.*, 5: 80-82.

Wright, R. 1969. Deep water movement in the Western Atlantic as determined by use of a box model. *Deep-Sea Res. 16(Supplement)*: 433-446.

Wyrtki, K. 1962. The oxygen minima in relation to ocean circulation. *Deep-Sea Res.*, 9: 11-23.

Wyrtki, K. 1964. Upwelling in the Costa Rica Dome. *Fishery Bull.*, 63(2): 355-372.

Marine Areas of Anomalous Chemistry Resulting from Oxygen Deficiencies

Francis A. Richards

University of Washington

ABSTRACT

Oxygen deficient environments arise when the circulatory replacement of dissolved oxygen to the deeper waters is outstripped by biochemical oxygen consumption. Such conditions characterize the Black Sea, Cariaco Trench, a variety of fiords and other embayments, and enormous areas in the eastern tropical Pacific Ocean.

As more and more organic matter is decomposed in a marine system, the sequence is 1) oxidation at the expense of dissolved oxygen, 2) nitrate reduction and denitrification, 3) sulfate reduction, and 4) anaerobic fermentation. Each step is accompanied by the production of CO_2 and phosphate ions. Step 1 eventually introduces nitrate ions, but free nitrogen is the end product of denitrification. Toxic sulfides and ammonia accumulate during Step 3, and Step 4 produces methane. Step 1 characterizes almost all the marine environment, but the subsequent steps result in "unusual" environments that may be biologically hostile.

INTRODUCTION

In most marine environments the water column contains dissolved oxygen, has a positive redox potential, and supports respiration at the expense of dissolved oxygen. In this oxidizing climate those elements easily oxidized or reduced are generally in the oxidized state; for example, Fe^{+3}, Mn^{+4}, NO_3^-, and SO_4^{-2}. In all but a few marine areas oxygen tensions are above the required threshold for the respiratory activity of most marine organisms, although recent work indicates that skipjack tuna avoid waters with less than about 3 ml/l dissolved oxygen (R. A. Barkley, personal communication). In these ventilated environments, photosynthesis, advection, and diffusion supply dissolved oxygen more rapidly than it is consumed by respiration or exchange across the sea surface, or both. Where this balance does not prevail, and respiratory consumption outstrips the supply mechanisms, oxygen deficient and anoxic environments arise. These conditions occur fairly commonly in bottom sediments, which are often reducing. However, they are much less common in the water column. The two largest of the latter systems are the Black Sea and the Cariaco Trench (*Richards*, 1965b). Smaller fjords and embayments that contain anoxic and sulfide-bearing waters exist in Norway, British Columbia, Canada, Costa Rica, Venezuela, Indonesia, etc. and have been discussed by the author (*Richards and Vaccaro*, 1956; *Richards*, 1965a, b, 1975; *Richards, Cline, Broenkow and Atkinson*, 1965).

The unusual environments to be considered in this paper include enormous oxygen-deficient areas in the eastern tropical Pacific Ocean in which oxygen concentrations approach zero, nitrate reduction and denitrification take place, but no sulfate reduction has been observed. Thus, we will consider oxygen deficient areas characterized by oxygen concentrations sufficiently low for nitrate reduction and denitrification to take place; systems that may seasonally or occasionally become anoxic and sulfide bearing; and systems that appear to be permanently anoxic and sulfide bearing. There are also some landlocked fjords that contain relict seawater in which sulfate reduction has gone to completion (*Strøm*, 1957, 1961; *Williams, Mathews, and Pickard*, 1961).

PROCESSES

In addition to nitrate reduction, denitrification, and sulfate reduction, anaerobic fermentation with the production of methane has been demonstrated in these systems (*Atkinson and Richards*, 1967) and, where sulfide-bearing waters are present in the photic zone, photosynthesis can be carried out by sulfur bacteria. Free sulfur, arising from the oxidation of sulfides, is produced during this process.

CHEMICAL CONSEQUENCES

Essentially all of the oxygen demand in the ocean is attributable to the organic matter ultimately formed photosynthetically in the photic zone. The oxygen demand of such reduced substances as Mn^{++}, Fe^{++}, and H_2S is practically negligible in most of the oceans. But, the chemical oxidation of these materials may assume importance at the sulfide-oxygen interface of sulfide-bearing systems, where reduced substances mix with oxygen-bearing water (*Cline and Richards*, 1969). An approximation of the oxygen demand of marine organic matter is stated in the stoichiometric model developed by *Richards* (1965a, b) and Redfield, Ketchum, and Richards (1963). The model organic matter is based on the C:N:P ratios in plankton organisms (Redfield, 1934). It assumes carbon in the oxidation state of CH_2O and wholly reduced nitrogen being wholly oxidized to CO_2 and NO_3^-:

$$(CH_2O)_{106}(NH_3)_{16}H_3PO_4 + 138\ O_2 = 106\ CO_2 + 106\ H_2O + 16\ HNO_3 + H_3PO_4 . \qquad (1)$$

This overly simplistic statement conceals many complexities, specifically the steps in the nitrogen cycle. For example, the oxidative process can be assumed to release simple amines and ammonia, which are then biochemically oxidized, presumably via nitrite, to nitrate.

In the environments of interest in this paper, the feeble circulation, or lack thereof, of the deep waters of many fjords and inlets, the Black Sea, and the Cariaco Trench, appears to be the major factor responsible for their becoming anoxic. We know too little about the balances in the eastern tropical Pacific to attempt to evaluate the relative importance of oxygen supplying and oxygen consuming processes responsible for the formation, maintenance, and decay of the highly developed oxygen minimum zone there.

The most obvious chemical consequence in these systems is the accumulation of the products of organic decomposition to higher levels relative to those observed in otherwise comparable marine environments. High concentrations of phosphate and silicate are readily demonstrable (Table 1, Richards, 1965a, b). The concentrations of CO_2 produced are also unusually high, and these can increase the carbonate alkalinity of the system by reactions with carbonates:

$$MeCO_3 + H_2O + CO_2 = Me(HCO_3)_2 = Me^{+2} + 2\ HCO_3^- . \qquad (2)$$

However, even larger increases in the titration alkalinity appear to be attributable to the formation and ionization of H_2S (Knull and Richards, 1969). It has also been speculated that accumulations of the anions of organic acids make significant contributions to the alkalinity of these systems (Gripenberg, 1960).

One can assume a stepwise sequence of the biochemical processes responsible for the decomposition of organic matter:
1) oxidation at the expense of dissolved oxygen
2) nitrate reduction and denitrification
3) sulfate reduction
4) anaerobic fermentation.

The reduction of CO_2 according to the equation $CO_2 + 8\ H = CH_4 + 2\ H_2O$ might also be considered a possibility. However, evidence of this process is uncertain. Evidence is also lacking for the use of phosphate and arsenate as the inorganic substrates, which are possibilities to be considered.

During step 2), which begins when oxygen concentrations reach low levels, nitrate reduction and denitrification result in lower-than-expected nitrate concentrations (nitrate anomalies or deficits),

TABLE 1. Approximate Concentrations of Phosphate and Reactive Silicate in Some Sulfide-Bearing and Otherwise Comparable, Nearby Marine Environments. (From *Richards*, 1965a)

	Sulfide-Bearing System	Otherwise Comparable Oxygenated System
Phosphate Concentrations		
Cariaco Basin, 900 m	∼2.6 µg-at./ℓ	
Nearby open Caribbean, 900 m		∼2.0 µg-at./ℓ
Saanich Inlet, 145 m	∼5.8 µg-at./ℓ	
Strait of Juan de Fuca, 145 m		∼2.6 µg-at./ℓ
Black Sea, 1600 m	∼8 µg-at./ℓ	
Mediterranean Sea, 1600 m		∼0.2 µg-at./ℓ
Reactive Silicate Concentrations		
Cariaco Basin, 900 m	∼60 µg-at./ℓ	
Nearby open Caribbean, 900 m		∼25 µg-at./ℓ
Black Sea, 1600 m	∼130 µg-at./ℓ	
Mediterranean Sea, 1600 m		∼15 µg-at./ℓ

the accumulation of larger-than-usual nitrite concentrations, and an increase in free nitrogen. Deficits of up to 15 µg-at./ℓ and nitrite concentrations of over 3 µg-at./ℓ have been observed in the eastern tropical North Pacific (*Codispoti*, 1973). The amount of free N_2 produced during denitrification is, even in the most extreme case, a small fraction of the N_2 present as a result of the solution of air in seawater. Ordinary analytical methods and estimates of the amount of N_2 dissolved from the air are too imprecise to evaluate with certainty by direct measurement the contribution of nitrogen produced during denitrification to the total amounts of dissolved free nitrogen.

The concentrations of H_2S that develop in these environments constitute a significant chemical characteristic. If natural seawater acted as an inorganic solution having an ionic strength of about 0.7, one might expect the concentrations of heavy metal ions that form highly insoluble sulfides to be undetectably low. Contrary to this expectation, soluble zinc occurs in much higher concentrations in Lake Nitinat, a sulfide-bearing fjord on Vancouver Island, British Columbia, than would be expected from the solubility of zinc sulfide (*Zirino*, 1970). On the other hand, *Richards, Cline, Broenkow, and Atkinson* (1965) demonstrated that the sulfide-bearing waters of lake Nitinat are probably a series of saturated solutions of a compound in which the ratio of iron to sulfide atoms is 1:1. The apparent solubility constant of the compound is 2.3×10^{-17}, compared to the solubility of FeS in distilled water of 3.7×10^{-19} (Hodgman, 1959, p. 1740). It is probable that zinc is kept in solution by organic chelating (complexing) agents. The specific nature of these compounds is almost wholly unknown, but experiments designed to characterize organic materials in natural seawater in terms of their ability to complex heavy metals and to determine the stability constants of these complexes are being carried out in our laboratory. Anodic stripping voltammetry yields estimates of the concentrations of free or loosely bound metal ions, while gel filtration and high speed liquid chromatography give some indication of the molecular weight ranges and functionality of the organic moieties.

The drastic biological consequences of the onset of sulfide production tend to eliminate from such systems the ordinary processes considered in biological oceanography; virtually all the biota and biological processes characteristic of the oxygenated ocean are eliminated or disrupted. Although the ability of the codlet *Bregmaceros nectabanus* to migrate into and out of the sulfide-bearing zone of the Cariaco Trench has been documented (*Mead*, 1963; *Wilson*, 1972; *Baird, Wilson, and Milliken*, 1973), this is the only example that I have found of organisms higher than bacteria surviving the toxic environment. Of more biological interest seem to be those zones in which sulfate reducing conditions are only approached, or in which the products of the mixing of sulfide-bearing and oxygenated or nitrate and nitrite-bearing waters exist. These may be particularly interesting in systems in which circulatory pulses replace sulfide-bearing waters annually, occasionally, or during rare catastrophic events.

The most obvious biological sequence to the replacement of sulfide-bearing waters is the mass killing of organisms by the toxic effects of such displacement. Mass mortalities of this sort were documented by *Brongersma-Sanders* (1957), and two fish kills in Lake Nitinat have been studied by *Ozretich* (1975). Regular alternations of oxidizing and reducing conditions are geologically recorded by varved sediments. The varves from Saanich Inlet have been shown to be annual (*Gross, Gucluer, Creager, and Dawson*, 1963); presumably the sediments respond to alternating oxidizing and reducing conditions by color and probably other changes, and because of the toxicity of the sulfides, benthic organisms do not destroy the layering by bioturbation.

An extreme and opposite case of reversing oxidizing conditions in a basin is recorded in the sediments of the Cariaco Basin, which were apparently laid down under oxidizing conditions during the Pleistocene. A change in sediment type corresponding to the advent of reducing conditions has been estimated to have taken place some 10,900 years b.p. by radiocarbon dating of the sedimentary interface (*Richards*, in press; *Heezen, Menzies, Broecker, and Ewing*, 1958).

More subtle biological consequences may characterize those regions of extreme oxygen deficiency where all or nearly all of the oxygen has been consumed, nitrate reduction and denitrification take place, and yet no sulfate reduction is evident. Nitrate reduction and denitrification are presumably the key microbiological processes. There are two main chemical expressions of these processes: 1) the biogenic production of free molecular nitrogen, although the quantities produced are practically insignificant in relation to the amounts of nitrogen present from the solution of air, and 2) the well-known "secondary" or deep nitrite maximum (*Brandhorst*, 1959; *Fiadeiro and Strickland*, 1968; *Codispoti*, 1973). In comparison to nitrite concentrations elsewhere in the oceans these concentrations are spectacular: concentrations over six times those occurring in the upper layers of the ocean (*Vaccaro*, 1965) are observed. No unusual concentrations of ammonia or other nitrogen compounds have been observed.

Perhaps a more basic property of the water column in the oxygen-deficient zone of the eastern tropical North Pacific is an enzymatic expression---the electron transport system activity (ETSA) discussed by *Packard* (1969, 1971), *Packard, Healy, and Richards* (1971), *and Devol* (1975), among others. *Devol* consistently observed a maximum in the ETSA at or near the oxygen minimum zone. From studies of the distribution of ETSA and other biochemical properties, he concluded that at oxygen concentrations of 15 µg-at./ℓ and above, respiration could be assumed to proceed at the expense of dissolved oxygen. At oxygen concentrations between 2 and 15 µg-at./ℓ, although the above process remains dominant, cytochrome levels may be elevated. At oxygen concentrations below 2 µg-at./ℓ denitrification becomes the dominant process. *Ozretich* (personal communication) has observed a threshold oxygen concentration for nitrate reduction of some 2 to 6 µg-at./ℓ in chemostat experiments with facultatively anaerobic bacteria, the concentration varying somewhat with the specific organisms.

An obvious consequence of nitrate reduction and denitrification is the occurrence of smaller concentrations of nitrate than would otherwise be present---so-called nitrate anomalies. Different methods have been used to estimate these quantities (*Goering, Richards, Codispoti and Dugdale*, 1973; *Codispoti*, 1973; *Fiadeiro and Strickland*, 1968; *Cline and Richards*, 1972; and *Codispoti and Richards*, 1976). Maximum anomalies in the eastern tropical Pacific are on the order of 15 µg-at. of N/ℓ. A more significant number in terms of biological cycles is the rate of denitrification, which has also been estimated in a variety of ways (*Goering et al.*, 1973; *Richards*, 1971; *Cline and Richards*, 1972; *Codispoti*, 1973). The estimate by *Cline and Richards* (23×10^{13} g/yr) appears, by their own statement and in comparison with other estimates, to be high; otherwise there is remarkable (fortuitous?) agreement that on the order of 1 to 3×10^{13}g of N_2 are formed annually by denitrification in the eastern tropical North Pacific, representing a significant loss to the bank of combined nitrogen in this part of the biosphere.

The above discussion has considered processes and consequences related to progressively increased organic decomposition in marine waters. Eventually, after all, or almost all of the dissolved oxygen, nitrite, and nitrate are reduced, sulfate reduction will begin. As it proceeds, the following will take place:

1) The redox potential will drop dramatically from positive to negative values.

2) Sulfides will be introduced into the system, altering metal solubilities and the toxicity of the environment.

3) The valance state of reducible elements, such as iron and manganese, will be reduced. This will result in greater solubilities for these two elements, because the solubilities of FeS and MnS are greater than those of such compounds as $Fe(OH)_3$ and MnO_2.

4) NO_3^- and NO_2^- will disappear from the system.

5) NH_3, PO_4^{-3}, and CO_2 will accumulate in proportion to the amount of H_2S formed, except that the excess CO_2 will react according to the equation

6) $MeCO_3 + CO_2 + H_2O = Me^{++} + 2\ HCO_3^-$, which will

7) increase the carbonate and titration alkalinity.

8) Sulfides, produced by the bacterial reduction of sulfate, are probably introduced as sulfide ions ($S^=$) and as such will increase the alkalinity by 2 meq./liter for each mg-at./liter of sulfide-sulfur added. This adds to the titration alkalinity but not to the carbonate alkalinity. Finally,

9) anaerobic fermentation ensues with the production of methane.

The accumulation of organic degradation products, such as CO_2 and PO_4^{3-}, presumably accompanies the concurrent accumulation in sulfide-bearing marine systems of other constituents of living organic matter, such as trace elements that are more concentrated in marine organisms than they are in the water in which they grow. This also applies to a large number of trace metals, and, perhaps even more importantly from the viewpoint of modern ecology and pollution problems, persistent organic compounds such as DDT and its derivatives. In addition, the polychlorinated biphenyls may be so concentrated. If marine organisms also concentrate petroleum hydrocarbons, they too might be accumulated to excessive concentrations in sulfide-bearing, stagnant marine environments. No evidence is known, positive or negative, for this speculation.

The foregoing is intended to emphasize the concentrating effect operating in sulfide-bearing marine environments. They are a special and extreme case of what *Redfield* (1958) called "nutrient traps." But, he concerned himself only with the accumulation of phosphorus, nitrogen, and carbon compounds.

Were there no processes by which these accumulated materials are returned to the circulating systems of the seas, they would represent chemical cul-de-sacs, important only as nutrient sinks and unusual geochemical environments and of little interest in the biological cycles of the sea. However, this is not the case, and the accumulated materials may be re-introduced at nonuniform rates into the surrounding waters in such a way as to produce unusual biological events. The catastrophic introduction of sulfides into the upper layers of fjords and inlets with resultant mass mortalities of marine organisms is an example. *Barlow* (1965) pointed out that the mixing of such waters with oxygen-bearing waters might have long-reaching subtle effects, even though the effects of the reoxidation of the sulfides might be trivial. In all the sulfide-bearing environments the combined nitrogen ends up as free nitrogen or ammonia, thus altering both the nitrate:phosphate and the combined nitrogen:phosphate ratios in waters to which an oxidizing condition has been restored by mixing. In a system as extensive as the eastern tropical Pacific, this cannot be a trivial change, but we have no idea of its biological consequences, if any. So far, we know of no extraordinary concentration of persistent organic chemicals in stagnant, oxygen-deficient or sulfide-bearing environments. Nevertheless, it is easy to imagine that if such accumulation were to take place, subsequent mixing or diffusion of the laden waters might well have serious biological consequences on the surrounding waters.

Altered nitrate:phosphate ratios are only one example of the effects of mixing oxygenated waters with those in which nitrate reduction, denitrification, sulfate reduction, and anaerobic fermentation have taken place. The resulting oxygen-bearing mixtures might be expected to be characterized by anomalies such as:
1) high phosphate:nitrate and phosphate:combined nitrogen ratios
2) high absolute concentrations of phosphate and silicate
3) the presence of sulfite, tetrathionate, and free sulfur (*Cline and Richards*, 1969)
4) methane
5) excess free nitrogen
6) low concentrations of heavy metals with insoluble sulfides
7) high concentrations of iron and manganese.

The biological consequences of none of these anomalies is understood.

A striking example of the effects of mixing oxygenated and sulfide-bearing waters is the marked changes it brings about in the vertical distribution of manganese and its partition between soluble and particulate forms. In general, soluble and particulate manganese are both low in the upper layers. Near, but below, the O_2-H_2S interface there is a dramatic increase in the concentrations of dissolved manganese. *Spencer and Brewer* (1971) observed increases from almost nil just above the interface to a peak of some 450 µg/kg just below the interface in the Black Sea. A strikingly similar peak in particulate manganese just <u>above</u> the interface was observed by *Flouriê*, (1971). High concentrations of soluble manganese (but significantly smaller than the peak) and negligible concentrations of particulate manganese characterize the sulfide-bearing zones (Spencer and Brewer, 1971; Flourié, 1971). It seems evident that mixing across the interface introduces soluble (reduced) manganese into an oxidizing medium, where it is precipitated, probably as MnO_2. As the particles re-enter the sulfide zone, they are reduced and re-solubilized. The biological effects of this probably cyclic process are unknown. The particles could present a surface favorable to bacterial growth. Richards and Devol (1973) observed a marked minimum in the concentration of dissolved organic carbon near the O_2-H_2S interface in the Black Sea. They suggested that the minimum was the result of the sorption of organic carbon on the particulate manganese (and presumably iron) compounds observed above the interface. Such a mechanism would tend to concentrate organic material and perhaps make it more available to organisms.

In conclusion, oxygen deficient and sulfide-bearing marine environments present biological environments that are characteristically different from most of the world's ocean. The presence of sulfides makes the environment toxic to most organisms. Denitrification alters the nitrate:phosphate ratio, and the mixing of waters in which sulfides have developed with oxygenated water produces mixtures that are anomalous in their biogeochemical properties.

ACKNOWLEDGEMENTS

This work was supported by a series of grants from the National Science Foundation, including GA-41349 and GA-24875 and by contract with the Office of Naval Research, including contract numbers N-00014-67-A-193-0014, N-00014-75-A-0026 and N-00014-75-C-0502. Contribution No. 878 from the Department of Oceanography, University of Washington, Seattle, Washington 98195

REFERENCES

Atkinson, L. P. and F. A. Richards. 1967. The occurrence and distribution of methane in the marine environment. Deep-Sea Res. 14(6): 673-684.

Baird, R. C., D. F. Wilson and D. M. Milliken. 1973. Observations on Bregmaceros nectabanus Whitley in the anoxic, sulfurous water of the Cariaco Trench. Deep-Sea Res. 20: 503-504.

Barkley, R. A. 1975. Personal ommunication, National Marine Fisheries Laboratory, Honolulu, Hawaii.

Barlow, J. P. 1965. Formal discussions of chemical observations in some anoxic, sulfide-bearing basins and fjords. P. 233-234. In Proceedings of the Second International Water Pollution Research Conference, Tokyo, 1964. Pergamon Press, New York.

Brandhorst, W. 1959. Nitrification and denitrification in the eastern tropical north Pacific. J. Cons., Cons. Perm. Int. Explor. Mer 25: 3-20.

Brongersma-Sanders, M. 1957. Mass mortality in the sea. *Geol. Soc. America Mem.* 67, 1: 941-1010.

Cline, J. D. and F. A. Richards. 1969. Oxygenation of hydrogen sulfide in seawater at constant salinity, temperature, and pH. *Environmental Science and Technology* 3: 838-843.

Cline, J. D. and F. A. Richards. 1972. Oxygen deficient conditions and nitrate reduction in the eastern tropical North Pacific Ocean. *Limnol. Oceanogr.* 7: 885-900.

Codispoti, L. A. 1973. Denitrification in the eastern tropical North Pacific Ocean. Ph.D. thesis, University of Washington, 118 p.

Codispoti, L. A. and F. A. Richards. 1976. An analysis of the horizontal regime of denitrification in the eastern tropical North Pacific. *Limnol. Oceanogr.* 21 (3): 379-388.

Devol, A. H. 1975. Biological oxidations in oxic and anoxic marine environments: rates and processes. Ph.D. thesis, University of Washington, 208 p.

Fiadeiro, M. and J. D. H. Strickland. 1968. Nitrate reduction and the occurrence of a deep nitrite maximum in the ocean off the west coast of South America. *Deep-Sea Res.* 26: 187-201.

Fluorié, E. J. 1971. Particulate manganese in seawater stressing regimes in marine anoxic basins. Ph.D. thesis, University of Washington, 151 p.

Gripenberg, S. 1960. On the alkalinity of Baltic waters. *J. Cons. Perm. Int. Explor. Mer* 26(1): 5-20.

Goering, J. J., F. A. Richards, L. A. Codispoti, and R. C. Dugdale. 1973. Nitrogen fixation and denitrification in the ocean: biogeochemical budgets. P. 12-27 In *Proc. Intn'l. Smp. Hydrogeochem. and Biogeochem.*, Vol. 2, Biogeochemistry (E. Ingerson, ed.), The Clarke Co., Washington, D. C.

Gross, M. G., S. M. Gucluer, J. S. Creager, and W. A. Dawson. 1963. Varved marine sediments in a stagnant fjord. *Science* 141: 918-919.

Heezen, B. C., R. J. Menzines, W. S. Broecker, and M. Ewing. 1958. Date of stagnation of the Cariaco Trench, southeast Caribbean. *Bull. Geol. Soc. Am.* 69: 1579.

Hodgman, C. D. (ed.) 1959. Handbook of chemistry and physics. 40th ed. Chemical Rubber Publishing Co., Cleveland, Ohio: 3456 p.

Knull, J. R. and F. A. Richards. 1969. A note on the sources of excess alkalinity in anoxic waters. *Deep-Sea Res.* 16: 205-212.

Mead, G. W. 1963. Observations on fishes caught over the anoxic waters of the Cariaco Trench, Venezuela. *Deep-Sea Res.* 10: 251-257.

Ozretich, R. J. 1975. Mechanisms for deep water renewal in Lake Nitinat, a permanently anoxic fjord. *Estuarine and Coastal Marine Sci.* 3: 189-200.

Ozretich, R. J. 1975. Personal communication, University of Washington, Seattle, Washington.

Packard, T. T. 1969. The estimation of the oxygen utilization rate in seawater from the activity of the respiratory electron transport system in plankton. Ph.D. thesis, University of Washington, 115 p.

Packard, T. T. 1971. The measurement of respiratory electron transport activity in marine phytoplankton. J. Mar. Res. 29: 235-244.

Packard, T. T., M. L. Healy, and F. A. Richards. 1971. The vertical distribution of the activity of the respiratory electron transport system in marine plankton. Limnol. Oceanogr. 16: 60-70.

Redfield, A. C. 1934. On the proportions of organic derivatives in seawater and their relation to the composition of plankton. P. 176-192 In James Johnstone Memorial Volume, University of Liverpool Press, Liverpool.

Redfield, A. C. 1958. The biological control of chemical factors in the environment. Amer. Scientist 46: 205-221.

Redfield, A. C., B. H. Ketchum and F. A. Richards. 1963. The influence of organisms on the composition of sea-water. Chapter 2 (p. 26-77) In The Seas, Vol. II (M. N. Hill, ed.). Interscience (Wiley), New York.

Richards, F. A. 1965a. Chemical observations on some anoxic, sulfide-bearing basins and fjords. P. 215-243 In Proceedings of the Second International Water Pollution Research Conference, Tokyo, 1964. Pergamon Press, New York.

Richards, F. A. 1965b. Anoxic basins and fjords. Chapter 13 (p. 611-645) In Chemical Oceanography (J. R. Riley and G. Skirrow, eds.). Academic Press, London.

Richards, F. A. 1971. Comments on the effects of denitrification on the budget of combined nitrogen in the ocean. La Mer (Bulletin de la Société franco-japonaise d'oceanographie) 9: 68-77.

Richards, F. A. 1975. The Cariaco Basin (Trench). In: Oceanography and Marine Biology: An Annual Review (H. Barnes, ed.), 13:11-67.

Richards, F. A., J. D. Cline, W. W. Broenkow, and L. P. Atkinson. 1965. Some consequences of the decomposition of organic matter in Lake Nitinat, an anoxic fjord. Limnol. Oceanogr. 10 (Supplement): R185-R201.

Richards, F. A. and A. H. Devol. 1973. Organic matter in anoxic environments. P. 480-490 In Proc. Intn'l. Symp. Hydrogeochem. and Biogeochem., Vol. 2, Biogeochemistry (E. Ingerson, ed.), The Clarke Co., Washington, D. C.

Richards, F. A. and R. F. Vaccaro. 1956. The Cariaco Trench, an anaerobic basin in the Caribbean Sea. Deep-Sea Res. 3: 214-228.

Strøm, K. 1957. A lake with trapped sea-water? Nature 180: 982-983.

Strøm, K. 1961. A second lake with old sea-water at its bottom. Nature 189: 913.

Spencer, D. W. and P. G. Brewer. 1971. Vertical advection diffusion and redox potentials as controls on the distribution of manganese and other trace metals dissolved in waters of the Black Sea. J. Geophys. Res. 70: 5877-5892.

Vaccaro, R. F. 1965. Inorganic nitrogen in seawater. P. 365-408 In *Chemical Oceanography, Vol. I* (J. P. Riley and G. Skirrow, eds). Academic Press, London and New York.

Williams, P. M., W. H. Mathews, and G. L. Pickard. 1961. A lake in British Columbia containing old sea-water. *Nature 191*: 830-832.

Wilson, D. F. 1972. Diel migration of sound scatterers into, and out of, the Cariaco Trench anoxic water. *J. Mar. Res. 30*: 168-176.

Zirino. A. 1970. Voltammetric measurement, speciation and distribution of zinc in ocean water. Ph.D. thesis, University of Washington, 205 p.

Turbulence as a Factor in Sound Scattering in the Upper Ocean

J. D. Woods

Southampton University, Southampton, England

ABSTRACT

The principal reasons for sound scattering in the interior of the upper ocean, namely, (1) small-scale spatial fluctuations in sound speed and (2) fish, are affected directly or indirectly by turbulent fluctuations in the velocity field. Recent theoretical developments in the study of oceanic turbulence are reviewed and their relevance to sound scattering discussed. Particular attention is paid to the intermittent nature of velocity fluctuations in the ocean and to the effects of density stratification and the earth's rotation on the structure of individual eddies. The role of fronts as areas of enhanced billow turbulence and nutrient upwelling, and hence, of enhanced sound scattering is discussed. A brief outline of a new model is presented for turbulent mixing designed to predict the fluxes of a dye, nutrients, etc. in a way that will be appropriate for coupled physical-biological models of primary production.

INTRODUCTION

Turbulent velocity fluctuations in the current systems of the upper ocean directly control the space-time distributions of the chemical constituents of the seawater and its temperature. These fluctuations are thought to play a significant role in determining the space-time distribution of plankton and, therefore, of larger biological sound scatterers.

The statistical properties of the turbulent velocity fluctuations are, as yet, poorly understood. This is due to the lack of suitable instruments to measure the velocity distribution in the ocean and because theories of geophysical turbulence are still in a primitive state of development. This paper will review some recent work on oceanic turbulence and indicate some of the simplications for sound scattering in the upper ocean.

THE PROBLEM

Velocity fluctuations in the upper ocean (0-1000 m in depth) have the spatial and temporal ranges shown in Table 1.

If it were possible to fully describe the velocity field in the ocean at some instant in time, with a resolution of 1 mm, together with the external forces (wind, tide) active on it and the effects of heat and moisture balance near the surface, then in principle it would be possible to compute the future evolution of the velocity field using the primitive equations, following the methods first developed by L. F. Richardson (1922) fifty years ago, and universally adopted by meteorologists for weather forecasting. The problem is that present-day computers take roughly real time (i.e., they take a day of computer time to forecast changes taking a day in the simulated ocean or atmosphere) for a model that contains only a limited range of all the possible scales of motion.

The range of scales depends upon the individual model. However, it is at best approximately $10^2 \times 10^2 \times 10$ for $L_x \times L_y \times L_z$. These ranges may be expected to increase slowly as computer technology advances. Nevertheless, it is inconceivable that any computer model will ever be able to give predictions (i.e. faster than real time) of more than a very limited part of the whole range of natural variability. (See Table 2).

TABLE 1. Velocity Fluctuations in the Upper Ocean.

Dimension	Maximum	Minimum	Range
Horizontal length, L_x	10,000 km	1 mm	10^{10}
Horizontal length, L_y	10,000 km	1 mm	10^{10}
Vertical length, L_z	1,000 m	1 mm	10^6
Time scale, T	10 years	1 second	10^9

TABLE 2. Comparison of the ranges of spatial and temporal fluctuations of the velocity field in the ocean and in computer models.

Dimension	Range of Fluctuations	
	In the Ocean	In Computer Models
L_x	10^{10}	10^2
L_y	10^{10}	10^2
L_z	10^6	10
T	10^9	10^2

Implications for Forecasting Sound Scattering

This inherent limitation in computer models of turbulent motions based on integrating the primitive equations á la *Richardson*, does not mean that such models can have no place in forecasting changing distributions of sound scattering. Such models can be effective if the application (in this case sound scattering) can be usefully served by predictions which (i) only contain structure with a limited range of space-time scales (normally called a spectral window) and (ii) do not extend so far into the future that the errors leaking into the forecast from neglected scales of motion outside the spectral window of the model become unacceptably large.

Spectral windows could possibly be specified within which *Richardson*-type models might be employed to predict variation in the structure of sound scattering in the upper ocean. However, it should be remembered that such computer models would have to be initiated using a set of observations with a similar spectral window; with the feasibility of collecting such observations being an integral part of specifying the model.

THE STATISTICAL APPROACH

Aims and Definitions

If it is impossible to achieve a deterministic description of the changing structure of the oceanic velocity field (and therefore hopefully of the sound scattering field) because of the great ranges of scales occurring in nature, it can still be hoped that progress will be made with a statistical approach in which a wide spectral window is retained at the expense of forecasting precise events. This statistical approach to describing flows with large ranges of scales (and hence large *Reynolds* numbers) forms the basis for turbulence theory and leads to the following definition. "Turbulence is a class of statistics used to describe the properties of large Reynolds number flows. A turbulent flow is one that has large *Reynolds* number." By "large", greater than a few thousands is normally meant. For the upper ocean, taking a typical dimension (L_z = 1 km) and speed (U = 10 cm/s), the *Reynolds* number, defined by $Re = UL/\nu$ (where ν is the kinematic viscosity = 0.01 cm^2/s for seawater), is 10^8. As a result, the ocean is always "turbulent" everywhere.

Specifications

Accepting that the velocity fluctuations in the upper ocean influence the space-time distribution of sound scattering, what can be learned from turbulence theory, remembering that it can only offer statistical relationships? The most elementary parameters are the mean, variance, and kurtosis of the velocity field. It might be asked how these statistical parameters vary in space and time and with scale, and whether it is possible to develop hypotheses about their distributions and correlations between them on the basis of simplified theoretical models of how turbulent fluids behave. If successful, the implications of such relationships for the distribution of scattering in the upper ocean could be explored. If the implications appear to be of service to the general problem of forecasting the distribution of sound scattering in the ocean, it will then be necessary to go back to the turbulence theory and to test its hypotheses against observations of the ocean. In practice, of course, the turbulence theories were originally stimulated by field observations, such as flow visualization.

Homogeneous, Isotropic Turbulence

The simplest statistical parameter to begin with is the variance of the velocity fluctuation. A naive first guess would be that the variance of the three Cartesian components of velocity, U_x U_y, and U_z, are each uniformly distributed throughout physical space (i.e. the turbulence is *homogeneous*) and that the magnitudes of σ_x, σ_y, and σ_z are equal (i.e. the turbulence is isotropic). These assumptions permit attractive mathematical simplification and have, therefore, formed the basis for many early developments in the theory of turbulence. Building on these assumptions (of homogeneity and isotropy) *Kolmogorof* (1941) *et al.* developed the first theories for the spectral distribution of the velocity variances (or turbulent kinetic energy, as the variance is often called); that is, their distribution in Fourier space (sometimes called wavenumber space). The concept of a cascade of turbulent kinetic energy through Fourier space, with a net flux from large scales to small scales, first described by L. F. Richardson (1920), was crystallized by Kolmogorof who introduced a new parameter, ε, the net rate at which energy was cascading from larger to smaller scales. Kolmogorof made two hypotheses concerning ε. First, that it was homogeneous in physical space, and second, that it was homogeneous in Fourier space. In later developments of the theories, alternative ε hypotheses have been explored. For the purposes of the present discussion it is important to remember that any theory of turbulence designed to explore the distributions of velocity variance in Physical and Fourier space must introduce some hypotheses about ε, supported by a physical model of the cascade. The cascade model implicit in most early theories of turbulence is that the flux of turbulent kinetic energy passes smoothly and continuously through Fourier space from some large scale where the variance is first created down to the tiniest scales, where molecular viscosity converts it to heat.

Intermittency

The theories of homogeneous, isotopic turbulence were developed primarily in the context of flows in wind-tunnels and in the wakes of objects in fluids of uniform density. It was shown experimentally (Batchelor and Townsend, 1949) that real flows did not exhibit *local* homogeneity even under controlled laboratory conditions. At any instant, the distribution of variance is uneven in both physical and Fourier space. These observations led to the abandonment of homogeneous ε parameterizations of the turbulent cascade, which is now thought to be effected in a series of spectral leaps rather than as a smooth continuous flow through the spectrum (Woods, 1974a; Leslie, 1973).

This switch of emphasis from the mean distribution, which may be statistically homogeneous, to the instantaneous pattern of local deviations from the mean, is an important step in the development of turbulence theories, because it permits the introduction of a dynamical model of the cascade, rather than a similarity model á la Kolmogorof. It is important to the analysis of sound scattering in the ocean, because the local deviations of velocity variance from the mean may be expected to produce corresponding variations of sound scattering properties. And, because the relationship between the fields of velocity fluctuations and sound scattering is non-linear, the ensemble mean statics of sound scattering as calculated on an intermittent local model are likely to differ significantly from those calculated from a homogeneous model.

Local Deviations in the Spectrum of Velocity Variance

Once the assumption of an energy cascade with a constant, ε, which is homogeneous in physical and Fourier space, has been relaxed (in order to be consistent with observations of intermittency) the assumption of a homogeneous spectral intensity is also lost. Moreover, at any instant the form of the spectrum will vary locally because the energy cascade is no longer smooth and continuous. So the original problem of describing the distribution of velocity variance at any instant in the three dimensions of physical space (x, y, z), has added to it the local distribution in the three wavenumber dimensions of Fourier space (k_x, k_y, k_z). Therefore,

$$\sigma_\chi = \sigma_\chi (x, y, z; k_x, k_y, k_z) \tag{1}$$

Furthermore, the local spectral distribution will change with time. As a result, the final analysis of variance has eight dimensions, four in physical space-time (x, y, z, t) and four in Fourier space-time (k_x, k_y, k_z, ω).

$$\sigma_\chi = \sigma_\chi (x, y, z, t; k_x, k_y, k_z, \omega) \tag{2}$$

It follows that the distribution of sound scattering variance should also be analyzed into these eight dimensions if its properties are to be related to the velocity field.

$$\sigma_s = \sigma_s (x, y, z, t; k_x, k_y, k_z, \omega) \tag{3}$$

Geophysical Effects

Velocity fluctuations at the smallest time scales in the ocean may be expected to behave rather like those in the laboratory. However, at larger time scales, two geophysical influences become important; namely *density stratification* and the *Earth's rotation*. As has been pointed out previously, the introduction of a local (i.e. eight dimensional) distribution of velocity variance allows the introduction of a dynamical model of the turbulent energy cascade and hence, an introduction of the effects of density stratification and the Earth's rotation. They enter into the momentum equation of fluid dynamics in the form of two frequencies

(i) the stability frequency, $N = \left(\dfrac{-g\partial\rho}{\rho\partial z}\right)^{\frac{1}{2}}$ (4)

(ii) the Coriolis frequency, $f = 2\Omega \sin\theta$ (5)

where g is the gravitational acceleration, ρ is the density, z is the depth, Ω is the Earth's speed of rotation, and θ is the latitude.

Using these two frequencies, geophysical fluid motions may be classified into three categories, according to the time scale T of the motion; so that, to a good approximation,

(i) If $\frac{1}{T} < f$, the flow is dominated by the Earth's rotation.

(ii) If $f < \frac{1}{T} < N$, the flow is dominated by the density gradient.

(iii) If $\frac{1}{T} > N$, the flow is independent of the effects of rotation and buoyancy.

Provided that $\omega = 1/T$ is carefully defined, it is appropriate to divide the spectrum of turbulence into the same three ranges, which are called the *rotation, buoyancy,* and *inertial* ranges, respectively (Woods, 1974a).

Definition of Frequency, ω

In order to define ω in a way that can be related to observations, it is necessary to introduce a hypothesis concerning the qualitative form of the instantaneous six-dimensional variance distribution

$$\sigma (x, y, z; k_x, k_y, k_z) \tag{6}$$

It is assumed that the distribution consists of an ensemble of discrete variance concentrations, distributed in $(x, y, z; k_x, k_y, k_z)$ like individual droplets in a cloud. This "condensed variance" model replaces the "continuous variance" model of homogeneous turbulence. It is appropriate to give the name "eddies" to these condensed droplets of variance. Without going into a detailed justification of this model, (which does appear to be consistent with recent observations in the upper ocean), it is worth drawing attention to the similarity existing between it and the meteorologist's description of the atmosphere in terms of discrete eddies, called cyclones and anticyclones and, on a much smaller scale, patches of clear air turbulence.

The characteristic time scale of an individual eddy is its lifetime. In terms of an energy cascade, the lifetime is the residence time of energy at a particular location in six dimensional space, it being injected there by a larger eddy and subsequently being lost to smaller eddies, or vice versa, or both. It is, of course, a Lagrangian time scale. The frequency is then defined as the reciprocal of the eddy lifetime (Woods, 1974a; Woods and Moen, 1975).

Relationships Between Length Scales

For an eddy of characteristic speed U and dimensions L_x and L_z

$$\frac{L_z}{L_x} = \frac{f}{N} Ro \cdot Ri^{\frac{1}{2}} \tag{7}$$

where the *Richardson* number, $Ri = N^2/(du/dz)^2 = N L_z^2/U^2$ and the Rossby number, $Ro = U/L_x f$.

Eddies in the inertial range ($\omega > N$) are independent of buoyancy or rotation effects and therefore likely to be statistically isotropic (i.e., $L_z = L_y = L_x$). The largest such eddy has $T = 1/N$. For these eddies $Ri = 1$. For the longer lived eddies, $Ri > 1$, but the product $Ro \cdot Ri^{\frac{1}{2}}$ is less than f/N, so that the eddies are no longer isotropic, but $L_z < L_x$. As T increases the ratio L_z/L_x remains small, but $L_x = L_y$, so the eddies in the buoyancy and rotation ranges are (statistically) *isotropic in the horizontal*,

i.e. $L_x = L_y > L_z$. (8)

Relationships Between Lifetime and Length Scales

By parameterizing the turbulent cascade in terms of the energy flux through Fourier space, ε, *Kolmogorof* was able to relate the length and time scales of eddies in the inertial subrange (where buoyancy and rotation do not affect the cascade dynamics) by the simple relationship $L = \text{constant} \times (\varepsilon/\omega^3)^{\frac{1}{2}}$. For the largest eddy in the inertial range $\omega = N$, and $L_N = \text{constant} \times (\varepsilon/N^3)^{\frac{1}{2}}$. This length scale is usually named after *Ozmidov* (1965) who originally pointed out the importance of the transition in cascade dynamics at this scale. *Panchev* (1971, p. 322) has suggested that the length scale of the transition between buoyancy and rotation ranges ($T = 1/f$) can be similarly calculated. Thus, $L_f = (\varepsilon/f^3)^{\frac{1}{2}}$.

If this scale analysis is accepted, then by introducing a hypothesis about the spectrum of ε, it is possible to calculate the relationships between L_x, L_y, L_z, and T within the uncertainty of the magnitude of the constants in the scale analysis above. *Woods* (1974a) investigated the hypothesis that $\varepsilon = \text{constant}$ (i.e. that there is no flux divergence of scales of order 100 km to viscous dissipation at scales of order 1 mm.) By analogy with the atmosphere, *Woods and Moen* (1975) argued that in practice the magnitude of ε was unlikely to vary by more than a factor of two from the mean value calculated from oceanic measurements in the inertial range. This leads to the not unexpected conclusion that all the eddies lie clustered in a relatively narrow band through Fourier space-time. Substituting typical values for the upper ocean leads to the following values for $L_f \sim 1$ km and $L_N \sim 1$ m.

Billow Turbulence

$L_N \sim 1$ m is the upper limit for isotropic turbulence in the upper ocean. Flow visualization studies (*Woods*, 1968; *Woods and Wiley*, 1972) show that there is a dramatic change in the flow field at this scale. Larger motions are effectively constrained to run along density surfaces which are more or less horizontally stratified; smaller motions are more characteristic of turbulence in the laboratory. The transition between the two types of turbulence (two-dimensionally isotopic when $L > L_N$, three-dimensionally isotopic when $L < L_N$) is effected by Kelvin-Helmholtz instability giving billows. These billows, some 20-30 cm high, are the largest overturning motion in the interior of the sea (i.e., below the near surface convectively-mixed layer). Flow visualization studies have shown that billow turbulence occurs in quasi-horizontal patches some 10 m or so in horizontal dimension and one or more billow height in vertical extent. Such patches occupy a tiny fraction (~5%) of the flow at any instant. These small-scale turbulent motions whose variance lies in a spectral band from about 30 cm to 1 mm, mix scalar properties of the sea to give a corresponding spectrum of variances of temperature and salinity. Because these fluctuations in temperature and salinity produce fluctuations in speed of sound with the same spectral form, sound waves with wavelengths of less than twice the billow height propagating through the patch of billow turbulence will be scattered by Bragg-scattering (*Munk and Garrett*, 1973). The patch of small scale sound speed fluctuations generated by the billow turbulence event will tend to persist rather longer than the lifetime of the billows ($T \simeq 5$ minutes), because molecular conduction acts more slowly than molecular viscosity. However, the patches of "fossil-turbulence" (*Woods*, 1969) left after velocity fluctuations have decayed, will be distorted by the shear in the larger scale ($L > L_N$) turbulence, so that the spectrum of (fossil) sound speed variance will slowly depart from its original form. This is usually assumed in sound scattering calculations á la *Tatarski* (1961) to be that of the spectrum of a passive scalar in equilibrium with homogeneous isotopic turbulence. It is worth noting that this assumption, while satisfactory for calculations of radio wave scattering by atmospheric billow turbulence (where Re is large), is probably not satisfactory for calculations of sound wave scattering in the ocean where billow turbulence has *Reynolds* numbers typically in the range 100 to 1000. This is too low for the development of an equilibrium velocity variance spectrum, within individual billows, or even for an ensemble of such billows.

Turbulence in the Rotation and Buoyancy Ranges

Although velocity fluctuations associated with eddies in the rotation and buoyancy ranges contribute to the time series collected from moored current meters, they are often masked by more energetic contributions from internal waves. While in principle it is possible to discriminate between eddies and waves on the basis of their dimensional relationships (i.e., between L_x, L_y, L_z, and ω), the elaborate arrays needed to achieve this have only recently been used and the results of preliminary experiments are not yet available. There is, however, indirect evidence from measurements of temperature (and sound speed) patterns on scales larger than L_N that rotation and buoyancy range eddies must exist everywhere and that their distribution in $(x, y, z, t; k_x, k_y, k_z, \omega)$ is probably highly intermittent, as postulated in the turbulence model described above (*Berkerle*, 1969; *Saunders*, 1972). In particular, this is revealed in maximum resolution spectra (*Woods and Moen*, 1975) of temperature fluctuations at a depth of 10 cm during calm weather when the interior motions outcrop at the surface without being masked by wind-driven convection (*Moen*, 1974).

FRONTS

It seems a reasonable hypothesis that eddies in the rotation and buoyancy ranges in the ocean will have many of the statistical characteristics of eddies with similar lifetimes in the atmosphere, allowing for absence of mixing by clouds in the ocean, and that the cascade mechanism is essentially the same in both cases. In the atmosphere, fronts are a common feature in both the rotation and buoyancy ranges and play a major role in transferring energy in great leaps from large scales to small (*Woods*, 1974a). There is ample evidence for the existence of fronts in the ocean, but the measurements are not yet sufficiently detailed to be sure of the role they play in oceanic turbulence. However, *MacVean* (1976) using a mathematical model has recently shown that fronts form rapidly in the deformation field of rotation range eddies. He is now attempting to model their instabilities which generate smaller eddies.

Importance of Fronts to Sound Scattering

Fronts affect sound propagation in a variety of ways. Sound waves impinging on the front are *refracted* in both the vertical and horizontal planes. As a result the distribution of sound intensity on the acoustic lee of the front is often markedly altered. In regions of strong frontal intensity, sonar range may be limited by these effects and the best prediction of sonar range may be the distance to the nearest front.

Scattering by small-scale fluctuations of sound speed created by billow turbulence is generally at least an order of magnitude more intense at fronts than elsewhere for two reasons. First, because the shear across the front is enhanced by the geostrophic (thermal wind) shear, which is much stronger at the front, reducing the *Richardson* number and increasing the frequency of occurrence of billow turbulence events. Second, because the temperature (and hence the sound speed) difference is usually larger across a front than elsewhere, so that a billow turbulence event generates larger temperature (sound speed) variance.

Scattering by fish is likely to be more intense at fronts, where strong upwelling raises nutrient-rich water towards the surface. Upwelling speeds approaching 1 mm/s have been estimated for fronts in rotation range eddies in the Western Ionian Sea (*Woods and Moen*, 1975).

DIFFUSION BY TURBULENCE IN THE ROTATION AND BUOYANCY RANGES

Motions with time scales longer than $2\pi/N$ (i.e., $L > L_N$) do not overturn, but are constrained to potential density surfaces. Thus, diffusion by eddies in the rotation and buoyancy subrange occurs only along potential density surfaces. If, as is the case at fronts, the surfaces are inclined to the horizontal, the eddy diffusion will have a vertical component. For motions in the rotation range (where geostrophy is assumed) the ratio of the horizontal to the vertical mixing rates, parameterized in terms of eddy transport coefficients, is given by *Woods* (1974b) as $K_x/K_z = (N/f)^2 \cdot Ri$. This describes the local transport rates in a single eddy in the rotation range. The instantaneous local transport rate at a given geographical location will depend on all the eddies present at the location at that time. According to the condensed variance model proposed earlier, only a limited number of eddies of different sizes are occupying a given location in physical space-time. In addition, the eddy inventory is a function of physical space-time.

If the turbulent transport due to a single eddy is parameterized in terms of an eddy transport coefficient K, then the magnitude of K will depend on the energy density in the eddy ($\frac{1}{2}\rho\sigma$) and its size (L). Thus $K \propto L\sigma^{\frac{1}{2}}$. K will increase with L for a red spectrum (i.e., one in which larger eddies tend to have greater energy density), which is the normal situation in the ocean. A number of empirical and theoretical relationships that have been proposed for $K(L)$ are reviewed by *Okubo* (1971). Because K varies with L for individual eddies, the total transport rate will depend upon the detailed inventory of eddies present at each location in physical space-time (i.e. $K = K(x, y, z, t)$). For the condensed variance model of turbulence, this will be a very irregular distribution. As a result, a passive scalar (e.g. dye) dropped into the sea will spread rather irregularly as eddies come and go, but with a trend that is equivalent to $\bar{K}$ increasing with time.

MIXING BY EDDIES IN THE ROTATION AND BUOYANCY RANGES

While it is often convenient to parameterize the turbulent spreading of passive tracers in the sea in terms of an eddy diffusivity, K, models based on this parameterization fail to reproduce the fine structure generated by the eddies while they are spreading the tracer. Yet, the existence of fine structure in dye being spread by oceanic turbulence is a striking feature of *Kullenberg's* (1974) tracer experiments, and the existence of a fine structure of temperature and salinity in the ocean has been revealed by modern profiling instruments. Furthermore, it has been suggested (*Steele*, 1975) that the generation of fine structure in nutrient and plankton concentrations may play a significant role in determining the interaction of physical and biological processes during plankton blooms. And as has already been presented out, the generation of sound speed microstructures by billow turbulence is an essential factor in sound scattering in the ocean.

Is it possible to devise an alternative parameterization of the effect of an individual eddy on a scalar contaminant, which creates microstructure in the way that is observed, rather than destroys it, as K theory does? In principle, it is quite easy to do so. However, the development of a mixing model incorporating the new parameterization has not yet been completed. In broad outline such a model would seek to follow the distributions of velocity U (x, y, z, t) and of velocity variance σ $(x, y, z, t; k_x, k_y, k_z, \omega)$ and of the scalar concentration $C(x, y, z, t)$ and of its variance σ_c $(x, y, z, t; k_x, k_y, k_z, \omega)$. If the model uses the condensed variance model of turbulence described in this paper, the effect of individual eddies can be treated separately. The action of an individual eddy may then be parameterized as follows. Action only occurs where velocity variance and scalar concentration variances overlap. At points of overlap the scalar variance is displaced to a slightly higher wavenumber location in Fourier space. In addition to this parameterization of the mixing effect, the model should also include advection of the dye concentration field by the velocity field. A mixing model of this type is currently being developed in the Oceanography Department at Southampton University, as a preliminary step towards developing a turbulent primary production model.

NUTRIENT SUPPLY TO THE EUPHOTIC ZONE

The classical description (c.f. *Dietrich*, 1975) of nutrient supply from the deep ocean to the euphotic zone is based on horizontally homogeneous vertical diffusion at a rate of about 1 cm^2/s, which is ascribed to "turbulence" because it is several orders of magnitude faster than the molecular diffusion rate for the nutrients. The new intermittent models of velocity fluctuations reviewed in this paper suggest that the nutrient supply may also be rather intermittent. The scale analysis, which leads to a classification of the eddies into rotation, buoyancy, and inertial (or billow turbulence) ranges, each with distinct physical characteristics and shapes (i.e. isotropy), helps to identify which motions effect the net upward diffusion of nutrients that is known to occur. At present, continuous sampling techniques do not have sufficient resolution to permit exploration of the microstructure of nutrient concentrations in the way that the microstructure of chlorophyll and temperature have been been studied by *Platt and Denman* (1976). It seems probable that the nutrient microstructure may be controlled to a far greater extent by biological processes than by turbulent motions in the sea. We must always bear in mind the fact that nutrients are nonconservative constituents of seawater. They are consumed in the euphotic zone by plankton and redistributed by the zooplankton and larger herbivores and carnivores, which migrate from near the surface at night to below

the euphotic zone during the day. This biological extraction and
redistribution can exploit even temporary (reversible) excursions of
nutrient enriched water into the euphotic zone, which would yield no
net vertical flux for a conservative constituent such as heat at
night.[1] It seems, therefore, attention should be concentrated on the
components of the turbulent motions in the thermocline, which produce
the largest vertical displacements, without worrying at this stage
about whether these displacements would achieve a net vertical flux
of a conservative constituent of seawater. The hypothesis is that
because of biological extraction in the euphotic zone, turbulent
motions may yield a faster vertical flux of nutrients than of any
other conservative constituent present in the same water mass at that
time. In a word, UPWELLING regions created by the turbulent motions
inside the thermocline should be sought.

Turbulent Upwelling

At this stage the statistics of the turbulent motion in the thermocline are temporarily put aside and, following the method of meteorologists, certain features of the flow within a succession of relatively narrow spectral windows are concentrated upon. If the flow field approximates to the "condensed variance" model proposed earlier, then the method adopted here introduces negligible errors provided that the boundaries of the spectral window do not intersect any of the variance droplets (i.e. the boundaries coincide with the spectral gaps between variance concentrations). Variance concentrations (eddies) lying to the high wavenumber-frequency side of the spectral window can be parameterized in terms of either the *diffusion* or the *mixing* method described before, while those within the spectral window are treated deterministically, just as a meteorologist describes weather systems such as depressions, anticyclones, fronts, and jet streams.

Billow Turbulence

To begin, the spectral window is located so that it overlaps the Kelvin-Helmholtz billows, which cause the transition from the two-dimensionally isotropic eddies of the buoyancy and rotation subranges to the three-dimensionally isotropic eddies of the inertial subrange.[2] As can be seen above, scale analysis based on the con-

---

[1] During the day, of course, short wave solar radiation absorbed near the surface can lead to an irreversibility in the heat flux.

[2] *Orlanski and Bryan* (1969) and *McEwan* (1971) have shown that internal wave breaking without Kelvin-Helmholtz instability may also effect the transition to isotropic turbulence, but this does not alter the conclusions reached here.

densed variance model and flow visualization in the thermocline both show that the vertical displacements achieved by these billows are typically 30 cm, and only very rarely as large as one meter. This upwelling is quite insignificant compared with the motions to be described later. Billow turbulence can safely be ignored as a direct factor in nutrient supply. It does, of course, play an indirect role by breaking down inhomogeneities in the nutrient field originally having scales comparable with the billow height into smaller fragments so that molecular diffusion can smooth them away. This process of scalar variance cascading was described before in the context of the mixing model. It does not control the rate of vertical transport of nutrients.

Quasi-Geostrophic Eddies

These eddies are two-dimensionally isotropic, with most of their kinetic energy consisting of motions along density surfaces. A quasi-geostrophic balance exists between the vertical displacements of the density surfaces and the flow along them. Idealizations of those patterns are shown in standard text books (e.g. *Defant*, 1961, Ch. 14). They are like miniature versions of the large ocean gyres in which the density surfaces are displaced vertically to reach geostrophic equilibrium with wind driven currents, with the kinetic energy coming from larger scale oceanic motions rather than the wind. So upward displacements of density surfaces (and hence of nutrient rich deep water) on the horizontal scales and lifetimes of rotation subrange eddies are created patchily by the intermittent flux of kinetic energy from (presumably) the MODE scale eddies described by *McWilliams* (1977). The vertical displacements achieved by individual eddies will depend on their mean kinetic energy density and their circulation pattern, which must be predicted by theory.

Taking a kinematic rather than a dynamical approach, one can seek to estimate the vertical displacement field from the patterns of convergence-divergence density surfaces. Such patterns occur in the (non-geostrophic) buoyancy range of eddies as well as the rotation range and the associated vertical motions appear to explain the observed vertical heat fluxes during calm weather (*Woods and Moen*, 1975).

Fronts

Areas of deformation along density surfaces enhance gradients and lead to frontogenesis which accelerates as the geostrophic flow increases, creating thin fronts in two to three days for typical deformation fields created by eddies with horizontal scales of about 100 km (i.e. MOED type eddies). Because these eddies and their associated deformation field persists for much longer than the time taken for frontogenesis, fronts are likely to be present in all such eddies for most of the time. During this time (the order of one month) instabilities on the fronts cause large amplitude waves of typically 10 km wavelength to distort them rather like Gulf Stream meanders. Horizontal accelerations associated with those waves produce alternating upwelling and downwelling tongues of water along the inclined density surface at the front. Observations in the western Ionian Sea (Woods, 1972) have shown that the vertical displacements of water in these fronts can exceed ±50 m and a parameterization of the vertical heat flux associated with this frontal upwelling process (Woods, 1974b) suggests that it can supply the monthly mean heat flux through the interior of the seasonal thermocline.

It is tempting to postulate that the vertical nutrient flux is also carried by frontal upwelling. If this were the case, then the supply of nutrients to the euphotic zone would occur mainly in the tiny fraction of the ocean where fronts are present. If the ocean were to be filled with MODE type eddies with a front occurring on the average in every one of them (a maximum likely packing), then the spacing between fronts would be given by the baroclinic Rossby radius of deformation (i.e. ~100 km). The horizontal thickness of the upwelling tongues is typically 2 km; and it can be assumed that the upwelling displacement is sinusoidal along the fronts. Taking these values, it can be calculated that the regions of strong upwelling occupy less than 1% of the surface area. This may well be a high estimate, since not all the eddies will contain active fronts. As a result, this model implies that nutrient supply to the euphotic zone will be better described in terms of an ensemble of rare events rather than the uniform vertical eddy diffusion of the classical picture. It is possible that even in regions where production is low on the average (e.g. central oceanic gyres), there may be locally productive regions around fronts. The frequency of occurrence and spatial concentration of such fronts may be a controlling factor in biological productivity in the ocean. In general, frontal activity is more vigorous in regions of topographic constraint such as the Malta channel, and relatively weak in oceanic gyres, but fronts have been found in the Sargasso Sea (e.g. Katz, 1969).

CONCLUSIONS

Scattering of sound by both physical and biological processes in the ocean is strongly dependent on the turbulent fluctuations of the flow. Improved models of oceanic turbulence in which the effects of intermittency play a central role offer hope for a better understanding of the observed space-time variations in billow turbulence and primary production, and therefore in sound scattering. However, many aspects of these models have not yet been worked out in detail, nor have they been fully tested against experimental data.

REFERENCES

Batchelor, G. K. and A. A. Townsend. 1949. The nature of turbulent motions at large wavenumbers. *Proc. Roy. Soc.*, A199: 238-255.

Berkele, J. C. 1969. Eddy circulation patterns in the Sargasso Sea. *Variability in the North Atlantic*. (Ed. Charnock and Lee) Proc. ICES meeting, Dublin.

Defant, A. 1969. *Physical Oceanography*. Pergamon, Oxford.

Dietrich, G. 1975. *General Oceanography*. Second Edition, Interscience, London.

Katz, E. J. 1969. Further study of a front in the Sargasso Sea. *Tellus*, 21: 259-269.

Kolmogorof, A. N. 1941a. The local structure of turbulence in incompressible viscous fluid for very large Reynolds numbers. *C. R. Acad. Sci. URSS*, 30: 301.

Kolmogorof, A. N. 1941b. On degeneration of isotropic turbulence in an incompressible viscous liquid. *C. R. Acad. Sci. URSS*, 31: 538.

Kullenberg, G. 1974. An experimental and theoretical investigation of the turbulent diffusion in the upper layer of the sea. *Report No. 25*, University of Copenhagen.

Leslie, D. C. 1973. *Developments in the theory of turbulence*. Clarendon Press, Oxford.

McEwan, A. D. 1971. Degeneration of resonantly-excited standing internal gravity waves. *J. Fluid Mech.* 50, part 3: 431-448.

MacVean, M. 1976. Numerical upper ocean frontogenesis models. Unpublished manuscript.

McWilliams, J. 1977. On the Large Scale Circulation of the Ocean: A Discussion for the Unfamiliar. In: "*Oceanic Sound Scattering Prediction*", N. R. Andersen and B. J. Zahuranec (Eds.), Plenum, N.Y., (this volume).

Moen, J. 1974. The spectrum of temperature variability in the Diurnal thermocline. M. Sc. *dissertation*, University of Southampton.

Munk, W. H. and C. J. R. Garrett. 1973. Internal wave breaking and microstructure. *Boundary Layer Met.*, 4: 37-45.

Okubo, A. 1971. Horizontal and Vertical Mixing in the Sea. In *Impingement of Man on the Oceans*. (Ed. D. W. Hood) Wiley-Interscience, New York-London: 89-168.

Orlanski, L. and K. Bryan. 1969. Formation of the thermocline step structure by large-amplitude internal gravity waves. *J. Geophys. Res.*, 74, (28): 6975-6983.

Ozmidov, R. V. 1965. On the turbulent exchange in a stably stratified ocean. *Atmos. & Oceanic Phys.*, 1, (8): 493-497.

Panchev, S. 1971. *Random functions and turbulence*. Pergamon, London.

Platt, T. and K. L. Denman. 1976. Unpublished manuscript (to be published in *Proc. NATO ASI*, Urbino, Sept., 1975.)

Richardson, L. F. 1922. *Weather prediction by numerical process*. Cambridge University Press, London.

Saunders, P. M. 1972. Space and time variability of temperature in the upper ocean. *Deep-Sea Research*, 19: 467-480.

Steele, J. 1975. Personal communication.

Tatarski, V. I. 1961. *Wave propagation in a turbulent medium.* McGraw-Hill, NY.

Woods, J. D. 1968. Wave-induced shear instability in the summer thermocline. *J. Fluid Mech.*, 32, (4): 791-800.

Woods, J. D. (Editor) 1969. Fossil turbulence. *Radio Science,* 4: 1365-1367.

Woods, J. D. 1972. The structure of fronts in the seasonal thermocline. *Proc. Conf. "Strait of Sicily"*, La Spezia: 144-152.

Woods, J. D. 1974a. Space-time characteristics of turbulence in the seasonal thermocline. *Memoires Société Royal des Sciences de Liège, 6e série,* tome VI: 109-130.

Woods, J. D. 1974b. Diffusion due to fronts in the rotation subrange of turbulence in the seasonal thermocline. *La Houille Blanche* No. 7/8: 589-597.

Woods, J. D. and R. L. Wiley. 1972. Billow turbulence and ocean microstructure. *Deep-Sea Research, 19:* 87-121.

Woods, J. D. and J. Moen. 1975. The diurnal thermocline. Presented at IAPSO Symposium on Ocean Microstructure, Grenoble, September, 1975.

Optical Parameters that may Affect Vertical Movement of Animals in the Ocean

J. Ronald V. Zaneveld

Oregon State University

ABSTRACT

The parameters that may affect vertical movement of animals in the ocean are discussed. These parameters are: total energy flux; spectral characteristics of the energy flux; shape of the lightfield; degree of polarization and direction of the $\bar{e}$ vector. Of these, only the total energy flux shows appreciable variation with depth beneath the euphatic zone. Turbidity may also influence animal migration.

INTRODUCTION

The physical property in the ocean to which marine animals respond visually is the lightfield. The lightfield is a vector field of radiances $L(\lambda,\theta,\phi,z)$ where λ is the wavelength, θ and ϕ denote the direction from which the light comes, and z is the depth. Radiance is defined as radiant flux per unit solid angle per unit projected area of a surface. (For definitions of this and related quantities, see *Jerlov*, 1968.) Each radiance is only fully characterized by the four Stokes parameters (*Van de Hulst*, 1957; *Beardsley*, 1968; *Lundgren*, 1971), which describe the state of polarization of the radiance.

The underwater lightfield at a given location is a function of the incident light at the surface of the ocean (which in turn depends on the entire process of atmospheric radiative transfer), the air-sea-interaction of light, and radiative transfer in the ocean itself. Radiative transfer in the ocean depends on the attenuation and scattering properties of the water and the materials dissolved and suspended therein. If the lightfield at the boundary of the ocean is given as well as the scattering and attenuation properties, the lightfield may be calculated at any depth. (For example, see *Chandrasekhar*, 1950; *Preisendorfer*, 1965; *Plass and Kattawar*, 1972).

Rather than analyzing the entire lightfield it can be assumed that animals respond to certain aspects of the lightfield: total energy flux; spectral characteristics of the light flux; shape of the lightfield; degree of polarization and direction of the $\vec{e}$ vector. The problem of underwater light and the orientation of animals has recently been reviewed in an excellent article by *Waterman* (1974). As a result, only the major aspects will be presented here.

TOTAL ENERGY FLUX

When considering the energy flux at a given point it is necessary to distinguish between scalar and vector irradiance. Scalar irradiance for a given wavelength λ is defined by

$$E_0(\lambda,z) = \int_0^{2\pi} \int_0^{\pi} L(\lambda,\theta,\phi, z) \sin\theta \, d\theta \, d\phi \quad (1)$$

and represents the total energy flux through a point. The vector irradiance for a given wavelength λ is defined by

$$E(\lambda,z) = \int_0^{2\pi} \int_0^{\pi} L(\lambda,\theta,\phi, z) \cos\theta \sin\theta \, d\theta \, d\phi \quad (2)$$

and represents the energy flux through a horizontal plane. The total energy fluxes $E_0(z)$ and $E(z)$ are obtained by integrating $E_0(\lambda,z)$ and $E(\lambda,z)$ over wavelength. The vector irradiance* is more commonly measured, although the scalar irradiance represents the total energy flux. The animal itself, of course, sees neither the vector nor the scalar irradiance, but some integral over the lightfield that depends on his physiology and his orientation. Figure 1 (*Clarke*, 1971; *Waterman*, 1974) shows the vector irradiance as a function of depth. The attenuation of vector irradiance is given by the parameter $k(z) = \frac{-1}{E(z)} \frac{dE(z)}{dz}$. The irradiance attentuation

---

*EDITORS' NOTE: Vector irradiance is also E_d-E_u, the difference between downwelling irradiance and upwelling irradiance. Biologists measure mainly E_d in routine work. On the other hand, E_u always remains small compared with E_d.

OPTICAL PARAMETERS

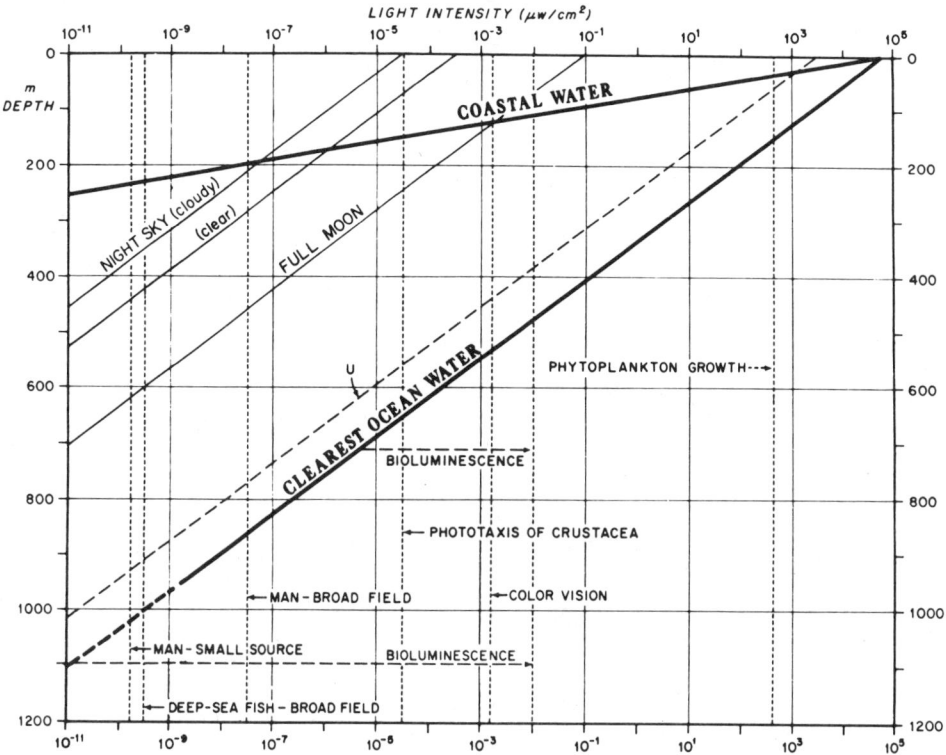

Fig. 1. Irradiance as a function of depth in the sea. The heavy lines diagrammatically indicate the penetration of sunlight into clear coastal water (k assumed to be 0.15 and constant throughout the water column) and into the clearest ocean water (k for the water column taken to be 0.033). In the latter case the sun's reflected upwelling irradiance is shown by the broken line (U). The penetration of moon and night sky light are also indicated. Estimates of relevant biological thresholds of boundaries are shown by the vertical broken lines (from Clarke, 1971).

coefficient for all wavelengths is strongly depth dependent near the surface as the red and infrared light is attenuated rapidly near the surface. Ocean water has been classified in a scheme developed by Jerlov (1968) (see Figure 2). Water of types between I and II are representative of most ocean waters. The attentuation of total light energy by the various water types are shown on Table 1 and Figure 3.

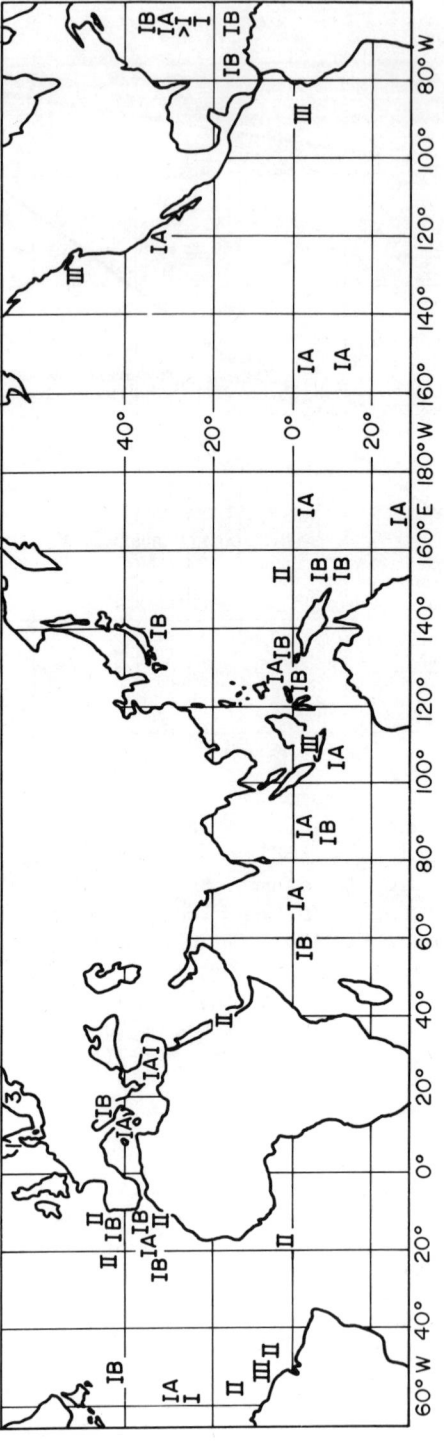

Fig. 2. Regional distribution of optical water types (from Jerlov, 1968).

OPTICAL PARAMETERS

TABLE 1. Percentage of Total Irradiance (300-2,500 nm) from Sun and Sky. (from Jerlov, 1968)

Depth (m)	Oceanic water					Coastal water				
	I	IA	IB	II	III	1	3	5	7	9
0	100	100	100	100	100	100	100	100	100	100
1	44.5	44.1	42.9	42.0	39.4	36.9	33.0	27.8	22.6	17.6
2	38.5	37.9	36.0	34.7	30.3	27.1	22.5	16.4	11.3	7.5
5	30.2	29.0	25.8	23.4	16.8	14.2	9.3	4.6	2.1	1.0
10	22.2	20.8	16.9	14.2	7.6	5.9	2.7	0.69	0.17	0.052
20						1.3	0.29	0.020		
25	13.2	11.1	7.7	4.2	0.97					
50	5.3	3.3	1.8	0.70	0.041	0.022				
75	1.68	0.95	0.42	0.124	0.0018					
100	0.53	0.28	0.10	0.0228						
150	0.056			0.00080						
200	0.0062									

[1]For oceanic water the solar altitude is 90°; for coastal water 45°.

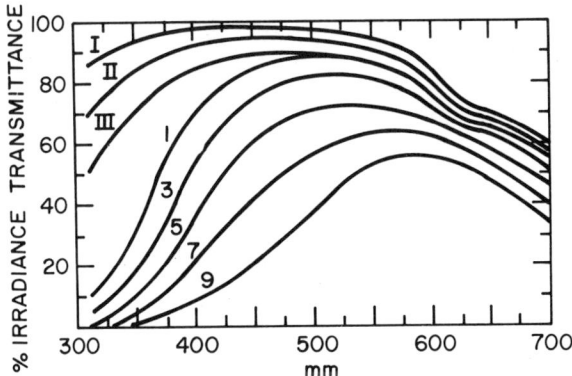

Fig. 3. Transmittance per meter of downward irradiance in the surface layer for optical water types. Oceanic types I, II, III and coastal types 1, 3, 5, 7, 9. (from Jerlov, 1968)

SPECTRAL ENERGY FLUX

The energy flux as a function of wavelength for the various water types is shown on Figure 4 and Table II (Jerlov, 1968). Coastal waters have peak transmissions at wavelengths from 500 to 575 nm, while open ocean waters have peak transmissions at about 475 nm. An example of spectral irradiance as a function of depth is shown on Figure 5 (Kampa, 1961; Jerlov, 1968). The light typically becomes more monochromatic as depth increases.

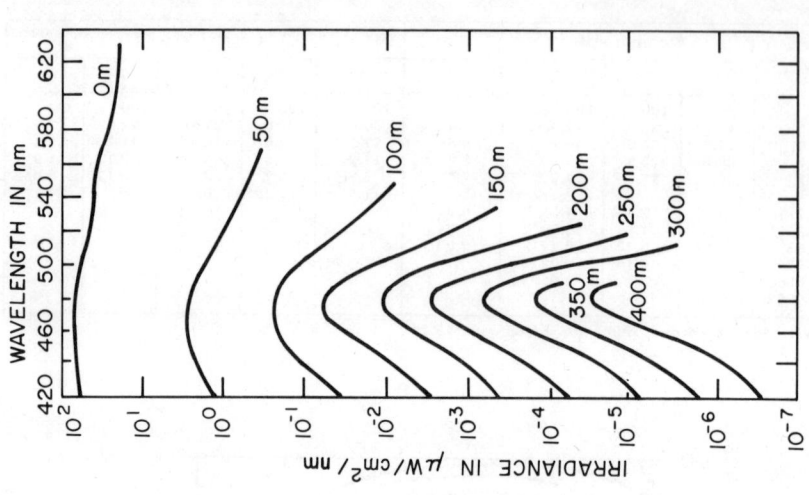

Fig. 5. Spectral distribution of downward irradiance in Golfe du Lion. (after Kampa, 1961)

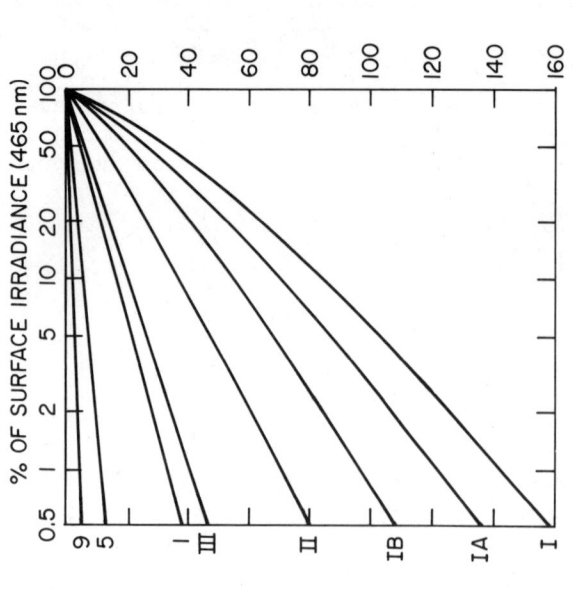

Fig. 4. Depth profiles of downward irradiance in per cent of surface radiance for defined optical water types. Oceanic types I, IA, IB, II, III and coastal types 1, 5, 9. (from Jerlov, 1968)

OPTICAL PARAMETERS

TABLE 2. Irradiance Transmittance for Surface Water of Different Water types. (from Jerlov, 1968)

Water Type	Irradiance transmittance (%/m) Wavelength (nm)															
	310	350	375	400	425	450	475	500	525	550	575	600	625	650	675	700
I	86	94	96.3	97.2	97.8	98.1	98.2	97.2	96.1	94.2	92	85	74	70	66	59
IA	83	92.5	95.1	96.3	97.1	97.4	97.5	96.6	95.5	93.6	91	84	73.5	69.5	65.5	58.5
IB	80	90.5	94	95.5	96.4	96.7	96.8	96.0	95.0	93.0	90.5	83	73	69	65	58.0
II	69	84	89	92	93.5	94	94	93.5	92.5	90.5	87.5	80	71	67.5	63.5	56
III	50	71	79	84	87	88.5	89	89	88.5	86.5	82.5	75	68	65	61	54
1	16	32	54	69	79	84	87.5	88.8	88.5	86.5	82.5	75	68	65	61	54
3	9	19	34	53	66	75	80	82	82	81	78	71	65	62	57	51
5	3	10	21	36	50	60	67	71	73	72	70	67	62	58	52	45
7		5.0	12	22	32	42	50	56	61	63	63	62	58	53	46	40
9		1.5	4.7	9	15	21	29	37	46	53	56	55	52	47	40	33

Water Type	Irradiance transmittance (%/10 m) Wavelength (nm)															
	310	350	375	400	425	450	475	500	525	550	575	600	625	650	675	700
I	22	54	69	75	80	82.5	83.5	75	67	55	43	20	4.9	2.8	1.6	0.5
IA	16	45	60.5	68.5	74.5	77	78	71	63	52	39	17	4.6	2.6	1.4	0.5
IB	11	37	54	63	69	71.5	72	66.5	60	48	37	15	4.3	2.4	1.3	0.4
II	2	18	31	43	51	54	54	51	46	37	26	11	3.5	1.9	1.0	0.3
III	0.1	3	10	18	25	29	31	31	29	23	15	5.6	2.1	1.3	0.8	0.2
1			0.3	2.7	9	18	26	30	29	23	14	5.7	2.0	1.3	0.8	0.2
3				0.2	1.8	5.9	10	14	15	12	7.9	3.7	1.6	0.9	0.4	0.1
5					0.1	0.7	1.8	3.3	4.3	4.0	3.1	1.9	0.9	0.4	0.2	
7							0.1	0.3	0.7	1.1	1.1	0.8	0.4	0.2		
9										0.2	0.3	0.3	0.2	0.1		

SHAPE OF THE LIGHTFIELD

Near the ocean surface the lightfield is peaked in the direction of the refracted image of the sun. As depth increases, the peak shifts towards the vertical as a result of multiple scattering. At great depths, the asymptotic condition is reached at which the lightfield is symmetrical about the vertical (Preisendorfer, 1959; Zaneveld and Pak, 1972; Højerslev and Zaneveld, 1975). Figure 6 illustrates this process (Lundgren and Højerslev, 1971; Ivanoff, 1975).

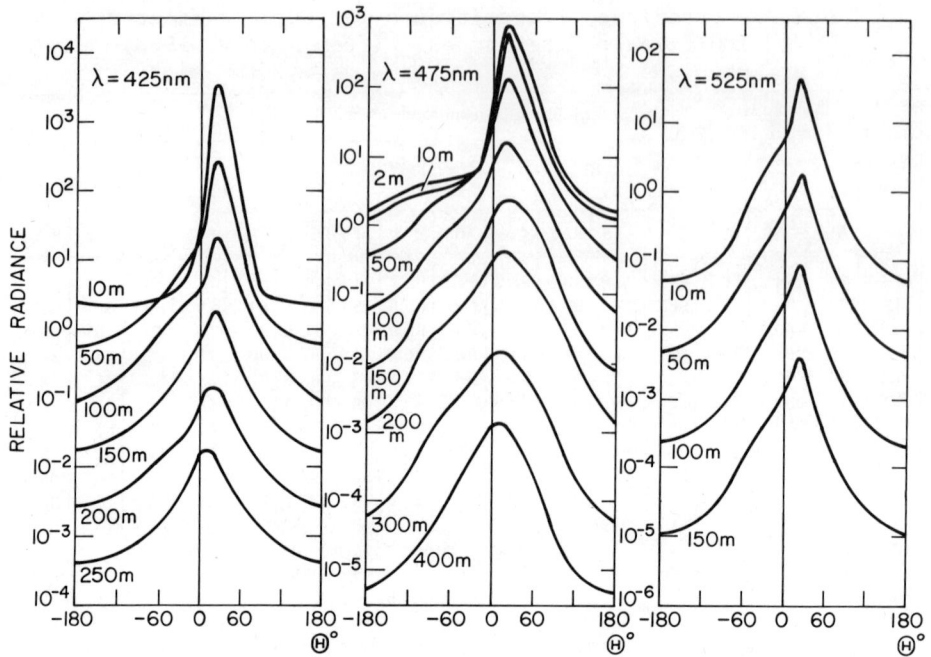

Fig. 6. Relative radiance as a function of vertical angle and depth for various wavelengths. (after Lundgren and Højerslev, 1971)

DEGREE OF POLARIZATION AND DIRECTION OF THE $\vec{e}$ VECTOR

A complete review of this subject may be found in Ivanoff (1974). Unpolarized light has no preferred direction for the plane of oscillation of the electric vector of the electromagnetic wave. Elliptically polarized light has an electric vector whose terminus describes an ellipse in the plane of oscillation. Light encountered in the ocean is partially polarized, i.e., it is a mixture of polarized and unpolarized light. The degree of polarization (p) indicates the proportion of the light that is polarized. The degee of polarization underwater depends on solar zenith angle, direction of the observation underwater and depth, as well as the optical properties of the water. Maximum polarization occurs perpendicular to the sun beam. In general, the degree of polarization decreases with more diffuse atmospheric light and increased turbidity (Waterman and Westell, 1956). The depth dependence of the degree of polarization is illustrated in Figure 7 (Ivanoff, 1974). The degree of polarization initially changes rapidly due to the increased scattered light and decreased direct skylight. At asymptotic depths the degree of polarization reaches a constant for all directions of sight.

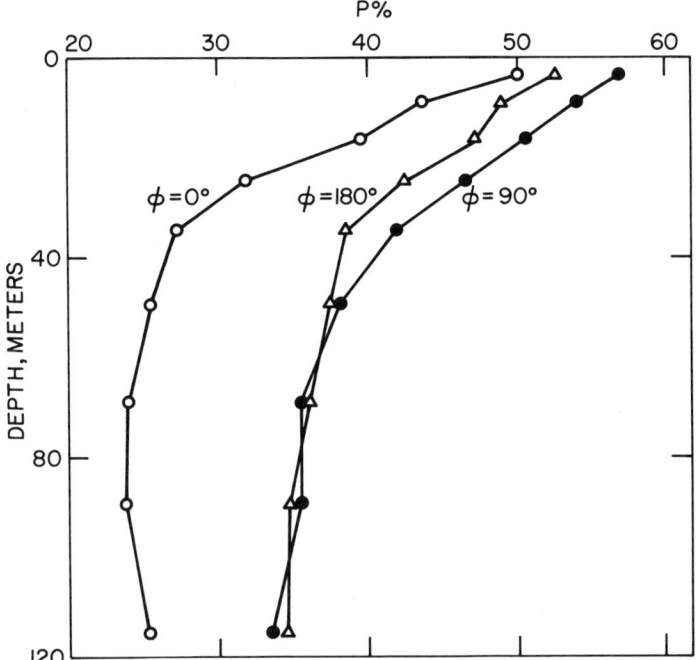

Fig. 7. Degree of polarization as a function of depth for horizontal lines of sight. φ is the azimuth relative to the direction of the sun. (from Ivanoff, 1975)

The preferred direction of the electric ($\vec{e}$) vector also depends on the solar zenith angle, direction of sight underwater, depth and optical properties of the ocean (Timofeeva, 1974). Near the surface the angle of the $\vec{e}$ vector with a horizontal plane for a line of sight perpendicular to the sun's bearing is equal to the angle of refraction of the direct sunlight. At greater depths, this angle decreases and at asymptotic dpeths it is zero. No detailed measurements of this parameter have been made as yet in the ocean.

DISCUSSION

The parameters described briefly above have several points in common. Great variations occur near the surface, whereas beneath the euphotic zone (about 80 m in clear ocean water and 20 m in coastal waters) the only parameter of the lightfield that can be used is the total energy flux (spherical or vector irradiance) As a result, this parameter is by far the most important in the pre-

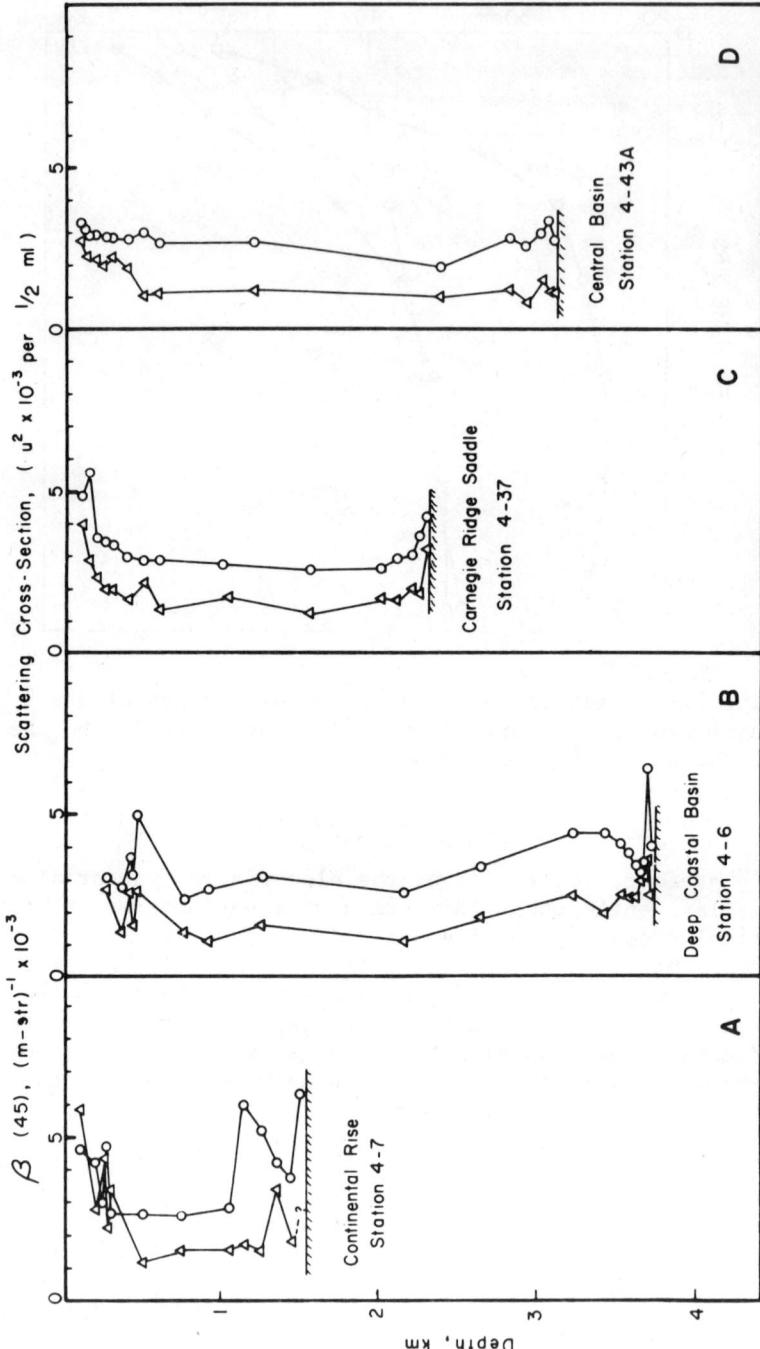

Fig. 8. Profiles of light scattering at 45° and scattering cross-section of suspended particles for various locations in the Panama Basin. (after *Plank et al.*, 1972)

diction of animal movement. This parameter is fortuitously also the most readily measured (for a review, see *Tyler and Smith*, 1970). Much more work needs to be done in order to finish the optical classification of ocean waters. One may predict underwater irradiance with some accuracy if the optical classification of the water is known in combination with irradiance at the earth surface (for example, see *Plass and Kattawar*, 1972) as a function of solar attitude.

Such an approach is, in fact, taken when the irradiance profile is measured at a given time, and then is assumed to remain constant over periods of the order of days. To obtain the irradiance at a given depth, one then multiplies the irradiance profile (normalized to 1.0 at the surface) by the incident irradiance measured on board the ship. This approach may lead to errors of up to an order of magnitude at about 100 m depth, due to changes in the attenuation of the intervening seawater. Particle "clouds", surface wave conditions and solar altitude all influence the irradiance at a given depth. It is strongly suggested that wherever possible the irradiance is measured at the same time and depth as the biological parameters. Inferring irradiance from one profile, or from an optical classification would yield useful "average" light conditions, but if animal migration is being studied at a given location, much potentially useful information is lost if the irradiance is not measured simultaneously.

For purposes of camouflage or feeding, animals may prefer a certain degree of turbidity. Turbidity is accurately specified by the beam attenuation coefficient or the volume scattering function (for definitions, see *Jerlov*, 1968). Little is known about this preference by animals. Turbidity, furthermore, varies greatly with space and time. Turbid layers generally exist near the surface and the bottom, but also may exist at any depth in between (see Figure 8). Due to the great time and space variability of turbidity (*Plank et al.*, 1972) prediction other than gross averages are very difficult.

ACKNOWLEDGEMENTS

This work was supported by the Office of Naval Research through contract N000-14-67A-0369-0007 under project NR 083-102.

REFERENCES

Beardsley, G. F. 1968. Mueller scattering matrix of seawater. *J. Opt. Oc. Am*, 58: 52-57.

Chandrasekhar, S. 1950. *Radiative Transfer*. Oxford Univ. Press, London. 393 pp.

Clarke, G. L. 1971. In: "*Proceedings of the International Symposium on Biological Sound Scattering in the Ocean, 1970.*" (G. B. Farquhar, ed.). Maury Center for Ocean Science, Dept. of the Navy, Washington, D.C. 41-50.

Højerslev, N., and J. R. V. Zaneveld. 1975. A theoretical proof of the existence of the submarine asymptotic daylight field. Submitted to J. Geophys. Res.

Ivanoff, A. 1974. In: *Optical Aspects of Oceanography*. (N. G. Jerlov and E. Steemann Nielsen, Eds.) Academic Press, London. 151-176.

Ivanoff, A. 1975. *Introduction a l'Oceanographie (Tome II)*. Vuibert, Paris.

Jerlov, N. G. 1968. *Optical Oceanography*. Elsevier, Amsterdam. 194 pp.

Kampa, E. M. 1961. Daylight penetration measurements in three oceans. *Union Géod. Géophys. Intern. Monographie*, 10: 91-96.

Lundgren, B. 1971. On the polarization of the daylight in the sea. Report #17. University of Copenhagen. Inst. for Fysisk Oceanogr.

Lundgren, B., and N. Højerslev. 1971. Daylight measurements in the Sargasso Sea. Results from the "Dana" expedition January-April 1966. Report #14. University of Copenhagen Inst. for Fysisk Oceanogr.

Plank, W. S., J. R. V. Zaneveld, and H. Pak. 1973. Distribution of suspended matter in the Panama Basin. J. Geophys. Res., 78(30): 7113-7121.

Plass, G. N., and G. W. Kattawar. 1972. Effect of aerosol variation on radiance in the earth and atmosphere-ocean system. App. Optics 11: 1598:1604.

Preisendorfer, R. W. 1959. Theoretical proof of the existence of characteristic diffuse light in natural waters. J. Marine Res., 18: 1-9.

Preisendorfer, R. W. 1965. *Radiative Transfer on Discrete Spaces*. Pergamon, New York. 462 pp.

Timofeeva, V. A. 1974. In: *Optical Aspects of Oceanography*. (N. G. Jerlov and E. Steemann Nielsen, Eds.) Academic Press, London. 177-220.

Tyler, J. E. and R. C. Smith. 1970. *Measurements of Spectral Irradiance Underwater*. Gordon and Breach, New York.

Van de Hulst, H. C. 1957. *Light Scattering by Small Particles*. Wiley, New York. 470 pp.

Waterman, T. H. and W. E. Westell. 1956. Quantitative effect of the sun's position on submarine light polarization. J. Marine Res., 15: 149-169.

Waterman, T. H. 1974. In: *Optical Aspects of Oceanography*. (N. G. Jerlov and E. Steemann Nielsen, Eds.) Academic Press, London. 415-444.

Zaneveld, J. R. V. and H. Pak. 1972. Some aspects of the axially symmetric submarine daylight field. J. Geophys. Res., 77: 2677-2680.

II

FOOD CHAINS

Recommendations of the Working Group on Food Chains

J. Mauchline, Chairman

Dunstaffnage Marine Research Laboratory

INTRODUCTION

Several subject areas of importance in examining the feasibility of predicting sound scattering in the oceans from physical/chemical/biological information were identified prior to the symposium. The proceedings of the Airlie House symposium of 1970 on "Biological Sound Scattering in the Oceans", the 1972 symposium on "Barobiology and the Experimental Biology of the Deep Sea", and the 1975 workshop on "Problems of Assessing Populations of Nekton" were used as baselines. Participants in Section II prepared statements on:
  a. The interpretation of data obtained from sampling pelagic organisms (Angel).
  b. Distribution of and trophic structure within the planktonic biomass (Blackburn).
  c. Regulation of and succession within pelagic community structure (Lewis, McGowan).
  d. Feeding ecology and growth of crustaceans (Mauchline, Nemoto).
  e. Distribution and ecology of mesopelagic fish (Badcock & Merrett, Hopkins & Baird, Karnella & Gibbs, Kinzer).
  f. Physiology of pelagic organisms (Childress).

Following delivery of the plenary papers, biologists met with acousticians and exchanged viewpoints, identified problems within the respective fields and discussed areas of mutual interest. Meetings with physicists, chemists and modellers further contributed to identifying the following areas of oceanic biology as important to the aims of the meeting:
  a. Structure of midwater communities
  b. Microscale distribution of mesopelagic nekton

c. Megascale distribution of mesopelagic nekton
d. Biomass of micronekton in upwelling situations
e. Feeding ecology of micronekton
f. Use of existing data bases in above studies
g. Physiological approaches
h. Technological aspects of physiological studies

Convenors appointed for discussions of these topics were: (a) M. V. Angel, (b) T. A. Clarke, (c) J. A. McGowan in collaboration with R. Backus, (d) M. Blackburn, (e) J. Kinzer, (f) R. H. Gibbs, Jr., (g) J. Badcock, (h) J. Childress.

These subjects are all interrelated and cannot be discussed in isolation, either from each other or from topics discussed in other sections of this symposium. In some cases, the physicists, chemists, acousticians and modelers consider closely related, if not identical, problems but from a different perspective. Consequently, this report should be examined in conjunction with the other section reports.

STRUCTURE OF MIDWATER COMMUNITIES

Vertical Structure

A detailed understanding of the vertical structure of oceanic pelagic communities will provide an insight into some of the processes governing the formation and behavior of deep scattering layers. The vertical zonation of plankton and micronekton has been described and is well documented. The boundaries of the zones relate to ecologically important features such as:
a. Bottom of the wind-mixed layer,
b. The depth at which the light field becomes symmetrical,
c. The depth at which daylight becomes dim relative to the intensity of bioluminescence,
d. The depth at which the scarcity of food leads the animals to adopt a life style of parsimony.

These zones approximate the classical bathymetric categories of epipelagic, shallow mesopelagic, deep mesopelagic, and bathypelagic.

Vertical migration by species of micronekton is usually performed by the majority of the population. Vertical migration of species of plankton, however, is often performed by only 30-50% of the population; this leads to a spreading of the population through the water column at night. Migration tends to be interzonal: epipelagic species migrate from the thermocline region into the middle of the wind-mixed layer; shallow mesopelagic species migrate into the epipelagic zone; deep mesopelagic species migrate into the shallow mesopelagic zone. In general, the deepest living planktonic species in a zone are the last to arrive and the first to leave.

FOOD CHAINS RECOMMENDATIONS

The daytime vertical layering of species tends to be maintained with their distributions more restricted vertically. Diel vertical migrations of plankton are restricted to the upper 700 m of the oceans; whereas migrations of micronekton occur, in a few species, from depths below the range 1000-1250 m. Overtaking of one species by another is more frequent in micronekton than plankton, *e.g.*, *Ceratoscopelus warmingi* migrating from 1250 m depth to the surface at 30°N 23°W.

Availability of data. Data from extensive sampling of various localities are available although not necessarily completely worked up, *e.g.*, Ocean Acre. Working up of material from such thoroughly sampled areas is recommended. A central data bank might be useful, although the quality control of the data may prove to be a serious problem.

Studies required. Vertical structuring of communities varies latitudinally both in species composition and in the distribution of biomass. Studies of microbial biomass should be included. The extent of vertical migrations also varies latitudinally. At any single station there will be seasonal variations that will be more or less extensive. Key stations should be sampled with opening/closing nets. Layers of 100 m thickness should be sampled through the surface to 1000 m depth water column by day and by night. The volume of water filtered by the sampler should be measured. This volume should be about 5-10 thousand cubic meters for plankton and an order of magnitude higher for nekton. Data from such stations would provide information on the vertical distributions and migrations of species, an assessment of the important species present, and reasonable estimates of their abundance. Simultaneous acoustical observations would allow correlation of movements of dominant species with changes in sonic reverberations.

Ideally, analyses of the vertical distributions and migrations need to be assessed in each of the major zoogeographic zones. A network of oblique hauls, preferably 0-1000 m, to sample the complete depth range of mesopelagic and epipelagic species by day and night could link the more detailed results from the vertical stations. A routine procedure, rigorously followed, should be designed for such vertical stations. Uniformity throughout the sampling program is essential to the meaningful interpretation of the data and this sampling should be conducted in conjunction with acoustic investigations. An initial pilot study of some given locality should be made to assess how far the vertical structuring of the midwater community can be related to the observed DSL structure. A 50% explanation of scattering phenomena would be acceptable in this initial study because as the data base increases so will the proportion of explainable DSL phenomena. One approach might be to fit profiles of acoustic energy scattered at a range of frequencies to the profile properties of the midwater communities. These properties could include measures such as biomass, species groupings as derived from "factor" analysis or some other objective assessment of the data, *e.g.*, size grouping of fish with swimbladders.

The oblique tows will locate and identify important faunal boundaries. As many taxonomic groups as possible should be studied because the boundaries may be more apparent in some groups than others, but not necessarily always the same groups. Changes in scattering phenomena, both in their profile structures and their variability, can be expected to coincide with faunal boundaries. Identification of faunal boundaries along a transect of oblique samples is probably best analyzed, at present, by principal components analysis so long as rotation to simple structure is carried out. Improved techniques of analyses are continually being developed so that the most appropriate methods should be selected for the interpretation of the data, as the data become available.

Microscale Structure

Non-random spatial distributions--usually in the direction of aggregation--are probably the rule rather than the exception among pelagic organisms. The following discussion is primarily of patchiness in the distributions of micronekton.

Angel's analysis of the sampling problems arising from patchiness of plankton is equally applicable to the micronekton. Likewise, aspects of mesoscale patchiness, discussed elsewhere at this symposium, are also pertinent to the micronekton. However, micronekton on the microscale, might be expected to behave differently from plankton. Micronekton are relatively good swimmers, possess good sensing capabilities, and are larger in body size. The same coactive factors that might produce an aggregation of copepods several meters in size may result in an aggregation of myctophids of the order of tens of meters in size. Consequently, it seems redundant to define in detail the need to study and recognize the effects of patchiness on sampling of micronekton or to give similar treatment to mesoscale distributions. It is more appropriate to consider the detection of microscale patchiness and its significance in understanding the ecology of micronekton.

Microscale patchiness can be related to many aspects of micronekton ecology. Several theoretical studies suggest that the adaptive values of schooling or aggregation lie in reduction of predation. Many epipelagic fish school during the day and disperse at night to feed; the behavior pattern may, in part, be correlated with reduced predation pressure at night. Diel variations in degrees of aggregation of populations of mesopelagic fish will probably reflect a similar situation, even though day and night light levels may differ little or not at all. Seasonal aggregation or aggregation at the onset of sexual maturity suggests an adaptive value related to reproductive success. Aggregations of like-sized individuals may reduce cannibalism. Aggregations that appear to be correlated with feeding chronology or with aggregations of principal prey organisms suggest that the species has evolved the behavior in response to aggregation of its prey--just as wolf packs of submarines evolved in response to their prey being aggregated in convoys.

In theory, it is by no means necessary that all departures from randomness be in the direction of aggregation; avoidance of conspecifics might be favored by the low food resources of the deep sea and lead to an overdispersed or even distribution.

Detection of patchiness of micronektonic species is made difficult by the generally low densities. Tows must be of the order of kilometers or tens of kilometers in length to obtain an average number of individuals per sample sufficient for statistical analysis. Replicate series of such samples are expensive in ship time. Consequently, the few previous attempts to consider patchiness of micronekton are based on inadequate sampling. Nevertheless, these analyses have generally indicated patchiness in the distributions of micronekton. Analysis of any replicate series already collected should be encouraged to expand the meagre data base.

A device similar to the Longhurst-Hardy Plankton Recorder (LHPR) that could collect shorter, contiguous serial samples would be helpful. Unfortunately, the probability of equipment similar to the LHPR functioning effectively in sampling micronekton is not high. The major problem encountered when sampling plankton--a non-systematic time lag between capture and entry into the cod-end gauzes--will almost certainly be magnified when sampling micronekton. Not only are the sustainable swimming speeds of the latter even more comparable to the towing speeds of the nets used, but there is also a tendency for many species to become entangled in the meshes. Consequently, a device that can sequentially fish a series of entirely separate nets offers the best prospect.

A "spot" sample, that is one whose spatial dimensions are more nearly equal to each other, would be far more appropriate for the detection of microscale patchiness than the elongate cylinders or prisms sampled by towed nets. Thus, development of a device that could sample a large sphere or cube of about 10^4 or 10^5 m^3 of water at depth might constitute a major advance. Acoustic devices presently available, in particular at-depth sonar using a range of frequencies, could provide information on the spatial distribution of individual scatterers. Densities of organisms, even within aggregations, are probably so low that packing is not likely to be a problem. A major limitation is that acoustic devices cannot discriminate between species with similar acoustic cross sections and consequently interpretation of sonar data from areas of high species diversity is difficult. A sonar might be mounted on a trawl or DSRV; the trawl collection or visual observations along with simultaneous scanning at more than one frequency might allow identification of aggregated scatterers or at least a narrowing of choices.

Megascale Distribution Patterns

Considerable progress has been made during the last 15 years in determining the shape of the faunal provinces of the Pacific and

Atlantic Oceans. Each province is of the order of 10^6 km^2 and is internally reasonably homogeneous. A remarkable agreement in basic patterns is observed between the Pacific and Atlantic Oceans despite the fact that zooplankton in the former and mesopelagic fish in the latter were used to determine the provinces. These faunal provinces are few in number and their locations correlate with the major circulational and climatic features of the oceans. The temporal persistence of faunal provinces is thought to be great.

The Indian and Southern Oceans are less well known. The Indian Ocean will probably show similar basic patterns to those of the Atlantic and Pacific even although it has no temperate region north of the equator and may show effects of monsoons. The Southern Ocean is faunally very distinct from the other oceans to the north but has a strong circumglobal continuity.

Some obvious objectives of future research are to advance the work in the Indian and Southern Oceans and to extend the observations in the Pacific to mesopelagic fish and those in the Atlantic to zooplankton.

Distribution of Biomass

Many species of oceanic micronekton, as previously discussed, make vertical diel migrations. Measurements or estimates of the biomass of micronekton by area and season would be useful in DSL studies, especially if partitioned according to taxa, migratory habits, trophic levels, etc. Existing measurements are spatially scattered and the data obtained in a variable manner. The biomass in a sea area is the product of the integrated parameters (physical, chemical and biological) of that area. It therefore forms a useful general parameter of that area for comparison with other large areas --on the scale of faunal provinces.

In general, the biomass of an area should be partitioned and the proportions of the components compared with each other and with comparable components of biomasses from other areas. There is evidence of pattern. Biomass of total micronekton over a given depth range is higher in the tropics and subarctic than in the subtropics of the central Pacific. This is very broadly similar to the distributions of zooplankton and phytoplankton. The biomass of fish and cephalopods combined in the micronekton of the eastern tropical Pacific is significantly related, with a time lag, to that of zooplankton. Another significant relation exists between biomasses of zooplankton and phytoplankton. More data are required to define and quantify these relationships and so make them useful.

Additional data sets may best be acquired in coastal upwelling situations. Strong spatial gradients of phytoplankton and zooplankton occur during seasons of upwelling but are relatively short lived. Patches and plumes of plankton may appear and disappear within an upwelling season. Thus, several processes whose rates are slow in other areas may be speeded up in upwelling situations and be amenable to study.

CONCLUSIONS

Various programs under the separate headings above are proposed because they would lead to an understanding of the spatial and temporal processes of oceanic faunal provinces and so to the governing parameters of the DSLs. Several of these programs could have a multiple purpose. For example, occasional closely spaced sets of oblique tows could provide measures of coherence within communities, and estimates of partitioned components of biomass. Simultaneous measurements of temperature/salinity, oxygen, light, nutrients, etc. would allow correlations to be made between the biological data, including microbial and phytoplankton biomasses, and the physical-chemical parameters of the environment. Data for megascale mapping, an important area of investigation, would be simultaneously derived.

There are, however, sampling problems. The greatest promise lies in development of methods for study of somewhat smaller scales. These include serial samples, the "spot" samples discussed above, and acoustic devices. The latter should be integrated with any studies using collection methods for the greatest probability of reliable data (or at least, knowledge of biases). In general, the success of such programs will be greater if the interactions between the biological, physical-chemical and acoustical processes are investigated. This approach will lead ultimately to increased understanding of biological phenomena and a better delineation of variability of sonic scattering in the open ocean.

FEEDING ECOLOGY OF MICRONEKTON

The feeding ecology of pelagic organisms is an important parameter in the vertical structuring of the biocoenosis and so of sonic scattering layers. Consequently, a concensus of opinion from the meeting may serve as a guideline for future investigations.

Most migrations of micronektonic organisms are probably associated with feeding (Vinogradov, 1970). The principal taxonomic components of the micronekton are fish, decapods, euphausiids and, of secondary importance, cephalopods. Determination of the proportions and abundance of the taxonomic components of the micronekton, including their diel vertical distributions were discussed in the previous part of this report. In addition, the following aspects of nutritional ecology should be investigated in certain important micronektonic species deemed representative of general classes of organisms:
    a. Taxonomic composition of diet
    b. Selectivity in feeding
    c. Ontogenetic changes of diet
    d. Feeding chronology
    e. Availability of prey (prey migrations, etc.)
    f. Functional morphology

g. Daily ration requirement
h. Resource partitioning
i. Aspects of life history--growth, fecundity, fat storage, etc.
j. Models of energy utilization
k. Integration of energy budget over life history
l. Migrations of population to assess impact on food resources
m. Assessment of trophic efficiency
n. Sensory physiology, especially vision and olfaction

Crustaceans may be either filter feeders or carnivores and, unlike most midwater fish, many species triturate, crush, or otherwise damage prey rather than swallowing it whole. It is therefore difficult to confidently define their diet, daily ration and selectivity of feeding as is possible for the midwater fish. Examination of feeding in nets, especially by invertebrate micronekton, must be evaluated, as must be the impact on dietary analysis of secondary stomach contents; that is, of portions of the stomach contents that are derived from the digestive tracts of prey.

These studies should not only be conducted on the micronektonic species in question but also on the respective dominant prey organisms. This is required to provide an understanding of the feeding ecology and vertical distributions of micronektonic organisms, including the primary components of DSLs.

EXISTING DATA BASES

It was considered necessary to discuss existing data bases in some detail. However, in reviewing the recommendations formulated here, it should not be inferred that new programs and new directions are not needed.

Prediction of unusual sound-scattering phenomena would be beneficial to anyone dependent upon sonic equipment for accomplishing underwater missions, and knowledge of normal scattering and its variability is basic to any such operation, no matter what the geographic location. Therefore, it is important to describe patterns of scattering in any given area as functions of space and time, and to determine the causes of their variation.

Studies relating vertical patterns of sound scattering with associated organisms have been few and relatively recent. Fewer still have attempted to study these relationships over time, and very few have described their correlation with physical and chemical factors. Prediction of any one kind of information--biological, acoustical, physical, or chemical--from the other is currently impossible, yet the materials to provide the basis for prediction are at hand.

While acoustic, physical, and chemical data are relatively easy to obtain and process (they can often be available in real time), this does not hold true for appropriate biological data. Even when sampling gear is adequate, collections are time-consuming and difficult to make, process, and analyze. Thus, the biological data are frequently regarded as the limiting factors for studies related to prediction.

The biological collections should be made in such a way as to permit a sufficiently accurate determination of the vertical distribution and vertical migration of each species or component of a species in a given locality on a seasonal or shorter-term basis. Seasonal or shorter-term coverage is needed to provide data for determining the life cycles of the species, because different life stages not only differ in size (therefore scattering contributions), but also in their vertical distribution patterns and abundance, and often in the development of morphological attributes (swimbladders) that affect sound scattering. A single sampling is not sufficient as the basis for prediction, even for that exact location.

Collections have been made that would provide basic seasonal data on vertical distribution and life cycles from several areas. They are not many, and they differ in their sampling methodology. Nevertheless, some among them should be sufficient as a starting point by providing the primary biological data necessary to make comparisons between and among locations.

Of particular importance would be the comparison of both life cycles and vertical distributions of the same species or equivalent species. If those characteristics are similar between locations, similarities in hydrologic and acoustic features can be sought. If they are different, differences in hydrologic features might explain them and different acoustic patterns might be explained by them. Differences in species not common to both locations may need to be invoked in discussing differences between locations.

A single location with adequate vertical and seasonal sampling could be compared with one-time comprehensive samples from other locations (for example, the IOS latitudinal series from 20°W) if there are shared species at the same life-history stage. Then hydrologic and acoustic data can be compared with the organisms' vertical distributions.

Significant correlations between vertical distribution of biological organisms and/or scattering intensities and any physical-chemical parameter can lead to a hypothesized relationship. This could then be tested by combining discrete samples of data from other locations chosen for the presence of specific biological, acoustical, or physico-chemical features.

The appropriate biological data are obviously the most difficult to provide. Such data are only available from one or two areas, though they are approaching availability from others. Completion of the necessary studies should be encouraged. Where comparisons between two or more areas are possible, biologists, acousticians, physical oceanographers, and chemical oceanographers should combine their talents in the determination of correlated factors that could form the basis of predictions for other areas or abnormal situations.

In summary, we urge that existing biological materials be analyzed and necessary data be made available. For this, the basic necessity is manpower in the laboratory, not ship time and gear. Then, the hydrologic and acoustic data must be located (and they probably exist for many of the locations of concern), and colleagues from the three involved disciplines will need to work together in the same place.

PHYSIOLOGICAL APPROACHES

Childress, in his paper tabled at this meeting, has provided the background to the present state of the art, its relevance and importance to the theme of this symposium. Further comment regarding that background is not required, pending presentation of the following recommendations.

1. Oxygen Consumption. Detailed studies should be made by direct methods on a relatively small number of species taken from a variety of depth distributions and life habits. It is essential that the animals used be kept alive for long periods to permit reasonable measurements of standard and maximum metabolic rate and activity. The effects of temperature, pH, nutrition, oxygen concentration, and pressure should be investigated. Similar, but less detailed studies of oxygen consumption should be carried out on a wide range of species that are less easily maintained in the laboratory. The interpretation of the results obtained should be compared to extrapolations made from the data of the detailed studies. Difficulties arise with midwater fish because few individuals live in captivity for more than a few hours. Crustaceans, on the other hand, can be kept successfully for considerably longer periods. It may be possible, with some reservations, to interpret fish data by extrapolating data obtained from detailed studies of crustaceans. Where possible, *in situ* observations of oxygen consumption may be of considerable value to confirm more detailed observations at the surface. Though this method has the advantage that traumatic responses are reduced, the conditions are poorly controlled. Where submersibles are used, the capture of specimens is difficult because of the handling capabilities of the submersible. Additional problems arise because the rates of oxygen consumption of many of these organisms are depressed and, further, only small volumes of water can be used to contain the animal.

2. Chemical Analyses. These indirect methods should be used extensively to gain insight into adaptations, habits, etc. For example, seasonal as well as other variations in chemical composition and caloric value should be assessed with regard to growth and reproductive effort. Growth should also be examined in the laboratory from the standpoint of variations in the diet, temperature, and other variables. Detailed studies of the assimilation of natural food should include analyses of the stomach contents

of specimens taken in nets, with specific reference to growth stages. Evaluation of the excretion of nitrogenous components and amino acid leakage should also be made. Studies of amino acid leakage have not been made on bacteria-free organisms; bacteria in and on the organisms must be considered.

3. Morphology. Studies of the morphology and structure of the swimbladder of mesopelagic fish are urgently required since present studies are far from comprehensive and, in general, lack necessary detail. In particular, the extent of intra-specific variation caused by ontogeny, the range of variation between populations, and seasonal variations must be considered.

THE PROBLEMS OF OBTAINING PHYSIOLOGICAL AND BEHAVIORAL INFORMATION

Unlike investigations in shallow aquatic or terrestrial habitats, it is impossible to study open ocean midwater animals in an undisturbed state in their natural environment. This lack of direct observation means that study of the biology of midwater organisms must rely largely on relatively indirect observations, each with appreciable associated errors. As many methods as possible must be used both independently and concurrently to infer aspects of the biology of these organisms. For example, direct studies of physiology and behavior in the laboratory provide insight into the organism's life in the natural environment. These types of observations can often be made on small numbers of animals and thus "sampling error" does not affect them.

One obvious field for study are those aspects of physiology relevant to studies of energetics as discussed in various of the tabled papers. Studies of sensory physiology, including bioluminescence can define capacities of organisms to respond to various parameters of their environment, necessary for understanding the behavior of the animals in that environment. Also useful is the direct observation of animals at depth from DSRVs.

For many physiological studies, the condition of recovered organisms is of prime importance. This problem can be divided into a consideration of mechanical damage and other forms of damage.

Mechanical damage results from abrasion of the surface of the organism in nets or from the expansion of gas-filled spaces within the organisms during retrieval. The abrasion can be greatly reduced by the use of low towing speeds, long trawl nets with low filtering velocities through the side panels, and specialized cod ends (RMT codend or Childress-Barnes thermally protecting cod end). Abrasion, nevertheless, remains a problem for those organisms with fragile surfaces (fishes, jellyfishes, etc.). On the other hand, the abovementioned techniques result in the recovery of most crustaceans with virtually no abrasion damage. For those species of fishes or other organisms which migrate to the surface, capture with virtually no abrasion is possible by dip-net collecting. Midwater traps might also be used in an attempt to avoid abrasion.

The expansion of closed gas-filled spaces results in embolism and gross damage to the internal structure of organisms, being especially acute for non-migratory, swimbladder containing fishes (hatchetfish for example). It is probably not a very important problem in the recovery of fishes which undergo vertical migrations to near the surface. A pressure retaining cod end could possibly be devised (with considerable effort) to bring the relatively few affected species to the surface at depth pressure. It is unlikely that this would be generally useful however, since abrasion problems would still be severe. In addition, any pressure recovery cod end would probably have to be specifically designed for a given depth range and purpose. General purpose units would probably not be very useful in practice.

The other kinds of damage relevant to the recovery of midwater animals alive are thermal and pressure (more subtle enzyme and membrane level effects). Temperature is obviously extremely important in this context. In most regions, shallow vertically migrating species can be brought to the surface alive in net tows, if other factors are not lethal. Deeper living species can be brought to the surface alive only in regions with a surface temperature below about 18°C. Survival of these species is greatly enhanced if the individuals are placed in cold water as soon as they are recovered. This temperature problem for the recovery of deeper living species and species from areas with higher surface temperatures has now been almost entirely solved. The Childress-Barnes thermally protecting cod end brings up midwater species at temperatures near those at the capture depth. This is a reliable, compact, and relatively inexpensive device which has been used routinely on two extended cruises (over 100 trawls) with a variety of trawl nets (RMT-8, IKMT, and a special 10-ft Tucker trawl). This device brings most of the captured individuals, except those killed by swimbladder expansion or by abrasion, to the surface alive. It has both demonstrated the importance of temperature as the decisive short-term lethal factor in recovery and greatly increased our recovery abilities, most dramatically with crustaceans. For fishes this device has considerably extended our short-term recovery abilities, but abrasion remains the factor which severely limits survival. The success of this cod end demonstrates that the more subtle pressure effects apparently do not greatly affect short-term survival of these organisms and therefore a pressure recovery system is not necessary to prevent this kind of damage. New pressure recovery systems would probably be of little value since even those midwater fish species which come to the surface and are captured in perfect condition cannot be maintained.

An equally important part of this problem is that of maintenance of the animals once recovered. Considerable success has been achieved by workers on copepods, ctenophores, and other shallow living forms. For deeper living forms, the longest survivals are about 2.5 years for the crustacean *Gnathophausia ingens* and 8 months for the fish *Melanostigma pammelos*. Maintenance systems generally need to be devised for particular species. The hope of a general advance in maintenance technology seems unfounded. Those aspects of maintenance which might be investigated for particular species are food, pressure maintenance, reduction of abrasion in captivity, and the use of very large aquaria.

Given the above difficulties, many have found the idea of *in situ* studies appealing. This has particularly involved the use of DSRVs to capture and to make measurements at depth on aminals which cannot otherwise be studied. This approach has many severe disadvantages. First is the fact that an animal enclosed in a container at depth is not *in situ*. It is, in addition, only possible to carry out crude experiments under these conditions since the precise control and observation of experimental conditions is extremely difficult. Further, the noise, vibration and light associated with the submersible is a severely disturbing factor. The enclosure of myctophids in a small container at depth would be as likely to cause death there as at the surface. Next, because of the rapid swimming speeds of most organisms of interest, catching them would be very difficult and capturing them without damage to their fragile surfaces very unlikely. Owing to the time and expense involved in this approach, the collection of a significant amount of data is virtually impossible. This is not to say that this approach should be entirely ruled out *a priori*, but that it is a crude method whose attempted use is justifiable only to obtain specific information obtainable in no other way to support information gained in other ways.

A different use of DSRVs is as platforms for observation of the distribution and behavior of midwater organisms. This kind of information has proven difficult to quantitate and therefore relatively little has been published. However, this anecdotal information is of tremendous value in understanding many aspects of the biology of midwater organisms. Some form of encouragement to individuals to publish it would be desirable. It might be necessary to publish a special volume to circumvent the problems of publishing anecdotal material in journals. Midwater workers who have made submersible dives have been generally surprised at how different many aspects were from what they expected. Because of this, all those studying biological and acoustical aspects of midwater animals should be given at least one midwater dive to get a more direct feel for this environment.

The potential usefulness of submersibles for further observations on behavior, distribution, etc. of midwater organisms is great. The realization of this potential depends on scientists obtaining sufficient experience in submersibles to allow them to formulate questions which can be answered using this equipment. There is also a potential for important work where submersible observations can be related to other studies on the organisms involved (laboratory behavior and sensory physiology, net avoidance estimates, acoustic observations, etc.).

Summary

A number of specific recommendations emerged from these discussions:
  a. A wide variety of physiological and biochemical approaches are appropriate.
  b. Present recovery methodology is functional for many midwater crustaceans. Physiological studies of fish should utilize vertically migrating species.
  c. The design and construction of expensive pressure recovery systems should be encouraged only for very specific projects.
  d. Thermally protecting recovery systems should be used extensively.
  e. Extensive work is required on the maintenance of particular species. Attention to food, large volume systems, anaesthetics, maintenance under pressure, and chemical quality of the water used is required. Use of vertically migrating fish is recommended.
  f. Easily maintained species should be used; extrapolation from their data may be feasible.
  g. *In situ* capture studies are regarded as crude and expensive, and the resultant data of limited value even in supporting indirect estimates of parameters of species that are impossible to maintain in the laboratory.
  h. Scientists working on midwater regimes should be given the opportunity to make at least one observation dive in a DSRV.
  i. Anecdotal type observations made from DSRVs should be published because of the insight they give into the possible behavioral characteristics of the animals.
  j. DSRVs would be most effective in this research area if combined with simultaneous observations of the following nature: laboratory behavioral and physiological studies, net sampling, acoustic observations, functional morphology, buoyancy measurements, biochemistry, etc. These studies are facilitated in regions where midwater biomasses are high.

GENERAL CONCLUSIONS

Various general recommendations can be selected, for reiteration, from the above discussions. The aim of the symposium was to examine the feasibility of predicting sonic scattering layers from physical/chemical/biological information. This report has examined the biological information required in conjunction with physical-chemical information to allow modelers to attempt formulation of the problem.

Studies of the structure of pelagic communities in the micro-, meso-, and megascales horizontally and vertically in the context of the physical and chemical parameters of the environment require much further development. These are quantitative studies that define detailed distributions of biomass but with the biomass partitioned into size and species components. Such partitioning allows empirical estimations of scattering phenomena to be made in the vertical and horizontal planes. In addition, delineation of faunal provinces, and so of sound scattering provinces, would also be derived. Provinces of the Southern, Indian, and regions of the Atlantic Oceans remain to be described although several can be inferred from existing data bases. The provinces vary from each other in their dynamic ecology, some individual or groups of provinces being more suitable than others for studies of DSLs.

There is a lack of information on the feeding ecology of many important species although new information is being currently obtained. Recovery techniques that allow maintenance of midwater crustaceans for laboratory observations are available and successful. This is not true for many mesopelagic fish, especially myctophids. Many of these suffer severe skin abrasion during catching and recovery; even if recovered in reasonable condition, they damage themselves in the laboratory container. There are, however, several groups of mesopelagic fish that perform diel vertical migrations to the surface and these can, at times, be caught by dip-netting. These are the primary fish species that should be used for physiological studies. Such studies should be concerned with feeding, metabolism, growth, breeding, biochemistry, microbiology, sensory physiology, and behavior.

Studies of the structure of pelagic communities combined with investigations of the feeding ecology and physiology of the species together provide the essential data for building a general ecosystem model. Such a model can be framed in biological and/or acoustical terms.

Finally, the strongest recommendation that emerged from the discussions was that there should be more encouragement of dialogue between the different disciplines contributing to the central problem. This should take place at all stages of the investigations--inception, planning, data collection, analysis, and synthesis.

Estimating Production of Midwater Organisms

J. Mauchline

Dunstaffnage Marine Research Laboratory

ABSTRACT

In an effort to estimate the production of midwater organisms, various approaches have been attempted which are assessed in the present paper. Before it is possible to study the productivity of oceanic ecosystems by estimating the production of midwater organisms, it is necessary to delimit the biocoenoses and define the food chains involved. In this review, emphasis is placed on the biology of the individual animal because it is measurements made on individuals within a population of a species that are used to produce generalities frequently considered to have wider application and relevance to the entire population. While data on growth, breeding, longevity, feeding, metabolism and behavior of the individuals in a population allow the construction of models of variation in the parameters within the animal, the development of population models requires information on population dynamics including information on the distribution of age groups within the population and generation times.

The difficulties in acquiring data on the production of oceanic species and ecosystems are a natural consequence of the difficulties of the environment: high pressure, low temperature, and great distances in a fluid environment with the consequent problems in identifying discrete populations of species and sampling them effectively. Thus, basic studies such as functional morphology and histology of individual species can provide data for use in the analyses of ecosystems and should be an integral part of large multidisciplinary programs.

INTRODUCTION

The prediction of sound scattering in the oceans from physical, chemical and biological information requires some amplification. Sonic scattering layers have been described in many regions and over large areas of the oceans as was instanced in the various papers of the Maury Center Symposium of 1970 (*Farquhar*, 1971). The acousticians are examining the reverberation properties of several types of organisms. The biologists are attempting to identify the organisms responsible for and associated with the layers. Barham, in the final discussion at the 1970 Symposium, talked about "...the friends of the euphausiids, the friends of the myctophids, or the friends of the physonects..." He suggested that this partisanship for various taxa as causative agents was being overcome by a realization that complex populations are being examined and considered as such. The identification of organisms primarily responsible for sonic scattering layers is still necessary. The prediction of sound scattering in the oceans in time and space can probably be approached in two ways.

The first would be the production of synoptic charts of layers in the world oceans following the general methods and forecasting principals of the meteorologists. The analogy can be developed, and undoubtedly has been, but the continuous effort required to maintain the flow of the basic data for such a prediction system is large. Further, it remains basically empirical and probably subject to more error than comparable meteorological forecasts.

The second approach is the study of biological production of organisms that generate the biological scattering layers. This, however, begs the question----which organisms? Biological production can be studied at the systems level. Models are constructed from simplified representations of real relations between variables; many variables are either smoothed or omitted. This leads to an understanding of how the system might work. The organisms producing sonic scattering layers are part of these systems. Refinement of the systems analyses requires identification of the organisms. Measurements or tested models of production of individual species allow more realistic estimates of parameters to be entered in the systems model, thus producing more realistic information on the behavior of the systems.

Many comprehensive review papers on different aspects of this broad field have been produced in recent years. *Parsons and Takahashi* (1973) provided a general introduction to the whole field. *Mann* (1969) and *Steele* (1974) examined the structure of marine ecosystems while *Vinogradov* (1968) and *Angel* (1977) looked at the organisms and the communities in the vertical plane of the oceans. *Steele* (1970) and *Strickland* (1972) provided information on marine food chains, both detailed and general, and *Mullin* (1969) studied the production processes of the zooplankton. *Marshall* (1973) reviewed information on the respiration and feeding of copepods, *Mauchline* and *Fisher* (1969) described the biology of euphausiids, *Omori* (1974) the biology of pelagic decapod crustaceans, and *Phillips* (1969) and *Cowey and Sargent* (1972) reviewed information on the nutrition of fish. Many of the general and particular difficulties of adequately sampling the different organisms in the water column are discussed by *Angel* (1977) and **Badcock and Merrett** (1977). These **reviews**, and their appended reference lists, reflect the vast diversity of information and approaches required for studying the production of oceanic organisms.

The present paper is concerned primarily with assessing the various approaches to estimating production in oceanic species. Emphasis is placed on the biology of the individual animal because it is measurements made on individuals within a population of a species that are used to produce generalities frequently considered to have wider applications and relevance.

BIOCOENOSIS

The sonic scattering layers exist in the 0-1000 m layer, but predominantly in the upper 750 m. The influence of downwelling daylight is present throughout this latter column in oceanic areas. *Roe* (1974), referring to his own and previous work, concludes that the vertical distributions of some species of copepods, euphausiids, decapods and fish are ordered with reference to isolumes while those of others are not. Many current investigations are involved with various aspects of vertical distribution and migration of organisms. The approaches to the subject are varied as discussed by *Angel and Fasham* (1973), *Angel* (1977), **Badcock and Merrett** (1977), *Clarke* (1970), *Deevey* (1971), *Deevey and Brooks* (1971), *Digby* (1972), *Donaldson* (1975), *Foxton* (1972), *Herring* (1972, 1973), *Hure and Scotto di Carlo* (1974), *Kinzer* (1977), **Legand et al.** (1972), *Macdonald et al.* (1972), *Marshall* (1972), *Moraitou-Apostolopolou* (1971), *Packard et al.* (1971), *Tranter* (1973), *Wilson* (1972).

Blackburn (1977) discussed biomass levels in the vertical column and referred to previous work. The biomass in the deeper layers of the oceans reflects the biomass occurring at the surface (Figure 1). A linear relationship often exists between log biomass and the depth at horizons deeper than 1000 m. There are major sampling problems, as well as interpretative problems, when large pelagic individuals are occasionally caught (or thought to have escaped). The important point is the correspondence between the biomass at depth and that at the surface. The biomass at 6000-8000 m depth is approximately one thousandth of that in the surface layers. This correspondence reflects the dependence of the deep biomass upon the surface biomass for its nutrition.

The zooplankton is divided on the basis of size into the mesoplankton (< *3-4 cm*) and the macroplankton or micronekton (> *3-4 cm*). *Vinogradov* stated that the mesoplankton extends from surface to

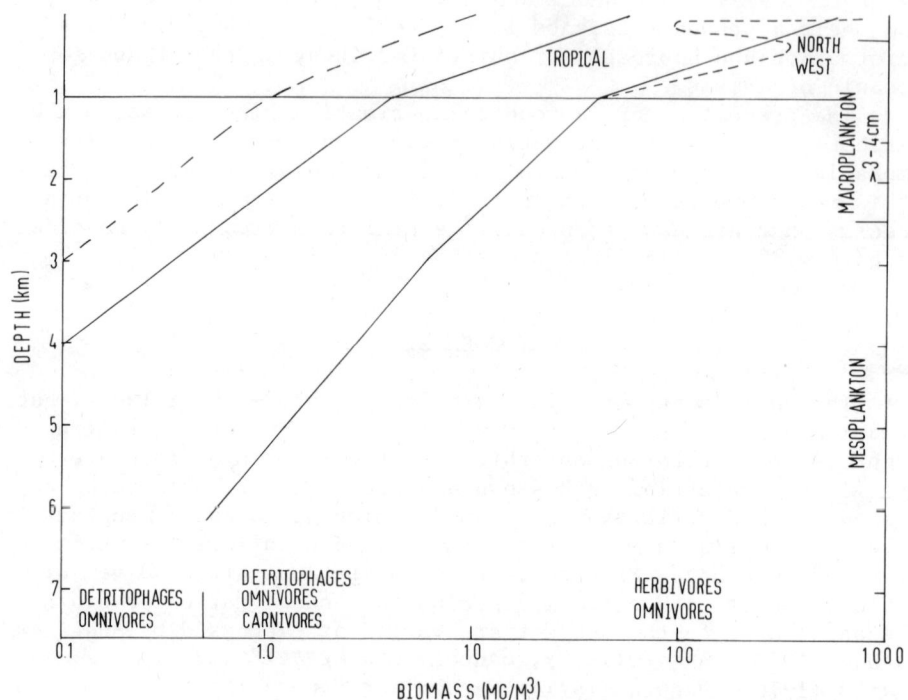

Fig. 1. The changing biomass of plankton related to depth in tropical and northwest Pacific. The biomass in tropical regions can be as low as that indicated by the broken line to the left. The biomass in the upper 1000 m does not decrease uniformly as shown by the solid lines but usually follows the type of decrease described by the broken line for the northwest Pacific. Micronekton (macroplankton) occurs predominantly in the 200-2500 m layer, the animals below 2500 m being predominantly mesoplanktonic. The deep communities at biomass less than 0.75 mg/m^3 consist of detritophages and mixed feeders. Carnivores probably become increasingly common at the higher biomasses of shallower depths while the corresponding numbers of detritophages decrease. (After *Vinogradov*, 1968).

bottom while the micronekton tends to be concentrated in the 200-2500 m depth layer, the depth and extent of the zone varying with latitude (see Blackburn, 1977). The organisms responsible for the sound reverberations belong to the micronekton. Blackburn (1977) reviewed the proportion of micronekton present at various depths and their biomasses. Some 10-25% of the total biomass of plankton in the 500-1000 m layer may consist of fish.

Vinogradov (1968), Legand et al. (1972) and Blackburn (1977) examined the proportions of various taxonomic groups of organisms in the vertical profile. There are several major investigations in progress that will amplify and modify their results. It is clear, however, that the mid-water populations of decapods and fish, occurring in the 200-2500 m layer are intimately associated with the formation of biological sonic scattering layers in the 0-1000 m column. These animals are probably divided vertically into several sub-populations of the same and different species. The 0-1000 m column can often be divided into two general areas (Figure 2). Legand et al. (1972) identified a superficial system occupying the 0-450 m layer. This consisted of the phytoplankton, mesoplankton and the shallower living micronektonic and nektonic species. A second system existed below 450 m depth. It had, of course, no phytoplankton, was relatively poor in mesoplankton and had the mesopelagic and intrusive bathypelagic micronekton and nekton. The ver-

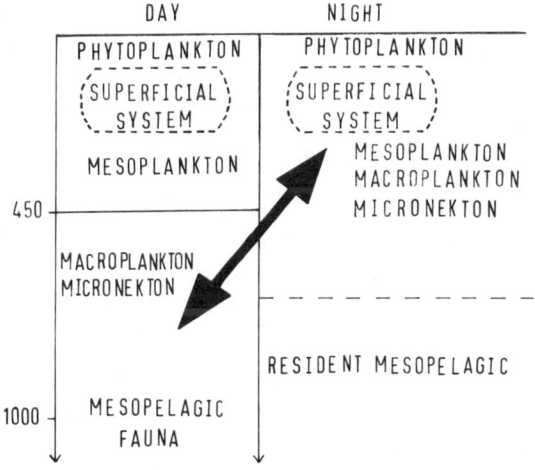

Fig. 2. Partitioning of the biomass in the 0-1000 m water column. There is an upper superficial system and a deeper mesopelagic system. The diurnal migrants form an interzonal system moving between the two. (After Legand et al., 1972).

tically migrating or interzonal species were in this deeper system during the day and in the superficial system at night. *Angel and Fasham* (1973) demonstrated as many as five biologically distinct zones in the 0-1000 m water column. Their analytical techniques had a higher resolution, (the degree of resolution being discussed by *Angel* (1977), than those of *Legand et al.* (1972). However, Legand's conception, in general terms, is useful. The intrusive elements of the lower mesopelagic and the bathypelagic associations suggest that study of the events in the 0-1000 m water column requires sampling to depths of about 2500 m, the lower region of maximum density of micronektonic species.

DEFINING FOOD CHAINS

Most investigations of food chains fall into three main categories. The first is primarily qualitative, being an attempt to define what a single species eats and to place its diet in the context of its biological and physical environment. In the second category are attempts to estimate production of a species, usually several species, and often devolving into a systems analysis. There are few examples of simple food chains of the type alga/copepod/small fish/large fish. The most profitable approach through systems analysis was the study of the food webs of the Black Sea by *Petipa et al.* (1970). They group the animals as:

PS	producers and saprophagous organisms
H	herbivores
CM	omnivores
CI	primary carnivores
CII	secondary carnivores
CIII	tertiary carnivores

Quantitative measurements can be attempted, in a biocoenosis, for each of these. Estimates or "guestimates" of transfer coefficients etc. can then be made.

A third category of investigation includes all the experimental approaches to gaining information on food chains. These are usually carried out in the laboratory, attempt to be quantitative, and attempt to relate to the natural situation.

The aims of all these approaches might be listed as follows:

1. Qualitative assessment of the diet.
2. Quantitative estimates of the diet.
3. Food requirements of the animal.
4. Conversion of food into growth of the animal and reproductive products.
5. Estimates of growth and reproduction of populations.
6. Descriptions of food chains or webs.
7. Analysis of the functioning of ecosystems.

Interpretation of Stomach Contents

How is an organism to be ascribed to one of the trophic classes of *Petipa et al.* (1970)? Food webs are extremely complex. Analyses of the stomach contents of animals show a high diversity of material in them (*Aizawa*, 1974; *Donaldson*, 1975; *Foxton and Roe*, 1974; *Gjosaeter* 1973b; *Harding*, 1974; *Hopkins and Baird*, 1973, 1975; *Isaacs*, 1973; *Itoh*, 1970; *Judkins and Fleminger*, 1972; *Kinzer*, 1977; *Mauchline and Fisher*, 1969; *Merrett and Roe*, 1974; *Omori*, 1974; *Raymont*, 1970; *Repelin*, 1972; *Tyler and Pearcy*, 1975). The present means of classification is extremely arbitrary and subjective. *Nemoto* (1968, 1972), using fluorometric techniques, examined stomach and intestinal contents and faecal pellets for chlorophyll a and phaeo-type pigments, with some surprising results. This type of approach could probably be developed for other "tracer substances".

An animal may be a herbivore during the spring increase of phytoplankton or when it locates a concentration or patch. The Antarctic euphausiid, *Euphausia superba*, is frequently quoted as a herbivore, as are many copepods, and yet there is little phytoplankton in its environment during the greater part of the year. The nutrition of copepods in temperate regions during winter minima of phytoplankton causes conjecture. *Corner et al.* (1974) have recently shown that *Calanus helgolandicus* is capable of carnivorous feeding on barnacle nauplii and could therefore satisfy its nutritional requirements during the winter through a carnivorous diet. Nutritional studies of copepods have concentrated on plant diets (*Marshall*, 1973) and little is known of the benefits to them of mixed or carnivorous diets.

The usual method of recording the different items of the diet in a stomach contents analysis is to indicate either presence or absence of the item or the percentage of stomachs that contain the item. This certainly provides an estimate of the commonest types of food eaten by many species but not necessarily of all species. *Ponomareva* (1954) described how euphausiids hold copepods in the "food basket" formed by the thoracic legs eat out of the soft parts and reject the empty chitinous husk with its oral appendages, head and abdomen. Thus, a diet of copepods can leave little identifiable trace in the stomachs of euphausiids. *Itoh* (1970) suggests that copepods with a high edge index (defined later) may take bites of prey organisms. Amphipods certainly take bites of mysids and animals feeding on carcasses in mid-water probably bite. Conventional analyses of stomach contents are valid when it is known that the animal swallows all of its prey. A habit of taking bites of this and that is almost impossible to demonstrate in a stomach analysis of preserved organisms.

The rate of digestion of the food in the stomach can influence the results of the analysis. Diatoms with strongly silicified frustules are more commonly reported from stomachs than those with less strongly silicified frustules. *Hart* (1934) stated that the stomach contents of *Euphausia superba* resemble green porridge in areas where the poorly silicified oceanic species of *Chaetoceros* are dominant. Many authors remark on debris, green mush, and detritus (flocculent and otherwise) among stomach contents. This material is often the dominant item but its importance is often eclipsed by an impressive list of scientific names of identified organisms or fragments of organisms.

There is still a further problem in interpreting contents. Some animals probably regurgitate stomach contents during the process of being caught. Others feed on organisms in the codend of trawls or in the buckets of plankton nets. This was demonstrated by *Judkins and Fleminger* (1972) in a study of the stomach contents of the shrimp *Sergestes similis*. They compared the stomach contents of shrimps caught in nets with those of shrimps taken from the stomachs of albacores. They found differences that could be explained by assuming that the shrimps were feeding in the plankton buckets. *Hopkins and Baird* (1977) discuss this aspect of the analysis of stomach contents of mesopelagic fish. Many organisms may only be eaten in the crowded and unnatural conditions of the plankton bucket, some of them rarely being available to the species under study because of different distributional patterns. The plankton bucket factor has to be borne in mind.

Functional Morphology

The particle size of food consumed by an organism often bears a direct relationship with the size of the organisms. This leads to the attractive idea of describing the structure of plankton communities in terms of particle size spectra (*Parsons*, 1969). This assumed relationship between size of food particle and size of feeding organism may be misleading. For instance, the filtering apparatus of crustaceans is examined and an optimal range of sizes of particles that it will extract from the water determined. The range of size so defined is usually indicative of the smallest particle that the animal can filter. It does not, however, necessarily define the maximum size taken. Most crustaceans have two or more methods of feeding, these methods being adapted for different foods. Many copepods, mysids and euphausiids can filter-feed on phytoplankton, catch and eat live prey, and scoop up sediment from the sea bed. Feeding on live prey usually involves maceration or biting and so presumably dead prey can be fed on with equal facility.

Hirota (1972) found that the optimal ratio of the size of food particles to the size of the body of the ctenophore *Pleurobrachia bachei* is 1/3 to 1/5, successful growth of the ctenophore being the criterion used in the assessment. The majority of other such assessments have been made from examinations of the morphology of the animal, especially of its mouthparts and stomach. *Omori* (1974) reviewed estimates of the particle sizes of the diets of pelagic decapods derived from studies of stomach contents and mouthparts. The structure of euphausiid mouthparts and their effect on diet was discussed by Mauchline and Fisher (1969). Salps, using their continuously renewed mucous nets, filter particles ranging from about 1 mm to less than 1 μm in size (Madin, 1974). Siphonophores feed on micro- and mesoplanktonic crustaceans (Mackie and Boag, 1963). Itõh (1970) has developed an "edge index" to describe the cutting edge of the mandibles of copepods. Changes in this index are correlated with the feeding habits of the copepods. Copepods with an index greater than 900 probably do not obtain food by filtration but seize prey and take bites of them. Marshall (1973) and Sekiguchi (1974) discussed other features of species of copepods that are specially adapted for trophic functions. Ebeling and Cailliet (1974) related mouth size and gill morphology of mesopelagic fish to their diets and reviewed previous information on this subject. The mouth size indicates the largest prey that can be taken whole while the structure of the gills indicates the smallest particles that can be retained. Hopkins and Baird (1977) discussed further morphological features related to feeding of mesopelagic fish.

The study of the functional morphology of the mouthparts and associated structures contributes significantly to the interpretation of stomach contents and so to the defining of trophic relations of a species. Extension of this study to the functional morphology of the foregut is also profitable (Nemoto, 1977).

Quantifying Stomach Contents

Is it possible to quantify items in the stomach contents? Legand et al. (1972) found that the weight of stomach contents varies as the log of body weight in the fish *Sternoptyx diaphana* and *Diplospinus multistriatus*. Nemoto (1977) found that the log weight of stomach contents varied as the log of body weight in the three species of euphausiids in the genus *Thysanopoda* that he examined. A relationship between log dry weight stomach contents and log body length was found in the amphipod *Parathemisto gaudichaudi* by Sheader and Evans (1975). The stomach contents of the euphausiids were approximately 1% of the body weight. The stomach contents in young *S. diaphana* approximated 8% of the body weight (100 mg) but they represented only about 1% of the body weight of a 10,000 mg fish. Hopkins and Baird (1977) stated that the food in the full stomach of *Diaphus taaningi* represents about 0.8% of the body weight. These analyses suggest that it may be possible to find such relationships in other single or groups of species. A subjective estimate of the degree of fullness of a stomach could then be converted to a measure of weight.

The routine estimation of weight of stomach contents relative to the animal's weight and to the time of day, season of the year, or other parameters such as depth of occurrence would be a major advance. Knowledge of the rate of digestion, or disappearance of material from the stomach is equally important. Sheader and Evans (1975) found that the time taken for ingested food to pass through the gut of

Parathemisto gaudichaudi is related to temperature----72 hours at 4°C, 36 hours at 7°C, 18 hours at 10°C and 3-4 hours at 12-14°C. This allows assessment of the significance of the percentage of empty or nearly empty stomachs recorded during the analysis. Estimates of weight of food consumed per unit time can then be made.

Nutritional Value of Food

The weight of food eaten per unit time has to be assessed against its nutritional value. The basic nutritional requirements of marine invertebrates are almost completely unknown. What minerals and vitamins are necessary for growth and development and are any of these limiting on the populations at any time or place? Nutritional requirements of fish have been studied primarily in salmon and trout. These studies are predominantly of fish being cultured and fed artificial foods (*Phillips*, 1969; *Cowey and Sargent*, 1972).

Comparable work on invertebrates is done by axenic culturing in the laboratory (*Dougherty et al.*, 1961; *Provasoli et al.*, 1959; *Provasoli and d'Agostino*, 1969; *Scott*, 1974). *Lewis* (1977) discussed other relevant aspects of this problem.

The assessment of the comparative nutritional values of different single and mixed diets requires chemical analyses of the species involved. The analyses of most value in production studies are estimates of carbon, nitrogen, phosphorus, protein, lipid, carbohydrate, caloric value and wet, dry and ash weights of the animals. *Beers* (1966) examined carbon, nitrogen, phosphorus and carbohydrate content of taxonomic groups of plankton while *Ikeda* (1972), and authors quoted by him, have examined the chemical composition of individual species. *Hopkins and Baird* (1977) and *Tyler* (1973) discussed caloric values of different organisms.

Recent attention has been centered on lipid, especially wax ester, contents of pelagic organisms (*Bottino*, 1974; *Herring*, 1973; *Kayama and Nevenzel*, 1974; *Morris*, 1973; *Morris and Sargent*, 1973; *Sargent et al.*, 1974). Triglyceride stores are usually built up during periods when food is plentiful. Temperate and high latitude organisms store triglyceride during the summer so that the stores reach their maximum size in the early autumn. These stores are then depleted during the winter, being almost non-existent by the spring when the last traces are found incorporated in the eggs. Lipids, triglycerides, and wax esters in mesopelagic organisms may be stored as insurance against periods of low food and/or for the development of the gonads. Stores of lipid have to be taken into account in the energy budgets of the animals. They also add to the nutritional quality of the organism as a food.

Diets in the Laboratory

Maintenance of pelagic animals alive under laboratory conditions is difficult (*Baker*, 1963; *Komaki*, 1966; *Lasker and Theilacker*, 1975). Considerable success has been achieved in recent years with calanoid copepods (*Omori*, 1973), the ctenophore *Pleurobrachia bachei* (*Hirota*, 1972) and a few other types of organisms including euphausiids and chaetognaths (*Forster and Beard*, 1973; *Fowler et al.*, 1971a,b; *Lasker*, 1966; *Mackie and Boag*, 1963; *Steele*, 1970).

One of the most interesting studies was that of *Conover and Lalli* (1974) on the pteropod *Clione limacina*. The pteropod feeds exclusively throughout its post-veliger life on two other pteropods, *Spiratella retroversa* and *S. helicina* and it is this complete natural diet that Conover and Lalli used in the laboratory. Such dietary specialization is very rare and being able to use the complete natural diet of an animal in the laboratory creates the most realistic feeding conditions possible. It should allow seasonal changes in the nutritional characteristics of the food organisms to be assessed against variations in the experimental parameters of the feeding animals.

The problems involved in replicating the natural diet of most pelagic organisms in the laboratory cannot be minimized. *Omori* (1973), reviewing information on the cultivation of marine copepods, lists 19 species of calanoids that have been successfully maintained. The diets of many of the cultures are mixed, as opposed to monospecific, yet only 18 species of organisms, primarily species of phytoplankton, are used as food. This is a function of the number of species of phytoplankton available in continuous culture in the different laboratories. The advent of more species of zooplankton available in continuous culture will help. Some detailed comparisons, however, of the chemical composition of cultured species with that of their wild counterparts will be needed. These studies should be related to season and life history of the species.

The use of artificial diets, taken to the extreme form of axenic culturing, requires more attention. The plumbing problems in such experiments probably restrict investigations to the study of small organisms. Rotifers (*Dougherty et al.*, 1961; *Scott*, 1974), *Artemia salina* and *Tigriopus* species (*Provasoli et al.*, 1959; *Provasoli and d'Agostino*, 1969) have been cultured in monoxenic conditions.

ESTIMATING FOOD REQUIREMENTS

An animal feeds to survive, grow and reproduce. Growth results from an excess of energy intake over expenditure. The production of energy stores and reproductive products are included as growth. Total metabolism is the sum of all the metabolic processes taking place in the animal for body maintenance as well as for growth. Natural diets contain growth and energy foods (carbohydrates, fats, proteins) and non-energy foods (minerals, vitamins, water and oxygen). The non-energy foods are essential to the well-being of the animal although many of them may only be required in very small quantities. A variable supply of many of these non-energy foods can therefore, potentially at least, act as a limiting factor on growth of an organism or a population.

Energy in nutrition studies is measured in calories. On the average, proteins contain 5.65 cal/mg, carbohydrates 4.1 cal/mg and fats 9.45 cal/mg of energy. However, the total energy in these groups of foods is not necessarily available to the animal. The quantity available is dependent upon the form in which the food is presented and the capabilities of the digestive system of the organism. This gives rise to digestibility factors and so to physiological calorific equivalents of these groups of food applicable to the various species of feeding organisms. Thus, there is variation between and within species in the values of these factors.

The statement that growth results from an excess of energy intake over expenditure can be modified and expressed as:

$$\textit{net efficiency of food conversion} = \frac{\textit{growth}}{\textit{food intake} - \textit{maintenance ration}}$$

Maintenance ration is that required for the basic metabolism of the animal, the maintenance of life without active growth----referred to as the routine rate of metabolism by *Childress* (1977). There are inherent problems in defining and measuring rates of metabolism. This is discussed in the context of fish nutrition by *Cowey and Sargent* (1972) and in a general context by *Childress* (1977). Using the data of *Brett et al.* (1969), the former suggested that a non-growing fish does not maintain its tissues in the same way as a growing fish. A fish fed a ration such that the body weight is maintained at a constant value shows significant alterations in the proportions of water, protein and fat in its body tissues. They concluded that the maintenance requirements of a growing fish may differ from that of a non-growing fish and so "it is tenuous to equate the food required to maintain a fish at constant weight with the portion of its food which a growing fish uses for basal metabolism, tissue wear-and-tear, and other forms of maintenance". The same argument may be applicable to the estimations of rates of metabolism made on starving invertebrates although the necessary evidence is not available at present. The amount of evidence of the situation in fish is minimal but the argument sounds correct. The life of the fish or invertebrate is, unless there is hibernation or a diapause, a continuous process that tends to assume a continuous supply of the required nutrients. Regulation of the physiological/metabolic processes enable the animal to survive, in the most effective manner, extreme events in the environment that are different from the modal conditions to which the organism's systems are adapted. Starvation, or the less severe situation of a temporary break in the food supply, or even a limitation to the normal ration, may elicit a fairly quick physiological/metabolic response. An energy store, lipid or protein, within the organism could be immediately mobilized resulting in the water content of the tissues being increased relative to lipid or protein. The specific responses from the organisms to these situations may vary considerably. Ontogenetic and phylogenetic variation in the morphology and behavior of marine species is large and this diversity is probably reflected by comparable degrees of variation in the adaptations of the physiological processes of the organisms. Superimposed on these sources of variation is the variation originating from the individual animal's physiological state at any one time. The values of various parameters are governed to a greater or lesser extent by historical events in the animal's life. A pelagic organism has no crystal ball; it is a compound of its present and past. Some of the genetically controlled adaptations of the physiological and biochemical systems may have developed in response to regular cyclic (seasonal) changes in the environment or to a progressively changing habitat preference as instanced in an ontogenetic migration. Consequently, in this context, given that these cyclic or progressive changes of the environment are more or less constant from one generation to the next, the animal can to some extent foresee the future. The past defines its present state, the present defines the processes that are continuing and those that are being initiated.

Ideally, the behavior of one parameter of an animal has to be placed in the context of the behavior of all the other parameters inside and outside the animal, not simply in the present context but also in the historical context.

The food requirements of an animal can be expressed as:

$$R = G + T + (E + e)$$

where R is the ration of food consumed, G the portion of the food used for growth and reproduction, T the portion used for general metabolism within the body, E the portion eaten but not assimilated, and e the portion lost through urinary excretion.

Growth (G)

The growth rates of several oceanic pelagic organisms responsible for or associated with sonic scattering layers have been estimated. The commonest expression of growth rate is increase in body length, or a linear equivalent of body length, against time. *Smoker and Pearcy* (1970) provide data on increase in weight in a lanternfish. Weight, either wet or dry, is a more useful parameter in comparative studies with "smoothed" conversions provided for obtaining the respective body dimensions.

Growth of organisms living in warmer waters is frequently continuous between egg and adult. Growth in areas at higher latitudes is often discontinuous, there being seasonal periods when little or no growth in body size takes place. This was instanced for pelagic decapods (*Omori*, 1974) and euphausiids (*Mauchline and Fisher*, 1969).

Gjösaeter (1973a) found that growth of the myctophid fish, *Benthosema glaciale* (Reinhart) seems to be retarded in the spring and summer in fjords of Norway. Little or no growth in body size does not necessarily mean that no growth of the animal is taking place; ovaries may be developing and this is active growth not necessarily reflected by an increase in the external dimensions of the body.

Euphausiids are members of the oceanic communities of pelagic organisms. Small species, of maximum body length about 25 mm, living in tropical, sub-tropical and warm temperate regions mature at a maximum age of one year, breed and usually die soon thereafter. Larger species, of maximum body length about 35 mm, living in cooler temperate regions or in the mesopelagic regime, mature at an age of about one year, breed and may often survive to be two years old when they breed a second time. A few large species, 30-66 mm total length, living in high latitudes or in the lower mesopelagic regime, may not mature until about two years old, at which time they breed and may then survive to be three years old when they breed for a second time (*Mauchline and Fisher*, 1969).

The sizes and ages at maturity of several other decapods are shown in Table 1. Size is measured as total length in the smaller species but usually as carapace length in the larger species. The ages at maturity of these decapods are comparable with those of the euphausiids, that is 1-3 years depending upon body size and latitude of occurrence.

Tåning (1918) examined body lengths of several species of mesopelagic fish but his samples were frequently small and his conclusions tentative. Information, however, is available for a few species of myctophids and gonostomatids (Table 2). The data suggest a shortening of the life history in lower latitudes relative to higher latitudes. The fish mature sexually at an age of 1-3 years, 4 years in *Stenobrachius leucopsarus*. They may survive to breed a second or third time but many of the smaller species maturing at one year old probably do not. The greatest sizes quoted in Table 2 for myctophids described by *Clarke* (1973) are the maximum sizes that he reported. Clarke gives no estimate for longevity but it is probably not more than 3 years, even in *Diaphus brachycephalus*. Tåning (1918) describes two size groups of *D. doffleini* and *Lampanyctus maderensis* and *Badcock* (1970) also found two size groups of the gonostomatid *Cyclothone braueri*. These species probably have a life-span of two or three years, if these size groups represent year groups.

Growth rates and life-spans of species of oceanic copepods are variable and being investigated. Small epipelagic species may have "resting periods"; they may breed in the spring and summer, producing two or three generations, and then over-winter as stage IV or V copepodites. This has been found in populations of *Rhincalanus gigas*, *Calanus finmarchicus*, *C. hyperboreus*, *Neocalanus tonsus* and *Pareuchaeta japonica* (*Ommaney*, 1936), and in *Calanoides acutus* (*Andrews*, 1966). All these species occur in higher latitudes. Little information is available on the seasonal cycles of oceanic

TABLE 1. Growth of Pelagic Decapods

Species	Area	Age at maturity (months)	Length at maturity (mm)	Average age	Greatest length	Authority
Sergestes similis	N. Pacific	ca 12	35	?	45	Omori et al., 1972
Sergestes similis	Oregon	ca 12	11(CL)*			Pearcy & Forss, 1969
Sergia lucens	Japan	10-12	> 37	15	40	Omori, 1969
Sergestes arcticus	Norway	12	15(CL)	< 24	17(CL)	Matthews and Pinnoi, 1973
Pasiphaea multidentata	Norway	18	18(CL)	24	25(CL)	" " " "
P. sivado	Norway	15	15(CL)			" " " "
Acetes japonicus						
winter generation	Ariake Sea	9-10	22-25			Ikematsu, 1953
summer generation	Ariake Sea	2.5-3	14-20			Ikematsu, 1953
Lucifer chacei	Hawaii	0.7-0.8	> 8			Zimmerman, 1973
Acanthephyra purpurea	N. Atlantic	24	12-17(CL)			Sivertsen-Holthuis, 1956
A. quadrispinosa	Japan	24	> 56	30	65	Aizawa, 1974
A. pelagica	N. Atlantic	30-36?	17-23(CL)			Sivertsen-Holthuis, 1956

*Carapace length

TABLE 2. Growth of Mesopelagic Fish

Species	Area	Age at maturity (months)	Length at maturity (mm)	Average age	Greatest length	Authority
Gonostomatidae						
Gonostoma gracile	Japan	24	90	36	120	Kawaguchi and Marumo, 1967
Vinciguerria nimbaria	Hawaii	12	27	–	49	Clarke, 1974
Myctophidae						
Benthosema glaciale	Norway			48	60	Gjøsaeter, 1973a
	Nova Scotia	36	46	48	55	Halliday, 1970
	Mediterranean	18	47	18	47	Tåning, 1918
B. simile	E. Indian O.	12	30	12	33	Legand and Rivaton, 1970
Myctophum affine	Japan	36	78	36	78	see Halliday, 1970
Stenobrachius leucopsarus	Oregon	48	65	48	80	Smoker and Pearcy, 1970
Notolychnus valdiviae	E. Indian O.	12	18–20	12	28	Legand, 1967
Benthosema suborbitale	Hawaii	12	25	–	38	Clarke, 1973
Diaphus schmidti	Hawaii	12	31	–	47	Clarke, 1973
D. elucens	Hawaii	12	48	–	65	Clarke, 1973
D. brachycephalus	Hawaii	12	28	–	61	Clarke, 1973
Lampanyctus steinbecki	Hawaii	12	43	–	56	Clarke, 1973
Scopelopsis multipunctatus	E. Indian O.	12	?	12	75	Legand, 1967

species in lower latitudes. However, it is probable that those in the 0-1000 m water column have a cycle of maximum length one year while strictly epipelagic species have generation times as short as 2-3 months.

The growth rates and life-spans of oceanic siphonophores are unknown (Totton, 1954, 1956; Alvarino, 1971; Pugh, 1974). Populations develop quickly under suitable environmental conditions. This presumably indicates relatively fast growth rates possibly punctuated with semi-resting stages resident in deeper water. Foxton (1966) estimated the rates of growth in Salpa thompsoni and S. gerlachei and inferred slower rates of growth at depths greater than 100 m than in the immediate surface layer. Life-spans of chaetognaths range from 3.5 months to 2 years, dependent upon the size of the species and latitude of occurrence (Mauchline, 1972).

Seldom have estimates of growth been expressed as gain in wet or dry weight per unit time. This information will undoubtedly become increasingly available. Growth rates are strongly affected by temperature regimes, either directly or indirectly. The 0-1000 m water column usually contains a marked gradient of temperature (Figure 3). In low latitudes the high surface temperature, often greater than $25°C$, decreases to $4-7°C$ at 1000 m depth. In higher latitudes, the surface temperatures are depressed while those at 1000 m depth remain in the range $4-7°C$. Differences in the extents of the gradients of temperature in different sea areas probably modify patterns of production.

It is instructive to compare several parameters of growth of an organism throughout its life. Euphausiids are representative of the general type of mesopelagic micronektonic organism of primary interest in this study. The growth of the euphausiid, Euphausia pacifica has been studied in some detail (Mauchline, 1977). This species lives for about one year off Oregon and grows continuously throughout its life. The growth, expressed as percentage increase in body length or dry weight, decreases at successive moults (Figure 4). The intermoult period increases logarithmically at successive moults which results in a direct correlation between duration of intermoult period and age of the animal. The result of decreasing growth factors and increasing intermoult period is a Von Bertalanffy type curve relating growth in body length against time. The corresponding curve for increase in dry weight against time is also given. The data conform approximately to a straight line over much of the crustacean's lifetime if wet weight (or dry weight) is plotted against age on a log/log scale. The increment to the body dry weight per day increases to a maximum value and then decreases. These are the general shapes of curves representative of species that grow continuously between birth and death.

Species in higher latitudes have periods of little or no growth in body size in winter. The growth curves are then interrupted at these seasons but the general pattern is the same (Figure 5). It is misleading to state that there is little or no growth in such

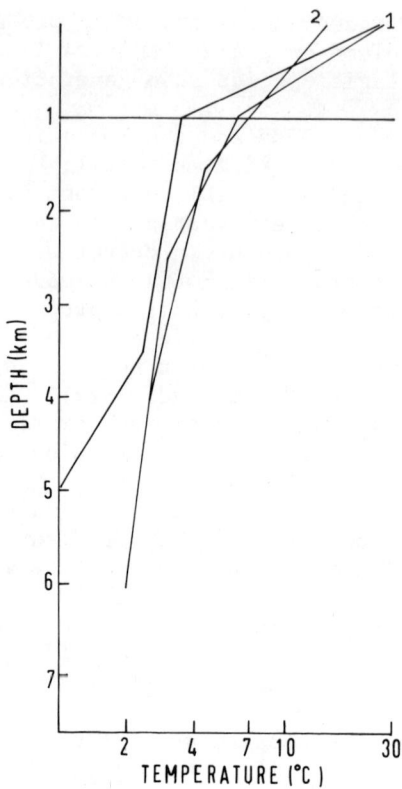

Fig. 3. Relationship of temperature to depth. 1. Range of temperature in lower latitudes. 2. Range of temperature at higher latitudes. (After Fuglister, 1960).

animals during the winter. Generally, these organisms breed in the spring in loose or strict association with the spring phytoplankton bloom. The ovary develops during the winter so that the eggs are ready for laying in the spring. The ovary of temperate euphausiids grows over a period of about three months, and represents about 10% of the body weight of the mature animal. The maximum daily increment of dry weight for E. pacifica is about 0.070 mg per day in an animal of about 14 mm total length and a dry weight of about 5 mg. The size of the daily increment then decreases (Figure 4). The growth of the ovary as increased dry weight of the animal has not been identified through weighing of the animals. It is not accounted for in the curves in Figures 4 and 5. The curves describe weight increments to general body tissue. If the total increment (Figure 4) were maintained at about 0.070 mg per day for about 100 days then the excess of the increment over that required for general body growth would produce an ovary equivalent to 10% of the weight of the animal. *Meganyctiphanes norvegica*, having a winter depression of general tissue growth, would only require to switch part of its growth increment of 0.38 mg per day to production of the ovary (Figure 5). Production of eggs takes place when the animal is, meta-

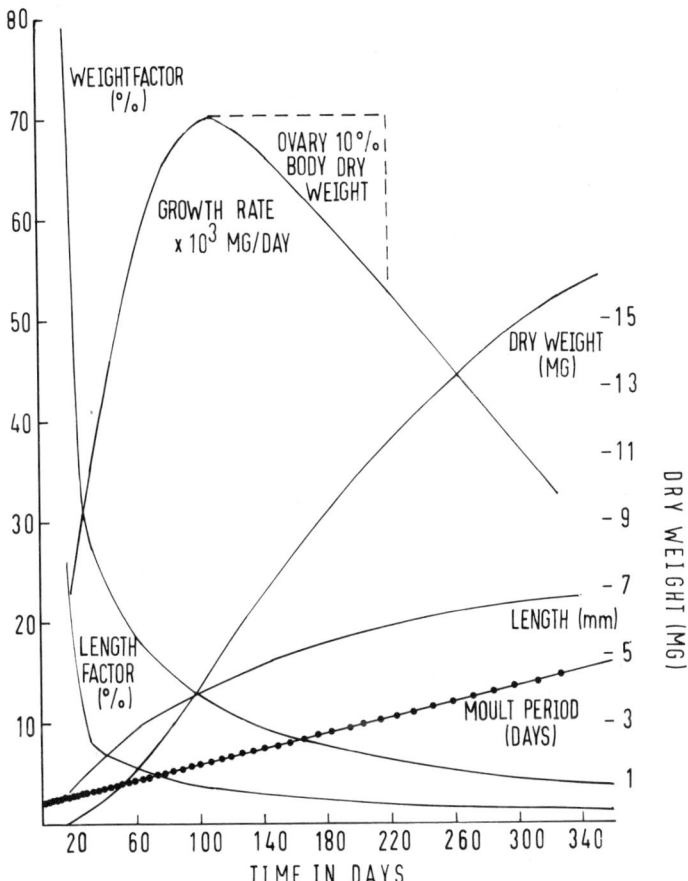

Fig. 4. *Euphausia pacifica*. Curves of selected growth parameters throughout the life of the animal. Length factor shows the decline of percentage increase in body length at successive moults. Weight factor shows the decline of percentage increase in body dry weight at successive moults. Moult period shows the increasing duration (in days) of successive intermoult periods. Length shows body length against time. Dry weight shows increase in dry weight of body against time. Growth rate shows actual daily increment to body dry weight; broken line indicates period over which ovary would be developed if daily growth increment remained at 0.071 mg/day.

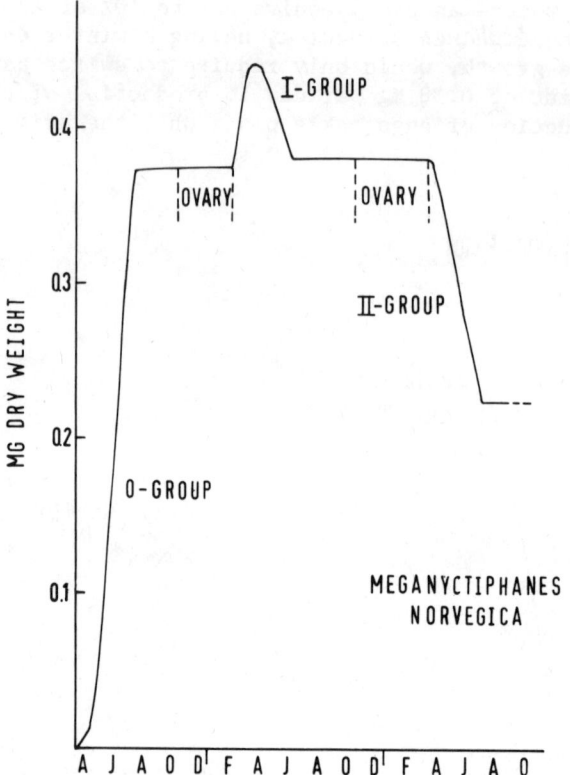

Fig. 5. *Meganyctiphanes norvegica*. The actual daily increment to body weight interrupted by the first winter period, August-February, and the second winter period July-March, of little or no growth. The broken lines indicate periods required at indicated daily increments for the production of ripe ovaries within the animals.

bolically speaking, in high gear.

Some organisms produce one brood of young in each of several successive years, others produce a single brood in their lifetime, while yet others produce several successive broods during one breeding season. Brood size is not necessarily related to fecundity, the total number of young produced by the animal. Brood size can often be measured directly by counting in species that carry their broods of eggs, for instance many crustaceans, or indirectly by counting ripe eggs in the ovaries and genital ducts of crustaceans, fish, chaetognaths etc. (*Chiba*, 1956; *Jensen*, 1958; *Mauchline*, 1968, 1972, 1973; *Nemoto et al.*, 1972; *Omori*, 1974; *Hopkins and Baird*, 1977). The determination of the numbers of such broods, and hence of fecundity, is much more difficult in many organisms, especially small tropical species. Brood size is related to body size and frequently varies seasonally (with temperature). In general, the volume of eggs produced in a single brood is approximately 5-10% of the body volume of the animal.

General Metabolism (T)

The problems involved in measuring the rates of metabolism have been discussed in the introduction to this section and more fully by *Childress* (1977). In general, the quantity of food used for metabolism is a function of the body weight (W) of the animal and can be measured in terms of respiration (T) as µl oxygen/animal/hour:

$$\log T = \log \alpha + y \log W$$

where α and y are constants. The theoretical value of y is 0.667 but considerable deviation from this value has been reported (*Parsons and Takahashi*, 1973; *Marshall*, 1973). Respiration measured in terms of µl O_2/mg wet weight of the animal/hour, known as weight specific respiration, decreases with increasing weight of the animal. *Marshall* (1973) has shown that the metabolic rate of copepods is affected by body size, body shape, reproductive state and also by the animal's feeding habits and the seasonal parameters of the environment. *Ikeda* (1970) found a correlation between latitude and changing values of y. *Conover and Lalli* (1974) found that the respiratory rate of the pteropod *Clione limacina* varies with locality and season but y remains constant. (Figure 6). *Paranjape* (1967) also found y remains constant, at a value close to 1.0 for *Euphausia pacifica* maintained at various temperatures (Figure 7). These latter measurements of the constancy of y are made in studies of single species. Ikeda examined a variety of organisms and pooled his data at the different latitudes and consequently it is not surprising that he obtained changing values of y.

A curve describing the general change in the rate of metabolism relative to body weight or age of *Euphausia pacifica* can be drawn in the context of the other curves in Figure 4. The rate of respiration per unit body weight decreases as the animal's body weight increases (Figure 8). The curve in Figure 8 is the one derived from the data obtained at 15°C by Paranjape.

Marshall (1973) examined the weight specific respiration rate of adult and earlier copepodite stages of a variety of species of copepods (Figure 6). The scatter of values about the calculated regression line is large enough to detract from the usefulness of any predictive capability. *Childress* (1971) found that the weight specific respiration rate in twelve species of pelagic crustaceans and fish is correlated with the depth at which the individual species

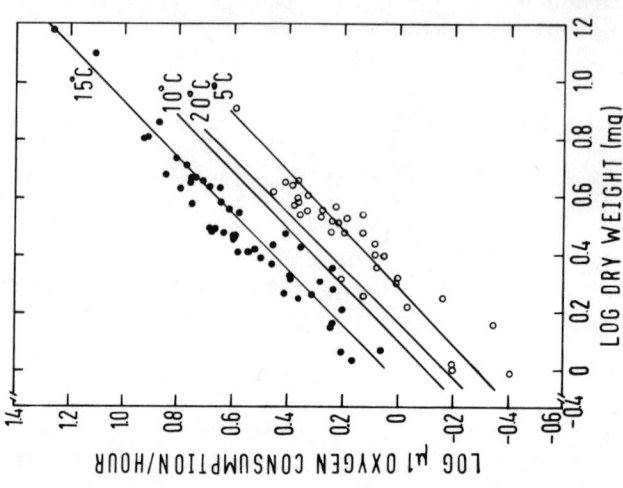

Fig. 7. *Euphausia pacifica*. Respiration rate related to body weight at various temperatures. Open circles are actual values at 5°C; solid dots are those at 15°C. (*Paranjape, 1967*).

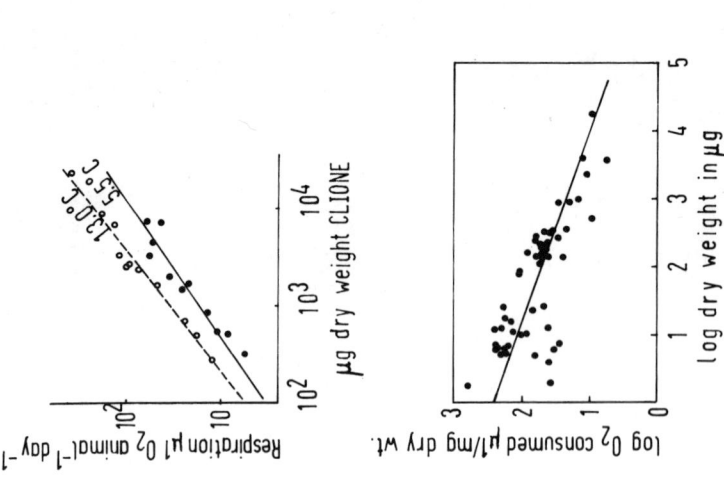

Fig. 6. Respiration rate. Upper figure: respiration of *Clione limacina* at two temperatures (*Conover and Lalli, 1974*). Lower figure: weight specific respiration related to body weight of a variety of copepods (*Marshall, 1973*).

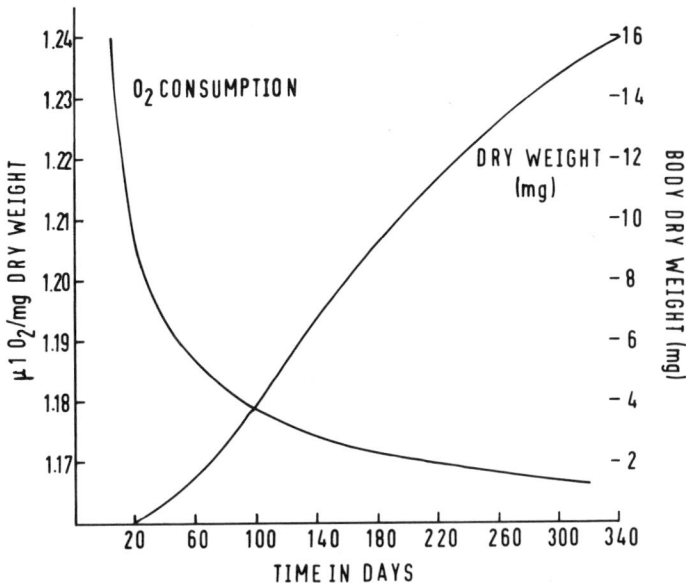

Fig. 8. *Euphausia pacifica*. Weight specific respiration rate, measured in µl O_2/mg dry weight of animal/hour, related to age of animal and to the curve for increase in dry weight as it changes with the age of the animal.

live. Those living in the 0-400 m layer consume ten times more oxygen per unit dry weight than those living in the 900-1300 m layer. *Smith and Hessler* (1974) found evidence of low respiration rates in benthopelagic fish, suggesting that the fish may be quiescent between meals. An alteration of quiescent periods with bursts of activity may be a fairly widespread behavioral trait of micronektonic and nektonic organisms in the deeper layers. Several workers, for example *Vinogradov et al.* (1970) and *Childress* (1977) have found evidence suggesting that deep water micronektonic animals have relatively less muscle (protein) and skeletal components and relatively more water in their bodies than shallower living species. This change in composition would produce an effect of reduced rates of metabolism in the deeper living animals.

More exploration of the variations of rates of metabolism in relation to the internal physiological and biochemical parameters of the animal and to the external physical/chemical/biological parameters of the environment are required. The rates of metabolism should be measured not only against the parameters existent at the time of the measurements but also in a historical context. Such studies should help to distinguish experimental errors in the measurements from other accountable sources of variation. A more realistic basis for the derivation of generalities describing metabolic rates will then exist.

Excretion (E + e)

Excretion in zooplankton, excluding fish, is generally measured as the quantity of food passing as feces and any equivalent urinary type excretion is considered to be small in comparison. *Cowey and Sargent* (1972) state that the quantity of nitrogen excreted in mg N/day is proportional to the body weight (W) in g of the fish:

$$\log N = b \log W - a$$

where a and b are constants.

Corner and Davies (1971) point out that the first attempt to measure nitrogen excretion in a planktonic organism seems to be that of *Harris* (1959) using *Acartia clausii*. They review the published information on the excretion of nitrogen in planktonic species. The number of species so far investigated is small----the copepods, *Acartia* spp., *Pseudocalanus minutus*, *Temora longicornis*, *Calanus finmarchicus*, *C. helgolandicus*, *C. hyperboreus* and *Oithona similis*; the mysids, *Neomysis integer* and *N. rayii*; the euphausiid *Euphausia pacifica*; and the chaetognath, *Sagitta hispida*. *Harris* (1973) estimated the rate of nitrogen excretion in the harpacticoid copepod, *Tigriopus brevicornis*. *Jawed* (1973) measured ammonia-nitrogen excretion of the euphausiids *Thysanoessa longipes* and *E. pacifica*, the medusa *Aequorea aequorea* and of mixed zooplankton. *Taguchi and Ishii* (1972) estimated ammonia-nitrogen excretion in the copepods *Neocalanus cristatus* and *N. plumchrus*.

The greater portion, approximately 80%, of the nitrogen excreted is in the form of ammonia-nitrogen. The remaining 20% may consist of about 19% amino-nitrogen and about 1% urea. Quantities excreted are in the range 0.5-43 μg N/mg dry weight of the animal/day.

Corner and Davies (1971) also reviewed the little information available on excretion rates of phosphorus by zooplankton. The greater portion of the phosphorus is probably excreted in an "organic" form.

The rates of nitrogen and phosphorus excretion are influenced by the physiological state of the organism, its body size, availability of food, and environmental temperature. Consequently, there are seasonal patterns of excretion.

Gross and Net Growth Efficiencies

Gross growth efficiency, K_1, is defined as the percentage of the captured food that is converted into new tissue while net growth efficiency, K_2, is defined as the percentage of the assimilated food that is converted into new tissue. *Corner and Davies* (1971) and *Parsons and Takahashi* (1973) list the different values of K_1 and K_2 so far determined. The concepts K_1 and K_2 are attractive but, as emphasized by Corner and Davies, much more information is required. How do values of K_1 and K_2 vary? There is evidence suggesting that K_1 diminishes at higher food concentrations and apparently equally good evidence that neither temperature nor quality of food affect K_1.

PRODUCTION OF POPULATIONS

Measurement of production in a population of a species depends upon having data on growth, breeding, longevity, feeding, metabolism and behavior of the individuals in the population. These data allow construction of models of variation in the parameters within the animal relative to each other and to parameters of the environment. Development of models of the population requires further information. This consists primarily of data describing the distribution of age groups within the population at successive intervals of time. The period described by the data should be related to the longevity of the individual, the duration of a generation being a primary parameter. Generation times may vary seasonally and sampling should take account of seasonal changes in population size and composition.

Acquisition of these data depends essentially upon being able to sample the same population often over an extended period of time. There must be reasonable confidence in the assumptions that exchange with neighboring populations of the species and/or loss to other areas is negligible or accounted for. Seasonal sampling of populations, biometrics and subsequent statistical analyses allow quantitative measurements of various parameters of the population to be made. The use of such methods to study populations of oceanic organisms present problems because of the difficulties of monitoring the same population at successive intervals of time. A general picture of the changing composition of the population can be obtained provided that samples of the population large enough to discern the age structure can be taken at successive intervals of time. These intervals in practice may be extremely irregular and spread over several years. The experience of the investigator is important in the interpretation of such data because subjective impressions are often important in determining the conclusions drawn from the data.

Assuming, however, that reliable data on changing structure of the population are added to the data discussed in previous sections of this paper, then various parameters of the population can be investigated. It is at this stage that the modelers start work. Reviews of theoretical and practical analyses of population data are given by and quoted in *Mann* (1969), *Krebs* (1972), *Williamson* (1972), *Parsons and Takahashi* (1973), and *Steele* (1974).

Mann (1969) discussed the possible usefulness of calculating the ratio of annual production (P) to the mean annual biomass (B). This ratio certainly seems to be worth exploring and such calculations were made for the two euphausiids, *Euphausia pacifica* and *Meganyctiphanes norvegica*. The data on growth of these species are given in Mauchline (1977).

Smiles and Pearcy (1971) analyzed the changing age composition of the population of *E. pacifica* off Oregon. The histograms represent a period on either side of the date of sampling, these periods being assessed in Table 3. Their samples are presented as number of animals per 1000 m^3 of water. The numbers for each period are converted to the equivalent dry weight. Growth, or weight addition to the population, was calculated by integrating weight addition to the different size classes of animals during each period to estimate total growth increment for the period. The mean biomass was calculated by multiplying the dry weight biomass by the duration of its respective period. The products were summed and divided by the total number of days, the sum of the periods. The ratio is:

$$\frac{P}{B} = \frac{113549}{13079} = 8.7$$

Similar calculations were made from the data on *M. norvegica* (Table 4). This species does not grow continuously but growth in body size is interrupted in the winter. The maximum growing season is 180 days, the minimum about 125 days. A season of 175 days is given in Table 4. The ratio is:

$$\frac{P}{B} = \frac{217990}{95918} = 2.3$$

The March period would be 30 days, the May period 60 days and the July period 35 days in a minimum growing season of 125 days. The ratio would then be:

$$\frac{P}{B} = \frac{152495}{95918} = 1.6$$

Comparable ratios for the small copepod *Acartia clausii* are 13 in the Black Sea and 8.7 in the English Channel. Greze (1966, 1970) modified this ratio to P/b, the daily increment. The concept of daily increments introduces seasonal ranges because P/B is dependent upon the size structure of the population, the growing season and probably several other parameters.

GENERAL CONCLUSIONS

The acquisition of accurate data on the production of oceanic species and ecosystems is a major challenge. The results from studies

TABLE 3. Production in the Population of *Euphausia pacifica* Off Oregon (Smiles and Pearcy, 1971)

	Numbers/ 1000 m^3	Dry weight (mg)/ 1000 m^3	Growth (mg)/ day/1000 m^3	Days Represented	Total production in period (mg)
October	8712	17103	400	90	36023
December	11340	18552	567	60	34028
February	7409	18951	446	60	26789
April	1260	8595	61	60	3676
June	3787	4482	145	90	13033
					113549

TABLE 4. Production in the population of *Meganyctiphanes norvegica* in Loch Fyne

	Numbers/ 1000 m³	Dry weight (mg)/ 1000 m³	Growth (mg)/ day/1000 m³	Days Represented	Total production in period (mg)
March	2000	89947	777	40	31071
May	10000	55351	1311	60	78645
July	3716	115555	1444	75	108274
					217990

species living in the epipelagic and coastal regimes cannot be extrapolated to mesopelagic and bathypelagic organisms without qualifications. Variation in the physical/chemical/biological parameters is greatest in the epipelagic and coastal environments, less in the mesopelagic, and least in the bathypelagic environment. The parameters become less variable with increasing depth but this stabilization is towards values that are extreme. The intensity of downwelling ambient daylight is negligible in the lower mesopelagic layer. Hydrostatic pressure increases from a range of 0-10 kg/cm^2 in the 0-100 m layer to 100 kg/cm^2 at 1000 m depth. Temperature decreases from 15-24°C at the surface to 4-8°C at 1000 m depth and to 3-4°C at 2000 m depth, in middle latitudes. Thus, with several variables simultaneously showing extreme values, the mesopelagic and bathypelagic environments elicit responses and adaptations rarely found in the epipelagic regime, where often only one variable is extreme.

The dominant properties of the deep layers are:
  a. low environmental temperature
  b. low light intensity
  c. high hydrostatic pressure
  d. low availability of food
  e. relative constancy of the values of these parameters.

The organism in its environment is a partially distinct subsystem within a system. The response of one area of an organism's system to a parameter of the environmental system may involve concomitant responses from other areas of the organism's system to other parameters of the environmental system. For instance, some species have been shown to respond to change in ambient light intensity by migrating vertically in association with an isolume. Such a response means that the animal moves into a different temperature regime and this may elicit concomitant changes in its rate of metabolism. The quantity and quality of food available and the density of the water also alter, probably eliciting further changed responses from the animal. The vertical range of water density is often appreciable and Bone (1973) suggested that a functional classification of myctophids based upon their density, lipid contents and the presence or absence of a gas-filled swimbladder is practicable.

Further to these changing parameters, is that of the hydrostatic pressure to which the organism is subjected. The *in situ* effects of or adaptations to the high hydrostatic pressures that meso- and bathypelagic animals experience is a subject of conjecture (Brauer, 1972). There are technical problems associated with investigations in this field and these were defined by various authors included in Brauer (1972) and by Gordon and Thomas (1974). Blaxter et al. (1971) examined dead and dying mesopelagic fish, measuring water content, heart weight, freezing point depression of body fluids, packed red blood cell volume, and blood viscosity. They also carried out histological and structural investigations of the musculature and of the lymph spaces and related their results to the bathymetric distributions of the fish. This type of study and the information presented at the workshop on barobiology (*Brauer*, 1972) suggest that the more classical approach through studies of functional morphology and histology combined with modern analytical techniques may be a major route for advancement in this field. Such studies made on selected deeper living organisms in combination with parallel studies of phylogenetically related shallow living species would probably be the most profitable approach. This comparative approach not only benefits the field of barobiology but also behavioral, physiological, biochemical and general production studies.

The simultaneous alteration of several parameters of the environment is not, of course, peculiar to a vertically migrating deep living organism. What is peculiar to this animal is that the values of several of these parameters are approaching extreme levels, in the biological sense. The activity of various enzymes is restricted to a limited range of values of various parameters. How much do we know about the enzymology of mesopelagic organisms? Multivariate analyses of experimental situations are rare and development of this approach to production studies is required. For instance, *Droop* (1974) has studied the relationships in chemostats between growth rate, rate of nutrient uptake and the internal and external concentrations of two nutrients, phosphorus and vitamin B_{12}, simultaneously in *Monochrysis lutheri*. *Gilfillan* (1972) designed experiments with the euphausiid *Euphausia pacifica* to examine both seasonal and latitudinal responses to simultaneous changes of temperature and salinity.

Many of the problems of production studies of offshore pelagic systems devolve from the difficulties of identifying discrete populations of species that can be sampled frequently enough and in sufficient numbers. More information relative to the numbers of specimens in the samples has to be obtained, often by approaches, including deductive methods, that are not normally applied to species in shallower environments. One result of this is that more basic studies such as functional morphology and histology assume greater importance within the framework of production studies. Such investigations of the biology of single species and of individual animals must be accented to provide data for use in the analyses of the ecosystems. These small programs, often carried out by single investigators, are an integral part of the large multidisciplinary program designed to examine production in small or large geographical areas.

ACKNOWLEDGEMENTS

I wish to thank the following colleagues at the Dunstaffnage Marine Research Laboratory for critically reading the manuscript; Mr. J. Blackstock, Mr. R. I. Currie, Dr. R. H. Millar and Dr. P. B. Tett. I am indebted to Miss C. M. Stewart for drawing the final text figures.

REFERENCES

Aizawa, Y. 1974. Ecological studies of micronektonic shrimps (Crustacea, Decapoda) in the western north Pacific. *Bull. Ocean Res. Inst.*, (6), Univ. Tokyo. 1-84.

Alvarino, A. 1971. Siphonophores of the Pacific with a review of the world distribution. *Bull. Scripps Inst. Oceanogr.*, 16: 1-432.

Andrews, K. J. H. 1966. The distribution of life-history of *Calanoides acutus* (Giesbrecht). *"Discovery" Rep.*, 34: 117-162.

Angel, M. V. 1977. Windows into a sea of confusion: sampling limitations to the measurement of ecological parameters in oceanic mid-water environments. In: *"Oceanic Sound Scattering Prediction"*, N. R. Andersen and B. J. Zahuranec (Eds.), Plenum, N.Y., (this volume).

Angel, M. V. and M. J. R. Fasham. 1973. Sond Cruise 1965: factor and cluster analyses of the plankton results, a general summary. *J. mar. biol. Ass. U.K.*, 53: 188-231.

Badcock, J. 1970. The vertical distribution of mesopelagic fishes collected on the SOND cruise. *J. mar. biol. Ass. U.K.*, 50: 1001-1044.

Badcock, J. and N. R. Merrett. 1977. On the distribution of midwater fishes in the eastern North Atlantic. In: *"Oceanic Sound Scattering Prediction"*, N. R. Andersen and B. J. Zahuranec (Eds.), Plenum, N.Y., (this volume).

Baker, A. de C. 1963. The problem of keeping planktonic animals alive in the laboratory. *J. mar. biol. Ass. U.K.*, 43: 291-294.

Beers, J. R. 1966. Studies on the chemical composition of the major zooplankton groups in the Sargasso Sea off Bermuda. *Limnol. Oceanogr.*, 11: 520-528.

Blackburn, M. 1977. Studies on pelagic animal biomasses. In: *"Oceanic Sound Scattering Prediction"*, N. R. Andersen and B. J. Zahuranec (Eds.), Plenum, N.Y., (this volume).

Blaxter, J. H., C. S. Wardle and B. L. Roberts. 1971. Aspects of the circulatory physiology and muscle systems of deep-sea fish. *J. mar. biol. Ass. U.K.*, 51: 991-1006.

Bone, Q. 1973. A note on the buoyancy of some lanternfishes (Myctophidae). *J. mar. biol. Ass. U.K.*, 53: 619-633.

Bottino, N. R. 1974. The fatty acids of Antarctic phytoplankton and euphausiids. Fatty acid exchange among trophic levels of the Ross Sea. *Mar. Biol.*, 27: 197-204.

Brett, J. R., J. E. Shelbourn, and C. T. Shoop. 1969. Growth rate and body composition of fingerling sockeye salmon, *Oncorhynchus nerka*, in relation to temperature and ration size. *J. Fish. Res. Bd. Can.*, 26: 2363-2394.

Brauer, R. W. (Editor). 1972. Barobiology and the experimental biology of the deep sea. North Carolina Sea Grant Program, University of N. Carolina. 428 pp.

Chiba, T. 1956. Studies on the development and systematics of Copepoda. *J. Shimonoseki Coll. Fish.*, 6: 1-90.

Childress, J. J. 1971. Respiratory rate and depth of occurrence of midwater animals. *Limnol. Oceanogr.*, 16: 104-106.

Childress, J. J. 1977. Physiological approaches to the biology of midwater organisms. In:*"Oceanic Sound Scattering Prediction"*, N. R. Andersen and B.J. Zahuranec (Eds.), Plenum, N.Y. (this volume).

Clarke, G. L. 1970. Light conditions in the sea in relation to the diurnal vertical migrations of animals. In: *"Biological Sound Scattering in the Oceans"*, G. B. Farquhar (Ed.), Maury Center for Ocean Science, Department of the Navy, Washington, D. C. 41-50.

Clarke, T. S. 1973. Some aspects of the ecology of lanternfishes (Myctophidae) in the Pacific Ocean near Hawaii. *Fish. Bull., U. S.*, 71: 401-434.

Clarke, T. S. 1974. Some aspects of the ecology of stomiatoid fishes in the Pacific Ocean near Hawaii. *Fish. Bull., U. S.*, 72: 337-351.

Conover, R. J. and C. M. Lalli. 1974. Feeding and growth in *Clione limacina* (Phipps), a pteropod mollusc. II. Assimilation, metabolism, and growth efficiency. *J. exp. mar. Biol. Ecol.*, 16: 131-154.

Corner, E. D. S. and A. G. Davies. 1971. Plankton as a factor in the nitrogen and phosphorus cycles in the sea. *Adv. Mar. Biol.*, 9: 101-204.

Corner, E. D. S., R. N. Head, C. C. Kilvington, and S. M. Marshall. 1974. On the nutrition and metabolism of zooplankton. IX. Studies relating to the nutrition of over-wintering *Calanus*. *J. mar. biol. Ass. U.K.*, 54: 319-331.

Cowey, C. B. and J. R. Sargent. 1972. Fish nutrition. *Adv. Mar. Biol.*, 10: 383-492.

Deevey, G. B. 1971. The annual cycle in quantity and composition of the zooplankton of the Sargasso Sea off Bermuda. I. The upper 500 m. *Limnol. Oceanogr.*, 16: 219-240.

Deevey, G. B. and A. L. Brooks. 1971. The annual cycle in quantity and composition of the zooplankton of the Sargasso Sea off Bermuda. II. The surface to 2000 m. *Limnol. Oceanogr.*, 16: 927-943.

Digby, P. S. B. 1972. Detection of small changes in hydrostatic pressure by Crustacea and its relation to electrode action in the cuticle. *Soc. exp. Biol. Symp.*, 26: 445-471.

Donaldson, H. A. 1975. Vertical distribution and feeding of sergestid shrimps (Decapoda:Natantia) collected near Bermuda. *Mar. Biol.*, 31: 37-50.

Dougherty, E. C., B. Solberg, and D. J. Ferral. 1961. The first axenic cultivation of a rotifer species. *Experientia*, 17: 131-132.

Droop, M. R. 1974. The nutrient status of algal cells in continuous culture. *J. mar. biol. Ass. U.K.*, 54: 825-855.

Ebeling, A. E. and G. M. Cailliet. 1974. Mouth size and predator strategy in midwater fishes. *Deep-Sea Res.*, *21*: 959-968.

Farquhar, G. B. (Editor). 1971. *Proceedings of an International Symposium on Biological Sound Scattering in the Ocean*. Maury Center for Ocean Science, Department of the Navy, Washington, D. C. 629 pp.

Forster, J. R. M. and T. W. Beard. 1973. Growth experiments with the prawn *Palaemon serratus* Pennant fed with fresh and compounded foods. *Fish. Invest.*, *Lond.*, *Ser 2*, *27*, (16):

Fowler, S. W., G. Benayoun, and L. F. Small. 1971a. Experimental studies on feeding, growth and assimilation in a Mediterranean euphausiid. *Thalassia Jugoslavica*, *7*: 35-47.

Fowler, S. W., L. F. Small, and S. Keskes. 1971b. Effects of temperature and size on moulting of euphausiid crustaceans. *Mar. Biol.*, *11*: 45-51.

Foxton, P. 1966. The distribution and life-history of *Salpa thompsoni* Foxton with observations on a related species, *Salpa gerlachei* Foxton. *"Discovery" Rep.*, *34*: 1-116.

Foxton, P. 1972. Observations on the vertical distribution of the genus *Acanthephyra* (Crustacea:Decapoda) in the eastern North Atlantic with particular reference to the species of the 'purpurea' group. *Proc. R. Soc. Edinb.*, *(B)*, *73*: 301-313.

Foxton, P. and H. S. J. Roe. 1974. Observations on the nocturnal feeding of some mesopelagic decapod Crustacea. *Mar. Biol.*, *28*: 37-49.

Fuglister, F. C. 1960. Atlantic Ocean Atlas. Woods Hole Oceanographic Institution, Woods Hole, Mass. 209 pp.

Gilfillan, E. 1972. Seasonal and latitudinal effects on the responses of *Euphausia pacifica* Hansen (Crustacea) to experimental changes of temperature and salinity. In: *"Biological Oceanography of the Northern North Pacific Ocean"*, A. Y. Takenouti et al. (Editors), Idemitsu Shoten, Tokyo. 443-463.

Gjösaeter, J. 1973a. Age, growth and mortality of the myctophid fish, *Benthosema glaciale* (Reinhart), from western Norway. *Sarsia*, *52*: 1-14.

Gjösaeter, J. 1973b. The food of the myctophid fish, *Benthosema glaciale* (Reinhart), from western Norway. *Sarsia*, *52*: 53-58.

Gordon, M. S. and T. J. Thomas. 1974. Comparative studies on the metabolism of shallow-water and deep-sea marine fishes. III. Apparatus for studies of tissue preparations under high hydrostatic pressures. *Mar. Biol.*, *28*: 73-77.

Greze, V. N. 1966. The rate of production in populations of heterotrophic marine organisms. *2nd Int. oceanogr. Congr.*, *Moscow*. 144-145.

Greze, V. N. 1970. The biomass and production of different trophic levels in the pelagic communities of south seas. In: *"Marine Food Chains"*, J. H. Steele (Ed.), Oliver and Boyd, Edinburgh. 458-467.

Halliday, R. G. 1970. Growth and vertical distribution of the glacier lantern fish, *Benthosema glaciale*, in the northwestern Atlantic. *J. Fish. Res. Bd. Can.*, 27: 105-116.

Harding, G. C. H. 1974. The food of deep-sea copepods. *J. mar. biol. Ass. U.K.*, 54: 141-155.

Harris, E. 1959. The nitrogen cycle in Long Island Sound. *Bull. Bingham oceanogr. Coll.*, 17: 31-65.

Harris, R. P. 1973. Fedding, growth, reproduction and nitrogen utilization by the harpacticoid copepod, *Tigriopus brevicornis*. *J. mar. biol. Ass. U.K.*, 53: 785-800.

Hart, T. J. 1934. On the phytoplankton of the south west Atlantic and the Bellingshausen Sea, 1929-1931. *"Discovery" Rep.*, 8: 1-268.

Herring, P. J. 1972. Depth distribution of the carotenoid pigments and lipids of some oceanic animals. 1. Mixed zooplankton, copepods and euphausiids. *J. mar. biol. Ass. U.K.*, 52: 179-189.

Herring, P. J. 1973. Depth distribution of the carotenoid pigments and lipids of some oceanic animals. 2. Decapod crustaceans. *J. mar. biol. Ass. U.K.*, 53: 539-562.

Hirota, J. 1972. Laboratory culture and metabolism of the planktonic ctenophore *Pleurobrachia bachei* A. Agassiz. In: *"Biological Oceanography of the Northern North Pacific Ocean"*. A. Y. Takenouti et al. (Editors), Idemitsu Shoten, Tokyo. 466-484.

Hopkins, T. L. and R. C. Baird. 1973. Diet of the hatchet fish *Sternoptyx diaphana*. *Mar. Biol.*, 21: 34-46.

Hopkins, T. L. and R. C. Baird. 1977. Aspects of the feeding ecology of oceanic midwater fishes. In: *"Oceanic Sound Scattering Prediction"*, N. R. Andersen and B. J. Zahuranec (Eds.), Plenum, N.Y., (this volume).

Hure, J. and B. Scotto di Carlo. 1974. New patterns of diurnal vertical migration of some deep-water copepods in the Tyrrhenian and Adriatic Seas. *Mar. Biol.*, 28: 179-184.

Ikeda, T. 1970. Relationship between respiration rate and body size in marine plankton animals as a function of the temperature of the habitat. *Bull. Fac. Fish. Hokkaido Univ.*, 21: 91-112.

Ikeda, T. 1972. Chemical composition and nutrition of zooplankton in the Bering Sea. In: *"Biological Oceanography of the Northern North Pacific Ocean"*. A. Y. Takenouti et al. (Editors), Idemitsu Shoten, Tokyo. 433-442.

Ikematsu, W. 1953. On the life-history of *Acetes japonicus* Kishinouye in Ariake Sea. *Bull. Jap. Soc. Sci. Fish.*, 19: 771-780.

Isaacs, J. D. 1973. Potential trophic biomasses and trace substance concentrations in unstructured marine food webs. *Mar. Biol.*, 22: 97-104.

Itoh, K. 1970. A consideration on feeding habits of planktonic copepods in relation to the structure of their oral parts. *Bull. Planktol. Soc. Jap.*, 17: 1-10.

Jawed, M. 1973. Ammonia excretion by zooplankton and its significance to primary productivity during summer. Mar. Biol., 23: 115-120.

Jensen, J. P. 1958. The relation between body size and number of eggs in marine Malacostrakes. Medd. Komm. Danm. Fisk.-og. Havunders., N.S. 2: 1-25.

Judkins, D. C. and A. Fleminger. 1972. Comparison of foregut contents of Sergestes similis obtained from net collections and albacore stomachs. Fish. Bull., U.S., 70: 217-223.

Kawaguchi, K. and R. Marumo. 1967. Biology of Gonostoma gracile (Gonostomatidae). 1. Morphology, life history and sex reversal. Inf. Bull. Planktol Jap., Commem. Number of Dr. Y. Matsue. 53-70.

Kayama, M. and J. C. Nevenzel. 1974. Wax ester biosynthesis by midwater marine animals. Mar. Biol., 24: 279-285.

Kinzer, J. 1977. Observations on the feeding habits of the mesopelagic fish Benthosema glaciale (Myctophidae) off NW Africa. In: "Oceanic Sound Scattering Prediction", N. R. Andersen and B. J. Zahuranec (Eds.), Plenum, N.Y., (this volume).

Komaki, Y. 1966. Technical notes on keeping euphausiids live in the laboratory, with a review of experimental studies on euphausiids. Inf. Bull. Planktol. Jap., 13: 95-105.

Krebs, C. J. 1972. Ecology. The experimental analysis of distribution and abundance. Harper and Row, New York. 694 pp.

Lasker, R. 1966. Feeding, growth, respiration and carbon utilization of a euphausiid crustacean. J. Fish. Res. Bd. Can., 23: 1291-1317.

Lasker, R. and G. H. Theilacker. 1965. Maintenance of euphausiid shrimps in the laboratory. Limnol. Oceanogr., 10: 287-288.

Legand, M. 1967. Cycles biologiques des poissons mésopélagiques dans l'est de l'Ocean Indien. Premiere note: Scopelopsis multipunctatus Brauer, Gonostoma sp., Notolychnus valdiviae Brauer. Cah. O.R.S.T.O.M., Ser. Oceanogr., 5: 47-71.

Legand, M., P. Bourret, P. Fourmanoir, R. Grandperrin, J. A. Gueredrat, A. Michel, P. Rancurel, R. Repelin, and C. Roger. 1972. Relations trophiques et distributions verticales en milieu pélagique dans l'Ocean Pacifique intertropical. Cah. O.R.S.T.O.M., Ser. Oceanogr., 10: 303-393.

Legand, M. and J. Rivaton. 1970. Cycles biologiques des poissons mésopélagiques dans l'est de l'Ocean Indien. Quatrième note: synthese des divers cycles décrits. Cah. O.R.S.T.O.M., Ser. Oceanogr., 8: 59-79.

Lewis, A. G. 1977. Possible factors affecting succession in plankton communites. In: "Oceanic Sound Scattering Prediction", N. R. Andersen and B. J. Zahuranec (Eds.), Plenum, N.Y., (this volume).

Macdonald, A. G., I. Gilchrist, and J. M. Teal. 1972. Some observations on the tolerance of oceanic plankton to high hydrostatic pressure. J. mar. biol. Ass. U.K., 52: 213-223.

Mackie, G. O. and D. A. Boag. 1963. Fishing, feeding and digestion in siphonophores. *Pubbl. Staz. Zool. Napoli.*, 33: 178-196.

Madin, L. P. 1974. Field observations on the feeding behavior of salps (Tunicata:Thaliacea). *Mar. Biol.*, 25: 143-147.

Mann, K. H. 1969. The dynamics of aquatic ecosystems. *Adv. Ecol. Res.*, 6: 1-81.

Marshall, N. B. 1972. Swimbladder organization and depth ranges of deep-sea teleosts. *Soc. exp. Biol. Symp.*, 26: 261-272.

Marshall, S. M. 1973. Respiration and feeding in copepods. *Adv. Mar. Biol.*, 11: 57-120.

Matthews, J. B. L. and S. Pinnoi. 1973. Ecological studies on the deep-water pelagic community of Korsfjorden, western Norway. The species of *Pasiphaea* and *Sergestes* (Crustacea, Decapoda) recorded in 1968 and 1969. *Sarsia*, 52: 123-144.

Mauchline, J. 1968. The development of the eggs in the ovaries of euphausiids and estimation of fecundity. *Crustaceana*, 14: 155-163.

Mauchline, J. 1972. The biology of bathypelagic organisms, especially Crustacea. *Deep-Sea Res.*, 19: 753-780.

Mauchline, J. 1973. The broods of British Mysidacea (Crustacea). *J. mar. biol. Ass. U.K.*, 53: 801-817.

Mauchline, J. 1977. Growth and moulting of Crustacea, especially euphausiids. In: *"Oceanic Sound Scattering Prediction"*, N. R. Andersen and B. J. Zahuranec (Eds.), Plenum, N.Y., (this volume).

Mauchline, J. and L. R. Fisher. 1969. The biology of euphausiids. *Adv. Mar. Biol.*, 7: 1-454.

Merrett, N. R. and H. S. J. Roe. 1974. Patterns and seclectivity in the feeding of certain mesopelagic fish. *Mar. Biol.*, 28: 115-126.

Moraitou-Apostolopolou, M. 1971. Vertical distribution, diurnal and seasonal migration of copepods in Saronic Bay, Greece. *Mar. Biol.*, 9: 92-98.

Morris, R. J. 1973. Relationships between the sex and degree of maturity of marine crustaceans and their lipid compositions. *J. mar. biol. Ass. U.K.*, 53: 27-37.

Morris, R. J. and J. R. Sargent. 1973. Studies on the lipid metabolism of some oceanic crustaceans. *Mar. Biol.*, 22: 77-83.

Mullin, M. M. 1969. Production of zooplankton in the ocean: the present status and problems. *Oceanogr. Mar. Biol. Ann. Rev.*, 7: 293-314.

Nemoto, T. 1968. Trace of chlorophyll pigments in stomachs of deep sea zoo-plankton. *J. oceanogr. Soc. Jap.*, 24: 46-48.

Nemoto, T. 1972. Chlorophyll pigments in the stomach and gut of some macrozooplankton species. In: *"Biological Oceanography of the Northern Pacific Ocean"*, A. Y. Takenouti et al. (Editors), Idemitsu Shoten, Tokyo. 411-418.

Nemoto, T. 1977. Food and feeding structures of deep sea *Thysanopoda* euphausiids. In: *"Oceanic Sound Scattering Prediction"*, N. R. Andersen and B.J. Zahuranec (Eds.), Plenum, N.Y., (this volume).

Nemoto, T., K. Kamada, and K. Hara. 1972. Fecundity of a euphausiid crustacean, *Nematoscelis difficilis*, in the North Pacific Ocean. *Mar. Biol.*, 14: 41-47.

Ommaney, F. D. 1936. *Rhincalanus gigas* (Brady), a copepod of the southern macroplankton. *"Discovery" Rep.*, 13: 277-384.

Omori, M. 1969. The biology of a sergestid shrimp *Sergestes lucens* Hansen. *Bull. Ocean. Res. Inst.*, Univ. Tokyo, (4): 1-83.

Omori, M. 1973. Cultivation of marine copepods. *Bull. Planktol. Soc. Jap.*, 20: 3-11.

Omori, M. 1974. The biology of pelagic shrimps in the ocean. *Adv. Mar. Biol.*, 12: 233-324.

Omori, M., A. Kawamura, and Y. Aizawa. 1972. *Sergestes similis* Hansen, its distribution and importance as food of fin and sei whales in the North Pacific Ocean. In: *Biological Oceanography of the Northern Pacific Ocean*, A. Y. Takenouti et al. (Editors), Idemitsu Shoten, Tokyo. 373-391.

Packard, T. T., M. L. Healy, and F. A. Richards. 1971. Vertical distribution of the activity of the respiratory electron transport system in marine plankton. *Limnol. Oceanogr.*, 16: 60-70.

Paranjape, M. A. 1967. Moulting and respiration of euphausiids. *J. Fish. Res. Bd. Can.*, 24: 1229-1240.

Parsons, T. R. 1969. The use of particle size spectra in determining the structure of a plankton community. *J. oceanogr. Soc. Jap.*, 25: 172-181.

Parsons, T. R. and M. Takahashi. 1973. Biological Oceanographic Processes. Pergamon Press, New York. 186 pp.

Pearcy, W. G. and C. A. Forss. 1969. The oceanic shrimp *Sergestes similis* off the Oregon coast. *Limnol. Oceanogr.*, 14: 755-765.

Petipa, T. S., E. V. Pavlova, and G. N. Mironov. 1970. The food web structure, utilization and transport of energy by trophic levels in the planktonic communities. In: *"Marine Food Chains"*, J. H. Steele (Ed.), Oliver and Boyd, Edinburgh. 142-167.

Phillips, A. M. 1969. Nutrition, digestion and energy utilization. In: *"Fish Physiology"*, W. S. Hoar and D. J. Randall (Editors), Academic Press, New York and London. 391-432.

Ponomareva, L. A. 1954. Copepods as food of euphausiids of the Sea of Japan. *Dokl. Akad. Nauk.*, S.S.S.R., 98: 153-154.

Provasoli, L. and A. D'Agostino. 1969. Development of artificial media for *Artemia salina*. *Biol. Bull.*, Woods Hole, 135: 434-453.

Provasoli, L., K. Shiraishi, and J. R. Lance. 1959. Nutritional idiosyncrasies of *Artemia* and *Tigriopus* in monoxenic culture. *Ann. New York Acad. Sci.*, 77: 250-261.

Pugh, P. R. 1974. The vertical distribution of the siphonophores collected during the SOND cruise, 1965. *J. mar. biol. Ass. U.K.*, 54: 25-90.

Raymont, J. E. G. 1970. Problems of the feeding of zooplankton in the deep sea. In: *"Biological Sound Scattering in the Oceans"*, G. B. Farquhar (Ed.), Maury Center for Ocean Science, Department of the Navy, Washington D. C. 134-146.

Repelin, R. 1972. Étude préliminaire des amphipodes du bol alimentaire de poissons pélagiques provenant de pêches a la longue ligne. *Cah. O.R.S.T.O.M., Ser. Oceanogr.*, 10: 47-55.

Roe, H. S. J. 1974. Observations on the diurnal vertical migrations of an oceanic animal community. *Mar. Biol.*, *28*: 99-113.

Sargent, J. R., R. R. Gatten, and R. McIntosh. 1974. Biosynthesis of wax esters in cell-free preparations of *Euchaeta norvegica*. *Comp., Biochem. Physiol.*, *47B*: 217-227.

Scott, J. M. 1974. A new marine rotifer of the genus *Encentrum*. Its morphology and cultivation. *Zool. J. Linn. Soc.*, *54*: 247-251.

Sekiguchi, H. 1974. Relation between the ontogenetic vertical migration and the mandibular gnathobase in pelagic copepods. *Bull. Fac. Fish. Mie Univ.*, *(1)*: 1-10.

Sheader, M. and F. Evans. 1975. Feeding and gut structure of *Parathemisto gaudichaudi* (Guerin) (Amphipoda, Hyperiidea). *J. mar. biol. Ass. U.K.*, *55*: 641-656.

Sivertsen, E. and L. B. Holthuis. 1956. Crustacea Decapoda. (The Penaeidae and Stenopodidae excepted). *Rep. Sci. Res. "Michael Sars" N. Atl. Deep Sea Exped. 1910*, *5, (12)*: 1-54.

Smiles, M. C. and W. G. Pearcy. 1971. Size structure and growth rate of *Euphausia pacifica* off the Oregon coast. *Fish. Bull., U.S.*, *69*: 79-86.

Smith, K. L. and R. R. Hessler. 1974. Respiration of benthopelagic fishes: *in situ* measurements at 1230 metres. *Science, N.Y.*, *184*: 72-73.

Smoker, W. and W. G. Pearcy. 1970. Growth and reproduction of the lantern fish *Stenobrachius leucopsarus*. *J. Fish. Res. Bd. Can.*, *27*: 1265-1275.

Steele, J. H. (Editor). 1970. Marine Food Chains. Oliver and Boyd, Edinburgh. 552 pp.

Steele, J. H. 1974. The structure of marine ecosystems. *Blackwell Scientific Publications*, Oxford. 128 pp.

Strickland, J. D. H. 1972. Research on the marine planktonic food web at the Institute of Marine Resources: a review of the past seven years of work. *Oceanogr. Mar. Biol. Ann. Rev.*, *10*: 349-414.

Taguchi, S. and H. Ishii. 1972. Shipboard experiments on respiration, excretion and grazing of *Calanus cristatus* and *C. plumchrus* (Copepoda) in the northern North Pacific. In: "*Biological Oceanography of the Northern North Pacific Ocean*", A. Y. Takenouti et al. (Editors), Idemitsu Shoten, Tokyo. 419-431.

Tåning, A. V. 1918. Mediterranean Scopelidae (*Saurus, Aulopus Chlorophthalmus* and *Myctophum*). *Rep. Dan. oceanogr. Exped. Mediterr.*, *2 (Biol)*: 1-154.

Totton, A. K. 1954. Siphonophora of the Indian Ocean together with systematic and biological notes on related specimens from other oceans. *"Discovery" Rep.*, *27*: 1-162.

Totton, A. K. 1965. A synopsis of the Siphonophora. *British Museum (Natural History)*, London. 230 pp.

Tranter, D. J. 1973. Seasonal studies of a pelagic ecosystem (Meridian 110°E). In: "*The Biology of the Indian Ocean*", B. Zeitschel (Ed.), Chapman and Hall Ltd., London. 487-520.

Tyler, A. V. 1973. Caloric values of some North Atlantic invertebrates. *Mar. Biol.*, *19*: 258-261.

Tyler, H. R. and W. G. Pearcy. 1975. The feeding habits of three species of lantern fishes (family Myctophidae) off Oregon, USA. *Mar. Biol.*, *37*: 7-11.

Vinogradov, M. E. 1968. Vertical distribution of the oceanic zooplankton. *Israel Program for Scientific Translations*, Jerusalem, 1970.

Vinogradov, M. E., O. K. Bordovskii, and E. A. Akhmet'eva. 1970. Biochemistry of oceanic plankton. Chemical composition of plankton from different depths of the northwestern Pacific. *Fish. Res. Bd. Can. Transl. Ser.*, (1833), 1971.

Williamson, M. 1972. The analysis of biological populations. Edward Arnold, London. 180 pp.

Wilson, D. F. 1972. Diel migration of sound scatterers into, and out of, the Cariaco Trench anoxic water. *J. mar. Res.*, *30*: 168-176.

Zimmerman, S. T. 1973. The transformation of energy by *Lucifer chacei* (Crustacea, Decapoda). *Pacific Science*, *27*: 247-259.

Windows into a Sea of Confusion: Sampling Limitations to the Measurement of Ecological Parameters in Oceanic Mid-Water Environments

Martin V. Angel

Institute of Oceanographic Sciences, United Kingdom

ABSTRACT

Oceanic midwater communities present broad spectra of variability in time, in space, and in size distributions among other ecologically important phenomena. Each biological sampler, even if properly and effectively deployed, only provides material from a limited band of each spectrum. These limited bands, which are specific to each sampler, are the "windows" of discrimination, within which sampling error and bias does not distort the natural patterns beyond recognition.

The windows have bounds that vary even at the species level from the smallest size of organism retained to the largest size limit that is effectively captured. Studies on patchiness show that it occurs at all scales. Physical oceanographic investigations indicate that the longevity of eddies is proportional to their size. Patches of organisms will be equally subject to turbulence, internal waves and diffusion even if they are actively maintained, so there are temporal limits to our ability to follow them. Thus in ecological studies for which long time series are required, there may be insuperable temporal limits to processes that can be observed in situ.

Field observations and sampling programs should be designed to stay within the bounds of this window. Any observations made outside the windows will be too distorted ever to be meaningfully interpreted. The interpretations of all previous and future sampling programs need critical examination to ascertain if the conclusions drawn are likely to be at all valid.

INTRODUCTION

Our knowledge and understanding of the distribution and structure of plankton and nekton communities is based on samples collected by a broad spectrum of samplers. Each sampler in itself has specific sampling characteristics which bias or distort the estimates of the community parameters. These distortions are produced by both the inter- and intraspecific variations in the catchability of organisms caused by mesh selection, avoidance, speed of tow etc., and also by the inherent variability of the planktonic population. Each sampler offers the investigator a series of limited "windows" into the spectra of variability. Beyond the bounds of these "windows" the degree of variability and sampling error obscures the ecological processes being investigated. The range of ecological parameters of interest to investigators include the specific composition of the community, its size structure, the delineation of community boundaries both vertically and horizontally, the trophic relationships within the community, and time series to provide date on diel and ontogenic vertical migrations and *in situ* growth rates. The limits of these "windows" in time and space will be discussed in relation to the present mid-water biological program at the Institute of Oceanographic Sciences (IOS).

SPECIFIC COMPOSITION

Specific composition is one of the basic properties of any community. An initial problem is to define the limits of the community. It may be defined by the broad ecological divisions of neustonic, epipelagic, mesopelagic and bathypelagic zones. The boundaries of these zones cannot be precisely defined but they have a real ecological significance (e.g. *Vinogradov*, 1968; *Angel and Fasham*, 1973, 1974). The community limits may be defined either by physical parameters such as depth, isolumes or isotherms, or by biological phenomena such as the depth of the deep scattering layer (*Backus et al*, 1970). Communities have been sampled with nets towed obliquely or vertically through the depth range. The depth ranges have been divided up into strata that are either sampled with nets towed horizontally (e.g. *Gibbs and Roper*, 1970; *Foxton*, 1972; *Angel and Fasham*, 1973; *Pugh*, 1974) or obliquely (e.g. *McGowan and Williams*, 1973). The sampling strategy used will determine the type of community boundary that can be discriminated.

Sample and sampler sizes are critical. *McGowan and Fraundorf* (1966), in comparing samplers ranging in size from 20-140 cm diameter, found that the larger sampler caught a greater diversity of species. *Wiebe* (1971) simulated the sampling with nets of four sizes with the organisms distributed in patches of several different sizes. He concluded that the sampling precision, based on the reduction in the coefficient of variation, increased with net size and tow length. When the patches were greater in size than 25m, the probability of hitting a patch was independent of net size. But the larger nets gave more precise estimates than samller ones, since the variability of the estimates was a function of the numbers caught. If the patch size is much greater than the net diameter, extending the tow length (up to a limit) is more effective in estimating population sizes than increasing the net diameter, assuming that the spatial structure of the patches is homogeneous. *Chanut* (1975) also used simulation to investigate sampling errors and how these were affected by the dispersion of the organisms. He found that for animals that were randomly distributed, increasing the mouth area of the sampler improved the accuracy of abundance estimates. As aggregation of the organisms was progressively increased, so length of tow and size of patch became more important factors in determining the errors. He suggested that simulations carried out prior to the design of sampling programs based on previous field evidence, could optimize the effectiveness of sampling.

The use of the Longhurst Hardy Plankton Recorder (LHPR) (*Longhurst et al*, 1966) resulted in field evidence for the length of tow needed to establish the specific composition of a midwater community. *Fasham et al* (1974) described the results of three LHPR tows at 550m at two positions in the N.E. Atlantic. Figure 1 shows the accumulation curves for species and specimens plotted against real time and log time. Such plots are familiar to botanists for determining the size of quadrat needed to study a plant community. The log time plot gives a simple extrapolation of the additional sampling needed either to add further species or take a sample of a given size. Additional sampling can be done either by extending the duration of the tow, or by carrying out repeated tows.

McGowan (1971) presented data from a series of tows taken with a 1m^2 net which increased in length so that the volume filtered increased by increments of about 100m^3. Between the filtered volumes of 128-1510m^3, the numbers of species of fish larvae and molluscs increased steadily whereas the number of euphausiid species was unchanged.

The species retained by a sampler is limited at both ends of the size spectrum. The upper limit is determined by a combination of increasing avoidance and decreasing density with increasing size of organisms. *Barkley* (1972) studied the effects of avoidance on the size distribution of the Hawaiian anchovy *Stolephorus purpureus* caught by a 1m^2 ring net and by a plankton purse seine (*Murphy and Clutter*, 1972). The plankton purse seine was assumed to catch everything it encircled, whereas the probability of catching any particular size class of organism in a towed ring net (P_c) was described mathematically as:

$$P_c = 1 - \left(\frac{x_0 \, u_e}{R \sqrt{u^2 - u_e^2}} \right)^2 \qquad (1)$$

where: x_0 is the organisms' mean reaction distance

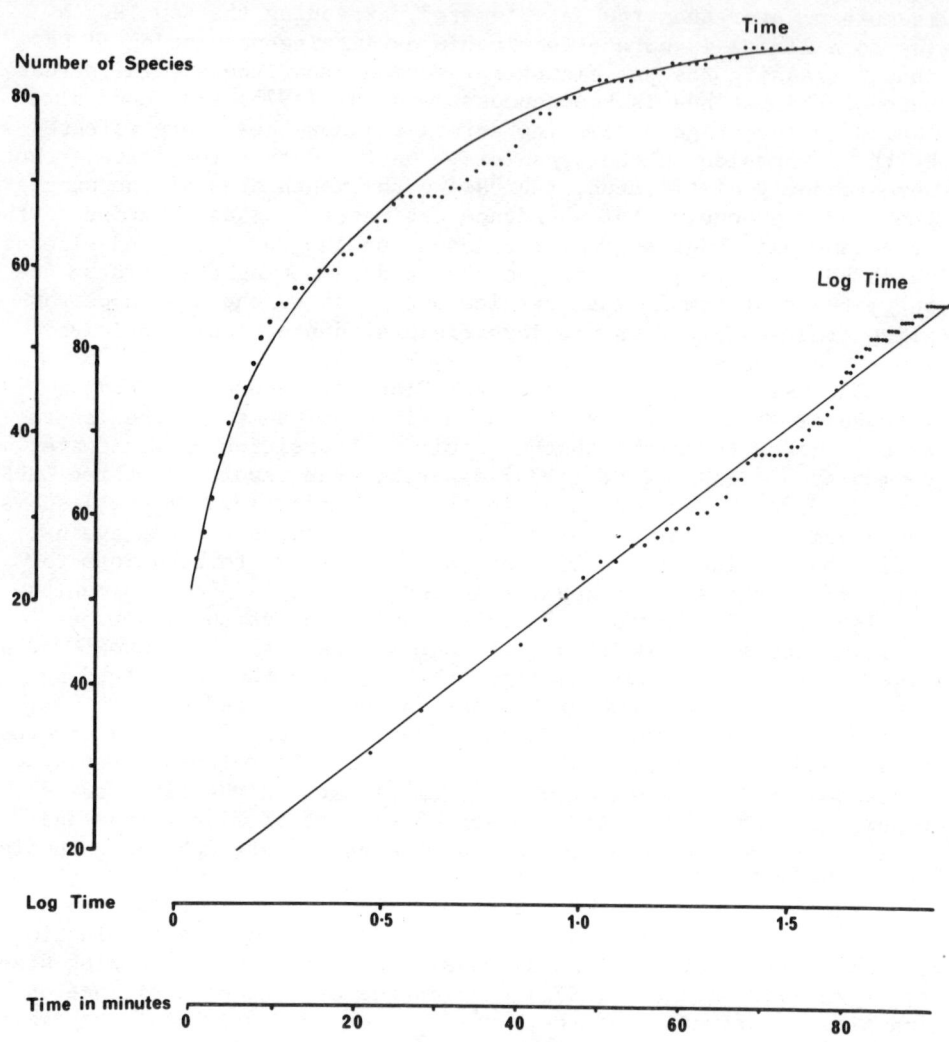

Fig. 1. Species accumulation curves against time and log time of the adult copepods captured in a Longhurst Hardy Plankton Recorder fitted in the codend of a 1m² net; towing speed 2 kts; depth 530-580m 'Discovery' station 7711 haul 49. Position 53°01'N 19°57'W. Time 1724-1924 GMT. Date 25 May 1971. (Fasham et al, 1974).

R is the net's radius

u_e is the organisms' mean escape speed

U is the speed of the net through the water

By making various assumptions, such as an organism's escape speed is proportional to 10 times its length, Barkley adjusted the $1m^2$ net results to account for avoidance. He concluded that the size range could be divided into four categories: (1) anchovies 1.5-2.5mm long which were partially extruded through the meshes but were unable to avoid the net effectively, (2) anchovies approximately 3.5mm long which were adequately retained and unable to avoid, (3) anchovies 4.5-11.5mm long reacting to the net at 3-4m and avoiding the net in proportion to their increasing length, and (4) anchovies 12.5-14.5mm long which were caught more effectively than was predicted. This may have been the result of either their reaction distances being reduced, or a breakdown in the assumption of a linear relationship between body length and escape speed.

Comparisons of ten foot Isaacs Kidd Midwater Trawl (IKMT) catches with the results of the $1m^2$ nets, showed that the IKMT performed less well than expected. The mean reaction distance to the $1m^2$ net was estimated to be 3.3m, in comparison to the 8.2m for the IKMT. Barkley concluded that 98-99% of the reduction in the catch of larger anchovies was the result of avoidance; this was assuming the size frequency was unrelated to the population size. *Aron and Collard* (1969) investigating the effects of net speed on the size distribution of fish caught by a six foot IKMT, found that small specimens (< 42mm) were caught more effectively at lower speeds while large specimens (> 47mm) were caught better at higher speeds. Barkley re-analyzing this data found that the length-frequency curves for the most abundant species *Bathylagus stilbius* was not determined by avoidance as was the case with the anchovies. However, avoidance still accounted for the fact that the faster towed hauls caught more animals overall than the slower. He further analyzed the effectiveness of a mixed mesh Cobb trawl in catching skipjack *Katsuwonus pelamis* and concluded that the fine mesh and coarse mesh parts acted as two nets fishing in tandem. Thus nets of mixed mesh cannot be used to analyze size frequency distributions without appropriate adjustment.

The limitations of mesh size had earlier been extensively reviewed by the SCOR Working Party review of zooplankton sampling (*Tranter*, 1968), in which Clutter summarized all the results on avoidance by zooplankton. Much of it appeared to be contradictory. *Le Brasseur et al.* (1967) found that dark nets caught more than light nets, otherwise most investigations suggest that the groups of small zooplankton such as copepods show little evidence of avoidance; euphausiids show some and fish and decapods much more. *Pearcy and Laurs*, (1966) found marked differences in six foot IKMT catches of fish by day and by night.

In the IOS investigations into zoogeographic variations in vertical distribution of plankton and micronekton using the opening-closing RMT 1+8 system, smaller zooplankton species such as ostracods (Angel, 1969; Angel and Fasham, 1975) and copepods (Roe, 1972) were caught in greater overall numbers by day. In contrast euphausiids (James, personal communication) and decapods (Foxton, 1972, personal communication) are more abundant in night catches. Fish present a more complex situation. At 30°N 23°W (Badcock and Merrett, 1976) the day : night ratios for non-migrant species for which the fishing effort was consistent were as follows Cyclothone braueri 3820:2659, Cyclothone microdon 1786:924, Valenciennellus tripunctulatus 177:459, Argyropelecus hemigymnus 406:203. Valenciennellus was a daytime feeder, but the ratio does not necessarily relate to feeding activity since most migrant species were caught more abundantly at night, when presumably most were feeding. This specific variability occurs in other groups as well. For example, ostracod species caught by day below 500m at 10°30'N 20°W had night catches containing more specimens than day catches: the usual ratio for the group is reversed. These differences can be size related. For example, Badcock and Merrett's raw data for juvenile Macrorhamphosus caught in the RMT 1 give day : night ratios of 1894:1340. When this is adjusted to equalize the fishing effort in the surface layers, the ratio becomes 522:775, the disparity resulting mainly from the smaller daytime catches of the larger size groups, a probable avoidance effect. One aspect that requires investigation is the effect of bioluminescence. Diurnal variations in the in situ luminescence and in the background light intensity may alter the visibility of the net and so change the length of the escape path.

Lower limits of retention are determined by the cross-section area of each individual species. Figure 2 shows the total size spectrum of most of the ostracod population at 50-100m at Ocean Acre close to Bermuda and illustrates how even closely related species vary in the size range retained by the mesh. The population consisted predominantly of two species with narrow, almost non-overlapping, size-frequency distributions, each peak representing different larval stages. Halocypria globosa is nearly spherical and is retained at much smaller size and consequently in larger numbers than the more cylindrical Conchoecia spinirostris.

Size frequency data from repeated tows might be expected to give information on in situ growth rates. Such studies on oceanic organisms have been largely restricted to fish, relying on the immense pool of fishery data. A few have been carried out on other nektonic groups (e. g. Pearcy and Forss, 1969). Figure 3 shows the results of such a study on the decapod Systellaspis debilis. The data is derived from seven repeated series of tows of 2 hours duration with the opening-closing RMT 1+8 from 100m strata at 300-600m from a 1° square centered on 44°N 13°W. Despite the use of tows standardized for length, speed and depth, the variations in total number caught were large. There is evidence of considerable changes in depth

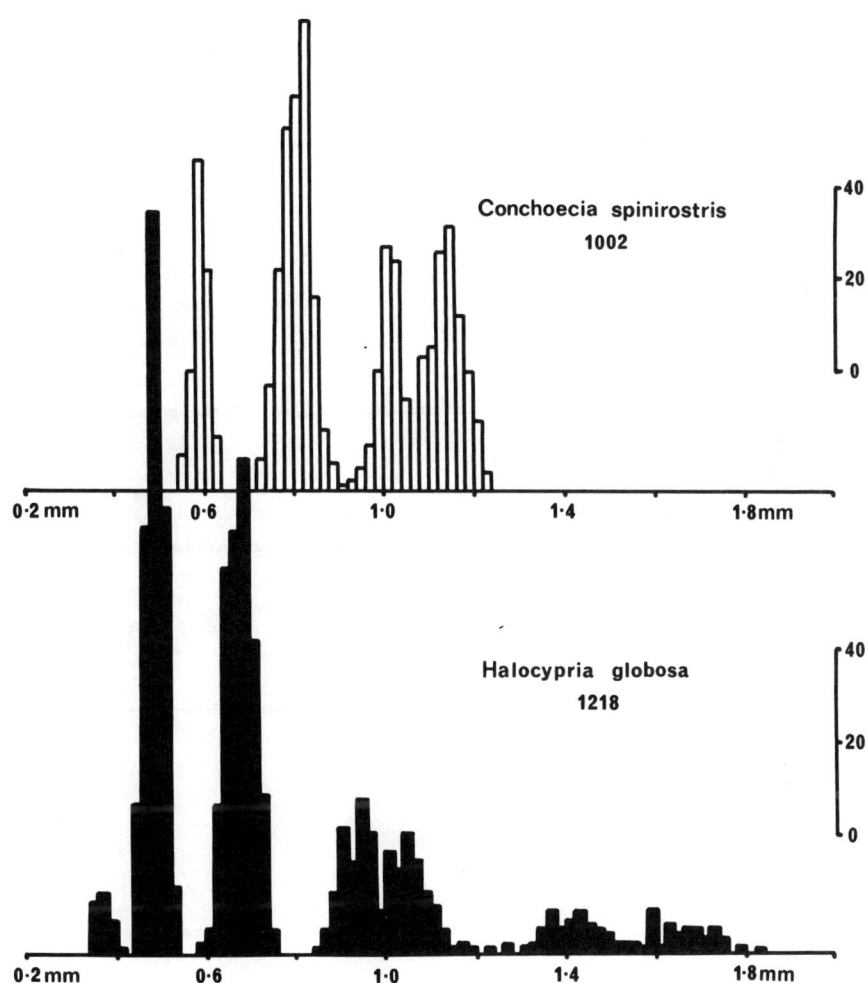

Fig. 2. Size spectrum of the two predominant ostracods (only six specimens belonged to other species) of a 1/8 subsample of an RMT 1 daytime haul taken in the region of Ocean Acre at 52-100m. 'Discovery' station 8281 haul 26. Position 31° 41.9'N 63° 39.5'W. Time 1904-2104 hours GMT. (Local time was GMT - 4 hours). Date 16 March 1973.

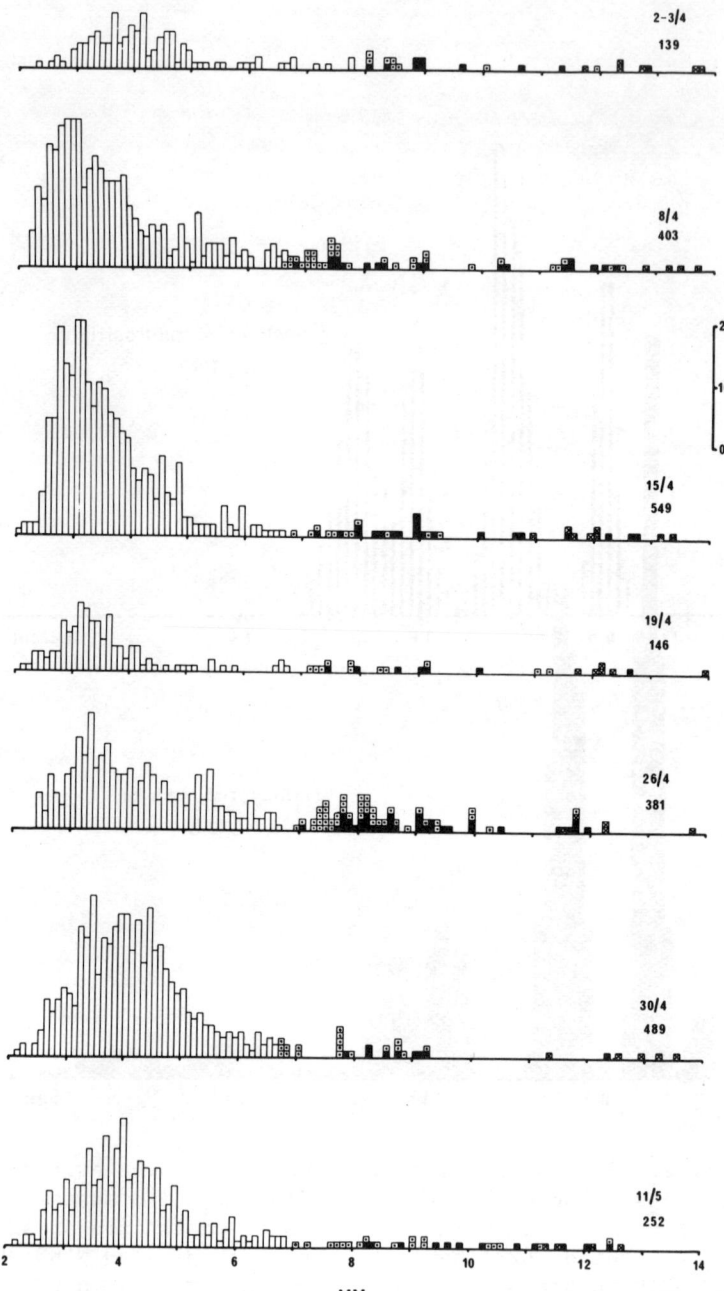

Fig. 3. Size distributions of the carapace lengths of the decapod *Systellaspis debilis* in daytime samples in the seven repeat series of tows from 100 meter horizons at 300-600m taken by R.R.S. Discovery in the region of 44°N 13°W in April - May 1974 in an attempt to study *in situ* growth rates of the mesopelagic fauna. Squares marked with a dot represent immature female species, marked with an x egg-carrying females, and solid squares represent males.

distribution that are size related. Thus, in some of the repeat series, the depth horizon of maximum abundance of some of the size classes may not have been sampled and abundances would have been substantially underestimated. Oblique tows taken immediately after the repeat samples captured too few animals to check the validity of the pooled size spectra. Although the timing of the hauls was kept as consistent as was practical, errors may have been introduced by sampling at different stages of the light cycle. Even variations in cloud cover at the surface may alter the day to day depth distributions of species (see p. 11). In addition, passive horizontal diffusion and accelerated horizontal diffusion produced by vertical migration through a stratified water column with horizontal shears between different levels would result in a constant change in the community sampled (e.g., *Miller*, 1970).

VERTICAL MIGRATION

The space-time relationships of plankton communities have an important effect on their stability (*Steele*, 1974). Vertical migration is an important phenomenon in pelagic communities; *Vinogradov* (1968) has reviewed the recent literature. Work in the N.E. Atlantic (*Angel and Fasham*, 1973, 1974, 1975; *Fasham and Angel*, 1975) suggests that below 700m planktonic species do not perform diel migrations, although many nektonic species do so (*Badcock and Merrett*, 1976; *Foxton*, 1972). *Roe* (1974) studied the planktonic and nektonic community at 30°N 23°W by sampling intensively with eleven repeated tows at 250m over a 24 hour period. He was able to interpret the data in the context of a comprehensive series of hauls taken immediately before, to examine the vertical structure of these communities at the same position. The populations of nektonic species tended to migrate as a unit, whereas the planktonic species tended to smear up through the water column. Planktonic species will therefore become more mixed by the current stress between different depth horizons than will the nektonic ones. The numbers of animals, even when unstandardized for water filtered (based on a flow meter mounted on the net monitor which was used to regulate the speed of the tow), showed remarkably smooth fluctuations in abundance. If patchiness had been a serious source of error, the numbers caught would have been far more erratic. Cyclic variations in avoidance could have produced the smooth fluctuations in abundance but in general the changes appear to have been real and not unduly distorted by sampling error.

From a sampling standpoint Roe's results show several important features. There are continuous and steady fluctuations in the numbers caught. The rate of change varies throughout the day, but as might be expected is generally greatest at dawn and dusk. Thus, in planning sampling programs in which replicated sampling is to be attempted over periods of several days, the comparative hauls should be fished as close to the same time of day as possible. Even so, the population structure of individual species is liable to show bias produced by the differential migration pattern between juvenile stages and males and females, as for example, in a species of halocyprid ostracod (Figure 4). Replicate samples of this species taken at midday would give an inverse sex ratio to samples taken two hours earlier. No matter how free of sampling errors the results can be made, the dynamic nature of the community will introduce bias.

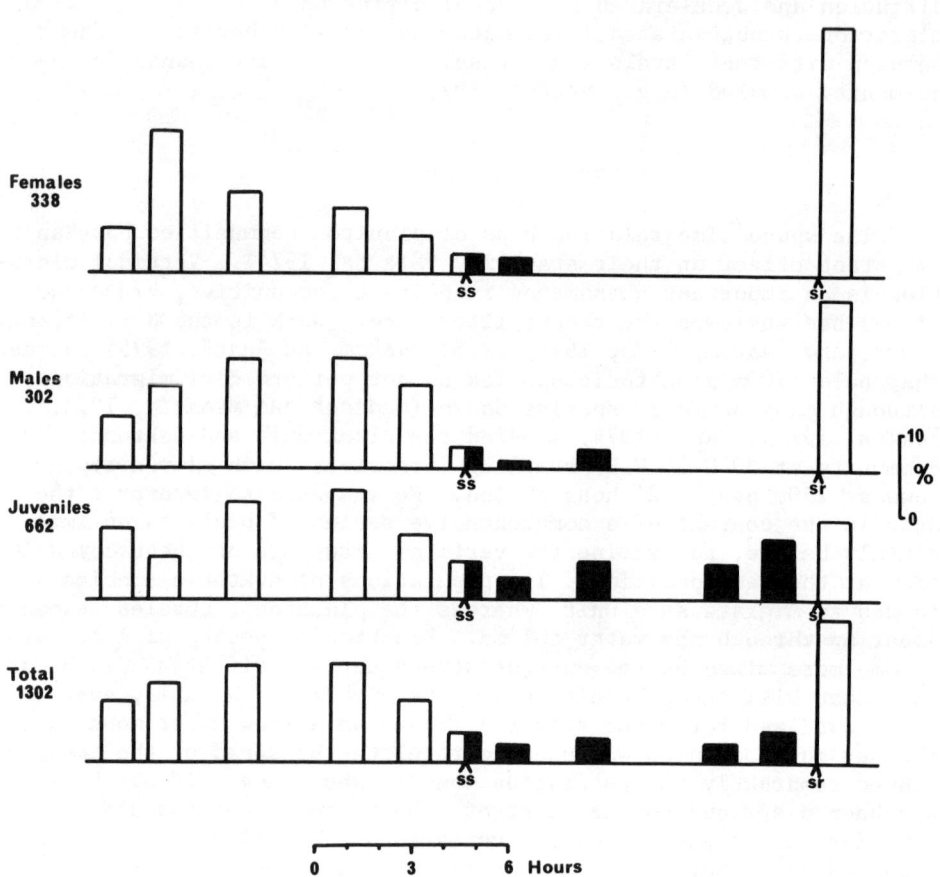

Fig. 4. Percentage occurrence of females, males and juveniles instars of the planktonic ostracod *Conchoecia oblonga* form B in the 24 hour series of repeated tows at a depth of 250m in the region of 30°N 23°W taken by Discovery on 7/8 April 1972 (*Roe*, 1974), showing intraspecific variation in the timing of vertical movements.

An attempt was made to establish some of the trophic relationships within the community sampled during this 24 hour series. Merrett and Roe (1974) analyzed the stomach contents of the seven species of fishes caught in reasonable numbers in the series. There was marked evidence of selectivity in the feeding of several species: *Argyropelecus aculeatus* selected for ostracods, *Valenciennellus tripunctulatus* for calanoid copepods, and *Lampanyctus cuprarius* for amphipods and possibly euphausiids. *Argyropelecus hemigymnus* and *Lobianchia dofleini* were more random in their feeding. Comparison of the time each species was present within the strata showed that only a single fish species was actively exploiting any one resource at any one time. Foxton and Roe (1974) analyzed the decapod stomach contents and also found evidence of some dietary differences between species and in timing of feeding. *Gennadas valens* was the only highly selective predator and fed extensively on Foraminifera. They concluded that many of the identifiable remains of small plankton in the stomach contents were secondarily derived from the diets of the prey. The lower degree of resource partitioning in the Decapoda may be a contributory causal factor of their lower species richness in pelagic ecosystems compared with fishes. Roger (1973) investigated the feeding of euphausiids in the Pacific and noted clear cyclic changes in the intensity of feeding. Vertical migrants fed most intensively at night in the epipelagic zone. The non-migrants tended to have feeding cycles unsynchronized with the light cycles. Thus of the two largest species of *Thysanopoda*, *T. cristata* was a non-migrant and showed maximum incidence of half-full stomachs from 0930-2130h, whereas *T. monacantha*, a migrant, showed evidence of having fed between 0130-0930h. The herbivorous species must have all been migrants, since by day all the species occurred below the euphotic zone. Angel and Fasham (1974) pointed out the spread in size of adult euphausiids occurring in one of their species groupings from the Canary Island region, and suggested this could be associated with resource partitioning.

In April 1974 further experiments were carried out from R.R.S. Discovery to investigate the timing of vertical migration. This time four depths, 100, 250, 450 and 600m were sampled repeatedly for 48 hours in an area 18km in diameter south of the Bay of Biscay. Though the material from the 96 samples is not completely analyzed, the unadjusted data for the ostracods show variations in the numbers caught at 100 and 250m caused by diural migrations (Figure 5). Other factors may have contributed to the variations in overall numbers at the deeper depths. For example there were marked differences in cloud cover and incident light at the surface between the first and second daylight periods of the 600m series. The bright clear conditions during the first daylight period may have induced the observed increase in ostracods caught, compared with the numbers caught on the second day.

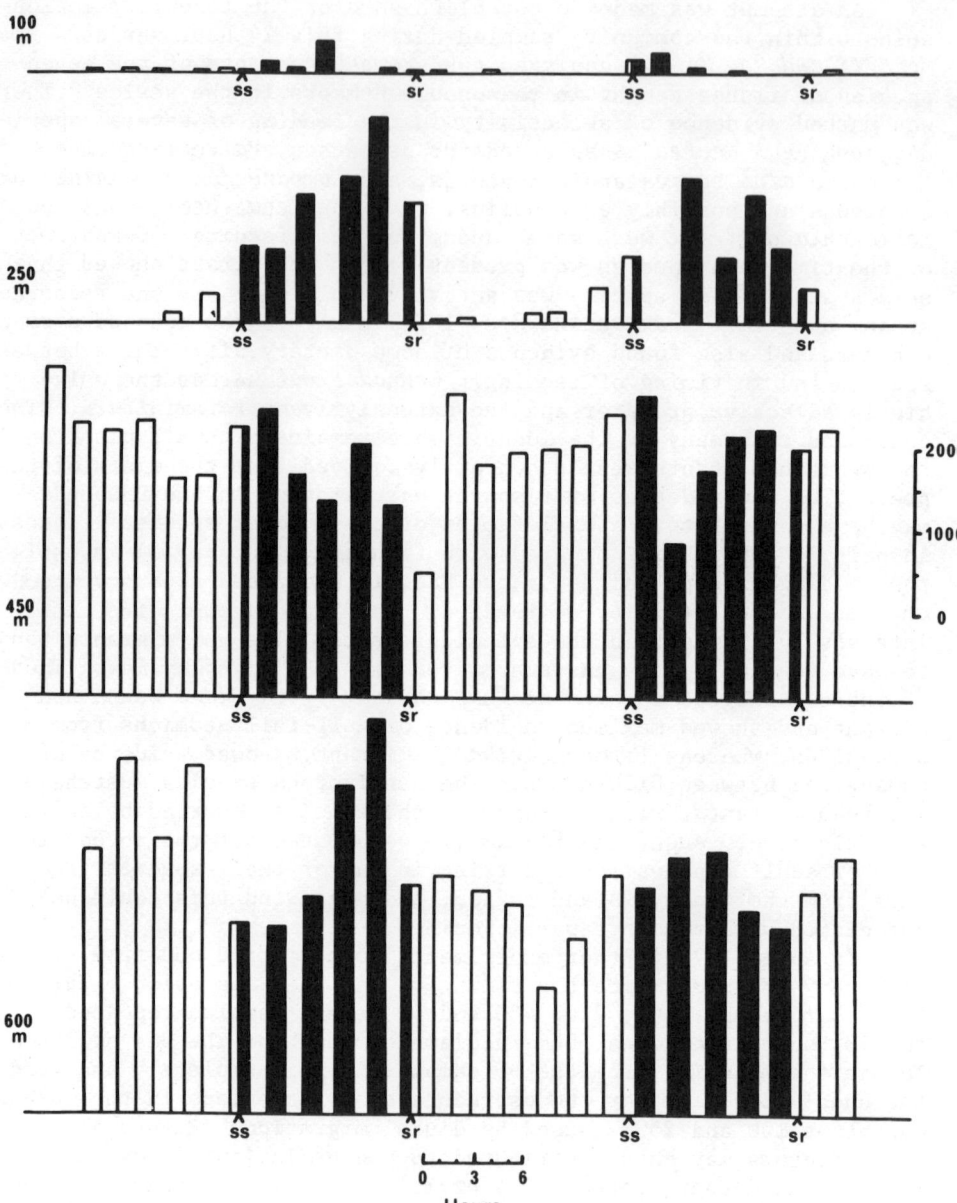

Fig. 5. Histograms of the numbers per haul of planktonic ostracods taken in 48 hours series of one hour tows with an RMT 1 at four different depths 100m, 250m, 450m and 600m, taken in the region of 44°N 13°W between 6-19 April 1974. The figures have not been adjusted for variations in the flow past the net.

ZOOGEOGRAPHIC DISTRIBUTIONS

McGowan (1974) has listed the objectives of biogeographers as
1) to determine what species are present; 2) to describe, quantitatively, their patterns of abundance; 3) to understand what maintains the patterns; 4) to determine how and why the patterns developed; 5) to describe and delineate the communities; and 6) to determine how these community-ecosystems are structured and how they function. Most zoogeographic studies have concentrated on the first two objectives. The aim has been to achieve maximum geographical coverage which has logistically necessitated compromises in the precision of the sampling. Large samples are generally taken which smooth out variability due to small scale patchiness and variations in patterns of depth distribution. The samples are usually sub-sampled and analyzed by "coarse" techniques often based merely on presence and absence of species (e.g., *Fager and McGowan*, 1963).

In studying nektonic groups there is the problem of making adequate collections. This is illustrated by the results of *Backus et al.* (1965) from the Central North Atlantic. They fished open Isaacs-Kidd Midwater Trawls within the zone of the sound-scattering maximum in order to mazimize their fish catches. Along the length of their transect the maximum fishing depth ranged from 42-495m. Using an index based on the first and last occurrence of the fish species, they identified a faunal boundary as occurring in the region of three hauls taken at 70m, 430m and 290m, containing 629 specimens of 31 species, 30 specimens of 13 species and 30 specimens of 15 species respectively. As the hauls were taken at night, *Fasham and Angel* (1975) have pointed out the apparently distinctive fish fauna of the South Atlantic Central Water could on this evidence have been produced by the migration of species from the underlying Antarctic Intermediate Water from below depths of 500-600m. The sample regime permitted the identification of the faunal boundary but not the elucidation of its underlying causes.

The value and sensitivity of such investigations is directly in proportion to the quality of the derived data. Data can be improved by standardizing the techniques employed and by increasing the number of samples. The design of a program is a compromise between the need for comprehensive coverage and logistic limitations. Ideally, the full depth distribution of the dominant species needs to be known at each position. This approach was adopted in the Institute of Oceanographic Sciences investigations of vertical distribution in the N.E. Atlantic with consequent deterioration in geographical coverage. In 1968-72 six stations were worked between 10°30'N and 60°N in the vicinity of the 20°W meridian. Opening and closing nets (RMT 1+8, *Baker et al.*, 1973) were fished to 2000m; the surface 100m subdivided into 0-10m, 10-25m, 25-50m and 50-100m horizons. Hundred meter horizons were fished to 1000m and below 1000m three broad horizons were sampled: 1000-1250m, 1250-1500m, and 1500-2000m. A day series and night series were completed at each position in which the assumption was made that major vertical movements were completed within an hour either side of sunrise and sunset. Data are available for some decapods (Foxton, 1972) and fish (Badcock and Merrett, 1976), and a complete data set is available for planktonic ostracods (Angel and Fasham, 1975; Fasham and Angel, 1975). The planktonic ostracod data has been subjected to a factor analysis to extract objectively the species associations and to group the hauls into those containing similar communities. In Fig. 6 the zonation of the grouping of the hauls is shown for this meridional section. In Fig. 7 the group number of each haul is plotted against the mean TS value for the sampling horizon derived from a STD probe. There is a close but by no means exact relationship between the ostracod fauna and the water masses (Fasham and Angel, 1975). The vertical zonation of the hauls is dependent on whether day or night samples are analyzed, but most of the zoogeographic boundaries are consistent between the day and the night series. The zonation of the hauls is based on changes in the relative abundances of the various species, rather than straight presence of absence.

A similar method of analysis was used to investigate the ecological grouping of deep-sea animals off Southern California by Ebeling et al. (1970) in a transitional faunal zone. Their analyses were based on open IKMT hauls fished at various depths for standard lengths of time. The standard duration of a deep tow was two hours at depth, and yet the longest total fishing time was over six hours. Thus, since the deeper the tow the more impoverished the fauna, the greater is the degree of contamination with species from shallower horizons. Ebeling et al. extracted rather ill-defined groupings which are probably in reality valid, but the inevitable degree of contamination caused by the use of non-open-closing nets introduced a large measure of uncertainty. No matter how sophisticated the statistical techniques used, if the initial sampling techniques could have been improved, the final results can be improved. Open-closing nets should be used in the future whenever logistically possible in all such studies.

A more satisfactory compromise between the need for extensive geographical coverage versus extensive vertical coverage was probably achieved by McGowan and Williams (1973). Working in the North Pacific they subdivided the near-surface water column (0-400m) into seven depth horizons through which they towed paired Bongo nets obliquely. They found that there was a pronounced latitudinal boundary for planktonic species between $38°N$ and $43°N$ with an associated steep gradient in both zooplankton and phytoplankton biomass, i.e., between the Central and Subarctic water masses. This boundary is correlated with pronounced differences in a number of habitat variables. Some of these variables showed seasonality in one water mass but not the other and may have been affected by fundamentally different processes. In the Central water mass, zooplankton biomass, phosphate concentrations and dissolved plus particulate organic carbon showed no seasonal variation in the surface 300m, whereas phytoplankton biomass and nitrite concentrations varied seasonally. In

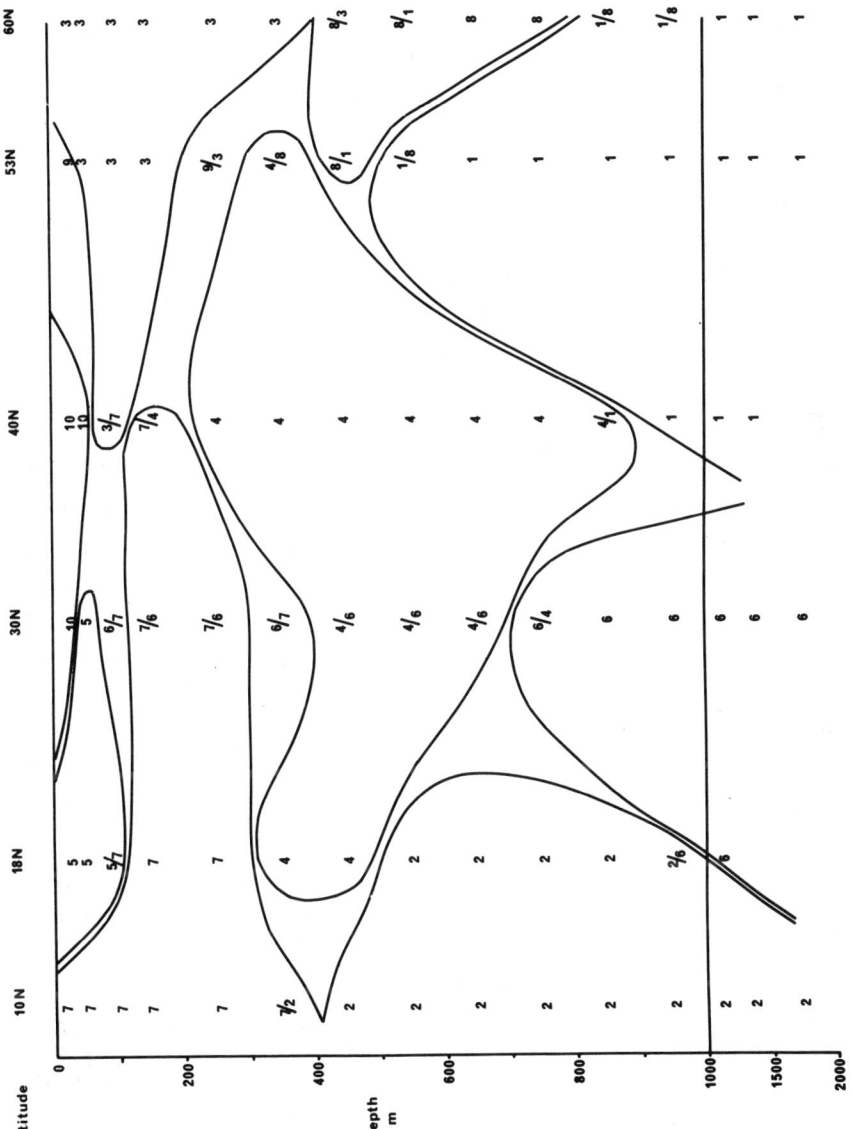

Fig 6. Zoogeographic and day depth zonation of the planktonic ostracod fauna in the N.E. Atlantic derived from a factor analysis of the data set of 63 species in 93 hauls from six stations close to the 20°W meridian at approximately 10° latitude intervals. The analysis grouped together those hauls which contain similar species in similar abundances (Fasham and Angel, 1975).

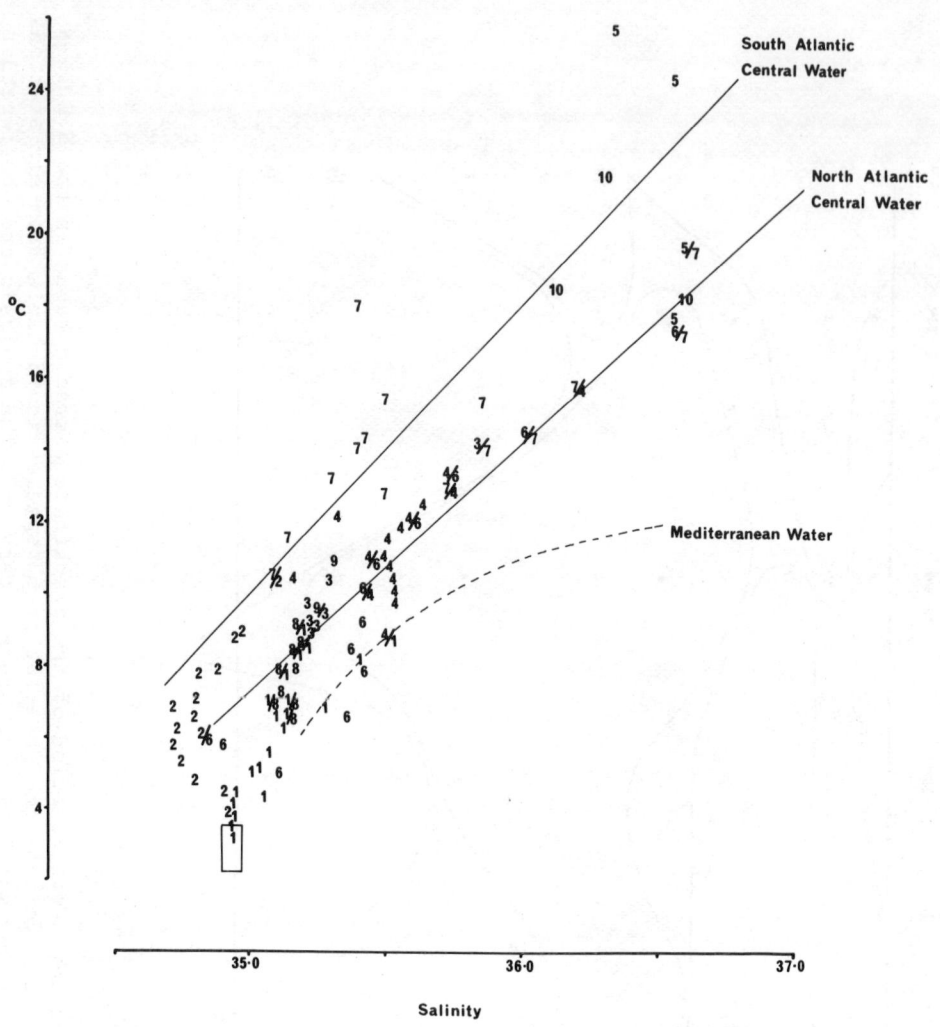

Fig. 7. The relationship between the zonation of the N.E. Atlantic planktonic ostracod fauna and the water masses. The group numbers as shown in Fig. 6 are plotted on the T.S. diagram derived from meansurements by a Bisset Berman STD probe taken during the course of each of the stations.

the Subarctic water mass the phytoplankton standing crop is high and unvarying throughout the year but plant nutrients and zooplankton biomass show strong seasonality. In summer the abundance of the zooplankton correlated with the phytoplankton abundance, but not in winter when it correlated with phosphate concentrations.

McGowan and Williams (1973) analyzed their data by multiple linear regression analysis. This type of analysis and the factor (or principal components) analyses used by others are far more sensitive than analysis by similarity indices based purely on presence or absence. However, the data has to be of high quality and complexity so that sacrifices have to be made in the extent of geographic coverage, degree of synchronization of the sampling and the speed at which samples can be processed. Such compromises may be achieved by using obliquely towed nets to give an integrated sample of the population in the water column (e.g., *Fager and McGowan*, 1963; *Pugh*, 1975). In the continuous plankton recorder survey (*Edinburgh Oceanographic Laboratory*, 1973) only a single depth is sampled; the logistic problems are further reduced by taking small samples, and by saving ship-time through the use of ships of convenience. The danger inherent in such programs is the deterioration in the accuracy of the estimates of species abundances (*Wiebe*, 1971) and in the chances of catching abundant but patchily distributed organisms (*Wiebe and Holland*, 1968, c.f. Fig. 1). No one sampling program can hope to fulfill all of McGowan's objectives listed at the beginning of this section. The types of questions that need to be answered when designing future sampling programs are: 1) what taxonomic groups will provide the optimum information for the zoogeographic problem being investigated? 2) what sampler gives the optimum catches of these groups? 3) how extensive is the geographical coverage to be? 4) what should be the extent of vertical range covered? 5) how much detail of the vertical structure of the community is needed? 6) is seasonal coverage required? 7) what environmental parameters need to be measured in association with the samples? and 8) what statistical analyses will be most appropriate for interpreting the data?

PATCHINESS

Patchiness of plankton is a well documented phenomenon (e.g., *Bainbridge*, 1957; *Cassie*, 1963; *Vinogradov*, 1968; *Fasham et al.*, 1974). It has extremely important ecological connotations. *Adams and Steele* (1966); *Parsons et al.* (1967) and *Paffenhoffer* (1970) have pointed out that the lower feeding threshold of many organisms is higher than the average density of prey organisms, so that aggregations of the prey are essential for the survival of the consumers, let alone their growth and reproduction. One clear example is the dependence of baleen whales on dense shoals of krill or other planktonic or micro-nektonic species (*Nemoto*, 1970). *Steele* (1974) on the basis of a mathematical model of a planktonic ecosystem has suggested that the threshold response coupled with patchiness and other types of heterogeneity of the community can result in "ecological stability" in certain conditions. Patchiness studies are important for understanding both sampling variability and the functioning of the ecosystem.

Bainbridge (1957) summarized many observations on patch size, and found reports of patches ranging from a few centimeters to over

100km. Large patches with dimensions of greater than tens of kilometers can be studied with conventional horizontal towed nets. The Longhurst-Hardy Plankton Recorder (LHPR) (*Longhurst et al.*, 1966) can be used for studying patch sizes of the order of 10-1000m (*Wiebe*, 1970; *Fasham, et al* 1974). One major source of potential error for the LHPR is the hang up of organisms in the netting in front of the recorder (*Haury*, 1973). Haury tested the LHPR by injecting plastic pellets or stained preserved plankton into the mouth of the sampler. Tests of the IOS sampler were carried out by injecting both preserved plankton and living organisms (otherwise not present in the water being fished). The preserved plankton was held up by the water pressure against the netting in front of the recorder, and was then slowly shaken down causing contamination for over 30 minutes of towing. The live organisms passed almost as quickly as expected down into the recorder. So the ability of the live animals to swim off the netting of the concentrater seems to be an important factor in the proper functioning of the device. The net should be designed to have a uniform distribution of pressure and particular attention should be paid to the region where the back pressure from the recorder is liable to increase the tendency for animals to hang up. Similarly, cross-seams in the netting will cause small localized regions of increased water pressure and hence hang up. Towing speeds should be kept comparatively low (speeds of 2 knots, i.e., $\sim$1m/sec have been generally used).

An alternative approach is by using a towed electronic counter similar to a Coulter Counter fitted as a cod-end device (*Boyd*, 1973). Boyd's counter is sensitive to particles with volumes of 0.078-8.69mm^3 equivalent to spheres of radii 0.531-2.55mm. It can discriminate particles separated by more than 15 millisec. at a towing speed of about 4 knots ($\sim$ 2m/sec) i.e. distances of $\sim$ 3cm. Initial results from oblique tows showed clear peaks in abundance at scales equivalent to a second of sampling time i.e. distances of 2m; so hang up does not appear to be a serious problem. The peaks often coincided with physical microstructure of the water. An important question is what are the horizontal dimensions of such peaks, do they represent layers of animals or schools with more spherical dimensions? At present Boyd's device has no collector, so the identity of his particles is unknown.

Pump samples are now being used more frequently in patchiness studies (e.g., *Platt*, 1972, 1974; *Fasham and Pugh*, 1976). Pump systems have been described by *Aron* (1958), *Beers et al.* (1967), and *Icanberry and Richardson* (1973). Fluorometers and particle counters tend to be utilized for quantification of the phytoplankton biomass and the microzooplankton in the water. Volumes of water processed are usually of the order of 100-500 liters per minute; sufficient for adequate samples of microplankton, but not macroplankton. Macroplankton may also be able to respond to the disturbance ahead of the intake and escape, although both *Beers et al.* (1967) and *Icanberry and Richardson* (1973) suggest that pumps are as efficient as nets at capturing organisms returned by nets with 303μ mesh.

Towed fluorometers mounted on undulators are now being used to investigate the horizontal and vertical distribution of chlorophyll and hence phytoplankton biomass (*Platt*, personal communication). This sort of application is most likely to resolve many of the problems involved in sampling the microscale structure of plankton communities. The use of pumps from a drifting ship has provided useful insights (e.g., *Platt*, 1972, 1974; *Fasham and Pugh*, 1976) but comes hard up against our understanding of physical processes within the euphotic zone. So far, patchiness studies using pumps have been almost entirely restricted to the surface 100-200m. However, before considering the possibility of limitations by physical processes on our ability to study biological patchiness, the causal mechanism of the patchiness needs to be discussed.

The dimensions of patches can give an indication of the causal mechanism of patch generation. Causal factors of spatial pattern are: a) vectorial patterns caused by physico-chemical gradients; b) stochastic-vectorial patterns caused by mass transport of water which may be subject to stochastic variations; c) reproductive patterns; d) social patterns, e.g. schooling; e) coactive patterns produced by competition or predation (*Hutchinson*, 1953; *Stavn*, 1971). Vectorial patterns produced by gradients would result in layers of organisms at discontinuities. If the discontinuity is then subjected to internal waves, a horizontally towed sampler will suggest a patchiness related to the wavelength of the internal waves. Turbulence will cause stochastic variations, as will coactive patterns such as shoals of fish eating holes in a layer of plankton. Social patterns are more likely to be limited to size by the attenuation of the signal used to maintain the shoal (sight, chemical messengers, sound or low frequency vibrations). Reproductive patterns will have to be large to persist for long periods under the dispersive influence of horizontal diffusion. *Steele* (1974) developed a mathematical model to investigate the critical patch size of growing phytoplankton. He concluded that a patch had to have a dimension of 10-100km to persist in the face of dispersion by diffusion. He pointed out that in areas like the Sargasso Sea where there is greater seasonal uniformity and the phytoplankton biomass and growth rates are low, these large patches may not be generated. When Steele applied non-linear effects to his models he found that perturbations could pass their effects both up and down the scale of dimension. Thus, small scale patchiness can generate larger scale patches and *vice versa*. A perturbation in this context could be provided by the change in grazing or diet of a developing zooplankton species. *Wroblewski et al.* (1975) elaborated the *Kierstead and Slobodkin* (1953) model which also predicts critical patch sizes. Their model includes the effect of herbivore grazing by including an Ivlev type grazing function, phytoplankton growth rates, and eddy diffusivity. They conclude that a realistic dynamic model also needs to include effects of nutrient limitation.

The importance of turbulent diffusion in the distribution of phytoplankton has only recently been recognized. *Platt* (1972) using a pump sampler and a fluorometer, examined local changes in phytoplankton abundance; spectral analysis of the variations showed a minus five thirds power relationship which is similar to the law deduced on theoretical grounds by *Kolmogorov* (1941) for the spectral shape of homogeneous and isotropic turbulence in three dimensions. *Ozmidov* (1965) has shown that spectral peaks of turbulent energy will be expected at dimensions of 10m, 10km and 1000km corresponding to the main energy input of waves, tidal oscillations and atmospheric cyclonic disturbances respectively.

Okubo (1971, 1974) has usefully summarized present knowledge on vertical and horizontal mixing. Dye diffusion experiments have yielded information about the rates of diffusion. For example, the Rhino experiment used 200kg of dye which was tracked for three weeks when the dye patch had reached an area of 3000km^2. Diffusion velocities were 0.2 - 0.4cm/sec. *Schuert* (1970) followed a dye patch at a depth of 300m and found after two days σ = 550m for the dye patch, a slower rate of horizontal diffusion than at the surface (which if it was extrapolated to 1000m would give σ = 200m after 2 days);

σ is the root mean square of the distribution of the dye in the patch. At the surface, dye patches become elongate with the "head" of the patch being dye that remains at the surface, and the tail becoming subsurface, so that the elongation is a function of the vertical shear. Phytoplankton patches will be similarly affected by passive diffusion, as will zooplankton patches in which the animals are not actively schooling or the distance between animals is out of range of their senses used in communication (e.g., sight, pheromones, or low frequency vibration). Patches with an extensive vertical dimension will rapidly break up. *Webster* (1969) analyzed long series of current records at 6 levels at 10-2000m and found the flows were highly variable and showed low coherence with depth. However, at frequencies with periodicities below 20 hours the energy density followed the minus five thirds power relationship.

The periodicity and persistence of turbulent features increase with size and reports of large mesoscale features are proliferating. For example, *Koshlyakov and Grachov* (1973) describing the Russian polygon experiment reported that two eddies, one 90km in dimension, the other 225km passed through the area during the sampling period. The eddies were obscured in the top 100m. There was a 15km displacement of the axis of the eddies at 1000m compared with 300m. Current eddies spawned from meanders in the Western Sargasso Sea have been tracked for up to 1½ years (*Parker*, 1971). These cold water "rings" are typically 100-180km in diameter at a depth of 200m. *Parker's* (1971) study just south of the Gulf Stream between 60° and 70°W of 62 rings showed that the 17° isotherm subsided at a rate of 0.6m/day, so that the isotherm's depth can be used to age a ring. The "rings" tend to move southwards and once south of about 29°N, they could not be recognized from surface temperature anomalies. *Richardson et al.* (1973) tracked one eddy which moved southwards at a speed of 1 nautical mile per day for two years and finally coalesced with the Gulf Stream again off the coast of Florida. Initially the diameter of the 15°C isotherm at a depth of 500m was 150km, and ringed isotherms at the surface had a diameter of 300km. Temperature anomalies were measured to a depth of 3000m. Barrett (1971) estimated from their energy content that the rings could have a duration of 5 years. Each year about five eddies are spawned off the Gulf Stream.

Lambert et al. (1973) observed a much more elaborate oxygen concentration microstructure within an eddy than outside it when its estimated age was 1½ years. Similar indirect evidence for such features having biological significance is given by Kester et al. (1973) who reported on oxygen minimum layer 0.4ml/l lower than the surrounding water at a depth of 350m, which was 80m thick in the vicinity of 22°N 35°30'W. Its north-south dimension was 37km and its east-west 15km. It moved at a speed of about 5km/day. Such a minimum must have been produced and maintained biologically, and may have been part of an eddy carrying an enriched fauna.

There is no overall agreed physical theory of the dispersal of eddies that allows estimates to be made of their persistence. Sanders (1972) studied sea-surface temperatures and found features with dimensions of 25km persisted for 2-5 days, 50km for 4-12 days and 100km for 8-16 days. Woods (1973) on the basis of a model of turbulence in the thermocline suggested persistences for eddies with characteristic vertical and horizontal dimensions (TABLE 1). Platt et al. (1975) fitted Okubo's (1971) diffusion coefficients to Robinson's (1971) equation for atmospheric predictability, predicting the 50% dilution of an eddy by horizontal diffusion. The disagreement between the values in Table 1 is considerable.

TABLE 1. Characteristic horizontal dimensions and persistence of eddies of various horizontal dimensions from Woods (1973), compared with values of predictability on times for 50% dilution of an eddy by horizontal diffusion derived by extrapolating values of Platt et al. (1975) obtained by fitting Okubo's (1971) diffusion coefficients to Robinson's (1971) equation for atmospheric predictability.

Horizontal Scale	Vertical Scale	Persistence	Predictability (50% dilution)
11.5km	115m	1 month	3 days
1.8km	18m	1 week	17 hours
200m	2m	1 day	2½ hours

It is tempting to relate the results of *Fasham et al.* (1974) in identifying 200m patches (σ = 50m) of mesopelagic plankton to Wood's estimates of the persistence of turbulent eddies. The patches would have a persistence of about a day, an appropriate time scale for a population largely consisting of diurnal migrants. However, *Fasham and Pugh* (1976) have shown such patch sizes can equally well be caused by internal wave effects.

The densities of the organisms were likely to have been too sparse for the patches to be generated by active schooling or any other behavioral mechanism. Further observations on whether the patches change in size or density during the 24 hour cycle might provide a useful insight into the causal mechanism. An additional observation that could be related is the occurrence of columnar scattering phenomena in the wind-mixed layer during very calm weather conditions. Figure 8 shows the scattering above the thermocline on 26 February 1973 from R.R.S. Discovery using a 10 kHz echosounder. The column of scattering appearing at the 0125 time marker in trace A (32°11.7'N, 20°28.2'W), with the ship's speed at 2 knots (∼1m/sec) took about four minutes to pass (i.e., ∼240m across). At 0520 with the ship traveling at 10 knots (∼5m/sec) the columnar scattering still occurred and still showed a maximum dimension of the order of 200m. The columns extended to the seasonal thermocline at 100 fms which was also marked with scattering typical of fish shoals. The columns disappeared with the downward migration which was well underway by 0730 (32°39.9'W) when the ship's speed was still 10 kts. This disappearance may have been caused by the dissipation of the column, or the movement of the organisms to depths where they no longer resonated at 10 kHz. The scattering organisms were either distributed in columns approximately circular in cross-section or in vertical bands. Subjectively the spacing of these features had a regularity; obvious hypotheses are that the scatterers are either responding to convection cells or to internal waves. An appropriate sampler towed through this community would probably yield patch sizes of 200m, either because both the zooplankton and the scatterers responded similarly to the physical processes occurring above the thermocline, or through coactive effects like the predation of the scatterers. The primary cause of the patches would be difficult if not impossible to elucidate. The patches generated at the surface might well persist in downwardly migrating (active or passive) organisms.

There are changes in the phenomena which appear to be the dominant influence in determining the patchiness at different size scales. *Platt* (1974) showed that at dimensions of 0.1 to 10km chlorophyll a (and hence the phytoplankton) behaves as a passive scalar, but at scales of 10 to 1000km it behaves as an active scalar with growth generating patches. Below scales of 0.1km Platt found low coherence between temperature and chlorophyll distribution and so suggests that active processes may be generating the patchiness at small scales. *Fasham and Pugh* (1976) obtained a similar lack of

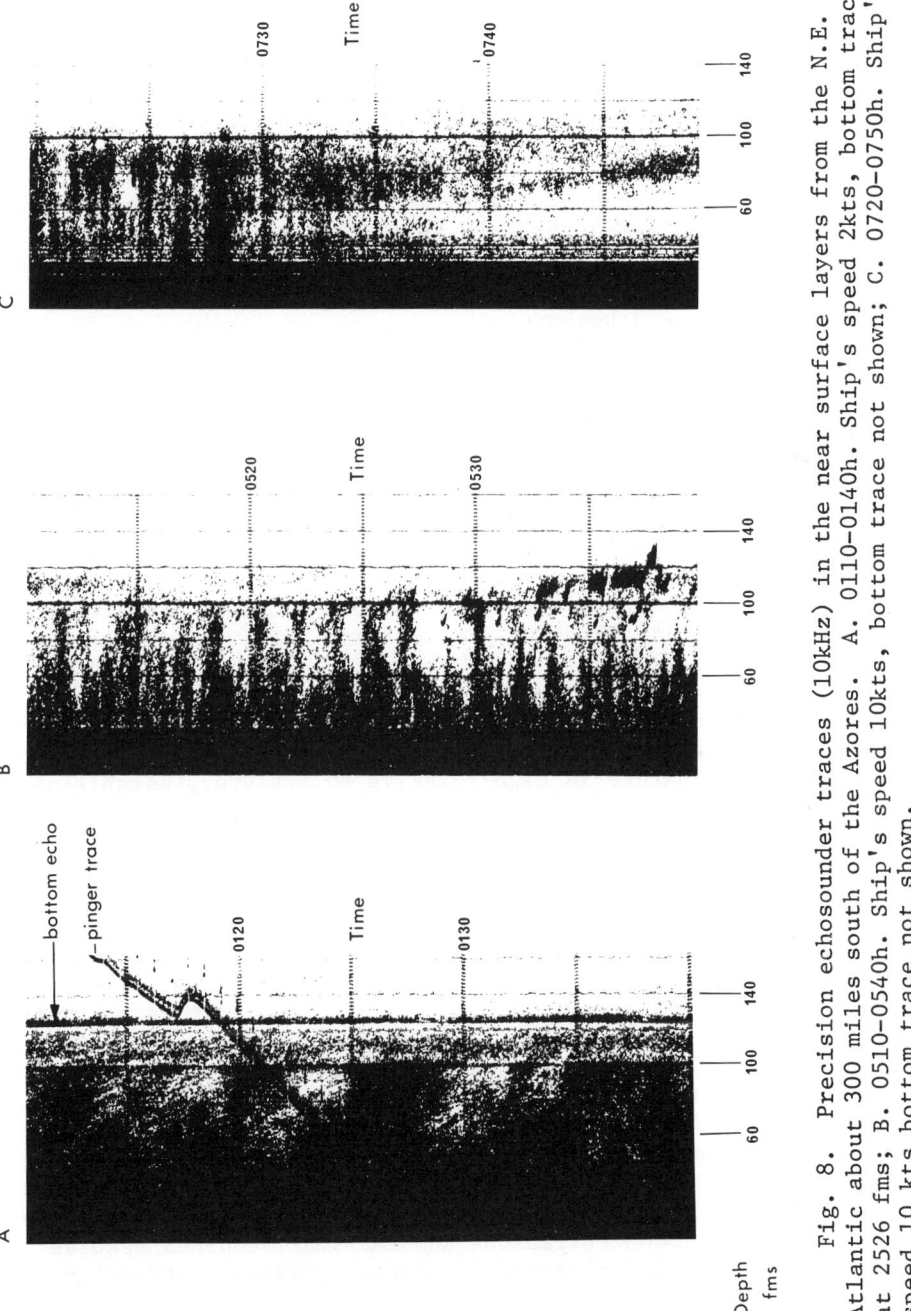

Fig. 8. Precision echosounder traces (10kHz) in the near surface layers from the N.E. Atlantic about 300 miles south of the Azores. A. 0110-0140h. Ship's speed 2kts, bottom trace at 2526 fms; B. 0510-0540h. Ship's speed 10kts. Bottom trace not shown; C. 0720-0750h. Ship's speed 10 kts bottom trace not shown.

coherence at small scales in their pump experiments at 8°45'N 23°W off the N.W. African coast. However, they found that the coherence was maintained during the period of the spring bloom and incipient stratification in the Bay of Biscay down to dimensions of 40m. Even more interesting was the maintenance of the coherence at length scales in excess of 1km. This gives support to *Platt and Denman's* (1974) first order approximation for the critical length scale below which population fluctuations will be damped out by horizontal turbulence. This suggests that when the phytoplankton growth rate balances or exceeds the herbivore grazing rate the critical length scale becomes very large. Hence the coherence between phytoplankton and temperature is maintained during bloom conditions in the Bay of Biscay at scales greater than 1km. In areas of high tidal mixing there may be no coherence between the microstructure of phytoplankton and temperature.

Fasham and Pugh (1976) further point out that peaks in their chlorophyll spectra occurred at length scales of 400m off the N.W. African Coast and 200m off N.W. Spain. These are consistent with internal waves being the dominant feature of the observed spatial heterogeneity and producing peaks at the local Brunt-Väisälä frequencies. One of the technical problems that are important in determining the causality of microscale patchiness, i.e. at scales of less than 500m, is in distinguishing the possible effects of turbulence and internal waves. *Fasham and Pugh* (1976) point out that there are two possible approaches, either to tow two samplers simultaneously separated vertically by about 10m, or else to tow a sampler that follows an isotherm.

Okubo (1974) points out that although diffusion is usually considered to result in great homogeneity, "diffusion-induced-instability" is a possible explanation of some patchiness. If the diffusion rate of predators is high relative to its prey, patchiness will result. This is precisely the situation with herbivore populations which vertically migrate through current shears, grazing on phytoplankton in the wind mixed layer. *Isaacs et al.* (1974) have pointed out that the diurnal vertical movement of DSL's is related to light intensity and so they move deeper in clearer than in turbid waters. Differential current shears increase with depth so *Isaacs et al.* point out that statistically the organisms of the DSL will tend to move into more turbid and hence more productive regions. It will also mean that diffusion-induced-instability will tend to be more important in poorly productive areas, precisely in those which Steele points out will not favor growth induced patchiness.

Steele (1974) considered that investigations into the correlations between spatial patterns of organisms of different trophic levels are critical to the validity of his concepts. *Fasham and Pugh* (1976) present some preliminary results showing coherence between chlorophyll concentrations and the small copepod species *Clausocalanus* sp. and *Paracalanus parvus*. Extending these observations to larger organisms will involve problems with the windows in the size spectrum of the samplers; either pump samplers will need to b designed to catch larger organisms, or zooplankton samplers such as the LHPR or the Boyd counter will need to sample smaller organisms, so that there is no gap in the size spectrum adequately sampled by the systems. Platt (1974) points out that there is a serious lack in our knowledge about longevity of patchiness at the critical mesoscale of 0.1 to 10km. Once the longevity of patches approaches the generation time of zooplankton it will have important connotations in secondary production. Breeding strategies of zooplankton species may well be related to the scale of heterogeneity of the environment, with rapid breeders favored in regions where the longevity of patches is about a month, but slower breeders in homogeneous or regions of large scale long-lived patches.

Off the coast of Peru, Beers et al. (1971) were able to follow two patches, detected by higher nutrient levels, higher chlorophyll concentrations and lower surface temperatures. Despite the actively growing phytoplankton populations (mostly small flagellates with doubling times of 1.4 days) and only about 25% of the production being removed by zooplankton grazing, the patches lost their identities in only 2-3 days. The first patch, when initially observed, was elliptical with an east-west dimension of 10km and a north-south one of 5km. The chlorophyll concentration was elliptical but with the axes of about 6 x 5 km rotated clockwise 45°. After four days the patch was again elliptical with axes of about 33½ x 20 km, an increase in area from about 100km^2 to 2100km^2.

In contrast, Cushing (1963) in summarizing the results of following a patch of Calanus in the North Sea for 66 days concluded that the reduction in density observed due to diffusion was very low. Grazing proved to be the dominant cause of phytoplankton mortality during the Spring bloom, although losses by sinking and diffusion were probably important even if low. So zooplankton patchiness may prove to be less influenced by physical processes than phytoplankton patchiness.

Miller (1970) attempted to follow a migratory zooplankton population by following a vertically migrating drogue. He found there was closer similarity between the populations sampled round the drogue than at a fixed point, but even so there was a progressive diminution in the similarities throughout the course of the observations. Any such attempts at in situ measurements of ecological parameters by taking repeated samples will yield samples that are representative of an ever increasing area as the sampling progresses. Using the more optimistic estimates of the persistence of turbulent eddies (TABLE 1), samples taken a month apart in the same body of water would represent a community from a patch 11.5km across, even supposing passive diffusion is the only dissipating mechanism. Thus the growth results presented in Figure 3 collected over a seven week period will only provide sensible estimates if no boundary conditions impinged into the total area from which diffusion could have occurred. Woods' estimate might set this critical area at around 500km^2. However, Folsom and Vine (1957) described the horizontal spread of a radioactive tracer over an area of 40,000km^2 in 40 days. Extrapolating the rate of spread of a patch observed by Beers et al. (1971) off Peru which increased in area from 100km^2 to about 2000km^2 in 4 days, an area of 40,000km^2 would have been covered in about 25 days. The ecologist has little chance of obtaining field data of any parameter with a long time scale, unless his samples can be considered to be representative of a vast area.

There seems no way that a circumscribed population can be followed for any length of time, in the oceanic environment, because it is dispersed and diluted too rapidly by diffusive processes. These temporal limitations are likely to make it impossible to improve field data.

Ecological parameters with long time scales will have to be measured in controlled laboratory conditions, or in enclosures where the diffusive effects are limited to the extent of the enclosure. Field data is probably best utilized in this context to test mathematical models. Models can be used to generate hypotheses which can then be tested within the "windows" of the various sampling devices. In their review, Platt et al. (1976) critically examine the approach to modeling primary production and emphasize the importance of defining the aims of a model as to whether it is descriptive, predictive or explanatory. Models are more useful if the parameters used have a possible ecological interpretation rather than a purely abstract one. They concluded that attempts to predict biomass in the sea by continuous integrations of a comprehensive production model over extended time periods cannot be expected to yield reliable results, due to fundamental theoretical obstacles. In atmospheric predictions this time limit is possibly in the region of two weeks. Even so, Dubois (1975a,b) using a non linear model of horizontal distribution of prey-predator plankton populations and taking into account residual currents and eddy diffusivity and assuming Lotka-Voltera type interactions has simulated patch changes that closely parallel observed distributions in the southern North Sea. Phytoplankton blooming starts as a steadily increasing circular disc. Once the time lag for the start of the response of the herbivores is completed, the center is eaten out, converting the phytoplankton patch into a ring structure with patches of high density around the circumference. The model appears to be a close analogue of the natural events (Wyatt, 1973).

So modeling, like any other technique, has boundaries to its discrimination, but this does not detract from the power of its proper application. Models such as those of Platt (1974) and Steele (1974) are already generating problems to be solved and ideas to be tested.

There are unresolved questions of the persistence of patches: how does patchiness vary with depth and with latitude? how long will expatriate populations entrained in an eddy maintain an ecological identity? will they disperse more rapidly than the eddy dissipates? do vertical migrants disperse more rapidly than non-migrants? if such expatriates do maintain themselves, by what mechanism? how are faunal boundaries maintained? Field sampling programs must be designed to answer such problems within the limitations of the samples available, or new samplers will need to be designed with different characteristics. However, added sophistication of techniques, for example, by adding extra sensors to a sampler, will only be useful if the additional information is relevant to the sampler's windows of discrimination.

Elton (1966, p. 31) considered that because plankton has no cover structure in which predators hunt for prey (i.e., there is no habitat structure in the terrestrial sense) planktonic communities are profoundly different in organization from terrestrial and "most aquatic communities". Possibly a more important influence is the uncertainty introduced by turbulent mixing and horizontal diffusion.

REFERENCES

Adams, J. A. and J. H. Steele. 1966. Shipboard experiments on the feeding of *Calanus finmarchicus (Gunnerus)*. Pp. 19-35 In *Some Contempory Studies in Marine Science*, Ed. H. Barnes. Allen and Unwin London.

Angel, M. V. 1969. Planktonic ostracods from the Canary Island region: Their depth distribution, diurnal migrations and community organization. *J. mar. biol. Ass. U.K.* 49: 515-553.

Angel, M. V. and M. J. R. Fasham. 1973. SOND Cruise 1965: Factor and cluster analyses of the plankton results, a general summary. *J. mar. biol. Ass. U.K.* 53: 185-231.

Angel, M. V. and M. J. R. Fasham. 1974. SOND Cruise 1965. Further factor analyses of the plankton data. *J. mar. biol. Ass. U.K.* 54: 879-894.

Angel, M. V. and M. J. R. Fasham. 1975. Analysis of the vertical and geographic distributions of the abundant species of planktonic ostracods in the Northeast Atlantic. *J. mar. biol. Ass. U.K.* 55: 709-737.

Aron, W. 1958. The use of a large capacity portable pump for plankton community sampling with notes on plankton patchiness. *J. Mar. Res.* 16: 158-173.

Aron, W. and S. Collard. 1969. A study of the influence of net speed on catch. *Limnol. Oceanogr.* 14 (2): 242-249.

Backus, R. H., G. W. Mead, R. L. Haedrich, and A. W. Ebeling. 1965. The mesopelagic fishes collected during cruise 17 of the R/V CHAIN with a method for analyzing faunal transects. *Bull. Mus. Comp. Zool.* 134: 139-158.

Backus, R. H., J. E. Craddock, R. L. Haedrich, and D. L. Shores. 1970. The distribution of mesopelagic fishes in the Equatorial and Western North Atlantic Ocean. *J. Mar. Res.* 28: 179-201.

Badcock, J. R. and N. R. Merrett. 1976. Midwater fishes in the eastern North Atlantic. 1. Vertical distribution and associated biology in 30°N 23°W, with developmental notes on certain myctophids. *Progress in Oceanography*, 7: 3-58.

Bainbridge, R. 1957. The size, shape and density of marine phytoplankton concentrations. *Biol. Rev. 32:* 91-115.

Baker, A. de C., M. R. Clarke, and M. J. Harris. 1973. The N.I.O. combination net (RMT 1+8) and further developments of rectangular midewater trawls. *J. mar. biol. Ass. U.K. 53:* 167-184.

Barkley, R. 1972. Selectivity of towed net samplers. *Fishery Bull. 70:* 799-820.

Barrett, J. R. 1971. Available potential energy of gulf stream rings. *Deep-Sea Res. 18:* 1221-1231.

Beers, J. R., M. R. Stevenson, R. W. Eppley, and E. R. Brookes. 1971. Plankton populations and upwelling off the coast of Peru, June 1969. *Fishery Bull. 69:* 859-876.

Beers, J. R., G. L. Stewart, and J. D. H. Strickland. 1967. A pumping system for sampling small plankton. *J. Fish. Res. Bd Can. 24:* 1811-8.

Boyd, C. M. 1973. Small scale spatial patterns of marine zooplankton examined by an electronic *in situ* zooplankton detecting device. *Netherlands J. Sea Res. 7:* 103-111.

Cassie, R. M. 1963. Microdistribution of plankton. In: *Oceanography and marine biology,* Pp. 223-252, Ed. H. Barnes. Allen and Unwin, London. Vol. 1.

Chanut, J. P. 1975. Simulation d'un plan d'experience en oceanographie: effets de la dispersion des organismes sur l'erreur d'echantillonage. *J. exp. Mar. Biol. Ecol. 17:* 239-260.

Cushing, D. H. 1963. Studies on a *Calanus* patch. V. The production cruises in 1954: summary and conclusions. *J. mar. biol. Ass. U.K. 43:* 387-389.

Dubois, D. M. 1975a. Simulation of the spatial structuration of a patch of prey-predator plankton populations in the Southern Bight of the North Sea. *Mem. Soc. Roy. Sci. Liege, 6th Series, 7:* 75-82.

Dubois, D. M. 1975b. A model of patchiness for prey-predatory plankton populations. *Ecol. Modelling, 1:* 67-80.

Ebeling, A. W., R. M. Ibara, R. J. Lavenberg, and F. J. Rohlf. 1970. Ecological groups of deep sea animals off southern California. *Science, 154:* 1-43.

Edinburgh Oceanographic Laboratory. 1973. Continuous plankton records: a plankton atlas of the North Atlantic and the North Sea. *Bull. Mar. Ecol., 7:* 1-174.

Elton, C. S. 1966. The pattern of animal communities. Methuen, London. 432 pp.

Fager, E. W. and J. A. McGowan. 1963. Zooplankton species groups in the North Pacific. *Science, 140:* 453-460.

Fasham, M. J. R., M. V. Angel, and H. S. J. Roe. 1974. An investigation of the spatial pattern of zooplankton using the Longhurst-Hardy Plankton Recorder. *J. exp. mar. Biol. Ecol. 16:* 93-112.

Fasham, M. J. R. and M. V. Angel. 1975. The relationship of the zoogeographic distributions of the planktonic ostracods in the Northeast Atlantic to the water masses. *J. mar. biol. Ass. U.K. 55:* 739-757.

Fasham, M. J. R. and P. R. Pugh. 1976. Observations on the horizontal coherence of chlorophyll a and temperature. *Deep-Sea Research* 23: 527-538.

Folsom, T. R. and A. V. Vine. 1957. On the tagging of water masses for the study of physical processes in the ocean. *Natl. Acad. Sci. - Nat. Res. Council, Publ.* 551: 121-132.

Foxton, P. 1972. Observations on the vertical distribution of the genus *Acanthephyra* (Crustacea: Decapoda) in the eastern North Atlantic, with particular reference to species of the "*purpurea*" group. *Proc. Roy. Soc. Edinb.* B, 73: 301-313.

Foxton, P. and H. S. J. Roe. 1974. Observations on the nocturnal feeding of some mesopelagic decapod Crustacea. *Mar. Biol.* 28: 37-49.

Gibbs, R. H. and C. F. E. Roper. 1970. Ocean Acre: Preliminary report on vertical distribution of fishes and cephalopods. In: *Proceedings of an International Symposium on Biological Sound Scattering in the Ocean.* Ed. G. Brooke Farquhar. Maury Center for Ocean Science, Washington, 1970. Pp. 119-133.

Haury, L. 1973. Sampling bias of a Longhurst-Hardy Plankton Recorder. *Limnol. Oceanogr.* 18: 500-506.

Hutchinson, G. E. 1953. The concept of pattern in ecology. *Proc. Acad. Nat. Sci. Phila.* 105: 1-12.

Icanberry, J. W. and R. W. Richardson. 1973. Quantitative sampling of live zooplankton with a filter pump system. *Limnol. Oceanogr.* 18: 333-5.

Isaacs, J. D., S. A. Tont, and G. L. Wick. 1974. Deep scattering layers: vertical migration as a tactic for finding food. *Deep-Sea Res.* 21: 651-656.

Kester, D., K. T. Crocker, and G. R. Miller. 1973. Small scale oxygen variation in the thermocline. *Deep-Sea Res.* 20: 409-412.

Kierstead, H. and L. B. Slobodkin. 1953. The size of water masses containing plankton blooms. *J. Mar. Res.* 12: 141-147.

Kolmogorov, A. N. 1941. The local structure of turbulence in an incompressible viscous fluid for very large Reynolds numbers. *Dokl. Akad. Nauk U.S.S.R.* 30: 299-303.

Koshlyakov, M. N. and Y. M. Grachov. 1973. Meso-scale current at a hydrophysical polygon in the tropical Atlantic. *Deep-Sea Res.* 20: 507-526.

Lambert, R. B. 1974. Small scale dissolved oxygen variations and the dynamics of gulf stream eddies. *Deep-Sea Res.* 21: 529-546.

LeBrasseur, R. I, C. D. McAllister, J. D. Fulton, and O. D. Kennedy. 1967. Selection of a zooplankton net for coastal observations. *Fish. Res. Bd. Can. Tech. Rep.* 37: 13pp.

Longhurst, A. R., A. D. Reith, R. E. Bower, and D. L. R. Seibert. 1966. A new system for the collection of multiple serial plankton samples. *Deep-Sea Res.* 13: 213-222.

McGowan, J. A. 1971. Oceanic biogeography of the Pacific. In: *The Micropaleontology of Oceans*, pp. 3-74. Eds. B. M. Funnell and W. R. Riedel, Cambridge Univ., Press, Cambridge.

McGowan, J. A. 1974. The nature of oceanic ecosystems. In: *The Biology of the Oceanic Pacific*, pp. 9-28. Ed. J. A. McGowan, Oregon State University Press.

McGowan, J. A. and V. J. Farundorf. 1966. The relationship between size of net used and estimates of zooplankton diversity. *Limnol. Oceanogr.* 11: 456-469.

McGowan, J. A. and P. M. Williams. 1973. Oceanic habitat differences in the North Pacific. *J. exp. mar. Biol. Ecol.* 12: 187-217.

Merrett, N. R. and H. S. J. Roe. 1974. Patterns and selectivity in the feeding of certain mesopelagic fishes. *Mar. Biol.* 28: 115-126.

Miller, C. B. 1970. Some environmental consequences of vertical migration in marine zooplankton. *Limnol. Oceanogr.* 15: 727-741.

Murphy, G. I. and R. I. Clutter. 1972. Sampling anchovy larvae with a plankton purse seine. *Fishery Bull.* 70: 789-798.

Nemoto, T. 1970. Feeding pattern of baleen whales in the ocean. In: *Marine Food Chains*, pp. 241-252. Ed. J. H. Steele. Oliver and Boyd, Edinburgh, 552 pp.

Okubo, A. 1971. Horizontal and vertical mixing in the sea. In: *Impingement of Man on the Oceans*, pp. 89-168. Ed. D. W. Hood, Wiley-Interscience New York, 1971.

Okubo, A. 1974. Diffusion-induced instability in model ecosystems: another possible explanation of patchiness. *Chesapeake Bay Institute, Johns Hopkins University, Technical Report*, 86: 17pp.

Ozmidov, R. V. 1965. Energy distributions between oceanic motions of different scales. *Izvestiya Atmospheric and Oceanic Physics*, 1: 257-261.

Paffenhofer, G. A. 1970. Cultivation of *Calanus helgolandicus* under controlled conditions. *Helgolander wiss. Meeresunters.* 20: 346-359.

Parker, C. E. 1971. Gulf Stream rings in the Sargasso Sea. *Deep-Sea Res.* 18: 981-994.

Parsons, T. R., R. J. LeBrasseur, and J. D. Fulton. 1967. Some observations on the dependence of zooplankton grazing on the cell size and concentration of phytoplankton blooms. *J. Oceanogr. Soc. Japan*, 23: 10-17.

Pearcy, W. G. and R. M. Laurs. 1966. Vertical migration and distribution of mesopelagic fishes off Oregon. *Deep-Sea Res.* 13: 153-165.

Pearcy, W. G. and C. A. Forss. 1969. The oceanic shrimp *Sergestes similis* off the Oregon Coast. *Limnol. Oceanogr.* 14: 755-765.

Platt, T. 1972. Local phytoplankton abundance and turbulence. *Deep-Sea Res.* 19: 183-187.

Platt, T. 1974. The physical environment and spatial structure of phytoplankton populations. *Mem. Soc. Roy. Sci. Leige, 6th Series* 8: 9-17.

Platt, T., K. L. Denman, and A. D. Jassby. In press. Modeling the productivity of phytoplankton. In: *Sea, Vol. 6*, (Eds.) E. D. Goldberg, J. J. O'Brien, I. McCave, J. Steele. Hill, 807-856.

Wiley-Interscience, New York (in press).

Pugh, P. R. 1974. The vertical distribution of the siphonophores collected during the SOND Cruise 1965. *J. mar. biol. Ass. U.K.* 54: 25-90.

Pugh, P. R. 1975. The distribution of siphonophores in a transect across the North Atlantic Ocean at 32°N. *J. exp. Mar. Biol. Ecol.* 20: 77-97.

Richardson, P. L., A. E. Strong, and J. A. Knauss. 1973. Gulf streams eddies: Recent observations in the western Sargasso Sea. *J. Phys. Oceanogr.* 3: 297-301.

Robinson, G. D. 1971. The predictability of a dissipative flow. *Quart. J. Roy. Meteorol. Soc.* 97: 300-312.

Roe, H. S. J. 1972. The vertical distribution and diurnal migrations of calanoid copepods collected on the SOND Cruise 1965. 1. The total population and general discussion. *J. mar. biol. Ass. U.K.* 52: 277-314.

Roe, H. S. J. 1974. Observations on the diurnal vertical migrations of an oceanic animal community. *Mar. Biol.* 28: 99-113.

Roger, C. 1973. Recherches sur la situation trophique d'un groupe d'organismes pélagiques (Euphausiacea). II. Compartements nutritionnels. *Mar. Biol.* 18: 317-320.

Sanders, P. M. 1972. Space and time variability of temperature in the upper ocean. *Deep-Sea Res.* 19: 467-480.

Schuert, A. E. 1970. Turbulent diffusion in the intermediate waters of the North Pacific Ocean. *J. Geophys. Res.* 75: 673-682.

Stavn, R. H. 1971. The horizontal-vertical distribution hypothesis: Langmuir circulations and *Daphnia* distributions. *Limnol. Oceanogr.* 16: 453-466.

Steele, J. H. 1974. Stability of plankton ecosystems. In: *Ecological Stability*, Pp. 179-191. Ed M. B. Usher and M. H. Williamson. Chapman Hall Ltd. London, 1974.

Tranter, D. J. (Ed.). 1968. Reviews on zooplankton sampling methods. Pt. 1. of "Zooplankton Sampling". UNESCO Paris 1968.

Vinogradov, M. E. 1968. Vertical distribution of the oceanic zooplankton. 339pp. Moscow: Nauka (Translation from the Russian. Jerusalem: Israel Program for Scientific Translation, 1970).

Webster, F. 1969. Vertical profiles of horizontal ocean currents. *Deep-Sea Res.* 16: 85-98.

Wiebe, P. H. 1970. Small scale spatial distribution in oceanic zooplankton. *Limnol. Oceanogr.* 15: 205-217.

Wiebe, P. H. 1971. A computer model study of zooplankton patchiness and its effects on sampling error. *Limnol. Oceanogr.* 16: 29-38.

Wiebe, P. H. and W. R. Holland. 1968. Plankton patchiness: effect on repeated net tows. *Limnol. Oceanogr.* 13: 315-321.

Woods, J. D. 1973. Space-time characteristics of turbulence in the seasonal thermocline. *Mem. Soc. Roy. Sci. Liege, 6th series*, 6: 109-130.

Wroblewskii, J. S., J. J. O'Brien, and T. Platt. 1975. On the physical and biological scales of phytoplankton patchiness in the ocean. *Mem. Soc. Roy. Sci. Liege, 6th series,* 7: 43–57.

Wyatt, T. 1973. The biology of *Oikopleura dioica* and *Fritillaria borealis* in the Southern Bight. *Mar. Biol.* 22: 137–158.

On the Distribution of Midwater Fishes in the Eastern North Atlantic

Julian Badcock and N. R. Merrett

Institute of Oceanographic Sciences, United Kingdom

ABSTRACT

From 1968-72 investigations into the vertical distribution of planktonic and micronektonic animals in the eastern North Atlantic were carried out at approximately 10° intervals of latitude from 11°N to 60°N near the 20°W meridian. A standard strategy of sampling contiguous depth strata was developed by using an acoustically operated and monitored, mouth opening/closing, combination net, the RMT 8, to permit certain intra- and inter-station comparisons. Preliminarily, the midwater fishes from the upper 1000m of the water column are broadly categorized into those with and those without gas-filled swimbladders. Species diversity and relative abundance of the two groups show bathymetric and latitudinal variations which are interpreted by indicating the influence of several ecological factors on selected species. The present results (relative to net selection) may be interpreted in more reliable detail than hitherto possible, but consideration of small and large scale temporal effects is paramount to the success of any future truly quantitative approach. Thus, while past studies have provided a superficial, but valuable, view of oceanic midwater fish ecology, a higher precision of sampling, coupled with a more detailed knowledge of oceanic ecology is required before correlations between species compositions and sound scattering layers become meaningful.

INTRODUCTION

In recent years several papers have been published pertinent to the vertical distribution of midwater fishes. Studies have been based upon collections made by open nets (e.g., Pearcy, 1964; Paxton,

1967a; *Clarke*, 1973, 1974) or nets designed to collect samples at discrete depths (e.g., *Pearcy and Laurs*, 1966; *Badcock*, 1970; *Gibbs and Roper*, 1970; *Robison*, 1972; *Badcock and Merrett*, 1976). Despite the wide variation in methods, the general characteristics of a midwater fish community portrayed by any one study show remarkable conformity with those of another. Thus some of the finer details, such as animal size stratification with depth shown by *Clarke* (1973, 1974) from open net collections, are confirmed by those obtained with the RMT 1+8 opening/closing net system (*Badcock and Merrett*, 1976). The development of reliable systems for collecting discrete depth samples has endowed the general interpretation of distributions with greater credibility, but perhaps the most significant gear developments made in recent years have been those in monitoring equipment (*Aron et al.*, 1964; *Bourbeau et al.*, 1966; *Clarke*, 1969; *Pearcy and Mesecar*, 1970; *Baker et al.*, 1973). Such systems can relay information on selected parameters to the ship during trawling, making accurately controlled sampling possible. This in turn has allowed studies to examine new perspectives in understanding the complex relations of sampling bias, introduced by both sampling procedure and animal behavior (see *Roe*, 1974).

Since 1968, an extensive program of investigations into various aspects of zooplanktonic and micronektonic distribution in the eastern North Atlantic has been carried out by workers of the Institute of Oceanographic Sciences (*Foxton*, 1971/72; *Clarke and Lu*, 1974, 1975; *Foxton and Roe*, 1974; *Merrett and Roe*, 1974; *Roe*, 1974; *Angel and Fasham*, 1975; *Fasham and Angel*, 1975; *Lu and Clarke*, 1975a and b; *Badcock and Merrett*, 1976). The main tools of investigation have been the mouth opening/closing rectangular midwater trawls, RMT 8 (*Clarke*, 1969) and RMT 1+8 combination net (*Baker et al.*, 1973). Throughout the program emphasis has been placed upon the use of standard sampling procedure to give directly comparable samples; the facility to monitor the net's performance directly has therefore been essential. This paper presents a preliminary survey of the vertical distribution of midwater fishes along a meridional section. The collection as a whole is being treated in greater detail elsewhere (e.g., *Badcock and Merrett*, 1976). In contrast to the "Discovery" SOND Cruise (*Currie et al.*, 1969), these investigations were not intended to correlate animal distributions and sound scattering layers. Nevertheless, the results do provide information on the trends in vertical distribution and migration of potential sound scatterers. To this end, the latitudinal and depth distributions of fishes with and without gas-filled swimbladders are discussed here in general terms. Certain species are examined in detail to show the fine structure of specific populations. The need for a more refined approach than used hitherto, for correlating deep sound scattering and sample composition, is discussed in the light of these and other results. For instance, the 24-hr study of a single depth (*Roe*, 1974) and subsequent 48-hr sampling at other selected depths (see *Angel*, 1977) have indicated the dynamic nature of the faunal composition at any one depth and the migratory differences between species. Similarly, feeding studies (e.g., Holton, 1969; Merrett and Roe, 1974) have emphasized the variability of behavioral patterns occurring among fish species. Such variability presents considerable problems in the interpretation of samples, especially if relations to sound scattering layers are to be demonstrated. Aspects of these problems are considered in the hope that ideas for their solution may be forthcoming.

MATERIALS AND METHODS

Collections were made at six positions from 11° N - 60° N, approximately along the meridian of 20° W, during the period 1968-1972 (Figure 1 and Table 1). Sampling in 1968 was carried out with an NIO (now IOS) mouth opening/closing RMT 8 (Clarke, 1969), while subsequent collections were made with a combination net (RMT 1+8) fished in a standard manner (Baker et al., 1973) during daylight and darkness, avoiding dawn and dusk periods. Aspects of this program have been described by Foxton (1971/72) and Angel and Fasham (1975), while Badcock and Merrett (1976) describe in detail the vertical distribution series of collections made at 30° N, 23° W and discuss sampling limitations with special reference to fish catches.

Briefly, the standard procedure was to fish the upper 1000m in 100m strata, which are the horizons of Foxton (1971/72) and Angel and Fasham (1975). However, we use the term "horizon" where single depths rather than strata have been fished. After the 1968 collections, the top 100m was subdivided into 3 strata (10-25, 25-50, 50-100m) for RMT 8 sampling (Table 1). Sampling in wider strata was also carried out below 1000m, generally down to 2000m, but only data in the upper 1000m have been used in this paper. The net was fished systematically through each stratum; opened at the lower horizon and closed at the upper. The tows were generally of 2 hour duration in the upper 1000m and, until 1972, fished to a ship speed of 2 knots. Because of bias due to currents affecting the net speed relative to that of the ship, an attempt was made to standardize the direction of tow at each station, so as to allow an estimate of relative abundance between day and night catches at any depth. During the 1972 series of collections, a flowmeter just above the net was used to keep the speed of the net as close to 2 knots as possible (see Badcock and Merrett, 1976).

Catches, excluding larvae and unidentified specimens of Cyclothone and Melamphaidae, are expressed as numbers per standard tow of 2 hours for each depth interval. The catches from the strata within the top 100m from stations north of 11° N have been adjusted and integrated to express them in terms equivalent in fishing effort to a single 100m stratum collection. These particular samples were normally taken consecutively during the same day or night period, which offsets day to day variation, but the bias introduced by

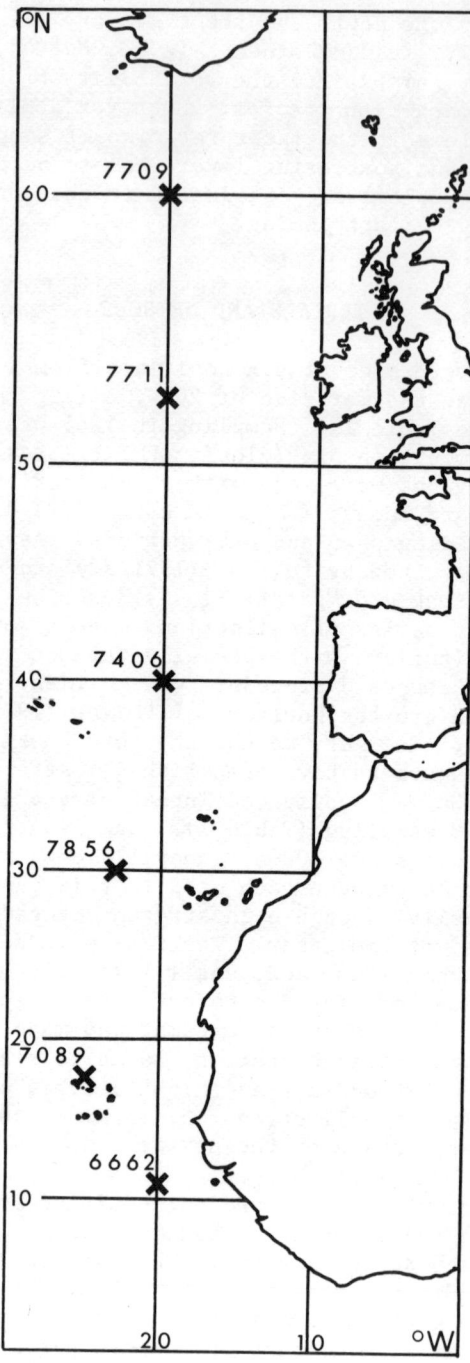

Fig. 1. Chart of the eastern North Atlantic showing the positions at which collections were made.

TABLE 1. Synopsis of the Collections Made by the RMT 8 from which Data were Utilized.

Localities of Observations

N	W	Station	Date	Nominal depth strata sampled (m)
11°	20°	6662	Feb 1968	0-100, et seq. to 1000.
18°	25°	7089	Nov 1969	10-25; 25-50; 50-100; then 100-200 et seq. to 1000.
30°	23°	7856	April 1972	As Stn 7089
40°	20°	7406	Oct 1970	As Stn 7089
53°	20°	7711	May 1971	As Stn 7089
60°	20°	7709	Apr/May 1971	As Stn 7089

migratory movements occurring during the hours of day or night could not be avoided (see Roe, 1974). At depths greater than 100m, catches are referred to nominal depth intervals, with the shallower figure quoted first for convenience, as the actual depths of opening and closing rarely varied by more than ± 10m (e.g., 295-205m would be quoted as 200-300m).

Hydrographic casts were made to 1200 or 2000m with a salinity/temperature/depth recorder, usually at the beginning, the middle and the end of the sampling program at each station. These data sets provided a hydrographic background to the catches and monitored any major hydrographic changes occurring during the sampling period. Only one major change was noted: at 60°N one hydrographic profile showed the movement of an oceanographic front across the position (Angel and Fasham, 1975). Composite meridional profiles of temperature (Figure 2) and salinity (Figure 3) have been synthesized from the records at each station. These represent the prevailing environmental conditions during each sampling series and therefore, together with the catches, must be regarded as spot observations from which only large-scale variations and trends can be observed.

The data presented here are drawn mainly from detailed laboratory analyses, but do contain some shipboard identifications and counts which may require minor modifications after more detailed examination. The total numbers of species and specimens used are

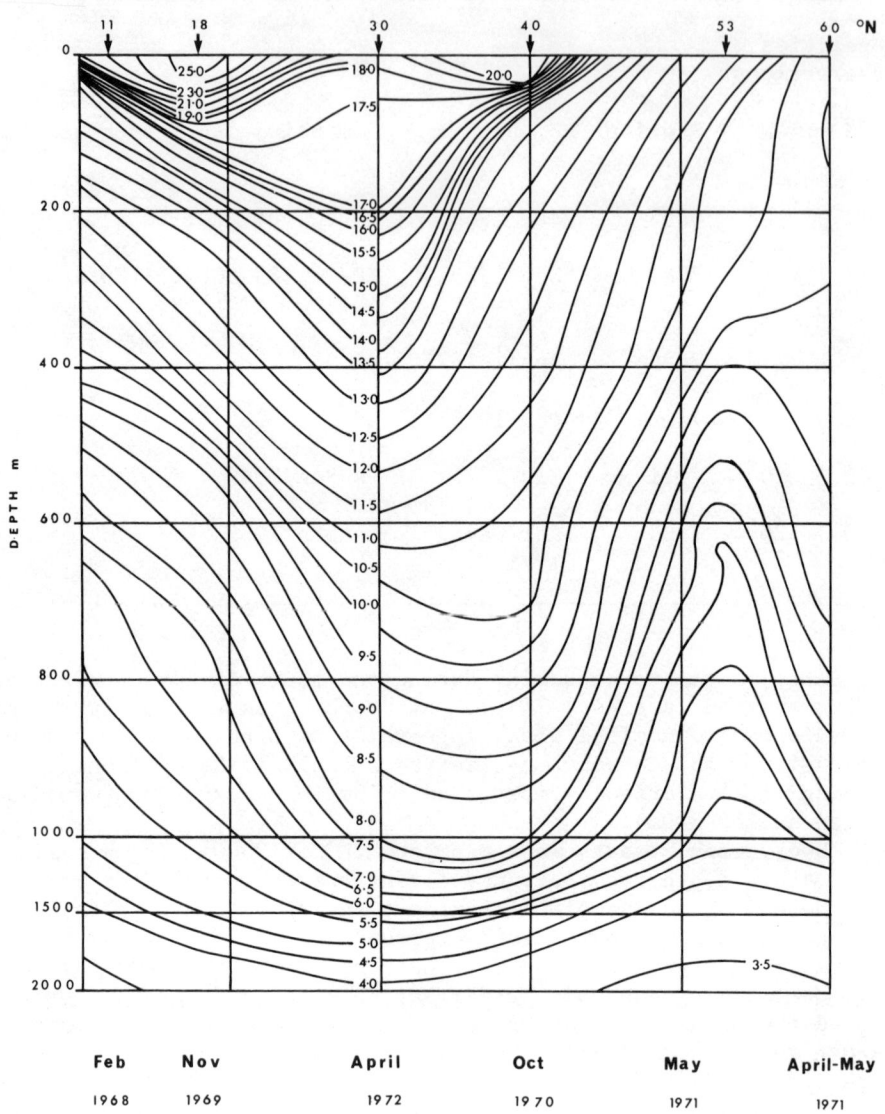

Fig. 2. Composite isotherms (°c) for the meridional section along 20°W derived from TSD measurements taken during the course of the collections at each station.

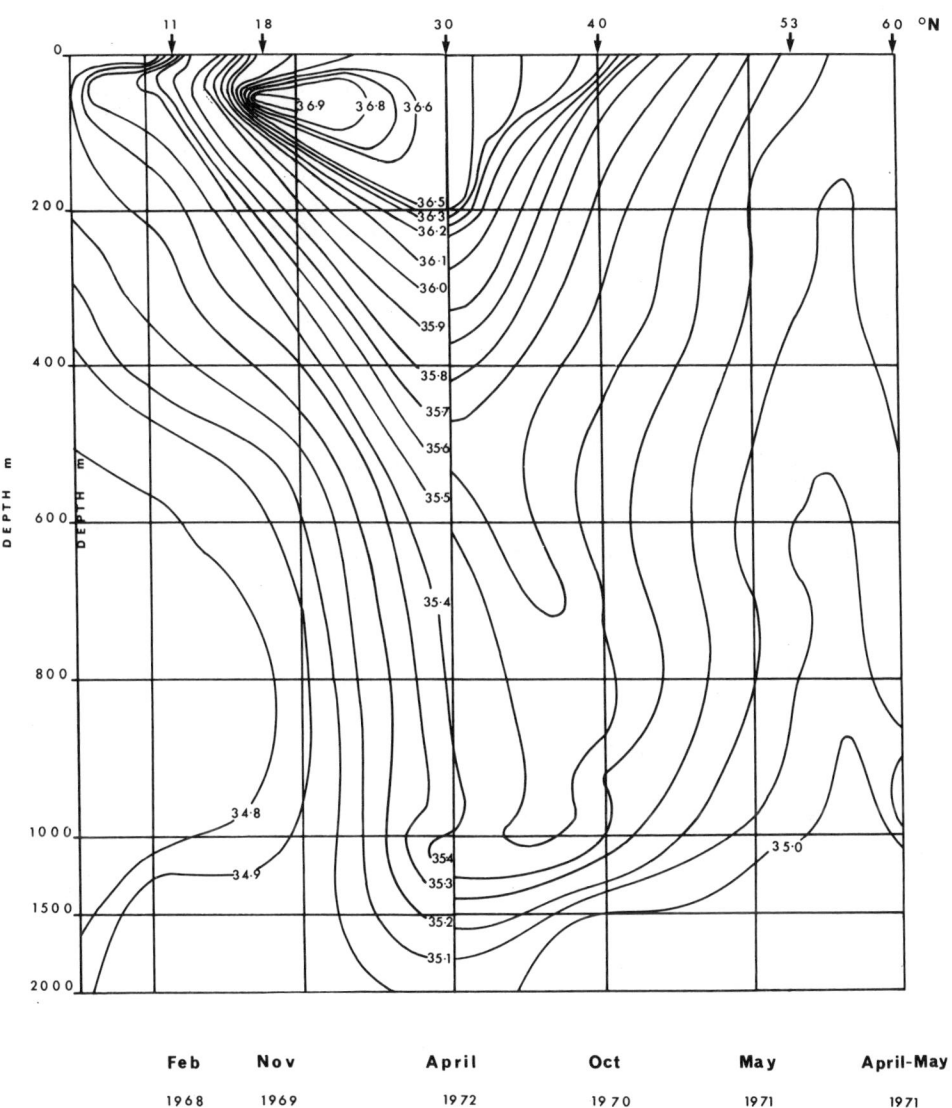

Fig. 3. Composite isohalines (°/oo) for the meridional section along 20°W, synthesized in the same manner as for Fig. 2.

given in Table 2. Species were identified as being swimbladdered or swimbladderless, either from the literature (e.g., Marshall, 1960) or from examination of fresh or preserved specimens. A size range of specimens was examined whenever possible to ascertain whether or not swimbladder regression occurred with growth. Specimen size was accounted for in swimbladder assessment for species where regression was known to take place. Even so, the division into swimbladdered and swimbladderless forms is not entirely satisfactory. Within many species there is a considerable variation between individuals in the degree of lumen occlusion, especially among lantern fishes (e.g., see Zahuranec and Pugh, 1970; Butler and Pearcy, 1972; Kleckner and Gibbs, 1972; Bone, 1973). Thus Benthosema glaciale (Reinhardt) has been regarded here as swimbladdered, although fat investing is known to occur in varying degrees (Marshall, 1960; Zahuranec and Pugh, 1970). The swimbladder of the Cyclothone species, pseudopallida Mukhacheva, pallida Brauer, acclinidens Garman, livida Brauer, and microdon (Günther) cease to be potential sound reflectors due to fat investment or regression at small size and, although the size range over which this occurs is variable (Marshall, 1960), these species are considered here to be swimbladderless. On the other hand the swimbladders of C. alba Brauer, and C. braueri Jespersen and Tåning, are probably gas-filled up to the later stages of gonad maturity (see Badcock and Merrett, 1976).

TABLE 2. Day, Night and Overall Totals of Swimbladdered and Swimbladderless Species and Specimens from the Meridional Collections.

		DAY	NIGHT	TOTAL
SPECIES	Swimbladdered	95	96	<u>112</u>
	Swimbladderless	112	98	<u>130</u>
	TOTAL	207	194	<u>242</u>
SPECIMENS	Swimbladdered	14791	16682	<u>31473</u>
	Swimbladderless	19196	16118	<u>35314</u>
	TOTAL	33987	32800	<u>66787</u>

Throughout the paper "species diversity" is used in the broad sense to describe the number of species present which is equivalent to the "species richness" of some workers.

RESULTS

Species Diversity and Abundance

The total numbers of species and animals contained in the collections are given in Table 2, subdivided into swimbladdered and swimbladderless classes. The diversity of species and the numbers of animals caught in the upper 1000m along the meridian are shown in Figure 4. As a consequence of the uniformity of fishing effort, direct area-to-area comparisons of both species diversity and relative abundance are possible, but the evaluation of these is limited by the sampling methods and the lack of seasonal approach. Nevertheless, consistent geographic trends are apparent in diversity and abundance patterns.

Two hundred and forty-two species (excluding the majority of the melamphaids) were identified and of these 130 were swimbladderless (Table 2). Overall more species occurred in day catches (207 v 194), due to a reduction in the number of swimbladderless species taken by night (Table 2). Since species were often common to catches made in two or more areas these figures are not directly related to the situation occurring along the meridian.

Overall, more swimbladdered and fewer swimbladderless individuals were taken by night (numerically and proportionately), but such features were not characteristic (Table 2, Figure 4). Variations as these may reflect the changes in relative abundance of particular species from area to area, as well as the inherent limitations of the study (e.g., the sampling of depths once each by day and night per station, and the reliance, at most stations, upon ship's speed as a net speed indicator). Even so, the geographical changes in relative abundance (and of species diversity) apparent are indicated not only in the combined data but also in the individual day and night data, and this strengthens the case for their reality.

Diversity differences along the meridional section show a conformity of latitudinal pattern between swimbladdered and swimbladderless forms (Figure 4). Thus diversity in both groups peaked at 18°N and thereafter decreased to the north and south. Moreover, the division into swimbladdered and swimbladderless species was fairly even at most stations (Figure 4). Day/night differences in species diversity were not excessive at any one position but those between day or night and the total diversity shown in the various categories plotted in Figure 4 indicate that a moderate number of species were not common to day and night catches. Such discrepancies are due mainly to intermittent occurrences of some species and also partly to the presence of those migrating into the upper 1000m from deeper in the water column.

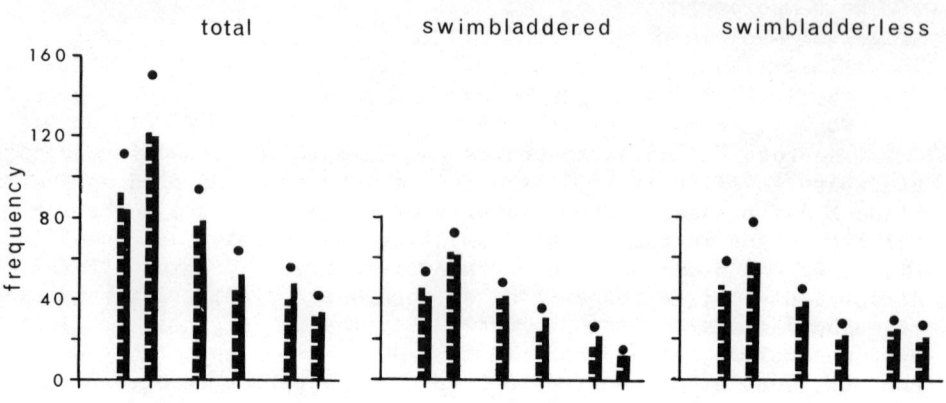
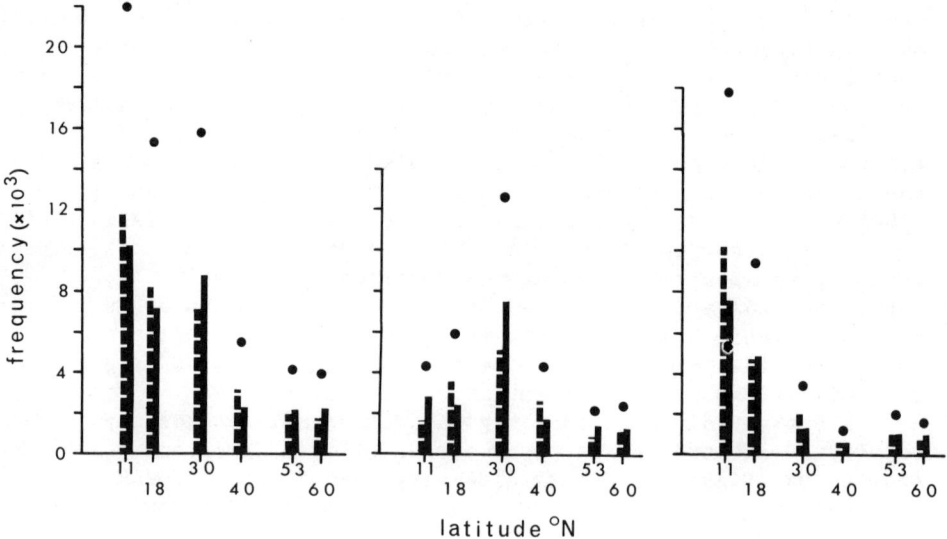

Fig. 4. Latitudinal distribution of species diversity and specimen abundance (•) from day (- - -) and night (———) catches at each position. In each case the totals are subdivided into values for swimbladdered and swimbladderless forms.

Total fish abundance was greatest at 11°N. Numbers of animals caught were similar, if fewer, at 18°N and 30°N, but greatly diminished at 40°N and northwards (Figure 4). The similarity between abundance at 18°N and 30°N is misleading, however, and due entirely to the temporary effect of the presence of the epipelagic life stages of *Macrorhamphosus scolopax* (*Linnaeus*) (4085 specimens). The relative abundance of mesopelagic species in general, then, declined south to north. Swimbladderless animals followed a similar pattern but, as with the species diversity of this group, showed a minimum at 40°N (Figure 4). Swimbladdered animals, on the other hand, displayed a very different pattern of abundance, the peak occurring at 30°N (even with *Macrorhamphosus* excluded) (Figure 4).

A consideration of the proportional contribution of swimbladdered and swimbladderless forms in each area gives a slightly different insight into the geographic changes. The proportion of swimbladdered animals increased sharply from a minimum of 19% of the population at 11°N to peaks of 79% at 30°N, or 71% excluding *Macrorhamphosus*, and 78% at 40°N. Thereafter the proportion decreased to 53% and 57%, respectively at 53°N and 60°N. Despite the evenness of division between swimbladdered and swimbladderless forms in species diversity at any given position, numberical dominance altered latitudinally in favor of swimbladdered species northwards. Even so, the numbers of swimbladdered animals caught at 11°N and 18°N were similar to or greater than those taken at 40°-60°N. Thus, although swimbladdered animals in high and low latitudes were apparently similar in absolute abundance, despite the seasonal spread of samples, their rôle and influence in the ecosystem may be different.

The data derived from any one of the investigated positions represent, in effect, spot checks and do not necessarily reflect a stable situation. In addition, the observations are based upon samples collected by a single net design, and as such are automatically, but consistently, biased. Factors which may affect relative abundance such as seasonal and annual fluctuations, the interrelations and influences of breeding cycles etc. have not been ascertained. Though these are likely to have an effect on abundance, especially in the northern sector, there is no evidence to suggest that they would alter the major diversity trends shown above.

The proportional relationship between swimbladdered and swimbladderless animals found at each station bathymetrically (Figure 5) showed the following characteristics common to most positions:

1) diurnally and nocturnally, peak abundances of swimbladdered forms tended to lie shallower than those of swimbladderless animals.

2) diurnally, 300m marked the approximate upper limit of the adult midwater fish community except at 53°N. The lower limits of peak abundances of swimbladdered species varied from area to area in an undulating pattern across the latitudes (Figure 5). The upper distributional limits of swimbladderless fish undulated similarly (c.f. species account below).

3) nocturnally, swimbladdered animals constituted the major component of the ichthyofauna of the upper 200m. At most positions a

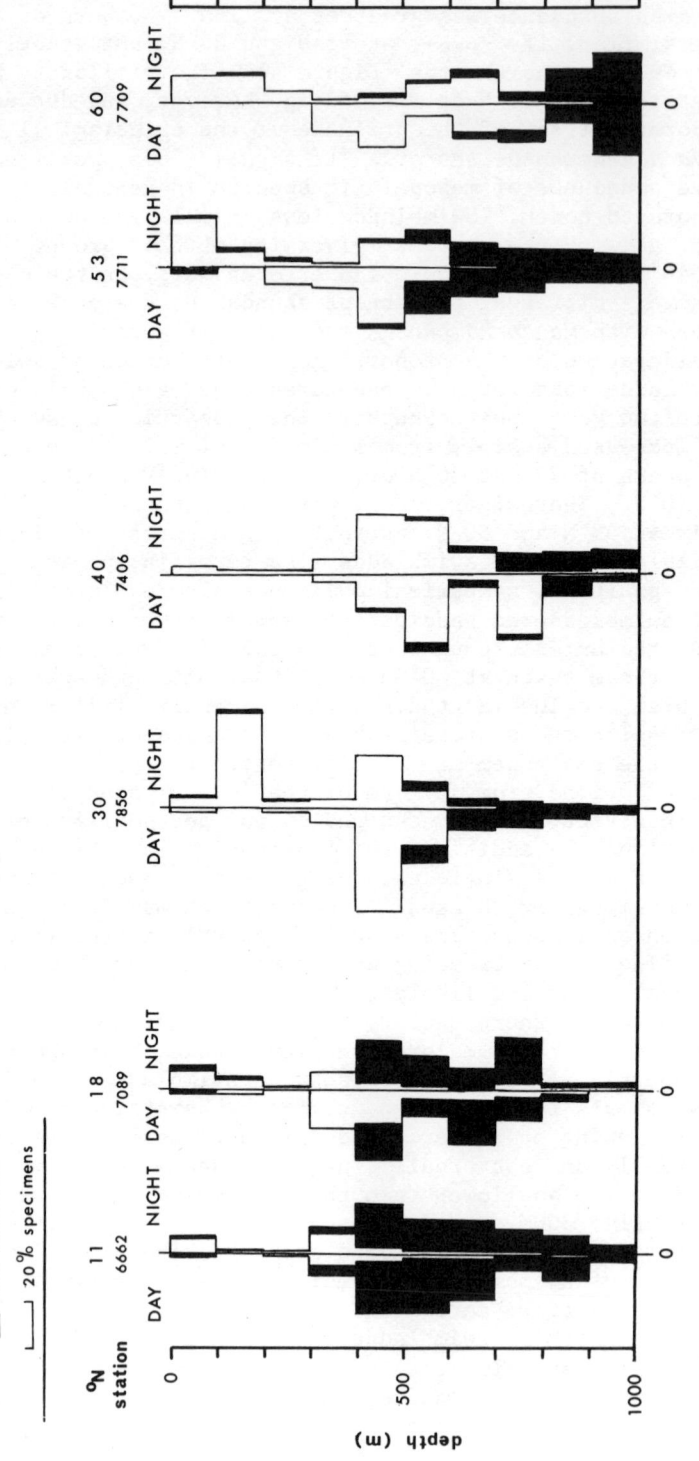

Fig. 5. Percentage distribution of specimens caught per 100m stratum from the total catches by day and by night in each position. The percentages of swimbladderless specimens are indicated in black.

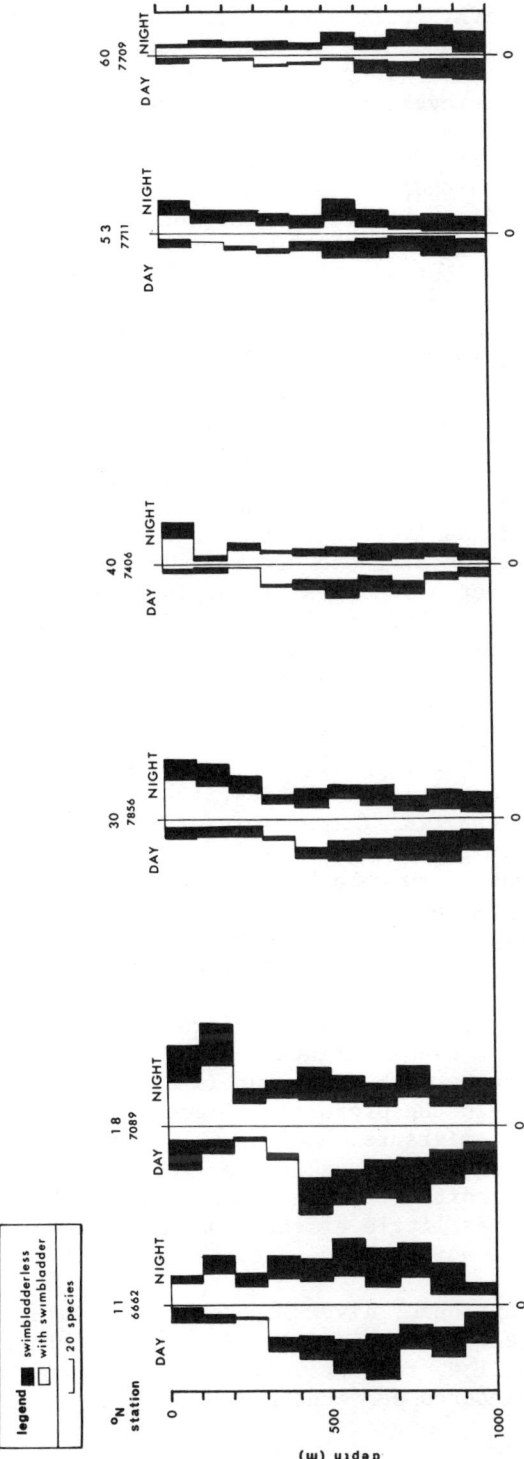

Fig. 6. Day and night distribution of the total number of species caught per 100m stratum from each position, with the numbers of swimbladderless species indicated in black.

layer, 100-200m thick, containing a low animal density lay between this fauna and peak abundances of swimbladdered and swimbladderless forms at greater depths. In general, the increased abundance of animals in the upper layers nocturnally were due to migrations of animals from greater, diurnally occupied depths. At 30°N, however, the nocturnal peak in 100-200m depth was the product of a reverse migration by *Macrorhamphosus scolopax*. The *Macrorhamphosus* size groups composing this peak were not sampled by RMT 8 during the day, since they then occupy the upper 10m of the water column (*Badcock and Merrett*, 1976). Thus at 30°N this species obscures the zone of minimum abundance nocturnally characteristic of other areas.

4) in general, not more than 50% of the swimbladdered population migrated into the near surface layers, and at 40°N a particularly low proportion did so.

In terms of species diversity, a nocturnal increase in the upper 200m was marked at all positions, although more so for swimbladdered than swimbladderless types, but day-night diversity changes at greater depths were less clear in most areas (Figure 6). The low density zones apparent nocturnally (Figure 5) are not reflected by differences in species diversities at these depths. Thus to a certain extent the bathymetric aspect of changes in species diversity by day and night is unrelated to abundance changes. The smallness of day/night changes in the diversity of species in depths greater than 200m is due partly to replacement by migratory species from greater depths and partly to populations of habitually migrant species containing non-migrant elements (see *Badcock and Merrett*, 1976). The nocturnal abundance of swimbladdered species in these depths, however, results from the predominance of a few species, e.g., *Cyclothone alba, C. Braueri, Valenciennellus tripunctulatus* (Esmark), *Argyropelecus hemigymnus* Cocco, in particular areas.

The proportions of swimbladdered and swimbladderless individuals per stratum are expressed as percentage of total individuals per stratum, day and night (Figure 7). While there are area to area variations apparent, this shows very clearly the consistency of population *proportions* at depths greater than 400m over the diel period and also a marked increase in swimbladdered proportions in the upper 200m by night. Taken alone, Figure 7 gives no indication of day/night abundance changes with depth but in conjunction with Figure 5 it emphasizes, through the proportional consistency shown below 400m, the dominance of non-migrants, both swimbladdered and swimbladderless, in the midwater fish fauna in these depths. Thus the vertical spread of migrant density diurnally is such that migration out of a particular stratum has little effect upon the swimbladdered/swimbladderless proportions, or else is compensated by migrants from greater depths.

As has been mentioned, diversity peaks of swimbladdered and swimbladderless species coincided at 18 N; yet in abundance they peaked at 30°N and 11°N, respectively. Doubts as to the accuracy of the abundance patterns depicted here must be expressed, however,

THE DISTRIBUTION OF MIDWATER FISHES

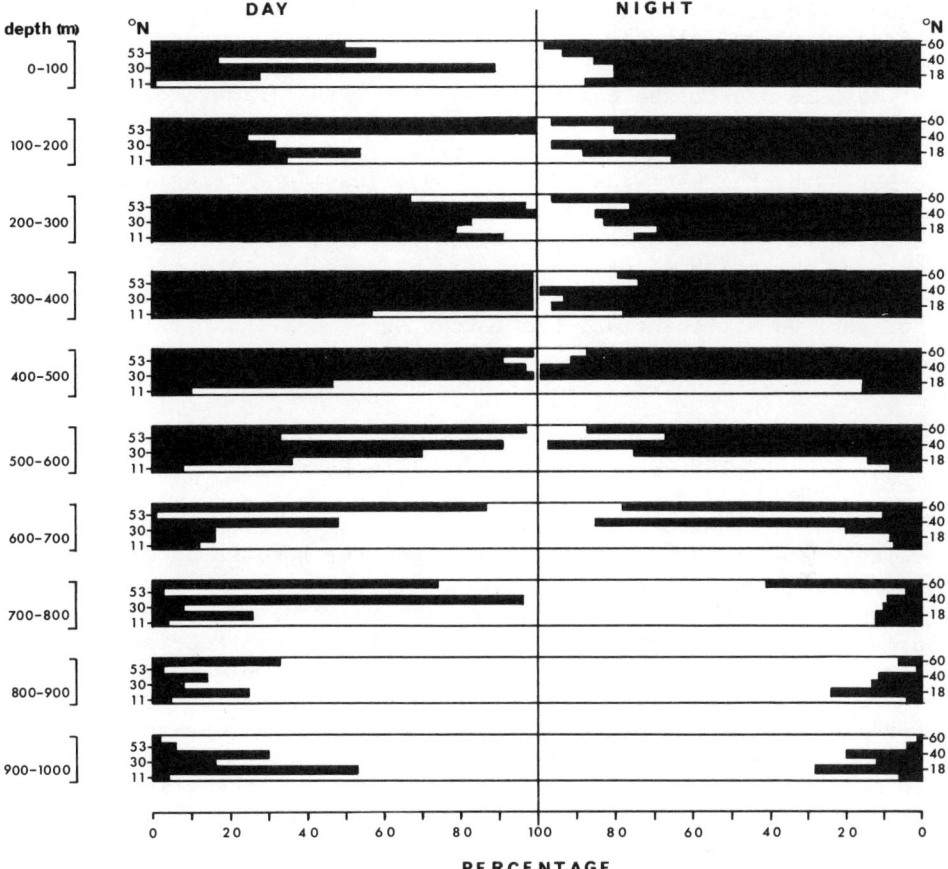

Fig. 7. Bathymetric distribution of the percentage abundance of swimbladdered (black) and swimbladderless (white) specimens per catch per position both by day and night.

since the collections were made by a single net (fishing mouth area $8m^2$). In nature, the shaping of abundance patterns geographically is probably greatly influenced by relatively few species. Abundance per station expressed by the present data should, ideally, represent an integration of the relative abundance of the various species concerned. Of the species caught, however, it is highly unlikely that they were taken in their correct interspecific proportions, because of net selectivity. The patterns of abundance indicated for swimbladdered and swimbladderless forms (Figure 4) therefore may reflect those of a few species selected by the net rather than depict the actual situation. The most abundantly sampled genus was *Cyclothone* (44331 specimens) and the influence of its species on the patterns of geographic abundance are clearly shown in Figure 8. Among swimbladderless animals the total abundance curve is ordered by *Cyclothone* species; with swimbladdered forms, the apparent abundance peak at 30°N is due to the dominance of this genus (Figure 8). The abundance patterns demonstrated geographically (Figure 4), then, reflect for the most part the situation with regard to *Cyclothone*. On the other hand, the patterns of swimbladderless *Cyclothone* and the remaining swimbladderless species correspond in showing a northward decline to a minimum at 40°N, which suggests this to be the basic pattern for swimbladderless species (Figure 8). The patterns for swimbladdered *Cyclothone* and the remaining swimbladdered species, however, do not coincide, the latter showing a peak at 18°N and a minimum abundance at 40°N. The overall pattern for swimbladdered animals (Figure 4) may, or may not, be distorted by biased sampling.

Interpretation of Specific Examples

The examination of overall latitudinal species diversity and relative catch abundance provides a general account of the situation and major trends in the ichthyofauna of the upper 1000m. Examples at specific level are necessary to interpret detail. While a comprehensive review of all species involved is inappropriate here, a few examples have been chosen to demonstrate some general trends in distributions, including 7 species of the genus *Cyclothone* (which contributed 70% and 62% respectively of the total day and night catches), 4 species of *Argyropelecus* and 2 species of *Benthosema*. In each case the indicated depth ranges portray the principal distribution layer, by plotting only strata containing 10% or more of the total day or night catch of the species at that station. Also discussed are water temperatures related to these depths, obtained during the sampling program at each position (cf. Figure 2). In all species other than *C. microdon* the principal distribution layer occurred within the upper 1000m.

No *Cyclothone* species were vertical migrators. Together they dominated the lower mesopelagic and upper bathypelagic catches all along the meridian. Latitudinally, a southerly and a northerly group of species emerge. *Cyclothone alba, pseudopallida, pallida, livida* and *acclinidens* were all most abundant at 11° and 18°N (Figures 9a and b). On the other hand, *C. braueri* and *C. microdon* are more widespread from 30°-60°N (Figure 9a). Any comparison of relative abundance among northern species here may be influenced by the different seasons sampled (see p. 189). Several species were breeding at 30°N in April and the continuance of breeding was still plain at 40°N in October. The population sampled at 53° and 60°N in April and May were composed entirely of adults in pre-breeding condition. Nevertheless, our data show *C. microdon* to be separated geographically from the group of southerly species. Although not isolated from the southerly group of species the proportion of the *C. braueri* caught to the north of 18°N was so great

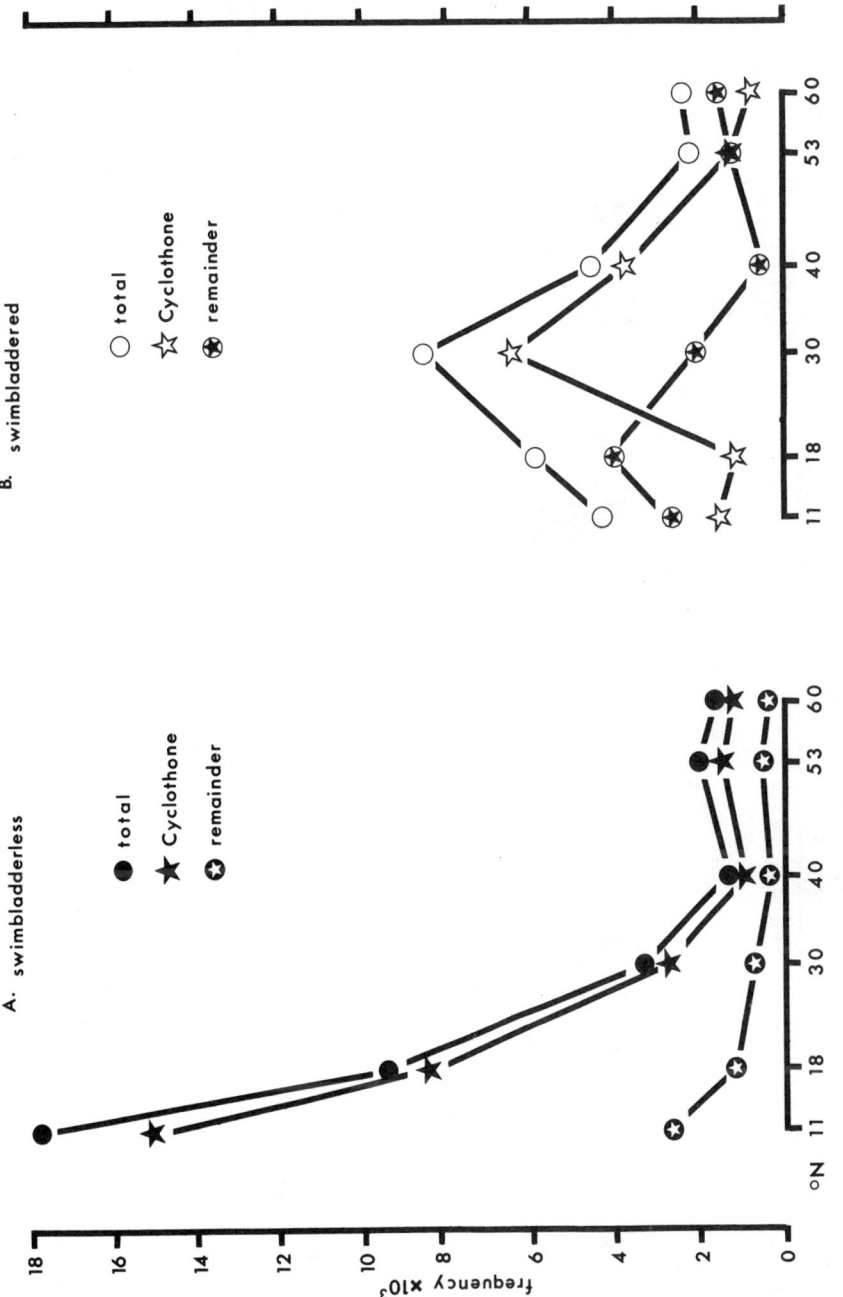

Fig. 8. Latitudinal distribution of total abundance of A, swimbladderless and B, swimbladdered specimens (excluding *Macrorhamphosus*).

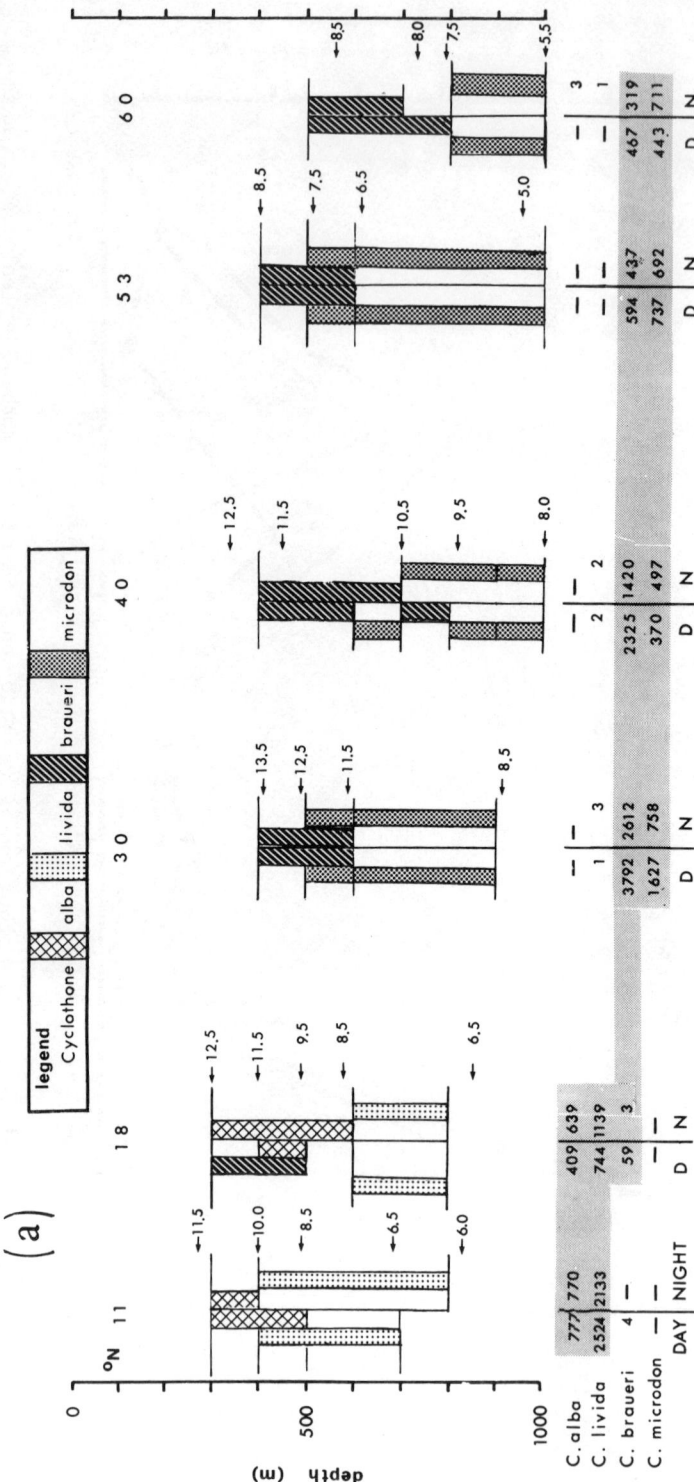

Fig. 9a. Principal distribution layers of the *Cyclothone* species, *alba*, *livida*, *braueri*, and *microdon*, from the day and night catches at various latitudes along the 20°W meridian with certain isotherms indicated. Day and night total catch per position for all species are given and areas of peak abundance marked.

THE DISTRIBUTION OF MIDWATER FISHES 267

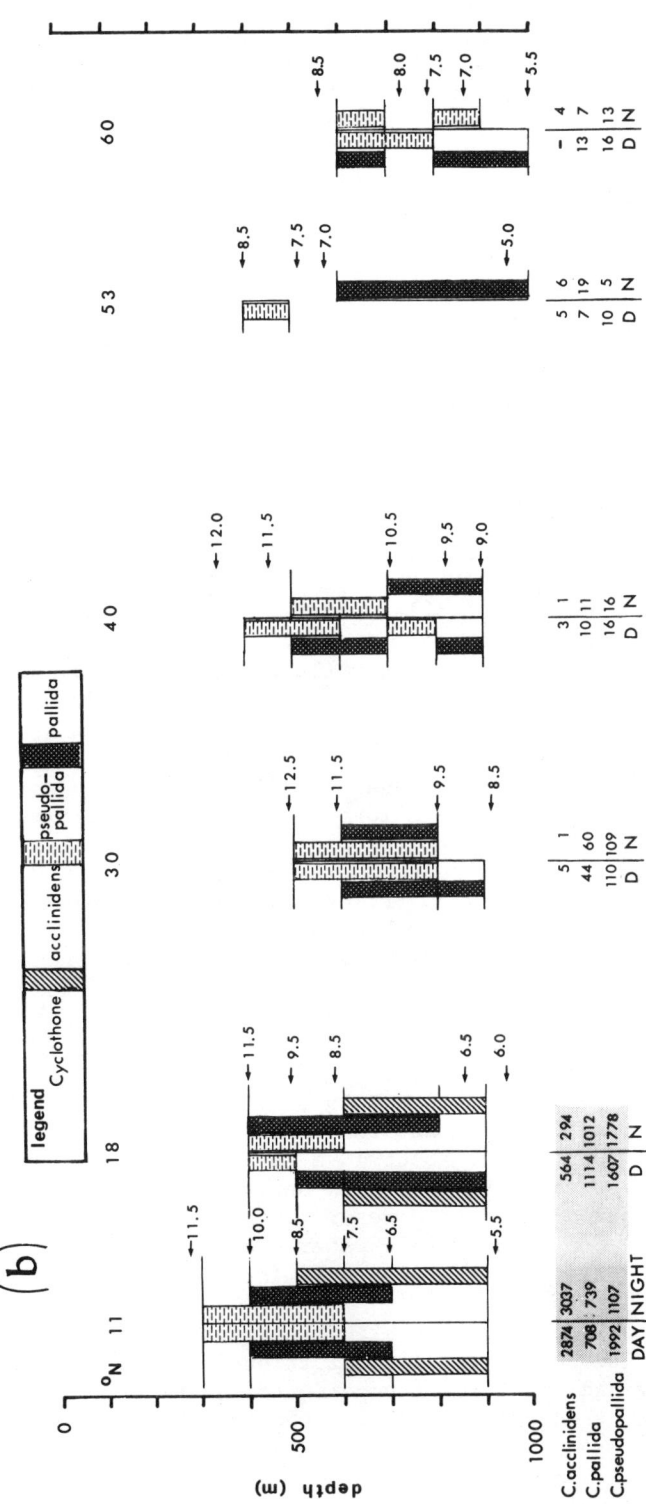

Fig. 9b. Principal distribution layers of the *Cyclothone* species, *acclinidens*, *pseudopallida* and *pallida*, from the day and night catches at various latitudes along the 20°W meridian, with certain isotherms indicated. Day and night total catches per position for each species are given and areas of peak abundance marked.

that this also is considered a northerly inhabitant.

The horizontal distributions of the species of *Argyropelecus* (in which only small migratory movements were observed) are in some instances similar to those of *Cyclothone*. *Argyropelecus sladeni* Regan, was found to be a southerly species which was caught only 11°N and 18°N. *Argyropelecus hemigymnus* had the broadest geographic range, extending from 18° to 60°N, with peak abundance in the catches at 30°N where breeding had begun (Figure 10). *Argyropelecus olfersi* (Cuvier) was restricted to the more northerly samples (40-60°N, Figure 10), while *A. aculeatus* Valenciennes was present only in the catches at 30°N (Figure 10). These distributions conform with those reported by Baird (1971) in his zoogeographic study of this family.

Benthosema suborbitale (Gilbert) and *B. glaciale* have an apparent north-south replacement, with the northern *B. glaciale* being superseded south of 40°N (Figure 11). *B. suborbitale* extends to about 14°30'N near the 20°W meridian (IOS unpublished), although it has been recorded further south from more westerly and less productive waters (Backus et al., 1965). *B. glaciale* reappears in the tropical Atlantic, occurring from about 22°N to 17°30'N at 20-25°W (IOS unpublished). Between 30°-18°N, collections show the species to be moderately abundant along the continental shelf to at least 23°N. This pattern of distribution was paralleled by *C. livida* and *C. microdon*. Along the 20°W meridian the catches of *C. livida* were small at 30°N (Figure 9a) while *C. microdon* was at its most abundant in the upper 1000m. From 28°N, 14°W, however, Badcock (1970) reported *C. livida* in moderate abundance while *C. microdon* was absent. Thus, these species tend to be allopatric. The surface temperature along the shelf at the 23°N station was considerably cooler (17°C in February) than was found further offshore along the 20°W meridian to somewhere northwards of 40°N (Figure 2). The presence of *B. glaciale* so far south thus is in keeping with Halliday's (1970) conclusion that the limit of its geographic range corresponded closely with the 15°C isotherm from average annual temperatures at 200m.

The results for these species of *Cyclothone* and *A. sladeni* and *A. hemigymnus* suggest a faunal change around 18°N. This was first observed by Backus et al. (1965 and 1970) among mesopelagic fishes. In addition, Foxton (1971/72) and Fasham and Angel (1975) observed a similar faunal boundary at about 18°N for mesopelagic decapods and ostracods, respectively, from data derived from the same series of collections that are reported here. In each case these authors related the change to identifiable water masses, such as the South Atlantic Central Water and the Antarctic Intermediate Water (e.g., Fasham and Angel, 1975).

Environmental factors which affect horizontal distribution may equally well influence the vertical distribution of a species within its geographic range. Before attempting to identify such factors we should examine the vertical distribution of the species under

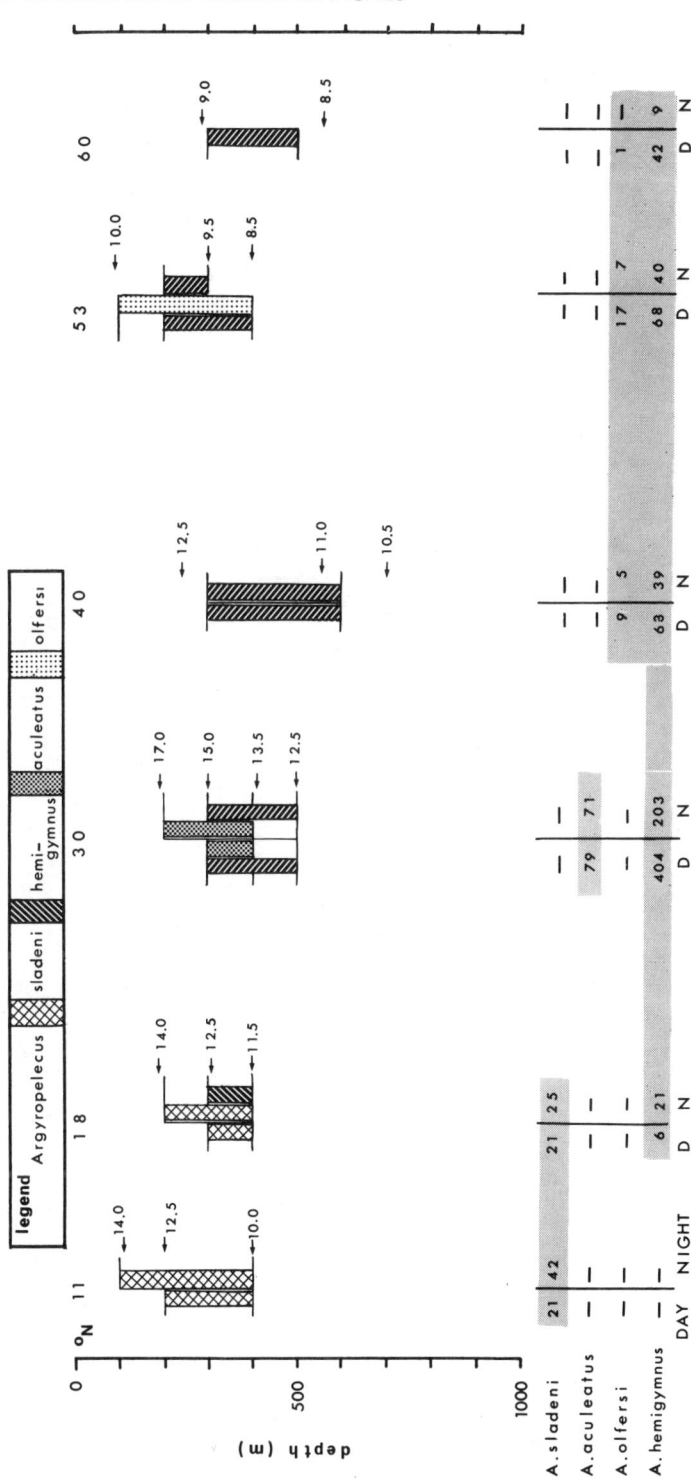

Fig. 10. Principal distribution layers of the *Argyropelecus* species, *sladeni*, *hemigymnus*, *aculeatus* and *olfersi*, from the day and night catches at various latitudes along the 20°W meridian, with certain isotherms indicated. Day and night total catches per position for each species are given and areas of peak abundance marked.

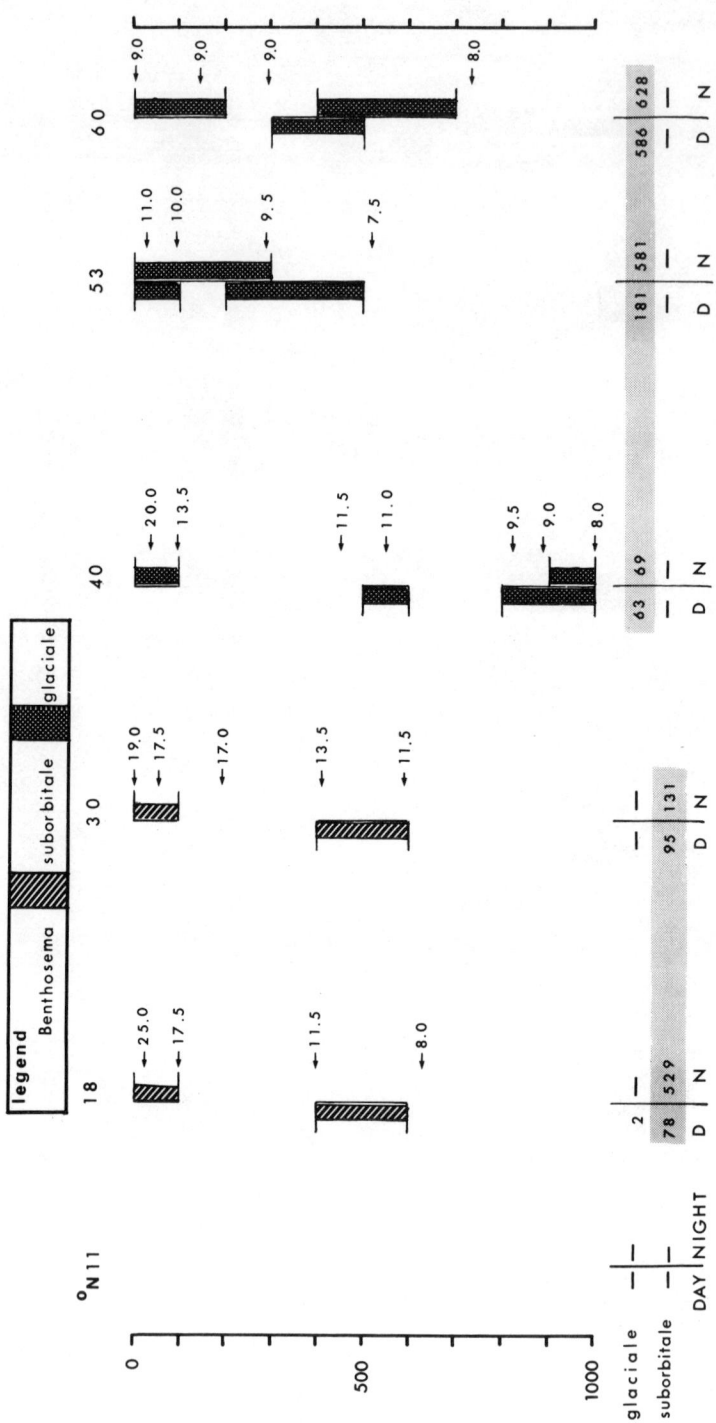

Fig. 11. Principal distribution layers of the *Benthosema* species, *suborbitale* and *glaciale*, from the day and night catches at various latitudes along the 20°W meridian, with certain isotherms indicated. Day and night total catches per position for each species are given and areas of peak abundance marked.

consideration. Despite their non-migratory habitude, the species of *Cyclothone* do display trends of meridional variation in the depths of their principal distribution layers (Figures 9a and b). In the south, *C. alba* and *C. acclinidens* roughly maintained their depth levels at 11° N and 18° N whereas the upper limits of *C. livida* and *C. pseudopallida*, occurred deeper in the water column at 18° N than at 11° N. While no definite trend is clear for *C. pallida* at 11° and 18° N, the upper limits of this species and *C. pseudopallida* are depressed at 30° N, but are located deeper at 60° N than 53° N (although few specimens were caught at these two northerly positions). The northerly species, *C. braueri* and *C. microdon*, occupied different levels in the water column. *Cyclothone braueri* was found to be shallowest at the edge of its range at 18° N and, while the upper limit of its principal distribution layer remained constant from 30°-53° N, the lower limits deepened considerably at 40° N. At 60° N the layer was depressed, by 100m, in common with its congeners. The principal distribution layer of *C. microdon* undulated from 30° to 60° N (Figure 9a).

The population structures of *C. braueri* and *C. microdon* from 30° N have been studied in detail (Badcock and Merrett, 1976). Adult males of these and other species of the genus are smaller than adult females, and a clear size/depth relation was observed when the sexes were separated, with larger animals occurring deeper in the principal distribution layer. In addition, adult female *C. braueri* were found to be stratified according to their gonad maturity; those closest to spawning were caught deepest. Moreover, in both species, proportionally more males were caught at shallower depths and proportionally more females were caught at deeper depths. Observations confirming these trends have been made from other stations along the meridian and elsewhere (IOS unpublished). Similar distributions were suggested by the data from 30° N for the less abundant species, *C. pseudopallida* and *C. pallida* (Badcock and Merrett, 1976). A positive size-depth relation for *C. acclinidens* has already been demonstrated (Talbot, 1973). Stratification due to size, sex and gonad maturity therefore, will greatly influence the vertical distribution of these and probably many other species within the range of tolerance imposed by environmental factors. Thus the depth of principal distribution layers may well be under considerable seasonal influence. For instance, the shallowing of the *C. microdon* population at 30° N is quite possibly due more to the presence of mainly adolescent fish than to any response to environmental effects. At 40° N, the shallower sections of the principal distribution layer of *C. microdon* were also composed mainly of adolescent specimens, effectively extending the layer into shallower water than if an entirely adult population had been sampled. At 53° N and 60° N where only adult *C. braueri* were sampled, a shallower principal distribution layer is to be expected after breeding as juvenile stages are recruited to the population.

The vertical distribution patterns of species of the genus *Ar-*

gryopelecus (Figure 10) are similar to those of several species of *Cyclothone*. The southern species, *A. sladeni*, is located rather deeper at 18°N than 11°N, while *A. hemigymnus* penetrates deepest at 40°N and is found shallower at 53°N than 60°N.

The diurnal depths of occurrence of the lanternfish *Benthosema suborbitale* vary little throughout the observed geographic range (18-30°N, Figure 11). The principal distribution layer of its congener *B. glaciale* by day was deepest at 40°N and was split, while at 53°N it was shallower than at either 40°N or 60°N. At night the bulk of the *B. suborbitale* population migrated into the upper 100m. The non-migratory portion of the *B. glaciale* population was large at 40°N and 60°N, but small at 53°N.

Many environmental factors (e.g., temperature, salinity, light, density, oxygen, pressure, primary productivity) have been evoked to explain the horizontal and vertical distribution of mesopelagic organisms (e.g., *Cushing*, 1951; *Moore and Foyo*, 1953; *Moore*, 1955; *Ebeling*, 1962; *Backus et al.*, 1965; *Clarke*, 1966; *Paxton*, 1967a; *Vinogradov*, 1968; *Backus et al.*, 1970; *Foxton*, 1971/72; *Robison*, 1972; *Fasham and Angel*, 1975). Singly or in combination they have been correlated with various patterns of distributions, with different factors having greater importance in different regions. Thus *Backus et al.* (1965) pointed out the significance of temperature in the North Atlantic saying that "The upper part of the mesopelagial in the North Atlantic does not cool from the equator polewards; rather it cools both equatorwards and polewards from a line connecting Guatemala in the west (ca. 15°N) with the Canary Islands in the east (ca. 28°N) (*Schroeder*, 1963)". Subsequently, *Foxton* (1971/72) showed that northwards of 28°N the vertical distribution of four mesopelagic species of the decapod *Acanthephyra* closely matched the geographical gradients in temperature. He also compared these distributions with the known gradient in solar energy latitudinally (i.e. decreasing from the equator northwards with seasonal variations becoming more important than diurnal at high latitudes). The distributions followed the light gradient from 28° to 53°N but ran counter to it to the south and north. Within the limitations imposed by the lack of seasonal sampling and despite some replacement of species along the meridian, the results presented here (Figures 9-11) indicate that the mesopelagic fishes follow similar trends with respect to temperature (cf. Figure 2) and light. We thus agree with *Foxton* (1971/72): "light is probably a significant local factor in determining depth distribution and diurnal migration of mesopelagic organisms but that environmental temperature, which normally changes gradually in the broad horizontal plane, can considerably modify the distribution." In addition, it is clear that the selective effect of ambient light on form, anatomy, and coloration of mesopelagic organisms is considerable, and is closely related to the preferred daytime living depth. An effect so influential on a species at an evolutionary level will be equally important behaviorally among individuals. Indeed, in the only direct study comparing the distribution of animals with the *in situ* light regimes, *Clarke* (1966) showed that the depth distributions of major luminous migrating organisms off California have a strong correlation with specific light levels. Foxton (1971/72) goes on to suggest: "the geographical limit to the horizontal distribution of a species may thus occur in those regions where the optimum isotherms become shallower in depth than the optimum isolumes."

Paxton (1967a) proposed a model to explain the interaction of temperature and light on the horizontal and vertical distributions of vertically migrating myctophids along the coast of western North America. He assumed that a vertically migrating species is restricted diurnally to a zone of optimal light intensity and nocturnally must migrate to a specific minimum depth for survival. In his model the southerly part of the range occurs where the isotherm of maximum temperature tolerance intercepts the depth of minimum migration. Thus, at the northern limit, temperature will only be restricting at diurnal depths while, in the south, it will only be limiting at nocturnal depths. Since in the eastern North Atlantic the isotherms of the mesopelagial tend to rise north and south of about $30°N$ (Figure 2), for Paxton's model to be applied in the southerly section the restricting temperature factors have to be reversed latitudinally. Thus the southern limits of a distribution would be reached when the minimum temperature tolerated lay shallower than the optimum isolume. The level of detail examined here is not great enough to make proper evaluation of the model but while certain species (e.g., the partial migrants, *Argyropelecus* spp) arguably fit the model, the northern curtailment of the full migrant *Benthosema suborbitale* appears not to fall within its terms. Correlation of salinity (Figure 3) with the geographic ranges of *B. suborbitale* and *B. glaciale*, including the latter's southward extension in the east, suggest that this may signal the separation of the two species. At nocturnal depths, *B. suborbitale* was captured in more saline waters than was *B. glaciale*. It is possible that the salinity and temperature gradients occurring around $40°N$ at $20°W$, on either side of which *B. suborbitale* and *B. glaciale* occurred, are related to a faunal change of the sort examined by McGowan and Williams (1973) in the North Pacific at about $40°N$. They show that the habitats, and seasonal influences affecting them, vary considerably on either side of this boundary and argue that evolutionary adaptation has led to very different complexes of species. They suggest that standing crop levels and yearly productivity are quite different on either side of the boundary, which, if such a situation occurred in a similar position in the North Atlantic, could explain the steep gradient in species diversity and abundance between our catches at $30°N$ and $40°N$ (Figures 4 and 8).

Hydrographic features alone, however, are unlikely to be wholly responsible for the horizontal and vertical distribution of species, for biological features such as interspecific competition may be equally, if not more, important. Such biological influences may be important also in separating the species mentioned in which apparent replacement occurs.

To sum up, the principal feature of the horizontal distributions

of the species cited is that several were abundant either to the
north or south of a faunal boundary around 18°N. This accords with
the reported distributions of other taxa from the same collections
(*Foxton*, 1971/72; *Fasham and Angel*, 1975) and appears to be related
to hydrographic changes. The separation of the horizontal ranges of
Benthosema suborbitale and *B. glaciale* around 40°N along the 20°W
meridian highlights the possibility of a further faunal boundary
analogous with the one investigated about this latitude in the North
Pacific (*McGowan and Williams*, 1973). In the vertical plane, the
diurnal depth distributions of several species were apparently influenced by temperature, but any links with ambient illumination
levels seem more tenuous. While this conforms with *Foxton's* (1971/
72) findings on decapod distributions, it is recognized that biological effects such as interspecific competition may be of considerable importance. The vertical distributions of some, if not all,
fishes, however, were further complicated by stratifications related
to size, sex and gonad maturity. Yet, to be more detailed in relating distributions to these or other environmental factors is outside the scope of this preliminary study.

DISCUSSION

We have presented a partial account of the distributional features of midwater fishes in the eastern North Atlantic, describing
the vertical and latitudinal changes in the populations. The results clearly show that any truly quantitative interpretation is
unwarranted, because of the lack of seasonal sampling. Nevertheless
there is no reason to doubt the reality of general trends in the context of the size of net used. As we have suggested elsewhere
(*Badcock and Merrett*, 1976), the results should give a better understanding of the factors that combine to stabilize the specific composition of the fish fauna both in depth and latitudinally in the
eastern North Atlantic. We have reviewed some of the parameters
possibly influencing distributions, but the resolution is too coarse
for detailed examination. The phenomena of size/depth and maturity/
depth stratifications imply that the factors limiting vertical distributions may be related to stage of life. Tolerances to temperature and light, for instance, may alter and thus depth distributions
of animals in differing developmental stages in different areas cannot
be compared directly.

The present study differs radically from many recent works on
vertical distribution in the approach adopted to sampling. It has
been common practice (e.g., *Paxton*, 1967a; *Badcock*, 1970; *Gibbs and
Roper*, 1970; *Clarke*, 1973, 1974) to identify distributions from a
series of samples collected at particular depth horizons which are
often well separated vertically, rather than from contiguous strata
as here. Despite differences in methods, these reports show good
qualitative agreement in their descriptions of the general features
of the fish community such as its general vertical structure, the
changes in composition that occur with depth and time, and the inter-

relations of species within it. In general, adults and juveniles of species occupy relatively narrow strata, and consequently considerable changes occur between samples collected from different depth horizons or strata.

Samples collected from specific and vertically separated depth horizons cannot be used for quantitative analysis. Nevertheless, estimates of abundance and biomass have been made from such collections (e.g., *Clarke,* 1973; *Vinogradov and Parin,* 1973). In the light of our own studies, this approach incorporates questionable assumptions and the validity of the derived estimates is, therefore, uncertain. Whereas in stratum fishing the catch number represents an integration of the density gradients that occur across the stratum, in horizon fishing the gradients between adjacent sampling depths have to be generated mathematically assuming linear relationships. It is assumed that the catches made are representative (within the context of the net fished) of the species composition and abundance of a broad band of the water column fished, and that samples from widely separated depths are directly comparable. For these assumptions to be reasonable, sampling must be consistent and comprehensive, and it is implicit that the biological situation at each depth is static per day and per night. In the studies already cited (*Clarke,* 1973; *Vinogradov and Parin,* 1973), ship speed was used as the primary indicator of the speed of the net. Currents, however, can modify the speed of a net through the water, with the result that the computation of distance fished, and hence volume of water filtered, is biased by an unknown amount. In addition, changes in net speed can affect catch composition (see *Angel,* 1977). As no measure of error is provided where ship speed is used as a net speed indicator, the sampling comparability, at a quantitative level, between different horizons must be in question.

Roe (1974) has outlined some of the biological and procedural phenomena that can influence the catch composition of samples taken from depth horizons (or indeed, narrow strata). One cannot assume that a migrant species returns to an identical depth each day (see *Boden and Kampa,* 1967) and, in addition, as *Roe* (1974) has shown, at a given depth species composition and abundance change continuously over the 24-hr period. Time is a very important sampling factor, and consequently samples from different horizons do not necessarily bear true comparison. The fishing of relatively broad depth strata (i.e. 100m) to a certain extent accommodates such fluctuations; variability due to mesoscale (10-1000m) patchiness remains a problem in the absence of repeat hauls, but it can be avoided by taking long tows.

Also implicit in the horizon sampling approach is that an even density gradient occurs across two sampled horizons, irrespective of the extent of vertical separation between them. The complexity of population structure in the upper 100m nocturnally is qualitatively well documented (e.g., *Pearcy,* 1964; *Clarke,* 1973, 1974; *Badcock and Merrett,* 1976). In view of the dynamic changes of species composition and abundance, and the observed species stratification it would be unreal to estimate nocturnal abundance and biomass within this layer by the interpolation of catch numbers from samples made at fixed depths of, for example, 5m and 100m. The reliability of a density gradient prediction between two horizons is clearly dependent upon the measure of their vertical separation, and the selection of the latter must be based upon prior knowledge of the population fine structure in the depths under consideration. What should the separation between horizons be, to ensure reliable estimates? Our knowledge of the population fine structure anywhere in depths greater than 100m is exceptionally poor. Foxton has examined the population fine structure of certain decapod crustaceans within a single 100m band (500-600m) by day at 11°N, 23°W by comparing samples collected in 25m contiguous strata with those obtained from an equivalent 100m stratum (Foxton, pers. comm.). His results showed considerable quantitative and qualitative variations between the 25m strata of the observed decapod population. Furthermore variations in sex ratios enhanced the complexity of the fine structure. Many species of fish diurnally exhibit a size stratification which in some cases can be related to maturity stages (e.g., Nafpaktitis, 1968; Krueger, 1972; Badcock and Merrett, 1976). Cyclothone braueri demonstrated such stratification at 30°N, 23°W (p.189) and other areas, as well as considerable changes in sex ratio with depth, which were similar by day and night (Badcock and Merrett, 1976). Clearly population fine structure is complex at any depth. This, combined with the relatively narrow depth ranges known for many species, implies that the vertical separation of fishing horizons is equally important below 100m as above it, if reliable estimates of abundance and biomass are required. Thus, for abundance and biomass to be reasonably estimated by horizon sampling, it is necessary to sample the water column at closely equidistant intervals. Logistic and biological factors (i.e. time spread of observations) make this strategy impractical so that the contiguous strata approach is the most effective compromise.

Samples obtained from a stratum should theoretically represent an integration of the various density gradients occurring within it. Thus a series of contiguous strata, each of equal breadth, provides an integrated estimate of the faunal composition in the water column. The success of the integration depends upon the accuracy of sampling and it is of paramount importance that strata should be fished evenly across their depth ranges. In practical terms, how good is the integration of the individual stratum? The comparisons by Foxton (pers. comm.) of the 25m strata catches with that from a 100m stratum, indicate that for some decapod species the integration in the 100m stratum sample was good, while in others it was poor. The 25m strata were fished on two consecutive days, and the problems of interpretation of narrow strata samples are similar to those for fixed horizon sampling, in that catch variability may be caused by different species behavior both from day to day and from hour to hour (Roe, 1974). Clearly further work is required but the problems involved are not easily overcome, since the temporal aspects of distribution and migration, as well as patchiness, need to be rationalized.

Pearcy and Laurs (1966) gave abundance and biomass estimates for midwater fishes in the 1000m water column off Oregon. Their net system included a flowmeter and a codend device that allowed the collection of discrete samples from the contiguous strata 0-150m, 150-500m and 500-1000m. Direct comparison between their estimates and any future estimates based upon the IOS procedure is complicated, firstly by the different nets and net sizes utilized, and secondly by the very different fishing procedures involved. They fished their nets downwards in an oblique manner, with the ultimate net closing at 1000m depth. Institute of Oceanographic Sciences procedure fishes the net effectively horizontally since the distance fished is great compared with the width of the stratum. The oblique sampling approach is perhaps more susceptible to variability caused by patchiness as the net fishes any distribution layer for a shorter time than if horizontally fished, but such variability may be normalized by the many repeat tows possible within a relatively short time period. On the other hand, oblique fishing may be more selective to certain types of fishes than the horizontal approach. Barham (1970) discussed various orientations adopted by fishes in the water and expanded the speculations of Harrisson (1967) upon the effects these may have in terms of net avoidance. He noted that lethargic, vertically oriented fishes dived downwards when approached by submersibles and suggested their escape reaction was more conducive to the avoidance of horizontally towed nets than was that of horizontally orientated fishes. Therefore, different procedures may select for different animals, and if quantitative estimates are to be comprehensive, the value of different techniques is indisputable.

Animals apparently residing in the deep sound scattering layers are generally reported as being "associated" with them. Indeed, convincing correlations between deep sound scattering layers and particular species or species groups are few and far between and generally not derived from net data (e.g., Barham, 1963, 1966; Backus et al., 1968). The identification of the principal components of a sound scattering layer from net data is fraught with interpretive difficulties. While some attempt has been made to name specifically "main scatterers" (e.g., Kashkin and Chindonova, 1971; Kashkin, 1974) the exact contribution such species make to the composition of deep sound scattering layers is uncertain.

The obtaining of reliable abundance estimates is prerequisite to the testing of correlations between species compositions and sound scattering layers. The uncertainties in providing these are symptomatic of poor sampling techniques giving an unbalanced understanding of the complex relations within the ecosystem. An accurate and comprehensive sampling procedure is essential if the fine details of these relations are to be examined in a meaningful way. Naturally, any estimate is biased by the type of net used and its size, but within the context of a consistent sampling procedure, catch variations or biases are of biological origin, due to differential species behavior. These features need to be examined and assessed. Recent studies have emphasized how differently mesopelagic species behave at any given time (e.g., Backus et al., 1968; Barham, 1970;

Roe, 1974; Merrett and Roe, 1974). Thus from submersibles, some species have been observed by day to be in general states of lethargy, either in aggregations (Backus et al., 1968) or occurring singly and adopting various orientations in the water (Barham, 1970). Yet other species have been seen in highly active states, horizontally oriented (Barham, 1970). Nocturnally species may behave in a different manner (Barham, 1970). Similarly, net data show that species differ in their food selection and feeding patterns (Paxton, 1967a; Merrett and Roe, 1974), and migratory behavior and its timings (Roe, 1974). With such apparent behavioral heterogeneity, it is inevitable that a truly representative sample cannot be obtained. Observations such as those above, however, give some insight into the biological causes of catch variability and perhaps in the future similar studies may provide a weighting on catch numbers based upon the influence of species' behavior. In turn, this may then give us reasonable estimates of species' composition and abundance with which to assess the importance of each species in generating the deep sound scattering layers. At the same time, however, biologically and acoustically our knowledge and understanding of the swimbladders of mesopelagic fishes is far from comprehensive and much work is required to unravel the details. Concerning the "problems and enigmas" of the deep scattering layer, Barham (1970) concluded "that to fully understand the complexities of this phenomenon, we must first understand the biology of mesopelagic fishes" - an apt sentiment.

ACKNOWLEDGEMENTS

We would like to thank Mssrs P. M. David, A. de C. Baker and Dr. M. V. Angel (I.O.S.) for their useful criticisms and discussions of the manuscript. Thanks also go to Miss R. Larcombe who made the figures.

REFERENCES

Angel, M. V. and Fasham, M. J. R. 1975. Analysis of the vertical and geographic distribution of the abundant species of planktonic ostracods in the North-East Atlantic. J. mar. biol. Ass. U.K. 55: 709-737.

Angel, M. V. 1977. Windows into a sea of confusion: Sampling limitations to the measurement of ecological parameters in oceanic midwater environments. In: "Oceanic Sound Scattering Prediction", N. R. Andersen and B. J. Zahuranec (Eds.), Plenum, N.Y., (this volume).

Aron, W., Raxter, N., Noel, R., and Andrews, W. 1964. A description of a discrete depth plankton sampler with some notes on the towing behavior of a 6-foot Isaacs-Kidd mid-water trawl and a one-meter ring net. Limnol. Oceanogr., 9 (3): 324-333.

Backus, R. H., Mead, G. W., Haedrich, R. L., and Ebeling, A. W. 1965. The mesopelagic fishes collected during Cruise 17 of the R/V Chain, with a method for analyzing faunal transects. *Bull. Mus. Comp. Zool. Harv.* 134 (5): 139-158.

Backus, R. H., Craddock, J. E., Haedrich, R. L., Shores, D. L., Teal, J. M., Wing, A. S., Mead, G. W., and Clarke, W. D. 1968. *Ceratoscopelus maderensis*: peculiar sound-scattering layer identified with this myctophid fish. *Science, 160,* (3831): 991-993.

Backus, R. H., Craddock, J. E., Haedrich, R. L., and Shores, D. L. 1970. The distribution of mesopelagic fishes in the equatorial and western North Atlantic Ocean. *J. mar. Res.* 28, (2): 179-201.

Badcock, J. 1970. The vertical distribution of mesopelagic fishes collected on the SOND cruise. *J. mar. biol. Ass. U.K.* 50: 1001-1044.

Badcock, J. and Merrett, N. R. 1976. Midwater fishes in the eastern North Atlantic. I. Vertical distribution and associated biology in 30°N, 23°W, with developmental notes on certain myctophids. *Progress in Oceanography.* 7: 3-58.

Baird, R. C. 1971. The systematics, distribution, and zoogeography of the marine hatchetfishes (Family Sternoptychidae). *Bull. Mus. Comp. Zool. Harv.* 142 (1): 1-128.

Baker, A. de C., Clarke, M. R. and Harris, M. J. 1973. The NIO combination net (RMT 1+8) and further developments of rectangular midwater trawls. *J. mar. biol. Ass. U.K.* 53: 167-184.

Barham, E. G. 1963. Siphonophores and the deep scattering layer. *Science.* 140 (3568): 826-828.

Barham, E. G. 1966. Deep scattering layer migration and composition: observations from a diving saucer. *Science* 151 (3716): 1399-1403.

Barham, E. G. 1970. Deep-sea fishes: lethargy and vertical orientation. *Proc. Internat. Symp. Biological Sound Scattering in the Ocean,* Warrenton, Va., 31 March - 2 April 1970. pp. 101-109.

Boden, B. P. and Kampa, E. M. 1967. The influence of natural light on vertical migrations of an animal community in the sea. *Symp. Zool. Soc. Lond.,* (19): 15-26.

Bone, Q. 1973. A note on the buoyancy of some lantern-fishes (Myctophoidei). *J. mar. biol. Ass. U.K.* 53: 619:633.

Bourbeau, F., Clarke, W. D., and Aron, W. 1966. Improvements in the discrete depth plankton sampler system. *Limnol. Oceanogr.,* 11 (3): 422-426.

Butler, J. L. and Pearcy, W. G. 1972. Swimbladder morphology and specific gravity of myctophids off Oregon. *J. Fish. Res. Bd Canada* 29: 1145-1150.

Clarke, M. R. 1969. A new midwater trawl for sampling discrete depth horizons. *J. mar. biol. Ass. U.K.* 49: 945-960.

Clarke, M. R. and Lu, C. C. 1974. Vertical distribution of cephalopods at 30°N 23°W in the North Atlantic. *J. mar. biol. Ass. U.K.* 54: 969-984.

Clarke, M. R. and Lu, C. C. 1975. Vertical distribution of cephalopods at 18°N 25°W in the North Atlantic. *J. mar. biol. Ass. U.K.* 55: 165-182.

Clarke, T. A. 1973. Some aspects of the ecology of lanternfishes (Myctophidae) in the Pacific Ocean near Hawaii. *Fish. Bull.* 71 (2): 401-434.

Clarke, T. A. 1974. Some aspects of the ecology of stomiatoid fishes in the Pacific Ocean near Hawaii. *Fish. Bull.* 72 (2): 337-351.

Clarke, W. D. 1966. Bathyphotometric studies of the light regime of organisms of the deep scattering layers. *Gen. Mot. Defense Research Laboratories, Final Report Contract* AT (04-3)-584, *Santa Barbara, California*.

Currie, R. I., Boden, B. P., and Kampa, E. M. 1969. An investigation on sonic-scattering layers: the R.R.S. "Discovery" SOND Cruise, 1965. *J. mar. biol. Ass. U.K.* 55: 165-182.

Cushing, D. H. 1951. The vertical migration of planktonic Crustacea. *Biol. Rev.*, 26: 158-192.

Ebeling, A. W. 1962. Melamphaidae I. Systematics and zoogeography of the species in the bathypelagic fish genus *Melamphaes* Günther. *Dana Rep.*, (58): 1-164.

Fasham, M. R. and Angel, M. V. 1975. The relationship of the zoogeographic distributions of the planktonic ostracods in the North-East Atlantic to the water masses. *J. mar. biol. Ass. U.K.* 55: 739-757.

Foxton, P. 1971/72. Observations on the vertical distribution of the genus *Acanthephyra* (Crustacea: Decapoda) in the eastern North Atlantic, with particular reference to species of the "purpurea" group. *Proc. R. Soc. Edinb.* (B) 73: 301-313.

Foxton, P. and Roe, H. S. J. 1974. Observations on the nocturnal feeding of some mesopelagic decapod crustaceans. *Mar. Biol.* 28: 37-49.

Gibbs, R. H. Jr. and Roper, C. F. E. 1970. Ocean Acre: preliminary report on vertical distribution of fishes and cephalopods. *Proc. Internat. Symp. Biological Sound Scattering in the Ocean,* Warrenton Va., 31 March - 2 April 1970. pp. 120-135.

Halliday, R. G. 1970. Growth and vertical distribution of the glacier lanternfish, *Benthosema glaciale*, in the northwestern Atlantic. *J. Fish. Res. Bd Canada,* 27: 105-116.

Harrisson, C. M. H. 1967. On methods for sampling mesopelagic fishes. *Symp. zool. Soc. Lond.* (19): 71-126.

Holton, A. A. 1969. Feeding behavior of a vertically migrating lanternfish. *Pacif. Sci.* 23: 325-331.

Kashkin, N. I. and Chindonova, Yu.G. 1971. Mesopelagic fishes as resonance scatterers in the deep-scattering layers of the Atlantic Ocean. *Oceanology,* 11 (3): 404-413.

Kashkin, N. I. 1974. Ichthyofauna of the sound-scattering layers in the northeastern Atlantic. *Oceanology,* 14 (3): 446:450.

Kleckner, R. C. and Gibbs, R. H. Jr. 1972. Swimbladder structure of Mediterranean midwater fishes and a method of comparing swimbladder data with acoustic profiles. *Mediterranean Biological Studies: Final Report ONR Contract N00014-67-A-0399-007 U. S. Govt. 1 (4)*: 230-281.

Krueger, W. H. 1972. Biological studies of the Bermuda Ocean Acre. IV. Life history, vertical distribution and sound scattering in the gonostomatid fish *Valenciennellus tripunctulatus* (Esmark). *Report to U. S. Navy Underwater System Center, Contract No. N00140-72-C-0315*: 1-37.

Lu, C. C. and Clarke, M. R. 1975a. Vertical distribution of cephalopods at 40°N, 53°N and 60°N at 20°W in the North Atlantic. *J. mar. biol. Ass. U.K.* 55: 143-163.

Lu, C. C. and Clarke, M. R. 1975b. Vertical distribution of cephalopods at 11°N, 20°W in the North Atlantic. *J. mar. biol. Ass. U.K.*, 55: 369-389.

Marshall, N. B. 1960. Swimbladder structure of deep-sea fishes in relation to their systematics and biology. *Discovery Rep.* 31: 1-122.

McGowan, J. A. and Williams, P. M. 1973. Oceanic habitat differences in the North Pacific. *J. exp. mar. Biol. and Ecol.*, 12: 187-217.

Merrett, N. R. and Roe, H. S. J. 1974. Patterns and selectivity in the feeding of certain mesopelagic fishes. *Mar. Biol.* 28: 115-126.

Moore, H. B. 1955. Variations in temperature and light response within a plankton population. *Biol. Bull. mar. biol. Lab., Woods Hole, 108*: 175-181.

Moore, H. B. and Foyo, M. 1953. A study of the temperature factor in twelve species of oceanic copepods. *Bull. Mar. Sci. Gulf Caribb.*, 13: 502-515.

Nafpaktitis, B. G. 1968. Taxonomy and distribution of the lanternfishes, genera *Lobianchia* and *Diaphus*, in the North Atlantic. *Dana Rep.* (73): 1-131.

Paxton, J. R. 1967a. A distributional analysis for the lanternfishes (Family Myctophidae) of the San Pedro Basin, California. *Copeia, 1967* (2): 422-440.

Paxton, J. R. 1967b. Biological notes on Southern California lanternfishes (Family Myctophidae). *Calif. Fish and Game,* 53 (3): 214-217.

Pearcy, W. G. 1964. Some distributional features of mesopelagic fishes off Oregon. *J. mar. Res.* 22 (1): 83-102.

Pearcy, W. G. and Laurs, R. M. 1966. Vertical migration and distribution of mesopelagic fishes off Oregon. *Deep-Sea Res.* 13: 153-166.

Pearcy, W. G. and Mesecar, R. S. 1970. Scattering layers and vertical distribution of oceanic animals off Oregon. *Proc. Internat. Symp. Biological Sound Scattering in the Ocean, Warrenton, Va., 31 March - 2 April 1970.* pp. 381-394.

Robison, B. H. 1972. Distribution of the midwater fishes of the Gulf of California. *Copeia, 1972,* (3): 448-461.

Roe, H. S. J. 1974. Observations on the diurnal vertical migrations of an oceanic animal community. *Mar. Biol.* 28: 99-113.

Schroeder, E. H. 1963. North Atlantic temperatures at a depth of 200 meters. *Serial Atlas Mar. Environment Folio 2,* Am. Geogr. Soc., *1-11,* 9 pls.

Talbot, J. J. 1973. Some aspects of the ecology and bathymetric distribution of *Cyclothone acclinidens* in the Santa Catalina Basin. *Copeia, 1973,* (3): 600-601.

Vinogradov, M. E. 1968. Vertical distribution of the oceanic zooplankton. pp. 1-320. *Moscow: Nauka.* (Translated from Russian, Israel Program for Scientific Translations, Jerusalem, 1970).

Vinogradov, M. E. and Parin, N. V. 1973. On the vertical distribution of macroplankton in the tropical Pacific. *Oceanology, 13,* (1): 104-113.

Zahuranec, B. J. and Pugh, W. L. 1970. Biological results from scattering layer investigations in the Norwegian Sea. *Proc. Internat. Symp. Biological Sound Scattering in the Ocean,* Warrenton, Va., 31 March - 2 April 1970, pp. 360-380.

Studies on Pelagic Animal Biomasses

Maurice Blackburn

*Institute of Marine Resources,
University of California, San Diego*

ABSTRACT

Selected data on biomasses of pelagic micronekton and large nekton, together with zooplankton, are given for different latitudes in the open North Pacific, and for a coastal upwelling area in the Atlantic.

The Pacific data, taken from the literature, show that several kinds of micronekton occur down to about 1000 m in the subtropics, 2000 m in the tropics, and 4000 m in the subarctic. These animals are principally mesopelagic fish, euphausiids and decapods. Areal distribution of their total biomass is very broadly like that of zooplankton, although the relation between the two biomasses is not linear. Comparable biomass data for large nekton are not available. In the upwelling study, on the other hand, good estimates of biomass of large nekton (epipelagic fish) were made acoustically.

INTRODUCTION

There is general agreement that pelagic sound scatterers are mostly animals 1 cm or more in size, i.e. macroplankton, micronekton and large nekton. Macroplankton and micronekton are more or less synonymous and refer to organisms about 1 to 10 cm long. Since they are all relatively mobile animals, apart from some tunicates and coelenterates, micronekton seems the better name. Large nekton are over 10 cm and include cetaceans, fishes, elasomobranchs, cephalopods and a few reptiles. They are mainly epipelagic, except possibly for some cephalopods and various fish. Both micronekton and large nekton feed on other nekton, small zooplankton (<1 cm), phytoplankton or detritus.

Large nekton of coastal waters, such as clupeids and carangids, frequently aggregate in dense schools or layers that can be detected in the water column by sonic methods. Those of the open ocean, such as tunas, seem not to form such aggregations at depth, although some of them school at the surface. Micronekton occur in sound-scattering aggregations in coastal and oceanic waters.

Measurement and mapping of biomasses of micronekton and large nekton is more difficult than for zooplankton biomass. Biomass of micronekton can be estimated from catches made by research midwater trawls, subject to various assumptions about their ability to avoid such nets. This work is time-consuming in the ocean, because of the great depth ranges that must be sampled. Quantitative sampling of large nekton with research tools is generally impossible because the animals have high avoidance. Where commercial fisheries exist the catches can be mapped, but it is extremely difficult to estimate biomass from them. The best methods of estimating biomass of large nekton seem to be acoustic, although sampling is still necessary in order to identify the species taken (*Thorne, Mathisen, Trumble, and Blackburn*, in press).

As a result, biomass data (*e.g.*, in g/m^2) are sparse for both micronekton and large nekton. This paper presents two sets of those data, one for an open ocean and one for a small coastal upwelling area, together with estimates of biomass of small zooplankton (sometimes called mesoplankton). The main purpose of the paper is to show in very broad terms what kinds of nekton biomass data are available with existing methods. Relations between nekton and zooplankton biomasses are suggested.

LARGE-SCALE DISTRIBUTION IN THE PACIFIC

This part of the paper reviews information on distribution of micronekton and summarizes information on their trophic relations from selected oceanic studies made in the Pacific. Micronekton are generally higher in the trophic pyramid than zooplankton, although there is overlap between them and exact trophic levels cannot be specified. It is convenient to discuss micronekton biomass in relation to zooplankton biomass. Zooplankton in all these studies were taken with nets of mesh-size 0.31 to 0.38 mm, so biomasses are comparable. Micronekton biomasses also appear broadly comparable, except that Soviet data refer to larger animals (≥ 3 cm) than the French and U.S. (≥ 1 cm). U. S. data on zooplankton and micronekton biomasses are displacement volumes, here converted to wet weight (1 ml = 1 g). Charts of actual and potential commercial catches of large nekton appear in F.A.O. (1972), but data on their biomass are not available.

Distribution of Total Micronekton Biomass

Figure 1 shows the general distribution of zooplankton biomass at 0-100 m in the Pacific. It is from F.A.O. (1972), redrawn from

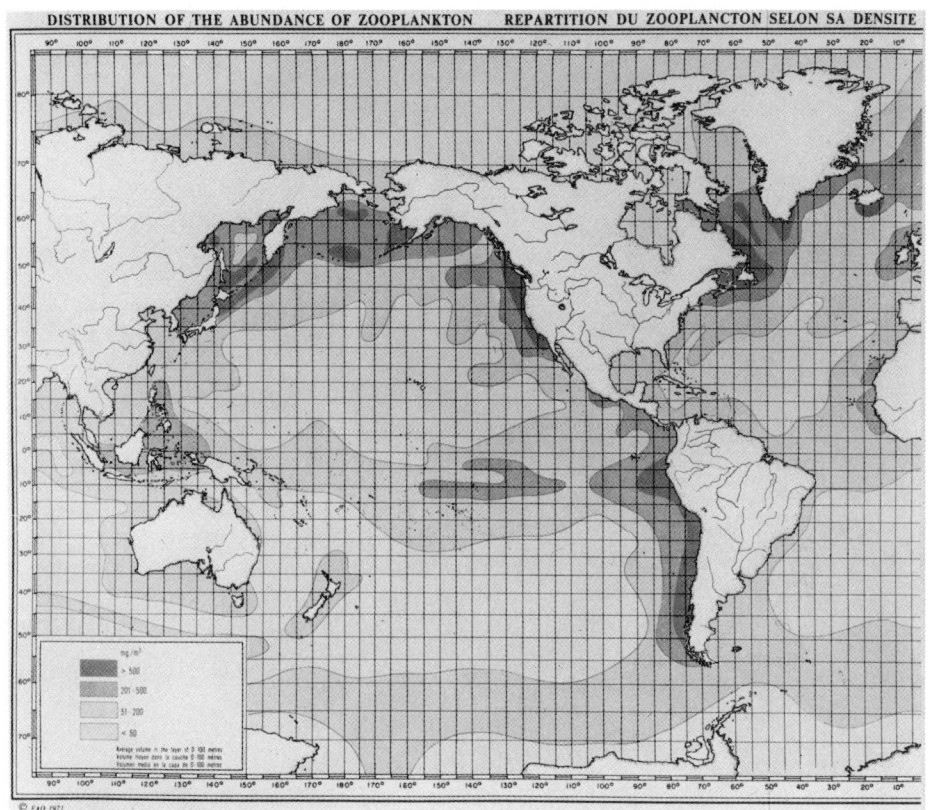

Fig. 1. Distribution of abundance of zooplankton at 0-100 m in the Pacific. From F.A.O. (1972), redrawn from Bogorov et al. (1968).

Bogorov et al. (1968). Reid (1962) published a similar map of zooplankton biomass at 0-150 m. There are maxima in the high latitudes and tropics and minima in the subtropics while high-latitude values average higher than tropical. The same situation appears in Pacific-wide charts of primary production (Koblentz-Mishke et al., 1970) and phosphate-phosphorus at 0-100 m (Reid, 1962). According to Vinogradov (1968) the ratio between biomasses of near-surface zooplankton (0-500 m) and total subsurface zooplankton (500-4000 m^2) is fairly constant over the ocean, at 2:1.

Figure 2 shows areas of the Pacific where distribution of oceanic micronekton was studied with research midwater trawls and large plankton nets. The four investigations reviewed are: Soviet cruises of VITYAZ and OB (Vinogradov, 1968); a series of French cruises called CARIDE (Legand et al., 1972); a series of U. S. cruises called EASTROPAC (Blackburn et al., 1970; Blackburn, unpublished data); and a series of cruises made by the Pacific Oceanic Fishery Investigations (POFI)

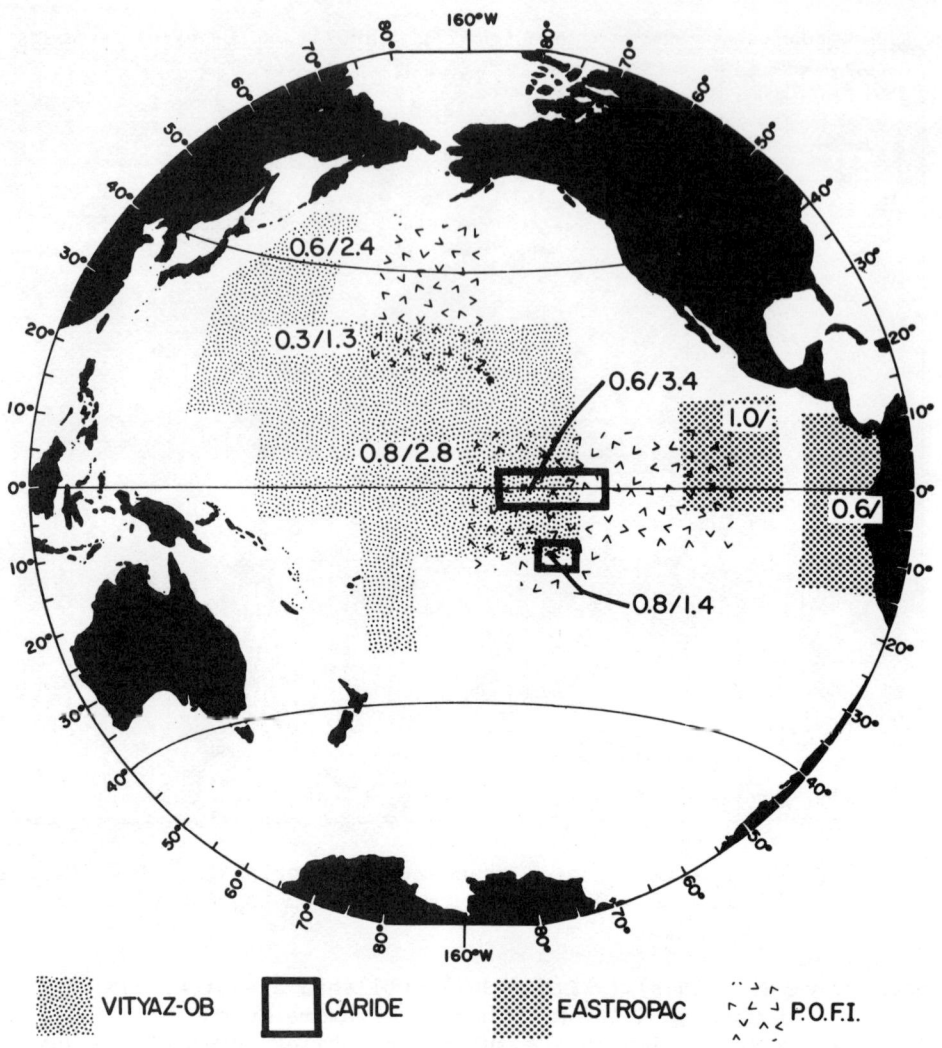

Fig. 2. Approximate areas of principal investigations on micronekton in the Pacific. Numbers are mean biomasses of micronekton in g/m^2 for the areas. Numbers to the left of the solidus are for the layer 0-200 m; numbers to the right of the solidus, where given, are for 0-1000 m. For further explanation, see Table 1 and text.

of the U.S. Bureau of Commercial Fisheries (*King and Iversen*, 1962). The first three yield comparable estimates of absolute biomass (Table 1). For the POFI study, estimates of absolute biomass are low and suspect but values of relative biomass are useful. In Table 1, mean biomasses of zooplankton and micronekton from CARIDE and EASTROPAC are equally weighted for day and night observations. The Soviet data

Table 1. Mean biomass (g/m^2) for specified areas and ranges of depth in the Pacific. Z is zooplankton, M is micronekton. See Figure 1 for location of areas.

A. Vityaz - Ob Data (original)

Depth Range (m)	12°N - 12°S			12° - 40° (N & S)			40°N - 50°N		
	Z	M	M/Z	Z	M	M/Z	Z	M	M/Z
0-100	5.79	0.76	0.13	2.68	0.00	0.00	34.80	0.00	0.00
0-200	7.67	0.79	0.10	4.14	0.00	0.00	45.70	0.00	0.00
0-1000	12.61	2.80	0.22	8.58	1.04	0.12	156.75	1.79	0.01
0-4000	14.27	8.74	0.61	11.18	1.34	0.12	191.32	4.50	0.02

B. CARIDE Data

Depth Range (m)	0°			10°S		
	Z	M	M/Z	Z	M	M/Z
0-110	5.90	0.37	0.06	5.70	0.75	0.13
0-225	7.10	0.61	0.09	6.32	0.76	0.12
0-950	8.34	3.41	0.41	8.13	1.43	0.18

C. EASTROPAC Data

Depth Range (m)	Western Area			Eastern Area		
	Z	M	M/Z	Z	M	M/Z
0-200	27.70	1.02	0.04	42.44	0.62	0.01

D. Vityaz - Ob Data (modified; see text)

Depth Range (m)	12°N - 12°S			12° - 40° (N & S)			40°N - 50°N		
	Z	M	M/Z	Z	M	M/Z	Z	M	M/Z
0-200	7.67	0.79	0.10	4.14	0.30	0.07	45.70	0.60	0.01
0-1000	12.61	2.80	0.22	8.58	1.34	0.16	156.75	2.39	0.02
0-4000	14.27	8.74	0.61	11.18	1.64	0.15	191.32	5.10	0.03

are as given in the literature, and assumed to be based on both day and night observations. Some data of Table 1 are shown in Figure 2. The data of *Omori* (1974) for the western Pacific are of interest as shown later, but are not comparable with those mentioned above. Micronekton and zooplankton were not clearly distinguished by *Omori*, and his zooplankton were taken with nets of 1.0 mm mesh.

Figure 3 from *Vinogradov* (1968) is a schematic picture of some relations between micronekton and zooplankton (respectively termed macroplankton and mesoplankton) in a longitudinal section in the

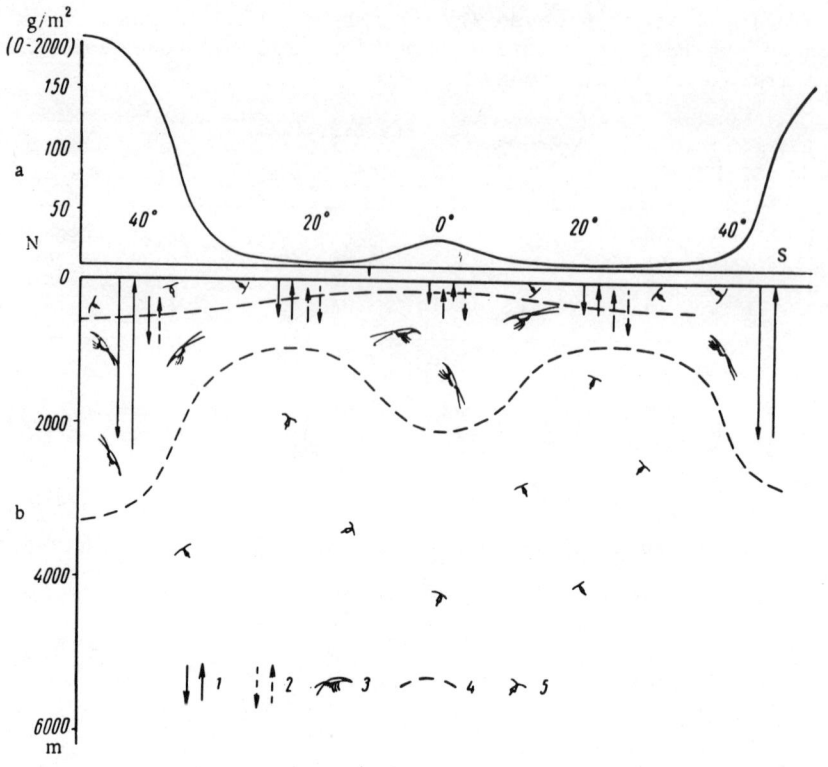

Fig. 3. Scheme of distribution of zooplankton (or mesoplankton) and micronekton (or macroplankton) in a longitudinal section of the Pacific Ocean. Above: biomass of both kinds combined in g/m², 0-2000 m. Below: showing vertical location of zone of greatest concentration of large organisms, between dashed lines. From *Vinogradov* (1968, Fig. 55).

Pacific. The upper part shows how the total biomass of both varies with latitutde in the top 2000 m. The lower part indicates the sections of water column in which micronekton are concentrated, between dashed lines. It suggests few micronekton in the top 200 m at high latitudes, but this is probably not true if organisms down to 1 cm are included. The great thickness of the layer in which micronekton are most abundant in high latitudes and the tropics, relative to that in the subtropics is most interesting.

In Table 1 and Figure 2, the Soviet, French and U. S. estimates of micronekton biomass agree fairly well as far as the tropics are concerned. Values at 0-200 m range from 0.6 to 1.0 g/m². Values at 0-1000 m (Soviet and French only) range from 1.4 to 3.4 g/m². The Soviet data show that the increase continues to 4000 m, proportionately more than the corresponding increase in zooplankton, although the actual concentrations (g/m) of both decline.

Vinogradov's data in Table 1 (part A) show less micronekton at all depths in subarctic waters, and much less in subtropical waters, than in tropical. They show none at all at 0-200 m in subtropical and subarctic waters. This is probably because only animals $\geq$ 3 cm were considered, which would exclude many euphausiids and other animals known to occur significantly at those depths and latitudes. POFI data on relative biomass (ml/hr) are available over a similar range of the latitudes covered by the VITYAZ and OB. They show approximately equal values in equatorial and subarctic waters, with subtropical values about half as high (Figure 4). These data refer to various depth ranges between 0-100 and 0-400 m, the mean range being 0-245 m. The best estimate of absolute biomass of micronekton at that depth at the Equator is from the CARIDE data, approximately 0.6 g/m^2. Then the Soviet zero biomasses at 0-200 m should be raised by about 0.6 at 40°-50°N and 0.3 at 12°-30°, and those increases added to the totals at 0-1000 and 0-4000 m. These modifications to the VITYAZ-OB data are shown in part D of Table 1.

Mean biomasses from parts B, C and D of Table 1 are shown in Figure 2 for the 0-200 and 0-1000 m layers. Together with Figure 3 they give a summary of what is known about areal and vertical distribution of total biomass of micronekton $\geq$ 1 cm, in the open waters of the Pacific. The areal distribution is very broadly like the distributions of zooplankton, primary production and phosphate, with subtropical biomass lower than tropical and subarctic. The tropical biomass of micronekton may, however, be higher than the subarctic, which is the reverse of the situation with the other properties. It is quite likely that some kinds of micronekton are undersampled because they can avoid nets. Thus the estimates in Table 1 and Figure 2 are minima.

The micronekton/zooplankton ratios in Table 1 are only approximate, but are of interest in view of predator/prey relations that must be involved. At 0-200 m they are 7 to 12% where zooplankton is 4 to 8 g/m^2, 4% where zooplankton is 27 g/m^2, and 1% where zooplankton is 42 to 46 g/m^2. Most of these data are for tropical areas where the ratio between biomass and production might be fairly uniform for zooplankton, and similarly for micronekton. Perhaps the transfer of material from zooplankton to micronekton is more efficient when zooplankton is relatively scarce. Similarly, *Blackburn* (1973) showed that biomass of fish-cephalopod micronekton varied with biomass of zooplankton to a power about 0.2 to 0.3, at 0-200 m in the western EASTROPAC area. It is doubtful if a similar conclusion can be drawn from the fewer micronekton/zooplankton ratios at 0-1000 m. The main interest in those ratios is that they are higher than at 0-200 m in the same areas. In the tropics they reach 20 to 40% at 0-1000 m, and more at deeper ranges. According to *Vinogradov* (1968) biomass of micronekton is 500% of that of zooplankton between 1000 and 2000 m, in the tropics.

Seasonal differences in micronekton biomass have been little studied, except at 0-200 m in the western area of EASTROPAC (*Blackburn et al.*, 1970). They are small there except for crustacean

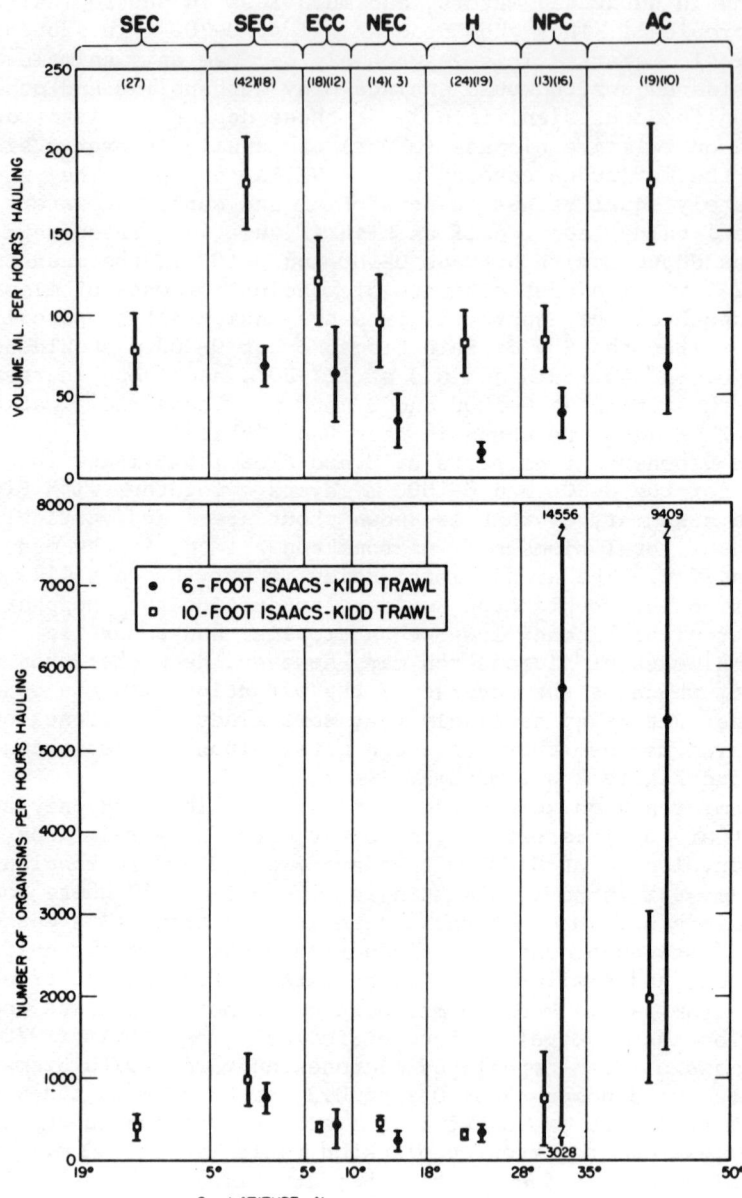

Fig. 4. Variation with latitude in average volumes (above) and numbers of organisms (below) captured per hour with Isaacs-Kidd midwater trawls in the Pacific. From *King and Iversen* (1962, Figure 13).

micronekton (mostly euphausiids) at the Equator, and probably do not invalidate comparisons already made for tropical waters.

Kinds of Micronekton and Their Trophic Relations

Micronekton in the 0-200 m layer had the following approximate composition by biomass on EASTROPAC cruises at night (Blackburn et al., 1970):

Western area:
- 36% mesopelagic fish
- 33% euphausiids
- 17% decapod crustaceans
- 5% epipelagic fish (mostly leptocephali)
- 5% cephalopods
- 4% other crustaceans

Eastern area:
- 57% mesopelagic fish
- 16% decapod crustaceans
- 11% euphausiids
- 7% other crustaceans
- 5% epipelagic fish (mostly leptocephali)
- 4% cephalopods

It is not clear why the percentages of mesopelagic fish and euphausiids are so different in these two areas, but clearly those groups together with decapod crustaceans are the three major components, totalling 84 to 86%, in each case. The fish are principally myctophids and gonostomatids, the decapods principally sergestids and penaeids. Similar data on composition of daytime biomass, which is much lower at 0-200 m, are not readily available. The lists of organisms given by King and Iversen (1962) cannot be expressed as biomass in the same way, but mesopelagic fish, euphausiids and decapods were the principal forms taken at night (about 0-245 m) in all latitudes. Euphausiids were most abundant and decapods least abundant north of 28°N.

Table 2 shows the data of Legand et al. (1972) on percentage composition of biomass by taxa of tropical Pacific micronekton at different depths, by day and night down to 1200 m. The day and night data are broadly comparable below 450 m, but quite different above that level with fishes, euphausiids and decapod crustaceans being much more abundant at night. They are the principal groups at night. On the basis of such data, the authors recognize the following three faunas within the micronekton:

Superficial fauna - animals that live entirely in the 0-450 m layer, whether or not they migrate diurnally: phronimid amphipods, certain euphausiids, heteropods, pteropods, juvenile cephalopods; larvae of fish and decapods, and adults of a few fish (especially Vinciguerria).

Table 2. Distribution of biomass of CARIDE micronekton by taxa, in percent of total micronekton biomass at each depth range; from Legand et al. (1972) Table 11.

Depth Range (m)	Time	Fishes	Euphausiids	Carids	Sergestids	Peneids	Mysids	Heteropods	Fish Larvae	Cephalopods	Amphipods	Phronimids	Pteropods
0-110	Day	6.4	0	0	0	0	0	28.8	1.7	35.9	16.2	7.4	3.6
110-225		0	0	0	0	0	0	3.9	2.1	75.0	4.2	5.5	9.4
225-450		0	5.6	3.8	0	0	0	0	3.5	54.3	17.6	0	15.2
450-700		70.9	9.6	9.9	8.2	1.2	0	0	0.1	0	0	0	0
700-950		62.8	0	10.1	4.2	8.6	8.9	0	0.5	4.0	0.9	0	0
950-1200		69.2	0	19.2	0	4.6	6.3	0	0.3	0	0.5	0	0
0-110	Night	29.8	20.1	8.9	15.3	0.5	0	5.3	1.0	13.2	3.4	0.9	1.5
110-225		33.0	14.5	14.3	18.5	4.7	0	1.3	1.7	9.6	1.5	0.2	0.7
225-450		21.8	21.1	23.7	2.6	16.3	0	0	0	13.1	1.4	0	0
450-700		80.8	2.7	7.7	3.8	4.7	0.1	0	0.1	0	0	0	0
700-950		77.2	0	4.9	3.9	1.9	9.1	0	0.4	2.1	0.4	0	0.1
950-1200		67.6	0	21.9	1.1	2.6	6.7	0	0	0	0	0	0

Deep fauna — animals that live entirely below 450 m, whether or not they migrate diurnally: fish (*Cyclothone* and *Sternoptyx*), mysids, some decapod crustaceans (penaeids, sergestids and carids), a few euphausiids, adult cephalopods, some amphipods, coelenterates, and some fish larvae.

Interzonal fauna — animals that live below 450 m by day and migrate above that level at night: most mesopelagic fish (myctophids, gonostomatids, etc.), most euphausiids, and most decapod crustaceans (as above).

The diet of micronektonic fish and cephalopods was studied from stomach contents by *Legand et al.* (1972). Copepods tend to be predominant in the diet of small individuals, euphausiids and other larger crustaceans in medium-sized individuals. For mesopelagic fish these results broadly agree with other studies (*Hopkins and Baird*, 1977). The deep fauna is assumed to derive most of its energy from the superficial flora and fauna, through the medium of the interzonal fauna. That is, the interzonal migrants obtain food near the surface at night, and are eaten by the deep fauna during the day.

Omori (1974) gave data on occurrence of fish, euphausiids and decapods over a range of depth similar to that of the French studies, at various latitudes (40°N to 0°) along 150°E longitude in the western Pacific. Above 500 m all three groups were well represented, but euphausiid biomass was generally minor below 500 m.

Omori estimated that euphausiids form 35 to 50% of the micronekton of the 0-1000 m layer, fish 20 to 45% and decapods 15 to 25%. He reviewed information on the feeding of decapods and concluded they are mainly carnivores (feeding especially on copepods and euphausiids), but also scavengers consuming dead organisms, fecal pellets, etc.

From various data given by *Vinogradov* (1968) it appears that the tropical micronekton biomass below 1000 m is mostly fishes, decapods and euphausiids, with the fishes and decapods disappearing below 2000 m. The fish are mainly *Cyclothone*, the decapods mainly Caridea, and the euphausiids mainly *Bentheuphausia*. The subarctic biomass differs in that decapods remain important to 4000 m. The subtropical biomass below 1000 m is minor, as shown in Table 1. *Vinogradov* considers most micronekton to be carnivores, except euphausiids which may also eat phytoplankton and detritus, as shown also by *Mauchline and Fisher* (1969). He envisages that these midwater carnivores obtain food by migrating towards the surface or by feeding on those that have returned from such migrations, and to a minor extent by scavenging. Thus little except detritus remains to be eaten below the layers in which these micronekton occur, whereby the zooplankton is extremely sparse at those depths.

RESULTS FROM A COASTAL UPWELLING AREA

Figure 5 shows an upwelling area off northwest Africa where an attempt was made to obtain data on the pelagic animal biomass. The

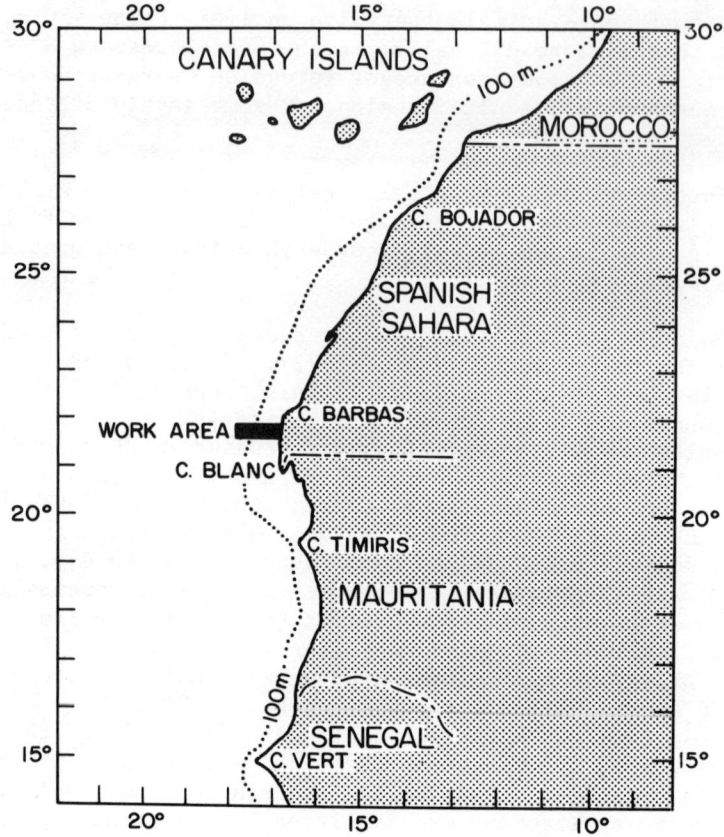

Fig. 5. Location of area of upwelling study.

work was done during expedition JOINT-I of the Coastal Upwelling Ecosystems Analysis (CUEA) program from March to May 1974, under Grant GX-33502 from the National Science Foundation. Preliminary results are available with regard to the analysis of collected material.

Zooplankton

The data given are unpublished. Net hauls for zooplankton on JOINT-I were made vertically from 200 m or the bottom, whichever was less, to the sea surface. Nets were of uniform mesh size, 102μ. One 1/4-aliquot of each catch was filtered through sieves in the laboratory, yielding subsamples in four size-ranges, approximately 100-200, 200-500, 500-1000 and >1000μ. These subsamples were weighed wet and standardized in g/m^2. Another aliquot was filtered in the same way to yield subsamples that were preserved for sorting. Another aliquot was generally used to determine the concentration of chlorophyll in

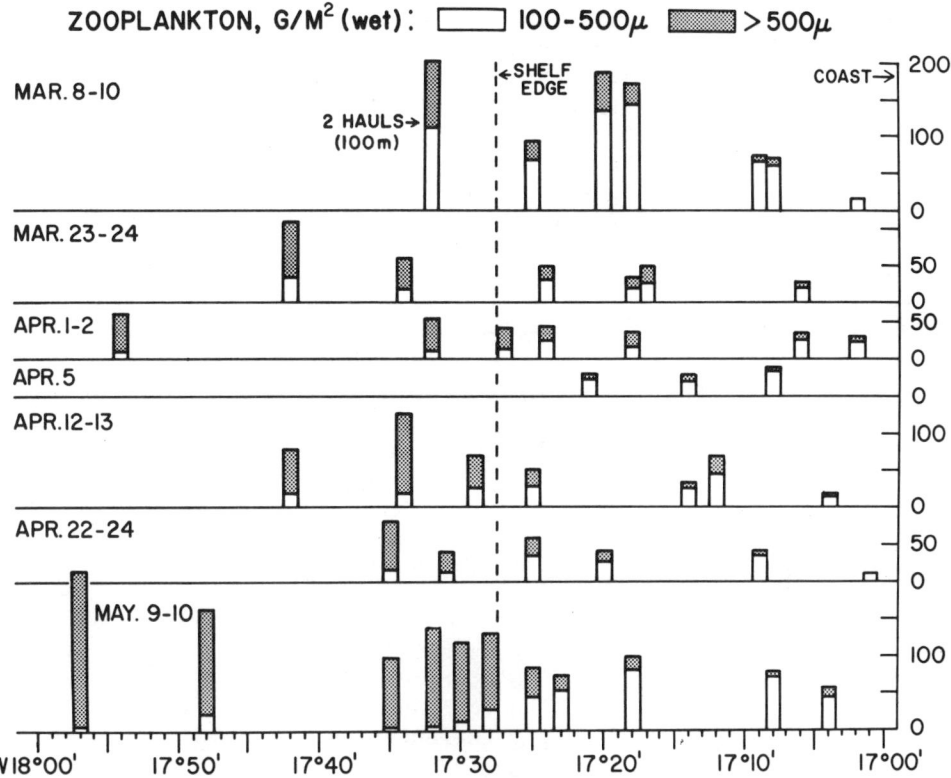

Fig. 6. Zooplankton biomass in the area of Figure 5.

the total sample. This was subsequently partitioned among the four size-fractions by inspection of the preserved samples, and converted to wet weight of phytoplankton contamination by ratios from *Lorenzen* (1968) and *Cushing et al.*, (1958). Sorting of preserved samples is not complete.

Figure 6 shows distribution of zooplankton biomass at 100-500μ and >500μ, in repetitive series of hauls across the area of study at about 21°40'N. Daytime hauls on the continental slope are omitted because some zooplankton might be below 200 m. These data have not been corrected for phytoplankton contamination but give a general idea of the variability of biomass. Most of the zooplankton on the shelf are small while most on the slope are large. Most of the large organisms are actually >1000μ. The change from predominantly small to large zooplankton occurs very close to the edge of the shelf. Within the shelf area there is a tendency for biomass to be least near the coast. Within the slope area (as far offshore as it was sampled) there is no clear spatial trend.

A three factor analysis of variance, by area (shelf vs. slope), period, and size of zooplankton, was performed on most of the data in

Figure 6 after subtracting the phytoplankton. It confirmed the area-size interaction just mentioned. No significant difference in biomass by period was detected.

The mean of 23 shelf hauls in the JOINT-I area, from March 23 to May 10, was 44 g/m^2 of zooplankton (free of phytoplankton): 28% at 100-200μ, 37% at 200-500μ, 12% at 500-1000μ, and 23% at >1000μ. The mean of 15 slope hauls was 102 g/m^2 (free of phytoplankton): respectively 6%, 8%, 7% and 79% in the above-mentioned size-ranges.

The approximate taxonomic composition by biomass in each size-fraction has been determined for a few hauls. From these and the preceding data are obtained the following very preliminary estimates of taxonomic composition of mean shelf and slope biomass:

```
Shelf:   27 g copepods
          5 g euphausiids
          2 g coelenterates
          2 g cladocerans
          2 g cephalochordate larvae
          1 g thaliaceans
          5 g others
         ―――
         44 g

Slope:   39 g euphausiids
         34 g copepods
         10 g thaliaceans
         10 g coelenterates
          1 g cladocerans
          1 g cephalochordate larvae
          7 g others
         ―――
        102 g
```

Thaliaceans and coelenterates are certainly herbivores and carnivores respectively, and are approximately equal in biomass. Chaetognaths, a minor component of "others", are also carnivores. The other groups probably eat both phytoplankton and zooplankton (including microzooplankton not sampled here) to some extent. It seems very probable that the shelf zooplankton is predominantly phytophagous, and the slope zooplankton significantly so. Phytoplankton biomass (measured as chlorophyll) was always relatively high on the shelf, except close inshore where zooplankton was also scarce. It was sometimes high on the slope as well.

Large Pelagic Nekton

Biomass was estimated from digital data obtained on frequent underway surveys with acoustic equipment (*Thorne et al.*, in press). The echosounders had frequencies of 120 and 38 kHz. Echo intensities were measured and integrated over time periods of 10 min and 10 m depth intervals. The integrated echo intensities were converted to fish densities in g/m^2 using the acoustic calibration data for the equipment and assuming specific target strengths (based on experience) for 1 kg of fish at the different frequencies.

There was much patchiness in space and time, but a general pattern was evident as follows: a rather discrete high concentration centered at the shelf edge, whose mean biomass fell from about 150 to 75 g/m^2 between April 6 and May 6; and a more diffuse concentration on the shelf, whose mean biomass rose from about 40 to 80 g/m^2 over the same period. This suggested the presence of two different kinds of pelagic fish.

Blackburn and Nellen (in press) showed, from studies of fish eggs and larvae taken in the JOINT-I zooplankton, that the fish at the shelf edge were probably horse mackerel, *Trachurus trachurus*; and those on the shelf predominantly sardine, *Sardina pilchardus*. This was confirmed later by the fishing results of a Polish research vessel. The decline in *Trachurus* biomass may have been due to fishing. The increase in biomass of *Sardina* was due to movement into the JOINT-I area, perhaps in response to an increase in phytoplankton biomass which occurred there.

The food of *Trachurus* in this area is known to be predominantly euphausiids and copepods (Blackburn, 1975). The food of *Sardina* in the area has not been studied. In Morocco and elsewhere *Sardina* eats both phytoplankton and small zooplankton. Blackburn and Nellen (in press) showed that its occurrence in the JOINT-I area was always at midshelf, where maximum biomass of both phytoplankton and small zooplankton (100-500µ) occurred. It could therefore have eaten either. Time changes in abundance of *Sardina* on JOINT-I appear roughly correlated with abundance of phytoplankton, but not zooplankton. It is therefore likely that phytoplankton was its main food.

From the data of Thorne et al. (in press) the mean biomass of pelagic nekton was very approximately 60 g/m^2 over the whole area and period of JOINT-I, about the same as for total zooplankton.

Micronekton

Mesopelagic fish were caught in micronekton nets just beyond the edge of the shelf. Eleven hauls were made there between 100 m and the surface on the same night. Assuming 90% water acceptance and no avoidance, the mean biomass of fish was 0.2 g/m^2. The mean biomass of total micronekton at 0-200 m was probably about 0.5 g/m^2, or about 0.5% of that of zooplankton on the continental slope.

REFERENCES

Blackburn, M. 1973. Regressions between biological oceanographic measurements in the eastern tropical Pacific and their significance to ecological efficiency. *Limnol. Oceanogr.* 18: 552-563.

Blackburn, M. 1975. Summary of existing information on nekton of Spanish Sahara and adjacent regions, northwest Africa. *Coastal Upwelling Ecosystems Analysis*, Tech. Rep. 8. 49 pp.

Blackburn, M., R. M. Laurs, R. W. Owen and B. Zeitzschel. 1970. Seasonal and areal changes in standing stocks of phytoplankton, zooplankton, and micronekton in the eastern tropical Pacific. *Mar. Biol.* 7: 14-31.

Blackburn, M. and W. Nellen. In press. Distribution and ecology of pelagic fish studied from eggs and larvae in an upwelling area off Spanish Sahara. *Fish. Bull.*

Bogorov, V. G., M. D. Vinogradov, N. M. Voronina, I. P. Kanaeva and I. A. Suetova. 1968. Distribution of zooplankton biomass within the superficial layer of the world ocean. *Dokl. Akad. Nauk SSSR* 182: 1205-1207. (in Russian).

Cushing, D. H., G. F. Humphrey, K. Banse and T. Laevastu. 1958. Report of the committee on terms and equivalents. *Rapp. Proc.-Verb. Cons. Perm. Internat. Explor. Mer.* 144: 15-16.

F.A.O. 1972. Atlas of the living resources of the seas. F.A.O. Fish. Circ. 126.

Hopkins, T. L. and R. C. Baird. 1977. Aspects of the feeding ecology of oceanic midwater fishes. In: *"Oceanic Sound Scattering Prediction"*, N. R. Andersen and B. J. Zahuranec (Eds.), Plenum, N.Y., (this volume).

King, J. E. and R. T. B. Iversen. 1962. Midwater trawling for forage organisms in the central Pacific 1951-1956. *U. S. Fish Wildl. Serv., Fish. Bull.* 62: 271-321.

Koblentz-Mishke, O. J., V. V. Volkovinsky and J. G. Kabanova. 1970. Plankton primary production of the world ocean. In: *Scientific Exploration of the South Pacific*: 183-193. (Wooster, W. S., Ed.) National Academy of Sciences, Washington.

Legand, M., P. Bourret, P. Fourmanoir, R. Grandperrin, J. A. Gueredrat, A. Michel, P. Rancurel, R. Repelin and C. Roger. 1972. Relations trophiques et distributions verticales en lilieu pélagigue dans l'Océan Pacifique intertropical. *Cah. O.R.S.T.O.M., sér. Océanogr.* 10: 303-393.

Lorenzen, C. J. 1968. Carbon/chlorophyll relationships in an upwelling area. *Limnol. Oceanog.* 13: 202-204.

Mauchline, J. and L. R. Fisher. 1969. The biology of euphausiids. *Adv. Mar. Biol.* 7: 1-454.

Omori, M. 1974. The biology of pelagic shrimps in the ocean. *Adv. Mar. Biol.* 12: 233-324.

Reid, J. L. 1962. On circulation, phosphate-phosphorus content, and zooplankton volumes in the upper part of the Pacific Ocean. *Limnol. Oceanogr.* 7: 287-306.

Thorne, R. E., O. A. Mathisen, R. J. Trumble and M. Blackburn. In press. Distribution and abundance of pelagic fish off Spanish Sahara during CUEA Expedition JOINT-I. *Deep-Sea Res.*

Vinogradov, M. E. 1968. Vertical distribution of the oceanic zooplankton. *Akad. Nauk SSSR, Moscow.* (in Russian). English translation 1970, Israel Program of Scientific Translations, 339 pp.

Blackburn, M. 1975. Summary of existing information on nekton of Spanish Sahara and adjacent regions, northwest Africa. *Coastal Upwelling Ecosystems Analysis, Tech. Rep. 8*, 49 p.

Blackburn, M., R. M. Laurs, R. W. Owen and B. Zeitzschel. 1970. Seasonal and areal changes in standing stocks of phytoplankton, zooplankton, and micronekton in the eastern tropical Pacific. *Mar. Biol.* 7: 14-31.

Blackburn, M. and W. Nellen. MS. Distribution and ecology of pelagic fish studied from eggs and larvae in an upwelling area off Spanish Sahara.

Bogorov, V. G., M. E. Vinogradov, N. M. Voronina, I. P. Kanaeva and I. A. Suetova. 1968. Distribution of zooplankton biomass within the superficial layer of the world ocean. *Dokl. Akad. Nauk SSSR 182*: 1205-1207. (in Russian).

Cushing, D. H., G. F. Humphrey, K. Banse and T. Laevastu. 1958. Report of the committee on terms and equivalents. *Rapp. Proc.-Verb. Cons. Perm. Internat. Explor. Mer. 144*: 15-16.

F.A.O. 1972. Atlas of the living resources of the seas. F.A.O. Fish. Circ. 126.

Hopkins, T. L. and R. C. Baird. 1976. Aspects of the feeding ecology of oceanic midwater fishes.

King, J. E. and R. T. B. Iversen. 1962. Midwater trawling for forage organisms in the central Pacific 1951-1956. *U. S. Fish Wildl. Serv., Fish. Bull. 62*: 271-321.

Koblentz-Mishke, O. J., V. V. Volkovinsky and J. G. Kabanova. 1970. Plankton primary production of the world ocean. In *Scientific Exploration of the South Pacific*: 183-193. (Wooster, W. S. ed.) National Academy of Sciences, Washington.

Legand, M., P. Bourret, P. Fourmanoir, R. Grandperrin, J. A. Gueredrat, A. Michel, P. Rancurel, R. Repelin and C. Roger. 1972. Relations trophiques et distributions verticales en milieu pelagique dans l'Ocean Pacifique intertropical. *Cah. O.R.S.T.O.M., ser. Oceanogr. 10*: 303-393.

Lorenzen, C. J. 1968. Carbon/chlorophyll relationships in an upwelling area. *Limnol. Oceanog. 13*: 202-204.

Mauchline, J. and L. R. Fisher. 1969. The biology of euphausiids. *Adv. Mar. Biol.* 7: 1-454.

Omori, M. 1974. The biology of pelagic shrimps in the ocean. *Adv. Mar. Biol. 12*: 233-324.

Reid, J. L. 1962. On circulation, phosphate-phosphorus content, and zooplankton volumes in the upper part of the Pacific Ocean. *Limnol. Oceanogr.* 7: 287-306.

Thorne, R. E., O. A. Mathisen, R. J. Trumble and M. Blackburn. MS. Distribution and abundance of pelagic fish off Spanish Sahara during CUEA Expedition JOINT-I.

Vinogradov, M. E. 1968. Vertical distribution of the oceanic zooplankton. *Akad. Nauk SSSR, Moscow*. (in Russian). English translation 1970, Israel Program of Scientific Translations, 339 p.

Physiological Approaches to the Biology of Midwater Organisms

J. J. Childress

University of California, Santa Barbara

ABSTRACT

It is impossible to see midwater animals in their natural environment in an undisturbed state. This makes investigations of the midwater environment very different from investigations in shallow aquatic habitats and terrestrial habitats. This lack of direct observation means that study of the biology of midwater organisms must rely largely on relatively indirect observations, each with appreciable associated errors. The strongest approach in this situation is to use as many methods as possible (independently and together) to infer aspects of the biology in these organisms. Among the newest approaches applied to midwater organisms are direct studies of the physiology and behavior of these organisms in the laboratory. From this sort of information one can describe many aspects of the organisms in the field. One especially valuable aspect of these types of observations is that they can often be carried out on small numbers of organisms and "sampling error" does not affect them.

The use of chemical composition and oxygen consumption measurements in this manner is considered. The great diversity of and patterns of compositional variation in midwater organisms are described. The need for more carefully controlled oxygen consumption experiments is emphasized.

INTRODUCTION

The difficulty of making direct observations, except near the surface, makes many kinds of knowledge about midwater organisms unobtainable. This problem makes physiological and biochemical approaches especially rewarding ways of obtaining information about these organisms. These disciplines study the individual organism and its parts without the need of a "normal environment" and without the complex problems of sampling. In particular, these approaches can provide information concerning the rates of biological processes in midwater animals--information which otherwise is virtually unobtainable due to the nature of the environment and the animals involved.

The study of midwater animals is also rewarding as a way of extending physiological knowledge and testing physiological ideas. Among the physiologically most interesting environmental characteristics of the midwaters are: 1) the impressive stability of physical and chemical parameters, 2) the hydrostatic pressure, 3) the very low oxygen in some regions, 4) the absence of sunlight at greater depths, 5) the way in which all of the properties are stratified with depth so that organisms undergo very significant environmental change with vertical migration, 6) the absence of microhabitats which means that organisms must deal physiologically, not behaviorally, with physiological problems. As a consequence of these factors a midwater animal's environment can be described to a much greater extent than can that of most benthic and terrestrial organisms. Furthermore, some environmental conditions are almost unique to midwater, $i.e.$, regions in which the oxygen at intermediate depths remains constantly low, but does not go to zero. Midwater organisms have proven to be especially favorable material for some physiological studies such as crustacean circulation, crustacean activity metabolism and bioluminescence. But rather than attempt to survey all of the physiology that has been done on midwater animals, I will confine my comments to those areas with which I am most familiar.

In the following discussion, I have tried more to be provocative than to be balanced in the hopes of stimulating more activity and dialogue at the interface between physiology and oceanography. I cannot emphasize too strongly, that to work effectively in this field, a physiologist must either obtain considerable knowledge of the methods and findings of oceanography or work very closely with oceanographers. Physiologists have had many problems in attempting to study midwater animals. Among these are the failure to be opportunistic in choice of animals and problems, the designing of experiments and apparatus without the necessary understanding of the animals to be used, the lack of knowledge of the natural history of midwater animals and poor adaptation to shipboard conditions. On the other hand, oceanographers generally have a point of view directed toward populations and communities with a limited knowledge of physiology. Consequently, they have tended to do physiologically unsophisticated studies and to ignore many physiological approaches of high potential for answering their questions.

Any physiological studies on midwater animals should very carefully take the animal's natural environment into account. For example, physiologists often make measurements at 5°C intervals of temperature at whole degree values. However, many of the deeper-living and non-migratory species live within temperature ranges of much less than 5°C. If one does not choose temperatures from a knowledge of the animal's distribution, it is very easy to choose temperatures of little meaning for the species under consideration and even to totally miss its characteristic temperature range. What is needed to properly plan physiological experiments and interpret much physiological data is knowledge of the vertical distribution, in the region of study, of the particular life stage being studied. This information allows one to define the environmental conditions to which the organism is exposed and thereby choose meaningful experimental conditions.

An important consideration relating to the organisms' environments is the matter of how to describe the vertical distribution of animals one is comparing. This is complex for these purposes: a depth description needs to be simple enough to be readily understandable yet usable for creating categories. One problem is that species often have different depth distributions in the day and in the night. Species which migrate to much shallower depths have a much greater potential food supply than do those species which remain within a narrow depth horizon. Another problem is that the same species may be found at rather different depths in different oceanographic regions, depending on the temperature regimes. Further, many species undergo ontogenetic changes of depth preference. Because of these facts, such terms as epipelagic, mesopelagic and bathypelagic are of little value because they are so loosely defined and fail to distinguish many obviously different life strategies. It may be better to create more finely divided categories which take into account day and night distributions. My preference has been for numerical expressions of the depth at which species occur, because numbers are easier to handle conceptually and they can be fitted into regression equations. However, numbers have the weakness of lacking a shorthand way of expressing depth distribution unless one wants to introduce the distortion of allowing one number to stand for the depth distribution of a species. Consequently, I consider the shallowest depth at which a species is commonly found as the most important one, since this generally represents the highest food level to which the species is exposed. To this end, the minimum depth of occurrence (MDO) has been defined as "that depth below which about 90% of the population lives" (*Childress* and *Nygaard*, 1974). The principal advantages of this parameter are that it is numerical, it separates strong migrators from deep-living non- or weakly-migrating forms and is more easily determined than most other depth parameters. Another numerical parameter in use is the depth of capture, but it fails to take into account the depth distribution of the species and of individuals over time.

A complicating aspect is the broad distribution of many midwater species. In many cases, single species are cosmopolitan with worldwide or circumglobal distributions: *Eucopia sculpticauda*, *Gnathophausia gigas*, *Bentheuphausia amblyobs*, *Gnathophausia ingens*. In fish taxonomy the tendency has been to describe cosmopolitan species as species complexes. In decapod and mysid crustacean taxonomy the tendency has been more often to consider these as widely distributed species with some local variations. At present, we cannot distinguish between these alternatives. We have almost no knowledge about the abilities of individual midwater organisms to modify their physiology and morphology under different hydrographic conditions (either through selection or acclimatization). Consequently, physiological, biochemical and ecological observations of a species should be assumed to apply to that species only at the site where it was studied. Wherever there are significant hydrographic differences or geographical distances between studied populations, the initial assumption should be that the populations and individuals are different until similarities are shown. This is not to say that we need more species names but only that with our present ignorance, geographical region is an important part of the identification of midwater species.

CHEMICAL COMPOSITION

The chemical composition of animals which live in midwater is probably more variable from species to species than is true of the inhabitants of any other environment. For example, midwater crustaceans and fishes may be as much as 95% water or as little as 63% water. With this variation in water content, measurements such as wet weight, volume or even dry weight are poor indicators of relative biomass. The structure of midwater communities and the productivity of their different species appear very different when considered in terms of caloric content. This wide variation probably results from different buoyancy strategies (most chemical components are either negatively or positively buoyant), the fact that the support of the water and absence of hard substrates and violent currents makes a wide range of mechanical strategies feasible, and the availability of very different habitats within a small vertical range. This variation in composition is useful for defining groups of species with different habits.

In making ecological or physiological generalizations from composition data, the more complete the analyses are, the more meaningful the interpretations can be. In addition, studying a wide range of species (in terms of depth and habits) provides a more comprehensive context for interpretation. The most important components in midwater animals are water, lipid, protein, ash, chitin (where applicable) and carbohydrate. Carbon and nitrogen are also often useful in building budgets for species and communities. Carbon can be used to estimate lipid content if the highest accuracy is not necessary. Nitrogen is also useful even if protein and chitin are known, since large fractions of the organic nitrogen are often tied up in forms such as amino acids. Because of the presence of chitin and other non-protein nitrogen, the common use of total nitrogen ($6.25N$ = protein) as an estimator of protein is not justified. One can, however, use empirical relations from the literature to make rough estimates of protein from nitrogen.

Another consideration is the form to use in expressing and discussing composition data. The different expressions have very different meanings and workers in this field have often been confused by them. Ideally, to compare the composition of different individuals or species, one would compare only individuals of the same size or weight. Then it would be possible to compare compositions directly without resorting to percentages. However, it is almost never possible to use animals all of the same size so composition data is usually expressed as percentages--the three most common of which are per cent of wet weight (live weight), per cent of ash-free dry weight (AFDW, organic matter) and per cent of dry weight. Confusion arises in that a numerical expression of a percentage only represents the partitioning of materials within the 100% entity, not the absolute amount of the material present. Because of this, to compare the relative amounts of a material in different individuals and species using percentages, the percentages must be in terms of the entire body weight of the organism (per cent of wet weight). This form of expression is most meaningful for ecological and physiological considerations, since this is the form in which the animals live. Percent of AFDW expresses the partitioning of organic matter and is thus extremely useful for understanding organisms' adaptive priorities concerning the growth component of energy usage. Per cent of dry weight, since it expresses the partitioning of materials of unlike kinds (organic and inorganic) and does not represent the entire organism, has little to recommend it except convenience. Confusion surrounding this issue is found in many papers. For example, *Lee, Hirota and Barnett* (1971) did lipid analyses on an extensive series of midwater copepods and expressed their data as per cent of dry weight. At one point they conclude, "most of the increase of lipid with increased depth was due to the addition of wax esters. Thus it is apparent that the lipid and wax esters are higher and triglycerides are lower in deeper living copepods." In fact, what their data shows is that a greater fraction of the total of ash weight and organic matter is in the form of lipids and wax esters in deeper living species. Their conclusions as stated can only be true if ash and water content were the same in all species. In another example, *Raymont et al.* (1971) determined wet weights on their specimens, but expressed the data as per cent of dry weight. Their figures show maximum protein and minimum lipid content in the spring and the reverse in the fall. They say "Figure 2 suggests that the higher lipid in autumn is apparently related to a fall in protein during the same period." Their figure really shows that since the major components of dry weight in this species are lipid and protein, an increase in per cent of dry weight of one results in a decrease in the other. Using their wet weight data, I recalculated protein and lipid on a wet weight basis. This clearly showed that protein, like lipid, is at its lowest in the spring (~12% wet weight) and like lipid rises slowly to a maximum (~13.3% wet weight) in the fall. Over this same time lipid rises from about 3% wet weight to about 7%. Since the water content declines by the same amounts, it is apparently being replaced by lipid and protein. This is a very different conclusion from that of the authors.

In summary, the use of both wet weight and AFDW expressions provides a powerful combination for examining the overall strategy and compositional nature of midwater organisms, but dry weight expressions are not very useful.

Since nearly all of the major components of midwater organisms affect their buoyancy relations, actual measurements of relative buoyancy or specific gravity are valuable for the interpretation of composition and physiological data. Buoyancy data are of course also useful by themselves.

Let us now consider the various major constituents of midwater organisms, particularly from the viewpoint of the range of compositions and the overall adaptive patterns.

Water

Water content is extremely variable in these species and is the major source of overall compositional variation. Midwater fishes range from about 62% water for some of the very high lipid content vertical migrators to about 94% for some of the very "flabby" deep living species (*Blaxter et al.*, 1971; *Childress and Nygaard*, 1973). Those species with less than about 80% water have displaced it with lipid while those species with more than about 85% water show a reduction in lipid, protein and skeletal material. In midwater crustaceans the pattern is more complex with the shallower living species having about 80% water while somewhat deeper living species (MDO = 400-700 m) have displaced the water with lipid and may reach water contents as low as 65%. Crustaceans which live even deeper sometimes show water contents as high as 95% (*Childress and Nygaard*, 1974).

The adaptive value of high water content for midwater fishes (*Denton and Marshall*, 1958), crustaceans of the genus *Notostomus* (*Herring*, 1973), and Cranchiid squids (*Denton et al.*, 1958) is that all or large fractions of their fluids are less dense than salt water and thus provide static lift. This hardly seems sufficient in itself, however, since lipids provide much more lift per unit volume and therefore require much less radical structural modifications in those organisms.

Lipids

The lipid content of midwater animals is perhaps the most studied aspect of their composition, with a focus on the chemistry of the lipids, especially wax esters, and their roles both in the individual organisms and in food chains. In midwater fishes, the highest lipid levels (up to 16% wet weight and 50% AFDW) appear to be found in that class of vertical migrators which *Barham* (1972) has characterized as "lethargic" (*Bone*, 1973; *Butler and Pearcy*, 1972; *Childress and Nygaard*, 1973; *Culkin and Morris*, 1969, 1970;

Nevenzell et al., 1969). The lowest levels (3% AFDW, .5% wet weight) are found in deeper living species and those species which have gas-filled swimbladders. The general trend is for lipid content to decrease with increasing depth of occurrence.

Lipids are also generally recognized to be the most variable component of midwater species (Conover and Corner, 1968; Fisher, 1962; Littlepage, 1964; Orr, 1934; Raymont et al., 1971), apparently because lipids are a major energy store. Their concentration therefore tends to reflect nutritional or reproductive status, age, etc. Increases or decreases in lipids in the whole animal generally involve displacing or taking in water (Childress and Nygaard, 1973, 1974).

Carbohydrates

As was pointed out by Raymont and Conover (1961), carbohydrate contents are very low in marine zooplankton (<2% AFDW). The same is also true of deeper living fishes and crustaceans (Childress and Nygaard, 1973, 1974).

Protein

Protein content tends to decline with increasing depth of occurrence in both midwater fishes and crustaceans, ranging from 3 to 15% wet weight for midwater fishes (Childress and Nygaard, 1973) and 0.6 to 12% for crustaceans (Childress and Nygaard, 1974). This decline with depth is apparently associated with the reduction of muscle tissue (Blaxter et al., 1971), oxygen consumption (Childress, 1971, 1975), and tissue density (Childress and Nygaard, 1974) found in deeper living species.

Chitin

Chitin shows no trend with increasing depth of occurrence. It accounts for about 2 to 16% of the wet weight of midwater crustaceans.

Carbon

Carbon content ranges from about 35 to 60% AFDW in midwater crustaceans and fishes (Omori, 1969; Childress and Nygaard, 1973, 1974). The variability in this parameter is positively related to lipid content.

Hydrogen

Hydrogen content ranges from 6 to 10% AFDW in midwater fishes and crustaceans and, like carbon, reflects the lipid content of the species.

Nitrogen

Nitrogen content ranges from 3 to 15% AFDW in midwater crustaceans and fishes (*Omori*, 1969; *Childress and Nygaard*, 1973, 1974). Much of the variability of this component is accounted for by protein and chitin content. However, a large fraction (up to 33% in the cases examined) of the nitrogen is in non-protein, non-chitin forms in both midwater fishes and crustaceans (*Cowey and Corner*, 1963; *Childress and Nygaard*, 1973, 1974; *Ferguson and Raymont*, 1974; *Raymont et al.*, 1971; *Srinivasagam et al.*, 1971).

Caloric Value

This is principally the result of the water, lipid and protein content of these species. It is possible to estimate this value directly using a calorimeter or indirectly by calculation from the composition. The two methods show reasonable agreement for midwater animals (*Childress and Nygaard*, 1974; *Tyler*, 1973). Calculation from composition has the advantage that one can calculate both the total caloric content of an animal (which the calorimeter also measures) as well as the assimilable caloric content taking into account the very different degrees of assimilation and energy yield from the different components. Midwater fishes examined range from 35 to 205 K cal/100 g wet weight while the crustaceans range from 4 to 235 K cal/100 g wet weight. In general, the lowest caloric values have been found in deeper-living species while the highest contents have been found in vertically migrating fishes and in 200-300 m MDO crustaceans (*Childress and Nygaard*, 1973, 1974). The very wide variation in caloric content makes consideration of this parameter essential for analyses of structural patterns in midwater communities. Volumes and wet weights have little meaning in terms of the dynamics of midwater communities, especially those of diverse composition like the nighttime near-surface groupings and the daytime scattering-depth groupings.

Buoyancy (Density)

Most deeper-living species are nearly neutrally buoyant with some apparently being positively buoyant (*Childress and Nygaard*, 1974). The deeper-living crustaceans and fishes generally approach neutral buoyancy by having reduced skeletons and increased water content (*Denton and Marshall*, 1958; *Childress and Nygaard*, 1973). Some also use lipids extensively (*Childress and Nygaard*, 1974) and others have replaced heavier cations with lighter ones (Denton et al., 1958). A greater diversity of buoyancy mechanisms is found at shallower depths where the crustaceans are considerably denser and often use lipids for buoyancy while the midwater fishes tend to remain near neutrality either by means of gas-filled swimbladders or lipids (Denton and Marshall, 1958; Butler and Pearcy, 1972; Capen, 1967). Composition and buoyancy are closely related although it appears that in some cases the buoyancy results from a composition selected for by other factors, while in other cases buoyancy itself appears to be a major selective factor. One intriguing aspect of this is that Childress and Nygaard (1974) have shown that buoyancy and oxygen consumption rate are apparently independent in the 400-700 m MDO crustaceans, indicating that neutral buoyancy is probably not an energy conservation mechanism in these species. This observation requires some rethinking of the adaptive value of neutral buoyancy in midwater species.

In addition to the above considered general aspects of composition (and composition as a function of depth), there are a number of other variables which affect it and have considerable ecological and physiological significance. None of these variables have been much studied in midwater organisms, and therefore more work is needed on them. One of the most important of these variables is age or size. The few studies relating this parameter to composition (Butler and Pearcy, 1972; Fisher, 1962; Raymont et al., 1971) generally show an increase in lipid content with age. This is generally interpreted as a reserve for reproduction. Another parameter related to reproduction and food availability is the effect of season on composition, but this has been studied only in shallow living species (Conover, 1968; Conover and Corner, 1968; Raymont et al., 1971). The lipid content is generally quite variable as a function of season in these species. Clearly, there are many interesting unstudied questions relating seasonal effects to the composition of deeper living species and on the effect of sex and reproductive status on composition. Morris (1973) has carried out a preliminary study showing some sexual differences but further study would prove useful for understanding the energy input into reproduction in midwater species.

The effect of hydrographic conditions on variations in composition may lead to a better understanding of their physiological and ecological significance. One approach of this kind involves studying latitudinal effects on composition. Conover (1960, 1964); Lee et al., (1971) and Littlepage (1964) have demonstrated higher lipid levels at higher latitudes and in some cases have related this to higher lipids in deeper species, concluding that the common causal factors may be low or sporadic food supply, nature of the food itself, or some unknown effect of low temperatures. Also largely unstudied but obviously important is the effect of diet on body composition (Lee et al., 1971; Morris et al., 1973; Morris and Sargent, 1973; Ackman et al., 1970).

The fact that species with common habits generally have similar compositions (Bone, 1973; Childress and Nygaard, 1973, 1974; Lee et al., 1971) can be used to define ecological groupings or better understand those defined in other ways, with or without other data such as stomach analyses. In addition, it is possible to predict the habits of species from composition data. This is perhaps one of the most potentially rewarding avenues for further research.

Future work needed in this field can be broken into two areas. First, investigation of the factors affecting composition of individual species of a variety of different habits is needed in order to understand the variability within individual species in a region. We now know most about certain copepods and euphausiids. As we develop comparable and even more complete data for species with other habits, our understanding and predictive ability will be greatly augmented. Second, more intensive investigation of variation in composition with depth and habit in particular areas is needed to understand the variability between species. An important aspect of this is the need for more detailed descriptions of vertical distribution, natural history, food habits, growth rates and metabolism of the species studied, coupled with differences in the composition of organisms living under different hydrographic conditions. This can best be studied in those areas where there are large variations in hydrography within relatively small areas (i.e., the Indonesian-Philippine area basins, the very well developed oxygen minima and sharply defined "transition" zones).

From the above, it is clear that the composition of midwater organisms is extremely variable and to a large extent poorly understood. Because of the magnitude of the variation, knowledge of the composition of the species is as important as data on abundance for understanding the ecology of midwater communities.

OXYGEN CONSUMPTION

Oxygen consumption studies on oceanic animals have a long history. In addition to being one of the easiest biological rates to measure, oxygen consumption is of ecological interest because it is an estimate of metabolism and metabolism generally accounts for the largest fraction of a zooplankter's energy use. Physiological interest in oxygen consumption has focused on understanding what factors determine the rate of consumption and how they do this. For physiologists, predictive ability in hypotheses is used largely to test the hypotheses and the underlying level of understanding. By contrast, ecologists want to predict metabolic rates of organisms in the field where they cannot be directly measured. These different goals have resulted in the collection of data of very different quality.

In measuring oxygen consumption, the rates are affected by so many factors that it is almost impossible to control all of the possible variables. Thus, the investigator's judgment about which variables to control becomes critical. Since oxygen consumption is one of the easiest characteristics of an animal to measure, yet one of the most difficult to measure in a meaningful way, it seems appropriate to consider the philosophy and methodology of oxygen consumption measurements on midwater animals.

The essence of a physiological approach to measuring metabolic rates is the careful separation of variables in experiments so that the influence of each can be quantified, as reviewed by F. E. J. *Fry* (1957, 1971) for fishes. The synthesis of detailed information from these kinds of studies to arrive at ecological conclusions is exemplified in the approach of J. R. *Brett*(1971). By contrast, oceanographers and others with primarily ecological interests have wanted to proceed from lab measurements to field estimates as quickly as possible. They have, as a result, tended to carry out studies in which many important variables are confounded. The usual approach has been to try to approximate the field conditions without really determining the critical factors in the field. Among the most flawed are the so-called "*in situ*" measurements. These are appropriate and valuable where the experimental entity cannot be recovered for laboratory study (*i.e.*, whole communities, *Smith and Teal*, 1973b; and individual deep-living organisms which have gas bladders, *Smith and Hessler*, 1974). However, even in these, and certainly in other "*in situ*" cases, the variables affecting metabolism are confounded and largely uncontrollable and unobservable. In reality, "*in situ*" is an inappropriate term since an organism completely enclosed in a container is not "*in situ*" and the location of this container makes little difference except in terms of experimental manipulation and ability to work with unrecoverable entities.

Even the laboratory studies of ecologically oriented workers have failed to separate many important variables, which as *Marshall* (1973) observed, has produced a mass of contradictory results. Clearly, the ecological advantage of a physiological approach is its powerful predictive ability, but it has the major disadvantage of the large amount of work needed to reach this position. The ecological approach has a much weaker predictive ability but has the often essential advantage of allowing one to reach conclusions in a reasonable time. Often, it is not the absolute quality of the data which is important but rather that the quality be sufficient for answering the questions asked. In this regard, we need detailed studies of a variety of midwater species, then the more superficial studies of other species will be more meaningful.

Methodology

In general, metabolic rate has most often been measured in marine animals in terms of oxygen consumption. The measurement of heat production is too difficult in small aquatic animals which is unfortunate since it could give the least ambiguous measurement of metabolic rate. Measurement of CO_2 production is potentially useful as an indicator (R.Q. when combined with O_2 consumption) of the type of metabolism occurring and the substrate being used. However, measurements of CO_2 production by marine animals have proven difficult due to the high solubility of CO_2 in seawater and the large amounts of CO_2, carbonate, and bicarbonate normally found there. The interpretation of CO_2 production data is also complicated by the fact that dissolution or formation of skeletal material as well as growth may affect CO_2 balance. In addition, high R.Q. values, which are often believed associated with anaerobic metabolism, may result not from the direct production of CO_2 by anaerobiosis but indirectly from the buffering of acidic end products of anaerobiosis. Further, the R.Q. in protein metabolism may be as low as .7 or as high as .95 depending on the nitrogenous end productions (*Elliot and Davison*, 1975). As a consequence of these problems, CO_2 production has been little studied in marine organisms and its interpretation is difficult. However, careful studies could yield interesting results. Nitrogen excretion has received moderate usage as a measure of protein metabolism. It has the potential when combined with O_2 consumption (assuming carbohydrate usage is low) of indicating the relative amounts of protein and lipid being metabolized. Although generally fruitful, the study of nitrogen metabolism has two major problems. The first is that nitrogen is excreted in several forms and most studies only measure about 80 to 90% of the total (there is also probably a non-excretory "leakage" component to N loss). The second problem is that it cannot be measured continuously but rather can only be analyzed destructively on relatively large samples.

Oxygen consumption has generally been the method of choice in metabolism experiments because it can be measured easily, cheaply, rapidly, continuously, in very small samples, non-destructively and by a large variety of methods. Although substrate metabolized and metabolic pathways used introduce some uncertainty into oxygen consumption data, their interpretation is generally less ambiguous than either CO_2 or N production results.

In examining oxygen consumption rates (or other metabolic rates), there are two basic patterns of measurement. The first of these is the method of continuous monitoring of oxygen uptake, most commonly by using the polarographic oxygen electrode. This has the advantage that an animal's actual consumption can be followed from moment to moment. If the experiment is long enough, one can exclude the initial high consumption rates due to handling. One can also estimate minimum and maximum rates and correlate consumption rates with short-term changes in behavior providing a great deal of information about each individual. The major disadvantage is the complexity and expense of the required equipment. The other major approach involves sampling the oxygen content at relatively long intervals (often only at the beginning and end of an experiment). In this sort of experiment one obtains an average rate over the measured time and there is no way to note variations in rate or to exclude initial periods of excitation. These effects can be minimized by using relatively long runs and placing the animal in the test chamber some time before the start of the run. One of the biggest advantages of this methodology is that it requires little specialized equipment, allowing more experiments to be conducted at a lower cost in time and money than by continuous methods.

A major consideration is how to express the results. Many zooplankton workers have now adopted μl O_2/mg weight/hr as the standard expression. However, it makes a considerable difference whether one expresses the weight as wet weight, dry weight, ash-free dry weight, protein, nitrogen, carbon or something else. To a large extent, the choice depends upon the use one wants for the data. In any case, observed differences in rate between different species (or the same species separated in space or time) may simply reflect compositional differences: more or less water if wet weight is used, or more or less lipid if dry weight or ash-free dry weight are used. Such differences might disappear if one chose another basis to express the rate. I prefer to express oxygen consumption at least in terms of wet weight, ash-free dry weight and protein (*Childress*, 1975). This allows one to separate out and examine the effects of the major constituents on rate. Dry weight rates seem less valuable since they combine organic and part of the inorganic components. Nitrogen seems less useful than protein because much nitrogen is in forms other than protein. In general, if one is using animals of the same species and size, collected at the same time and place and studied at the same time, the form of expression makes no difference. The farther one deviates from these conditions, the more important a consideration of compositional effects becomes.

Another aspect of the expression of results is the effect of size on metabolic rate, concerning which, many studies and generalizations have been made. Many authors have tried to find a single explanation and/or expression of the variation in rate with size but it is still not fully understood. In general, a species' oxygen consumption is directly proportional to weight, to surface area or to some intermediate value. There are exceptions and some species even show much less than surface proportional changes in rate with change in size. Many further factors confound size-metabolism comparisons, including the differences between life stages in terms of activity and composition. Clearly, size corrections of data to a standard size are most meaningful when one is working with a restricted size range of the species being considered. There is no particular reason to expect that a single relationship will characterize this variation over the whole life of a species.

For most animals, the activity level is the most important factor producing the observed metabolic rate. For example, in *Gnathophausia ingens* (a bathypelagic mysid, *Childress*, 1971), the oxygen consumption at maximum sustainable activity is 9.5 times that at minimum activity. Given this tremendous influence, it is obviously important to reduce activity as an uncontrolled variable in metabolism studies. This was appreciated long ago by terrestrial vertebrate physiologists and their concepts have been adapted to a variety of freshwater and shallow marine fishes over the past 25 years (Fry, 1971). This has resulted in three definitions of metabolic rates which correspond to particular activity levels and a considerable technology for simultaneously measuring oxygen consumption and activity. The "standard" rate applies to an animal which is minimally active and not digesting and absorbing food. This can be determined either by observation of minimal activity or by extrapolation of metabolism--activity relations to zero activity. The "routine" rate applies to organisms which are spontaneously active in the absence (as much as possible) of external stimuli. The metabolic rate corresponding to the maximum sustainable swimming speed is the "active" rate. Description of the relation between quantitative activity levels and metabolic rate has allowed very precise comparisons of the effects of other parameters on the activity level itself, and on the metabolic rate at defined activity levels (Fry, 1971). One of the most significant results has been the demonstration that the metabolic effects of the same set of conditions may be different at different activity levels. For example, while standard metabolism rises regularly over the thermal range of an organism, the active metabolism often reaches a maximum well below the upper lethal thermal limits and then declines with increasing temperature. In another example involving salinity effects on metabolism, the effects were found to be proportional to activity with little effect at low activity (and therefore metabolic) rates and large effects at high activity rates (Rao, 1968). The standard rate is generally thought of as the minimum possible rate set by the conditions to which the organism is exposed, while the active rate is the upper limit for metabolism and activity imposed by these same conditions and the organism's intrinsic abilities and conditioning. Thus, the effects of environmental factors on standard metabolism have a different meaning than their effects on active or intermediate metabolic rates. The extensive literature on these topics in fishes provides many examples for creating a more analytical approach to midwater animal respiration.

In most previous metabolic studies of zooplankton and micronekton, the usual assumption has been that the measured rates corresponded directly to field rates. These studies have made no attempt to separate activity and metabolism, the assumption being that routine metabolism was being measured and is a reasonable approximation of field metabolism. The work on fishes (Fry, 1971; Ware, 1975) supports the general idea that both routine and field metabolism tend to be about 2 to 3 times standard metabolism even in highly active species where active metabolism may exceed standard by 10-12 times. However, a critical comparison of the methodology used for midwater animal metabolism with the definition of routine rate (above and Fry, 1971) reveals that many of these studies are carried out under conditions inappropriate for "routine" metabolism determinations (protection from outside stimuli) and therefore represent metabolic rates corresponding to unmeasured, higher activity and excitement levels. High activity levels are strongly indicated by the fairly frequent observation of negative Q_{10} values at higher temperatures in these studies. These values are associated with high activity levels where oxygen becomes limiting at higher temperatures in the fish studies (*Brett*, 1964), which also seems reasonable for midwater animals.

The importance of carefully considering the activity levels of midwater animals in studies has often been neglected because of the belief that only rough estimates are desired and that sophisticated statistical design and analysis will suffice to demonstrate differences. Obviously, a range between standard and active rates of 10 to 12 times the standard rate and between routine and active of 5 to 6 times standard for many animals makes measurements which fail to account for activity exceedingly crude. Since responses to environmental factors can be different depending upon the activity level, no amount of statistical sophistication can retrieve meaningful data when activity is ignored. The two major shortcomings in measuring routine metabolic rates for midwater animals involve the failure to exclude the initial excitement phase (caused by handling the animals), and the agitation of the experimental vessel. For fish, several hours (as much as 6 or more) are often required for animals to "calm down" after handling and assume a routine rate. This excitement phase is characterized both by increased metabolism, expressed as increased activity, and by increased metabolism independent of activity (*Fry*, 1971). I have observed that many hours are also required for midwater crustaceans to "calm down" and assume a steady rate of oxygen consumption. This can be mitigated by using relatively long runs and frequent or continuous oxygen measurement, or by keeping the animals in the experimental chamber for an extended period before commencing the run. The first practice allows one to actually measure the initial phase and discard it. The second practice simply provides time for this phase to be passed before measurement commences. A third, moderately satisfactory method is to make runs quite long (12 hours or more) so that the initial phase does not dominate the run, but runs as short as 15 minutes to 2 hours are common even recently.

The second major failing of routine metabolism measurements of midwater animals has been the degree and method of agitation of the experimental medium. Agitation is required to keep oxygen partial pressure the same throughout the chamber, for proper operation of some sensors (oxygen electrodes), or equilibration of gases between phases in mixed phase systems (*Warburg, Gilson, Scholander*, etc.). Ideally, the only agitation should be that provided by the animal which is sufficient for active animals in small volumes, analyzing either by removing water samples or by using micro-cathode oxygen electrodes (which do not require stirring). When using electrodes which require more stirring, the animal should be shielded from all but a very gentle flow of water. Due to the need for considerable agitation and the extended time scale over which excitement can persist, none of the commonly used mixed-phase systems can be used except for measuring activity levels which depart rather far and to an unknown degree from routine rates. Furthermore, different animals may react very differently to different agitation levels. For example, excessively stimulated ostracods tend to withdraw their antennae and become quiescent, while excessively stimulated midwater fishes tend to become quite excited since, due to their body structure, it is difficult for them to stabilize themselves in a confining chamber. On the other hand, mysids and decapod crustaceans are less sensitive to agitation since they become lodged in a corner and expend little energy to keep from being tossed about.

While the above is an indictment of much previous work in this field, one cannot simply adopt the methodology and concepts used on shallow water fishes. These animals are generally neutrally buoyant and capable of very rapid swimming. They orient and swim in currents, are relatively large, have relatively high metabolic rates, and can maintain their positions in chambers by hovering in place. In contrast, midwater species are generally heavier than water, do not orient in currents, swim at rates so low as to make measurement difficult, cannot hover in confined spaces, are relatively small, and have low metabolic rates. Thus, most of the methodology used for shallow-water fish is either inapplicable or adapted only with difficulty to our needs. Especially difficult to deal with are the small shallow-living animals like copepods, with erratic activity patterns and food consumption rates so high that starvation may become a critical factor in a few hours. Clearly, much more experimental and conceptual work is needed to understand metabolic rates in these species.

Over the last six years, we have been working on systems for simultaneously measuring the activity and metabolism of euphausiid, decapod and mysid crustaceans. To get the equipment into a small enough volume of water, each size class of each species must have a separate system. Clearly, it will never be possible to extend this approach to very many species or size classes. Further, quantitative measurements of some taxa (for example copepods, small euphausiids, ctenophores) seem at the moment almost impossible even at a conceptual level. Studies are needed quantitatively relating metabolism and activity in a few suitable and varied species with more studies taking a middle path of considering activity in less quantitative ways.

Even without measuring activity quantitatively, one can determine standard metabolism simply by observing the experimental animals and noting the oxygen consumption rates which correspond to minimal activity. Similarly, one can estimate the active rate by stimulating the animal to the maximum rate which it can sustain and measuring oxygen consumption under these conditions. This approach has had considerable success when applied to intertidal snails, whose activity is difficult to quantitate (*Newell and Roy*, 1973). I have some reservations about the definition of routine rate for midwater species. The definition given earlier relates to organisms which can orient in relatively confining containers, are obviously aware of their confinement and are quite active. Midwater animals on the other hand are generally unable to orient in chambers and appear generally unaffected by their confinement (perhaps because such a thing is never experienced in their normal habitat). Further, my own and others' observations from submersibles suggest that the usual activity of these species is just sufficient to keep them from sinking. Support for the concept of very low activity in midwater animals also comes from observations that oxygen minimum layer animals can obtain enough oxygen for their needs only if they have quite low activity levels (*Childress*, 1971b) and that the midwater fishes which have been examined are capable only of very low sustained swimming speeds (*Gordon and Belman*, pers. comm.). Thus, it seems appropriate to try to estimate the metabolic rate which corresponds to just sufficient activity for the animal to support its weight in water, sometimes using *Zeuthen's* (1947) term "normal" for this rate. This is a very low level of activity and in practice has usually turned out to be virtually the same as the rates which most species reach after being in a chamber for 3 to 6 hours (about twice the standard metabolic rate). Therefore, the use of routine rate seems appropriate for these species even though its meaning may be a little different.

Another approach (*Childress*, 1975) is to make a continuous recording of oxygen uptake and then examine it for the maximum and minimum rates which are sustained for a few minutes in the interval after the initial thermal equilibration has occurred and before the P_c is reached. The minimum is probably a good estimate of standard metabolism, although the maximum rate will usually be somewhat less than the active rate (*Fry*, 1971; *Childress*, personal observation). It is also possible to examine the recording and take the relatively constant rate achieved after a period of hours and before the P_c is reached as an estimate of routine rate (40-70 mm HgO_2 rate, *Childress*, 1975). To summarize briefly, activity should be taken into account as much as possible in metabolic studies of midwater animals. In particular, 1) it is important to minimize initial handling effects, 2) where possible, measure activity quantitatively, 3) observe the animal's behavior in and out of the chambers and correlate this with metabolic rate, 4) manipulate the activity to get a maximum rate, 5) analyze continuous recordings not just for means but also for sustained maxima and minima, 6) take extreme care concerning the effects of the agitation used upon the animal's activity, 7) suspect very high activity levels when negative Q_{10} values occur. The above measures can greatly improve the quality and predictive value of midwater animal metabolism studies. The data should be sufficiently sophisticated for the intended use and conclusions drawn from it, not as sophisticated as possible.

Much previously collected metabolism data fails to meet these minimum standards which makes it difficult to discuss the effects of other factors on metabolism. Though I will attempt to outline those conclusions justified by the data, a relatively complete understanding will require more work relating metabolism and activity.

Temperature

Most of the data available suggest that oxygen consumption and activity increase with temperature with a Q_{10} of around 1.5 to 3 (*Marshall*, 1973; *Mauchline and Fisher*, 1969; *Teal*, 1971; *Smith and Teal*, 1973a). This data, when taken with the fish data (*Holeton*, 1974), makes it seem likely that there is no latitudinal temperature adaptation in these species. There has been very little work on acclimation of these species to different thermal regimes, although many species are found over very wide temperature ranges both geographically as a species and daily as individuals. The most thorough examination of this in a copepod (*Halcrow*, 1963) demonstrates that one copepod species has considerable capacity to acclimate to different temperatures. When interpreting temperature-metabolism data where activity is not measured, it should be remembered (as *Fry*, 1971, pointed out) that flat areas of rate temperature curves are more likely to indicate behavioral expressions of thermal preferences rather than temperature acclimation or metabolic regulation.

Pressure

This is one of those topics where the number of review papers almost exceeds the number of original works. In many cases of shallow living, vertically migrating forms, pressure appears to have virtually no effect on metabolism, so that temperature is a major determinant of metabolic rate over the species' vertical range (*Teal and Carey*, 1967a; *Smith and Teal*, 1973a; *Quetin and Childress*, in prep.). In the case of somewhat deeper living species which migrate to within about 100 m of the surface, increasing pressure appears to increase the metabolic rate, thus acting opposite to temperature with increasing depth (*Teal*, 1971). Two non-migratory, deeper living (~400-800 m) species appear to be little affected by pressure within their normal ranges (*Meek and Childress*, 1973; *Belman and Gordon*, personal communication). Until we know more about the effects of pressure on the activity of these species, this information is of relatively limited value.

Dissolved Gases

There has been little work outside my lab on the effect of dissolved gases on oxygen consumption of oceanic animals. *Calanus finmarchicus* is affected by oxygen concentrations of about 3 ml O_2/l (*Marshall et al.*, 1935). *Teal and Carey* (1967b) demonstrated that an oxygen minimum layer euphausiid was tolerant of low oxygen conditions. Off California, most midwater crustaceans can regulate their oxygen consumption rates down to the lowest oxygen concentrations at which they are found (as little as 0.1 ml O_2/l) and generally have very limited anaerobic abilities. (*Childress*, 1968, 1969, 1971b, 1975; *Quetin and Childress*, in prep.) These species are able to regulate their consumption and survive only if they maintain a low activity level in the minimum layer. Those species which cannot regulate their oxygen consumption to low levels have considerable anaerobic abilities. Oxygen tolerances and regulatory abilities of species in other regions should be examined in order to understand what oxygen levels will tend to limit which species and ecological groups. The relationship between oxygen consumption, activity, and P_c (*Childress*, 1971a) is also useful for field activity and oxygen consumption rates in these species. Investigation of the mechanisms for oxygen uptake and transport to tissues in these species could also help elucidate the metabolic strategies of species which cannot yet be collected alive.

The effect of the lower pH values (higher CO_2) found at greater depths has also not yet been investigated.

Food

There are two primary effects of food on metabolism. The first is associated with the initial consumption of the food (costs of digestion and assimilation and the specific dynamic effect) and the second is concerned with changes in metabolism of the organism depending upon its stored food levels. These aspects have been little studied but are clearly of importance for understanding energy relationships, perhaps especially for the small, short-lived, rapidly metabolizing surface forms.

Salinity

Salinity probably does not vary enough in oceanic environments to produce measurable effects on metabolism, even with very sophisticated measuring systems.

Behavioral Effects

Metabolic responses to light quantity and quality and to chemical substances have significance in that they indicate the operation of orienting responses to these stimuli. The metabolic responses are very sensitive ways of detecting and quantifying the behavioral responses. Few metabolic studies of midwater animals have taken this approach.

Composite Factor

Many of the environmental aspects ecologists want to predict do not affect metabolic rates directly. Instead, the observed changes related to them stem from a complex of covarying primary factors. Two examples here are variation in metabolism with season within a single species and with depth distribution among species. The unraveling of the causality in these situations is of considerable physiological interest. It can substantially increase predictive abilities and help in the understanding of other aspects of the organisms' lives.

Studies of seasonal effects on shallow living species show that highest oxygen consumption generally coincides with high availability of food (*Conover*, 1968). The available studies suggest the importance of temperature acclimation, food concentration, feeding activity, body composition, metabolic substrate, reproduction and growth in producing this pattern (*Conover*, 1968). More information on the relative roles of these component factors for animals of different latitudes, depths and habits, will greatly enhance our ability to understand and predict.

Studies on metabolic rates of deeper living species have shown that species with minimum depths of 100 meters may require only 1/20th of the oxygen that surface-living species need (*Childress*, 1971a, 1975). This finding of much lower rates at greater depths has been supported by studies on two deep-living individual fish (*Smith and Hessler*, 1974). This decline with depth has significant effects on considerations of the role of deeper living species in trophic interactions, biological processes, etc. The causes of this decline have been examined in detail by *Childress* (1975), but the examination of the size and pattern of this decline in a variety of hydrographically different regions offers the best way to gain a thorough understanding of its causes and consequences in oceanic communities.

SUMMARY

A rich potential of additional physiological methods exists for providing predictive information about oceanic communities including, for example, visual structure and function, bioluminescence, circulatory physiology, chemosensitivity and mechanosensitivity. Some of these approaches have already been studies extensively. For the future, more can be gained by studying some species very intensively and by using these findings as the basis for more superficial studies over wider varieties of species and hydrographic conditions. One important approach for understanding many phenomena in midwater communities will be to undertake broad physiological and ecological studies of midwater organisms where the vertical or horizontal distribution of hydrographic properties is especially appropriate for testing particular hypotheses.

REFERENCES

Ackman, R. G., C. A. Eaton, J. S. Sipos, S. H. Hooper, and J. D. Castell. 1970. Lipids and fatty acids of two species of North Atlantic krill and their role in the aquatic food web. J. Fish. Res. Bd. Canada 27: 513-533.

Barham, E. G. 1972. Deep-sea fishes: lethargy and vertical orientation, pp. 100-118. In G. B. Farquhar, ed., Proceedings of an International Symposium on Biological Sound Scattering in the Ocean. Maury Center for Ocean Science Rept. 005, 629 pp.

Blaxter, J. H. S., C. S. Wardle, and B. L. Roberts. 1971. Aspects of the circulatory physiology and muscle systems of deep-sea fish. J. Mar. Biol. Ass. U.K. 51: 991-1006.

Bone, Q. 1973. A note on the buoyancy of some lantern fishes (Myctophoidei). J. Mar. Biol. Ass. U.K. 53: 619-633.

Brett, J. R. 1964. The respiratory metabolism and swimming performance of young sockeye salmon. J. Fish. Res. Bd. Canada 21: 1183-1226.

Brett, J. R. 1971. Energetic responses of salmon to temperature. A study of some thermal relations in the physiology and freshwater ecology of sockeye salmon (Oncorhynchus nerka). Amer. Zool. 11: 99-113.

Butler, J. L., and W. G. Pearcy. 1972. Swimbladder morphology and specific gravity of myctophids off Oregon. J. Fish. Res. Bd. Canada 29: 1145-1150.

Capen, R. L. 1967. Swimbladder morphology of some mesopelagic fishes in relation to sound scattering. U.S. Navy Electronics Lab. Res. Rep., 29 pp.

Childress, J. J. 1968. Oxygen minimum layer: vertical distribution and respiration of the mysid Gnathophausia ingens. Science 160: 1242-1243.

Childress, J. J. 1969a. The respiratory physiology of the oxygen minimum layer mysid Gnathophausia ingens, 142 pp. Ph.D. Dissertation, Stanford University, Stanford.

Childress, J. J. 1969b. The respiration of deep-sea crustaceans as related to their depth of occurrence and the oxygen minimum layer. Amer. Zool. 9: 222.

Childress, J. J. 1971a. Respiratory rate and depth of occurrence of midwater animals. Limnol. Oceanog. 16: 104-106.

Childress, J. J. 1971b. Respiratory adaptations to the oxygen minimum layer in the bathypelagic mysid Gnathophausia ingens. Biol. Bull. Mar. Biol. Lab. Woods Hole 141: 109-121.

Childress, J. J. 1975. The respiratory rate of midwater crustaceans as a function of occurrence and relation to the oxygen minimum layer off Southern California. Comp. Biochem. Physiol. 50A: 787-799.

Childress, J. J., and M. H. Nygaard. 1973. The chemical composition of midwater fishes as a function of depth of occurrence off Southern California. Deep-Sea Res. 20(12): 1093-1109.

Childress, J. J., and M. H. Nygaard. 1974. The chemical composition of midwater crustaceans as a function of depth of occurrence off Southern California. *Mar. Biol.* 27: 225-238.

Conover, R. J. 1960. The feeding behavior and respiration of some marine planktonic Crustacea. *Biol. Bull.* 119: 399-415.

Conover, R. J. 1964. Food relations and nutrition of zooplankton, pp. 81-91. *Proc. Symposium of Experimental Ecology.* Univ. of Rhode Island, Occ. Pub. No. 2 (Oceanography).

Conover, J. R. 1968. Zooplankton - life in a nutritionally dilute environment. *Amer. Zool.* 8: 107-118.

Conover, R. J., and E. D. S. Corner. 1968. Respiration and nitrogen excretion by some marine zooplankton in relation to their life cycles. *J. Mar. Biol. Ass. U.K.* 48: 49-75.

Cowey, C. B., and E. D. S. Corner. 1963. Amino acids and some other nitrogenous compounds in *Calanus finmarchicus*. *J. Mar. Biol. Ass. U.K.* 43: 485-493.

Culkin, R., and R. J. Morris. 1969. The fatty acids of some marine crustaceans. *Deep-Sea Res.* 16: 109-116.

Culkin, R., and R. J. Morris. 1970. The fatty acids of some marine teleosts. *J. Fish. Biol.* 2: 107-112.

Denton, E. J., and N. B. Marshall. 1958. The buoyancy of bathypelagic fishes without a gas-filled swimbladder. *J. Mar. Biol. Ass. U.K.* 37: 753-767.

Elliot, J. M., and W. Davison. 1975. Energy equivalents of oxygen consumption in animal energetics. *Oecologia* (Berlin) 19: 195-201.

Ferguson, C. F., and J. K. B. Raymont. 1974. Biochemical studies on marine zooplankton. XII. Further investigations on *Euphausia superba* Dana. *J. Mar. Biol. Ass. J.K.* 54: 719-725.

Fisher, L. R. 1962. The total lipid material in some species of zooplankton. *Rapp. P.-v. Reun. Cons. Perm. Int. Explor. Mer.* 153: 129-136.

Fry, F. E. J. 1957. The aquatic respiration of fish. In M. E. Brown, ed., *The Physiology of Fishes*, Vol. I, 1-63. Academic Press, New York.

Fry, F. E. J. 1971. The effect of environmental factors on the physiology of fish. In W. S. Hoar and D. J. Randall, eds., *Fish Physiology*, Vol. VI, 1-98. Academic Press, New York.

Halcrow, K. 1963. Acclimation to temperature in the marine copepod *Calanus finmarchicus*. *Limnol. Oceanog.* 8: 1-8.

Herring, P. J. 1973. Depth distribution of the carotenoid pigments and lipids of some oceanic animals. 2. Decapod crustaceans. *J. Mar. Biol. Ass. U.K.* 53: 539-562.

Holeton, G. F. 1974. Metabolic cold adaptation of polar fish: Fact or artifact? *Physiol. Zool.* 47: 137-152.

Lee, R. F., J. Hirota, and A. M. Barnett. 1971. Distribution and importance of wax esters in marine copepods and other zooplankton. *Deep-Sea Res.* 18: 1147-1166.

Lee, R. F., J. C. Nevenzel, and G. A. Paffenhofer. 1971. Importance of wax esters and other lipids in the marine food chain: phytoplankton and copepods. *Mar. Biol.* 9: 99-108.

Littlepage, J. L. 1964. Seasonal variation in lipid content of two Antarctic marine crustacea. *Actual. Scient. Ind. 131:* 463-470.

Marshall, S. M. 1973. Respiration and feeding in copepods. *Adv. Mar. Biol. 11:* 57-120.

Marshall, S. M., A. G. Nicholls, and A. P. Orr. 1935. On the biology of *Calanus finmarchicus*. VI. Oxygen consumption in relation to environmental conditions. *J. Mar. Biol. Ass. U.K. 20:* 1-28.

Marshall, S. M., and A. P. Orr. 1958a. *The Biology of a Marine Copepod.* Oliver and Boyd, Edinburgh. 188 pp.

Marshall, S. M., and A. P. Orr. 1958b. On the biology of *Calanus finmarchicus*. X. Seasonal changes in oxygen consumption. *J. Mar. Biol. Ass. U.K. 37:* 459-472.

Mauchline, J. and L. R. Fisher. 1969. The biology of euphausiids. In F. S. Russel and M. Yonge, eds., *Advances in Marine Biology,* Vol. 7. Academic Press, New York, 454 pp.

Meek, R. P., and J. J. Childress. 1973. The effect of hydrostatic pressure on the respiratory rate of *Anoplogaster cornuta*. *Deep-Sea Res. 20:* 1111-1118.

Morris, R. J. 1973. Relationships between the sex and degree of maturity of marine crustaceans and their lipid compositions. *J. Mar. Biol. Ass. U.K. 53:* 27-37.

Morris, R. J., C. F. Ferguson, and J. E. G. Raymont. 1973. Preliminary studies on the lipid metabolism of *Neomysis integer* involving labelled feeding experiments. *J. Mar. Biol. Ass. U.K. 53:* 657-664.

Morris, R. J., and J. R. Sargent. 1973. Studies on the lipid metabolism of some oceanic crustaceans. *Mar. Biol. 22:* 77-83.

Nevenzell, J. C., W. Rodegker, J. S. Robinson, and M. Kayama. 1969. The lipids of some lantern fishes (Family Myctophidae). *Comp. Biochem. Physiol. 31:* 25-36.

Newell, R. C., and A. Roy. 1973. A statistical model relating the oxygen consumption of a mollusk (*Littorina littorea*) to activity, body size, and environmental conditions. *Physiolog. Zool. 46:* 253-275.

Omori, M. 1969. Weight and chemical composition of some important oceanic zooplankton in the North Pacific Ocean. *Mar. Biol. 3:* 4-10.

Orr, A. P. 1934. On the biology of *Calanus finmarchicus*. Part IV. Seasonal changes in the weight and chemical composition in Loch Fyne. *J. Mar. Biol. Ass. U.K. 19:* 613-632.

Rao, G. M. M. 1968. Oxygen consumption of rainbow trout (*Salmo gairdneri*) in relation to activity and salinity. *Can. J. Zool. 46:* 781-786.

Raymont, J. E. G., and R. J. Conover. 1961. Further investigations on the carbohydrate content of marine zooplankton. *Limnol. Oceanog. 6:* 154-164.

Raymont, J. E. G., R. T. Srinivasagam, and J. K. B. Raymont. 1971.

Biochemical studies on marine zooplankton. VIII. Further investigations on *Meganyctiphanes norvegica*. *Deep-Sea Res.* *12*: 1167-1178.

Smith, K. L., and R. R. Hessler. 1974. Respiration of benthopelagic fishes: In situ measurements at 1230 meters. *Science* *184*: 72-73.

Smith, K. L., and J. M. Teal. 1973a. Temperature and pressure effects on respirations of the costomatous pteropods. *Deep-Sea Res.* *20*: 853-858.

Smith, K. L., Jr., and J. M. Teal. 1973b. Deep-sea benthic community respiration: An in situ study at 1850 meters. *Science* *179*: 282.

Srinivasagam, R. T., J. E. G. Raymont, C. F. Moodie, and J. K. B. Raymont. 1971. Biochemical studies on marine zooplankton. X. The amino acid composition of *Euphausia superba*, *Meganyctiphanes norvegica* and *Neomysis integer*. *J. Mar. Biol. Ass. U.K.* *51*: 917-925.

Teal, J. M., and F. G. Carey. 1967a. Effects of pressure and temperature on the respiration of euphausiids. *Deep-Sea Res.* *14*: 725-733.

Teal, J. M., and F. G. Carey. 1967b. Respiration of a euphausiid from the oxygen minimum layer. *Limnol. Oceanog.* *12*: 548-550.

Teal, J. M. 1971. Pressure effects on the respiration of vertically migrating decapod crustacea. *Amer. Zool.* *11*: 571-576.

Tyler, A. V. 1973. Caloric values of some North Atlantic invertebrates. *Mar. Biol.* *19*: 258-261.

Ware, D. M. 1975. Growth, metabolism and optimal swimming speed of a pelagic fish. *Jour. Fish. Res. Bd. of Canada* *32*: 33-41.

Zeuthen, E. 1947. Body size and metabolic rate in the animal kingdom. *Compt. Rend. Trav. Lab. Carlsberg, Ser. Chimie* *26*: 15-61.

Aspects of the Feeding Ecology of Oceanic Midwater Fishes

T. L. Hopkins and R. C. Baird

University of South Florida

ABSTRACT

A review of the feeding ecology of oceanic midwater fishes is presented in which aspects of vertical distribution and migration, diet composition, feeding chronology, daily food ration, selective feeding, resource partitioning and bioenergetics are discussed. Recommendations are made of areas of emphasis for future research.

INTRODUCTION

Midwater fishes are major contributors to sound scattering at various frequencies in oceanic environments (Farquhar, 1970) and a predictive theory of sound scattering in the ocean must include considerable knowledge of the distribution and ecology of this fauna. An extensive literature on systematics, zoogeography, sound scattering pattern and more recently vertical distribution of midwater fishes reveals that in general, 1) species composition, vertical distribution, and sound scattering layer characteristics vary with hydrographic conditions, 2) fish populations are highly vertically stratified and 3) marked diurnal changes occur in vertical distribution among and within species. While such information is essential to understanding the biology of midwater fishes, little is known concerning the underlying factors for the observed distributions or their relation to oceanic primary production. Studies of functional morphology, feeding ecology (e.g., diet, feeding chronology, prey selectivity) and distribution, however, strongly suggest that patterns of vertical stratification result from behavior evolved to enable each species to efficiently utilize food resources (e.g.,

Marshall, 1954, 1971). Considering the apparent importance of feeding strategies to trophic dynamics and patterns of vertical stratification in oceanic ecosystems, the literature contains relatively sparse information on the feeding ecology of midwater organisms.

This paper is intended primarily to summarize the limited information available concerning various aspects of feeding ecology of mesopelagic fishes. Preliminary results of studies now in progress are included, as well as suggestions for future research emphasis. The concept of feeding strategy encompasses an extremely broad range of considerations which together impinge on the ecology of midwater organisms. A secondary but no less important goal is to reduce the complexity to a more manageable subset of topics from which meaningful inferences and hypotheses can be formulated. The first four subsections consider aspects of the diet and feeding of individual species. The topics are arranged in the following sequence: 1) what is eaten (Diet Composition), 2) when feeding occurs (Feeding Chronology), 3) how much is consumed (Daily Food Ration) and 4) prey choice (Selective Feeding). The fifth topic explores feeding in relation to competitive interactions within and among species (Resource Partitioning) while the sixth subsection considers the impact of a species on its environment in terms of energy utilization (Bioenergetics). Because of limited available data, a number of extrapolations have been made from the literature pertaining to predation, predation theory, and observations on non-midwater fish species. While considerable refinement and revision can be expected to occur with increasing knowledge, the discussion should provide a framework around which investigators of the trophic structure of oceanic ecosystems can direct their research.

DIET COMPOSITION

Net Feeding

A number of investigators have pointed to the potential bias in stomach analysis data because of fish feeding in trawl cod ends. Before considering diet composition, then, some discussion of net feeding is appropriate. Net feeding has been implicated as the probable cause for observed diet patterns in *Leuroglossus stilbius* (*Anderson*, 1967), *Pleuragramma antarcticum* (*DeWitt and Hopkins*, in press), *Argyropelecus aculeatus* (*Merrett and Roe*, 1974) and the ceratioids *Lasiognathus saccostoma* and *L. waltoni* (*Nolan and Rosenblatt*, 1975). *Judkins and Fleminger* (1972) also presented indirect evidence for net feeding by the shrimp *Sergestes similis*. They found significant differences in species composition in gut contents of shrimp taken in midwater trawls and from tuna stomachs. *Collard* (1970) and *Hopkins and Baird* (1975), however, mention the severe stress which must certainly occur to the relatively fragile midwater fishes commonly caught in midwater trawl cod ends. This would make it unlikely that extensive net-feeding occurs. Further, studies of feeding chronology (Holton, 1969; DeWitt and Cailliet, 1972; Baird, Hopkins and Wilson, 1975) presented convincing evidence (i.e., high incidence of empty or near empty stomachs) that at certain periods in the diel cycle, at least, a number of midwater fishes do not feed in trawl cod ends. Hopkins and Baird (1975) found no consistent statistical evidence for net feeding in 11 species of abundant midwater fishes from the eastern Gulf of Mexico in an experiment specifically designed to evaluate net feeding. In this experiment a paired trawl arrangement was used wherein a coarse mesh "catcher" prevented fish from mixing with plankton in the cod end of one trawl while the adjacent trawl was of conventional design without a catcher. Paxton (1967), Holton (1969) and Collard (1970) also found no evidence for net feeding in fishes off California. The evidence, then, while indicating that some degree of net-feeding may occur and that behavior in nets may vary among species, suggests that net feeding is probably not extensive for many midwater fishes.

Diet Components

Diet studies of midwater fishes have concentrated on the Myctophidae, Gonostomatidae and Sternoptychidae, the principal taxonomic components of the midwater fish fauna in most oceanic regions. Though there is considerable phyletic diversity in their diet, crustaceans appear to be the principal forage, copepods, euphausids, ostracods, amphipods and small decapods being the most important elements (Jespersen, 1915; Beebe and Van der Pyl, 1944; Günther and Deckert, 1953; Anderson, 1967; Paxton, 1967; Legand and Rivaton, 1969; Collard, 1970; Nakamura, 1970; Cailliet, 1972; DeWitt and Cailliet, 1972; Samyshev and Schetinkin, 1973; Hopkins and Baird, 1973; Gorelova, 1974; Merrett and Roe, 1974; Tyler and Pearcy, 1975). The dominance of crustaceans in diets is obvious in our data on 17 representatives from the 3 major groups of midwater fishes. In all but 2 species in Table 1 (Diaphus taaningi, Ceratoscopelus warmingi) crustaceans exceed 70% of stomach contents numerically. Also, the importance of copepods is apparent in that this group constituted over 50% of the diets of 8 of the 17 fishes listed and were the most abundant group in all but 5 species. It is apparent as well that the diets of all 17 species are taxonomically diverse, the median number of taxa being 23.

Fishes from other than the 3 dominant midwater fish groups mentioned above can be locally important in pelagic food webs, e.g., the deep-sea smelt (Leuroglossus stilbius: Bathylagidae) in coastal waters of California (Anderson, 1967; Cailliet, 1972) and the codlet (Bregmaceros nectabanus: Bregmacerotidae) in the Cariaco Trench (Baird et al., 1973). The former forages heavily on larvaceans and salps and to a lesser extent on small crustaceans while the latter appears to be a copepod-euphausid predator (Milliken, 1975). Deep

TABLE 1. Diet Composition of 17 Species of Oceanic Midwater Fishes. Values in body of table are percents of total number of prey items. Number in brackets after each fish species listed in heading represents area of collection. Area 1 = E. Gulf of Mexico; 2 = Cariaco Trench; 3 = Off Southern California; 4 = Gulf of Guinea; 5 = Off Baja California.

Stomach Contents	Benthosema suboribitale [1] (52 fish, 23-33mm SL; 168 prey)	Ceratoscopelus warmingi [1] (50 fish, 28-52mm SL; 464 prey)	Diaphus dumerili [1] (12 fish, 37-66mm SL; 175 prey)	Diaphus taaningi [2] (9 fish, 35-45mm SL; 1414 prey)	Lampanyctus alatus [1] (64 fish, 28-46mm SL; 420 prey)	Lepidophanes guentheri [1] (71 fish, 37-58mm SL; 490 prey)	Notolychnus valdiviae [1] (49 fish, 16-23mm SL; 130 prey)	Gonostoma elongatum [1] (52 fish, 68-117mm SL; 121 prey)	Valenciennellus tripunctulatus [1] (20 fish, 18-29mm SL; 219 prey)	Argyropelecus aculeatus [1] (14 fish, 22-38mm SL; 68 prey)	Argyropelecus affinis [3] (9 fish, 27-47mm SL; 141 prey)	Argyropelecus hemigymnus [3] (9 fish, 26-30mm SL; 129 prey)	Argyropelecus lychnus [5] (16 fish, 52-57mm SL; 71 prey)	Argyropelecus sladeni [3] (17 fish, 27-47mm SL; 384 prey)	Sternoptyx diaphana [4] (14 fish, 20-40mm SL; 267 prey)	Sternoptyx obscura [3] (17 fish, 20-40mm SL; 399 prey)	Sternoptyx pseudobscura [4] (7 fish, 20-40mm SL; 104 prey)
Copepoda																	
Acartia	—	—	—	0.1	0.2	—	0.8	—	—	—	—	—	—	—	—	—	—
Arietellus	0.6	—	—	—	—	—	—	—	—	—	—	0.8	1.4	2.9	—	0.8	—
Calanus	—	0.2	—	—	0.2	—	—	—	—	—	50.4	0.8	5.6	—	—	—	—
Calocalanus	4.8	0.4	0.6	0.2	—	3.9	0.8	0.8	0.9	—	—	—	—	—	—	—	—
Canadacia	—	—	20.0	—	3.3	—	—	—	—	—	—	—	—	1.0	20.2	4.3	3.8
Centropages	1.2	—	0.6	—	—	0.4	—	—	—	—	1.4	—	—	—	—	0.3	—
Clausocalanus	—	2.4	—	0.5	0.2	—	1.5	—	0.5	—	—	—	—	0.3	—	—	—
Euaetideus	—	—	—	—	0.5	—	—	—	—	—	—	—	—	—	—	—	1.0
Eucalanus	—	0.9	—	2.3	2.4	1.2	—	—	—	—	—	—	—	—	—	—	—
Euchaeta	1.2	0.4	1.7	—	—	1.0	6.9	0.8	17.8	—	0.7	—	1.4	0.5	13.9	0.8	10.6
Euchirella	1.2	0.2	—	—	0.2	0.8	0.8	0.8	9.1	1.5	4.3	—	1.4	0.3	7.5	0.5	2.9
Gaetanus	—	—	—	—	0.2	—	—	—	0.5	—	0.7	—	—	1.0	0.4	0.8	—
Haloptilus	—	—	—	—	—	—	—	—	1.0	1.5	—	0.8	—	—	—	—	—
Heterorhabdus	—	0.4	1.1	—	0.5	0.8	—	0.8	2.7	—	4.3	3.1	1.4	1.3	0.4	—	2.9
Labidocera	—	—	—	—	—	—	—	—	—	—	—	0.8	—	—	0.4	—	—
Lophothrix	—	—	—	—	0.2	—	—	—	—	—	—	—	—	—	—	—	—

ASPECTS OF THE FEEDING ECOLOGY

Taxon	1	2	3	4	5	6	7	8	9	10	11	12	13	14	15	16	17
Lucicutia	2.4	1.5	-	0.4	0.2	0.2	1.5	-	0.5	-	-	0.8	-	-	-	-	-
Metridia	-	-	4.6	-	1.0	2.9	-	-	-	-	-	9.3	-	0.5	1.1	-	-
Nannocalanus	0.6	1.7	0.6	-	2.1	0.6	3.8	-	1.8	-	-	-	-	-	6.7	1.0	-
Neocalanus	1.2	-	-	1.1	-	-	0.8	-	-	-	-	-	-	-	-	-	-
Paracalanus	-	0.2	1.7	-	2.1	1.6	2.3	-	-	-	-	-	-	-	1.5	-	-
Paracandacia	1.2	0.4	-	-	-	-	-	-	-	-	-	-	-	-	-	-	-
Phaenna	-	0.2	-	0.1	23.8	28.0	16.2	23.1	13.7	1.5	-	1.6	-	19.0	1.5	-	3.8
Pleuromamma	19.0	3.7	4.0	-	-	0.2	-	-	-	-	-	-	-	0.3	-	-	-
Pontellina	-	-	-	0.1	-	-	0.8	-	2.7	-	0.7	27.9	1.4	-	-	-	-
Rhincalanus	-	-	-	-	-	-	-	0.8	2.7	-	-	-	-	-	-	-	-
Scolecithricella	2.4	0.9	1.1	-	0.7	1.2	0.8	1.7	-	-	-	-	-	-	1.5	-	-
Scolecithrix	-	-	-	-	0.7	-	-	-	-	-	-	-	-	-	-	-	-
Scottocalanus	0.6	0.9	-	7.2	2.4	1.4	8.5	-	-	-	-	-	-	0.5	2.2	-	-
Temora	1.2	0.2	-	-	-	1.2	-	-	-	-	-	-	-	-	-	-	-
Undeuchaeta	-	0.2	-	-	-	-	-	-	-	-	-	-	-	-	-	-	-
Undinula	7.2	4.8	3.4	2.2	15.8	6.1	9.2	4.1	30.6	1.5	15.5	27.1	2.8	3.4	2.2	0.5	-
Xanthocalanus	-	-	-	-	-	-	-	-	0.5	-	-	-	-	-	-	-	-
Calanoida	-	-	-	0.1	1.7	1.2	0.8	1.7	-	4.4	-	-	-	-	1.1	0.3	-
Conaea	3.0	2.6	1.1	0.7	-	-	-	-	-	-	-	-	-	-	-	-	-
Copilia	-	-	-	0.1	-	0.8	-	-	0.5	-	-	0.8	-	-	-	-	-
Corycaeus	4.8	1.3	-	-	-	-	-	-	0.5	-	-	-	-	-	-	-	-
Farranula	23.2	6.3	0.6	0.2	3.1	3.9	20.8	0.8	11.0	1.5	-	3.9	-	-	0.7	0.3	1.9
Lubbockia	1.8	1.3	6.9	7.9	-	0.6	-	-	-	-	-	-	1.4	-	0.4	-	-
Oithona	0.6	0.9	0.6	-	-	-	2.3	-	-	-	-	-	-	-	-	-	-
Oncaea	-	0.2	-	-	-	-	-	-	1.8	-	-	4.7	-	-	-	-	-
Sapphirina	-	-	-	0.1	-	-	-	-	-	-	-	-	-	-	-	-	-
Cyclopoida	-	0.4	-	-	-	-	-	-	-	-	-	-	-	-	-	-	-
Aegisthus	-	-	-	3.1	-	-	-	-	-	-	-	-	-	-	-	-	-
Macrosetella	10.7	13.1	1.1	0.3	6.9	10.2	11.5	11.6	1.4	36.8	5.7	18.6	7.0	64.1	4.5	29.1	6.7
Microsetella	-	-	-	0.1	-	-	-	-	-	-	-	-	-	-	-	-	-
Miracia	-	-	-	-	-	-	-	-	-	1.5	-	-	-	-	0.7	-	2.9
Copepod nauplii	-	-	-	-	0.5	0.6	-	-	0.5	-	-	-	-	-	1.1	0.5	1.9
Ostracoda	-	0.2	-	-	0.5	0.2	-	-	0.5	-	-	-	1.4	-	2.6	1.3	2.9
Conchoecinae	-	-	-	-	-	-	-	-	11.0	2.9	-	-	-	0.3	-	-	-
Cladocera	-	-	-	-	0.7	0.6	-	0.8	-	-	-	-	-	-	-	-	-
Penilia	0.6	1.3	2.9	-	0.5	-	-	-	-	-	-	-	-	-	-	-	-
Amphipoda																	
Amphithyrus																	
Anchylomera																	
Brachyscelus																	
Eupronoe																	
Hyperia																	
Hyperioides																	

TABLE 1 (cont.)

Stomach Contents	B. suborbitale	C. warmingi	D. dumerili	D. taaningi	L. alatus	L. guentheri	N. valdiviae	G. elongatum	V. tripunctulatus	A. aculeatus	A. affinis	A. hemigymnus	A. lychnus	A. sladeni	S. diaphana	S. obscura	S. pseudobscura
Paraphronima	-	-	-	-	-	-	-	-	-	-	-	-	-	-	-	-	-
Parathemisto	-	-	-	-	-	-	-	-	-	-	-	-	-	-	-	0.5	-
Phronima	-	0.2	-	-	0.2	-	-	-	-	-	-	-	-	-	-	3.3	-
Phronimella	-	-	-	-	-	-	-	-	-	-	-	-	-	-	-	0.3	1.0
Phronimopsis	-	0.2	-	-	2.6	-	-	-	-	-	-	-	-	-	0.7	-	3.8
Phrosina	-	-	-	-	0.2	0.2	-	0.8	-	-	-	-	-	-	-	-	-
Platyscelus	0.6	-	-	-	0.2	0.2	-	0.8	-	-	-	-	-	-	-	-	-
Primno	-	-	-	-	1.2	-	-	-	-	5.9	-	-	-	-	3.0	-	5.8
Scina	-	-	-	-	-	0.4	-	0.8	-	-	-	-	-	-	1.1	8.8	1.9
Streetsia	-	-	-	-	-	-	-	-	-	-	-	-	-	-	-	-	-
Sympronoe	-	-	-	-	-	-	-	-	-	-	-	-	-	-	0.7	1.0	1.0
Vibilia	-	-	-	-	-	-	-	-	-	-	0.7	-	-	-	-	-	-
Hyperiidea	-	0.4	-	-	0.7	0.2	0.8	4.1	-	2.9	-	-	1.4	1.3	2.6	-	1.0
Gammaridea	0.6	-	-	-	0.5	-	-	0.8	-	-	-	-	-	-	0.4	-	-
Amphipoda	-	0.2	-	-	0.2	0.4	0.8	0.8	-	1.5	-	-	-	-	2.6	-	-
Euphausiacea	-	-	-	-	-	-	-	-	-	-	-	-	-	-	-	-	-
Euphausia	-	-	-	-	1.2	9.8	1.5	-	-	1.5	-	-	-	-	-	-	-
Nematoscelis	1.8	0.9	-	-	0.7	0.4	-	5.0	-	1.5	-	-	25.4	2.3	1.5	7.5	-
Stylocheiron	-	-	0.6	-	14.0	4.3	6.2	19.0	-	-	-	-	-	-	0.7	-	5.8
Thysanoessa	-	-	-	-	-	-	-	-	-	-	-	-	-	-	-	2.3	-
Thysanopoda	-	-	-	-	-	-	-	3.3	-	-	-	-	-	-	-	-	-
Euphausiacea	-	1.7	-	0.1	4.0	3.7	0.8	6.6	-	-	-	-	5.6	0.3	1.9	14.0	1.9
Decapoda	-	-	-	-	-	-	-	-	-	-	-	-	-	-	-	-	-
Lucifer	-	-	0.6	0.3	-	0.6	-	-	-	-	-	-	-	-	-	-	-
Sergestes	-	-	12.2	-	-	0.4	-	0.8	-	-	-	-	-	-	5.6	-	1.9
Decapoda	-	-	-	-	-	1.2	-	-	-	-	-	-	-	-	-	-	-
Larvae	-	0.9	17.1	0.4	0.5	-	-	4.1	-	16.2	-	-	28.2	-	-	2.3	4.8

ASPECTS OF THE FEEDING ECOLOGY

Prey	C1	C2	C3	C4	C5	C6	C7	C8	C9	C10	C11	C12	C13
Stomatopoda													
Larvae	–	–	0.6	–	–	–	–	–	–	0.4	–	–	–
Mollusca													
Cavolinia	–	0.4	–	–	0.2	–	–	–	–	–	–	–	–
Clio	–	0.2	–	–	–	–	–	–	–	–	–	–	–
Creseis	–	0.2	–	–	–	–	–	–	–	–	–	–	–
Diacria	–	0.6	–	–	–	–	–	–	–	–	–	–	–
Limacina	4.2	15.1	1.1	–	0.2	3.5	–	2.9	–	–	–	–	–
Styliola	–	2.6	–	–	–	0.8	–	2.9	–	–	–	–	–
Pteropoda	–	0.2	–	–	–	0.2	–	4.4	0.7	–	–	–	1.0
Atlantidae	–	1.7	–	0.1	0.2	–	–	1.5	–	–	–	–	–
Cephalopoda	–	–	–	0.1	–	–	–	–	–	0.4	0.3	–	–
Pelecypod larvae	–	0.2	–	–	–	–	–	–	–	0.4	–	–	–
Polychaeta													
Alciopidae	–	–	–	–	0.7	0.4	0.8	–	–	–	–	–	–
Tomopteris	–	0.2	–	–	–	–	–	–	–	–	–	–	–
Typhloscolecidae	–	–	–	–	–	–	–	–	–	–	–	–	–
Polychaeta	–	–	–	0.3	–	–	–	–	2.1	7.0	–	–	–
Siphonophora													
Chelophyes	–	–	–	–	–	–	–	–	–	–	–	–	–
Siphonophora	1.2	11.2	2.3	–	–	0.2	–	1.5	–	–	0.3	–	–
Chaetognatha													
Sagitta	–	–	–	0.3	–	–	–	2.9	1.4	–	–	3.0	1.8	8.7
Chaetognatha	–	–	3.4	0.1	–	0.2	–	–	7.1	0.8	1.4	–	6.4	14.8	7.7
Tunicata													
Oikopleura	–	–	–	71.3	–	–	–	–	–	–	–	–	–
Larvacea	–	2.2	4.0	–	0.2	0.2	–	–	–	–	–	–	–
Salpidae	–	4.3	1.1	–	–	0.2	–	–	–	–	–	–	–
Doliolum	–	0.2	–	0.1	–	–	–	–	0.7	–	–	–	–
Echinodermata													
Larvae	–	–	–	–	–	–	–	–	0.7	–	–	–	–
Protozoa													
Foraminifera	–	0.2	–	–	–	–	–	–	2.8	–	–	–	–
Radiolaria	–	0.2	–	–	–	–	–	–	–	–	–	–	–
Pisces													
Myctophidae	–	–	–	–	–	–	1.7	–	–	1.4	–	0.7	–
Cyclothone	–	–	1.7	0.1	0.5	–	–	–	–	4.2	–	–	–	2.3
Pisces	–	–	–	–	–	0.8	2.5	–	–	–	–	0.7	–	–	1.0
Phaeophyceae													
Sargassum	–	0.2	–	–	–	–	–	–	–	–	–	–	–
Unident. Prey	2.4	6.7	2.9	–	0.2	–	–	1.5	–	–	–	–	0.3	–	1.0

dwelling fishes such as larger stomiatoids and ceratioids which constitute minor fractions of trawl catches utilize a wide range of food including copepods, euphausids, decapods, squids and fishes (*Bertelsen*, 1951; *Tckernavin*, 1953; *Marshall*, 1954; *Legand and Rivaton*, 1969; *Merrett and Roe*, 1974; *Nolan and Rosenblatt*, 1975).

Diet Variability

Factors complicating investigations of diet composition are: 1) the influences of ontogeny on feeding behavior and prey selection, 2) seasonal changes in prey availability and 3) regional variations in feeding pattern.

Ontogenetic changes in diet have been documented in many non-midwater fishes (e.g., *Gerking*, 1954; *Reid*, 1954; *Seaburg and Moyle*, 1964; *Hartmann*, 1970; *Keast*, 1970; *Adams*, 1972; *Carr and Adams*, 1973) and have been observed as well in the diets of oceanic midwater fishes e.g., *Sternoptyx diaphana* (*Hopkins and Baird*, 1973) and *Diaphus theta* and *Tarlentonbeania crenularis* (*Tyler and Pearcy*, 1975). In our data on 6 midwater species in Table 2 a change occurs in both size and taxonomic composition in the diet with increasing fish size. In general, larger individuals appear to take a greater proportion of larger food items than do smaller fish. In each of the species listed, copepods play a major role in nutrition, particularly in the earlier stages of development. It should be cautioned, however, that numerical abundance does not necessarily equate to nutritional contribution. A single euphausiid, for example, is equivalent to many copepods in terms of biomass. Lack of coincidence of the most numerous prey taxon with that contributing most to food biomass in stomachs can be seen in our data on *S. diaphana* (*Hopkins and Baird*, 1973; Table 7) and that of *Tyler and Pearcy* (1975) from *Stenobrachius leucopsarus*, *Diaphus theta*, and *Tarlentonbeania crenularis*.

Seasonal changes have been observed in midwater fish diets as is apparent in *Cailliet's* (1972) study of *Leuroglossus stilbius* in the Santa Cruz Basin off southern California. In the period of relatively high primary productivity (January-June), *L. stilbius* feeds primarily on salps and larvaceans; during the less productive season (the thermocline period of August-December) when this prey is less abundant, *L. stilbius* forages more on small copepods. A seasonal change in diet of the myctophid *Benthosema glaciale* in fjord waters has also been observed by *Gjøsaeter* (1973b) who found that in summer this species feeds almost exclusively on copepods while in winter euphausids become an important component of the diet.

Regional variation in diet composition of midwater fishes is apparent from the comparative data in Table 3. Least inter-regional variation in diet components at higher taxonomic levels (i.e., copepods, ostracods, etc.) seems to occur in species which are relatively small as adults (i.e., *N. valdiviae*, *V. tripunctulatus*, *A. hemigymnus*) and in earlier stages of species which attain a greater size

TABLE 2. Ontogeny in the Diets of 6 Midwater Fishes as Indicated by Changes in Size and Taxonomic Composition of Food Items. Area of collection indicated in parentheses (for boundaries of these geographic areas see Baird, 1971).

	No. specimens	size range (SL, mm)	No. food items	% food <5mm	% food >5mm	Trend in diet Composition (increasing %)
Argyropelecus aculeatus (WN Central Atlantic)	14 19	10-20 50-60	198 213	99 12	1 88	ostracods, copepods → fish
Argyropelecus affinis (off Southern California)	39 32	21-34 52-68	259 209	94 68	6 33	copepods, ostracods → euphausiids, salps, chaetognaths
Argyropelecus sladeni (off Southern California)	12 13	10-20 30-50	151 138	66 24	34 76	ostracods, copepods, chaetognaths → euphausiids
Sternoptyx diaphana[1] (Pacific subantarctic)	12 12	10-24 40-80	104 110	100 60	0 40	copepods, ostracods → euphausiids
Sternoptyx obscura (off Southern California)	16 14	10-20 30-50	271 277	86 63	14 37	copepods → ostracods, amphipods, chaetognaths
Valenciennellus tripunctulatus (E. Gulf of Mexico)	21 30	15-20 30-35	116 109	85 58	15 42	small copepods → larger copepods; >%Pleuromamma

[1]These specimens belong to a species closely related to, but distinct from S. diaphana. The new form is being described by the second author.

TABLE 3. Diet Comparisons of Midwater Fish Species Taken from Several Different Oceanic Regions. Taxonomic composition expressed as % total number of food items in stomachs. First (heavy line) and second (light line) ranked categories underlined. Oceanic areas (for geographic boundaries see Baird, 1971): 1 = Indian Ocean Central; 2 = Caribbean – Gulf Central; 3 = EN Atlantic Central; 4 = E Pacific equatorial; 5 – N Pacific Transitional; 6 = WN Atlantic Central; 7 = Pacific Subantarctic; 8 = EN Pacific Central; 9 = SE Atlantic Transitional; 10 = Venezuelan-Caribbean; 11 = NW Atlantic pocket; 12 = Indian Ocean Equatorial. Diversity index = $\bar{D}'$, a modification of the Shannon index as defined by Travers, 1971.

	Oceanic area	No. specimens	Size range (SL, mm)	Diet Composition (% no. food items)										No. food items	Diversity index
				Copepods	Ostracods	Amphipods	Euphausiids	Other Crustaceans (mostly decapods)	Molluscs	Polychaetes	Gelatinous orgs. (siphon, tunicates etc.)	Chaetognaths	Fish	Other		
Ceratoscopelus warmingi	1a	60	ND	66	1	7	14	2	–	–	3	4	1	–	147	1·8
	2	50	28-52	34	13	3	3	1	21	<1	24	–	–	<1	464	2·3
Lampanyctus alatus	1a	59	ND	73	–	3	13	11	–	–	–	–	–	–	38	1·2
	2	64	28-46	62	7	8	20	1	1	1	<1	–	1	<1	420	1·7
Notolychnus valdiviae	1a	90	ND	77	–	14	–	6	–	–	–	3	–	–	35	1·1
	2	49	16-23	79	12	2	9	–	–	–	–	–	–	–	130	0·9
	3b	19	12-23	77	17	–	7	–	–	–	–	–	–	–	30	0·9

TABLE 3 (cont.)

Valenciennellus tripunctulatus	3[b]	85	8-19	83	16	<1	<1	1	–	–	–	–	–	354	0.7
	2	21	15-20	97	3	–	–	–	–	–	–	–	–	116	0.2
Argyropelecus aculeatus	3[b]	42	8-38	21	74	2	<1	–	1	1	1	–	–	483	1.1
	2	31	13-38	17	29	14	2	2	31	1	2	<1	1	371	2.4
	2	23	30-50	21	22	19	4	5	16	1	5	3	2	161	2.9
	3	13	30-60	8	5	11	7	1	41	3	7	2	2	98	2.8
	4	11	30-50	22	6	4	60	3	–	–	3	12	–	68	1.8
Argyropelecus hemigymnus	2	51	16-34	78	19	2	1	–	–	1	–	–	–	202	0.9
	5	21	10-30	77	22	–	1	–	–	1	–	1	–	210	0.8
	6	12	10-30	64	33	1	–	–	–	1	–	–	–	81	1.1
	7	28	10-30	68	14	–	–	–	2	–	–	15	2	59	1.4
	8	10	10-30	61	39	–	–	–	–	–	–	–	–	129	1.0
	3[b]	18	14-26	64	29	–	–	1	3	1	–	1	–	73	1.3
Argyropelecus sladeni	5	24	10-30	56	39	1	4	–	–	–	–	1	–	210	1.3
	9	14	10-30	52	35	7	4	–	–	–	–	3	–	95	1.6
Sternoptyx diaphana	6[c]	21	20-40	7	16	44	8	3	<1	<1	–	17	4	446	2.3
	9[c]	14	20-40	59	5	15	4	6	1	1	–	8	2	267	2.0
	10[c]	13	20-40	4	10	21	29	13	1	<1	–	17	5	115	2.6
	11[c]	13	20-40	18	11	18	10	4	3	1	–	4	32	72	2.6
Sternoptyx obscura	5	24	10-30	45	9	8	19	2	1	3	–	12	2	479	2.3
	12	14	10-30	70	8	6	4	1	–	4	–	8	–	253	1.6

[a]From Legand and Rivaton, 1969, Tables 1,4; their C. townsendi is probably C. warmingi (see Nafpaktitis, 1969).
[b]From Merrett and Roe, 1974, Table 1(A).
[c]From Hopkins and Baird, 1973, Tables 5,7.

(e.g., *A. sladeni*). These smaller sized fish also show less diversity in diet at higher taxonomic levels of prey than do the larger fish listed in the table, though prey diversity at the generic level even in smaller species can be considerable (see Table 1). Greater regional variation seems to occur in the intermediate to larger sized fishes and is especially apparent in the diets of *A. aculeatus* and *S. diaphana*. Data in Table 3 are also supportive of earlier statements on diet trends in Table 1 in that on a world wide basis crustaceans, copepods in particular, predominate in the diets of oceanic midwater fishes. Of theoretical interest here is that regional variations in diet may reflect not only differences in prey availability, but indicate, as well, species adaptability in utilization of locally available prey and the ability to respond to a range of different visual or other stimuli in obtaining food.

Summarizing, the available data indicate that midwater fish species generally consume a wide range of prey types and sizes which may vary seasonally and geographically. Most species can probably be classified as opportunistic predators (*sensu Holling*, 1968; *Schoener*, 1971) in which a significant proportion of the diet is a function of prey availability. The generally planktivorous diet of most species also suggests that more time is spent in search than pursuit and capture of individual prey items. This in part is related to the fact that prey size and mobility are small in comparison to the size and speed of midwater fish and because of presumed limits on visual detectability of prey in poorly illuminated midwater environments. Finally, the complexity of diets (as can be seen in Table 1) and the occurrence in stomachs of prey from several trophic levels reinforces the concept of a food web rather than food chain in oceanic ecosystems.

FEEDING CHRONOLOGY

Diurnal Patterns

If, as *Marshall* (1954) suggests, diurnal changes in patterns of vertical distribution are associated with feeding, then diurnal periodicity in rate of prey ingestion in migrating species might be expected. Investigations of feeding chronology in a number of midwater fishes are summarized in Table 4. Distinct diurnal feeding cycles are present in many strong migrators which ascend into the epipelagic zone (<200 m) at night. Only two of the myctophids listed in Table 4 (*Diaphus dumerili*, *Lepidophanes guentheri*), for instance, show little evidence of feeding periodicity. For species which remain below the epipelagic layer in the mesopelagic zone, feeding patterns are more varied and may include periodic feeding at various times or acyclic feeding behavior. The weakly migrating *Valenciennellus tripunctulatus* feeds primarily during the day, while the two sternoptychids listed, *Argyropelecus aculeatus*, a migrator,

TABLE 4. Feeding Chronology Data on Oceanic Midwater Fishes.

Principal Feeding Period	Species	Depth Zone (m) Day	Depth Zone (m) Night	Region of Study
Night	Myctophum spinosum[1]	500-1000	Near surface	W equatorial pacific
	Myctophum aurolaternatum[1]	500-1000	Near surface	W equatorial pacific
	Symbolophorus evermanni[1]	500-1000	Near surface	W equatorial pacific
	Centrobranchus andreae[1]	500-1000	Near surface	W equatorial pacific
	Benthosema glaciale[2]	>200	<200	Norwegian fjords
	Diaphus taaningi[3]	220-270	0-100	Cariaco Trench, Venezuela
	Lampanyctus cuprarius[4]	700-900	100-500	NE Atlantic 30°N 23°W
	Lampanyctus mexicanus[5]	600-700	~100-250	Southern Gulf of California
	Lampanyctus mexicanus[9]	300-700	50-300	Catalina Basin, California
	Lobianchia dofleini[4]	300-500	25-200	NE Atlantic 30°N 23°W
	Notolychnus valdiviae[4]	400-700	25-200	NE Atlantic 30°N 23°W
			300-500	NE Atlantic 30°N 23°W
			600-700	NE Atlantic 30°N 23°W
	Cyclothone acclinidens[6]	>600	>600	Off Santa Barbara, California
	Leuroglossus stilbius[6]	400-600	50-100	Catalina Basin, California
	Bregmaceros nectabanus[7]	~100	400-900	Cariaco Trench, Venezuela
Night-Morning	Stenobrachius leucopsarus[10]	300-500	<200	Off Oregon coast
	Diaphus theta[10]	300-400	<200	Off Oregon coast
	Tarletonbeania crenularis[10]	300-400	Surface	Off Oregon coast
Day	Valenciennellus tripunctulatus[4,7]	100-500	100-500	NE Atlantic 30°N 23°W
		200-600	200-600	E Gulf Mexico 27°N 86°W
Dusk	Argyropelecus aculeatus[4]	300-500	200-300	NE Atlantic 30°N 23°W
	Argyropelecus hemigymnus[4]	300-600	200-600	NE Atlantic 30°N 23°W
Acyclic	Diaphus dumerili[8]			NW Africa, Slope, 21°N 7-18°W
	Lepidophanes guentheri[8]			NW Africa, Slope, 21°N 7-18°W
	Maurolicus muelleri[8]			NW Africa, Slope, 21°N 7-18°W
	Cyclothone signata[6]	<600	<600	Off Santa Barbara, California
	Chauliodus danae[4]	>250	100-300	NE Atlantic 30°N 23°W
			400-900	

[1] Gorelova, 1974
[2] Gjösaeter, 1973a, b
[3] Baird et al., 1975
[4] Merrett & Roe, 1974
[5] Holton, 1969 (Triphoturus mexicanus)
[6] DeWitt & Cailliet, 1972
[7] Milliken, 1975; authors unpub. data
[8] Samyshev & Schetinkin, 1971
[9] Anderson, 1967 (=Triphoturus mexicanus)
[10] Tyler and Pearcy, 1975; depths represent peak abundances

and *A. hemigymnus*, a weak migrator, appear to feed primarily at dusk when downwelling light is apparently still sufficient for visual prey location. The gonostomatid *Cyclothone acclinidens*, a species which does not migrate to epipelagic depths, appears to be a nocturnal feeder. An acyclic pattern may predominate in species which ingest very large prey items, particularly those living in food poor bathypelagic regions, though evidence is generally lacking for these groups.

Because of lack of uniformity in field and lab methods and in data reporting in midwater nutritional investigations, variations in diurnal feeding patterns related to oceanic regions are difficult to assess. Additional feeding studies on a worldwide basis may reveal, for example, that in regions of upwelling, midwater fishes show less distinct feeding cycles than in central, less turbulent or enriched oceanic areas. *Samyshev and Schetinkin* (1973) found no diurnal feeding cycle in *Lepidophanes guentheri* and *Diaphus dumerili* in upwelling waters off the West African coast (see Table 4), and *Paxton* (1967) was unable to report cyclic feeding in the myctophids he examined from productive waters off southern California. While *Anderson* (1967) found most active feeding at night in *Triphoturus mexicanus* and *Bathylagus stilbius* in another study off southern California, his data were inconclusive as to the extent of daytime feeding. Also, *Tyler and Pearcy* (1975) though noting night and morning peaks in foraging, also had evidence for continuous feeding throughout the diurnal period in North Pacific transitional waters (see *Baird*, 1971) off the Oregon coast.

Digestion Rates

Rate of digestion has not been determined by direct means for any species of oceanic midwater fish. *Hochachka* (1974) suggested, however, on the basis of studies of enzymes extracted from *Stomias* livers that enzymes of deep sea fishes have unusually low activation energies and are highly efficient. Also, it seems probable that fish with a pronounced feeding cycle would clear their digestive tracts daily. The authors (*Baird et al.*, 1975a) found stomachs and intestines of *Diaphus taaningi* to be mostly empty by early morning. The number of prey items in stomachs and intestines of fish taken at different times during the night indicated that the rate of digestion appeared similar to that reported for the bluegill *Lepomis macrochirus*, a species with a digestion rate typical of many predaceous fishes (*Windell*, 1967). Of particular interest in the study of *D. taaningi* was the indication that most of the digestion also occurs at night at the highest diurnal temperatures encountered by the species. Whether these patterns are typical of many strong vertical migrators remains to be documented and the bio-energetic ramifications (e.g., see *McLaren*, 1963) need further investigation. There is evidence that midwater fishes ingesting relatively large prey have rates of digestion somewhat comparable to fishes feeding on small food items. Marshall (1954) and Merrett and Roe (1974) suggest that Chauliodus can digest most of its meal within a diel period.

DAILY RATION

Daily Ration and Fish Biomass

Little information exists on daily ration of oceanic midwater fishes in terms of biomass. In one of the few papers considering this aspect of diet, Legand and Rivaton (1969) found that myctophids with full stomachs contained approximately 1% of their body weight in prey. The species examined had fed primarily on herbivorous plankton or at Legand and Rivaton's designated trophic level "A". Their study also included the stomiatoid Chauliodus sloani, a predator feeding at a higher trophic position (level "B"). Full stomachs of Chauliodus contained proportionally more food on a body weight basis, though it is not known if this species feeds on a daily cycle.

Diaphus taaningi with full stomachs contained 0.8% of body wieght in food (Baird et al., 1975a) a percentage in Legand and Rivaton's (1969) range for myctophids feeding at trophic level "A". Taxonomic analysis of the diet revealed that this myctophid also fed on small herbivorous plankton. Similarity of daily ration percentages, however, may be fortuitous because of variations in methods of stomach analysis. Legand and Rivaton, for instance, weighed gut contents and consequently may have underestimated food biomass because of losses from digestion. Holton (1969), following Kow's (1950) procedures, gravimetrically estimated "% nutrition" for Lampanyctus mexicanus (=Triphoturus mexicanus; see Paxton, 1972). In this method the weight of the digestive tract is also included which restricts the use of these data in estimations of daily ration. The authors (Hopkins and Baird, 1973; Baird et al., 1975a) have attempted to determine weight of undigested ration through measuring food items and estimating biomass from size-dry weight curves. These curves, however, were derived from preserved plankton material and appropriate correction factors are needed (see Hopkins, 1968) for accurate determinations of dry weight of fresh prey. In our analyses both stomach and intestines are examined to determine daily ration, the intestinal contents also allowing some estimate of evacuation rates.

The above data imply that daily ration may be a constant proportion of body weight in midwater fishes. However, no adequate study of the relative weight of the daily ration with increasing size has been attempted. There may well be size-specific differences in energy requirements per unit of biomass which are not simply related to total increased metabolic demands of larger size. Several herpetological studies on lacertid and Anolis lizards suggest size-specific differences in energy requirements which are reflected in relative differences in daily ration (e.g., *Schoener and Gorman*, 1968; *Jackson and Telford*, 1975).

Daily Ration and Feeding Periodicity

A number of other factors related to fish behavior also complicate the estimation of daily intake as well. It is apparent, for instance, from most published results on midwater fish diets that the population of any given species is not feeding in perfect synchrony. In any trawl catch, even those made during the period of active feeding, fish stomachs range from being packed with food to containing no food at all. Further, as found in *Valenciennellus tripunctulatus* (*Merrett and Roe*, 1974), fresh food items can occur in fishes collected even during the least active feeding periods. It may be, then, as *Merrett and Roe* (1974) and *Tyler and Pearcy* (1975) suggested, that despite a daily peak in foraging activity, feeding at a low level can occur in some species at any time in the diel period. In the bluegill, for instance, *Windell* (1967) has shown that feeding temporarily ceases after satiation but resumes as the stomach empties. The relationships of feeding rate in midwater fishes to diurnal periodicities in metabolism and to degrees of satiation (i.e., amount of food in the stomach relative to stomach capacity) remain unexplored.

Daily ration should be easiest to assess in species with sharply defined feeding periodicities. Strongly migrating fishes such as myctophids which feed in the epipelagic zone at night should, then, provide some of the most reliable information on daily intake. Stomach analysis of *Diaphus taaningi* taken from the Cariaco Trench (*Baird et al.*, 1975) indicate, for example, that virtually no feeding occurs during the day. The same is apparently true for the gadoid *Bregmaceros nectabanus* which resides in the anoxic zone of the trench during daylight hours (*Baird et al.*, 1973; unpublished data). Conversely, daily ration is much more difficult to estimate in species which "snack" between main feeding bouts, e.g., *Valenciennellus tripunctulatus*, or feed continuously or randomly throughout the diel period.

Another factor potentially complicating the estimation of daily intake is regurgitation. *Holton* (1969) suggested that *Lampanyctus mexicanus* (=*Triphoturus mexicanus*) regurgitates its undigested stomach contents prior to downward migration at dawn. Also, as *Marshall* (1954) stated, regurgitation might be possible as well in some individuals under stress in the trawl. In analyses of thousands of midwater fish stomachs, we have on occasion observed hatchetfish stomachs which have been everted through the expansion of the swimbladder. We have not found this, however, in other fishes taken in midwater trawls. Whether or not midwater fishes can indeed regurgitate has not been reported, though it has been noted in fishes from other environments (*Markus*, 1932; *Molnár et al.*, 1967, and others).

Daily Ration and Caloric Equivalency

Daily intake in terms of calories has not been estimated for any oceanic midwater fish, though considerable information is available for freshwater species (e.g., *Phillips*, 1972; *Warren and Davis*, 1967). Adequate bomb calorimetry information on a unit weight basis is lacking on even the more common types of prey utilized by midwater fishes, though some data are available on a few calanoid copepods which are important forage species in boreal-temperate waters, e.g., *Calanus finmarchicus* and *C. helgolandicus*. The caloric equivalents for these copepods range from 5400 to 7400 cal./g ash free dry weight (AFDW), the value depending on sex and season (*Comita and Schindler*, 1963). In tropical-subtropical waters and bathy-abyssopelagic depths where seasonality is less pronounced, the range in caloric content for a given species or sex might be less. An average caloric value for crustaceans, probably the principal food of the bulk of the midwater fish population, falls in the range of 5200-5900 cal./g AFDW. This is based on data from copepods, caldocerans, amphipods, mysids and decapod shrimp summarized in *Cummins and Wuycheck's* (1971) survey of caloric equivalents. Caloric values for a variety of midwater fishes are listed in *Childress and Nygaard* (1973) who found on a unit wet weight basis, caloric value in general decreases with depth of habitat.

As previously indicated, midwater fish generally utilize a broad range of forage items both by size and taxonomic composition. The daily ration, then, is constituted by organic units of different caloric "worth" both in terms of absolute caloric equivalents and probably the percentage of this energy which can be digested and assimilated by the predator. In the case of the hatchetfish *Sternoptyx diaphana* most of the prey items found in 20-40 mm fish were in the <1 cm size fraction (full stomachs contained 30 to 40 items) whereas greater food biomass and presumably caloric content were in size classes larger than 1 cm. This species, which inhabits the deep mesopelagic zone, is foraging in a relatively food-poor environment and appears to be a highly opportunistic predator ingesting the nearest prey it can perceive and capture (*Hopkins and Baird*, 1973). In contrast is the pattern of the myctophid *Diaphus taaningi* which feeds in the upper 50 m at night in relatively plankton-rich waters (*Baird et al.*, 1975). This species ingests 150-250 small prey items in a 4 to 6 hours period generally prior to 0200 hours. Most of the forage is small, the number and biomass of food items (mostly larvaceans and copepods) being concentrated in the <4 mm size fraction. *D. taaningi*, then, is obtaining its daily ration from numerous small energy "packets".

SELECTIVE FEEDING

Characterizing most midwater fishes as opportunistic predators (i.e., ingesting a relatively large range of prey types and sizes) does not imply that prey are taken in direct proportion to their absolute abundance at the time of feeding. Several forms of prey selectivity appear to be operating which influence the size distribution and taxonomic composition of the diet in individual species. If one assumes that midwater fishes in general encounter food resources in a "fine grained" manner (i.e., in proportion to absolute abundance; *MacArthur and Wilson*, 1967) then major shifts in diversity and size of prey items from absolute abundances represent some form of selectivity.

Size Selectivity

Feeding preference can involve simply size selection in which certain prey size classes are eaten disproportionately to their absolute abundance regardless of taxonomic composition. A number of factors may be involved in this kind of selective predation. Though much of the evidence comes from studies of organisms other than fishes (e.g., lizards, insects), there can apparently be differences in feeding efficiencies on certain size classes by different predators as well as differences in functional response (*sensu Holling*, 1959) to various spectra of prey density and size distributions (see *Schoener*, 1969; *O'Connell*, 1972; *Frost*, 1974). Also involved are energy requirements of various predators which may vary considerably and consequently significantly affect prey size selection and feeding strategy (e.g., *Schoener and Gorman*, 1968; *Brocksen et al.*, 1970).

Size selectivity in midwater fishes is apparent in the diets of *Valenciennellus tripunctulatus* and *Lampanyctus alatus* (authors, unpublished). Both appear to select prey items in the 1-5 mm size range (particularly the 2-4 mm prey sizes) while feeding sparsely on the 0.5-0.9 mm size classes despite the latter's much greater relative abundance in surrounding waters. Size choice has been shown for other planktivorous fishes (*Parsons and Le Brasseur*, 1970) and prey selectivity, especially on the basis of size as *Frost* (1974) and other infer, has probably played a role in determining the present relative abundance and taxonomic composition of oceanic plankton. The effect of selective feeding has been convincingly demonstrated in fresh water lakes where zooplankton populations have been significantly altered by selective feeding of planktivorous fish (*Brooks and Dodson*, 1965; *Galbraith*, 1967).

Taxonomic Selectivity

Selectivity may also include preferential feeding on specific prey taxa. Such differences may result from complex attributes of both predator and prey such as avoidance capabilities of prey, differential abilities of various prey taxa to provide stimuli to elicit predator response, and factors concerned with palatability and assimilatable energy. *Hartmann and Weikert* (1969) present evidence for taxonomically selective feeding in the neustonic myctophid *Centrobranchus nigroocellatus* which, in collections from the tropical Atlantic, were found to contain only the molluscs *Atlanta peroni* and *Styiola subula*. *Merrett and Roe* (1974) in comparing taxonomic composition and abundance of stomach contents with that of plankton collected concurrently with midwater fishes found *Argyropelecus aculeatus* to prey selectively on ostracods, *Lampanyctus cuprarius* on amphipods, and *Valenciennellus tripunctulatus* on members of the copepod genus *Pleuromamma*. Though selective predation has not been confirmed, other midwater fishes such as *Lobianchia dolfleini*, *Lepidophanes guentheri*, *Lampancytus alatus*, *Notolychnus valdiviae*, *Benthosema suborbitale* and *Gonostoma elongatum* feed heavily on *Pleuromamma* as well (*Merrett and Roe*, 1974; *Baird et al.*, 1975b; and unpublished data). The frequent occurrence of species of this copepod genus in diets of midwater fishes may derive in part from factors other than their abundance in the plankton. *Zaret and Kerfoot* (1975), for instance, observed in aquarium experiments that *Melaniris chagresi*, a planktivorous atherinid, selected *Bosmina longirostris*, a cladoceran with heavy eye pigmentation, over larger forms with a less conspicuous eye. Similarly, the large pigment spot on the side of the metasome of *Pleuromamma* may make this genus relatively conspicuous to visually oriented predators.

Selectivity and Sampling Adequacy

Confirmation of selectivity in feeding behavior requires both adequate sample size for diet analysis and an accurate estimate of prey availability at the time and depth of feeding. Midwater fishes often show considerable variability both in stomach fullness and diet composition in any given sample (see our data on intrasample variability in *Sternoptyx diaphana*; *Hopkins and Baird*, 1973, Table 3); consequently, sufficient material should be available for analysis so that fish-to-fish variability does not obscure general diet patterns of a given species under study. Additionally, prey populations undergo considerable diurnal variation in depth as do fish predators. *Merrett and Roe's* (1974) data on the copepod genus *Pleuromamma* from 30°N 23°W, for example, indicate that *P. piseki* and *P. gracilis* populations are centered at 300-400 m during the day and 0-25 m at night. Such migrations, when coupled with specific feeding chronologies and possible patchiness in prey distribution (see *Collard*, 1970 and *Tyler and Pearcy*, 1975 regarding uniformity of stomach contents and plankton patchiness), requires large sample sizes taken over short periods of time in assessing prey availability. The determination of prey availability particularly demands accurate quantitative sampling of all size classes of prey found in fish diets. This necessitates a multiple net sampling approach since the use of a single mesh size often results in a biased picture

(Clutter and Anraku, 1968) of total array of food available. In
our investigations of the nutrition of *Valenciennellus tripunctulatus*, both 165μ and 333μ mesh nets, while appearing to adequately
sample the upper range of prey ingested by this species, seriously
underestimated the availability of the smaller (<1 mm) elements of
the diet. 30 l water bottle sievings (28μ gauze) from the 200-600 m
zone, for instance, averaged 3 times more copepods in the 0.5-0.9 mm
size class than 165μ mesh plankton nets. Inadequate collecting of
the small plankton by nets commonly used in oceanic sampling programs
is also apparent from the findings of *Beers and Stewart* (1967, 1969
a, b, 1971), who demonstrated the abundance of small plankton in the
epipelagic zone with a submersible pump system and series of fine
mesh sieves.

RESOURCE PARTITIONING

Given the absence of different structural microhabitats and
the generally food-poor environments characteristic of many pelagic
oceanic ecosystems, one would expect marked selective pressure to
reduce competition both within and among species for limited
available resources. The nature of competitive interactions among
individual species are not well understood or documented, and prey
selectivity and feeding efficiency may well be important in this
regard. It also follows that patterns of vertical distribution
could play a major role in competition phenomena. In general, physical parameters remain relatively constant within any depth stratum
over large areas of the ocean, particularly in interior oceanic
regions, the major heterogeneity being provided by marked vertical
differences in physical-chemical properties (*Sverdrup et al.*, 1942).
This is reflected in the generally observed pattern of decreasing
biomass with depth and the concentration of primary production in
the upper 100 to 150 m (*Vinogradov*, 1970). These conditions, then,
place certain constraints on midwater fishes regarding feeding
strategies and the partitioning of available resources among competing species.

Each midwater fish predator forages on a spectrum of prey, the
limits of which are assumed to be determined by the unique anatomical and behavioral characteristics of the species and by prey
availability. Within these constraints diet optimization in terms
of energy costs to obtain the daily ration would be predicted to
determine the actual observed diets. As a number of authors have
suggested (*Ivlev*, 1961; *MacArthur and Pianka*, 1966; *Emlen*, 1966;
Morse, 1971), a varied diet would be expected in food-poor environments where highly selective feeding might be too costly in terms
of the energy of search and capture. While midwater fishes can be
generally characterized as opportunistic feeders, various mechanisms
for resource partitioning can be predicted. Resource partitioning
would most likely involve time-space separations related to cyclic
migrations and feeding as well as reduction of competition through

physiological, anatomical, and behavioral specialization. This would especially be true among species competing for resources within the same depth stratum.

Resource partitioning is in fact readily apparent in much of the literature on life histories, vertical distribution and diets of midwater fishes. The most obvious and general form of resource partitioning is where species feed at different depths in the water column. Species living in different depth zones thus have available different food choices in their respective feeding layers. Further, in many taxa of midwater fishes, e.g., myctophids, gonostomatids, sternoptychids and ceratioids, (Tåning, 1918; Marshall, 1954; Ahlstrom, 1959) various life history stages inhabit different levels in the water column and thus probably encounter prey populations of varying composition. Larval ceratioids, as an example, develop in the epipelagic zone (Marshall, 1954) while the adults are bathypelagic. Also, different size classes of Chauliodus danae (Merrett and Roe, 1974), several species of Gonostoma (Kawaguchi, 1973; Krueger and Bond, 1972) and a number of myctophids and other stomiatoids (Merrett and Roe, 1974; Clarke, 1973, 1974) are concentrated at different depth levels. There is possible, then, intraspecific resource partitioning achieved through ontogeny and spatial dispersion.

Resource partitioning among species feeding within the same narrow depth zone has also been indicated. The diets of seven species taken from the same depth by Merrett and Roe (1974), despite considerable rodundancy, were species unique. The same is obvious from our stomach analyses of species collected in the same haul, though it is not known that all feeding occurs at the depths indicated. In all comparisons in Table 5 marked differences in percent composition of major food categories and/or prey size distribution are apparent. In cases where some similarity occurs in major taxa percentages (e.g., tow 157: A. hemigymnus; V. tripunctulatus), there are conspicuous differences at the generic level (see percentages for Pleuromamma and Oncaea). Much of the plankton in the mesopelagic layer also migrates vertically over considerable distances (Bainbridge, 1961; Moore and O'Berry, 1957; Vinogradov, 1970; Roe, 1974) which alters prey availability at each depth horizon over a diel period. Differences in feeding chronology already discussed above and the suggestion by Merrett and Roe (1974) that some myctophids may forage at dusk on ascent while others feed later at night imply, then, a potential time apportioning of resources in the water column.

A number of other mechanisms for partitioning resources among and within species may be operating as well. For instance, one derives partitioning from size variations both ontogenetically and among species in which a larger fish is able to ingest larger prey items than can a smaller individual. Additionally, Goodyear et al. (1972) have pointed to the marked differences in breeding season among a number of midwater fishes with epipelagic larvae in the Mediterranean. They suggest that non-synchronous breeding could be

TABLE 5. Resource Partitioning Among Species of Midwater Fishes Taken in the Same Haul Within a Relatively Narrow Depth Zone. Percentages listed for *Pleuromamma* and *Oncaea* (in addition to those for major plankton groups) as these are often the most abundant genera. All tows from E. Gulf of Mexico (27°N 86°W). First (heavy line) and second (light line) ranked categories underlined.

	No. specimens	Size range (SL, mm)	No. food items	Pleuromamma	Oncaea	Total copepods	Ostracods	Amphipods	Euphausiids	Other crustaceans	Molluscs	Polychaetes	Gelatinous orgs.	Chaetognaths	Fish	Other	<2 mm	2-4 mm	>4 mm
Tow 137; 90-100m																			
Benthosema suborbitale	52	23-33	168	(19)	(23)	<u>78</u>	<u>11</u>	2	2	-	4	-	-	-	-	2	<u>62</u>	32	6
Certescopelus warmingi	50	28-52	464	(4)	(6)	<u>34</u>	13	3	3	1	21	<1	1	-	-	<1	<u>65</u>	30	5
Lepidophanes guentheri	71	37-58	490	(28)	(4)	<u>58</u>	10	3	<u>18</u>	2	5	<1	<u>24</u>	<1	1	1	36	<u>37</u>	27
Tow 13; 60-80m																			
Ceratoscopelus warmingi	40	34-60	354	(8)	(3)	<u>35</u>	8	4	1	2	5	<1	<u>42</u>	1	-	2	40	<u>52</u>	7
Diaphus dumerili	18	26-56	230	(4)	(11)	<u>51</u>	9	3	1	2	11	1	<u>17</u>	2	<1	<1	<u>70</u>	24	7
Tow 141; 130-140m																			
Notolychnus valdiviae	49	16-23	130	(16)	(21)	<u>78</u>	<u>12</u>	2	9	-	-	-	-	-	-	-	48	38	15
Lampanyctus alatus	64	28-46	420	(24)	(3)	<u>62</u>	7	8	<u>20</u>	1	1	1	<1	-	1	<1	24	<u>52</u>	24
Tow 144; 155-200m																			
Gonostoma elongatum	52	68-117	121	(23)	(1)	<u>36</u>	12	9	<u>34</u>	5	-	1	-	-	4	-	10	36	<u>54</u>
Argyropelecus aculeatus	14	22-38	67	((12))	(2)	12	<u>37</u>	15	3	<u>16</u>	12	-	3	3	-	-	22	<u>53</u>	25
Tow 157; 280-320m																			
Valenciennellus tripunctulatus	12	20-29	45	(47)	(9)	<u>93</u>	7	-	-	-	-	-	-	-	-	-	33	<u>53</u>	13
Argyropelecus hemigymnus	16	16-34	44	(23)	(14)	<u>75</u>	<u>18</u>	2	-	2	-	2	-	-	-	-	<u>64</u>	34	2

a mechanism to reduce competition among larvae and juveniles of these species thereby partitioning resources seasonally. Studies on the distribution of *Diaphus taaningi* (Baird et al., 1974, 1975a) suggest that the spacing of individuals also has important implications concerning intraspecific resource partitioning. Presumably, if individuals with high grazing efficiencies are too closely spaced in regard to prey density then each may receive a reduced ration due to removal of prey in the vicinity by other fish. The apparent lack of marked horizontal patchiness in the distribution of individuals of the myctophid *D. taaningi* may be, then, a mechanism for reducing intraspecific competition.

BIOENERGETICS

Because of inadequacies of present sampling methods and difficulties in obtaining physiological data, an energy balance is not available for any midwater fish species. Much can be deduced, however, concerning energy requirements and constraints on a midwater species from information on population structure, feeding chronology, daily ration, and extrapolation from the literature. The equation below, developed in the course of studies on freshwater fish bioenergetics (Warren and Davis, 1967; Weatherly, 1972), can be applied to midwater fishes:

$$Q_c - Q_w = Q_g + Q_s + Q_d + Q_a$$

Where: Q_c = energy of ingested ration.

Q_w = energy lost as feces, urine and secretions through skin and gills.

Q_g = increase in potential energy through growth.

Q_s = energy of metabolism of unfed, resting fish (i.e., standard metabolism).

Q_d = energy cost of digestion, assimilation and storage.

Q_a = energy of swimming activity.

Q_c

There are few data on caloric values of marine plankton, though as previously mentioned, crustaceans, the principal food of many midwater fishes generally have a caloric equivalency in the range of 5200 to 5900 cal/g AFDW. Q_c, then, is estimated by determining biomass (dry weight) of individual diet components, converting to the appropriate caloric equivalent, and cumulating caloric values of individual food items for total caloric content of the daily ration.

$$Q_w$$

Energy lost as wastes is usually assumed to be 0.2Q_c (*Mann*, 1967), though recent work on the bleak, *Alburnus alburnus*, suggests a urine loss of approximately 2% and infers that other wastes account for 12% of ingested energy (*Ware*, 1975). As an initial estimate for midwater fishes, energy available for growth, maintenance and activity can be assumed to be in the range of 0.80-0.85 Q_c (i.e., Q_w = 0.15-0.20 Q_c).

$$Q_g$$

Calculation of potential energy accumulated in growth and its metabolic cost requires 1) determination of biomass and caloric content of various size classes from egg to adult, 2) information on the duration of each phase of the life history, and 3) an estimation of the efficiency of conversion of ingested energy to growth. The latter can be determined only subsequent to obtaining values for the other components of the bioenergetics equation (i.e., losses and metabolic costs, Q_w, Q_s, Q_d, Q_a). The increase in organic biomass throughout ontogeny can be determined gravimetrically and by ashing, or indirectly through extrapolation of size - weight curves. Caloric equivalency of each size class can be derived from a regression relationship between length or biomass (AFDW) and caloric content as determined from bomb calorimetry.

Knowledge of duration of each life history phase is difficult to obtain because of problems of age determination. However, information from studies of successfully reared clupeid fishes (assumed for purposes of initial estimates as having developmental patterns similar to midwater fishes), as well as observations on gonad development, seasonal changes in size classes, size at first reproduction and maximum size of various midwater species should allow for estimates of time requirements for each stage of maturation. The first or larval growth stage, that is, from hatching through metamorphosis, based on a number of clupeid species is perhaps from 10 to 30 days (*Blaxter and Holliday*, 1963; *Houde and Palko*, 1970). Growth in this period for many midwater fishes including myctophids, gonostomatids, sternoptychids and other stomiatoids, is from 2-4 mm to 8-25 mm (*Ahlstrom and Counts*, 1958; *Grey*, 1964; *Gibbs*, 1964, 1969; 1964a, b, c; *Moser and Ahlstrom*, 1970; *Krueger and Bond*, 1972; *Clarke*, 1973; *Ozawa*, 1973). Egg size in many midwater fishes is between 0.5 and 1.2 mm (unpublished data) and the caloric conversion of yolk to newly hatched larvae, has been estimated for non-midwater species to be as high as 65-70% (*Phillips*, 1972).

A second growth phase is from post metamorphosis to onset of sexual ripening. Midwater fish growth rates undoubtedly vary, but many species, especially in temperate - tropical latitudes, are thought to be annuals and reach sexual maturity at 25-50 mm (Gibbs et al., 1971; Clarke, 1973, 1974; Krueger and Bond, 1972; Baird et al 1974). Data on Benthosema glaciale (Halliday, 1970) and Stenobrachius leucopsarus (Smoker and Pearcy, 1970) indicate that as in zooplankton (Bogorov, 1958; Marr, 1962) at higher latitudes, more than one year may be required to reach sexual maturity. Also, larger midwater species such as Astronesthes, Chauliodus and Gonostoma may require more than one year to sexually mature (Kawaguchi and Marumo, 1967; Clarke, 1974).

Increase in length during the second growth phase may eventually prove divisible into two stages, a postlarval and a gonad maturation period. It is possible that the growth rate (in terms of length) may lessen to some extent near the end of the subadult period as food energy is diverted from somatic growth to reproduction (Bagenal, 1967).

Assuming a life span of one year the authors made a preliminary estimate of the duration of the growth phases of Valenciennellus tripunctulatus. The larval through metamorphosis stage (2-4 mm to 7-10 mm) possibly lasts 15-30 days, the juvenile through subadult stage (7-10 mm to 20-23 mm) 175-225 days and the mature adult stage (20-23 mm to 30-32 mm) 140-190 days. The number of copepods required to reach an adult female size of 32 mm (69 mg AFDW) is estimated at 3000 to 6000 (a range based on estimated average and maximum daily ration for each size class) approximately half of which are in the <2 mm size class and half >2 mm. Smaller copepods are most important in the nutrition of the early phases of growth whereas larger prey individuals contribute the bulk of caloric intake in the later growth stages.

Q_s

Data on standard metabolism of midwater fishes is virtually nonexistent. Childress (1971), however, was able to obtain valuable respiration data for midwater crustaceans and two midwater fishes (a zoarcid and a liparid) from California waters. He determined their protein content and using a protein-respiration rate regression, estimated respiration rate indirectly for a wide variety of midwater fishes by assaying protein content (Childress and Nygaard, 1973). Their data indicate a decrease in protein on a live weight basis with respect to depth and decreasing temperature. If their respiration-protein regression is valid, then commensurate with this depth-temperature decrease would be a decline in respiration rate per unit of fish biomass. It is not clearly stated in their paper, however, whether these indirectly estimated rates are closer to standard or normal "field" activity. Childress (personal communication) suggests their data relate more to the latter and estimates that in the relatively inactive mesopelagic animals they studied, standard metabolism possibly constitutes 1/2 to 2/3 field metabolism. Consequently, until correction factors become available, their respiration values multiplied by a factor of 0.5 to 0.7 can be used as initial estimates of standard metabolism in midwater fishes.

$$Q_d$$

Little known of the energy requirements, commonly referred to as the specific dynamic effect of food or "entropic tax", for digestion, assimilation and storage. Recent estimates, from non-midwater forms, however, range between 14 and 19% of Q_c (Ware, 1975).

$$Q_a$$

The estimation of active metabolism in midwater fishes is complicated by the marked diurnal differences in activity of many species as well as the differences in metabolic activity required by different feeding strategies. Observations from submersibles indicate that certain midwater species remain inactive during much of day, including migrating forms, and a number of species (e.g., *Chauliodus*, *Cyclothone*; Barham, 1970) appear to employ feeding tactics in which little energy is spent in active search. Studies based primarily on freshwater fishes suggest that active or "field" metabolism is 2-3 times standard metabolism (i.e., $Q_a + Q_s = 2-3Q_s$). For the present, as previously mentioned, a reasonable estimate of Q_a might stand at 0.5 to $1.5Q_s$. A factor of $1.5Q_s$ might be appropriate for relatively active fishes such as myctophids during diurnal periods of migration and searching for prey while $0.5Q_s$ perhaps would be more applicable to quiescent bathypelagic species and to migrating species in their period of inactivity. In the terms of the bioenergetics equation, then, $Q_s + Q_a = 1.5$ to $2.5Q_s$. Q_a as considered here, however, does not include burst speeds (i.e., specifically stimulated activity) required for pursuit, escape or possibly mating behavior. Such expenditures, while presumably expensive energetically and potentially exceeding standard metabolism by an order of magnitude (Ware, 1975), would, because of short duration, perhaps constitute only a small fraction of the daily Q_a. Error from this source might be more serious in energetics calculations for active forms such as *Gonichthys* and *Myctophum* which can be observed darting about while foraging in surface waters, than for more lethargic meso- and bathypelagic species.

In summary, to estimate the impact of a species of midwater fish on its environment in terms of energy utilization, the energetics equation discussed above must be integrated over all size classes of the population. This requires quantitative sampling of all size classes. Several estimates are available of standing crop of certain midwater fishes per unit area of sea surface (Pearcy and Laurs, 1966; Kashkin, 1967; Clarke, 1973; Baird et al., 1974). However, the reliability of these estimates is limited by the collecting methods (Harrison, 1967; Chapman, 1967), a number of factors operating to bias estimations of population abundance and size-frequency distributions (e.g., net avoidance, mesh size, patterns of spatial distribution, seasonality). By integrating catch data from an array of nets, plus perhaps using independent estimates from submersibles or acoustic data, it might then be possible to construct an accurate picture of the population abundance and size structure of a midwater fish species. Even with adequate sampling, however, there remains the formidable taxonomic problem of readily identifying and counting larval and early postlarval stages of midwater fishes. Continued research along the lines of Tåning (1918), Moser and Ahlstrom (1970) and Ozawa (1973) is essential for significant progress in this area.

Some approximation of energy utilization by larval and early post larval stages is possible if larger size classes can be adequately sampled and fecundity determined. Mean fecundity per female can be estimated from egg counts which also can be expressed in terms of calories. If the number of spawning females in the population over time can be approximated from gonad analysis, then total spawn can be estimated both in terms of number of eggs and their caloric worth. Knowing this in addition to the length of time spent in each stage prior to recruitment to sizes that are quantitatively sampled, one can construct a probable survival curve. The product of estimation of daily ration, time per developmental stage and probable numerical density of each stage would yield total energy consumed by a particular segment or species of the midwater community.

FUTURE RESEARCH

Research in virtually all phases of the trophic ecology of midwater fishes is in its initial stages and additional information of even the most descriptive kind (e.g., diet analysis) is needed. Certain lines of investigation, however, should perhaps receive special emphasis. One such line includes field studies of mesopelagic fish and their zooplankton prey (i.e., to assess prey availability). Consideration should be given, as well, to standardizing procedures both in the field and laboratory to maximize comparability. Effort should also be directed towards modeling energy budgets. This will require more accurate determination of population abundance and size structure through improved collecting procedures and will demand a better understanding of the metabolism of midwater fishes. The latter is made especially difficult because of the fragility of midwater species, though "breakthroughs" in physiology may be possible in the near future through technological advances in shipboard aquaria. Spherical chambers have been designed, for instance, which should minimize trauma to fishes returned to the deck alive (Robison, 1973). Further, as a result of Clarke's (1969) flume trails and the development of thermally protected buckets (*Childress*, personal communication), trawl cod-end designs are now available which can return the catch in good condition. Perhaps a better alternative to shipboard aquaria would be to acquire data *in situ*. Ultimately, it may be possible, for example, to obtain more realistic measures of metabolic rate with special chambers manipulated from a submersible.

In addition to data on abundance and size structure of populations, information is needed on aspects of life history such as fecundity, size at first reproduction, age and growth rate, larval development and mortality. The relationship of these parameters with primary production is of course critical to progress in predictability in deep sea trophic-dynamics. Consideration should also be given to the physiology and physics of sensory systems, particularly vision, in the sea. Research in this area would ultimately lead to a better understanding of the physical parameters and limitations involved in prey location and predator avoidance as well as diurnal vertical migration.

Another important area for emphasis is behavioral research. Information on diet, functional morphology and metabolism become much more meaningful if normal behavior patterns of midwater fishes can be documented. This, again, demands the return of material virtually undamaged to suitable aquaria, or observations *in situ* with television, possibly acoustic techniques, or from submersibles. Submersibles have been used for midwater observations since the 1930's (*Beebe*, 1935; *Bernard*, 1955; *Pérès*, 1958a, b; *Trégouboff*, 1958; *Backus et al.*, 1968; *Barham*, 1970), but their potential as a research tool has certainly not been fully exploited.

ACKNOWLEDGEMENTS

We wish to thank Dr. R. J. Lavenberg for providing specimens for stomach analysis from the Los Angeles County Museum fish collection. A considerable portion of the authors' research reported in this paper was supported by NSF grant DES 75-03845.

REFERENCES

Adams, C. A. 1972. Food habits of juvenile pinfish (*Lagodon rhomboides*), silver perch (*Bairdiella chrysura*), and spotted seatrout (*Cynoscion nebulosus*) of the estuarine zone near Crystal River, Florida. *Master's Thesis*, University of Florida. 147 p.

Ahlstom, E. H. 1959. Vertical distribution of pelagic fish eggs and larvae off California dn Baja California. *Fish. Bull. U.S.* 60: 107-146.

Ahlstrom, E. H. and R. C. Counts. 1958. Development and distribution of *Vinciguerria lucetia* and related species in the eastern Pacific. *Fish. Bull. U.S.* 58: 363-416.

Anderson, R. 1967. Feeding chronology in two deep-sea fishes off California. *Master's Thesis*, Univ. So. California. 22 p.
Backus, R. H., J. E. Craddock, R. L. Haedrich, D. L. Shores, J. M. Teal, A. S. Wing, G. W. Mead, and W. D. Clarke. 1968. *Certoscopelus maderensis*: peculiar sound-scattering layer identified with this myctophid fish. *Science 160*: 991-993.
Bagenal, T. B. 1967. A short review of fish fecundity, p. 89-111. In: S. D. Gerking (ed.), *The biological basis for freshwater fish production*. Wiley, New York.
Bainbridge, R. 1961. Problems of fish locomotion. *Symp. Zool. Soc. London 5*: 13-32.
Baird, R. C. 1971. Systematics, distribution, and zoogeography of the marine hatchetfishes (Family Sternoptychidae). *Bull. Mus. Comp. Zool., Harvard 142*: 1-128.
Baird, R. C., T. L. Hopkins, and D. F. Wilson. 1975a. Diet and feeding chronology of *Diaphus taaningi* (Myctophidae) in the Cariaco Trench. *Copeia 1975*: 356-365.
Baird, R. C., N. P. Thompson, T. L. Hopkins, and W. R. Weiss. 1975b. Chlorinated hydrocarbons in mesopelagic fishes of the eastern Bulf of Mexico. *Bull. Mar. Sci. 25*: 473-481.
Baird, R. C., D. F. Wilson, R. C. Beckett, and T. L. Hopkins. 1974. *Diaphus taaningi* Norman, the principal component of a shallow sound-scattering layer in the Cariaco Trench, Venezuela. *J. Mar. Res. 32*: 301-312.
Baird, R. C., D. F. Wilson, and D. M. Milliken. 1973. Observations on *Bregmaceros nectabanus* Whitley in the anoxic, sulfurous waters of the Cariaco Trench. *Deep-Sea Res. 20*: 503-504.
Barham, E. G. 1970. Deep-sea fishes: lethargy and vertical orientation, p. 100-118. In: G. B. Farguhar (ed.), *Proc. Int. Symp. Biological Sound Scattering in the Ocean*. Scient. Rep. Maury Center Ocean. V.5.
Beebe, W. 1935. Half mile down. John Lane, London. 343 p.
Beebe, W. and M. Van der Pyl. 1944. Eastern Pacific expeditions of the New York Zoological Society XXXIII, Pacific Myctophidae (fishes). *Zoologica, New York 29*: 59-95.
Beers, J. R. and G. L. Stewart. 1967. Micro-zooplankton in the euphotic zone at five locations across the California Current. *J. Fish Res. Bd. Canada 24*: 2053-2068.
Beers, J. R. and G. L. Stewart. 1969a. Micro-zooplankton and its abundance relative to the larger zooplankton and other seston components. *Mar. Biol. 4*: 182-189.
Beers, J. R. and G. L. Stewart. 1969b. The vertical distribution of micro-zooplankton and some ecological observations. *J. Cons. Int. Explor. Mer 33*: 30-44.
Beers, J. R. and G. L. Stewart. 1971. Micro-zooplankters in the plankton communities of the upper waters of the eastern tropical Pacific. *Deep-Sea Res. 18*: 861-883.
Bernard, F. 1955. Densite du plancton vu au large Toulon depuis le bathyscaphe F.N.R.S. III. *Bull. Inst. Oceanogr. 1063*: 1-16.

Bertelsen, E. 1951. The ceratioid fishes. *Dana Repts. 39:* 1-276.

Blaxter, J. H. S. and F. G. T. Holliday. 1963. The behavior and physiology of herring and other clupeids, p. 261-393. IN: F. S. Russell (ed.), *Advances in Marine Biology, V. 1.* Academic Press, New York.

Bogorov, B. G. 1958. Perspectives in the study of seasonal changes of plankton and the number of generations at different latitudes, p. 145-158. In: A. A. Buzzati - Traverso (ed.), *Perspectives in marine biology.* Univ. California Press, Berkeley.

Brocksen, R. W., G. E. Davis, and C. E. Warren. 1970. Analysis of trophic processes on the basis of density - dependent functions, p. 468-498. In: J. H. Steele (ed.), *Marine Food Chains.* Univ. California Press, Berkeley.

Brooks, J. L. and S. I. Dodson. 1965. Predation, body size, and composition of plankton. *Science 150:* 28-35.

Cailliet, G. M. 1972. The study of feeding habits of two marine fishes in relation to plankton ecology. *Trans. American Microscop. Soc. 91:* 88-89.

Carr, W. E. S. and C. A. Adams. 1973. Food habits of juvenile marine fishes occupying seagrass beds in the estuarine zone near Crystal River, Florida. *Trans. American Fish. Soc. 102:* 511-540.

Chapman, D. W. 1967. Production in fish populations, p. 3-29. In: S. D. Gerking (ed.), *The biological basis of freshwater fish production.* Wiley, New York.

Childress, J. J. 1971. Respiratory rate and depth of occurrence of midwater animals. *Limnol. Oceanogr. 16:* 104-106.

Childress, J. J. and M. H. Nygaard. 1973. The chemical composition of midwater fishes as a function of depth occurrence off Southern California. *Deep-Sea Res. 20:* 1093-1109.

Clarke, M. R. 1969. A new midwater trawl for sampling discrete depth horizons. *J. Mar. Biol. Assoc. United Kingdom 49:* 945-960.

Clarke, T. A. 1973. Some aspects of the ecology of lanternfishes (Myctophidae) in the Pacific Ocean near Hawaii. *Fish. Bull. U.S. 71:* 401-434.

Clarke, T. A. 1974. Some aspects of the ecology of stomiatoid fishes in the Pacific Ocean near Hawaii. *Fish. Bull. U.S. 72:* 337-351.

Clutter, R. I. and M. Anraku. 1968. Avoidance of samplers, p. 57-76. In: D. J. Tranter (ed.), Zooplankton Sampling. *Monographs on Oceanographic methodology, 2.* UNESCO.

Collard, S. B. 1970. Forage of some eastern Pacific midwater fishes. *Copeia 1970:* 348-354.

Comita, G. W. and D. W. Schindler. 1963. Caloric values of microcrustacea. *Science 140:* 1394-1395.

Cummins, K. W. and J. C. Wuycheck. 1971. Caloric equivalents for investigations in ecological energetics. *Int. Assoc. Theoret. Applied Limnol. Communications No. 18.* 158 p.

DeWitt, F. A. and G. M. Cailliet. 1972. Feeding habits of two bristlemouth fishes, *Cyclothone acclinidens* and *C. signata* (Gonostomatidae). *Copeia* 1972: 868-871.

DeWitt, H. H. and T. L. Hopkins. Aspects of the diet of the Antarctic herring *Pleuragramma antarcticum* Boulenger. *SCAR/IUBS Symposium on Antarctic Biology,* 1974. In press.

Emlen, J. 1966. The role of time and energy in food preference. *American Natur.* 100: 611-617.

Farquhar, C. B. (ed.). 1970. *Proc. Int. Symp. Biol. sound scattering in the ocean.* Maury Center for Ocean Science, Dept. Navy, Washington, D. C. MC Rep. 005. 642 p.

Frost, B. W. 1974. Feeding processes at lower trophic levels in pelagic communities. In: C. B. Miller (ed.), *The Biology of the oceanic Pacific,* p. 59-77. Proc. 33rd Ann. Biol. Colloq., Oregon State Univ. Press, Corvallis.

Galbraith, M. G. 1967. Size selective predation on *Daphnia* by rainbow trout and yellow perch. *Trans. American Fish. Soc.* 96: 1-10.

Gerking, S. D. 1954. The food turnover of a bluegill population. *Ecol.* 35: 490-498.

Gibbs, R. H., Jr. 1964. Family Astronesthidae, p. 311-350. In: Y. H. Olson (ed.), Fishes of the western North Atlantic. Pt. 4. *Mem. Sears Found. Mar. Res.,* Yale Univ.

Gibbs, R. H., Jr. 1969. Taxonomy, sexual dimorphism, vertical distribution, and evolutionary zoogeography of the bathypelagic fish genus *Stomias* (Stomiatidae). *Smithsonian Contrib. Zool.* 31: 25 p.

Gibbs, R. J., R. H. Goodyear, M. J. Keene, and C. W. Brown. 1971. Biological studies of the Bermuda Ocean Acre II. Vertical distribution and ecology of the lanternfishes (family Myctophidae). *Rept. to U.S. Navy Underwater Systems Center,* Contract No. N00140-70-C-0307 (processed), 141 p.

Gjøsaeter, J. 1973a. Age, growth, and mortality of the myctophid fish, *Benthosema glaciale* (Reinhardt), from Western Norway. *Sarsia* 52: 1-14.

Gjøsaeter, J. 1973b. The food of the myctophid fish, *Benthosema glaciale* (Reinhardt), from western Norway. *Sarsia* 52: 53-58.

Goodyear, R. H., B. J. Zahuranec, W. L. Pugh and R. H. Gibbs, Jr. 1972. Ecology and vertical distribution of Mediterranean midwater fishes. (ONR Contract N0014-67-A-0399-0007). V. 1: 91-230. Washington, D. C.

Gorelova, T. A. 1974. Zooplankton from the stomachs of juvenile lanternfish of the Family Myctophidae. *Oceanology* 14: 575-580 (AGU English trans).

Grey, M. 1964. Family Gonostomatidae. In: Y. H. Olsen (ed.), Fishes of the western North Atlantic. Pt. 4., p. 78-273. *Mem. Sears Found. Mar. Res.,* Yale Univ.

Günther, K. and K. Deckert. 1953. Morphologisch - Anatomische und vergleichend Ökologische untersuchungen über die Leistungen des Viszeralapparates bei Tiefseefishen der Gattung *Cyclothone* (Teleostei, Isospondyli). *Zeit. Morph. u. Ökol. Tiere* 42: 1-66.

Halliday, R. G. 1970. Growth and vertical distribution of the glacier lanterfish *Benthosema glaciale*, in the northwestern Atlantic. *J. Fish. Res. Bd. Canada* 27: 105-116.

Harrison, C. M. H. 1967. On methods for sampling mesopelagic fishes. *Symp. Zool. Soc. London* 19: 71-126.

Hartmann, J. 1970. Verteilung and Nahrung des Ichthyoneuston im subtropischen Nordostatlantik. *"Meteor" ForschErgebn (D)* 8: 1-60.

Hartmann, J. and H. Weikert. 1969. Tages gang eines Myctophiden (Pisces) and zweier von ihm getressener Mollusken des Neuston. *Kieler Meers-forsch.* 25: 328-330.

Hochachka, P. W. 1974. Enzymatic adaptations to deep sea life, p. 107-136. In: C. B. Miller (ed.), The Biology of the Oceanic Pacific. *Proc. 33rd Ann. Biol. Colloq.*, Oregon State Univ. Press, Corvallis.

Holling, C. S. 1959. Some characteristics of simple types of predation and parasitism. *Canadian Entomol.* 91: 385-398.

Holling, C. S. 1968. The tactics of a predator. *Symp. Roy. Entomol. Soc. London* 4: 47-58.

Holton, A. A. 1969. Feeding behavior of a vertically migrating lanternfish. *Pacific Sci.* 23: 325-331.

Hopkins, T. L. 1968. Carbon and nitrogen content of fresh and preserved *Nematoscelis difficilis*, a euphausiid crustacean. *J. Cons. Perm. Int. Explor. Mer* 31: 300-304.

Hopkins, T. L. and R. C. Baird. 1973. Diet of the hatchetfish *Sternoptyx diaphana*. *Mar. Biol.* 21: 34-46.

Hopkins, T. L. and R. C. Baird. 1975. Net feeding in mesopelagic fishes. *Fish. Bull. U.S.* 73: 908-914.

Houde, E. D. and B. Palko. 1970. Laboratory rearing of the clupeid fish *Harengula pensacolae* from fertilized eggs. *Mar. Biol.* 5: 354-358.

Ivlev, V. S. 1961. Experimental ecology of the feeding of fishes. Yale Univ. Press, New Haven (trans. from Russian by D. Scott). 302 p.

Jackson, D. R. and S. R. Telford, Jr. 1975. Food habits and predatory role of the Japanese lacertid *Takydromus tachydromoides*. *Copeia 1975*: 343-351.

Jespersen, P. 1915. Sternoptychidae (*Argyropelecus, Sternoptyx*). *Rep. Dan. Oceanogr. Exped. Mediterranean V. 2 (Biol.: A)*: 1-41.

Judkins, D. C. and A. Fleminger. 1972. Comparison of foregut contents of *Sergestes similis* obtained from net collections and albacore stomachs. *Fish. Bull. U.S.* 70: 217-223.

Kashkin, N. I. 1967. On the quantitative distribution of lanterfishes (Myctophidae S.L.) in the Atlantic Ocean, p. 206-264. In: T.S. Ross (ed.), The pelagic and bathypelagic fishes of the world oceans. (Transl. No. AD 686384, Nat. Tech. Info. Serv., Springfield, Va.).

Kawaguchi, K. 1973. Biology of *Gonostoma gracile* Gunther (Gonostomatidae) II. Geographical and vertical distribution. *J. Oceanogr. Soc. Japan* 29: 113-120.

Kawaguchi, K. and R. Marumo. 1967. Biology of *Gonostoma gracile* (Gonostomatidae) I. Morphology, life history and sex reversal. *Inform. Bull. Planktol. Japan Commemoration Number of Dr. Matsue.* p. 53-69.

Keast, A. 1970. Food specializations and bioenergetic interrelations in the fish faunas of some small Ontario waterways, p. 377-411. In: J. H. Steele (ed.), *Marine food chains*. Univ. California Press, Berkeley.

Kow, T. A. 1950. The food and feeding relationships of the fishes of Singapore Straits. *London, H. M. Stationery Office*. 35 p.

Krueger, W. H. and G. W. Bond. 1972. Biological studies of the Bermuda Ocean Acre III. Vertical distribution and ecology of the bristle-mouth fishes (family Gonostomatidae). Rept. to U.S. Navy Underwater Systems Center Contract No. N00140-72-C-0315. 50 p.

Legand, M. and J. Rivaton. 1969. Cycles biologiques des poissons mésopélagiques dans l'est de l'Ocean Indien. Troisimème note: action predatrice des poissons micronectoniques. *Cahiers, Office de la recherche scientifique et technique outre-mer, oceanographie* 7: 29-45.

MacArthur, R. H. and E. R. Pianka. 1966. On optimal use of a patchy environment. *American Natur.* 100: 603-609.

MacArthur, R. H. and E. O. Wilson. 1967. The theory of island biogeography. Princeton Univ. Press, Princeton. 203 p.

Mann, K. H. 1967. The cropping of the food supply, p. 243-257. In: S.D. Gerking (ed.), The biological basis for freshwater fish production. Wiley, New York.

Markus, H. C. 1932. The extent to which temperature changes influence food consumption in largemouth bass (*Huro floridana*). *Trans. American Fish. Soc.* 62: 202-210.

Marr, J. W. S. 1962. The natural history and geography of the Antarctic krill (*Euphausia superba* Dana), *Discovery Rep.* 32: 33-464.

Marshall, N. B. 1954. Aspects of deep-sea biology. Hutchinson, London. 380 p.

Marshall, N. B. 1971. Explorations in the life of fishes. Harvard Univ. Press, Cambridge, Mass. 204 p.

McLaren, I. A. 1963. Effects of temperature on growth of zooplankton and the adaptive value of vertical migration. *J. Fish. Res. Bd. Canada* 20: 685-727.

Merrett, N. R. and H. S. J. Roe. 1974. Patterns and selectivity in the feeding of certain mesopelagic fishes. *Mar. Biol.* 28: 115-126.

Milliken, D. M. 1975. Species of the genus *Bregmaceros* in the Gulf of Mexico and Caribbean Sea, including the description of a new species from the Cariaco Trench. *Master's thesis*, Univ. South Florida. 74 p.

Molnár, G., E. Tamássy, and I. Tölg. 1967. The gastric digestion of living predatory fish, p. 135-149. In: S.D. Gerking (ed.),

The biological basis of freshwater fish production. Wiley, New York.

Moore, H. B. and D. L. O'Berry. 1957. Plankton of the Florida Current IV. Factors influencing the vertical distribution of some common copepods. *Bull. Mar. Sci. Gulf and Caribbean* 7: 297-315.

Morrow, J. E., Jr. 1964a. Family Chauliodontidae, p. 274-289. In: Y.H. Olsen (ed.), Fishes of the western North Atlantic. Pt. 4. *Mem. Sears Found. Mar. Res.*, Yale Univ.

Morrow, J. E., Jr. 1964b. Family Stomiatidae, p. 290-310. In: Y.H. Olsen (ed.), Fishes of the western North Atlantic. Pt. 4. *Mem. Sears Found. Mar. Res.*, Yale Univ.

Morrow, J. E., Jr. 1964c. Family Malacosteidae, p. 523-549. In: Y.H. Olsen (ed.), Fishes of the western North Atlantic. Pt. 4. *Mem. Sears Found. Mar. Res.*, Yale Univ.

Morse, D. H. 1971. The insectivorous bird as an adaptive strategy. *Ann. Rev. Ecol. Systm.* 2: 177-200.

Moser, H. G. and E. H. Ahlstrom. 1970. Development of lanternfishes (family Myctophidae) in the California Current. Pt. 1. Species with narrow-eyed larvae. *Bull. Los Angeles Co. Mus. Nat. Hist. Sci.* 7: 1-145.

Nafpaktitis, B. G. and M. Nafpaktitis. 1969. Lanternfishes (family Myctophidae) collected during cruises 3 and 6 of the R/V ANTON BRUUN in the Indian Ocean. *Bull. Los Angeles Co. Mus. Nat. Hist., Sci.* 5: 1-79.

Nakamura, E. L. 1970. Observations on the biology of the myctophid, *Diaphus garmani*. *Copeia 1970*: 374-377.

Nolan, R. S. and R. H. Rosenblatt. 1975. A review of the deep-sea angler fish genus *Lasiognathus* (Pisces: Thaumatichthyidae). *Copeia 1975*: 60-66.

O'Connell, C. P. 1972. The interrelation of biting and filtering in the feeding activity of the northern anchovy (*Engraulis mordax* Girard) in the laboratory. *J. Fish. Res. Bd. Canada* 79: 285-293.

Ozawa, T. 1973. On the early life history of the gonostomatid fish, *Vinciguerria nimbaria* (Jordan and Williams), in the western North Pacific. *Mem. Fac. Fish. Kagoshima Univ.* 22: 127-141. (In Japanese).

Parsons, T. R. and R. J. Le Brasseur. 1970. The availability of food to different trophic levels in the marine food chain, p. 325-343. In: J.H. Steele (ed.), *Marine food chains*. Univ. California Press, Berkeley.

Paxton, J. R. 1967. Biological notes on southern California lanternfishes (family Myctophidae). *California Fish and Game* 53: 214-217.

Paxton, J. R. 1972. Osteology and relationships of the lanternfishes (family Myctophidae). *Bull. Los Angeles Co. Mus. Nat. Hist., Sci.* 13: 81 p.

Pearcy, W. G. and R. M. Laurs. 1966. Vertical migration and distribution of mesopelagic fishes off Oregon. *Deep-Sea Res.* 13: 153-167.

Pérès, J.-M. 1958a. Remarques générales sur un ensemble de quinze plongées effectuées avec le bathyscaphe F.N.R.S. III. p. 259-286. In: Resultats scientifiques des Campagnes du bathyscaphe, F.N.R.S. III 1954-1957. *Ann. Inst. Oceanogr.* 35: 235-341.

Pérès, J.-M. 1958b. Trois plongées dans le canyon du Cap Sicié, effectuées avec le bathyscaphe F.N.R.S. III de la Marine Nationale. *Bull. Inst. Oceanogr.* 1115: 1-21.

Phillips, A. M., Jr. 1972. Calorie and energy requirement, p. 1-28. In: J.E. Halver (ed.), Fish nutrition. Academic Press, New York.

Reid, G. K., Jr. 1954. An ecological study of the Gulf of Mexico fishes in the vicinity of Cedar Key, Florida. *Bull. Mar. Sci. Gulf and Caribbean* 4: 1-94.

Robison, B. H. 1973. A system for maintaining midwater fishes in captivity. *J. Fish. Res. Bd. Canada* 30: 126-128.

Roe, H. S. J. 1974. Observations on the diurnal vertical migrations of an oceanic animal community. *Mar. Biol.* 28: 99-113.

Samyshev, E. Z. and S. V. Schetinkin. 1973. Myctophids: feeding patterns of some species of Myctophidae and *Maurolicus muelleri* caught in the sound dispersing layers in the northwestern Africa area. *Annales Biologiques* 28: 212-215.

Schoener, T. W. 1969. Models of optimal size for solitary predators. *American Natur.* 103: 277-313.

Schoener, T. W. 1971. Theory of feeding strategies. *Ann. Rev. Ecol. Syst.* 2: 369-404.

Schoener, T. W. and G. C. Gorman. 1968. Some niche differences in three Lesser Antillean lizards of the genus *Anolis*. *Ecol.* 49: 819-830.

Seaburg, K. G. and J. B. Moyle. 1964. Feeding habits, digestive rates, and growth of some Minnesota warm water fishes. *Trans. American Fish. Soc.* 93: 269-285.

Smoker, W. and W. G. Pearcy. 1970. Growth and reproduction of the lanternfish *Stenobrachius leucopsarus*. *J. Fish. Res. Bd. Canada* 27: 1265-1275.

Sverdrup, H. U., N. W. Johnson and R. H. Fleming. 1942. The oceans. Prentice-Hall, Englewood Cliffs. 1087 p.

Tåning, A. V. 1918. Mediterranean Scopelidae (*Saurus, Aulopus, Chloropthalmus*, and *Myctophum*), p. 1-154. In: J. Schmidt (ed.), *Rep. Danish Oceanogr. Exped.* 1908-10. Host, Copenhagen, V. 2 (A7).

Tchernavin, V. V. 1953. The feeding mechanisms of a deep-sea fish *Chauliodus sloani* Schneider. *British Mus. (Nat. Hist.).* 101 p.

Travers, M. 1971. Diversité du microplancton du Golfe de Marseille en 1964. *Mar. Biol.* 8: 308-343.

Trégouboff, G. 1958. Prospection biologique sous-marine dans la région de Villefranche-sur-Mer au Cours de l'année 1951. I. Plongées en bathyscaphe. *Bull. Inst. Oceanogr.* 1117: 1-37.

Tyler, H. R. and W. G. Pearcy. 1975. The feeding habits of three species of lanternfishes (Family Myctophidae) off Oregon, USA. *Mar. Biol.* 32: 7-11.

Vinogradov, M. E. 1970. Vertical distribution of the oceanic zooplankton. IPST translation. U.S. Dept. Interior Document, TT 69-59015. 339 p.

Ware, D. M. 1975. Growth, metabolism, and optimal swimming speed of a pelagic fish. *J. Fish. Res. Bd. Canada* 32: 33-41.

Warren, C. E. and G. E. Davis. 1967. Laboratory studies in the feeding, bioenergetics and growth of fish, p. 175-214. In: S.D. Gerking (ed.), The biological basis of freshwater fish production. Wiley, New York.

Weatherley, A. H. 1972. Growth and ecology of fish populations. Academic Press, New York. 293 p.

Windell, J. T. 1967. Rates of digestion in fishes, p. 151-173. In: S.D. Gerking (ed.), The biological basis of freshwater fish production. Wiley, New York.

Zaret, T. M. and W. C. Kerfoot. 1975. Fish predation on *Bosmina longirostris*: body-size selection versus visibility selection. *Ecol.* 56: 232-237.

The Lanternfish *Lobianchia dofleini*: An Example of the Importance of Life-History Information in Prediction of Oceanic Sound Scattering

Charles Karnella and Robert H. Gibbs, Jr.

Smithsonian Institution

ABSTRACT

Lobianchia dofleini in waters off Bermuda reaches a maximum size of about 38 mm, spawns mainly in winter, and has a one year life cycle. Its vertical range is between 400-750 m during the day, 50-250 m at night. Size stratification is evident at all seasons, smaller and younger specimens usually occurring shallower than larger and older specimens. In winter, however, the stratification apparently is reversed by day. Upward migration appears to be about two hours in duration, nocturnal depths being reached within an hour after sunset. Descent begins at about sunrise and diurnal depths are reached in about two hours. Patchiness was indicated at 50-100 m at night in spring and summer, and was attributed to juveniles in spring and to juveniles and subadults in summer. Indications of daytime patchiness were not convincing, and none was indicated in winter. Catch rates in winter, when adults predominated, were higher during daytime than at night. In spring, with juveniles predominating, catch rates were much higher at night; in summer rates were only slightly higher at night. The species certainly breeds near Bermuda, and the hypothesis is rejected that populations in the slope water of the western North Atlantic are comprised of expatriates from the eastern North Atlantic. Available information suggests the existence of discrete populations in the eastern and western North Atlantic and also in the Mediterranean. The importance of knowing the genetic status, geographic distribution, life history, and behavior of components of oceanic communities is stressed as being essential to seasonal or geographic prediction of sound-scattering phenomena.

INTRODUCTION

Sound scattering intensity over a considerable range of frequencies in the open ocean is largely the product of the community of resonant organisms present. Communities are dynamic. Even when the same species are present throughout the year, the relative abundances and sizes of the species will change with the progress of the seasons. This account of the biology of one species of lanternfish provides an example of some of the many factors that may influence acoustic measurements and result in differences over a yearly time span. The data are not easily obtained or interpreted, but they are indispensable to prediction of sound scattering either in time or in space.

The species considered here, *Lobianchia dofleini*, is a medium-sized lanternfish with a maximum size of 38 mm standard length in our collections. It has a gas-filled swimbladder for at least part of its life (*Kleckner and Gibbs*, 1972) and has been implicated in sound scattering in the study area (*MacDonald*, 1972; *Brown and Brooks*, 1974).

This paper is part of a larger study of the biology of midwater fishes, based upon the material collected on the Ocean Acre program. The program was designed to investigate the acoustical and biological properties of the water column below the one-degree quadrangle of open ocean water centered at 32° N, 64° W, approximately 43 nautical miles ESE of Bermuda. *Roper et al.* (1970), *Gibbs et al.* (1971), and *Brown and Brooks* (1974), among others, have given descriptions of the program, its objectives and methodology, and *Brooks* (1972) has discussed the oceanography of the sampling site and adjacent areas.

SAMPLING METHODS

Samples were collected on 15 cruises to provide data concerning the biology and sound-scattering potential of the midwater fauna. One cruise involved the R/V DISCOVERY and was an adjunct to the program of the British Institute of Oceanographic Sciences, using their RMT 1+8 opening closing midwater trawl (*Baker et al.*, 1973). Fourteen cruises were made by the Ocean Acre program, using a 1-m (the first two cruises) or a 3-m (12 cruises) Isaacs-Kidd Midwater Trawl, fully lined with 3/8-inch stretch knotless nylon mesh. Eight of the 14 cruises made discrete-depth samples, using a cod-end sampler similar to that developed and described by *Aron et al.* (1964) with four chambers closed by doors that were activated from the ship by electrical signals passed through conductor cable.

The routine employed in discrete-depth sampling was to shoot the net with all chambers open and allow the net to stabilize at a selected depth, after which the chambers of the sampler were closed sequentially, usually at one-hour intervals. When the system functioned perfectly this procedure resulted in three discrete-depth samples from the selected depth and an oblique sample during retrieval from depth to surface. The depth at which the net was fished was at first determined by wire angle and the actual depth measured by a time-depth recorder attached to the net. During the last five cruises, an electronic sensor system mounted on the spreader bar allowed constant monitoring of depth from aboard ship (*Battista and Guliano*, 1971).

METHODS OF ANALYSIS

The primary objective of the program was to make discrete-depth samples at a selected series of depths from the surface to 1500 meters during both day and night on each cruise. This goal, however, was seldom fully attained. Because of this, we have chosen to combine data from three pairs of relatively successful cruises and to use the paired-cruise data as the basis for analyses. The three pairs of cruises provide good depth coverage, at least to 1000 m, both night and day for three different seasons: winter (cruise 11 in January 1971 and cruise 13 in February-early March 1972), late spring (cruises 10 and 14 in June 1970 and 1972), and late summer (cruise 4 in September 1968 and cruise 12 in August-September 1971) (Figures 1, 2 and 3).

Every specimen taken in the discrete-depth samples was measured (in mm SL) and sexed, and the developmental stage recorded. These data and the collection data were entered into computer storage to form the base that was processed by several programs to give sample statistics.

For the primary analyses, data were grouped by 50 m depth intervals for day and night separately. Day (= diurnal) is defined as from 1.5 hours after sunrise to 1.5 hours before sunset; night (= nocturnal) is from 1.5 hours after sunset to 1.5 hours before sunrise; crepuscular (= twilight) times are those within 1.5 hours of either sunrise or sunset. Only day and night samples were used to determine abundance and vertical distribution. Recourse to the original, ungrouped data (Figures 1-3) was made when there was reason to believe that the grouped data led to erroneous interpretations.

Total catch rates and catch rates for each size and stage were determined for all 50 m depth intervals by calculating the catch per hour for each sample (including negative samples) taken within the interval and using the arithmetic mean of these catch rates. Catch rates for unsampled intervals thought to be within the vertical range of the species were estimated by interpolation. The interpolated total catch and catch for each stage were taken as the mean of the catch rates of the two intervals immediately adjoining the unsampled one (cf. Table 4). Interpolated catch rates for each size were

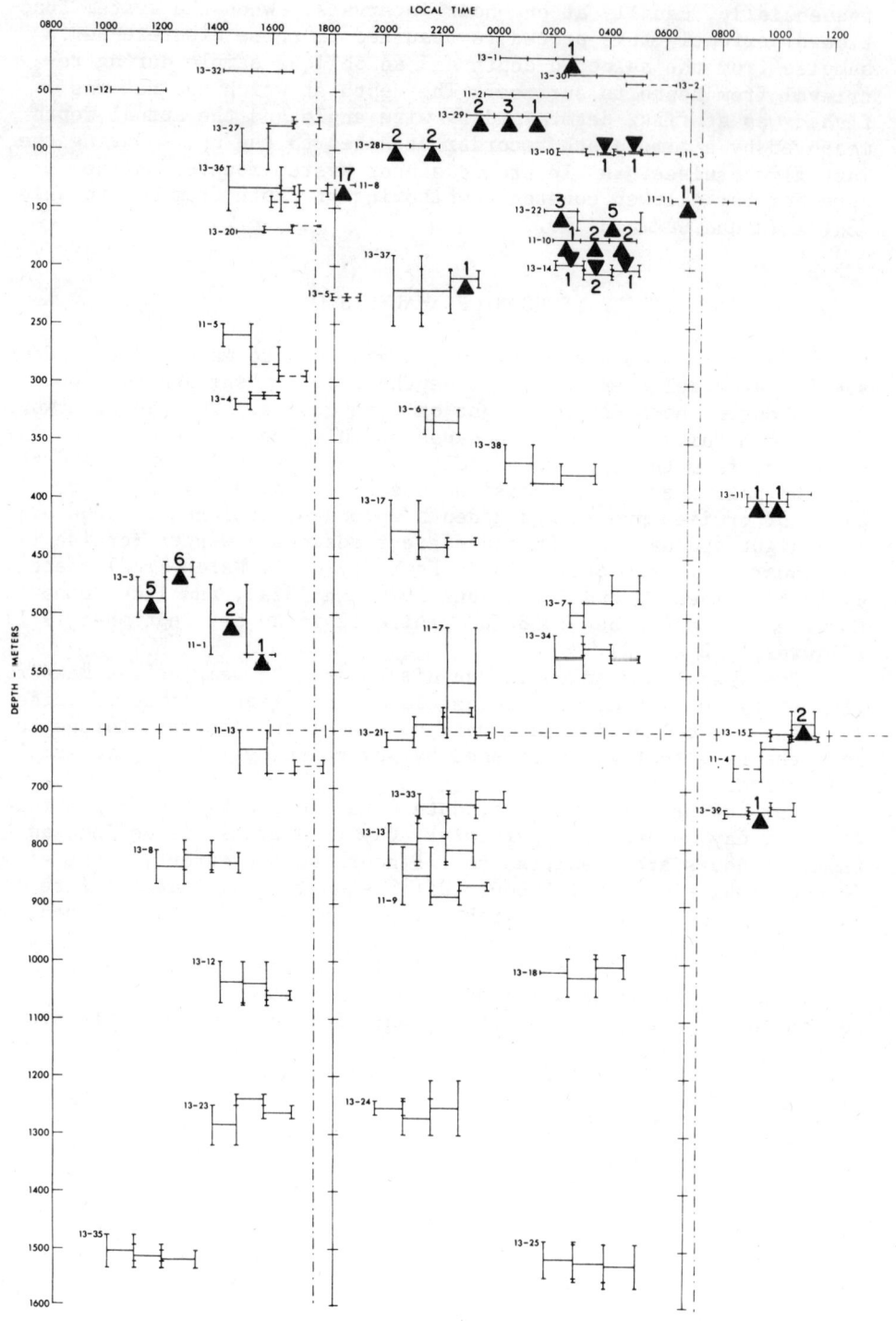

determined by using a mean size midway between the mean sizes for the two immediately adjoining depth intervals and by taking the average of the catch rates for each millimeter of size above and below each of the mean sizes (Table 2 is the summation of the results). If an unsampled interval was at the suspected shallow or deep depth limit, the total catch was estimated as half that of the adjoining positive interval, and the mean size was set equal to that of the positive interval. Total abundance for any stage of size was then taken as the sum of the catch rates for all depth intervals and includes interpolated values. No attempt has been made to convert these figures to absolute abundances.

The possibility that distributions were non-random (patchiness) was examined using the coefficient of dispersion (CD; see Pearcy, 1964). Clumping was said to occur only if the CD was significantly greater than 1.0 ($p = .05$; highly significant, $p = .01$). This analysis was applied only to samples within a single 50 m depth interval; when clumping is noted for a larger depth range, the CD for each of the included 50 m intervals was significant.

Ontogenetic Stages: Definition and Size

Ontogenetic stages were defined primarily on the basis of gonadal development, and the distinction between successive stages is somewhat arbitrary, especially at the transition between stages. Males were more difficult to categorize than females because development of the testes involves a less dramatic morphological change.

Postlarvae were not used in any of our analyses for this species. Juveniles were 10-24 mm and had thread-like gonads that showed no regional enlargement. Generally only the largest juveniles, 16 mm and larger, could be sexed under the dissecting microscope. Juvenile females had eggs mostly less than 0.05 mm in diameter. Subadults were 19-36 mm, and had large easily recognizable ovaries or testes that were cylindrical rather than thread-like and showed regional enlargement. Eggs showed little development and had a

---

Fig. 1. Time and depth excursion for all discrete-depth samples in winter (Cruises 11 and 13). The time of each sample is shown by the horizontal lines banded by short vertical lines; the dashed horizontal lines indicate samples made partly or entirely within 1.5 hours of sunrise or sunset. The horizontal time lines are at the main sampling depth; depth excursions are indicated by the vertical lines bounded by short horizontal marks. The approximate times of sunset and sunrise are given by vertical lines running the length of the figure and crossed by depth marks (Cruise 11 - broken lines, Cruise 13 - Solid lines). The dashed horizontal line at 600 m indicates a change in depth scale. The small numbers are the cruise and trawl number. The number of specimens of Lobianchia dofleini in each positive trawl is shown near the time line (usually above it).

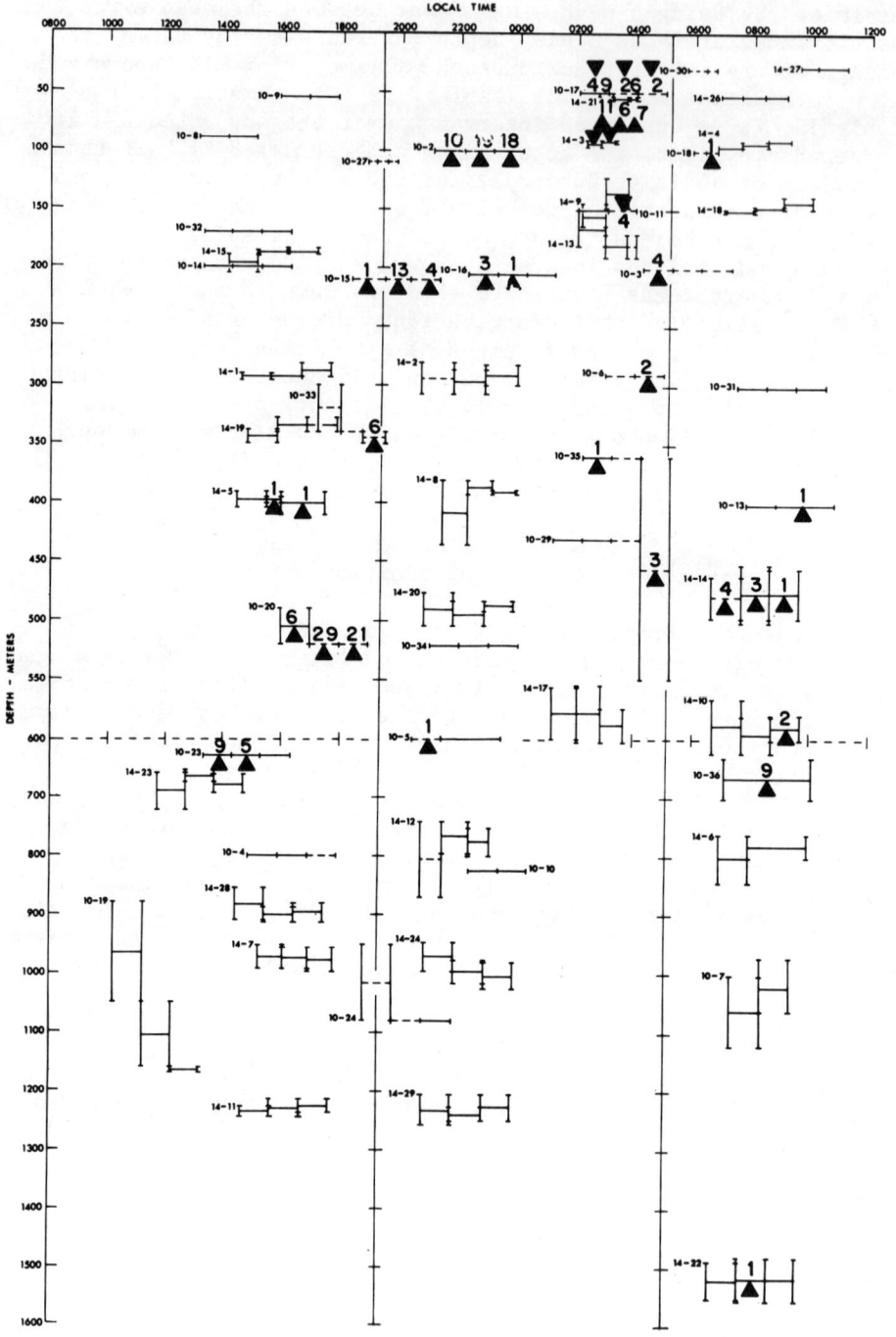

Fig. 2. Time and depth excursion for all discrete-depth samples in late spring (Cruises 10 and 14). See Fig. 1 for explanation of details.

THE LANTERNFISH *LOBIANCHIA DOFLEINI*

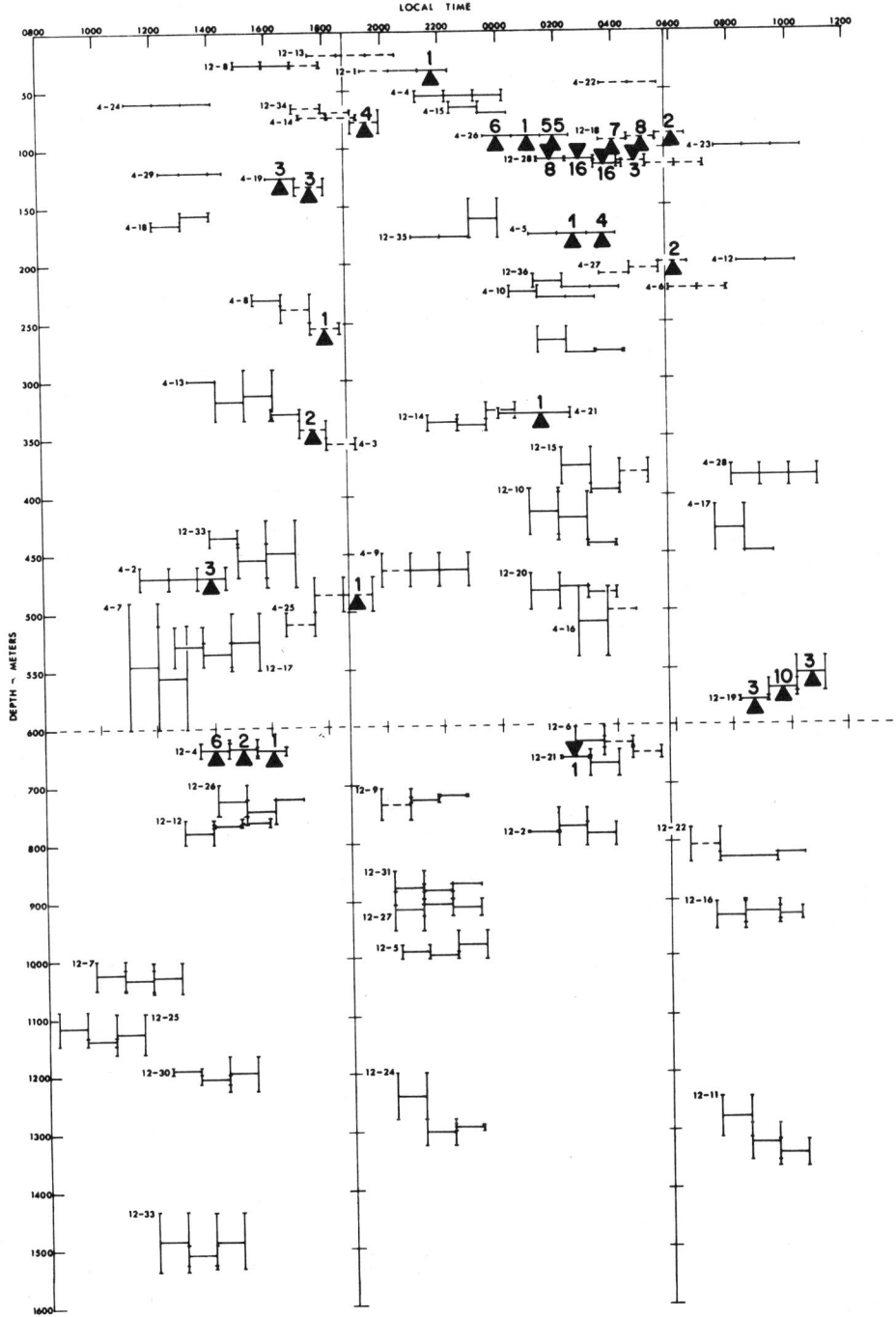

Fig. 3. Time and depth excursion for all discrete-depth samples in late summer (Cruises 4 and 12). See Fig. 1 for explanation of details.

maximum size of 0.1 mm. Adults were 24-34 mm and had testes that generally showed much regional enlargement or ovaries that were greatly enlarged, with eggs as large as 0.6 mm, but mostly 0.2-0.3 mm in diameter. Sexual dimorphism was apparent externally at 18-21 mm, with males developing supracaudal luminous tissue and females infracaudal luminous tissue (see Nafpaktitis, 1968, p. 14); males also have noticeably larger eyes than females (Tåning, 1918; Nafpaktitis, 1968).

The Biology of Lobianchia dofleini in the Ocean Acre Area

Lobianchia dofleini was one of the more abundant species in the Ocean Acre, 958 specimens being taken on the paired seasonal cruises. Discrete-depth samples included 597 specimens, of which 409 were collected during non-crepuscular times.

Life Cycle and Seasonal Abundance. The species has a one-year life cycle. It breeds from January (possibly December) to June, with a marked peak of spawning intensity in winter. It was most abundant in late spring, intermediate in late summer, and least abundant in winter (Table 1). The catch in each season was strongly dominated by a different developmental stage; juveniles were predominant in late spring, subadults in late summer, and adults in winter. More than 80% of the winter population consisted of adults (Table 1). Although adult-sized females were taken throughout the year, only in winter did many have eggs larger than 0.1 mm in diameter.

Small juveniles (10-15 mm) were taken from February (they were not taken in January collections) to September, but were most abundant in late spring, when they accounted for about 15% of the night catch of all stages (Table 2). In late spring large juveniles were the most abundant category, suggesting that recruitment was past its

TABLE 1. Lobianchia dofleini: Seasonal abundance and percent (in parentheses) of the given developmental stages and of all specimens combined. Abundance is the sum of catch rates for all 50 m intervals for day or night, whichever is greater, and includes interpolated values.

	Juveniles	Subadults	Adults	Total
Winter	1.3(10.4)	0.8(6.4)	10.4(83.2)	12.5
Late Spring	46.0(82.4)	9.2(16.5)	0.6(1.1)	55.8
Late Summer	3.3(14.0)	19.6(83.0)	0.7(3.0)	23.6

TABLE 2. *Lobianchia dofleini*: Abundance of each size by season. Numbers are total catch rates of each size from all 50 m intervals combined, those in parentheses include interpolated values.

SL	Winter		Late Spring		Late Summer	
mm	Day	Night	Day	Night	Day	Night
10	0.4	0.1		0.2		
11			0.1	0.7		
12	0.8			1.5		
13			1.0	1.0	0.3	
14		0.1	(0.5)	2.5	0.3	
15			(1.7)	2.5	0.3	
16			(1.0)	8.4		0.6
17			(1.6)	2.7		
18			(0.9)	1.7	0.5	
19			(0.6)	5.3	0.5	0.1
20			(2.1)	7.8		0.3
21			(0.5)	4.8	2.7	0.5
22			(0.8)	6.0	1.9	1.0
23			(0.4)	2.2	1.6	3.1
24			(1.1)	2.0	0.6	2.7
25			0.7	2.5	3.3	3.1
26	3.3	0.7		2.2	3.1	3.3
27	(2.3)	1.0	0.1	0.3	1.3	3.4
28	(1.7)	1.7			0.8	2.0
29	(2.4)	0.8		0.5	0.3	2.2
30	(1.3)	0.8			0.3	0.7
31						
32		0.2	0.3	0.2		
33		0.3	0.7	0.3		
34				0.3		0.2
35		0.1				
36		0.6				

peak. Subadults also increased in abundance from winter to late spring, but this increase was only partially due to growth of the earliest spawned recruits; some were of adult size and were probably spent during the recent spawn. Adults were at a minimum, suggesting that little additional spawning would take place. Post spawning mortality from winter to late spring was reflected by the decrease in abundance of adults generally and particularly of specimens greater than 25 mm (Tables 1 and 2).

By late summer there was little additional recruitment, as indicated by the low abundance of all individuals less than 22 mm (Table 2). Continued growth and development in the recruit class

TABLE 3. *Lobianchia dofleini*: Numbers of each sex for the given stages by season for all individuals sexed and staged. Significant differences from equality indicated by a Chi-square test are given by an asterisk (p < 0.05).

	Juveniles		Subadults		Adults		Totals	
	♂	♀	♂	♀	♂	♀	♂	♀
Winter	0	0	3	8	26	30	29	38
Late Spring	64	61	47	55	8	2	119	118
Late Summer	24	10*	71	93	4	0	99	103

resulted in an increase in subadult abundance and a decrease in juvenile abundance (Table 1). Adult abundance remained at its low late spring level. All adults were males, indicating that spawning activity had been completed by early to mid-summer.

Sex Ratios. Males and females are probably equally abundant at all seasons. Males and females were taken in about equal numbers (1.0:1) in late spring and late summer, while more females were taken than males in winter (1.3:1), but the difference is not significant (Table 3). Juvenile males were more numerous than juvenile females in late spring and late summer. Subadult females were more numerous than subadult males in all three seasons. Adult males were more numerous than adult females in late spring and late summer and less numerous than adult females in winter, when adults were most abundant. The only significant difference ($p<0.05$), however, was for juveniles in late summer, and this probably reflects sexual dimorphism in rate of development rather than a fundamental difference in the numbers of each sex.

Vertical Distribution. Diurnal vertical range in winter was 400-750 m with a maximum abundance at 451-500 m, in late spring 400-700 m with a maximum at 601-650, and in late summer 460-650 m with a maximum at 501-600m. Depth range at night in winter was 68-ca. 200 m (a single 10 mm juvenile was taken at 18-19 m) with no apparent concentration within that range, in late spring 50-ca. 200 m with a maximum abundance at 50 m, and in late summer 90-175 m (a single 16 mm juvenile was captured at 33 m) with a maximum at 90-120 m (Table 4, Figures 1, 2 and 3).

Stage and size stratification were evident at all seasons and, except for size stratification by day in late spring, were apparent both day and night. By night at all seasons juveniles were most abundant at shallower depths and had a shoaler upper depth limit than adults (Table 4). In terms of size, the largest individuals

TABLE 4. *Lobianchia dofleini*: Vertical distribution day and night for juveniles (Juv), subadults (Sa), adults (Ad), and all specimens (Tot) taken at each depth interval. Numbers are mean catch rates for each category at the indicated depths. Interpolated values are in parentheses. NS indicates no samples were taken from that interval and interpolation was not used.

DAY

Depth (m)	Winter				Late Spring				Late Summer			
	Juv	Sa	Ad	Tot	Juv	Sa	Ad	Tot	Juv	Sa	Ad	Tot
400	0	0	0.9	0.9	0.2	0	0	0.6	0	0	0	0
450	(0)	(0)	(3.2)	(3.2)	(1.4)	(0.4)	(0)	(2.2)	0	0	0	0
500	0	0	5.5	5.5	2.6	0.7	0	3.3	0	0.6	0	0.6
550	0	0.8	0.8	1.6	(1.5)	(0.4)	(0)	(2.0)	2.0	4.5	0	6.5
600	0.9	0	0	0.9	0.7	0	0	0.7	1.3	5.2	0.3	6.7
650	0	0	0	0	2.7	1.7	0.3	4.7	0	2.7	0.3	3.0
700	NS	NS	NS	NS	0.5	0.2	0	0.7	(0)	(1.4)	(0.2)	(1.5)
750	0.4	0	0	0.4	(0.2)	(0.1)	0	(0.3)	0	0	0	0

NIGHT

Depth (m)	Winter				Late Spring				Late Summer			
	Juv	Sa	Ad	Tot	Juv	Sa	Ad	Tot	Juv	Sa	Ad	Tot
50	0.1	0	0	0.1	34.0	1.5	0	35.5	0.5	0	0	0.5
100	0.1	0.5	0.6	1.3	11.5	6.7	0.3	18.5	2.1	5.6	0	7.7
150	0	0	2.4	2.4	0	0	0	0	0	12.7	0.7	13.3
200	0	0.2	1.3	1.5	0	0	0	0	0	0.8	0	0.8
250	0	0	0.3	0.3	0	1.0	0.3	1.3	0	0	0	0

TABLE 5. *Lobianchia dofleini*: Mean size in mm and (in parentheses) number and size range of specimens taken from the given 50 m depth intervals. Dashes indicate no sampling effort.

DAY

Depth (m)	Winter	Late Spring	Late Summer
351-400	28.0(2, 27-29)	19.4(3, 15-22)	0
401-450	--	--	0
451-500	27.2(11, 26-29)	19.2(10, 15-24)	21.0(1, 21-21)
501-550	28.0(2, 26-30)	--	22.9(26, 13-28)
551-600	10.5(2, 10-11)	16.5(2, 13-20)	23.9(27, 18-29)
601-650	0	22.1(14, 13-36)	24.8(9, 21-33)
651-700	--	21.3(9, 11-35)	--
701-750	11.0(1, 11-11)	--	0

NIGHT

Depth (m)	Winter	Late Spring	Late Summer
1-50	10.0(1, 10-10)	17.9(60, 11-26)	16.0(1, 16-16)
51-100	26.4(12, 12-29)	21.4(79, 10-32)	23.5(62, 16-27)
101-150	28.2(8, 26-30)	0	26.8(40, 22-30)
151-200	29.1(9, 26-33)	0	27.2(5, 24-34)
201-250	28.0(1, 28-28)	29.2(4, 24-34)	0

were taken only at the lower depth limit at all seasons, and in winter and late summer the smallest individuals were taken only at the shallower depth limit. Nocturnal stratification is indicated by the increase in mean standard length with depth (Table 5).

By day in late spring juveniles and subadults were most abundant at the same depth (601-650 m) and had similar depth ranges, but juveniles were more abundant above than below 600 m while subadults were more abundant below than above 600 m (Table 4). Only a single adult was taken at this season. Small (11-13 mm) and large (25-36 mm) specimens were taken only at and near the lower depth limit, while those of intermediate size were taken throughout the vertical range (Table 5).

In late summer juveniles were most abundant at a shallower depth during the day than subadults and adults and, with subadults, had a shoaler upper depth limit than adults. Individuals at the small extreme (13-15 mm) were taken only near the upper depth limit, those at the large extreme only at the lower limit, and there was a slight increase in the mean SL of the catch with depth (Table 5).

In winter, stratification by day apparently was reversed. Adults were most abundant at a shallower depth and had a shoaler upper depth limit than the few juveniles and subadults, and the juveniles were taken only below the deepest depth at which the more advanced stages were taken (Table 4).

Diel migrations occurred at all three seasons and apparently for all sizes and developmental stages. Only a single individual, a probable contaminant, was captured at daytime depths by night. Recently metamorphosed juveniles (10-11 mm) were taken only at the greatest day depths and at or near the shoalest night depths in winter and late spring, suggesting that these individuals may undergo more extensive diel migrations than the more advanced developmental stages.

Our sampling regime was not directed at studying actual migratory movements, hence, crepuscular trawls were generally avoided. There were, however, at each season, several depths fished during the morning and evening crepuscular periods, and the data gathered from these samples (see Figures 1, 2 and 3) are used to estimate the times and rates of diel migratory movements.

Morning and evening migrations were best covered in late spring. Evening migrations were on the order of 1.5-2.0 hours in late spring and perhaps as long as 3.0 hours in late summer and winter (but almost certainly they were of shorter duration, for the population at the latter two seasons consisted mostly of more advanced developmental stages and larger sizes than in late spring). Day depths were still occupied about an hour before sunset in late spring (520 m), about 2.5 hours before sunset in late summer (ca. 650 m), and about two hours before in winter (535 m). (In late summer a single individual was captured between 470 and 500 m at about sunset.) Nocturnal depths were reached within an hour after sunset in all three seasons (Figures 1, 2 and 3). From these data the estimated minimum rates of migration between day and night depths of maximum abundance were about 300 m/hr in late spring (ca. 625 to 50 m), about 140 m/hr in late summer (550 m to ca. 125 m), and about 120 m/hr in winter (451-500 m to 101-150 m).

Apparently the descent to day depths starts about the time of sunrise, for specimens were taken in the upper 200 m at or near that time in each season (Figures 1, 2 and 3). Diurnal depths were reached about two hours or less after sunrise in late spring and three hours or less after sunrise in late summer and winter, giving rates of descent similar to those for upward migration at each season.

Captures were made at several different depths at or near sunset in late spring and late summer and at or near sunrise in late spring, suggesting that different rates of migration may exist in various elements of the population.

Patchiness. Patchiness by night was indicated at 50-100 m in late spring and late summer. In late spring each of the three developmental stages occurred in maximum abundance in this interval, but probably only juveniles had a patchy distribution. They accounted for more than 95% of the catch at 50-56 m and, hence, were responsible for the observed variation in the catch at that depth. Between 90 and 100 m, where more than 60% of the catch consisted of juveniles, two samples from about 90 m, containing almost exclusively juveniles, suggested clumping, while three samples from 100 m, containing mostly subadults, did not. In late summer clumping was indicated for juveniles and subadults at 50-100 m. Juveniles were most abundant in this depth interval and, although subadults accounted for more than 70% of the catch from that depth, they were most abundant but showed no clumping at 101-150 m.

We did not accept all significant coefficients of dispersion (CD) as indicative of a clumped or patchy distribution. Significant CD's for 151-200 m at night in late summer resulted from a series of three negative samples from one year being tested with a series of three samples from the other year containing two positive samples (Figure 3). The positive samples, taken near the morning crepuscular period, may have captured early downward migrants or may represent year to year variation in vertical range. By day in late summer the CD's for 501-600 m were barely significant, and the combined samples suggest a random distribution. At 451-500 m only a single sample was positive, suggesting a low population density at the upper day depth limit. At 101-150 m, a depth much too shallow to be within the diurnal vertical range, only a single sample, taken near the evening crepuscular period (Figure 3), was positive, suggesting that the catch consisted of migrants. By day in late spring at 551-700 m the distribution may be patchy, but more likely is random, similar to that of 501-600 m by day in late summer, but at a lower population density.

In winter patchiness was not indicated for any stage at any depth day or night.

Night to Day Catch Ratios. Night to day catch ratios (including interpolated values) for discrete-depth captures were 0.5:1 in winter, 4.2:1 in late spring, and 1.3:1 in late summer (Table 6).

Seasonal differences in clumping, abundance, vertical distribution, discrete-depth coverage, and stage and size composition render it most unlikely that any single factor was the principal cause of the observed differences in diel catch rates.

Adults were most abundant in winter and accounted for most of the difference between day and night catches at that season. Juveniles had the greatest proportional difference between day and night catches, but they constituted little more than 10% of the day catch and less than 5% of the night catch. Increased net avoidance at night by adults, inadequate discrete-depth sampling within the vertical range by night, or both may have been responsible for the higher daytime catches in winter. Apparently *L. dofleini* feeds at night or during migrations (Merrett and Roe, 1974), and increased activity associated with feeding may result in enhanced net avoidance. Bond (1974) has proposed a similar behavior for *Cyclothone microdon* and *C. braueri*, the two most abundant fishes in our collections. The largest catches of *L. dofleini* in winter came from discrete-depth samples taken at 130 m and 140 m in the crepuscular period (Figure 1).

TABLE 6. *Lobianchia dofleini*: Seasonal night to day catch ratios for the given developmental stages and for all specimens combined. Ratios are based upon the sums of the catch rates for all 50 m intervals including interpolated values.

	Juveniles	Subadults	Adults	Total
Winter	0.2:1	0.9:1	0.5:1	0.6:1
Late Spring	4.8:1	2.6:1	2.0:1	4.2:1
Late Summer	0.8:1	1.4:1	1.2:1	1.3:1

Discrete-depth samples at night were taken at 100 m and 150 m, but not between, suggesting that the depth of maximum abundance of the species was not sampled, and as a result the night catch may have been artificially low.

Juveniles were responsible for most of the difference between late spring day and night catches and subadults for the difference in late summer. Patchiness was greater in these seasons and the vertical range more constricted by night than by day, and these differences may have resulted in increased night catches.

Impact of Life-history Information on Prediction of Sound Scattering

The species considered here, *Lobianchia dofleini*, has been shown to be a major contributor to sound scattering at certain frequencies by virtue of its abundance and its swimbladder size (MacDonald, 1972; Brown and Brooks, 1974). Its contribution, however, will not be the same at all seasons, for the population is composed of different numbers of different-sized fish at different times of the year. The depths inhabited by fish of different sizes have been shown to vary seasonally, as does the rate and duration of the vertical migration. Patchiness differs from night to day as well as seasonally and appears to be size dependent. Night versus day catch ratios are not seasonally consistent, suggesting changing behavioral attributes throughout the year's progression.

Lobianchia dofleini in Bermuda waters is an annual fish, with a one-year life cycle, as are many other fishes in the same area. Others, of course, have longer cycles. The integration of the behavior of all species in the area affects patterns of sound scattering. Studies such as these are sorely needed as the raw data for creating predictive models.

near Bermuda apparently does, it is quite possible that *O'Day and Nafpaktitis* found no gravid females because winter collections were not available to them. Normal but undeveloped oocytes, the presence of males with mature sperm, the large population density in slope water off New England, and the appearance of 12 mm juveniles off Cape Hatteras all suggest the presence of a reproductive population in or near slope water off eastern North America. In fact, one of us (Karnella) has examined winter specimens with large eggs from the slope water south of New England. We are thereby convinced that expatriation is not a factor in maintaining the populations of *L. dofleini*, either at Bermuda or in the slope water of the western North Atlantic, and that these populations are breeding populations.

Having rejected the expatriation hypothesis, we point out that *Bolin* (1959) and *Nafpaktitis* (1968) both noted that *L. dofleini* was variable in several morphological characters, and they related this variation to geographic distribution. Nafpaktitis demonstrated differences between eastern and western North Atlantic populations. Added to this, our data indicate that the Bermuda population matures at a smaller size and does not reach the maximum size of the eastern North Atlantic (*Nafpaktitis*, 1968) or Mediterranean (*Tåning*, 1918; *Goodyear et al.*, 1972) populations. Furthermore, there is a difference in breeding seasons: March-October in the eastern North Atlantic (*Nafpaktitis*, 1968), January-June at Bermuda and probably in the western North Atlantic generally, mainly February-June in the Mediterranean.

Taken together, the data suggest the existence of two distinct populations in the North Atlantic, one eastern and one western, and a third Mediterranean population. A detailed taxonomic evaluation of these populations could even show them to be different species. Whatever the level of differentiation, the possibility of differences in ecology and behavior is great and must be resolved in any geographically predictive model of sound scattering.

EPILOGUE

We have discussed *Lobianchia dofleini* only from the North Atlantic. The species occurs also in the eastern South Atlantic, across the southern South Atlantic and the southern Indian Ocean and in the eastern South Pacific Ocean. How many additional distinct populations (or species) may exist within the wide range of this nominal species is a matter for investigation.

We know little enough of the environmental factors that influence the life history and behavior of this or any other mesopelagic fish. Clearly, the genetic-systematic status of the populations must be understood for the other factors to be interpreted and used properly for predictive purposes. The need is great for intensive studies of different populations in different geographic areas of the world ocean, the kinds of studies that would provide information

Broader Implications of Geographic Variation and Expatriation

Even when the biology of a species is reasonably well known in a given area, there is no assurance that the same data will apply to populations of that species in another area. There must always be a question as to the degree of genetic differentiation even to the species level, between populations in widely separated regions, even though occurrence of the nominal species may appear to be continuous between regions. Conclusions of research scientists, too often are taken as the final word. Aspects of the geographic distribution and variation of *Lobianchia dofleini* in the North Atlantic and its implied expatriation in the western North Atlantic provide evidence of the potential hazards in projecting data from one geographical area to another or in accepting published conclusions uncritically.

Lobianchia dofleini has been taken across the North Atlantic between about 20° and 50° N. *O'Day and Nafpaktitis* (1967) and *Nafpaktitis* (1968) concluded the area of spawning in the North Atlantic was in the eastern part of the ocean, south of 48°N and east of 35°W. These authors found no gravid females and no juveniles smaller than 16 mm in the western North Atlantic north of Cape Hatteras. Histological evidence presented by *O'Day and Nafpaktitis* (1967) revealed that female *L. dofleini* in the western North Atlantic had only oogonia and oocytes, and no large, yolk-filled eggs in their ovaries. The oocytes appeared to be normal, but had not undergone vitellogenesis. These authors concluded that *L. dofleini* was expatriated and non-breeding in the western North Atlantic, with population numbers maintained by periodic recruitment from the spawning area in the eastern Atlantic. They noted that such a crossing would take about a year, yet they made no attempt to explain the presence of 11-12 mm juveniles off Cape Hatteras.

Since the species is known to transform to the juvenile stage at 11-13 mm (*Tåning*, 1918), the proposed system of recruitment would involve a larval stage nearly a year in duration. Such a prolonged period of slow development followed by a sudden resumption of growth is extremely unlikely. Moreover, if one considers the inherent mortality for these small recruits, truly astronomical numbers of this species must be transported, for the population density in the "expatriate" area is almost or quite as great as that in the spawning area (*O'Day and Nafpaktitis*, 1967).

According to their list of specimens *O'Day and Nafpaktitis* (1967) did not examine any western Atlantic females taken from November through February, and only a single female taken in March. Most specimens examined were taken from June through October. We have already indicated that Bermuda has a breeding, not expatriated population that spawns in winter. If the population of *L. dofleini* off the continental United States in slope water is a reproductive population, and that population spawns only in winter, as the population on life histories and data useful for relating behavior to environmental factors, while at the same time contributing material for determination of the genetic-systematic status of geographic populations.

ACKNOWLEDGEMENTS

We express our gratitude to R. C. Kleckner, J. F. Janosky, and, especially, M. J. Keene and W. H. Howell for sorting and initially identifying most of the collections used in this study; and to S. J. Karnella for preparing the figures. The Ocean Acre Program has been supported by the U. S. Naval Underwater Systems Center. This particular work has been supported in part by the Smithsonian Research Foundation and the Office of Naval Research.

REFERENCES

Aron, W., N. Raxter, R. Noel, and W. Andrews. 1964. A description of a discrete-depth plankton sampler with some notes on the towing behavior of a 6-foot Isaacs-Kidd midwater trawl and a one-meter ring net. *Limnol. Oceanogr.* 9(3): 324-333.

Baker, A. de C., M. R. Clarke, and M. J. Harris. 1973. The NIO combination net (RMT 1+8) and further developments of rectangular midwater trawls. *J. Mar. Biol. Ass. U.K.* 53: 167-184.

Battista, G. and D. F. Guiliano. 1971. Improved discrete-depth sampler. *Naval Underwater Systems Center Rept. No. 4083*: 1-7.

Bolin, R. 1959. Iniomi. Myctophidae from the "Michael Sars" North Atlantic deep-sea expedition 1910. *Rep. Sci. Res. "Michael Sars" N. Atl. Deep-Sea Exped. 1910.* 4, pt. 2(7): 1-45.

Bond, G. W. 1974. Vertical distribution and life histories of the gonostomatid fishes (Pisces: Gonostomatidae) off Bermuda. *Rep. to U. S. Navy Underwater Systems Center.* Contract No. N 00140-73-C-6304: 1-276.

Brooks, A. L. 1972. Ocean Acre: Dimensions and characteristics of the sampling site and adjacent areas. *Naval Underwater Systems Center Rept. No. 4211*: 1-16.

Brown, C. L. and A. L. Brooks. 1974. A summary report of progress in the Ocean Acre Program. *Naval Underwater Systems Center Tech. Rept. No. 4643*: 1-44.

Gibbs, R. H., Jr., C. F. E. Roper, D. W. Brown, and R. H. Goodyear. 1971. Biological studies of the Bermuda Ocean Acre I. Station data, methods and equipment for Cruises 1 through 11, October 1967-January 1971. *Rept. to U. S. Navy Underwater Systems Center.* Contract No. N00140-70-C-0307. Smithsonian Inst., Wash., D. C. 1-49.

Goodyear, R. H., B. J. Zahuranec, W. L. Pugh, and R. H. Gibbs, Jr. 1972. Ecology and vertical distribution of Mediterranean midwater fishes. In *Mediterranean Biological Studies Final Rept. Rept. to U. S. Navy Office of Naval Research,* Contract No. N00014-67-A-399-000-7: 91-229.

Kleckner, R. C. and R. H. Gibbs, Jr. 1972. Swimbladder structure of Mediterranean midwater fishes and a method of comparing swimbladder data with acoustic profiles. In *Mediterranean Biological Studies Final Rept. Rept. to U. S. Navy Office of Naval Research,* Contract No. N00014-67-A-0399-007: 230-281.

MacDonald, R. B. 1972. A comparison of acoustical and biological backscattering strengths at discrete-depth for Ocean Acre Cruise 6 (April, 1969) and 10 (June, 1970). *Naval Underwater Systems Center Tech. Mem. No. TA131-225-72.*

Merrett, N. R. and H. S. J. Roe. 1974. Patterns and selectivity in the feeding of certain mesopelagic fishes. *Marine Biol.,* 28: 115-126.

Nafpaktitis, B. G. 1968. Taxonomy and distribution of the lanternfishes, genera *Lobianchia* and *Diaphus,* in the North Atlantic. *Dana Rep.,* 73: 1-131.

O'Day, W. T. and B. Nafpaktitis. 1967. A study of the effects of expatriation on the gonads of two myctophid fishes in the North Atlantic Ocean. *Bull. Mus. Comp. Zool.,* Harvard Univ., 136(5): 77-89.

Pearcy, W. G. 1964. Some distributional features of mesopelagic fishes off Oregon. *J. Mar. Res.,* 22: 83-102.

Roper, C. F. E., R. H. Gibbs, Jr., and W. Aron. 1970. Ocean Acre: an interim report. *Rept. to U. S. Navy Underwater Sound Lab.,* Contract No. N00140-69-C-1066: 1-31.

Tåning, A. V. 1918. Mediterranean Scopelidae (*Saurus, Aulopus, Chlorophthalmus,* and *Myctophum*). *Report on the Danish Oceanographical Expeditions 1908-1910 to the Mediterranean and Adjacent Seas,* 2 (*Biology*) (5, A. 7): 1-154.

Observations on Feeding Habits of the Mesopelagic Fish Benthosema glaciale (Myctophidae) off NW Africa

Johannes Kinzer

University of Kiel

ABSTRACT

In the slope waters north of Cape Blanc, NW Africa, a total of 1273 Glacier Lanternfish Benthosema glaciale have been collected in 12 oblique hauls between 0-600 m depth. Using the RMT 1+8, essentially a combined trawl and plankton net of 1 and 8 m² opening with a frontal opening and closing device. In the area where upwelling occurs, B. glaciale migrated in a diel pattern from a daytime depth of 150-400 m to the surface (25-100 m) at night, considerably shallower than in non-upwelling oceanic areas. Analysis on 784 stomach contents and the degree of filling revealed, that B. glaciale feeds randomly on copepods and ostracods, occasionally also on chaetognaths, decapod larvae, and pteropods. At night near the surface, for specimens > 30 mm SL 45% consisted of larger prey such as euphausiids (Euphausia krohnii) and a few amphipods. Feeding activity dominated in the early night, but was continued during the day at all depths. The feeding pattern of B. glaciale off NW Africa is compared with other myctophid species there and from other areas.

INTRODUCTION

In recent years, investigations on the feeding ecology of mesopelagic fishes have increasingly contributed to our knowledge of the lower levels of the marine food web (DeWitt and Caillet, 1972; Hopkins and Baird, 1976; and Merrett and Roe, 1974). In particular, the myctophids, among the most common fish of the world ocean, play an important part in the vertical transport of organic matter (Paxton, 1967; Holton, 1969; Samyshev and Schetinkin, 1971; Baird et al., 1975; and Wörner, in print). Many myctophid species are known to migrate in a diurnal pattern (*Badcock*, 1970; *Pearcy and Laurs*, 1966; and *Clarke*, 1973). Since many species have gas-filled swimbladders, they have long been recognized as acoustic scatterers (*Marshall*, 1951; *Tucker*, 1951; and *Backus et al.*, 1968) and many deep scattering layer (DSL) studies have focused on myctophid fishes, furnishing additional information on their biology.

According to *Halliday* (1970) and *Giosaeter* (1973a), the Glacier Lanternfish *Benthosema glaciale* is the most common myctophid species in the Atlantic north of about 35°N. Off the west African coast, a few catches have been described from the Cape Verde Islands (*Norman*, 1939; cited from *Halliday*, 1970). The present investigation describes the feeding patterns of *B. glaciale* from off NW Africa, an area of coastal upwelling where this species is by far the most abundant mesopelagic fish. Previously, the feeding behavior of this myctophid had been described only from a fjord in western Norway (*Gjoesaeter*, 1973a). Other recent studies on feeding of *B. glaciale* sampled in the surface layers off NW Africa have not yet been published (*Wörner*, in print).

In further contributions, the feeding patterns of other mesopelagic fish species of the same area will be described.

MATERIALS AND METHODS

Sampling was done during the "Upwelling '75" Expedition, a joint cruse of RRS "Discovery" and RV "Meteor" from 14 January to 21 March 1975. The material for the present study was collected in 12 hauls of the Rectangular Midwater Trawl (RMT 1+8) aboard RRS "Discovery" over and beyond the continental shelf north of Cape Blanc at about 23° North (Figure 1). The gear, as developed by *Baker et al.* (1973), is essentially a twin net, consisting of an 8 m^2 trawl of 4.5 mm mesh size and a simultaneously fished 1 m^2 plankton net of 300 μ mesh size. The opening and closing of both nets are acoustically triggered. A net monitor telemeters depth, temperature, speed, and distance of the gear to the ship's laboratory. The RMT 1+8 is towed at a speed of 2 knots. During the expedition, the two vessels collected a total of 80 RMT 1+8 samples.

The RMT 1+8 was towed in a series of oblique stratified hauls, the strata covering an average depth level of 50 to 100 m (Table 1). The fish samples were first preserved in 4% formalin. After the cruise they were bottled in a preserving fluid as described by *Steedman* (1974).

A total of 784 stomachs of *B. glaciale* were dissected for food composition and degree of filling. From samples with more than 100 specimens, only 100 stomachs were dissected, except from Station 8776/3, where 150 stomach contents were analyzed. Each stomach content was preserved separately. The amount of stomach filling was classified according to the following scale: 0 = empty, 1 = nearly empty, 2 = half full, 3 = full, 4 = extended stomach.

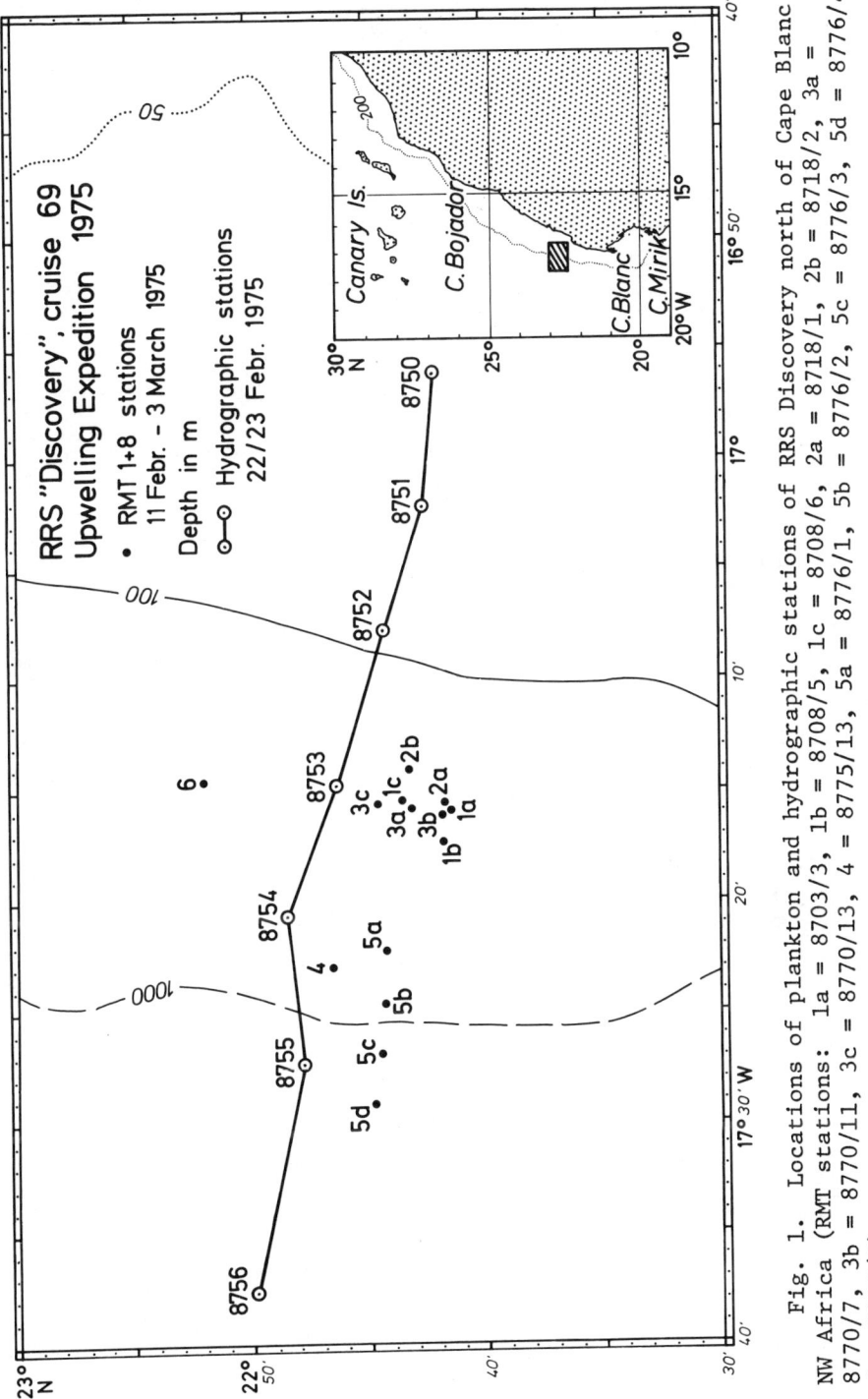

Fig. 1. Locations of plankton and hydrographic stations of RRS Discovery north of Cape Blanc, NW Africa (RMT stations: 1a = 8703/3, 1b = 8708/5, 1c = 8708/6, 2a = 8718/1, 2b = 8718/2, 3a = 8770/7, 3b = 8770/11, 3c = 8770/13, 4 = 8775/13, 5a = 8776/1, 5b = 8776/2, 5c = 8776/3, 5d = 8776/4, 6 = 8807/2).

Table 1. Vertical distribution of *Benthosema glaciale* and depth of the deep scattering layer (DSL) from six series of RMT 8 hauls (time in GMT).

Station:	8708
Date:	11/2/75
Time:	1148-1810 hrs
DSL:	160-200 m
Water Depth:	450-651 m

Depth (m)	No. Fish
50-75	0
75-100	0
100-128	0
125-150	12
150-200	202 DSL
200-250	42

Station:	8718
Date:	12/2/75
Time:	0941-1346 hrs
DSL:	120-150 m
Water Depth:	492-617 m

Depth (m)	No. Fish
18-50	0 DSL
240-300	182
300-400	230

Station:	8770
Date:	25/2/75
Time:	0906-1941 hrs
DSL:	160-185 m
Water Depth:	550-778 m

Depth (m)	No. Fish
0-15	0
25-55	0
100-150	0
195-220	4 DSL
270-300	23
300-500	29

Station:	8775
Date:	27/2/75
Time:	0835-1751 hrs
Water Depth:	720-1210 m

Depth (m)	No. Fish
0-12	0
13-50	0
53-75	0
75-100	0
100-150	0
155-200	0
200-250	0
250-300	33

Station:	8776
Date:	27/2/75
Time:	2004-2357 hrs
Water Depth:	912-1278 m

Depth (m)	No. Fish
12-25	0
25-50	116
50-100	309
100-150	3
505-600	32

Station:	8806
Date:	2/3-3/75
Time:	2246-0353
Water Depth:	114-323 m

Depth (m)	No. Fish
0-10	0
12-25	0
25-50	2
50-100	0
100-150	0
150-200	0

RESULTS AND DISCUSSION

The 12 samples contained a total of 1273 *B. glaciale*, comprising about 85% of the total catch of mesopelagic fish. In contrast to the higher latitudes, where *B. glaciale* reaches a standard length of about 68 mm (Halliday, 1970), the largest *B. glaciale* from our collection measured only 42 mm. The average was between 25 and 32 mm SL.

Most *B. glaciale* were adults, only 8 specimens were juveniles. The sex ratio was nearly 1.0.

Vertical Distribution and the Deep Scattering Layers

At six stations on the shelf and along the continental slope, with the depth of water varying between 114 and 1278 m, the RMT 1+8 was fished in oblique hauls at various depth ranges between 0 and 600 m (Table 1). The shallowest positive daytime sample came from 125-150 m depth with 12 specimens (Station 8708), while at night *B. glaciale* appeared at the 25-50 m depth level at Station 8776 (116 ind.) and Station 8806 (2 ind.). The low number of specimens at Station 8806 is probably due to the shallow location on the shelf with a sounded depth of only 114-323 m. The greatest depths sampled were from 505 to 600 m (Station 8776), where 32 *B. glaciale* were caught.

The few samples indicate that in the area of investigation *B. glaciale* is distributed along the slope of the continental shelf. In the daytime, *B. glaciale* aggregated mostly between 150 and 400 m depth, much shallower than in the slope waters off Nova Scotia, where during the day they dominate at depths below 500 m (Halliday, 1970). The small range of vertical migration in *B. glaciale* off the NW African coast is probably caused by the reduced light penetration in upwelling areas where increased primary production causes the low transparency of the surface water. As with many other vertical migrators, the maximum depth of distribution is primarily regulated by light responses.

Few RMT 8 catches from the "Upwelling '75" Expedition have been analyzed, so it is not yet possible to determine the extent to which *B. glaciale* contributes to the scattering of the sound. In the area of investigation, the depth of the DSL varied almost continuously during the day between 80 and about 200 m, probably closely correlated to the turbidity of the water. Only occasionally the DSL was as deep as 320 m. The echogram in Figure 2 shows the typical variations in depth of DSL, with depth-intervals of about 80 m and the ship going at speed 5 knots. During the day, *B. glaciale* seemed to have its maximum distribution somewhat deeper than the DSL. Only at Station 8708, where the DSL appeared at 160 to 200 m depth, did *B. glaciale* concentrate at about the same depth level (Table 1).

Food Composition

The results of the stomach-content analysis are given in Figure 3. Feeding in the cod-end of the net was unlikely to occur since the

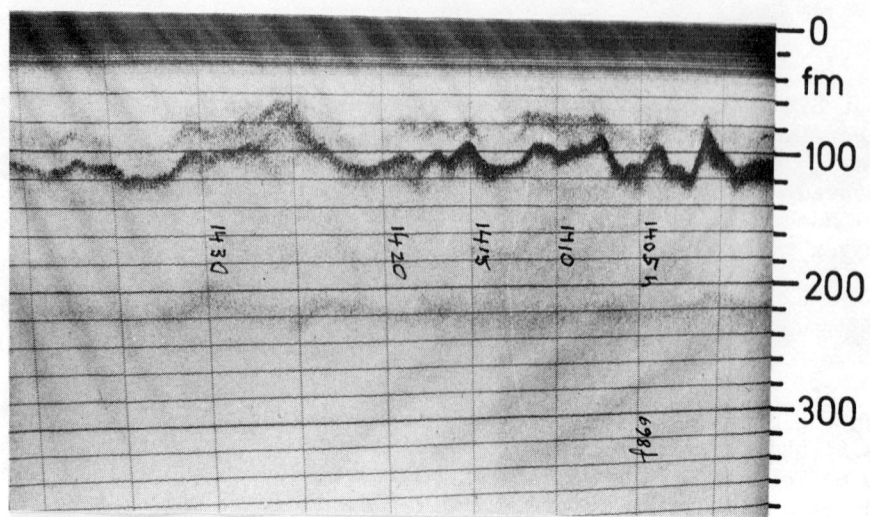

Fig. 2. Sound scattering at 12 kHz in the slope waters at 23°N off NW Africa. The echogram shows the typical depth variations of the DSL in areas of upwelling. (Mufax facsimile recorder, 27 February 1975, 1355-1440 hrs GMT, sounded depth 698-750 fm).

mesh size of the cod-end (0.7 mm) is too coarse to hold the smaller zooplankton, such as copepods and ostracods (see also Merrett and Roe, 1974), whereas the primary food at all depth levels consisted of copepods and conchoecid ostracods. However, at night, with B. glaciale feeding at 25-100 m depth, up to 45% of the prey were euphausiids (Station 8776/3). Occasionally (Station 8770/11), amphipods were also found in the stomachs. Chaetognaths, pteropods and decapod larvae were particularly rare in the diet. Only a small fraction of the stomach contents could be identified.

Usually the prey was only slightly digested, so that identification to genus of some prey organisms was possible. The copepod Pleuromamma was easily identifiable, as the dark spot on the thorax is very resistant to digestion. This copepod has an average size of 3 mm and was observed in the stomach contents from all depths (Figure 2). Pleuromamma was a particularly important constituent (45%) at Station 8775/13, from 250-300 m depth. Besides Pleuromamma, the copepod Rhincalanus was also identified. At Station 8807/3 from 128-150 m depth, at least 25% of all copepods eaten were Rhincalanus. Candacia and harpacticoids played only a minor role in the diet of B. glaciale. No trace of phytoplankton could be found in the stomachs.

From the data compiled in Figure 3, the large variations in the food composition at different depth levels are obvious. Apparently, B. glaciale is feeding randomly on a certain "suitable" size fraction of food organisms. Only B. glaciale > 30 mm SL were found feeding on

FEEDING HABITS OF THE MESOPELAGIC FISH

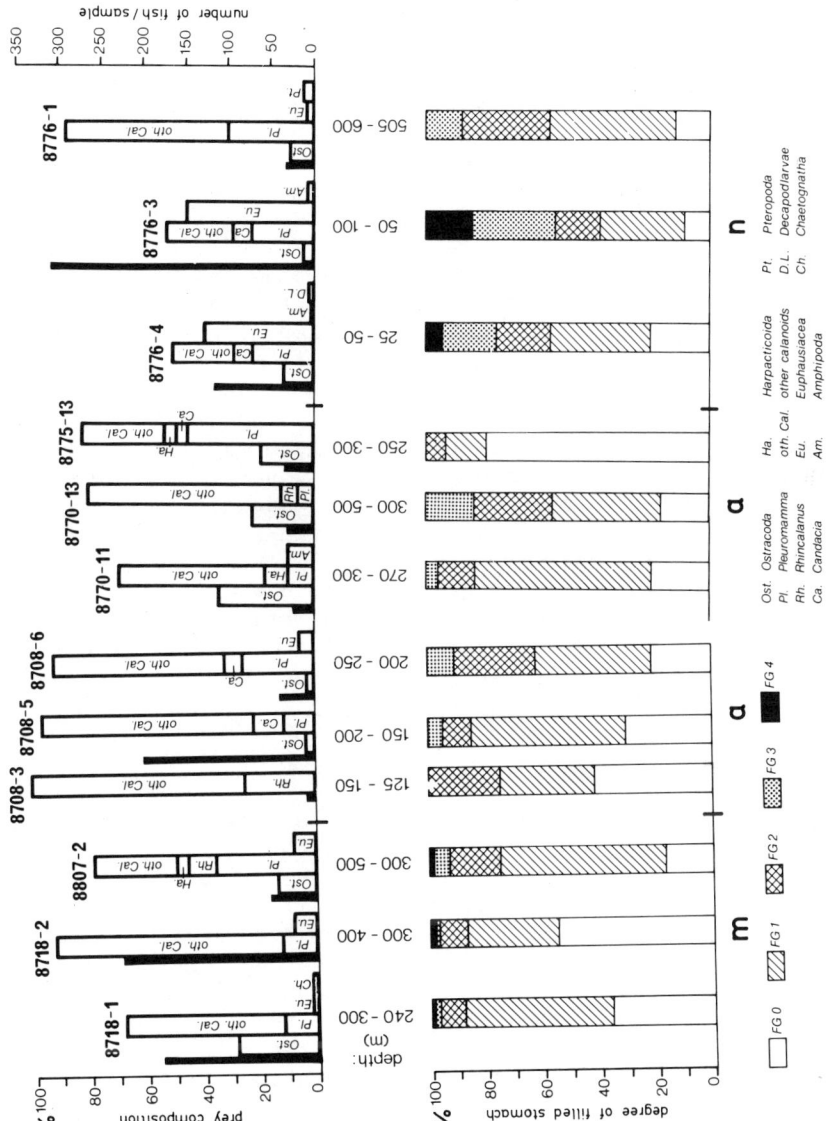

Fig. 3. Percentage composition of total stomach contents and degree of filled stomachs per sample. The black bars indicate the number of *B. glaciale* per haul. (m = morning, a = afternoon, n = night, FG 0 = empty, FG 1 = nearly empty, FG 2 = half full, FG 3 = full, FG 4 = extended stomach).

larger items, such as amphipods and particularly euphausiids. The myctophids sampled at night (2206-2323 hrs GMT) at Station 8776/4 and 8776/3 at 25-50 and 50-100 m depth had preyed on *Euphausia krohnii*, mostly adolescents, but partly also adult specimens. According to *Baker* (personal communication) *E. krohnii* has its maximum abundance in the slope waters off NW Africa at a daytime depth of only 50-150 m, while in oceanic water it lives in 300-700 m depth. At night they aggregate at about 25 m depth, or somewhat deeper. Future analysis of the RMT 1 sample might give evidence whether or not *B. glaciale* has been feeding in the maximum layer of *E. krohnii*.

In an investigation by *Gjösaeter* (1973a) on the feeding pattern of *B. glaciale* from a Norwegian fiord, the author found a similar food spectrum in the stomachs: copepods (*Calanus finmarchicus*, *Metridia sp.*, and *Pareuchaeta norvegica*) dominated and of the euphausiids species, the diet consisted mostly in *Thysanoessa* and occasionally *Meganytiphanes norvegicus*. With changing relative abundance of the various kinds of food organisms in the plankton, the composition of food also varied during the year.

Paxton (1967) analyzed about 300 stomach contents of 9 myctophid species from samples collected off southern California. Most of the species were larger than *B. glaciale*. Therefore euphausiids dominated as prey. Other food items were copepods, sergestids, and amphipods. Fish remains were found only in larger species (> 60 mm SL). At least 5 species are possibly highly specific feeders.

In a recent study *Merrett and Roe* (1974) investigated the stomach contents of 9 mesopelagic fish species in the eastern North Atlantic, among them 3 myctophid species. The samples were taken in a series of 11 RMT 1+8 hauls over a 24 h period from 250 m depth. All 3 myctophid species were found to be vertical migrators feeding in a nocturnal pattern on a similar diet as other myctophiids: mainly copepods, ostracods, amphipods, and euphausiids.

Selectivity and Feeding Chronology

B. glaciale from the samples described in this study has an average size of 25-32 mm SL, probably too small for selective feeding towards prey size larger than copepods. Only the larger specimens of > 30 mm SL preferred euphausiids to copepods and ostracods.

Merrett and Roe (1974) observed size selectivity in *Lampanyctus cuprarius* of 57-72 mm SL. The stomachs contained larger prey, mostly amphipods, but also euphausiids. Similar to the selectivity reported for *L. cuprarius*, *L. alatus* also has been described to select prey in particular size ranges (*Hopkins and Baird*, 1977). *Tyler and Pearcy* (1975), in a study on three common myctophids collected in the eastern North Pacific, observed a preference toward either euphausiids or copepods in a single feeding period.

Selectivity toward certain prey species has been described by *Hartmann and Weikert* (1969) from *Centrobranchus nigroocellatus*. This myctophid, as collected with a neuston net from the 0-10 cm surface layer, feeds during the night on the neuston gastropods *Atlanta peroni* and/or *Styliola subula*.

Next to the composition of the diet, the amount of food in the stomach is of particular interest. Together with the degree of digestion, it provides some information on the feeding chronology.

From Figure 3 (night stations, n) it is evident, that *B. glaciale* is feeding at night, as the specimens from the 25-50 and 50-100 m depth levels had the highest degree of filled stomachs, compared to the other samples in the histogram. The haul from 50-100 m depth yielded 309 specimens, the highest number of myctophids in this series of RMT 8 hauls. The night samples were collected from 2206 to 2323 hrs GMT. The 32 specimens from 505-600 m (Station 8776/1), collected from 2004 until 2237 hrs, had less filled stomachs and—quite in contrast to the shallow hauls—were feeding mostly on copepods and a few pteropods.

The morning samples, particularly those from 240 to 400 m depth, collected from 0941 to 1142 hrs revealed mostly empty and nearly empty stomachs. Later during the afternoon, a slightly greater amount of food was found in the stomachs, except for Station 8775/13 with 80% empty stomachs.

The analysis on the rate of stomach fillings gives evidence that *B. glaciale* feeds intensively in a nocturnal pattern. Probably most feeding occurs in the evening or early night in the surface water. This corresponds to the observations by *Gjösaeter* (1973a) on *B. glaciale* from the Norwegian fiord population, although from his data "diurnal variation is probably not very great."

The slight increase in stomach filling during the day, with only traces of digestion, suggests continuous feeding. In addition, the small population which did not rise to the surface at night (Figure 3, Station 8776/1) had been feeding at the depth of 505-600 m, since the stomach contents showed traces of digestion.

Relatively few specimens of *B. glaciale*, collected in deep samples at 300 to 500 m in the morning and afternoon (Station 8807/2 and 8770/13) had empty stomachs, but many had half-filled and even full stomachs. By comparison, in eastern Arabian Sea during daytime at scattering layer depth (300-500 m), *B. pterota* was feeding intensively on copepods and ostracods (*Kinzer*, 1969). The stomach contents showed only slight traces of digestion, indicating that they had been feeding recently. Stomachs of *Diaphus fulgens* and two other *Diaphus* species were empty.

It should be considered, that due to low temperatures at the depths of feeding and capture, the rate of digestion might be low. This holds true particularly for animals from the slope waters where upwelling occurred. Figure 4 shows the upwelling along the continental slope during the time of investigation: the cold water reaches some 100-200 m further up toward the surface than it does in the warmer oceanic region. On the other hand, there might be physiological adaptations towards an accelerated digestion rate, as discussed by *Merrett and Roe* (1974) for the chauliodontid *Chauliodus*.

Samyshev and Schetinkin (1971) did not observe any significant change of feeding intensity night or day in 3 myctophid species

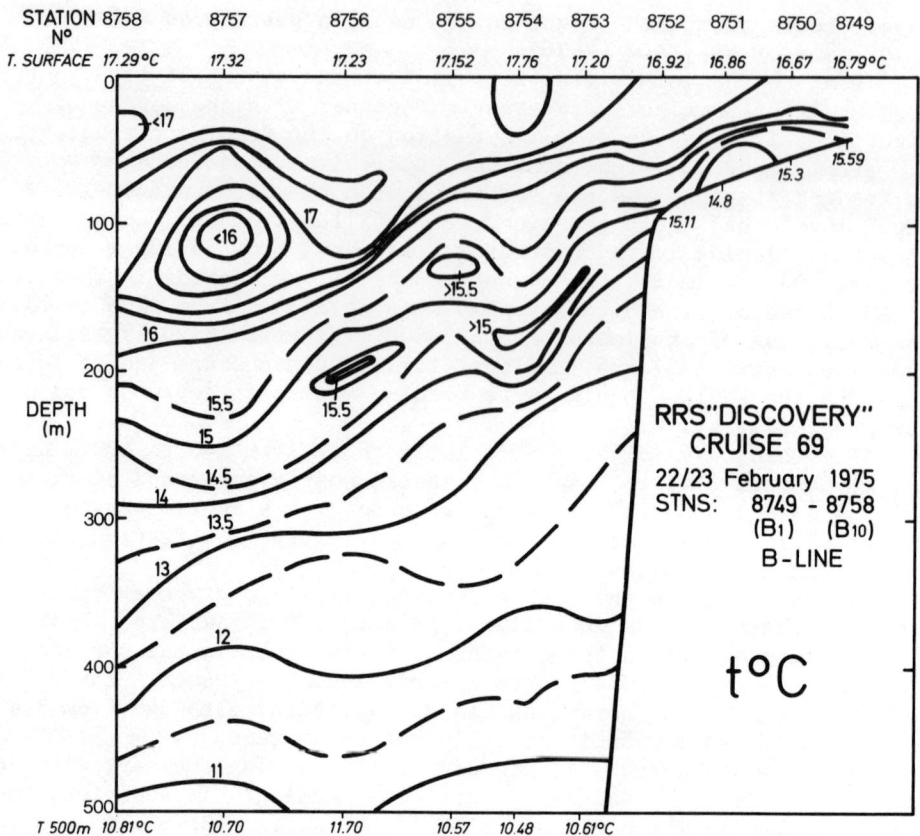

Fig. 4. Distribution of temperature along transect B, stations 8749 to 8758. (see Fig. 1 for location of hydrographic stations 8750-8756.)

collected from the DSL off NW Africa. Since their net hauls were also rich in zooplankton, the authors assumed that at the depth of sampling the predators found sufficient food.

Similar results were described by *Paxton* (1967) for 9 southern California myctophid species: 117 specimens caught during the day and 86 specimens caught at night had "identifiable" stomach contents. *Holton* (1969) who investigated the feeding pattern of *Lampanyctus mexicanus* off southern California, also observed continuous feeding in this vertically migrating myctophid species, although nocturnal feeding prevailed, "probably because of the greater amount of food in the surface waters."

Considering the migratory and feeding pattern of *B. glaciale* populations in the upwelling waters off NW Africa, it is obvious, that this abundant fish is an important link in the mesopelagic food web. Probably it also contributes considerably to an active vertical transport of organic matter in this area (in the sense of

Vinogradov, 1962). However, the range of vertical migrations is quite reduced compared with oceanic areas. B. glaciale in the area of upwelling migrates only to a depth of about 300 m, while in the less turbid oceanic waters it migrates to more than 500 m depth. Some zooplankton also have a much narrower vertical range where upwelling occurs, for instance Euphausia krohnii, an important prey organism and several siphonophore species (Kinzer, in preparation). Besides the different vertical migration patterns of prey and predator in upwelling areas, the reduced diversity in micronekton and zooplankton species should have a marked influence on the food chains.

ACKNOWLEDGEMENTS

This contribution was supported by the German Research Council (DFG). Dr. Renate Weigmann (Euphausiacea) and Dr. Sigrid Schnack (Copepoda) kindly identified some of the species from the stomach contents. The hydrographic data have kindly been contributed by Dr. Peter Hughes, Liverpool.

REFERENCES

Backus, R. H., et al. 1970. The distribution of mesopelagic fishes in the equatorial and western North Atlantic Ocean. In G. B. Farquhar, ed., *Proceedings of an International Symposium on Biological Sound Scattering in the Ocean*: 20-41. Washington, DC.

Badcock, J. 1970. The vertical distribution of mesopelagic fishes collected on the SOND Cruise. *J. Mar. Biol. Ass. U.K.* 50: 1001-1044.

Baird, R. C., T. L. Hopkins, and D. F. Wilson. 1975. Diet and feeding chronology of *Diaphus taaningi* (Myctophidae) in the Cariaco Trench. *Copeia 1975*: 356-365.

Baker, A. de C., M. R. Clarke, and M. J. Harris. 1973. The NIO combination net (RMT 1+8) and further developments of rectangular midwater trawls. *J. Mar. Biol. Ass. U.K.* 53: 167-184.

Clarke, T. A. 1973. Some aspects of the ecology of lanternfishes (Myctophidae) in the Pacific Ocean near Hawaii. *Fish. Bull. U.S.* 71: 401-434.

DeWitt, F. A. jr. and G. M. Caillet. 1972. Feeding habits of two bristlemouth fishes, *Cyclothone acclinidens* and *C. signata* (Gonostomatidae). *Copeia 1972(4)*: 868-871.

Gjösaeter, J. 1973a. The food of the myctophid fish, *Benthosema glaciale* (Reinhardt), from western Norway. *Sarsia* 52: 53-58.

Gjösaeter, J. 1973b. Age, growth, and mortality of the myctophid fish, *Benthosema glaciale* (Reinhardt), from western Norway. *Sarsia* 52: 1-14.

Halliday, R. G. 1970. Growth and vertical distribution of the Glacial Lanternfish, *Benthosema glaciale*, in the Northwestern Atlantic. *J. Fish. Res. Bd. Canada* 27: 105-116.

Hartmann, J. and H. Weikert. 1969. Tagesgang eines Myctophiden (Pisces) und zweier von ihm gefressener Mollusken des Neuston. *Kieler Meeresforsch* 25(2): 328-330.

Holton, A. A. 1969. Feeding behavior of a vertically migrating lanternfish. *Pac. Sci.* 23(3): 325-331.

Hopkins, T. L. and R. C. Baird. 1977. Aspects of the feeding ecology of oceanic midwater fishes. In: *"Oceanic Sound Scattering Prediction"*, N. R. Andersen and B. J. Zahuranec (Eds.), Plenum, N.Y., (this volume).

Kinzer, J. 1969. On the quantitative distribution of zooplankton in deep scattering layers. *Deep-Sea Res.* 16: 117-125.

Marshall, N. B. 1951. Bathypelagic fishes as sound scatterers in the ocean. *Mar. Res.* 10: 1-17.

Merrett, N. R. and H. S. J. Roe. 1974. Patterns and selectivity in the feeding of certain mesopelagic fishes. *Mar. Biol.* 28: 115-126.

Nafpaktitis, B. G. 1973. A review of the lanternfishes (family Myctophidae) described by A. Vedel Taning. *Dana-Rep.* 83: 1-46.

Paxton, J. R. 1967. Biological notes on southern California lanternfishes (Family Myctophidae). *Calif. Fish Game* 53: 214-217.

Pearcy, W. G. and R. M. Laurs. 1966. Vertical migration and distribution of mesopelagic fishes off Oregon. *Deep-Sea Res.* 13: 163-166.

Samyshev, E. B. and S. V. Schetinkin. 1971. Feeding patterns of some species of Myctophidae and *Maurolicus mülleri* caught in the sound-dispersing layers in the northwestern African area. *Ann. Biol., Cons. Int. Explor. Mer.* 28: 212-214.

Steedman, H. F. 1974. Laboratory methods in the study of marine zooplankton. *J. Cons. Int. Explor. Mer* 35(3)

Tåning, A. V. 1918. *Myctophum glaciale* (Reinhardt). In Mediterranean Scopelidae (*Saurus, Aulopus, Chlorophthalmus,* and *Myctophum*). *Rep. Dan. Oceanogr. Exped. Mediter. 2, Biol.* A7: 31-45.

Tucker, G. H. 1951. Relations of fishes and other organisms to the scattering of underwater sound. *J. Mar. Res.* 10: 215-238.

Tyler, H. R. and W. G. Pearcy. 1975. The feeding habits of three species of lanternfishes (Family Myctophidae) off Oregon, USA. *Mar. Biol.* 32: 7-11.

Vinogradov, M. E. 1962. Feeding of the deep-sea zooplankton. *J. Cons. Int. Explor. Mer* 153: 114-119.

Wörner, F. G. (in print) Untersuchungen an drei Myctophiden-Arten *Benthosema glaciale, Ceratoscopelus maderensis* und *Myctophum pumctatum* aus dem nordwestafrikanischen Auftriebsgebiet im Frühjahr 1972. Ph.D. thesis, Univ. Kiel, 136 pp.

Possible Factors Affecting Succession in Plankton Communities

A. G. Lewis

University of British Columbia

ABSTRACT

Succession is described as an orderly and directed change by Margalef (1958). In terrestrial plant communities the unique nature of the community is known to affect succession while in marine situations this is less well documented. The stability of the water controls the exchange occurring between near surface and deep water. This either directly or indirectly controls many of the events occurring during succession. Upwelled oceanic water is rich in nutrients and generally low in trace metals. The combination of low metal concentrations and reduced availability through the action of naturally occurring complexing agents may be biologically limiting. As upwelled water ages near the surface the effect of photo-oxidation and biologically produced complexing agents produces a unique situation. Comparison of the types of organisms occurring during succession suggests that the nature of the community may, in part, be controlled by these events. In contrast to the offshore upwelling situation, metal concentrations in inshore waters are generally higher and not as limiting. The effect of complexing agents in inshore waters would then be more like that of a buffering agent.

INTRODUCTION

It is known that different amounts of various trace metals are required by organisms (*e.g.*, Bowen, 1966). It is also suspected that the form or state of the metal is important (Steemann Nielsen and Wium-Andersen, 1970; Provasoli, 1963). Finally, there is evidence that the chemical nature of the water can affect the biological availability of metals (Whitfield and Lewis, 1976). If the chemical nature of the water does control the biological availability of trace metals, this could affect succession in plankton communities. The intention of this paper is to discuss evidence of the ability of natural water to control trace metal availability and to relate this to succession.

The change in composition and population density of plankton with time frequently occurs in an orderly fashion or succession. Margalef (1958, page 27) has defined succession as ". . . the occupation of an area by organisms involved in an incessant process of action and reaction which, in time, results in change in both the environment and the community, both undergoing reciprocal influence and adjustment." He goes on to say that "the important point is that . . . (succession) . . . is an orderly and directed change", and his work on the ria of Vigo, a marine nearshore environment, demonstrates this.

Succession infers that the community produces a change in the environment which in turn changes the community. In the terrestrial environment this is more apparent than in the marine. Examples of the unique nature of the community and its effect on succession are given in the papers of Knapp (1974), on mutual influences between plants, and Hanes (1971), on succession in the chaparral of Southern California. Influences originate from chemical substances exuded or leached from living parts of plants as well as from dead parts. The effect of these exudates may be direct, on neighboring plants, or indirect, through soil and microorganisms. In the marine environment, the biological uptake of nutrients such as reactive silicate has been shown to reduce the concentration of biologically important chemicals (e.g., Postma and van Bennekom, 1974). Changes in water quality have also been attributed to trace amounts of "biologically active" organic compounds (Lucas, 1958; Provasoli, 1963; Barber and Ryther, 1969; Lewis, et al., 1971). Examples suggest the ability of the organism to modify the environment in both the terrestrial and marine condition.

Succession can be discussed from the overly simplistic diagram in Figure 1. This shows upwelling as a means of bringing nutrient rich water to the euphotic zone. Phytoplankton and zooplankton biomass are seen to increase gradually at the beginning and then more rapidly as water column stability increases. The peak in biological activity occurs during maximum stability of the water column which is followed by downwelling. The figure could also be used to show temporal succession including the three stages discussed by Margalef (1958). Stage 1 would be the upwelling or initial phase, characterized by low biomass and small-celled phytoplankton (mainly diatoms: e.g., Skeletonema, Rhizosolenia, Leptocylindrus) with high surface to volume ratios, stage 2 with higher biomass and a mixed community of diatoms with bigger cells and smaller surface to volume ratio (Lauderia, Rhizosolenia, Thalassiosira), stage 3 with an increasing number of dinoflagellates. One interesting aspect pointed out by Margalef is that stage 1 organisms grow well in "rough" cultures suggesting tolerance to adverse conditions, stage 2 organisms may be cultured

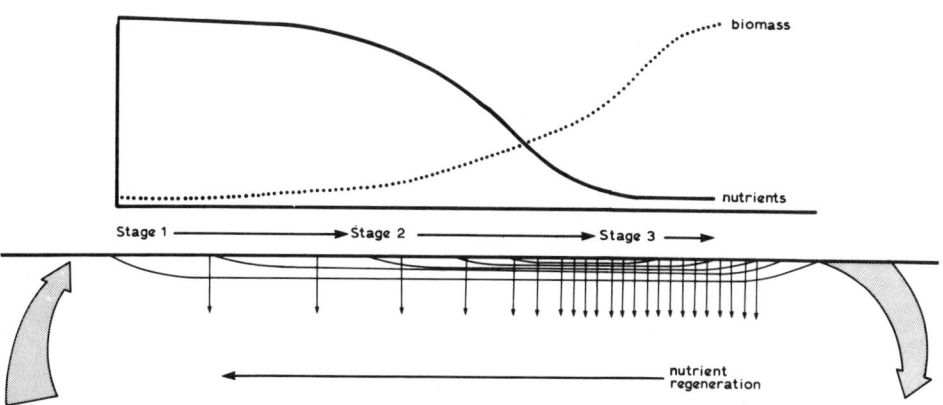

Fig. 1. Diagramatic representation of events occurring between upwelling (left) and downwelling (right). Stages are those of Margalef (1958), isopleths represent lines of equal density, vertical arrows represent partical fallout from nearsurface productivity. Upper figure represents hypothetical profiles of nutrients and biomass during the period.

well but not as easily as those in stage 1 suggesting reduced tolerance. Stage 3 organisms are the most difficult to culture suggesting intolerant organisms. Applying this to succession suggests that a change effected in the environment by early stages leads to environment specialization and organisms capable of living under these conditions but which are increasingly intolerant to change.

The organisms characteristic of the three stages appear to offer an explanation for succession without bringing in the problem of metabolites. Stage 1 organisms occur in well mixed water but as stability of the water column increases they are replaced by organisms capable of maintaining their position in the euphotic zone and surviving under increasingly unique near surface conditions. Further, it can be argued that changes in levels of inorganic nutrients, with the changing hydrographic conditions, cause the unique near surface conditions. But to eliminate metabolites as important factors is to eliminate such things as vitamins which are required by many planktonic organisms (e.g., Fogg, 1966). Additionally, the direct uptake of dissolved organics as a possible food source (Stephens and Schinske, 1961) may also make use of these materials. In addition to these there is still a considerable amount of dissolved material of biological origin which may include biologically active compounds.

Bioassay work in our laboratory (Lewis et al., 1971), using the prefeeding stages of the calanoid copepod Euchaeta japonica, indicated that water from an inshore station possessed what we termed "chelator" activity and that water from an oceanic station rarely possessed this activity. Part of this was suggested by the frequent improval to survival of the bioassay organism by addition of a chelating agent (ethylenediaminetetraacetic acid, EDTA). The beneficial effect of EDTA in oceanic water was compared with the results of *Barber and Ryther* (1969) who found that addition of EDTA increased phytoplankton growth in recently upwelled water. The indication was that the state of the metals was biologically limiting in both cases. In the case of *Barber and Ryther* (1969), the growth of phytoplankton was improved after the upwelled water had remained at the surface for some time or extracts of zooplankton were added. The suggestion by these authors was that natural organic chelators are produced by organisms as the water ages. Oceanic water for the *Lewis et al.* (1971) study was from 200-220 meters depth and resembled upwelled water in the effect of EDTA. Both of these studies suggested that the effect of EDTA or of the water remaining at the surface for some time is the addition of organic material which complexes metals, that the organometallic complex is not present in recently upwelled warter or water occuring at some depth. If this is true, organic complexing agents should be at a low level in upwelled or deeper oceanic water.

Photooxidation of water with a high intensity ultraviolet lamp has been found capable of oxidizing organics (*Beattie et al.*, 1961; *Armstrong et al.*, 1966) including organic complexing agents in sea water (*Strickland*, 1972). If biologically important complexing agents are at low levels in upwelled and deeper oceanic water ultraviolet photooxidation should not change survival of the bioassay organism unless it changes some other component of the water. Additionally, addition of EDTA to the ultraviolet treated water should improve survival in a manner similar to that shown in the 1971 study. In 1973 we were able to test this, with a 1200 watt ultraviolet source, on water obtained from the same general area and depth. Techniques used were those described in *Whitfield* (1974) and *Whitfield and Lewis* (1976).

The results of the bioassay with ultraviolet irradiated water June of 1973 are shown in Figure 2. This was during a period of time when upwelling normally occurs near the source of the water. The effect of the ultraviolet irradiation was to increase the survival of the bioassay organism while the addition of EDTA to the irradiated water decreased survival. This indicated that the water contained some agent or series of agents which reduced survival of the bioassay organism and whose effect could be reduced by ultraviolet photooxidation. Further, the addition of a chelating agent (EDTA) changed the state of the water to a condition which was not beneficial to the organism. It is of interest to note that the decrease in survival with increasing EDTA is not linear, that there is a sharp decrease between 0.2-0.4 um EDTA and only a small decrease between 0.0-0.2 and 0.4-0.6 um. When compared with the survival of the control this suggests that low levels of EDTA would increase survival while higher levels would not.

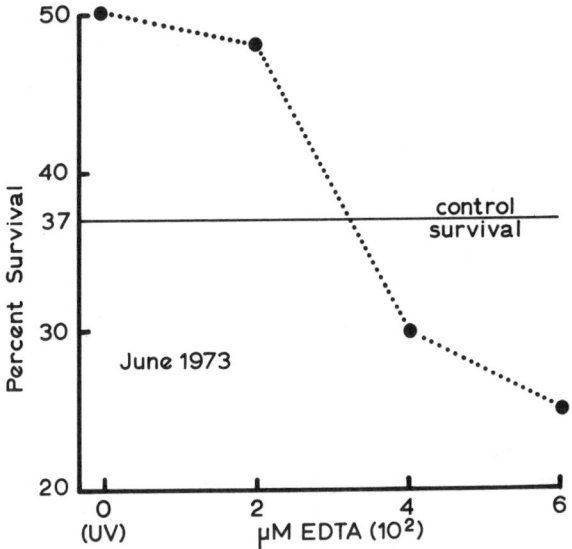

Fig. 2. Survival of Euchaeta japonica in ultraviolet irradiated water with increasing levels of EDTA. (Control survival is in untreated water.)

The case for June (Figure 2) is typical of the months studied in which upwelling was occurring in offshore waters. Ultraviolet irradiation of the water caused an increase in the survival of the bioassay organism, addition of EDTA to the irradiated water caused a decrease. The difference between months is the amount of decrease in survival with increasing EDTA, the decrease varying with time. In offering an interpretation for this it is accepted that the EDTA will chelate metals preferentially. With this assumption it is suggested that the concentrations of different metals may be fluctuating with time, that the biological availability of the metal is changing and is controlled by the concentration of the metal and the amount of added EDTA.

If an organism exists under higher than necessary levels of required metals, natural or added complexing agents serve as a buffer (e.g., Provasoli, 1963). If, however, the metals exist in very low concentrations the complexing agents may actually reduce the biological availability of the metal. The nature of the agent is important as the metal-organic associations will differ. Finally, the response of different organisms to different metal complexes will vary (INCRA project #177-B).

Metals tend to be higher in inshore waters than in offshore, which suggests that the buffering effect of natural complexing agents would be more important in inshore waters. It appears that in upwelling or deeper oceanic water the low concentration of required trace metals and an ultraviolet photooxidizable agent controls the biological availability of metals. (This is presumably some metal complexing agent.) The addition of EDTA to the water, through interaction with the ultraviolet photooxidizable agent, appears to produce an improvement in survival up to a point and then may decrease survival.

The poor growth of phytoplankton noted by *Barber and Ryther* (1969) in upwelled water may have been due to the low concentration of metals and the presence of complexing agents that bind the metals in such a manner that they are essentially unavailable to the organism. This state could be compared to that produced by addition of EDTA in both the *Barber and Ryther* (1969) and *Lewis et al.* (1971) studies.

The change of the metals into a more suitable state could be due to the breakdown of the complexing agent upon exposure to solar irradiation. *Kashiwada et al.* (1960), for example, noted that 50-81 percent of vitamin B_{12} is decomposed in the upper 20 m upon exposure to sunlight. *Spoehr et al.* (1949) found that photooxidation of unsaturated fatty acids from *Chlorella* extracts leads to antibacterial activity of the compounds. The change could also be due to equilibrium between complexing agents formed at the surface by biological activity (e.g., *Barber and Ryther*, 1969). More than likely it is a combination of the two. It is important to remember, however, that although the activity of any strong complexing agent in upwelled water may decrease due to solar irradiation, the effect is still present prior to upwelling, on subsurface populations of organisms.

It is interesting to try to fit the types of organisms one might find into a situation produced by an environment with low concentrations of trace metals which are further limited by strong complexing agents. It would be advantageous for the organism to have requirements for very low levels of metals, be able to use the organometallic complex, or be hardy enough to tolerate low levels of metals. A large surface to volume ratio might also be beneficial in accumulating metals at low levels. These requirements appear to be met by the types of organisms described by *Margalef* (1958) for stage 1 in plankton succession (Figure 1).

As the stability of the water increases the exchange with deep water decreases and there would be less loss of dissolved materials. Continuing productivity in this more stable system would cause the accumulation of dissolved organics, possibly including complexing agents. Equilibrium between existing organometallic compounds and new organic ligands could change the nature of the organometallics as time progressed. Organisms with rather specific requirements, including those for metals or organometallic compounds, might then find conditions more suitable (e.g., stages 2 and 3 of *Margalef*, 1958; Figure 1).

Since there is a net loss of dead and dying particles from the euphotic zone the breakdown of these particles would occur at some depth. Duursma (1960), for example, found that dissolved organic substances come more from dead cells than from living. Smayda (1963) suggests that the greatest amount of ectocrines are produced after death of the organism. The breakdown is known to be important for nutrient regeneration, it could also be important to generation of complexing agents. Presumably the longer lasting complexing agents would survive in deep water until upwelling occurred. It is important, however, to remember that organisms occurring at depth may be exposed to the effects of complexing agents produced during decomposition.

The array of metals available, the differences in the metal requirements of organisms, and the complexing ability of different agents for different metals all make specific explanations potentially difficult. The possible control over the composition and population density of marine organisms by natural complexing agents, their effect on community structure in different bodies of water, and the potential biological uniqueness produced by different agents in different areas does, however, offer a rather tempting though frustrating study. This study should help provide an idea of the sequence of events occurring in succession in plankton communities.

REFERENCES

Armstrong, F. A. J., P. M. Williams, and J. D. H. Strickland. 1966. Photo-oxidation of organic material in seawater by ultraviolet radiation, analytical and other applications. *Nature*, 211: 481-483.

Barber, R. T., and J. H. Ryther. 1969. Organic chelators: Factors affecting primary production in the Cromwell Current upwelling. *J. Exp. Mar. Biol. Ecol.*, 3: 191-199.

Beattie, J., C. Bricker, and D. Garvin. 1961. Photolytic determination of trace amounts of organic material in water. *Anal. Chem.* 30: 1890-1892.

Bowen, H. J. M. 1966. *Trace elements in Biochemistry*. Academic Press, London and New York. 241 pp.

Duursma, E. K. 1960. Dissolved organic carbon, nitrogen, and phosphorus in the sea. *Netherlands J. Mar. Res.*, 1: 1-148.

Fogg, G. E. 1966. *Algal Cultures and Phytoplankton Ecology*. Univ. Wisconsin Press, Madison. 126 pp.

Hanes, T. L. 1971. Succession after fire in the chaparral of Southern California. *Ecol. Mono.*, 41: 27-52.

INCRA project #177-B, Research and Development Report #103-74. Chemistry of Copper in Water. Submitted January 30, 1974.

Kashiwada, K., D. Kakimoto, and A. Kanazawa. 1960. Studies in Vitamin B_{12} in natural water. *Rec. Oceanogr., Wks. Japan*, 5: 51-76.

Knapp, R. 1974. Mutual Influences between Plants, Allelopathy, Competition, and Vegetation Changes. Handbook of Vegetation Science. *Vegetation Dynamics*. (Ed. R. Knapp): 111-122.

Lewis, A. G., A. Ramnarine, and M. S. Evans. 1971. Natural chelators—an indication of activity with the calanoid copepod *Euchaeta japonica*. *Mar. Biol.*, 11: 1-4

Lucas, C. E. 1958. External metabolities and productivity. *Cons. Explor. Mer. Rapp. Proc. Verb.*, 144: 155-158.

Margalef, R. 1958. Temporal succession and spatial heterogeneity in phytoplankton. *Perspectives in Marine Biology*. (Ed. A. A. Buzzati-Traverso), Univ. California Press, Berkeley.

Postma, H., and A. J. van Bennekom. 1974. Budget aspects of biologically important chemical compounds in the Dutch Wadden Sea. *Neth. Journ. Sea Research 8*: 312-318.

Provasoli, L. 1963. Organic regulation of phytoplankton fertility. (Ed. M. N. Hill). *Interscience*: 165-219. John Wiley and Sons, Inc., New York.

Smayda, T. J. 1963. Succession of phytoplankton, and the ocean as an holocoenotic environment. *Symposium on Marine Microbiology*. (Ed. C. H. Oppenheimer). C. C. Thomas, Publisher, Springfield: 260-274.

Spoehr, H. A., J. H. Smith, H. H. Strain, H. W. Milner, and G. J. Hardin. 1949. *Fatty acid antibacterials from plants*. Carn. Inst. Wash. Publ., 586: 1-67.

Steemann Nielson, E. and S. Wium-Anderson. 1970. Copper ions as poison in the sea and in freshwater. *Mar. Biol. 6*: 93-97.

Stephens, G. C. and R. A. Schinske. 1961. Uptake of amino acids by marine invertebrates. *Limnology and Oceanography 6*: 175-181.

Strickland, J. D. H. 1972. Research on the marine planktonic food web at the Institute of Marine Resources: A review of the past seven years of work. *Ocean. Mar. Biol. Ann. Rev. 10*: 349-414.

Whitfield, P. H. 1974. Seasonal changes in hydrographic and chemical properties of Indian Arm and their effect on the calanoid copepod *Euchaeta japonica*. M. Sc. thesis University of British Columbia, Vancouver, Canada. 79 p.

Whitfield, P. H., and A. G. Lewis. 1976. Control of the biological availability of trace metals to a calanoid copepod in a coastal fiord. *Estuarine and Coastal Marine Science 4*: 255-266.

Whitfield, P. H. and A. G. Lewis. 1974. The biological importance of copper in the sea—A literature review. Final Report; INCRA project No. 223. 132 p.

Growth and Moulting of Crustacea, Especially Euphausiids

J. Mauchline

Dunstaffnage Marine Research Laboratory

ABSTRACT

There are problems with determining growth rates of offshore pelagic crustaceans because it is generally impossible to sample a discrete population repeatedly. Thus, conventional methods of statistical analyses of seasonal samples of the same population cannot be applied. A completely different approach to the problem is outlined, using three species of euphausiids as examples. The percentage growth factors at successive moults decrease logarithmically while the durations of the intermoult periods increase logarithmically. The changing growth factors and intermoult periods can be described by two regression lines on body length or successive moult numbers. Methods of drawing these regression lines are given. Growth and moulting sequences throughout the life of the organism can then be derived and compared with information from other sources.

INTRODUCTION

Rates of growth of crustaceans are usually estimated by repeated sampling of natural populations. An identifiable population is found and an estimate of the life time of the animal made. A series of samples of the populations are then taken for a period equal to the life time of the animal or over a minimum period of one year if the animal lives one year or longer. Growth rates are then determined by statistical analyses of length or weight measurements made on the individuals in the samples. This method depends upon identifying a distinct population and being able to sample it repeatedly over an, often, extended period of time. This is usually impossible when studying species living offshore in the epipelagic and, especially, the mesopelagic and bathypelagic regimes.

A radically different approach is suggested by the studies of Kurata (1962). He examined moulting, growth, and age relationships in a range of species and defined various parameters acting throughout the lives of the animals. He postulated the following relationships within species:
1. The logarithm of the duration of the intermoult period is related to the body length or to successive moult numbers and increases as the animal grows.
2. The duration of the intermoult period is directly related to the wet or dry weight of the animal, increasing with increasing weight.
3. The logarithm of the duration of the intermoult period of individuals of a species of the same size decreases as the logarithm of the environmental temperature increases.

Kurata examined the increments of length or weight at moulting of crustaceans and plotted them on, what he termed, Hiatt growth diagrams. A Hiatt growth diagram (Hiatt, 1948), relates post-moult size to pre-moult size. Size in this context is usually defined by a linear measurement of length rather than weight. He then calculated linear regression equations to fit the points on the diagram, the equations being of the normal form:

$$y = a + bx$$

where x and y are the pre- and post-moult lengths respectively and a and b are constants.

Mauchline (1976a) criticized the use of the Hiatt growth diagram because it examines the increment at moulting as a measurement contained within the post-moult length. A more rigorous analysis of growth is obtained by expressing the increment at each moult as a percentage of pre-moult length, then equivalent to a growth factor at the moult. Growth factors at successive moults follow an exponential decrease such that log growth factor is linearly related to body length or successive moult number (Mauchline, 1976a, 1976b). Some of the best sets of data for testing these hypotheses are those on commercial species of shrimps, crabs, and lobsters. The following highly significant linear relationships were demonstrated with these data (Mauchline, 1976b).
1. Log intermoult period (days) on a measurement of body length or on successive moult moult number.
2. Log growth factor (%) on a measurement of body length or on successive moult number.
3. Log of the cubic function of a measurement of body length ($log\ L^3$) on log age (days).
4. Log intermoult period (days) on the cube root of body weight (g).
5. Long increment in weight (g) on log body weight (g).
6. Log body weight (g) on log age (days).

No attempt is made in this paper to examine the implications of these relationships in a general theory of growth of crustaceans. Lasker, in a personal communication concerning Zweifel and Lasker

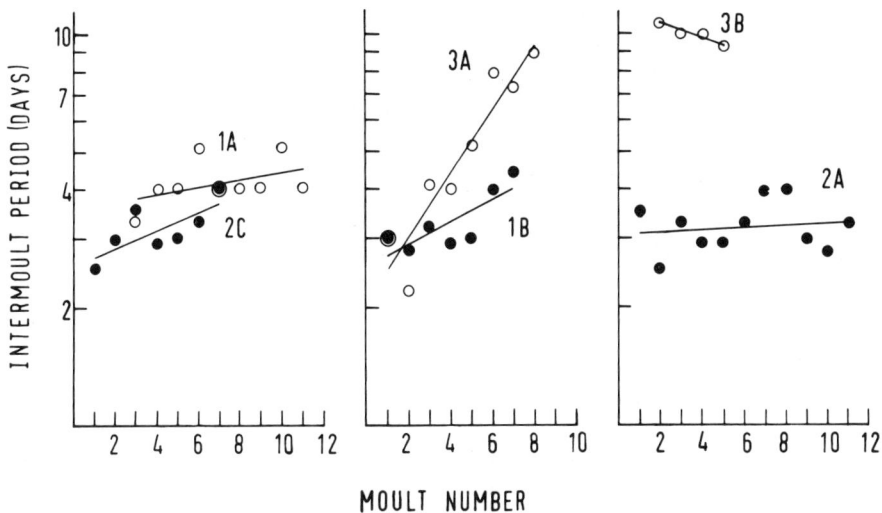

Fig. 1. *Nyctiphanes couchii*. Intermoult periods of successive furciliae in the series of experiments by Le Roux (1973). Closed circle inside open circle is point subscribed to by two series.

(1976), suggests that the growth and moulting characteristics of crustaceans, as defined by these relationships, may be described by a Gompertz type growth function with moults occurring at regular intervals along the decay of the specific growth rate. This certainly appears to be an appropriate model to be tested as further data become available.

Statistical analysis of populations of several species of euphausiids have provided estimates of growth rates (Mauchline and Fisher, 1969; Smiles and Pearcy, 1971; Mackintosh, 1972). Recent studies on moulting and growth have provided enough data to estimate growth rates using the first two of the six relationships defined above. These estimates can be compared with the estimates obtained by statistical analyses of the wild populations.

DURATION OF THE INTERMOULT PERIOD

Le Roux (1973) examined the intermoult periods of furciliae, and Fowler et al. (1971b) of adult *Nyctiphanes couchii*. Fowler et al. (1971b) also examined the intermoult periods of other species, and Gopalakrishnan (1973) observed moulting in *Nematoscelis difficilis*.

The intermoult periods of the furciliae of *N. couchii* at successive moults in a Series 1A to 3B of experiments made by Le Roux are shown in Figure 1, along with the calculated regression lines. The equations and correlation coefficients are given in Table 1.

TABLE 1. Log/Linear Regression Equations Relating Intermoult Period (y) to Successive Moult Number or Body Length (x). The correlation coefficients (r) are given and the intermoult period slope factors (f) are calculated for equations relating to successive moult numbers.

Species			Equations	r	f
Nyctiphanes couchii					
furciliae	Ser	1A	log y = 0.4059 + 0.0227x	0.7395	1.0537
		1B	log y = 0.4048 + 0.0279x	0.7955	1.0664
		2A	log y = 0.4848 + 0.0033x	0.1733	1.0076
		2C	log y = 0.5526 + 0.0091x	0.4250	1.0212
		3A	log y = 0.3140 + 0.0819x	0.9375	1.2075
		3B	log y = 1.0634 − 0.0183x	−0.9484	0.9588
Ser 1A, 1B and 2A			log y = 0.4608 + 0.0092x	0.3977	1.0214
Nyctiphanes couchii					
adults			log y = 0.6928 + 0.0058x	0.8928	1.0134
Nematoscelis difficilis			log y = 0.6006 + 0.0147x	0.5463	1.0344
Meganyctiphanes norvegica			log y = 0.6370 + 0.0122x	0.9638	
Meganyctiphanes norvegica					
Temperatures:	10°C		log y = 0.8655 + 0.0079x	−	
	13°C		log y = 0.6153 + 0.0128x	0.9791	
	15°C		log y = 0.6134 + 0.0096x	0.9994	
	18°C		log y = 0.5331 + 0.0109x	−	

Le Roux found the fastest growth rates in Series 1A, 1B, and 2A. The data for these three experiments were combined (Figure 2) and the regression equation calculated (Table 1). The regression lines in Figures 1 and 2 are termed intermoult period lines and have a slope defined in the regression equations. A more useful constant, however,

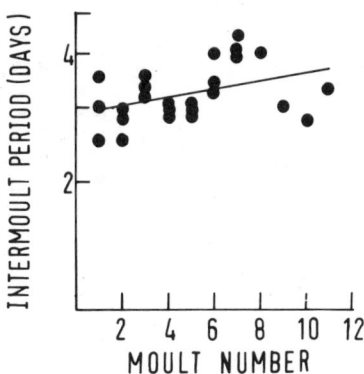

Fig. 2. *Nyctiphanes couchii*. Intermoult periods of successive furciliae in experimental Series 1A, 1B, and 2A of *Le Roux* (1973).

is intermoult period slope factor. This is the factor by which the intermoult period increases at each successive moult; these are also given in Table 1. The intermoult period increased at the slowest rate in Series 2A. This was the Series in which the fastest rate of growth was achieved not simply against time but also against moult number; that is, the successive furciliae in this Series were larger than comparable furciliae in the other series of experiments. *Le Roux* indicates in his text-figures that the third calyptopes in Series 2 were slightly larger than the third calyptopes in Series 1 and 3. Consequently, the experiments are not detailed enough to define possible relationships between intermoult period slope factors and growth of the larvae at successive moults.

Fowler et al. (1971b) measured the successive intermoult periods of a single *N. couchii* maintained for eleven months at 13 ± 0.3°C in the laboratory (Figure 3). The regression equation, correlation

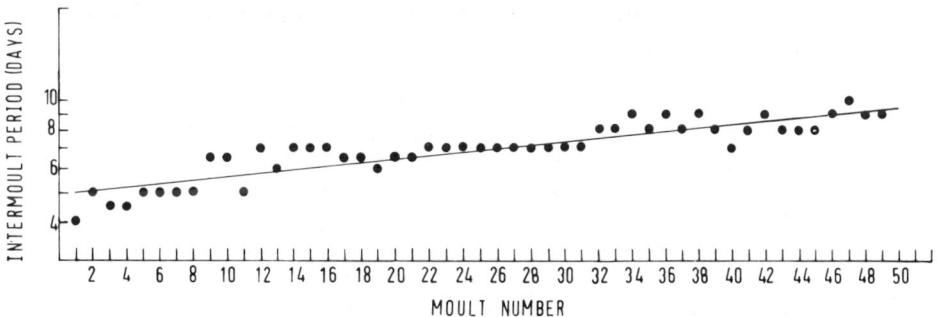

Fig. 3. *Nyctiphanes couchii*. Successive intermoult periods of a single individual maintained at 13°C (*Fowler et al.*, 1971b).

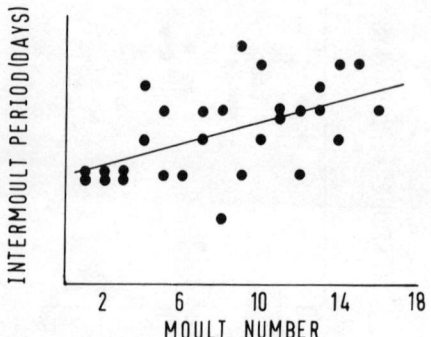

Fig. 4. *Nematoscelis difficilis*. Intermoult period of successive moults. Data from *Gopalakrishnan* (1973).

coefficient and moult slope factor are given in Table 1. The intermoult period slope factor is low at 1.0134. *Gopalakrishnan* (1973) examined the moulting frequency of three *Nematoscelis difficilis*, two of which were equal in initial size and attained the initial size of the larger third individual after ten moults. The data from the three animals are combined in Figure 4, the first moult of the larger animal being at moult 11. The regression equation, correlation coefficient, and intermoult period slope factor are given in Table 1.

Fowler et al. (1971b) describe the increase in the duration of the intermoult period in relation to body dry weight in *Meganyctiphanes norvegica*. Regression equations relating body length exclusive of uropods and the length of the uropods to dry weight are given by *Fowler et al.* (1971a). These equations were used to convert measurements of body dry weight to measurements of total length. The intermoult period was then related to total body length (Figure 5), the regression equation and correlation coefficient being given in Table 1. *Fowler et al.* (1971b) examined the duration of the intermoult period of individuals of this species of different body dry weights at different temperatures. The dry weights were again converted to equivalent measures of total length. The duration of the intermoult

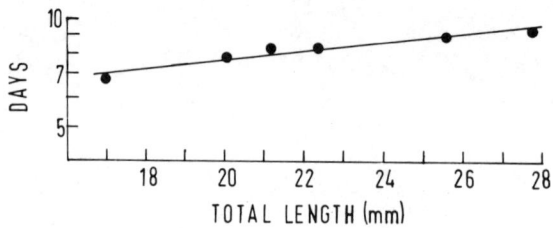

Fig. 5. *Meganyctiphanes norvegica*. Intermoult period related to total body length. Data derived from *Fowler et al.* (1971a, 1971b).

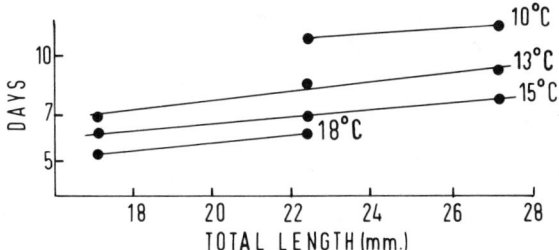

Fig. 6. *Meganyctiphanes norvegica*. Intermoult period at different temperatures related to total body length. Data derived from Fowler et al. (1971a, 1971b).

periods of animals at different temperatures are shown related to body length in Figure 6. Regression equations and their correlation coefficients are given in Table 1.

Fowler et al. (1971b) examined the effects of temperature on the duration of the intermoult period when the size of the experimental animals was kept approximately constant. They related the duration of the intermoult periods of *M. norvegica* and *Euphausia krohnii* linearly against temperature. Data from other crustaceans (Kurata, 1962) indicate that the logarithm of the intermoult period decreases with increasing logarithm of the temperature. (Figure 7).

Jerde and Lasker (1966) found that the intermoult periods of *Euphausia eximia* are 4 - 6 days, of *E. pacifica* and *Thysanoessa spinifera* 5 - 6 days, and of *Nyctiphanes simplex* 5 - 7 days. Temperature was not rigidly controlled in the experiments. Lasker (1966) maintained *E. pacifica* in a temperature regime that fluctuated between 6° and 17°C. Raising the temperature above 12°C did not further de-

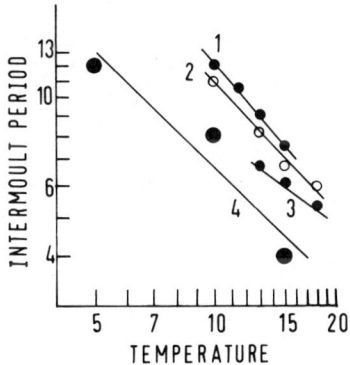

Fig. 7. Intermoult period related to temperature *Meganyctiphanes norvegica*: 1, body dry weight 39.6 mg; 2, body dry weight 21 mg; 3, body dry weight 8.6 mg. *Euphausia krohnii*: 4, large solid circles.

crease the duration of the intermoult period but temperature alterations between 7° and 12°C were quickly reflected by changes in the duration of the intermoult period. The periods, on average, ranged between 5 and 7 days for animals below 10 mm in total length.

Paranjape (1967) found that, at temperatures between 11° and 15°C, *E. pacifica* has an intermoult period of 4 - 6 days, *Thysanoessa spinifera* a period of about 5 days, and *T. raschii* 6 days. The large Antarctic species, *Euphausia superba* was found to have an intermoult period of about 14 days by Mackintosh (1967) who examined animals in the size range 20-30 mm body length.

Komaki (1966), Lasker (1966), and Fowler et al. (1971b) have shown that moulting of euphausiids takes place predominantly at night. Thus intermoult periods measured experimentally tend to conform to periods of whole days rather than fractions of days.

GROWTH FACTORS

Allometric growth of the abdomen takes place in the calyptopes; these three larval stages follow the metanauplius in the molt sequence. The third calytopis has body proportions that are essentially adult and little additional allometric growth takes place (Sheard, 1953). The growth factor for the moult CIII-FI (third calyptopis to first furcilia) is 1.26 in the species examined in the genera *Thysanopoda, Meganyctiphanes, Nyctiphanes, Pseudeuphausia, Euphausia,* and *Thysanoessa* with a few exceptions. The three species *Euphausia triacantha, E. vallentini,* and *E. frigida* have factors that range from 1.40 - 1.46 while *Thysanoessa inermis* has a factor of 1.17 (Mauchline, unpublished). Species in the genera *Nematoscelis* and *Stylocheiron* have factors ranging between 1.10 and 1.26.

The three *Euphausia* species along with *E. superba* and *E. crystallorophias* are those that live at highest latitudes in the Antarctic. The only other euphausiids at these latitudes in the Antarctic are *Thysanoessa vicina* and *T. macrura*. The larvae of *E. superba* have a growth factor of 1.26 at the CIII-FI moult. The larvae of the *Thysanoessa* species and *E. crystallorophias* have not been adequately described to assess the growth factors. These Antarctic species live in an environment with a maximum upper range of temperature of 2° - 8°C. It raises the possibility of low temperature regimes generating high growth factors in larvae that tend to be initially larger than comparable larvae produced at higher environmental temperatures.

Fowler et al. (1971a) determined length increments at moulting of *Meganyctiphanes norvegica* of 12 to 17.5 mm total length. Growth factors ranged from 1.053 to 1.260 (Figure 8).

ASSESSMENT OF GROWTH AND MOULTING SEQUENCES

The duration of the intermoult period increases logarithmically at successive moults. Moult slope factors varying from 1.0134 in juvenile and adult *Nyctiphanes couchii* to 1.0664 in furciliae of the

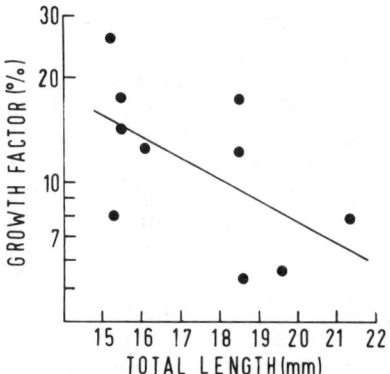

Fig. 8. *Meganyctiphanes norvegica*. Growth factors related to total body length. Data derived from *Fowler et al.* (1971a). The regression equation is: $\log y = 2.1051 - 0.0608x$; $r = -0.5722$.

same species were calculated in Table 1. The mid-point of this range is 1.04. An intermoult period line in which the period increases by a factor of 1.04 at each successive moult is drawn in Figure 9.

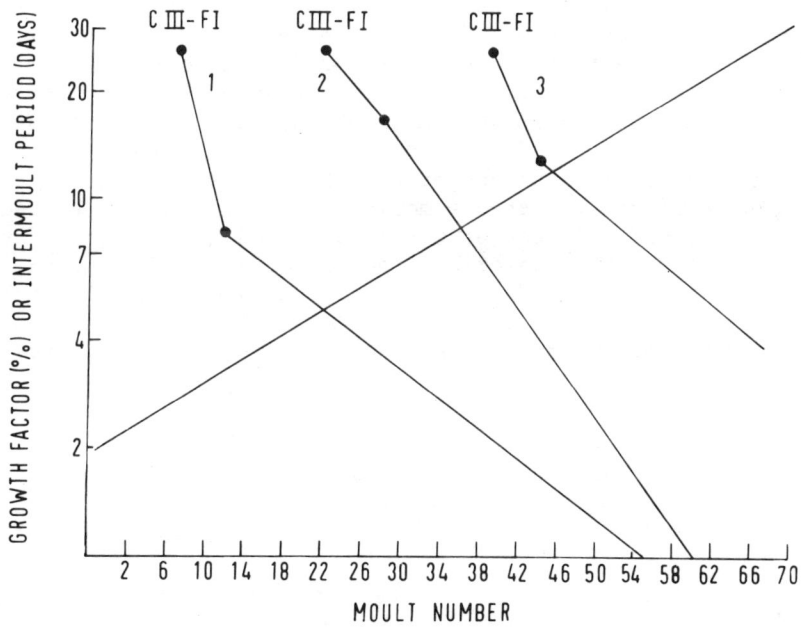

Fig. 9. Growth factor lines and intermoult Period lines for 1, *Euphausia pacifica*; 2, *Meganyctiphases norvegica*; and 3, *E. superba*.

Euphausia pacifica

Smiles and Pearcy (1971) describe the growth of E. pacifica in the coastal waters of Oregon. Their estimates are based on an analysis of a series of samples collected over a period of five years. Breeding takes place throughout the year, but is most intense in the autumn. Smiles and Pearcy, quoting Ponomareva and Brinton, accept that furciliae develop 16 to 18 days after the eggs are laid. This means that the first seven moults are passed through in about 17 days. Moults 1 - 7 described by the intermoult period line in Figure 9 have a cumulative period of 17 days. The calculated intermoult periods for these moults are given in Table 2. The larval development proceeds through two successive nauplii, a metanauplius, three calytopes, followed by six furciliae. The development is complete by moult 13 when the sixth furcilia moults to the first post-larva or juvenile. The following twelve moults, moults 13 - 25, have intermoult periods ranging from 3.4 - 5.5 days. These periods agree with those estimated by Lasker (1966) and Paranjape (1967) for juveniles up to a length of about 12 mm. Consequently, the slope of the intermoult period line in Figure 9, is approximately correct for this species.

Successive growth factors of the furciliae decrease logarithmically (Mauchline, unpublished). A moult slope factor of 0.79 decreases the growth factor at the CIII-FI moult of 1.260 to a factor of 1.080 at the last furcilia moult, moult 12 (Table 3, Figure 9). The moult slope factor then changes. The further application of the factor 0.79 would produce a series of growth factors that would not allow the animal to grow to the observed size at sexual maturity. Kurata (1962) has documented the evidence showing that the trend of the growth factors can change at the end of the larval development and produce an inflexion point on a Hiatt growth diagram. Various moult slope factors were fitted from moult 13 and a factor of 0.953 was found to provide a series of growth factors that described the animal's growth pattern better than any other. Growth factors, derived with this moult slope factor at successive moults are shown in Table 2 and Figure 9, along with the intermoult periods of the successive moults.

TABLE 2. *Euphausia pacifica*. Calculated Intermoult Periods and Growth Factors and Consequent Body Lengths and Dates.

Moult No.		Intermoult period (days)	Growth factor	Total length (mm)	Date
EGG					13.IX
1	NI	2.14			15.IX
2	NII	2.22			17.IX
3	Mn	2.31			20.IX
4	CI	2.40			22.IX
5	CII	2.50			25.IX
6	CIII	2.60		2.50	27.IX
7	FI	2.70	1.260	3.15	30.IX
8	FII	2.81	1.205	3.80	3. X
9	FIII	2.92	1.162	4.41	6. X
10	FIV	3.04	1.128	4.98	9. X
11	FV	3.16	1.101	5.48	12. X
12	FVI	3.29	1.080	5.92	15. X
13		3.42	1.076	6.37	19. X
14		3.56	1.073	6.83	22. X
15		3.70	1.069	7.31	26. X
16		3.85	1.066	7.79	30. X
17		4.00	1.063	8.28	3.XI
18		4.16	1.060	8.78	7.XI
19		4.33	1.057	9.28	11.XI
20		4.50	1.054	9.78	16.XI
21		4.68	1.052	10.28	20.XI
22		4.87	1.049	10.79	25.XI
23		5.06	1.047	11.30	30.XI
24		5.26	1.045	11.81	5.XII
25		5.47	1.043	12.31	11.XII

Table 2 (cont'd).

Moult No.	Intermoult Period (days)	Growth factor	Total length (mm)	Date
26	5.69	1.041	12.82	16.XII
27	5.92	1.039	13.32	23.XII
28	6.16	1.037	13.81	29.XII
29	6.40	1.035	14.29	4.I
30	6.66	1.034	14.78	11.I
31	6.93	1.032	15.25	18.I
32	7.20	1.031	15.73	25.I
33	7.49	1.029	16.18	1.II
34	7.79	1.028	16.63	9.II
35	8.10	1.026	17.07	17.II
36	8.43	1.025	17.49	26.II
37	8.76	1.024	17.91	6.III
38	9.12	1.023	18.33	16.III
39	9.48	1.022	18.73	25.III
40	9.86	1.021	19.12	4.IV
41	10.25	1.020	19.50	14.IV
42	10.66	1.019	19.88	25.IV
43	11.09	1.018	20.23	6.V
44	11.53	1.017	20.58	17.V
45	11.99	1.016	20.91	29.V
46	12.47	1.016	21.24	11.VI
47	12.97	1.015	21.56	24.VI
48	13.49	1.014	21.86	7.VII
49	14.03	1.013	22.15	21.VII
50	14.59	1.013	22.43	5.VIII

TABLE 3. *Meganyctiphanes norvegica*. Calculated Intermoult Periods and Growth Factors and Consequent Body Lengths and Dates.

Moult No.		Intermoult period (days)	Growth factor	Total length (mm)	Date
Egg					1.IV
16	NI	3.85			4.IV
17	NII	4.00			8.IV
18	Mn	4.16			12.IV
19	CI	4.33			16.IV
20	CII	4.50			21.IV
21	CIII	4.68		2.40	26.IV
22	FI	4.87	1.260	3.02	30.IV
23	FII	5.06	1.242	3.76	5.V
24	FIII	5.26	1.225	4.60	11..V
25	FIV	5.47	1.209	5.56	16.V
26	FV	5.69	1.194	6.64	22.V
27	FVI	5.92	1.181	7.84	28.V
28	FVII	6.16	1.168	9.16	3. VI
29		6.40	1.154	10.57	9.VI
30		6.66	1.141	12.06	16.VI
31		6.93	1.129	13.62	23.VI
32		7.20	1.118	15.22	30.VI
33		7.49	1.108	16.87	8.VII
34		7.79	1.099	18.54	15.VII
35		8.10	1.090	20.21	24.VII
36		8.43	1.083	21.88	1.VIII
		NO GROWTH IN WINTER			
37		8.76	1.076	23.55	9.III
38		9.12	1.069	25.34	18.III
39		9.48	1.063	26.93	27.III
40		9.86	1.058	28.49	6.IV
41		10.25	1.053	30.00	16.IV
42		10.66	1.048	31.44	27.IV

Table 3 (cont'd).

Moult No.	Intermoult Period (days)	Growth factor	Total length (mm)	Date
43	11.09	1.044	32.83	8.V
44	11.53	1.041	34.17	20.V
45	11.99	1.037	35.44	1.VI
NO GROWTH IN WINTER				
46	12.47	1.034	36.64	12.IV
47	12.97	1.031	37.78	25.IV
48	13.49	1.028	38.84	9.V
49	14.03	1.026	39.85	23.V
50	14.59	1.024	40.80	7.VI
51	15.18	1.022	41.70	22.VI
52	15.78	1.020	42.54	8.VII
53	16.42	1.018	43.30	24.VII
54	17.07	1.017	44.04	10.VIII

There are no notable seasonal changes in the environmental water temperatures in the coastal region of Oregon (Smiles and Pearcy, 1971). Active growth of this species therefore takes place throughout the year. Assuming that the egg is laid on 13 September, the dates and body lengths of subsequent moults can be calculated (Table 2). These are compared with the growth curve estimated by Smiles and Pearcy in Figure 10. The two curves correspond closely.

Meganyctiphanes norvegica

Attempts in the laboratory to determine the intermoult period of the larvae and young adolescents of M. norvegica produced limited results because of an inability to maintain the animals in healthy condition. The laboratory experiments were complimented by twice weekly sampling of the larvae in Loch Fyne, Scotland, during April and May.

The furciliae moulted in the laboratory once every 4 - 7 days, while adolescents of 10 - 12 mm total length had an intermoult period of 5 - 8 days. The eggs were first laid in Loch Fyne at the beginning of April and the first adolescents were found in the first week of June. These adolescents, during their development from the egg, had moulted 14 times in approximately 70 days. The same intermoult period line (Figure 9) is used to estimate the durations of

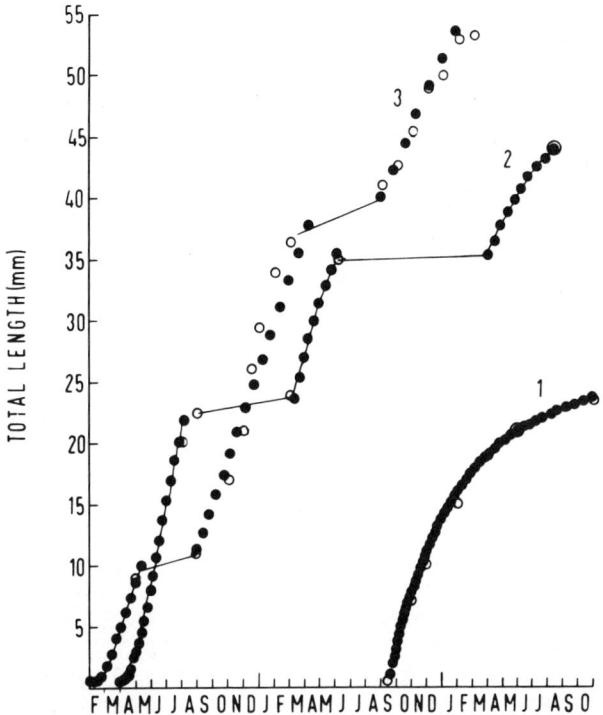

Fig. 10. Growth curves of 1, *Euphausia pacifica*; 2, *Meganyctiphanes norvegica*; and 3, *E. superba*. Solid circles, values calculated in Tables 3-5; open circles: 1, values from *Smiles and Pearcy*, 1971; 2, values derived from *Mauchline*, 1960; and 3, values from *Mackintosh*, 1972.

successive intermoult periods in this species. Moults 16 - 29 have a cumulative period of 70 days and these are used to define the intermoult periods of the larvae (Table 3, Figure 9).

The growth factor at the CIII-FI moult is 1.260, the same as that in *E. pacifica*. A moult slope factor of 0.93 decreases this growth factor of 1.26 at moult 22 to a factor of 1.168 at moult 28 (Table 3, Figure 9). On completion of the larval development, the growth factors at successive moults appear to decrease at an accelerated rate (Figure 9). A moult slope factor of 0.915 produced a series of growth factors that describe the growth of this species better than any other (Table 3). The growth factors agree with those determined experimentally by *Fowler et al.* (1971a) for animals in the size range 15 - 22 mm (Figure 8). The durations of the intermoult periods of moults 33 - 40 (Table 3) agree with those determined experimentally by *Fowler et al.* (1971b) for animals in the size range 17 - 28 mm (Figure 5).

Meganyctiphanes norvegica in Loch Fyne is subjected to seasonal changes in environmental temperature. Moulting appears to be inhibited during the autumn and winter because little growth in length of these animals was found through statistical analyses of the population (Mauchline, 1960). The eggs are laid at the beginning of April, develop to juveniles by June, and active growth in body length ceases in August. Growth in the second year takes place in the period March to June while growth during the third year occurs in April to August. The breeding season extends through April and May, occasionally into June. The lengths at and dates of successive moults are calculated (Table 3) and compared with the growth curve estimated by statistical analysis of the wild population in Figure 10. The two curves correspond closely.

Euphausia superba

Mackintosh (1972) has produced a detailed study of the rate of growth of the Antarctic E. superba. The most detailed data are presented for the populations in the North Weddell Drift. The water temperature in the Antarctic during the early development of this animal is 2° - 3°C. Comparable temperatures in Loch Fyne during the breeding season of M. norvegica are 6° - 8°C. The slopes of the lines in Figure 7 relating intermoult period to environmental temperatures suggest that the duration of the intermoult periods at Antarctic temperatures should be about twice those at the temperatures in Loch Fyne. Mackintosh suggests that the egg, and, in consequence, the succeeding larval stages, of E. superba may take twice as long to develop as those of M. norvegica. This is assumed here and the moult CIII-FI is identified as moult 39 in Figure 9.

Fraser (1936) measured the body lengths of successive larval stages of E. superba. The growth factor of the CIII-FI moult is 1.26 and decreases to 0.87 of its value at each successive moult of the furciliae (Table 4). The development of the six furciliae is not completed in this species before the onset of the Antarctic winter. Furciliae V and VI are usually passed through during the winter months, May to August (Mackintosh, 1972). Mackintosh shows that furciliae in the North Weddell Drift attain a length of approximately 8.5 mm by the end of April, spawning having taken place on 31 January. Moults 33 - 42 have intermoult periods (Figure 9) that conform to this time scale (Table 5). Furciliae V and VI develop within the period May to August. Mackintosh (1967) found that the intermoult periods of E. superba of body length 20 - 30 mm ranged from 12 - 17 days. The calculated intermoult periods of moults 50 - 55 in Table 4 correspond with his observations.

The larval development is completed by moult 44. A moult slope factor of 0.95 produces a series of growth factors describing the growth of this species from moult 45 onwards (Table 4, Figure 9). Mackintosh suggests that spawning of the North Weddell Drift population takes place at the end of January. Using January 31 as his

TABLE 4. *Euphausia superba*. Calculated Intermoult Periods and Growth Factors and Consequent Body Lengths and Dates.

Moult No.		Intermoult period (days)	Growth factor	Total length (mm)	Date
Egg					31.I
33	NI	7.49			7.II
34	NII	7.79			15.II
35	Mn	8.10			23.II
36	CI	8.43			4.III
37	CII	8.76			13.III
38	CIII	9.12		4.00	22.III
39	FI	9.48	1.260	5.04	31.III
40	FII	9.86	1.226	6.18	10.IV
41	FIII	10.25	1.197	7.40	20.IV
42	FIV	10.66	1.171	8.66	1.V
43	FV	11.09	1.149	9.95	11.VI
		NO GROWTH IN WINTER			
44	FVI	11.53	1.130	11.25	1.IX
45		11.99	1.124	12.65	12.IX
46		12.47	1.117	14.12	24.IX
47		12.97	1.111	15.69	7.X
48		13.49	1.106	17.36	21.X
49		14.03	1.101	19.11	4.XI
50		14.59	1.096	20.94	19.XI
51		15.18	1.091	22.85	4.XII
52		15.78	1.086	24.81	20.XII
53		16.42	1.082	26.85	5.I
54		17.07	1.078	28.94	22.I
55		17.76	1.074	31.08	9.II
56		18.47	1.070	33.26	27.II
57		19.20	1.067	35.49	18.III
58		19.97	1.063	37.72	7.IV

Table 4 (cont'd).

Moult No.	Intermoult period (days)	Growth factor	Total length (mm)	Date
		NO GROWTH IN WINTER		
59	20.77	1.060	39.99	1.IX
60	21.60	1.057	42.27	22.IX
61	22.47	1.054	44.55	14.X
62	23.36	1.052	46.87	6.XI
63	24.30	1.049	49.16	1.XII
64	25.27	1.047	51.47	26.XII
65	26.28	1.044	53.73	21.I

origin, he constructs a growth curve for *E. superba* which is compared with the one calculated here in Figure 10. The two curves correspond closely.

PRODUCTION OF CASTES

Lasker (1966) showed that castes of *Euphausia pacifica* potentially under still hydrographic conditions could sink at a rate of 300 m per day. Moulting takes place at night and will occur predominantly in the surface layers because many euphausiid species approach the surface at night in a diurnal vertical migration. He suggests, on experimental evidence, that *E. pacifica* may produce a dry weight of castes equal to its own initial body dry weight every 50 days.

Smiles and Pearcy (1971) quote numbers of *E. pacifica* per 1000 m^3 of water seasonally. They describe the general pattern of population ecology of this species; their raw data for October and December 1965, February and April 1967, and June 1966 seem most representative of the populations of these months of the year. Body dry weights for each size class were calculated and total dry weight, or biomass, of *E. pacifica* per 1000 m^3 water for each month determined (Figure 11). The numbers of *E. pacifica* equivalent to the biomass dry weight

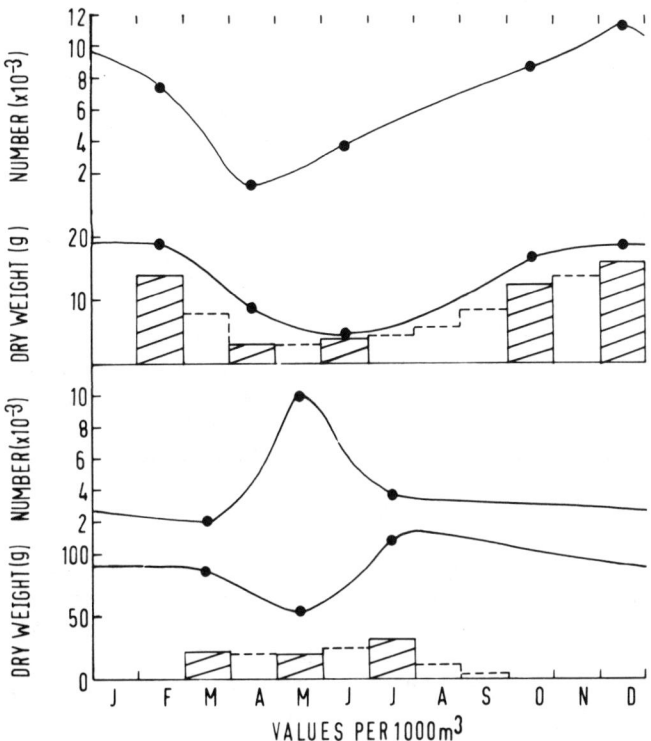

Fig. 11. *Euphausia pacifica*, upper part; *Meganyctiphanes norvegica*, lower part. In each part, the upper line shows changing numbers, in thousands per 1000 m^3 of water, of the species The center line hows changing dry biomass per 1000 m^3 of water. The lower histogram shows dry weight of castes produced in months specified by solid lines; interpolated weights per month are indicated by hatched lines.

are also shown. The numbers of castes produced per month by each size class was estimated from Table 3. The dry weight of a caste is approximately 10 percent of the body dry weight of the euphausiid (*Lasker*, 1964). The total weight of castes produced by each class in the month was determined and so the total weight produced by the population in 1000 m^3 of water (Figure 11). The total monthly weight of castes produced is closely related to the dry weight of the population, approximating to 75 percent of the dry biomass.

A similar analysis of the population of *Meganyctiphanes norvegica* in Loch Fyne was made (Figure 11). Moulting is restricted to the summer period, as is breeding. Seasonal increase in the population in April and May is the result of breeding and the consequent addition of larvae. The simultaneous decrease in the biomass is a result of mortality of adults at breeding. Growth and moulting increases the biomass to a peak in the early autumn. The weight of castes produced in the spring and summer remains relatively constant through the season at a monthly level approximating to 25 to 50 percent of the dry biomass (Figure 11).

DISCUSSION

The basic information required for this analysis of the growth and moulting sequences of crustaceans is:
1. The time in days required for a known number of successive or individual moults. This can be estimated in the laboratory or from sampling of natural populations, especially larvae, and early post-larvae.
2. An accurate estimate of the increments at moulting for a few successive or individual moults of post larval stages or an accurate estimate of the increase in size over a known number of moults.
3. An approximate growth curve for the animal throughout its life, derived from seasonal sampling of natural populations.

The growth curves of crustaceans all approach asymptotic sizes and this fact can be used to provide empirical estimates of growth and moulting sequences of the different species. The late larval and early post larval stages of many crustaceans can often be identified in samples. Estimates of the growth factors at these moults can be made and so determine the position of the upper end of the growth factor line of the juveniles and adults. The maximum or asymptotic size of the adults can also be estimated from the samples. A growth factor line can then be fitted empirically starting with the growth factors of the late larvae or early post larval stages. Different moult slope factors are applied until a range of them is found such that the derived asymptotic value of body length approximates that estimated from the samples. The range in the moult slope factors describing the growth from the young juvenile of known length to the maximum known size of the adult is limited; it frequently requires the moult slope factor to be defined to three decimal places.

An empirically fitted growth factor line provides an estimate of the number of moults required for the young animal to achieve its maximum size. Estimates of the duration of intermoult periods during any part or parts of the animal's life allows placement of an intermoult period line relative to the growth factor line on the graph. This then gives an estimate of the minimum duration of the life cycle. The next problem is to decide whether growth throughout the life cycle is continuous or whether there are seasonal or other periods when no increase in body length takes place. Discontinuous growth can be caused by inhibition of moulting during winter or through development of ovaries and carriage of eggs in female decapods. Estimates can be made of the duration of annual periods of active growth in body length. Annual sequences of growth and moulting, such as those in Tables 2 - 4 and Figure 9, can be estimated and compared with information on the species derived from other sources.

Small low latitude species grow continuously throughout the year. The populations contain individuals at all stages of development. There tend, however, to be seasonal periods of more intensive breeding, frequently giving rise to "generations" such as occur most clearly in populations of mysids (Mauchline, 1971). There must then be periods of the year when deaths of mature animals, after breeding, occur more commonly. Smiles and Pearcy (1971) suggest that this happens in E. pacifica off Oregon in the summer because large individuals have disappeared from their samples by the winter. This is also indicated by the analysis of their data in Figure 11; the numbers per 1000 m^3 begin increasing in June but the biomass is less in June than it was in April.

This is even more striking in the analysis of the population of M. norvegica (Figure 11). The adults, after breeding in March to April, suffer a consequent mortality (Mauchline, 1960). This is reflected by a decreased biomass in May. The large addition of young from the breeding period is reflected by the increased numbers per 1000 m^3. These young are small relative to the adults and, although growing and moulting at a faster rate than the adults, do not contribute markedly to the production of castes until June. By then their growth is also relfected by an increasing total biomass, even though mortality of these young through predation and other causes, has taken place throughout May and June. The population has two or three modal size ranges (year groups), the number being dependent upon the time of sampling relative to the breeding season. The discontinuous pattern of growth of the animals usually results in the modes being distinctly separated.

The populations of species such as E. superba in high latitudes show the same patterns of production as those of M. norvegica, except that the animals require two years to attain sexual maturity. The population has year groups that can be separated and which have different rates of growth and production of castes.

REFERENCES

Fowler, S. W., G. Benayoun, and L. F. Small. 1971a. Experimental studies on feeding, growth, and assimilation in a Mediterranean euphausiid. Thalassia Jugoslavica 7: 35-47.

Fowler, S. W., L. F. Small, and S. Keckes. 1971b. Effects of temperature and size on moulting of euphausiid crustaceans. Mar. Biol. 11: 45-51.

Fraser, F. C. 1936. On the development and distribution of the young stages of krill (Euphausia superba). 'Discovery' Rep. 14: 1-192.

Gopalakrishnan, K. 1973. Development and growth studies of the euphausiid Nematoscelis difficilis (Crustacea) based on rearing. Bull. Scripps. Inst. Oceanogr. 20: 1-87.

Hiatt, R. W. 1948. The biology of the lined shore crab, *Pachygrapsus crassipes*. *Randall. Pacific. Sci.* 2: 135-213.

Jerde, C. W., and R. Lasker. 1966. Moulting of euphausiid shrimps: shipboard observations. *Limnol Oceanogr.* 11: 120-124.

Komaki, Y. 1966. Technical notes on keeping euphausiids live in the laboratory, with a review of experimental studies on euphausiids. *Inform. Bull. Planktol. Japan* 13: 95-105.

Kurata, H. 1962. Studies on the age and growth of Crustacea. *Bull. Hokkaido Reg. Fish. Res. Lab.* 24: 1-115.

Lasker, R. 1964. Moulting frequency of a deep-sea crustacean, *Euphausia pacifica*. *Nature, Lond.* 203: 96.

Lasker, R. 1966. Feeding, growth, respiration, and carbon utilization of a euphausiid crustacean. *J. Fish Res. Bd. Can.* 23: 1291-1317.

Le Roux, A. 1973. Observations sur le développement larvaire de *Nyctiphanes couchii* (Crustacea: Euphausiacea) au laboratoire. *Mar. Biol.* 22: 159-166.

Mackintosh, N. A. 1967. Maintenance of living *Euphausia superba* and frequency of moults. *Norsk. Hvalfangst.-Tid.*, 5: 97-102.

Mackintosh, N. A. 1972. Life cycle of Antarctic krill in relation to ice and water conditions. *'Discovery' Rep.* 36: 1-94.

Mauchline, J. 1960. The biology of the euphausiid crustacean *Meganyctiphanes norvegica* (M. Sars.). *Proc. R. Soc. Edinb.*, B. (*Biol.*) 67: 141-179.

Mauchline, J. 1971. Seasonal occurrence of mysids (Crustacea) and evidence of social behavior. *J. Mar. Biol. Ass. U.K.* 51: 809-825.

Mauchline, J. 1976a. The Hiatt growth diagram for Crustacea. *Mar. Biol* in press.

Mauchline, J. 1976b. Growth of shrimps, crabs, and lobsters--an assessment. *J. Cons. Perm. Int. Explor. Mer.*: in press.

Mauchline, J., and L. R. Fisher. 1969. The biology of euphausiids. *Adv. Mar. Biol.* 7: 1-454.

Paranjape, M. A. 1967. Moulting and respiration of euphausiids. *J. Fish. Res. Bd. Can.* 24: 1229-1240.

Sheard, K. 1953. Taxonomy, distribution, and development of the Euphausiacea (Crustacea). *Rep. B.A.N.Z. Antarct. Res. Exped., Ser. B, (Zoology and Botany)* 8: 1-72.

Smiles, M. C. and W. G. Pearcy. 1971. Size structure and growth rate of *Euphausia pacifica* off the Oregon coast. *Fish. Bull, U.S.* 69: 79-86.

Zweifel, J. and R. Lasker. 1976. Pre- and post-hatch growth of fishes--a general model. *Fish. Bull., U.S.*: in press.

What Regulates Pelagic Community Structure in the Pacific?

John A. McGowan

Scripps Institution of Oceanography

ABSTRACT

Ecologists, biological oceanographers, and biogeographers wish to understand the factors that serve to influence the temporal and areal variations in abundance of all "important" functional groups and/or species within food webs. Works pertaining to this are referred to variously as community or ecosystem studies. However, the theoretical background to such studies usually includes the assumption of an essentially closed system. Apart from the input of solar energy and some losses due to sedimentation, the systems are thought of as being self regulating. There are large areas of the Pacific Ocean where it seems very likely that such variations in abundance are strongly influenced by large-scale horizontal advection of allochthonous nutrients and populations. In these areas the basic assumption of in situ processes predominating the regulation is not met. However, other areas appear to essentially self regulate and it is in these that studies of processes and events may lead us more quickly to an understanding of the nature of the regulation and enhance the development of predictive capability. One of the reasons for this is the great lateral homogeneity of these systems. Thus the results of sampling and measurement at limited locales are representative of much larger areas.

Biogeographic evidence helps us decide which are likely to be the in situ systems and which are the advective systems. However, there are additional tests that may be applied. The California Current (an advective system) and the North and South Pacific central gyres (in situ systems) are examples where these tests have been used.

INTRODUCTION

Ecosystems are generally conceived as consisting of a community of organisms plus the physical habitat of that community. The term ecosystem implies that aspects of their structure, function, and dynamics are in some sense systematic. Although the history of the use of the terms community and ecosystem is complex it is useful to attempt to summarize the concepts involved.

Community ecosystems contain food webs but not all food webs are ecosystems. For example: a hobbyist's aquarium containing plants indigenous to southern USA, Europe, and Africa; herbivores from northern USA, Central America, and Japan; carnivores from Borneo, the Amazon, and Thailand; and degraders endemic to the tap water of San Diego county, would contain species at all trophic levels which would form a food web and might eventually form some sort of quasi-stable equilibrium for a considerable period of time, especially with the occasional addition of an allochthonous nutrient source. The dynamics of the "community" can be studied and no doubt interesting results might be obtained. However, it is clearly a very unnatural assemblage and may be contrasted with a small, old, tropical lake where the components of the food web have undergone a period of evolutionary co-adaptation representing thousands of generations and where predator-prey and competitive interactions have been strong selective forces. In this latter case we might expect that the constituent species have evolved aspects of their physiology, behavior, and population dynamics to maximize fitness within the context of the limited spectrum of physical and biological variables of the habitat, and that these organisms might be far less fit to survive in a situation where the habitat variables were very different or were constantly changing. This is the sort of situation that *Hutchinson* (1967) was referring to when he said (p 232) that a lake may be ordinarily considered an ecosystem, in a single biotope (habitat) supporting a single biocoenosis (community). It is what *Holling* (1973) refers to as "self-contained" ecosystems. It is clear that the food web statics and dynamics in the case of the evolutionarily accommodated, old lake are likely to be quantitatively and qualitatively different than the hobbyist's aquarium.

Although the analogy is somewhat strained there are similarities in the oceanic Pacific.

BIOGEOGRAPHIC EVIDENCE AND RESULTANT HYPOTHESES

Large-scale sampling of the epipelagic zooplankton of the Pacific has revealed that a pattern of faunal provinces exist (*Johnson and Brinton*, 1963; *Beklemishev*, 1969; *Bogorov*, 1967; *McGowan*, 1971, 1974). A subsample of the data was used by *Fager and McGowan* (1963) to show that there are groupings of species that form a nearly constant part of each others' environment. The locations of these groups were consistent with patterns of the major provinces (Figure 1). An examination of the individual patterns of species within groups and provinces shows a low variability of relative abundances within the central portions of their ranges, thus indicating a great lateral homogeneity of what appears to be the carrying capacity of the habitat. The provinces may in some senses be considered habitats because of the lateral coherence of physical-chemical properties as exemplified by the close knit family of T-S curves, and because of the regularity of: nutrient concentration, standing crops of phytoplankton, zooplankton and of primary productivity (within seasons) (*Sverdrup et al.*, 1942; *Reid*, 1965; *NorPac Committee*, 1960; *Koblentz-Mishke et al.*, 1970). Most of the groups showed highly significant levels of large-scale, areal concordance of abundance indicating that they were reacting in a similar way to such habitat variations that occur.

All but one of the spatial patterns formed by the faunal provinces and groups occur in areas that may be characterized by either gyre-like or zonal, circulation-recirculation systems. Indeed these two mechanisms are in part responsible for the conservation of properties of the habitats and lead to the great lateral monotony of physical-chemical-biological variables. Each of these systems has its own climate, seasonality, and characteristic density structure. Primary productivity is regulated in each by slightly different processes. All of the systems are geologically old (*McGowan*, 1974).

Thus they are the oceanic analogues of *Hutchinson's* lakes except that they are very large and very old and apparently very stable (*Riedel and Funnell*, 1964). They also have the necessary prerequisites for climax ecosystems. However, these systems do not have closed boundaries as in lakes. The "boundaries" are there but they exist as rather broad areas of physical water mass and faunal mixing; that is, ecotones. These ecotones are the equivalent of the hobbyist's aquarium where species which evolved in radically different habitats with very different selection pressures, are mixed together; allochthonous flora, fauna, nutrients, and biomass are constantly being added through large-scale lateral mixing processes. Thus, while populations are responding to physical processes and interactions are occurring in terms of competition and predation in the ecotones, the results are seriously aliased by large-scale horizontal advection. Because of the much weaker rate of lateral transfer of allochthonous flora, fauna, nutrients, and biomass into the centers of the gyre-like and zonal systems, the interactions and feedback loops between the finely tuned (in terms of evolutionary fitness) components are the primary regulators of the state of the system. In other words the faunal provinces are *in situ* regulated ecosystems while the ecotones are regulated by both *in situ* processes and large-scale and variable horizontal advection.

HYPOTHESES

If the above summary is substantially correct, there are certain expectations about the structure of these two very different sorts of

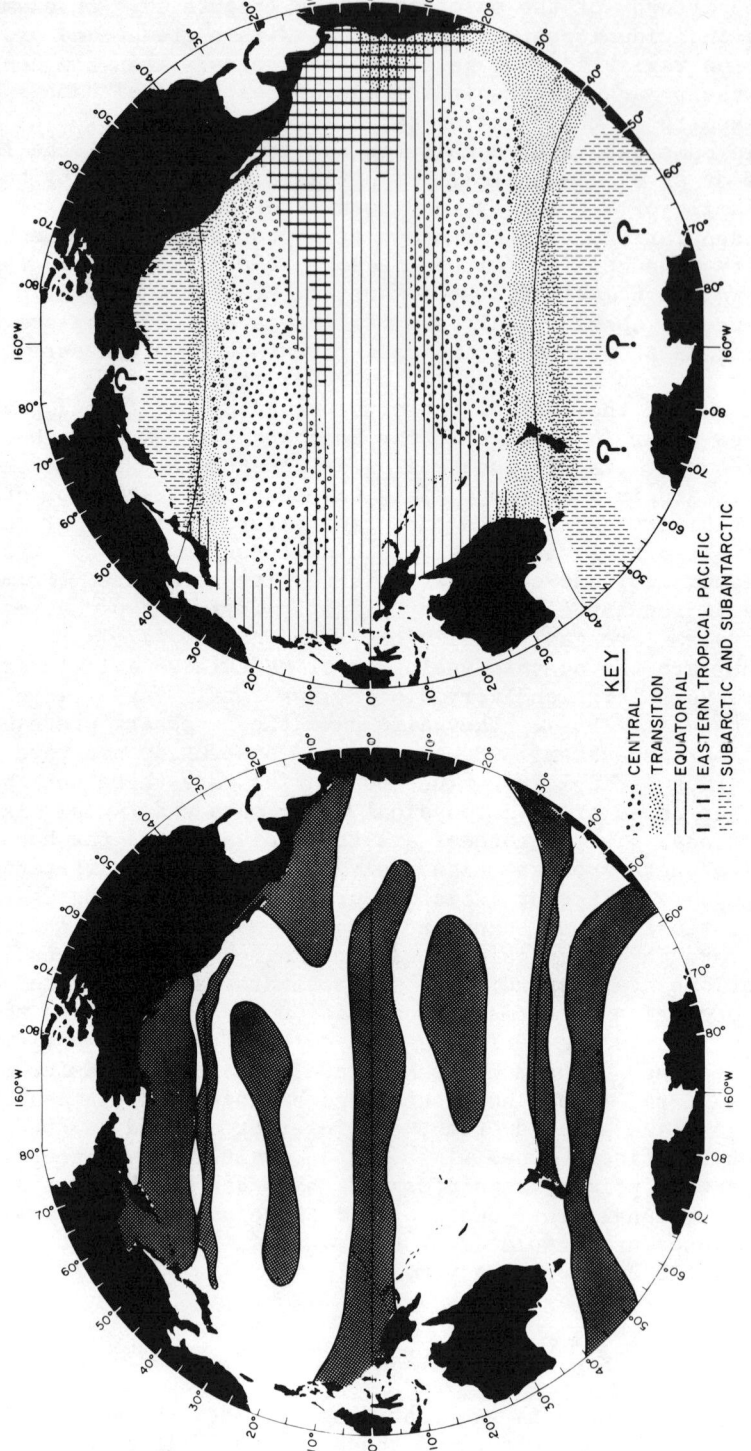

Fig. 1. Major faunal provinces of the oceanic Pacific. Left: the cores of the faunal provinces (100% of the species forming the faunal group are present). Right: extended habitats (≥60% of the species of a faunal group are present). Question marks indicate lack of adequate, quantitative sampling.

systems. The heterogeneity of biomass within the highly tuned gyre
systems should be low because everywhere the processes are similar.
However, in the ecotones, one might expect great heterogeneity because
mixing plus interaction can get out of phase, especially if the rate
of mixing is variable and the organisms being mixed have varying fitness depending on the unpredictability of the habitat. Thus in testable hypothesis I, the variability of biomass in a gyre is compared
to that in an ecotone. The ecotone should be more heterogeneous in
both space and time.

A further consequence of evolutionary adaptation in predictable
habitats is that of species equilibrium. *In situ* systems should have
a relatively constant species list and the species should show a relatively high frequency of occurrence in both space and time. This can
be easily measured as testable hypothesis II.

If competitive interaction works the way the theoreticians believe it does, resulting in competitive dominance and degrees of
specialization, two more testable hypotheses are generated. The *in
situ* systems should have relatively stable rank orders of dominance
within functional groups. A set of Spearman rank correlation coefficients of species relative abundances in the two areas is an appropriate test of hypothesis III. A more sensitive and slightly different test could use Whittaker percent similarity indices. A further
consequence of a stable equilibrium in interaction between species
and their environment and among the species is that of the equitability of diversity. That is, does the shape of the diversity curve
tend to vary spatially, more in one sort of habitat than the other?
This hypothesis, IV, is related to but not identical with number III.

Hypothesis V may be tested but with considerably more sampling
effort. In hypothesis V, one might see if closely related species
tend to divide the water column in the vertical so that there is a
tendency for a discrimination against the co-occurrence of congeners
(the one species to a niche hypothesis). Thus elements of small-scale spatial structure should show reproducible pattern at *in-situ*
locales as opposed to the advective ones.

These five testable hypotheses are by no means an exclusive list
of the possible ones. However, they do represent fairly easy, direct
measures that may be rather quickly carried out perhaps even with
existing samples. Examples of other attributes that are certainly of
importance are: the stability of the trophic pyramid and the temporal predictability of ecological "events". The first of these is
concerned with the shape of the trophic pyramids and it means that
on an average, over time, primary productivity should exceed secondary,
secondary should exceed tertiary, and so forth. Further in an evolutionary mature, climax system one might expect a higher efficiency of
transfer between trophic levels than in a system that has had no
chance to develop biological accommodation. The second concept has to
do with the degree to which organisms can afford to specialize to
"anticipate" the occurrence and magnitude of an event such as the
spring bloom. In the subarctic Pacific gyre there is almost no lag
between the spring productivity peak and zooplankton bloom (*Parsons*

and LeBrasseur, 1968). The main grazers have adjusted their life histories and population dynamics such that they can be thought of as anticipating the productivity peak. This must mean that from the organism's viewpoint it is a very predictable event and is the sort of fine tuning that might be expected in old climax ecosystems. Additional attributes of importance are the stability and resiliency (Holling, 1973). That is, how does the system respond to perturbations? For example, does the rank order of species abundance (the species structure) change radically or not in response to an environmental change such as a drop in the mean mixed layer temperature of 2°C? Do some species "crash" completely after such a perturbation?

While these last three questions about ecosystems are of great theoretical and practical significance, they require a large and long-term effort for resolution. Appropriate time series of samples and measurement are clearly necessary. These then probably cannot be put into the form of easily tested hypotheses but rather ones that take considerably more effort and time though they are very worthwhile.

METHODS

The evidence from biogeographic studies strongly implies that the gyres and circulation-recirculation systems are likely candidates for stable, climax in situ ecosystems in the strict sense. Similarly, the areas of mixing appear to be good candidates for ecotones or advective systems. Thus we may compare these areas to one another with regard to aspects of structure and stability. Hypothesis I may be tested by comparing the indices of dispersion of biomass ($S^2/\bar{x}$) within each. Hypothesis II may be looked at by examining the variance in length of species list and frequency within areas. Hypothesis III may be examined by constructing matrices of Spearman rank correlation coefficients of species abundances in each and comparing central tendencies. The variability in equitability of diversity curves may be compared for hypothesis IV. The degree of co-occurrence of congeneric species may be determined with sets of samples taken on a scale relevant to the species populations spatial ambit, for a test of hypothesis V.

The central gyre zooplankton, nekton, and chlorophyll a samples used in these tests all came from an area no larger than 20 by 80 nautical miles centered about the point 28°N 155°W (Figure 1). The California Current zooplankton samples were selected from a very large set. In making this selection an attempt was made to have the samples as comparable as possible in their spatial and temporal arrangement to those of the gyre. Not all of the types of statistical tests were done on the same California current sample sets. For example, the indices of dispersion for zooplankton biomass came from CalCOFI stations 80.70 off Pt. Conception, 100.60 off northern Baja California and 130.60 off southern Baja California (Figure 2). These are meant to represent three widely separated parts of the California

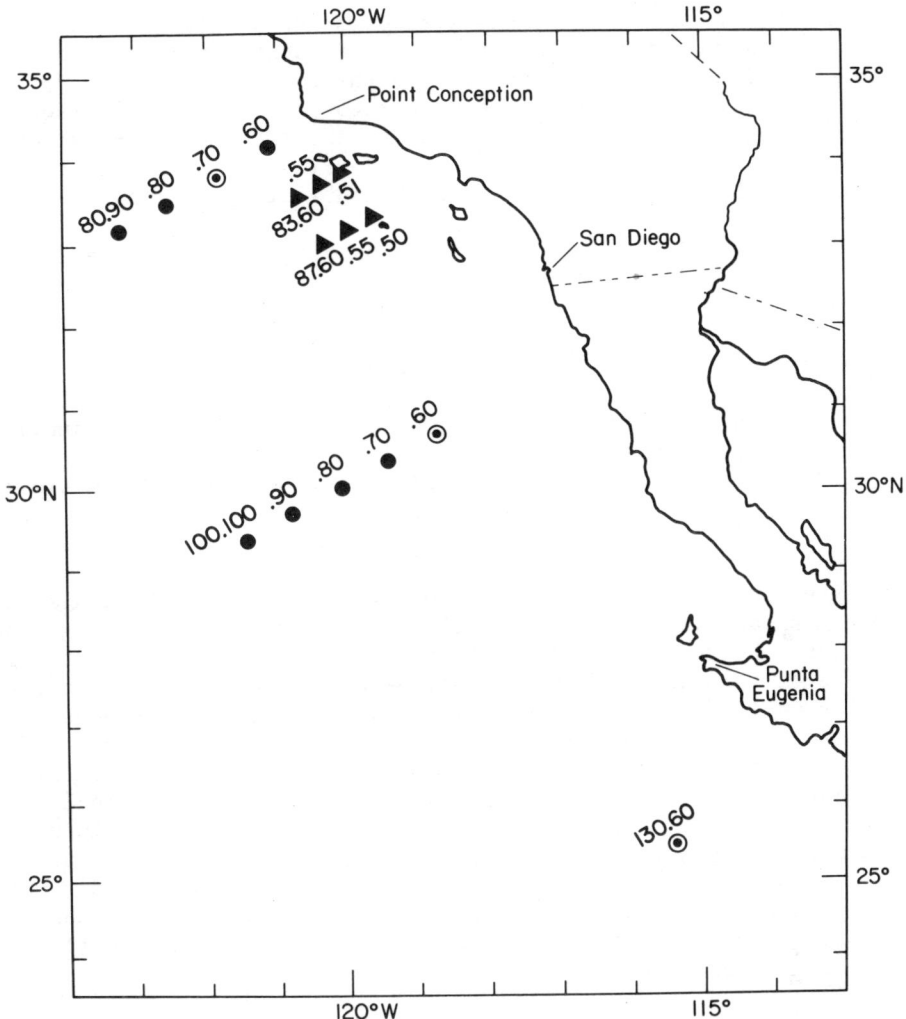

Fig. 2. Sampling locales within the California Current. The solid circles inside of open circles are stations where the long term estimates of variance of zooplankton biomass were made. The solid dots are stations from which samples were taken for species rank difference and PSI calculations. The solid triangles are stations where samples were taken for major category rank difference (Table 2) and PSI calculations (Table 3). See text for fuller explanations.

current. All of these stations are greater than 50 nautical miles offshore, well away from local neretic conditions. All seasons are represented and the tows were taken over a period of about eight years. The gyre samples were also taken seasonally but early winter is not represented and the time span is only four years.

The California current samples for copepod species intercomparisons by Spearman rank correlations and Percent Similarity Indices came from lines of stations normal to the coastline (Figure 2). The stations, within each set, ranged from 45 to 178 nautical miles apart and were done over a period of 2 days each. Thus the spatial and temporal array is similar to but not identical with the Climax II set of samples which ranged from 20 to 80 nautical miles apart over a ten day period.

The rank order comparisons of major planktonic taxa were based on a set of samples from the California current selected to be as similar as possible in spatial and temporal array to those of the gyre (triangles in Figure 2). Only night samples were used and only those categories counted in common were ranked and compared. However, the correlation coefficients of species of mesopelagic fish from a series of cruises in the gyre are not compared to the California current because there are no comparable sample sets available.

Although the temporal and spatial array of sampling is similar in the two areas being compared, the nets used and length of tow were not. The California current samples were made with a net of .785 m^2 mouth area, 505 µ mesh and filtered about 500 m^3 of water. The gyre plankton samples were made with a net 7.78 m^2 mouth area, 505 µ mesh and filtered about 10,000 m^3 of water. These large tows were taken because of the low standing crops of plankton in the gyre. Because of this the amount of zooplankton caught per tow in the gyre and the California current are comparable. These differences in gear and length of tow have probably not affected the results of this study. While it is conceivable that net size and tow length might influence the average rank of species or taxa, it is difficult to visualize that the constancy of rank order would be so influenced. This might happen with very small nets and very short tows but it seems unlikely with the type of tow taken in the California current. These are by some standards "large" net tows. This problem can, of course, be tested by making the direct comparisons, in the field but has not yet been done.

RESULTS

Figure 3 shows that both chlorophyll and macro-zooplankton biomass is much more variable spatially in the California Current than in the gyre. This figure also shows the ranges for Shannon diversity indices. The data used are counts of species of adult copepods. This index is sensitive to both the equitability of relative proportions and length of species list. In both cases the list includes those species necessary to make up 90% of the individuals present. The index in the California Current has a much broader range than in the gyre (log scale). Figure 4 shows the matrices of the Spearman rank difference correlation coefficients of copepod species counts at two locales in the California Current and at a

INDEX OF DISPERSION
BIOMASS

CHLOROPHYLL

		$S^2/\bar{x}$	n
CAL. CURRENT	10 m	0.3	18
	20 m	0.9	18
GYRE	25 m	0.012	18
	75 m	0.013	18

ZOOPLANKTON

		$S^2/\bar{x}$	n
CAL. CURRENT	north	1300.8	69
	central	73.3	78
	south	16.3	75
GYRE		3.6	52

DIVERSITY

GYRE	CAL. CURRENT
H : 4.5 - 4.9	H : 1.8 - 5.1

Fig. 3. Comparisons of the variability, as measured by the index of dispersion ($S^2/\bar{x}$ for n measurements), of chlorophyll and macrozooplankton biomass within the California Current and within the central North Pacific gyre at indicated depths or sub-areas (positions shown as circle-dots in Figure 2). Comparison of the variability of diversity as illustrated by the range of the Shannon index of diversity (H') for copepod species in the gyre and in the California Current.

locale in the center of the gyre. The mean and medians for the gyre are both positive and about half of the samples are significantly positively correlated while in two locales in the California Current the two measures of central tendency are both negative at each of the locales. Figures 5 and 6 are the same data expressed in terms of Whittaker percent similarity indices (PSIs). Within the California Current the relative proportions of species vary in a manner not distinguishable from random fluctuations, while in the gyre the mean PSIs are consistently much higher and show a great stability of relative proportions. However, counting species proportions of copepods is tedious, time consuming, and difficult, and only the one taxon is accounted for. Because of these problems, the tests were repeated using only major categories of planktonic organisms (Table 1). Here it is assumed that the total species proportions can be no less

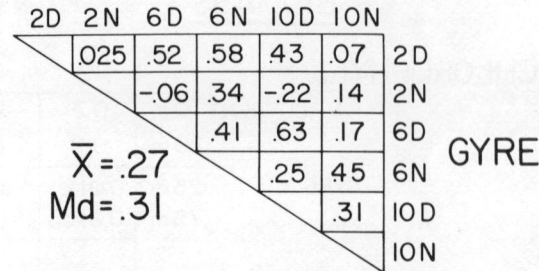
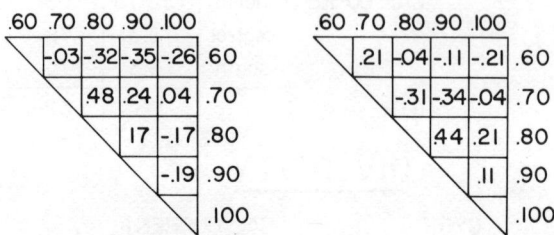

Fig. 4. Spearman rank difference correlation coefficients for copepod species in the central North Pacific gyre and in two areas of the California Current. Coefficients are based on data from cruise Climax II (for the gyre) day and night hauls from stations 2, 6, and 10, and from CalCOFI cruise 5804 (for the California Current) standard day and night stations of CalCOFI line 80 and line 100 (all solid circles in Figure 2). Within each of these data sets only the ten or so most abundant species in each sample were considered in the ranking procedure and subsequent calculations.

variable than that of the categories. Table 2 presents the Spearman rank difference coefficients compared for major categories of zooplankton, Table 3 shows PSIs for major categories, and Figure 7 gives an indication of spatial coherence of rank order in each system. Table 4 shows the ranges of percent importance of major planktonic taxa. Table 5, for the gyre only, shows rank correlations for species of mesopelagic fish and clearly resembles the copepod species rank order stability. Thus, two important and very different trophic levels are represented. Figures 8 through 10 are the distributions

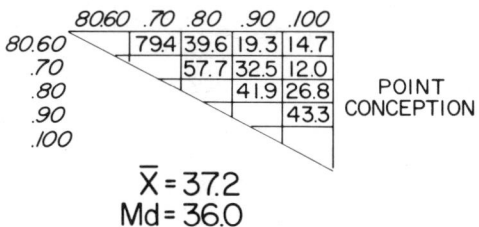
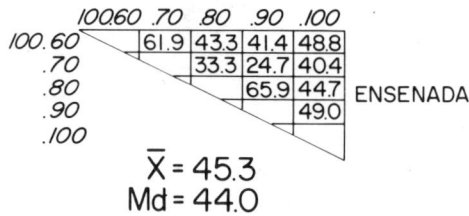

Fig. 5. Whittaker percent similarity indices (PSIs) between samples from two areas of the California Current. The relative proportion of all species of copepods, in pairs of samples, are compared. A PSI of 100 is perfect agreement. The samples are those of Figure 2.

Table 1. Major categories of planktonic organisms counted in the Central North Pacific Gyre and in the California Current. Categories were counted in both areas except those marked with * (gyre only) or † (California Current only). Gyre samples were night Isaacs-Kidd Plankton Trawls from Climax II (Sept. 1969, Stations A2, A4-10). California Current samples were from CalCOFI Cruise 5610 (Sept. 1956, standard Stations 83.51, 83.55, 83.60, 87.50, 87.55, 87.60; see Figure 2) and were selected by the following criteria: (1) grid spacing approximately the same as the gyre samples; (2) all night time samples; (3) samples were from an area not representing a pronounced faunal boundary.

Copepods
Chaetognaths
Euphausiids
Thaliacians
Ostracods
Amphipods
Pteropods
Decapods
Heteropods
"Jellies"

*Squid
*Larval Fish
*Polychaetes

†Larvaceans
†Crustacean Larvae

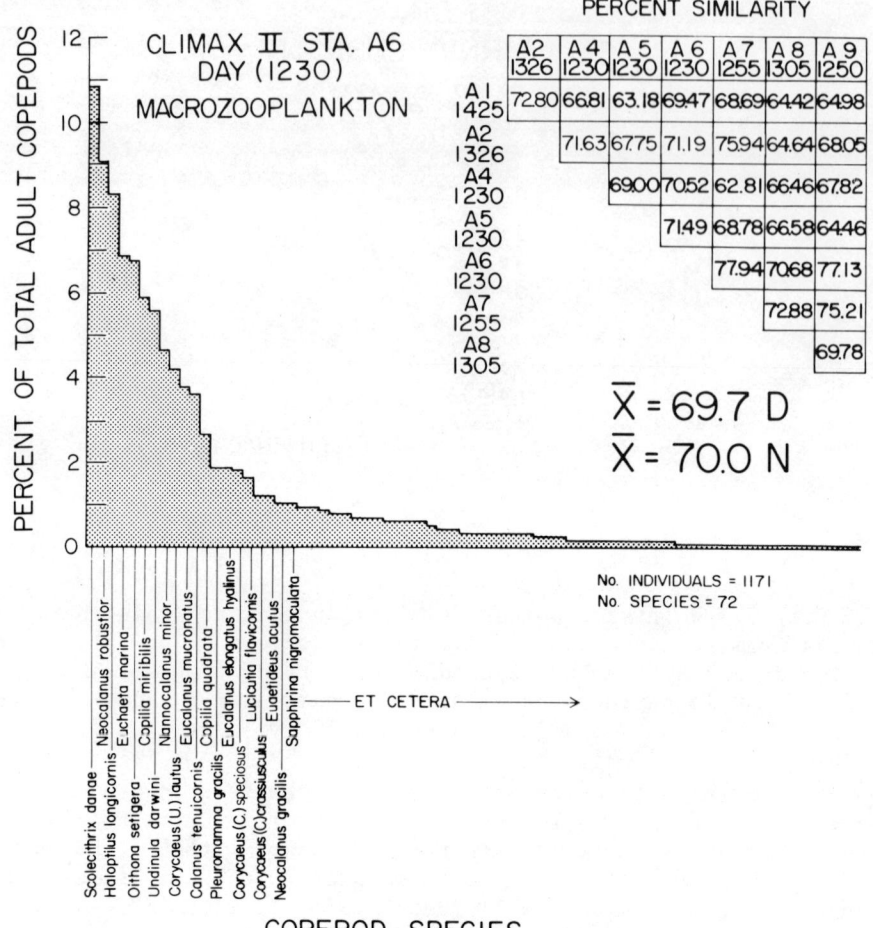

Fig. 6. Typical Whittaker percent similarity index matrix for all copepod species for the central North Pacific gyre. Histogram of relative proportions of copepod species in a typical gyre sample (A6 at 1230) is presented. Data are from day samples (A1 at 1425 local time, etc.) from Climax II. Mean PSI for both day and night samples from this cruise are given.

of various properties in the California Current and show strong north-south gradients. The probable effects of upwelling can also be seen in several of these, but horizontal advection is orders of magnitude greater than vertical. Thus, the rate of supply of phosphate (Figure 8) during the months of August to Septermber 1969, to the California Current was primarily from the north rather from the

Table 2. Spearman rank-difference correlation coefficients for major categories of planktonic organisms within the central North Pacific Gyre (Stations A2, etc.) and within the California Current (Stations 83.51, etc.).

A. Central North Pacific Gyre

A2	A4	A5	A6	A7	A8	A9	A10	
	.822	.817	.867	.935	.959	.953	.864	A2
		.994	.970	.911	.893	.911	.935	A4
			.976	.917	.876	.917	.953	A5
				.941	.917	.941	.953	A6
					.959	.964	.953	A7
						.947	.870	A8
							.953	A10

$\bar{x}$ = .924

md = .938

Range: .817 - .994

B. California Current

83.51	.55	.60	87.50	.55	.60	
	.573	.343	.580	.587	.413	83.51
		.720	.748	.818	.706	.55
			.455	.797	.811	.60
				.434	.336	87.50
					.881	.55
						.60

$\bar{x}$ = .613

md = .587

Range: .336 - .881

Table 3. Whittaker percent similarity indices (PSIs) between stations, using major taxonomic categories, within the Central North Pacific Gyre (Stations A2, etc.) and within the California Current (Stations 83.51, etc.).

A. PSIs: Central North Pacific Gyre

A2	A4	A5	A6	A7	A8	A9	A10	
	86.7	88.1	86.3	92.0	94.3	90.2	90.6	A2
		89.8	92.4	86.9	88.7	92.7	90.1	A4
			88.5	90.7	88.5	86.5	92.6	A5
				87.9	86.6	90.7	88.7	A6
					92.3	89.2	92.2	A7
						90.4	90.3	A8
							90.0	A9
								A10

$\bar{x}$ = 89.8

md = 90.3

Range: 86.3 - 94.3

B. PSIs: California Current

83.51	83.55	83.60	87.50	87.55	87.60	
	63.2	31.3	67.4	65.1	60.0	83.51
		64.6	62.2	76.1	65.0	83.55
			29.4	54.5	29.8	83.60
				65.4	85.7	87.50
					66.7	87.60

$\bar{x}$ = 59.3

md = 65.1

Range: 29.4 - 85.7

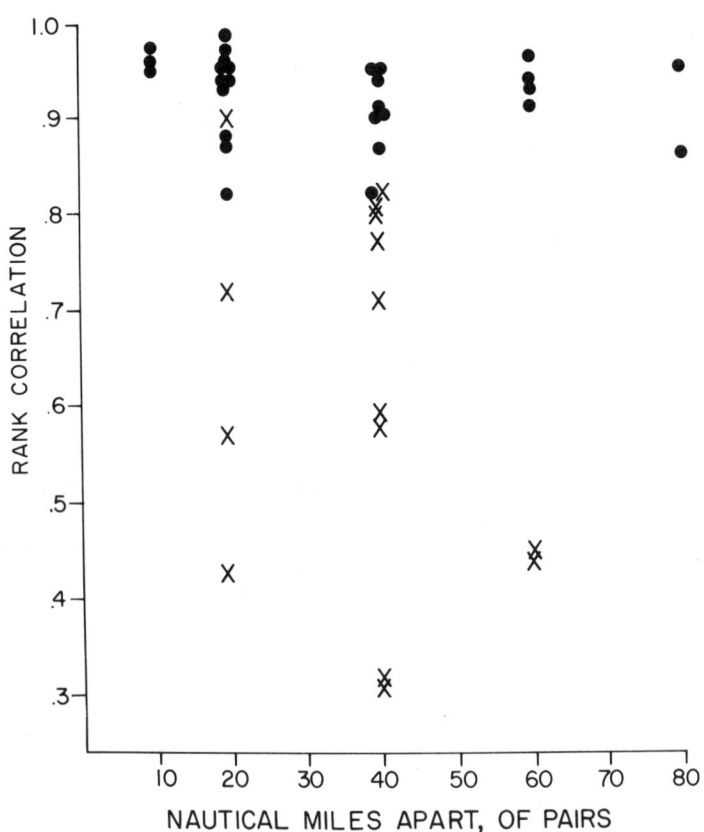

Fig. 7. Spatial coherence within the central North Pacific gyre and within the California Current. The Spearman rank difference correlation coefficients for major taxa are plotted against the distance separating the pair of samples on which a coefficient is based. The California Current samples are shown as solid triangles in Figure 2.

in situ process of upwelling. The advective trends illustrated in Figures 8 through 10 then give some insight into the reasons for the results of the tests of the hypotheses.

In every case where the data were appropriate for testing, the gyre shows much greater spatial stability and/or structure than does the California Current. Results of tests of the temporal

Table 4. The range of the percent (of total plankton counted) that each of the ten taxa counted in both areas contributes to samples in the Central North Pacific Gyre and in the California Current.

Taxon	Central Gyre	Calif. Current
Copepods	51.2 - 60.9%	15.9 - 72.2%
Chaetognaths	6.2 - 10.6	2.5 - 36.7
Euphausiids	7.1 - 13.9	6.3 - 34.0
Thaliacians	1.0 - 11.2	0.4 - 48.8
Ostracods	3.9 - 9.0	0.0 - 1.8
Amphipods	3.9 - 7.4	0.4 - 3.4
Pteropods	0.8 - 4.1	0.0 - 1.4
Decapods	0.7 - 2.5	0.0 - 5.4
Heteropods	0.2 - 0.8	0.0 - 0.3
"Jellies"	1.8 - 3.5	0.2 - 2.8

aspects of these hypotheses have not as yet been completed but the preliminary results indicate that a similar situation prevails. A reasonable interpretation is that large scale horizontal mixing has a pronounced effect on structure and, by implication, the function and dynamics of populations in the California Current. It appears that the north Pacific gyre is a well regulated system while the California Current is not.

However, these hypotheses should be tested in other suspected *in situ* systems and other suspected advective systems. The spatial coherence of structure and stability should be mapped, in order to find out where the communities are, and the shapes and sizes of their boundaries. A very careful design of sampling strategy will be required.

Table 5. Spearman rank-difference correlation coefficient (r_d) matrix for the top 20 mesopelagic fish species, between cruises in the Central North Pacific Gyre. Data is lumped within each cruise. Column (H') is Shannon-Wiener diversity index for each cruise. The within-cruise percent similarity indices (using all species) range from 42.9 to 81.7; mean is 63.9. Information primarily from *Barnett* (1975).

r_d Matrix

1	2	3	4	5	6	7	Cruise	(H')	Month/Year
	.288	.405	.223	.421	.545	.480	1 Aries IX	2.6	Sept.-Oct. 1971
		.614	.500	.605	.733	.535	2 Cato 1	2.7	June 1972
			.659	.498	.766	.661	3 SOTW XIII	2.7	Feb. 1973
				.703	.478	.518	4 TSDY I	3.0	June 1973
					.801	.562	5 TSDY II	2.8	July 1973
						.642	6 TSDY XI	2.6	Mar. 1974
							7 Climax II	2.3	Aug.-Sept. 1969

$r_d \ \bar{x} = .554$

$r_d \ md = .545$

r_d Range: .223 – .801

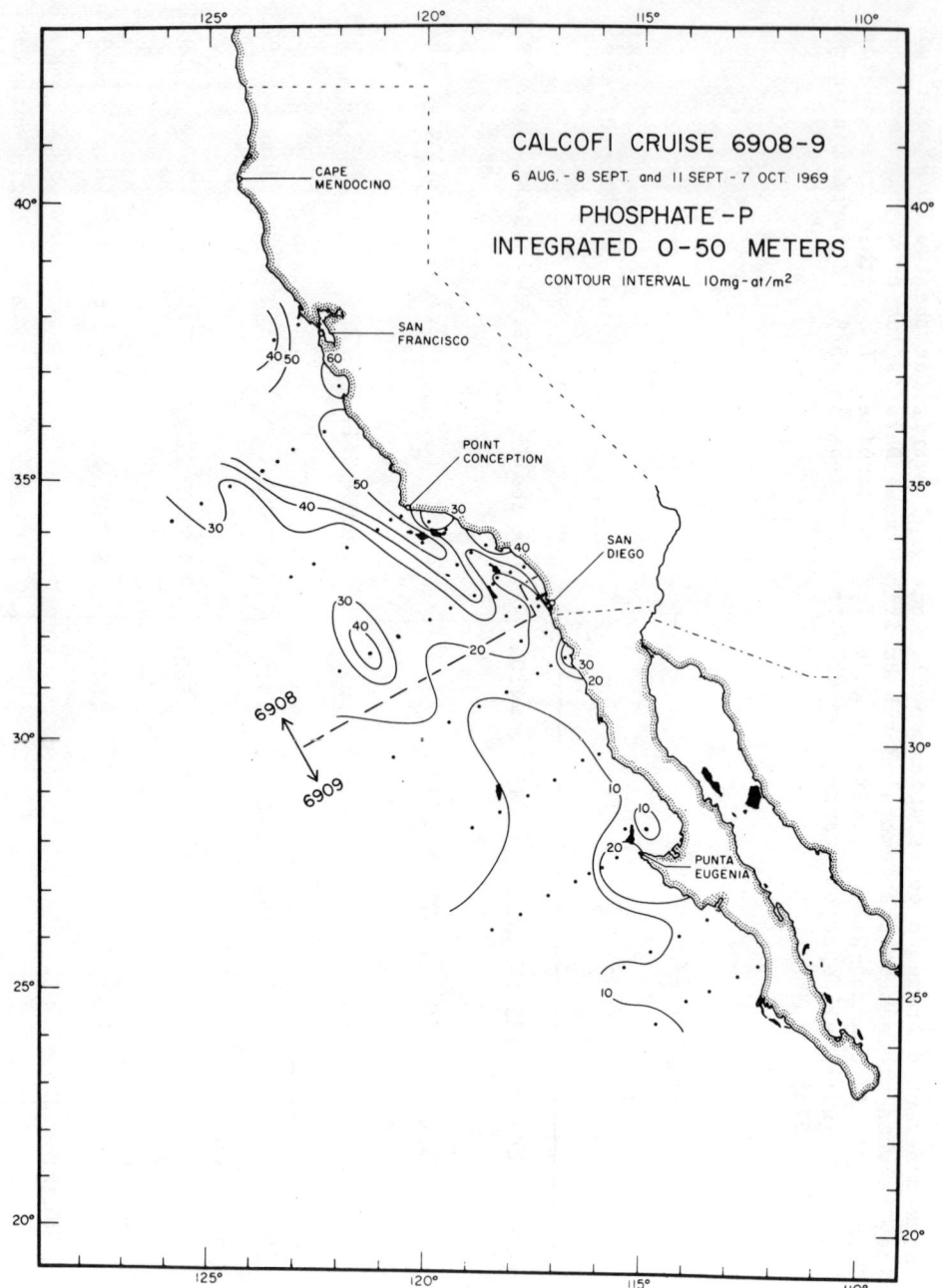

Fig. 8. Example of phosphate-phosphorous distribution in the California Current. Data are from CalCOFI cruises 6908 and 6909

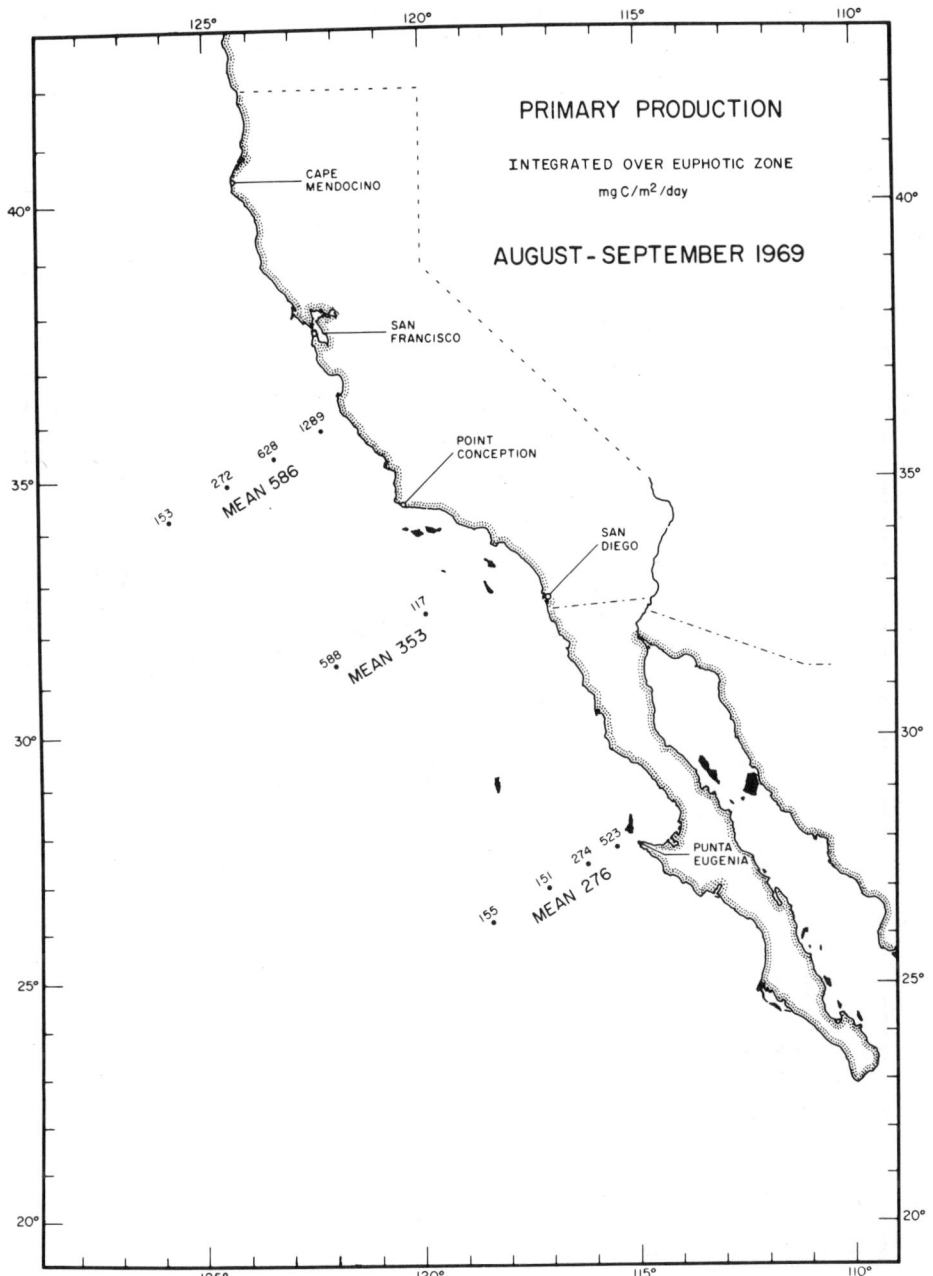

Fig. 9. Example of primary production in regions of the California Current. Source of data as in Figure 8.

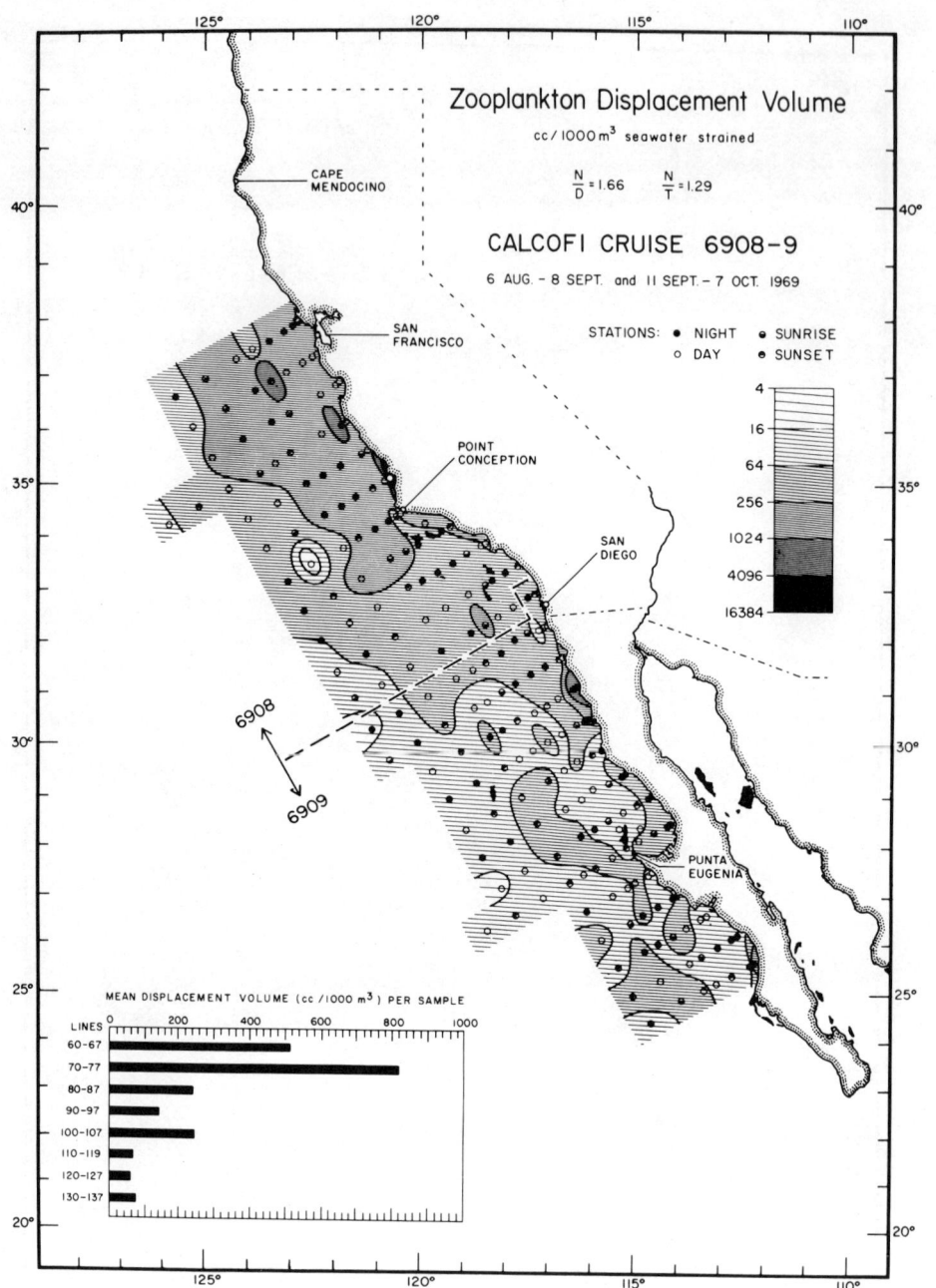

Fig. 10. Example of zooplankton biomass distributions in the California Current. N/D = night/day ratio; N/T = night/twilight ratio Source of data as in Figure 8.

REFERENCES

Barnett, M. A. 1975. Studies on the patterns of distribution of mesopelagic fish faunal assemblages in the Central Pacific and their temporal persistence in the gyres. Ph.D. Thesis. Scripps Institution of Oceanogr., Univ. Calif., San Diego.

Beklemishev, C. W. 1969. Ecology and Biogeography of the Open Ocean. Publishing House "Nauka", Moscow. 291 p.

Bogorov, V. G. [ed.] 1967. Biology of the Pacific Ocean; Book 1: Plankton. Publishing House "Nauka", Moscow. 266 p.

Fager, E. W., and J. A. McGowan. 1963. Zooplankton species groups in the North Pacific. *Science 140*: 453-460.

Holling, C. S. 1973. Resilience and stability of ecological systems. In R. J. Johnson, ed., *Annual Review of Ecology and Systematics 4*: 1-23. Annual Reviews, Inc., Palo Alto, California.

Hutchinson, G. E. 1967. *A Treatise on Limnology. Vol. II: Introduction to Lake Biology and the Limnoplankton.* John Wiley and Sons, Inc., New York.

Johnson, M. W., and E. Brinton. 1963. Biological species, water masses, and currents. In M. N. Hill, ed., *The Sea 2*: 381-414. Interscience, New York.

Koblentz-Mishke, O. J., V. V. Volkovinsky, and J. G. Kabanova. 1970. Plankton primary productivity of the world ocean. In. W. S. Wooster, ed., *Scientific Exploration of the South Pacific*: 183-193. Nat. Acad. of Sciences, Washington, DC.

McGowan, J. A. 1971. Oceanic biogeography of the Pacific. In B. M. Funnell and W. R. Riedel, ed., *The Micropaleontology of Oceans*: 3-74. Cambridge Univ. Press, Cambridge.

McGowan, J. A. 1974. The nature of oceanic ecosystems. In C. B. Miller, ed., *The Biology of the Oceanic Pacific*: 9-28. Oregon State Univ. Press, Corvallis.

NorPac Committee. 1960. *Oceanic observations of the Pacific: 1955, the NORPAC Atlas*: 123 plates. Univ. of Calif. Press, Berkeley.

Parsons, T. R., and R. J. LeBrasseur. 1968. A discussion of critical indices of primary and secondary production for large scale ocean surveys. *Calif. Coop Oceanic Fish. Invest. Reports XII*: 54-63.

Reid, J. L. 1965. *Intermediate waters of the Pacific Ocean. The Johns Hopkins Oceanographic Studies Number 2.* The Johns Hopkins Press, Baltimore. 85 p.

Riedel, W. R., and B. M. Funnell. 1964. Tertiary sediment cores and microfossils from the Pacific Ocean floor. *Quart. J. Geolog. Soc. Lond. 120*: 305-368.

Sverdrup, H. V., M. V. Johnson, and R. H. Fleming. 1942. *The Oceans: Their Physics, Chemistry, and General Biology.* Prentice-Hall, New York. 1087 p.

The Role of Microorganisms in the Marine Environment

Richard Y. Morita

Oregon State University

ABSTRACT

Although they are so small that they cannot be seen by the naked eye, because of their enormous abundance and physiological diversity, microorganisms play a crucial role in the marine environment. The biomass of the bacteria may approach that of the entire fauna. They are of prime importance as decomposers and thus, play a great role in the regeneration of nutrients and the production of various dissolved gases. Their other contributions to the various trophic levels in the sea include being a source of food for many filter feeders and producing ectocrine compounds. They are the prime cause of anoxic environments through their utilization of dissolved oxygen and play an important, though not clearly understood, role in the diagenesis of marine sediments.

INTRODUCTION

"The role of the infinitely small in nature is infinitely great."
--Louis Pasteur.

There is a tendency in oceanography to neglect the "infinitely small" organisms. The reasons for this situation may be attributed to (1) the lack of interaction between marine microbiologists and other types of oceanographers, (2) insufficient training of oceanographers in microbiology, and (3) a tendency to underestimate the microbes' role--mainly because they cannot be seen with the naked eye. However, marine microbiology plays an important role in the oceans because it interfaces not only with biology, but with the chemistry and geology of the marine environment.

The aim of this paper is to review, in a limited way, the functions of microbes in the marine environment.

BACTERIAL BIOMASS

There are many technical difficulties in trying to assess the bacterial biomass of any given aquatic system. The cultural procedures used in the past only permit us to determine a small fraction of the bacterial population. This is to be expected since there is such a diversity of microbial types. The reliability of the ATP method (*Holm-Hansen*, 1966 and 1969) for determining the bacterial cellular organic carbon is now being questioned by Jassby (1975). The most reliable method for determining the bacterial numbers and biomass is still the direct count method. The newer approaches for determining the bacterial biomass utilize epifluorescent microscopy (*Zimmerman and Meyer-Reil*, 1974, and *Daley and Hobbie*, 1975) or the Lipopolysaccharide method (*Dexter, Sullivan, Williams, and Watson*, 1975) which has now been introduced. The latter method gives a lower count than the epifluorescent method. The former method permits one to count microbes that are as small as 0.1 µ (see Figure 1). By the epifluorescent direct count method we are beginning to realize that many bacteria in the ocean are less than 0.5 µ. In our laboratory the method of *Zimmerman and Meyer-Reil* (1974) is an extremely good tool and it does not appear that the comparison between the methods of *Francisco, Mah, and Rabin* (1973) and *Zimmerman and Meyer-Reil* (1974) made by *Daley and Hobbie* (1975) is valid.

Although there exist in the Russian literature (*Kriss*, 1962) much data on the bacterial biomass in the oceans, the data are in dispute--mainly due to the method of collecting the water samples. The collected data in the older literature concerning microbial biomass in the water column are not very reliable but the current data are beginning to shed more light on the bacterial biomass in the ocean. According to Dale (1974) (employing direct count measurements) the bacterial biomass is about equal to the faunal standing crop. Likewise *Watson* (personal communication) states that the bacterial biomass is slightly greater than the biomass of other forms in nearshore waters.

Employing the epifluorescent method for direct counting of bacteria *Meyer-Reil* (personal communication) reported that the bacterial biomass (in terms of cellular matter) in surface water nearshore in the Baltic Sea was 100 to 500 mg/m^3 whereas offshore the range was 50 to 200 mg/m^3. The total bacterial organic biomass in nearshore areas can range from 0 to 180 g/m^3/year. Off the coast of Oregon, using epifluorescent counts, there were ca. 80 x 10^4 bacterial cells/ml at 20 m and 9.8 x 10^4 bacterial cells/ml at 700 m (unpublished data, *Meyer-Reil, Geesey and Morita*).

Standing crop bacterial biomass studies in themselves are not sufficient to understand the microbial contribution to the organic matter in the sea because of the ability of the bacterial cells to multiply in a short time, especially in nutrient rich waters. (The shortest generation time reported for a bacterial cell is slightly less than 10 minutes and the information was obtained using a marine bacterium.)

THE ROLE OF MICROORGANISMS

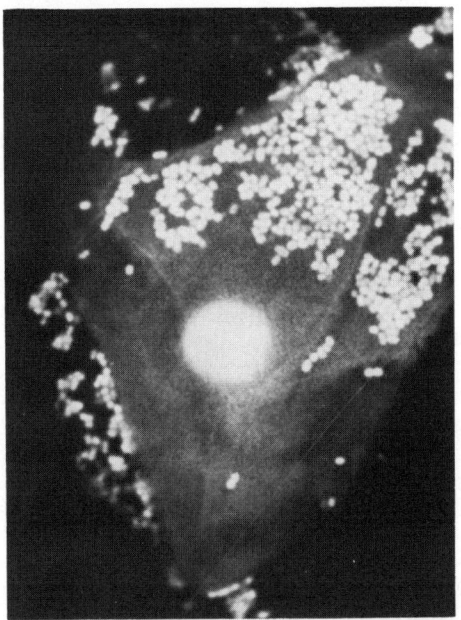

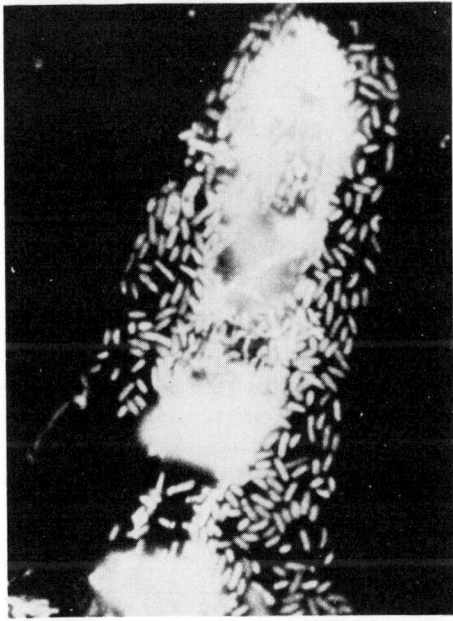

Fig. 1. Epiflourescent migrographs of bacteria on detritus (upper photo) and on phytoplankton undergoing decomposition. Microbial cells in the upper photo are approximately 0.1 µ in diameter while in the lower photo the length of the bacteria cells are 0.2 to 0.3 µ. Both materials were taken from the Baltic Sea. Courtesy of Dr. L.-A. Meyer-Reil.

MICROBIAL CONTRIBUTION TO THE TROPHIC LEVELS

Biological oceanographers, when dealing with the trophic levels relegate the bacteria to the role of decomposers in the sea. The assignment of microbes as decomposers is an important one since the amount of carbon dioxide fixed per year should approximately equal the amount of carbon dioxide regenerated. Because of the importance of microbes in the cycles of matter, *Russell-Hunter* (1970) stated that if the function of microbes ceased, then all forms of life on earth would expire in three months. Although the mineralization of organic matter is one of the primary functions of microbes, they have other important functions. The activities of microbes in the water column are of primary importance and their biomass mainly serves as a food source for other forms. In addition, one must always consider the microbes on and in various other forms of life. They are also the main reason for the fertility of sea water.

Microbes as Food

Since microbes form a large part of the biomass in the oceans, they are of greater importance in the trophic levels than to merely be assigned the role of decomposers.

As they fall though the water column, microbes that normally live in the surface waters, as well as terrestrial contaminants, are often not capable of further growth. By our current methods, we cannot detect the stages between life and death of the sedimenting microorganisms. Whether the cells are living, dead, or metabolically inactive, they will all constitute a source of food, especially for the mud eaters on the bottom. The low temperatures of the area below the thermocline as well as the increasing hydrostatic pressures tend to preserve cells in the intact form providing they are not psychrophilic, psychrotrophic, barophilic, or eurybaric. Naturally, the degree to which these types of microorganisms will function will depend on the temperature, hydrostatic pressure, as well as the availability of organic matter. Both low temperature and increased hydrostatic pressure are important environmental factors but they will not be discussed in detail here (see *Morita*, 1976).

Bacteria are known to be the source of food for protozoa. However, such filter feeders as coral (*Sorokin*, 1973) and sponges (*Reiswig*, 1971) are now known to utilize bacteria as their main source of food. Bacteria and allied microorganisms are known to colonize detritus making the detritus a more nutritive substance. Bacteria can also form detritus particles and are known to interact with bubbles to form organic aggregates (*Barber*, 1966). In addition, bacteria can utilize dissolved organic matter in sea water more efficiently than any other biological forms. Because of the ability of bacteria to colonize particles, various organisms such as cladocerans, sponges, sabellids, appendicularians, ascidians, copepods, euphausiids, polyps, and molluscs are capable of utilizing bacteria as food (for a complete list see *Sorokin*, 1974). The relationship between microbes and detritus is fully appreciated by many contributors to the IBP-UNESCO symposium held to discuss the role of detritus (*Melchiorri-Santolini and Hopton*, 1972).

Although the microorganisms in the lower depths of the sea may not constitute the major portion of the organic matter present, they are nutrient rich in proteins, phospholipids, and ectocrine compounds, etc.

Jannasch et al. (1971) reported slow rates of degradation of materials in the submersible ALVIN which had sunk to the bottom in 1540 m. Further work by *Jannasch and Wirsen* (1973) showed extremely slow rates of organic matter conversion by bacteria (*in situ* studies at 1830 m). Slow rates of microbial activity are to be expected when one looks at the data concerning the hydrostatic pressure effect on most organisms. If high rates of microbial activity should occur with most bacteria in the deeper portions of the sea, then there would not be sufficient organic matter to support the life of the higher forms. The residence time of organic matter in the sea as well as the time for the microbes to act upon it should be taken into consideration. Unfortunately, experiments that are designed in microbiology do not take into consideration geological time, but must in order to fully understand the function of microbes in the sea.

On the other hand, microbial activities in the deep sea (5200 m) have been reported to be sizable (*Seki*, 1972). *Schwartz, Vayanos, and Colwell* (1976) reported excellent growth of bacteria isolated from a deep-sea invertebrate (7200 m). We have also isolated such organisms. However, keeping in mind the source of the microorganisms and their habitat, it is the eurybiotic forms that do the best in the marine deep sea environment (*Morita*, 1976).

Coprophagy

Fecal pellets are utilizable sources of food for the macrobenthic organisms and many organisms practice coprophagy. *Johannes and Satomi* (1966) found that the organic matter of fecal pellets was, to a large degree, in the form of bacteria. Furthermore, fecal pellets are generally low in nitrogen when first excreted but the microorganisms grow and multiply on and in the fecal pellet thereby enriching it nutritionally in many ways.

Production of Ectocrine Compounds

Many organisms require ectocrine compounds for growth. Microbes, whether in the intestinal trace or free living in the water column, produce ectocrine compounds required by other forms. Although certain algae are known to produce ectocrine compounds (*Carlucci*, 1974), they cannot compete with the bacteria (*Carlucci*, personal communication). We are now beginning to realize that certain bases (purines and pyrimidines) enhance the growth of red tide organisms (*Iwasaki*, 1971). Reviews on vitamins in the sea have been written by *Belser* (1963), *Provasoli* (1958) and *Riley* (1965) while *Ohwada and Taga* (1972) present the most recent data on the distribution and seasonal variation of vitamin B_{12}, thiamine, and biotin in the sea.

Interaction of Microbes with other Forms in the Trophic Level

Muller and Lee (1969) found that bacteria were required for sustained reproduction of foraminifera. Their results indicate that the bacteria produced a nutritional factor or factors required by the foraminifera.

A hormonal-like compound produced by a bacterium was found to be essential for the normal development of *Ulva lactuca* (*Provasoli and Pinter*, 1972). Without the presence of the bacterial epiphyte or the presence of the cultural fluid in which the bacteria were grown, the *Ulva* would develop abnormally.

An *Azotobacter* (nitrogen fixing bacterium) was found to be associated with the marine macroalga *Codium fragile* (*Head and Carpenter*, 1975). In this relationship, the bacteria apparently obtain an energy source (glucose) from the alga and the *Azotobacter* furnishes the alga with nitrogenous material (nitrate or ammonia). Nitrogen-fixing bacteria have also been found in the gut of marine shipworms. The nitrogen fixation activity appeared to be inversely related to the ability of the marine shipworms to obtain combined nitrogen compounds in their diet (*Carpenter and Culliney*, 1975).

Data in our laboratory obtained by T. D. Goodrich indicate that chitinase, which is necessary to degrade chitin, in the intestinal tract of fish is microbial. No chitinase is produced by the fish. In view of the tremendous amounts of chitin produced yearly in the oceans, the chitinoclastic activity of bacteria assumes an important role not only in the release of a utilizable nitrogenous source for the fish but also in the nitrogen and carbon cycles.

MICROBES AND THE FERTILITY AND CHEMISTRY OF SEA WATER

Nitrite, ammonia, nitrate, sulfate, phosphate, etc., are primary nutrients for phytoplankton growth and their formation contributes to the fertility of sea water. Many textbooks in marine chemistry discuss these compounds (*Horne*, 1969; *Riley and Chester*, 1971; and *Riley and Skirrow*, 1965) and *ZoBell* (1946) discusses the role of microbes in the regeneration of these nutrients.

Many chemical oceanographers have made analyses of phosphate, silicate, oxygen, etc. in water masses. These are static measurements since they do not represent the rate at which these compounds are produced or utilized. The measurements simply represent values at that one particular time at that one location and depth. We do not know whether the values obtained are steady state values, or points in a decreasing reaction rate or an increasing reaction rate.

Harvey (1955) discussed the changes in seawater brought about by microbial activity. During the past generation, progress has been made in understanding the biochemistry and physiology of organisms that catalyze various reactions that can change the chemistry and fertility of sea water. Unfortunately, since Harvey's book appeared, the fields of chemistry of sea water and the microbiology of sea water appear to have gone their separate pathways. As a result, no recent intense study has addressed the microbial functions as they occur in sea water and how they affect the chemistry of the water.

With modern scientific apparatus plus the knowledge gained from other areas of research, the microbe's contribution to the chemistry and fertility of sea water should be re-investigated to produce data on the rates of activity. Since the residence time of specific bodies of water can be in terms of years or even a thousand years, low rates of reaction may be a factor. Team effort is called for in many of the problems mentioned above.

MICROBES AND THE CARBON DIOXIDE SYSTEM

Approximately 90% of the carbon dioxide produced is by microbial means (Stanier, Doudoroff, and Adelberg, 1970). Because of the ability of microbes to produce carbon dioxide, the evolution of the 'steady state' carbon dioxide system can be attributed to the microorganisms (Morita, 1975). The 'steady state' carbon dioxide system described by Pytkowicz (1971) occurred approximately 1 billion years ago and was maintained by the procaryotic cells for a long period of time before the evolution of the higher forms (Lovelock and Margulis, 1974 and Margulis and Lovelock, 1974). Thus the procaryotic cells assume a great importance in the carbon dioxide-carbonate system in the ocean.

The oxygen distribution patterns (Redfield, 1942; Pytkowicz, 1968) resulting in oxygen utilization in the biological production of carbon dioxide are probably due to past microbial events. Such distribution patterns never give us an insight as to the rate of the reactions creating such conditions because they are past events. There is a correlation between the pCO_2 values and microbial activity off the Oregon Coast during an upwelling situation (Morita, Geesey, and Goodrich, 1974) which could give rise to the areas of low oxygen distribution described by Pytkowicz (1968).

ANOXIC ENVIRONMENTS

The creation of anoxic areas in the ocean such as the Black Sea, various fiords, etc., is due mainly to the utilization of dissolved oxygen by microbial action faster than the oxygen can be replaced.

Chemically, these areas are well described by *Richards* (1965). Recent studies of the microbiology of the Black Sea have been done by *Jannasch, Truper, and Tuttle* (1974). The microbiological processes of sulfate reduction are well documented for geochemical processes (*Thode, Kleerekoper, and McElcheran*, 1951; *Jones*, 1956).

Once the anoxic environments are created, the anaerobic organism *Desulfovibrio* produces hydrogen sulfide which helps maintain the anoxic condition. Since hydrogen sulfide in the presence of oxygen can be oxidized abiologically, a constant input of oxygen into the anoxic environment would eventually revert the system back to the oxidized state again. However, if the amount of hydrogen sulfide generated is great enough to counter the constant input of oxygen into the system, then the environment will maintain its anoxic condition.

The reduced conditions in anoxic environments can also influence the oxidation-reduction state of many metals found in the environment. In anoxic environments, nitrate, nitrite, as well as sulfate mentioned above, can be utilized by microbial systems as the ultimate hydrogen and electron acceptor.

MICROBIAL PARTICIPATION IN GEOLOGICAL OCEANOGRAPHY

Microbes are the main catalytic agents affecting the transformation of sedimentary materials, especially when endothermic reactions are involved. The biogeochemistry of carbon and sulfur are also affected by microbial activity. The function of microbes in the diagenesis of sediments should be investigated jointly by microbiologists and geologists using modern techniques. The biogenic gas production in the ocean (*Mechalas*, 1974), especially in marine sediments, should be scrutinized further. The formation of these biogenic gases (carbon dioxide, oxygen, nitrogen, ammonia, hydrogen sulfide, methane, and hydrogen) is a complex interaction between chemical, physical, and biological processes. Many reactions, whether in the water column or in sedimentary material, are governed by the microbes' ability to change the oxidation-reduction potential and the pH. These two factors are of extreme importance since they determine, in many instances, the direction and the magnitude of chemical and biochemical reactions.

Eh and pH are known to be changed in sedimentary material that contain organic matter. The pH and Eh are extremely important in the diagenesis of nearshore sediments. Early work on the diagenesis of sediments by microbial action was performed mainly in the late 1940s and early 1950s in Dr. ZoBell's laboratory but interest in this aspect of microbiology has faded. Much is still unknown concerning the microbial processes in the diagenesis of sediments and a renewed effort should again be made.

REFERENCES

Barber, R. T. 1966. Interaction of bubbles and bacteria in the formation of aggregates in sea water. *Nature 111*: 257-259.

Besler, W. L. 1963. Bioassay of trace substances. In M. N. Hill, ed., *The Sea, 2:* 220-231. Interscience, New York.

Carlucci, A. F. 1974. Production and utilization of dissolved vitamins by marine phytoplankton. In R. R. Colwell and R. Y. Morita, eds., *Effect of the ocean environment on microbial activities:* 449-456. University Park Press, Baltimore.

Carpenter, E. J., and J. L. Culliney. 1975. Nitrogen fixation in marine shipworms. *Science 187:* 551-552.

Dale, N. G. 1974. Bacteria in intertidal sediments: Factors related to their distribution. *Limnol Oceanogr. 19:* 509-518.

Daley, R. J. and J. E. Hobbie. 1975. Direct counts of aquatic bacteria by a modified epifluorescence technique. *Limnol. Oceanogr. 20:* 875-882.

Dexter, S. C., J. D. Sullivan, Jr., J. Williams III, and S. W. Watson. 1975. Influence of substrate wettability on the attachment of marine bacteria to various substrates. *Appl. Microbiol. 30:* 298-308.

Francisco, D. E., R. A. Mah, and A. C. Rabin. 1973. Acridine orange epifluorescence technique for counting bacteria in natural waters. *Trans. Am. Microsc. Soc. 92:* 416-421.

Harvey, H. W. 1955. *The chemistry and fertility of sea waters.* Cambridge University Press. London. 224 p.

Head, W. D., and E. J. Carpenter. 1975. Nitrogen fixation associated with marine macroalga *Codium fragile. Limnol. Oceanogr. 20:* 815-823.

Holm-Hansen, O. 1969. Determination of microbial biomass in ocean profiles. *Limnol. Oceanogr. 14:* 740-747.

Holm-Hansen, O., and C. R. Booth. 1966. The measurement of adenosine triphosphate in the ocean and its ecological significance. *Limnol. Oceanogr. 11:* 510-519.

Horne, R. A. 1969. *Marine Chemistry.* Wiley-Interscience. New York, London, Sydney, and Toronto.

Iwasaki, H. 1971. Studies on red tide flagellates. *J. Oceanogr. Soc. Japan 27:* 152-157.

Jannasch, H. W., Eimhjellen, K., Wirsen, C. O., and Farmanfarmaian, A. 1971. Microbial degradation of organic matter in the deep sea. *Science, 171:* 672-5.

Jannasch, H. W., H. G. Truper, and J. H. Tuttle. 1974. Microbial sulfur cycle in Black Sea. *Amer. Assoc. Petrol. Geol. Memoir No. 20:* 419-425.

Jannasch, H. W., and Wirsen, C. O. 1973. Deep-sea microorganisms: *in situ* response to nutrient enrichment. *Science, 180:* 641-3.

Jassby, A. D. 1975. An evaluation of ATP estimations of bacterial biomass in the presence of phytoplankton. *Limnol. Oceanogr. 20:* 646-648.

Johannes, R. E., and M. Satomi. 1966. Composition and nutritive value of fecal pellets of a marine crustacena. *Limnol. Oceanogr. 11:* 191-197.

Jones, G. E. 1956. Fractionation of the stable isotopes of sulfur by microorganisms with particular reference to the sulfate-reducing bacteria. Ph.D. Thesis, Rutgers University.

Kriss, A. E. 1962. *Marine Microbiology (Deep Sea)*. Translated by J. M. Shewan and Z. Kabata. Edinburgh and London: Oliver and Boyd.

Lovelock, J. E. and L. Margulis. 1974. Atmospheric homeostasis by and for the biosphere: the gaia hypothesis. *Tellus 36*: 1-10.

Margulis, L. and J. E. Lovelock. 1974. Biological modulation of the Earth's atmosphere. *Icarus 21*: 471-489.

Mechalas, B. J. 1974. Pathways and environmental requirements for biogenic gas production in the ocean. In I. R. Kaplan, ed., *Natural gases in marine sediments*: 11-25. Plenum Publ. Co., New York.

Melchiorri-Santolini, U., and J. S. Hopton. 1972. Detritus and its role in aquatic ecosystems. *Proc. IBP-UNESCO Symp. Memorie Dell' Instituto Italiano di Idrobiologia 29*: Supplement.

Morita, R. Y. 1975. Microbial contribution to the evolution of the 'steady state' carbon dioxide system. *Origins of Life 6*: 37-44.

Morita, R. Y. 1976. Survival of bacteria in cold and moderate hydrostatic pressure with special reference to psychrophilic bacteria and barophilic bacteria. In T. G. R. Gray and J. Postgate, eds., *Survival of vegetative microbes*: 279-298. 26th Sym. Soc. Gen. Microbiol.

Morita, R. Y., G. G. Geesey, and T. D. Goodrich. 1974. Potential microbial contribution to the carbon dioxide system in the sea. In R. R. Colwell and R. Y. Morita, eds., *Effect of the ocean environment on microbial activities*: 386-391. Univ. Park Press, Baltimore.

Muller, W. A., and J. J. Lee. 1969. Apparent indispensability of bacteria in foraminiferan nutrition. *J. Protozool. 16*: 471-478.

Ohwada, K. and N. Taga. 1972. Distribution and seasonal variation of vitamin B_{12}, thiamine, and biotin in the sea. *Mar. Chem. 1*: 61-69.

Provasoli, L. 1958. Nutrition and ecology of protozoa and algae. *Ann. Rev. Microbiol. 12*: 279-308.

Provasoli, L., and I. J. Pintner. 1972. Effects of bacteria on seaweed morphology. *J. Phycol. (suppl.) 8*: 10.

Pytkowicz, R. M. 1968. Water masses and their properties at 160° W in Southern Ocean. *J. Oceanogr. Soc. Jap. 24*: 1-22.

Pytkowicz, R. M. 1971. The chemical stability of the oceans and the CO_2 system. In D. Dyrssen and D. Jagner, eds., *Nobel Sym, 20, The changing chemistry of the oceans*: 147-152. Wiley Interscience, New York, London, and Sydney.

Redfield, A. C. 1942. The processes determining the concentration of oxygen, phosphate, and other organic derivatives within the depths of the Atlantic Ocean. *Pap. Phys. Oceangr. 9*: 1-22.

Reiswig, H. M. 1971. Particle feed in natural populations of three demosponges. *Biol. Bull. 141*: 508-591.

Richards, F. A. 1965. Anoxic basins and fiords. In J. P. Riley and G. Skirrow, eds., *Chemical Oceanography 1*: 611-645. Academic Press, London and New York.

Riley, J. P. 1965. Analytical chemistry of sea water. In J. P. Riley and G. Skirrow, eds., *Chemical Oceanography*. Academic Press, New York.

Riley, J. P. and R. Chester. 1971. *Introduction to Marine Chemistry*. Academic Press, London and New York.

Riley, J. P. and G. Skirrow. 1965. *Chemical Oceanography 1 & 2*. Academic Press, London and New York.

Russell-Hunter, W. D. 1970. *Aquatic productivity*. Macmillan Co., New York.

Schwarz, J. R., A. A. Yayanos, and R. R. Colwell. 1976. Metabolic activities of the intestinal microflora of a deep-sea invertebrate. *Appl. and Environmental Microbiol. 31:* 1001-1002.

Seki, H. 1972. The role of microorganisms in the marine food chain with reference to organic aggregates. In U. Melchiorri-Santolini & J. W. Hopton, eds., *Detritus and its role in aquatic ecosystems* 245-259. *Proceeding of an IBP-UNESCO Symposium. Memorie Dell' Instituto di Idrobiologia 29:* Supplement.

Sorokin, Yu. I. 1973. Trophical role of bacteria in the ecosystem of the coral reef. *Nature 242:* 415-417.

Sorokin, Yu. I. 1974. Bacterial production in bodies of water. In Z. I. Kuznetsova, ed., *General Ecology biocenology hydrobiology 1:* 29-80. G. K. Hall & Co., Boston.

Stanier, R. Y., M. Doudoroff, and E. A. Adelberg. 1970. *The Microbial World*. Prentice-Hall, Inc., Englewood Cliff, New Jersey.

Thode, H. G., G. Kleerekoper, and D. McElcheran. 1951. Isotopic fractionation in the bacterial reduction of sulfate. *Research (London) 4:* 581-582.

Zimmerman, R. and L. A. Meyer-Reil. 1974. A new method for fluorescence staining of bacterial populations on membrane filters. *Kieler Meeresforschungen 30:* 24-27.

ZoBell, C. E. 1946. *Marine Microbiology*. Chronica Botanica Co., Waltham, Massachusetts.

Food and Feeding Structures of Deep-Sea Thysanopoda Euphausiids

Takahisa Nemoto

Ocean Research Institute, Tokyo

ABSTRACT

Food and feeding structures of three bathypelagic giant Thysanopoda have been examined. The stomach contents are dominated by copepods, chaetognaths being the next most important food, then euphausiids, amphipods, mysids, and pelagic shrimps present in smaller amounts. T.egregia is conisdered to be the most voracious euphausiid, taking micronectonic fishes and even oceanic squid larva. One T.egregia (48 mm) was estimated to have eaten two fishes. The thoracic legs of T.cornuta, T.egregia and T.spinicaudata have no developed filtering barbs along the setae in each joint. The mandibles of deep sea Thysanopoda are much reduced while the teeth of the pars incisiva are well developed. Stomachs of these giant Thysanopoda are large in volume but, become flat and folded when empty. Strong solitary spines and setae that are typical of carnivorous euphausiids are found along the inner side of the stomach wall. There are no clusters of spines such as those found in the stomach walls of typically herbivorous euphausiids. Bathypelagic Thysanopoda species are large and presumed to be very actively carnivorous. They are considered to occupy significant position in the food chain in the deep sea ecosystem.

INTRODUCTION

Among eighty-five described species of euphausiids (Mauchline & Fisher, 1969), three species of Thysanopoda and Bentheuphausia amblyops are recognized as existing in the deep sea. For example, adults of Thysanopoda, T.egregia, T.cornuta and T.spinicaudata; and B.amblyops are distributed widely in the ocean depths (Brinton, 1962; Vinogradov, 1968; Nemoto, 1967). The biomass of euphausiids in the deeper waters of the North-western Pacific Ocean has been discussed by *Vinogradov* (1968). He suggested that euphausiid biomass is about 10 percent of the total at a depth of 3,000 meters. The species comprising this biomass are *Thysanopoda* and *Bentheuphausia*.

Some aspects of the food and feeding structures of these deep sea euphausiids have been examined (*Marshall*, 1954; *Tchindonova*, 1959; *Mauchline*, 1967; *Nemoto*, 1967), and it appears that they have an important role in the food chains of the deep sea ecosystem. The present study is concerned with the food and feeding structures of the three deep sea *Thysanopoda* species to clarify their ecological characteristics and to compare them with those of certain epipelagic and mesopelagic euphausiids. The fourth bathypelagic species, *Bentheuphausia amblyops*, is also examined and compared with the other three.

MATERIALS AND METHODS

The specimens used in the present study are as follows:
1. Specimens collected by research vessels of the Scripps Institution of Oceanography, University of California, San Diego, and kept by the Marine Life Research Group.
2. Specimens collected on cruises KH 72-1 in 1972, KH 73-1, and KH 73-2 in the western North Pacific in 1973 by the research vessel "Hakuho-Maru" of the Ocean Research Institute, University of Tokyo.
3. Specimens collected during the International Indian Ocean Expedition (IIOE) by the training ship "Kagoshima-Maru" of Kagoshima University in 1963.
4. Specimens collected by the "Anton Bruun" off the western coast of South America.

All specimens were preserved in 10% solution of neutral formalin in sea water. The numbers of individuals of each species and the ranges in their body lengths are shown in Table 1. No larval stages were present in the samples. All specimens were adult and almost sexually mature except one specimen of *Thysanopoda spinicaudata*, having a body length of 13.9 mm.

1. The wet weight of the specimens was measured by weighing the animals on a Metler Balance, Types H14 and H15. Excess moisture was removed from each individual by blotting paper and the body weight measured to 1 mg, following *Nakai and Kudo* (1967).
2. Measurements of total length and the dimensions of various parts of the body, segments and appendages were made on a 0.1 mm scale of a micrometer eye-piece in a dissecting microscope. The parts measured are the same as those measured by *Nemoto* (1968) in other species of euphausiids.
3. The weight of the stomach contents was measured to 0.1 mg using the Metler Balance, Type H15 by: a) Measuring the weight of the stomach plus its contents, and b) measuring

TABLE 1. Number of *Thysanopoda* Specimens Treated, Cruises and Organization from which Material is Derived

Species	Number of Specimens examined	Marine Life Research group Specimen	Kagoshima-Maru Cruise (IIOE)	Anton Bruun Cruise	Hakuho-Maru Cruise			Extent of body length (mm)
					KH 72-1	KH 73-1	KH 73-2	
T. cornuta	28	16	-	-	2	2	8	36.0~116.0
T. egregia	12	9	1	-	1	-	1	42.0~57.0
T. spinicaudata	6	5	-	1*	-	-	-	13.9~150.0
T. cristata	4	4	-	-	-	-	-	38.9~56.0
T. acutifrons	6	6	-	-	-	-	-	35.2~48.0
T. tricuspidata	32	-	-	-	-	32	-	15.5~23.2

* by courtesy of Mr. T. Antezana

the weight of the contents and the stomach wall separately. The weight of the former should equal the total of the latter if there is no loss of weight, and generally they show good agreement.
4. Organisms among the stomach contents were identified using the dissecting microscope. These organisms were usually broken. Measurements of whole and parts of organisms such as lenses and vertebrae of fish and mouth spines of chaetognaths were made.
5. In the case of the micronectonic fish *Cyclothone*, the relative sizes of the eye lens and vertebrae to the body length of the fish are used to estimate the size of fish from which lenses and vertebrae in the stomachs of *Thysanopoda* originated.

RESULTS

Stomach Contents

The relation between body length and body weight is expressed as

$$\log y = 2.83 \log x - 4.76$$

where y is body weight and x is body length. This equation is nearly the same as that for euphausiids in general (Mauchline & Fisher, 1969) The genus *Thysanopoda* contains larger sized species than any other genus of euphausiids. Two specimens of *T. spinicaudata* examined in the study are the largest and the second-largest euphausiids described to date (Brinton, 1962). The stomach contents of three bathypelagic species of *Thysanopoda* are shown in Table 2. Food was present in the stomachs of 23 of the 31 individuals examined. The food consisted predominantly of crustaceans, such as copepods, euphausiids, mysids, and amphipods. Copepods were most common among the stomach contents, being found in the stomachs of 14 out of 19 specimens of *T. cornuta*, and 7 out of 10 specimens of *T. egregia*. Chaetognaths are second in importance, being found in 8 out of 19 specimens of *T. cornuta*, 5 out of 10 specimens of *T. egregia*, and 1 out of 2 specimens of *T. spinicaudata*. Ommatidia and spermatophores of crustaceans were found in *T. cornuta*. Chaetognaths and radiolarians were found in stomachs of *Thysanopoda cornuta* and *T. egregia*.

The lenses of fish eyes were present in the stomachs of three *T. egregia* and fish vertebrae were found in the stomachs of a *T. egregia* and a *T. cornuta*. Fish scales were also detected in 5 specimens. However, some of these fish scales may have been eaten after the euphausiids were caught in the net. One *T. spinicaudata* had only scales in an empty stomach suggesting this. One *T. egregia* 48 mm long had 45 fish vertebrae in its stomach. Size and shape of these fish vertebrae strongly indicate that this *T. egregia* caught two or more

FOOD AND FEEDING STRUCTURES

TABLE 2. Stomach Contents of Three *Thysanopoda* Euphausiids

Species	Pisces			Crustacea								Cephalopoda	Radiolaria	Chaetognatha	Oil globules**			Detrital matter	No. of specimens containing food	No. of specimens examined
	Vertebrae	Lens	Scales*	Copepoda	Euphausiacea	Mysidacea	Amphipoda	Decapoda	Other fragments	Spermatophore	Ommatidia				o	+	++			
T. cornuta	1	-	3	14	1	3	2	2	3	2	3	-	4	8	11	6	2	-	14	19
T. egregia	1	3	1	7	3	1	-	-	2	1	-	1	1	5	5	4	1	1	7	10
T. spinicaudata	-	-	1	-	-	-	-	-	1	1	-	-	-	1	1	1	-	-	2	2

* Some may be taken in the collecting net.
** Oil presence is judged through stomach wall.

small *Cyclothone* fish or a related species. Another *T. egregia* had a larva of a pelagic cephalopod, the species being unknown. Radiolarians are the only dominant protozoa found in the stomach contents of the deep sea *Thysanopoda* species. Fourteen specimens had oil globules in their stomachs that apparently originated from the body contents of deep sea crustaceans. In a fecal pellet (1.5 cm long) in the intestine of a *T. cornuta* (78.2 mm), 4 mandibles of copepods were observed.

In Figure 1 the wet weights of stomach contents are plotted against wet body weights of euphausiids. Three feeding conditions determined subjectively and expressed as r, rr, rrr (r: little; rr: moderate; rrr: heavy) correspond to the regressions between weights of contents against body weight in logarithmic scales. The maximum wet weight of food in the stomach is 173 mg or about 1.4% for *T. cornuta* of 12.870 g in body weight.

Feeding Structures and Internal Digestive Tracts

Mandible. The general shape of the mandibles of bathypelagic *Thysanopoda* species do not differ from that of other mesopelagic and epipelagic *Thysanopoda* species (Figure 2). The *pars molaris* of the mandibles of *Thysanopoda* species is smaller than those of surface herbivorous euphausiids such as *Euphausia superba*.

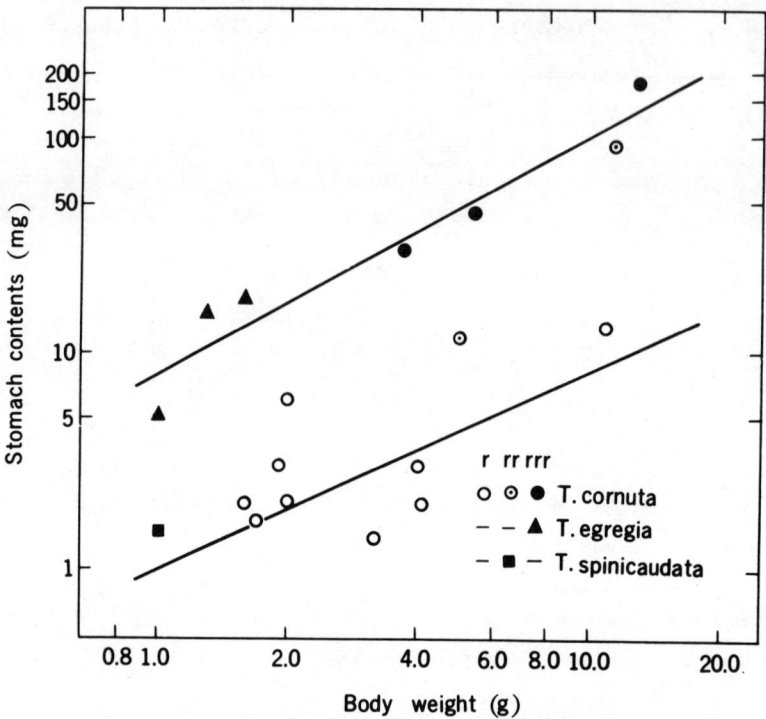

Fig. 1. Weight of stomach contents of deep sea *Thysanopoda* species. The condition of stomach fill is given as, r: little; rr: moderate; rrr: much.

Among *Thysanopoda* species, the pars molaris of the mandible is smaller in bathypelagic species and larger in the epipelagic *T. tricuspidata*. The fragments of food organisms found in the stomach of *T. spinicaudata* suggest that they are cut to this size by the pars molaris of the mandible.

The ratio of the margin length of the pars incisiva to the length of the pars molaris also varies according to the habitat of the species. *Thysanopoda egregia*, the most voracious euphausiid, shows the lowest ratio, 0.34. Values for other bathypelagic species are *T. cornuta*: 0.51; *T. spinicaudata*: 0.41. The ratio for the mesopelagic species *T. acutifrons* is 0.64, and that for *T. cristata* is 0.49. The bathypelagic, but presumed filter feeding, euphausiid *Bentheuphausia amblyops* has mandibles with a ratio of 0.71. The ratio for the presumed surface filter feeding euphausiid *Thysanopoda tricuspidata* is 0.78, which is higher still, but not yet comparable to that of the typical herbivorous filter feeding euphausiid in the surface waters, *Euphausia superba* which has a ratio of 1.09.

Maxilla and Maxillule. The general aspects of the feeding appendages of *Thysanopoda* species are described by Mauchline (1967). He also proposed clear trends in the development of maxillae and

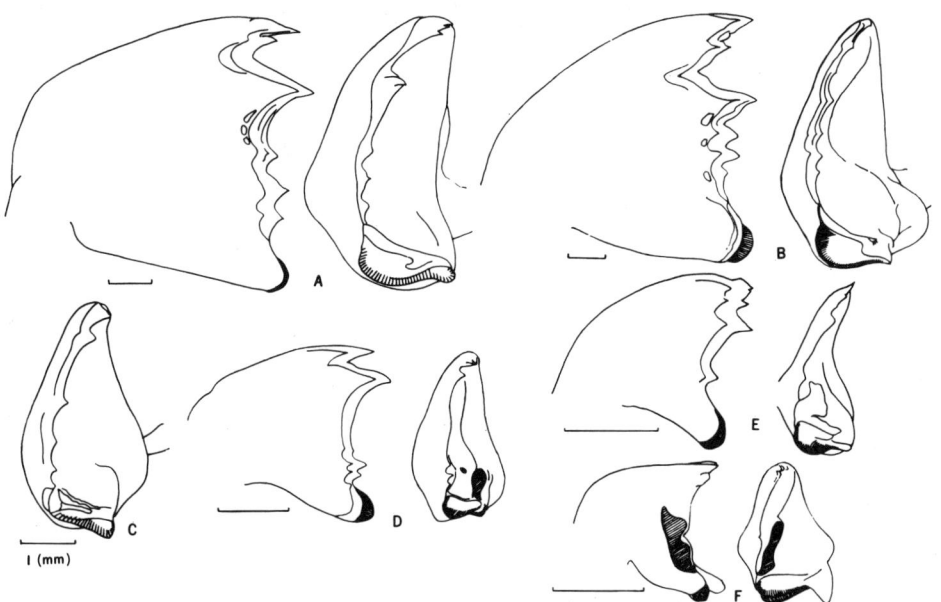

Fig. 2. Mandibles of deep sea *Thysanopoda* and *Bentheuphausia amblyops*. *Pars molaris* and effective parts for grinding are shaded. A: *Thysanopoda spinicaudata*. B: *T. cornuta*. C: *T. egregia*. D: *T. cristata*. E: *T. acutifrons*. F: *Bentheuphausia amblyops*. Each bar shows 1 mm.

maxillules. The development of the endopod and pseudexopod of the maxillule is the same as that described by Mauchline (1967). Three bathypelagic species have large endopods and small pseudexopods.

The maxilla of *T. cornuta* and *T. spinicaudata* have large and slender endopods as was shown by Brinton (1953). They bear plumose setae along the distal margin as was postulated by Mauchline (1967). The endopod of *T. egregia*, on the other hand, is rather small, perhaps smaller than that of *T. cornuta*. The three bathypelagic *Thysanopoda* species have fewer setae along the maxillar endopodite than along the maxillular endopodite. *T. egregia* in particular lacks all barbs on the margin of the lobe of the third joint. Only spines are found along that margin. The epipelagic *Thysanopoda tricuspidata* and presumed mesopelagic *T. cristata* have more and better developed setae along the margin of endopodite.

Thoracic Legs. The three bathypelagic *Thysanopoda* species have very few filtering setae and barbs along the thoracic legs. *T. egregia* completely lacks setae on the *ischium* and *merus* of the legs (Nemoto, 1967). The comparative development of the thoracic legs is different among *Thysanopoda* species. The epipelagic *T. tricuspidata* has functional filtering setae and barbs along the *ischium*, *merus*, and *carpus* of the endopodite of the legs. The allomorphs based on leg lengths

divide the three bathypelagic and one epipelagic *Thysanopoda* species
into two groups as is shown in Figure 3. *T.tricuspidata* has an elongated *carpus* relative to the *merus* of the second legs. The relations
between total length of the first legs relative to carapace margin
length, and the length of the *merus* relative to that of the *ischium*
also divide the species into two groups.

Shape and Internal Structures of Stomach. The lateral shapes of
stomachs of *Thysanopoda* species vary among species (Figure 4). Stomachs of bathypelagic *Thysanopoda* species and *T.cristata* are similar in
shape, the cardiac part being expanded. The mesopelagic species *T.
acutifrons* and *T.monacantha* have stomachs of somewhat different shapes.
T.acutifrons has a very swollen cardiac part, but the posterior part
is slender and typical of filter feeding euphausiids. General shape
of this species is similar to that of *Bentheuphausia amblyops*. The
shape of the stomach of *T.monacantha* is, on the other hand, related
to those of epipelagic species. *T.tricuspidata* also has stomach shape
typical of filter feeding euphausiids.

The stomachs of bathypelagic *Thysanopoda* euphausiids (Figure 5),
when they are empty, become folded up along the folds between stomach
wall plates. They are expanded when filled with food. The stomach
wall of surface filter feeding euphausiids such as *Euphausia* species
are very thick and do not have clear folds along the wall. The peculiar spines on the inner wall of deep sea *Thysanopoda* euphausiids are
illustrated in Figure 6. The spines are typical of carnivorous species. Along the inner wall of stomachs of herbivorous filter feeding
euphausiids such as *Euphausia superba*, there are well developed cluster spines on the plate for crushing hard shelled organisms. The
three *Thysanopoda* species have short spines in the upper part of side
plate (postero-lateral cardiac plates). However, their number is far
smaller and they do not form clusters.

Eye and Abdominal Segment. With respect to allomorphosis in body
parts, the three deep sea *Thysanopoda* species generally form one group
separated from the epipelagic *T.tricuspidata*. For example, the eyes
of *Thysanopoda* species are spherical and not constricted, but are far
smaller relative to body size than those of surface euphausiids (Nemoto, 1967). As shown in Figure 7, *T.spinicaudata* has relatively
smaller eyes than *T.egregia* and *T.cornuta*, but these three species
form a loose allometry group separate from *T.tricuspidata*. The relation between the length of the fifth and sixth abdominal segment is
shown in Figure 8. Deep sea *Thysanopoda* species have a shorter sixth
abdominal segment relative to the fifth segment.

DISCUSSION

The food of deep sea *Thysanopoda* species mainly consists of
zooplankton and micronecton, in which copepods usually dominate.
Other crustacean remains and jaws of chaetognaths are also common in
the stomach contents. Micronectonic fish and larvae of an oceanic

FOOD AND FEEDING STRUCTURES 465

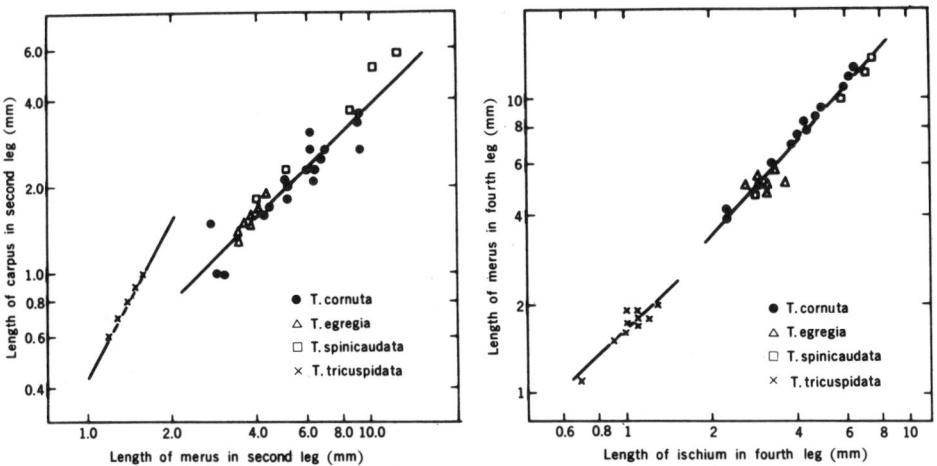

Fig. 3. Allomorphosis in the relation in the joints of the thoracic legs of *Thysanopoda* euphausiids. Left: Relation between the length of the merus of *Thysanopoda* and the lengths of the carpus in the second legs. Right: Relation between the length of the ischium and the length of the merus of the fourth leg.

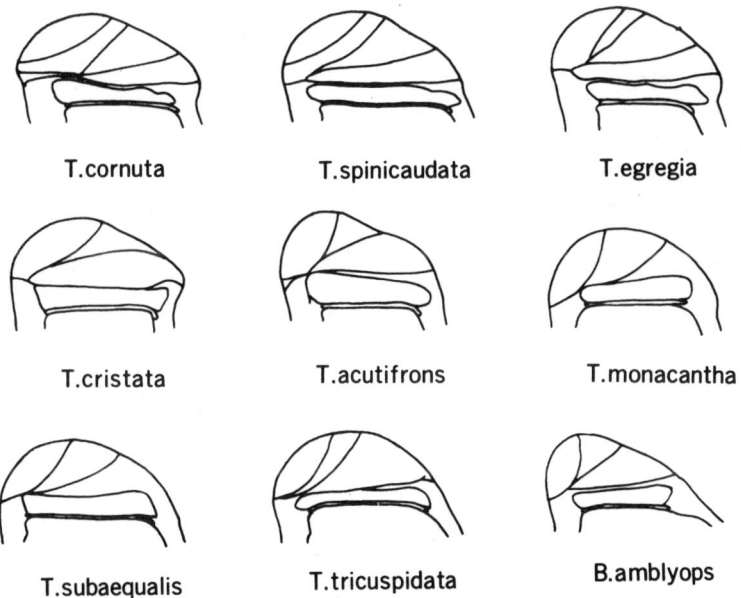

Fig. 4. Shapes of stomachs of *Thysanopoda* euphausiids and *Bentheuphausia amblyops*.

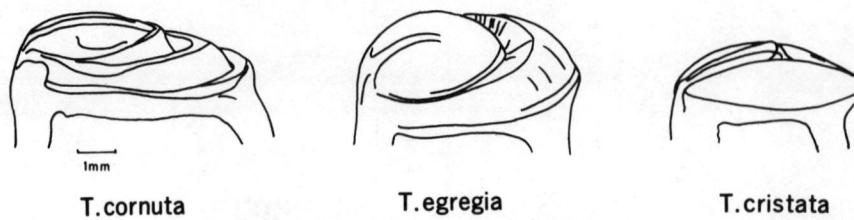

Fig. 5. Vacant stomachs of deep sea *Thysanopoda* showing the folding. Bar shows 1 mm.

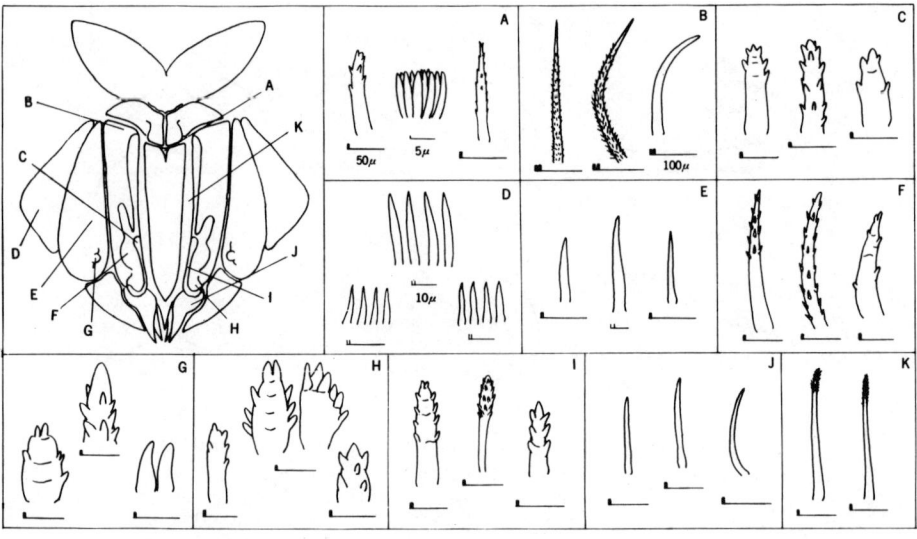

Fig. 6. Open cut view of the stomach of *T.cornuta* and shapes of spines along the inner side of the stomach walls of three deep sea *Thysanopoda* euphausiids. Left: *T.cornuta*. Middle: *T.spinicaudata*. Right: *T.egregia*. Only *T.cornuta* and *T.spinicaudata* are shown in K.

FOOD AND FEEDING STRUCTURES 467

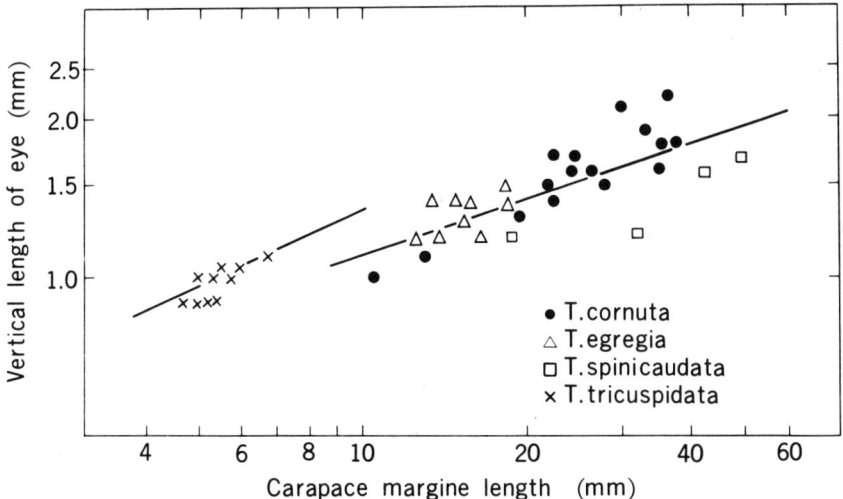

Fig. 7. Allomorphosis in the relation between the carapace margin length and the vertical length of the eyes in *Thysanopoda* euphausiids.

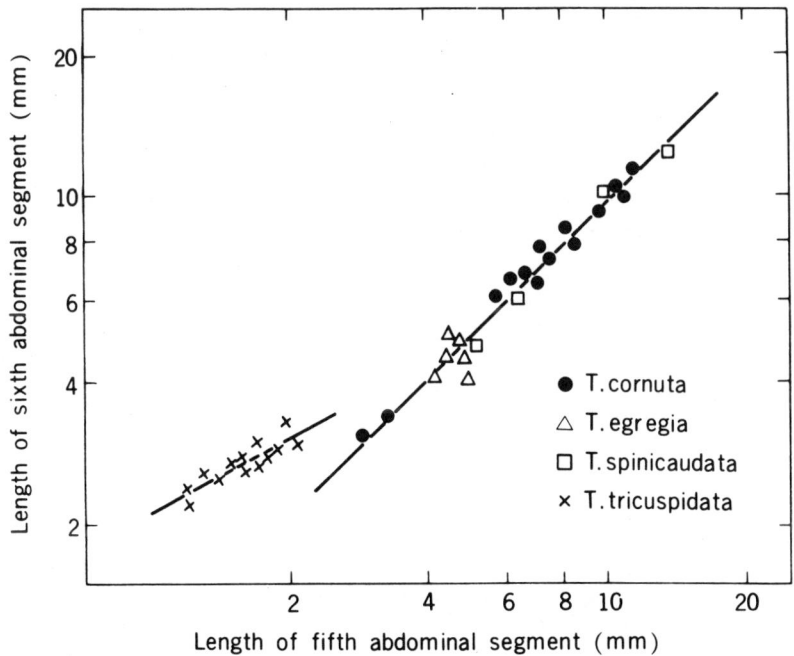

Fig. 8. Allomorphosis in the relation between the length of the fifth abdominal segment and the length of the sixth abdominal segment.

squid were also found in the stomachs of T.egregia and T.cornuta. These results suggest that the three bathypelagic Thysanopoda species are essentially carnivorous.

Tchindonova (1959) and Nemoto (1967) examined T.cornuta and T.egregia, and reported that they feed mainly on copepods. Some of the copepod fragments in stomachs of Thysanopoda euphausiids examined in this study are considered to be those of copepod detritus, as they are similar to specimens described by Wheeler (1967). Chase (1940) Pearcy & Forss (1969), Judkins & Fleminger (1972), and Aizawa (1974) also reported copepods as the main food for deep water oceanic shrimps

Chase (1940) considered that the pelagic prawns sometimes captured their prey after the specimens were in the net, and the mandibles did not crush the food into an amorphous mass before it is passed into the stomach. Judkins and Fleminger (1972) compared foregut contents of an oceanic shrimp Sergestes similis obtained from net collections and albacore stomachs. The diversity of calanoid species, and the presence of the numbers of fish scales only in shrimps caught by net tows suggest S.similis also feeds in collecting nets under tow. Calanoids and euphausiids were found to be appreciably commoner in the foreguts of net caught S.similis than in the guts of fish caught shrimps.

In this study, some stomach contents are also considered to have been taken in the net after the catch. Fish scales and some fresh euphausiids are considered to be eaten in the collecting net under tow. It is likely that Thysanopoda feed to a lesser extent in the collecting net because it is generally observed that the bathypelagic euphausiids T.cornuta and T.egregia become inactive sooner than Acanthephyra, Sergestes and other shrimps which are often still very active on board the research vessel after being caught.

To examine the possibility of feeding in the collecting nets, the plankton organisms in the food grooves of Thysanopoda were also examined. Ponomareva (1963) considers the organisms trapped in the food basket (food grooves) are important items of food. Mauchline (1967), on the other hand, states that the trapped organisms in the food basket are packed while in the collecting net and are not necessarily normal food items of euphausiids. Later, Mauchline (personal communication) agreed that the packed organisms express the size range of food of euphausiids to some extent since the organisms are subjected to the sorting activity of the thoracic legs in the collecting net. Judkins (Judkins & Fleminger, 1972) also observed that many netcaught Sergestes similis have fish scales, chaetognaths and small crustaceans packed into their mouth parts and sometimes gripped in their mandibles.

The comparison between stomach contents and organisms packed into the food basket is given in Table 3 for each of 5 specimens of T.cornuta and T.egregia. Fish scales and whole bodies of copepods, and other crustaceans, chaetognaths as well as other plankton organisms, are the main organisms caught. The fish scales found in the food basket are very large, ranging from 1.2 to 3.2 mm, and are never found in the stomachs of Thysanopoda. The lists of stomach contents

TABLE 3. Comparison of Category, Number and Size of Stomach Contents and Organisms Packed in Food Basket of *Thysanopoda cornuta* and *T. egregia**

	Body length(mm)	Stomach contents	Organisms in food basket
T. cornuta			
No. 1	88.8	Decapoda 1(0.6)	Pisces 1(6.4) Pisces scale 5(2.4) Copepoda 2(4.4) Mysidacea 1(6.4) Chaetognatha 1(2.4) Crustacea fragments many(2.4)
No. 2	54.5	Copepoda fragments many(2.4) Chaetognatha jaw 13(1.0)	Copepoda 5(2.4)
No. 3	75.5	Mysidacea 1(1.0) Amphipoda 1(0.8) Crustacea fragments (1.5)	Copepoda 7(2.0) Chaetognatha 1(8.1) Globigerina 1(0.8)
No. 4	78.2	Copepoda leg many(3.6)	Copepoda 5(2.4) Chaetognatha 1(7.3)
No. 5	79.5	Copepoda fragments many(5.7)	Copepoda 3(3.5) Chaetognatha 1(24.9)
T. egregia			
No. 1	53.3	Copepoda 2(4.0) Euphausiacea fragments Chaetognatha jaw (1.5) Radiolaria 2(0.4)	Pisces egg 2(0.8) Pisces scale 9(1.2) Copepoda 5(2.4) Mysidacea 1(6.4) Euphausiacea leg 1(3.2) Chaetognatha 2(4.0)
No. 2	—	Copepoda fragments 4(3.2) Crustacea fragments many Chaetognatha jaw 15(1.3)	Pisces scale 2(2.4) Copepoda 2(2.1) Decapoda egg 1(1.2)
No. 3	45.4	Chaetognatha jaw 10(1.3)	Copepoda 4(2.4)
No. 4	47.8	Copepoda 1(3.2)	Pisces scale 4(3.2) Chaetognatha 1(6.4) Globigerina 1(0.8)
No. 5	50.0	Crustacea fragments 25(0.2)	Pisces scale 4(1.7) Copepoda 2(1.7)

*The sizes of the organisms or the sizes of the largest fragments are in brackets in mm.

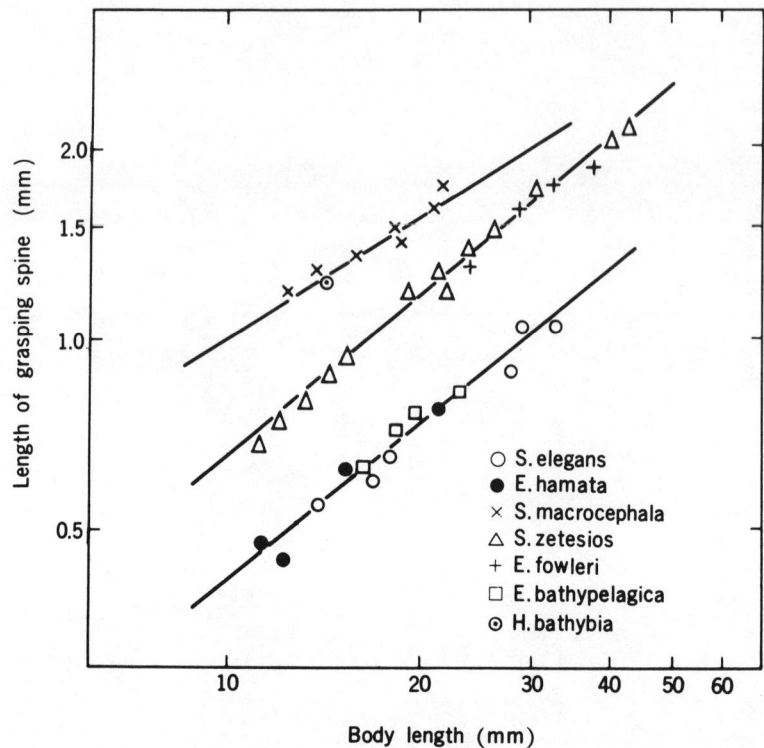

Fig. 9. Allomorphosis in the relation between the hook length of the grasping spines of chaetognaths and body length.

and organisms in the food basket show considerable differences. This suggests that large *Thysanopoda* species are less active in feeding in collecting net than other pelagic prawns of comparable size.

Almost all food is cut by the heavy mandible (pars incisiva). However, some fragments of food still remain undamaged. Jaws of chaetognaths, fish vertebrae and lens of eyes are good indicators of the body sizes of the original organisms.

The relative growth between jaws and the body length of chaetognaths is shown in Figure 9, that of vertebrae length and body length and diameter of lens of eye and body length of *Cyclothone* fish are indicated in Figure 10. The allomorphosis between jaws and body length in chaetognaths gives three groups, namely 1) *Sagitta elegans Eukrohnia hamata* (both epipelagic species) and *Eukrohnia bathypelagica*: (mesopelagic species), and 2) *Sagitta zetesios* and *Eukrohnia fowleri*: (bathypelagic species) and 3) *Sagitta macrocephala, Heterokrohnia bathybia*) both bathypelagic species). The body sizes of chaetognaths are estimated from the length of jaws found in the stomachs of *Thysanopoda*. The species of chaetognaths are also determined from the characteristics of jaws, as revealed by scanning electron microscope (Nagasawa & Marumo, 1973). Nagasawa & Marumo (1973)

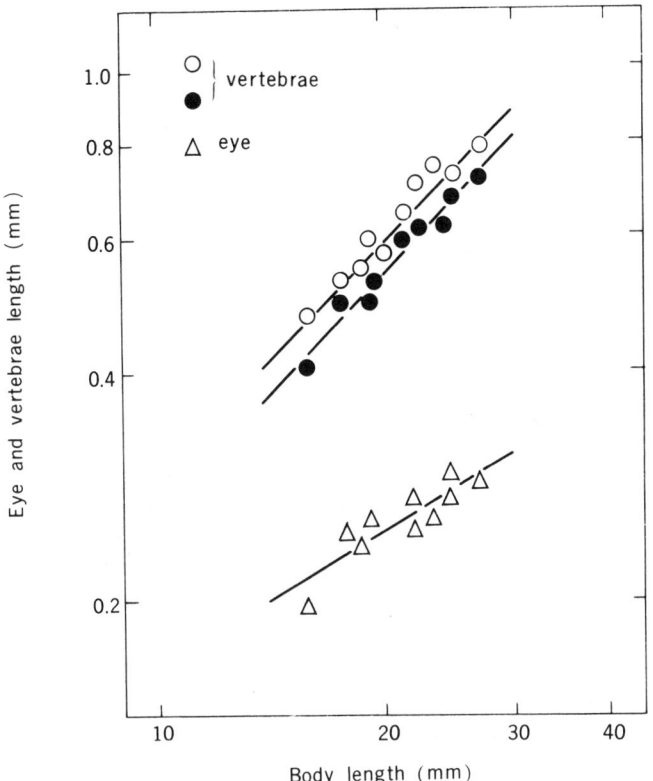

Fig. 10. Allomorphosis in the relation between the lateral length of vertebrae and the eyes and body length of *Cyclothone alba*. Open circle: 15th vertebrum. Closed circle: 5th vertebrum.

reported the characteristics of these jaws and reported one instance of *E. bathypelagica* being eaten by *T. cornuta*. The body size distribution of chaetognaths suggests that there is rarely size selection for chaetognaths by *Thysanopoda* euphausiids.

The sizes of grasping spines of chaetognaths suggest that most of the chaetognaths taken are large specimens. The small and fine spines of chaetognaths have rarely been observed in the stomachs.

The vertebrae and eye lens of fish are detected in the stomachs of *Thysanopoda cornuta* and *T. egregia*. These fish are considered to be mesopelagic *Cyclothone* species. The reconstruction of the size of this fish is based on the allomorphosis between vertebrae length and body length of one species of *Cyclothone alba*. These fish were eaten by a *T. egregia* of 48 mm (1,2666 mg) length. The ratio of combined prey fish weights against predator euphausiid is 6.1%: these fish were probably caught simultaneously, as their state of digestion was comparable. This value is high compared with the ratio estimated for

the surface feeding euphausiid (Parsons et al, 1967). Thus T.egregia
is considered to be one of the most voracious micronecton in the
bathypelagic community.

Some oceanic micronecton Crustacea also have been recorded as
feeding on micronecton fish. Chase (1940), Tchindonova (1959), and
Aizawa (1974) reported the presence of fish in the stomachs of oceanic
prawns. But, it is considered that such occurrences of fish are
comparatively few as compared with T.egregia in this report. Cyclothone fishes in turn take copepods, ostracods and chaetognaths together with unrecognizable materials (Dewitt & Cailliet, 1972). The
size range, pattern of vertical distribution (Badcock, 1970) and
the abundance of Cyclothone fish (Beebe & Vander Pyl, 1944; and
Kawaguchi, 1969) strongly suggest that they are playing a very
important role connecting smaller zooplankton, such as copepods, to
larger micronecton including giant bathypelagic euphausiids.

Some of the remains in the stomachs of Thysanopoda are considered to be eaten by smaller crustaceans such as copepods and
Bentheuphausia which are in turn fed on by Thysanopoda. This is
suggested as Thysanopoda lacks filtering setae to take radiolarians,
and by the presence of stomach walls of Bentheuphausia amblyops and
copepods appearing sometimes along with the body fragments of radiolarians. The presence of oil globules has also been observed in the
stomachs of Thysanopoda euphausiids. Oil globules are also common
in the stomachs of mesopelagic euphausiids. (Einarsson, 1945). The
origin of these oil globules is apparently copepods and other Thysanopoda and Bentheuphausia euphausiids.

The stomach samples of another bathypelagic euphausiid, Bentheuphausia amblyops, often contains degraded chlorophyll pigments in
vast quantity (Nemoto, 1968), and phytoplankton (Roger, 1974). As
B.amblyops bears functional filtering barbs on the setae along the
joints of thoracic legs (Nemoto, 1967), it is strongly suggested that
B.amblyops feeds on degraded pigments which form aggregates suitable
for filtration (Nemoto, 1968). Mauchline (1967) reported that the
mouth parts and thoracic limbs of B.amblyops suggest that this species
can feed efficiently by filtering food from the surrounding water.
Thus, there is a contrast between deep sea Thysanopoda species which
caught mainly zooplankton and micronecton, and another deep sea euphausiid Bentheuphausia amblyops which took degraded phytoplankton pigments (Nemoto, 1967, 1972).

Vinogradov (1968) considers that B.amblyops can rise to the euphotic zone and feed on phytoplankton. However, the degraded pigments
are here considered to be eaten far below the euphotic zone. Similar
degraded chlorophyll pigments were also found in the deep sea copepod,
Megacalanus princeps (Nemoto, 1972), which feeds by a steady forward
movement, filtering organisms and material from the water with the
setae of the maxillipedes (Wickstead, 1962). The typical filter
feeding euphausiids in the epipelagic layer such as Euphausia superba
in the Antarctic and Thysanopoda raschii in the northern polar seas
have generally well developed thoracic legs and filtering setae,
especially in the posterior part (Nemoto, 1966, 1967). On the other

hand, bathypelagic *Thysanopoda* species lack completely the filtering barbs along the setae in the thoracic legs (*Nemoto*, 1967). The mesopelagic *T.monacantha* has filtering barbs on the setae along the *ischium* and *merus* of the thoracic legs. Epipelagic filter feeding *Thysanopoda tricuspidata* has filtering barbs on the setae not only along *ischium* and *merus* but also along the *carpus*.

The allomorphosis relations among body portions of bathypelagic *Thysanopoda* show different groupings from the epipelagic *T.tricuspidata*, which correspond also to different feeding groups. *T.tricuspidata* has larger eyes, longer sixth abdominal segment against fifth abdominal segment, and longer *carpus* against *merus* in second legs. These characteristics apparently correspond to individuals in layers between the surface and waters of 700 depth and heavy diel vertical migration for feeding in the euphotic zone at night.

The shape of stomachs of euphausiids has been discussed by *Marshall* (1954) and *Nemoto* (1967). *Marshall* considered that the stomachs of *Euphausia* species such as *E.superba* in the high latitudes are relatively small and thick walled, and have powerful external muscles. *E.hanseni* in the tropical waters, on the other hand, has a stomach that is relatively larger and less thick walled. *Nemoto* (1967) described the swollen part observed in the pyrolic region in stomachs of carnivorous euphausiids, which is never observed in herbivorous euphausiids. The relative size of stomachs of euphausiids also differs among different feeding groups, being comparatively small in species with fine thoracic filtering setae and larger in species with coarse filters (*Marshall*, 1954).

Ponomareva (1963) reported cannibalism among the epipelagic euphausiids such as *Thysanoessa longipes* in the North Pacific. In this study, few fragments of deep sea *Thysanopoda* species have been found in the stomachs of three *Thysanopoda* species examined. Only four cases of euphausiids have been found, one of which is considered to have been taken in the collecting net. Three other traces are considered to be fragments of stomach walls and of other parts those of *Bentheuphausia amblyops*.

Predation of these *Thysanopoda* species by other deep sea organisms, has not been described. Characteristics of the hardest parts of euphausiids (i.e. mandible, spines on stomach walls and spines of thoracic legs) as revealed in this study, will be of help in identifying euphausiid fragments in stomachs of large deep sea micronecton organisms.

The large sized bathypelagic *Thysanopoda* produce fewer, but larger eggs, and possibly larger larvae after hatching (*Mauchline*, 1972). The food items of adolescents of *Thysanopods* examined here do not differ from those of adults, and one smaller *T.spinicaudata* (13.9 mm) also bears scarcely functional filtering setae along the thoracic leg joints. These results indicate that young stages of bathypelagic *Thysanopoda* adopt the diet of adults quickly (*Mauchline*, 1972), and possibly they feed on zooplankton although they live in general, in shallower waters (*Nemoto*, 1967).

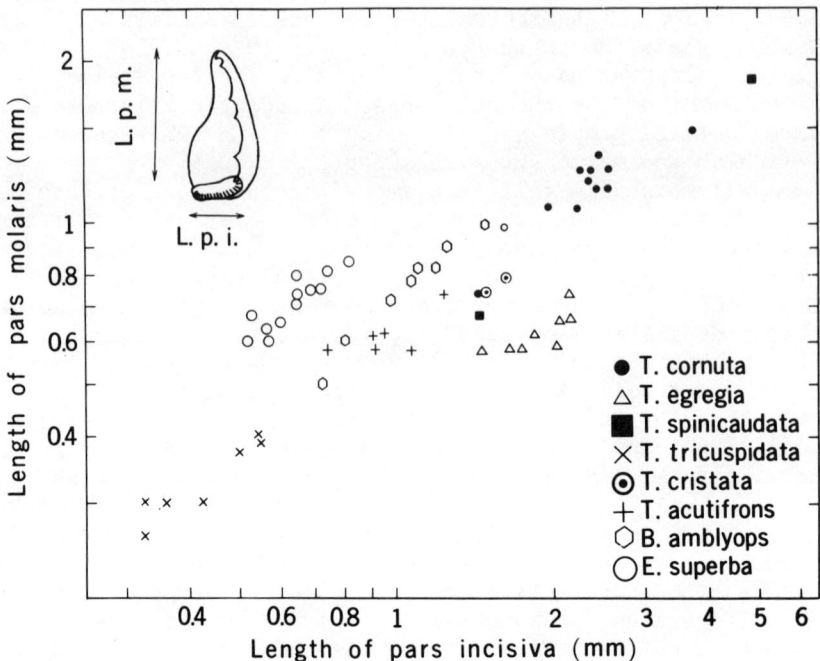

Fig. 11. Allomorphosis in the relation between the length of the *pars molaris* and the length of the *pars incisiva* in the left mandibles of euphausiids.

The allomorphosis between the length of *pars molaris* and *pars incisiva* is illustrated in Figure 11. Differences in allometry are clearly observed in *E. superba* and *Thysanopoda egregia*. These two species are typically herbivorous and carnivorous, respectively. Other euphausiids, including bathypelagic euphausiids, form one allometric group, although it is somewhat loose. The growth of the *pars molaris* shows less increase against that of *pars incisiva*, of all euphausiids described. However, in *B. amblyops*, *T. tricuspidata*, and *E. superba*, the *pars molaris* is relatively larger than in *T. egregia*, *T. acutifrons*, *T. cornuta*, and *T. spinicaudata*. This indicates that the *pars molaris* does generally not develop to the same extent as the *pars incisiva* part of the mandible.

The length of the sixth abdominal segment against the fifth abdominal segment shows a clear difference between *T. tricuspidata* and bathypelagic *Thysanopoda*. *T. tricuspidata* has a longer sixth abdominal segment (Figure 8), while the three deep sea *Thysanopoda* form the same allometric group. The longer sixth abdominal segments are possibly advantageous for the extended diurnal vertical migration in *T. tricuspidata*.

The pattern of vertical distribution of euphausiid genera in the Western North Pacific is given in Figure 12, which generally agrees with *Brinton* (1967) and *Baker* (1970). In this scheme the main euphau-

FOOD AND FEEDING STRUCTURES

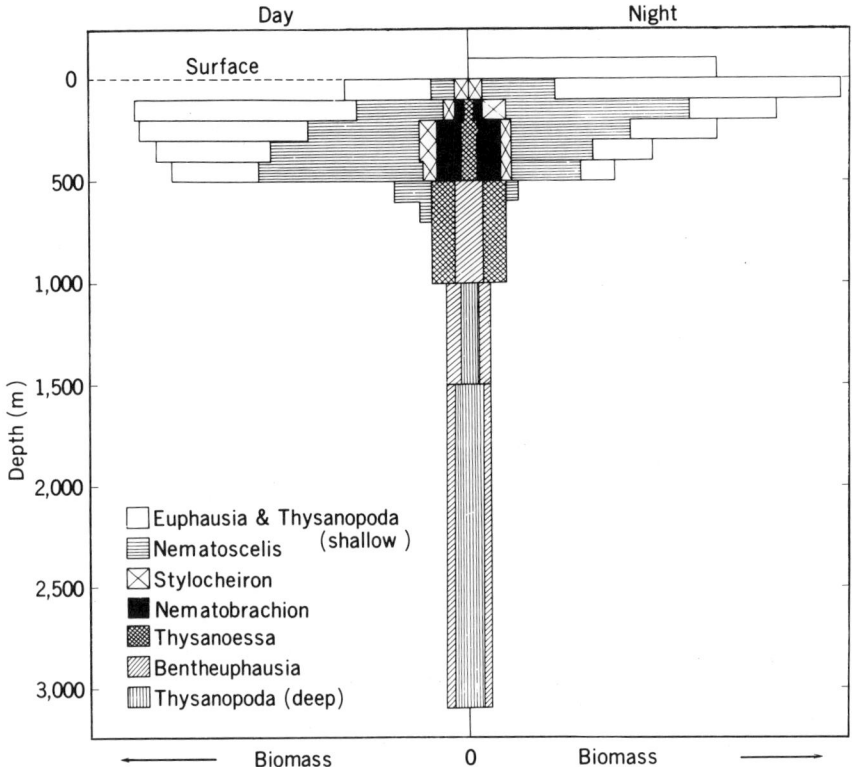

Fig. 12. Pattern of vertical distribution of euphausiids in the western North Pacific. The biomass is given in logarithmic scales.

siid species belonging to non-migrants are *Thysanoessa gregaria, Nematobrachion boopis, Stylocheiron longicorne, S.carinatum, Bentheuphausia amblyops*, and deep sea *Thysanopoda cornuta* and *T.egregia*. Migrant species are *Euphausia similis, E.nana*, and *Nematoscelis difficilis*. Epipelagic *Thysanopoda tricuspidata* (Brinton, 1962) is also a migrant. A clear diurnal migration is observed in *Euphausia* and *Nematoscelis* species. Five genera (*Thysanoessa, Stylocheiron, Nematobrachion*, and *Bentheuphausia*, and deep sea *Thysanopoda*) do not exhibit diurnal migration. Among them *Euphausia* (mainly *E.similis* and *E.nana*) and *B. amglyops* are mainly filter feeding euphausiids. The former comes up to the surface and takes phytoplankton (Nemoto, 1967), but the latter takes degraded phytoplankton pigments, possibly existing as aggregates and fecal pellets in the deep waters. Epipelagic *Thysanopoda tricuspidata* is included in this pattern of the *Euphausia* group, and comes up to the surface layer to feed at night. Weigmann (1970) described dinoflagellates and detritus along with Crustacea and other zooplankton in Arabian Sea, and Nemoto (unpublished) also observed phytoplankton pigments in stomachs of *T.tricuspidata*.

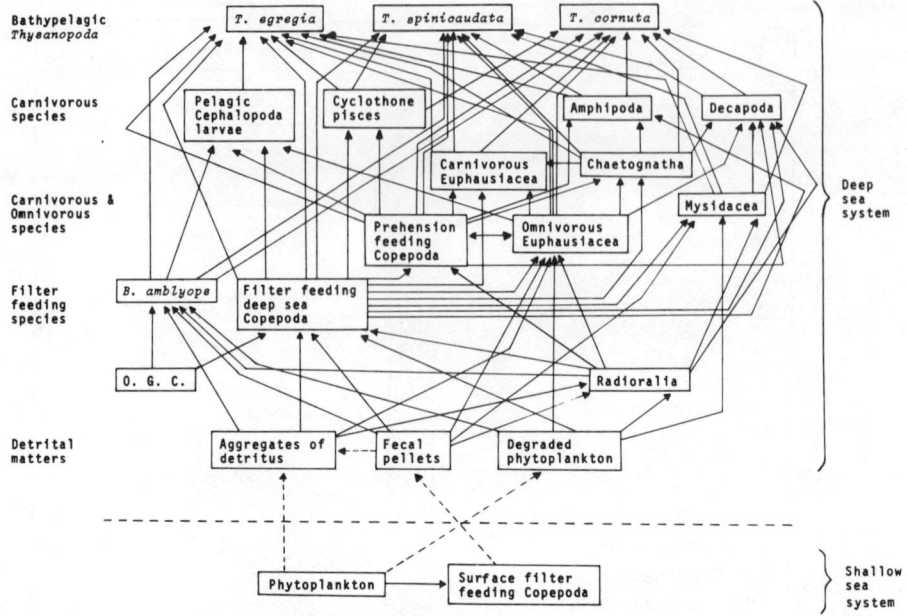

Fig. 13. Food web involving three bathypelagic *Thysanopoda* euphausiids and *Bentheuphausia amblyops*.

The size of the three bathypelagic *Thysanopoda* is very large, and they possibly live 3-7 times as long as epipelagic species (Mauchline, 1972). The large body size and increased mobility (Mauchline, 1972) and vast storing of food organisms at every chance may mean that they do not need to extend range by vertical migration. The food relations of deep sea *Thysanopoda* and *Bentheuphausia amblyops* are shown in Figure 13. Three *Thysanopoda* species are at least two steps higher than *Bentheuphausia*. *B.amblyops* ranks at the same level as filter feeding deep sea copepods such as *Megacalanus princeps*. The heavy concentration of arrows to *T.egregia* indicates its catholic list of food items. The food and feeding structures of other deep sea decapods and mysids which have body structures similar to euphausiids indicate that they may also be mainly carnivorous. However, some of them may possibly feed on detrital matter and degraded phytoplankton, their feeding processes being little known (Tchindonova, 1959). An examination should be made to divide them into separate trophic levels and estimate their biomass and activities in the deep sea ecosystem.

Perhaps one important criterion for pre-predator relationships around deep sea *Thysanopoda* is the size of organisms. *Thysanopoda* species take smaller decapods and mysids, but on the other hand, larger decapods and mysids will take larvae and adolescents of *Thysanopoda* species. The larger organisms such as macrourid fishes (*Coryphaenoides armatus*, *C.leptolepis*, and *C.filifer*) took *Thysanopoda*

(Pearcy & Ambler, 1974). Perhaps micronectonic fishes such as myctophids will not take fully grown *Thysanopoda* euphausiids, although those fishes commonly take smaller euphausiids (Paxton, 1967).

CONCLUSIONS

Studies of the food and feeding mechanisms have been made of the three deep sea euphausiid species of the genus *Thysanopoda*. The morphology of the mandible, first and second maxillae, and thoracic appendages have been described. The structure of the stomach and alimentary canal was also examined. The weight of food and type of food in the stomachs was determined.

1. The three bathypelagic species of *Thysanopoda* have mandibles of a distinctive form. The *pars molaris* is much reduced in area relative to the size of the mandible while the teeth of the *pars incisiva* are very strongly developed. The shallow living *T. tricuspidata* and *T. acutifrons* have a more developed *pars molaris* and reduced *pars incisiva* while the more predominantly herbivorous euphausiids, for example *Euphausia superba*, has a large *pars molaris* and much reduced *pars incisiva*. Another deep sea filter feeding euphausiid, *Bentheuphausia amblyops* also has moderate (not reduced) *pars molaris*.
2. The filtering barbs and setae are observed on the *carpus*, *merus*, and *ischium* in the epipelagic species, *T. tricuspidata*. These filtering barbs are also present on the *ischium* and *merus* of the mesopelagic species *T. monacantha*, but are rarely found in the three bathypelagic *Thysanopoda*.
3. The three bathypelagic species have large stomachs that become flat and folded when empty. They are distended when full. The number of spines in the cluster in the inner wall of the stomachs is smaller than in herbivorous species. There are no setae and spines as observed in more herbivorous species, the walls having the stray setae or spines characteristic of carnivorous species.
4. Copepods were the most frequently occurring organisms in the stomach contents. Other crustacea such as euphausiids, mysids, amphipods, and decapods were found along with lenses, scales, and vertebrae (probably of the genus *Cyclothone* or related species) of fish and remains of chaetognaths. Copepods appear to be the dominant and most important food organisms, followed by chaetognaths.
5. One *T. egregia* of 48 mm total body length had two fish of about 28 and 18 mm estimated body length of the genus *Cyclothone* in its stomach. The young of some oceanic squid were also found in a stomach of *T. egregia*. Thus, *T. egregia* may be the most active of the carnivorous euphausiids.
6. The weight of the stomach contents was directly related to the body weight of the *Thysanopoda* species. The maximum weight of stomach contents found was 173 mg in a *T. cornuta* of 12.870 g body weight. The stomach contents represented about 1.4% of the body weight.

7. Bathypelagic species of the euphausiids, *T.cornuta*, *T.egregia*, and *T.spinicaudata*, are all of large size. It has been shown that they feed on zooplankton, but also take small active micronekton such as fish and cephalopods. They are, therefore, very active carnivorous species. They are considered to play an important role in the ecosystem of the deep sea since they live long and occupy a high position in the food chain. Present estimates, however, suggest that their biomass is not great.

ACKNOWLEDGEMENTS

I wish to thank Miss K. Kamada and C. Takada, Ocean Research Institute, University of Tokyo, for their able collaboration in the course of this study. I also thank Dr. J. Mauchline, Dunstaffnage Marine Research Laboratory, Scotland, for his fruitful discussions of the problems. Part of this study was carried out while I was at Scripps Institution of Oceanography, University of California, San Diego, in 1971 and 1972 as a visiting scientist supported by the Japanese Ministry of Education, and as a Research Associate in the University of California, San Diego. I wish to thank Dr. E. Brinton and Prof. J. Isaacs, Program Director of the Marine Life Research Group, and members of the MLR group for their kind hospitality and stimulating discussions.

REFERENCES

Aizawa, Y. 1974. Ecological studies of micronektonic shrimps (Crustacea, Decapoda) in the western North Pacific. *Bull. Ocean. Res. Inst. Univ. Tokyo*, 6: 84 p.

Badcock, J. 1970. The vertical distribution of mesopelagic fishes collected on the Sond Cruise. *J. Mar. Bio. Assoc.*, *U.K.*, 50: 1001-1044.

Baker, A. de. 1970. The vertical distribution of euphausiids near Fuerteventura, Canary Islands (Discovery SOND cruise, 1965). *J. Mar. Bio. Assoc.*, *U.K.*, 50: 301-342.

Beebe, W. and M. Vander Pyl. 1944. Eastern Pacific Expeditions of the New York Zoological Society, XXXIII. Pacific Myctophidae (Fishes) *Zoologica*, N.Y. 29: 59-95.

Brinton, E. 1953. *Thysanopoda spinicaudata*, a new bathypelagic giant euphausiid crustacean, with comparative notes on *T.cornuta* and *T.egregia*. *J. Wash Acad. Sci.*, 43 (12): 408-412.

Brinton, E. 1962. The distribution of Pacific euphausiids. *Bull. Scripps Inst. Oceanogr.*, 8: 51-270.

Brinton, E. 1967. Vertical migration and avoidance capability of euphausiids in the California current. *Limnol. Oceanogr.*, 12: 451-483.

Chace, F. A., Jr. 1940. Plankton of the Bermuda oceanographic expeditions. *Zoologica.*, N.Y., 25: 117-209.

DeWitt, F. A. and G. M. Cailliet. 1972. Feeding habits of two bristlemouth fishes, *Cyclothone acclimidens* and *C. signata* (Gonostomatidae). *Copeia*: 868-871.

Einarsson, H. 1945. Euphausiacea, I. North Atlantic species. *Dana Rep.*, 27: 1-185.

Judkins, D. C. and A. Fleminger. 1972. Comparison of foregut contents of *Sergestes similis* obtained from net collections and albacore fishes. *Fisheries Bull.*, 70: 217-223.

Kawaguchi, K. 1969. *Ecological studies of micronectonic fishes in the Western-North Pacific*. Dissertation, University of Tokyo.

Marshall, N. B. 1954. *Aspects of deep sea biology*. Hutchinson, London. 380 pp.

Mauchline, J. 1967. Feeding appendages of the Euphausiacea (Crustacea). *J. Zool. Lond.*, 153: 1-43.

Mauchline, J. and Fisher. 1969. The biology of euphausiids. *Adv. in Mar. Biol.*, 7: 1-454.

Mauchline, J. 1972. The biology of bathypelagic organisms, especially Crustacea. *Deep-Sea Research*, 19: 753-780.

Nagasawa, S. and R. Marumo. 1973. Structure of gasping spines of six chaetognath species observed by scanning electron microscopy. *Bull. Plankton Society of Japan*, 19: 5-15.

Nakai, J. and Kudo. 1967. An error and its correction in the measurements of plankton volume and weight. *Inf. Bull. Planktology Japan. Commem., No. Dr. Y. Matsue*: 145-150.

Nemoto, T. 1966. *Thysanoessg* euphausiids, comparative morphology, allomorphosis and ecology. *Sci. Rep. Whales Res. Inst. Tokyo*, 20: 109-155.

Nemoto, T. 1967. Feeding pattern of euphausiids and differentiations in their body characters. *Inform. Bull. Plank. Japan, Commemoration, No. of Dr. T. Matsue*: 157-171. 3 pls.

Nemoto, T. 1968. Chlorophyll pigments in the stomachs of euphausiids. *J. Oceanogr. Soc. Japan*, 24: 253-260.

Nemoto, T. 1972. Chlorophyll pigments in the stomach and gut of some macrozooplankton species. *Biological Oceanography of the Northern North Pacific Ocean*: 411-418.

Parsons, T. R., R. J. LeBrasseur, and J. D. Fulton. 1967. Some observation on the dependence of zooplankton grazing on the cell and concentration of phytoplankton blooms. *J. Oceanogr. Soc. Japan*, 23: 10-17.

Paxton, J.R. 1967. A distributional analysis of the lantern fishes (Family Myctophidae) of the San Pedro Basin, California. *Copeia*: 422-440.

Pearcy, W. G. and Forss. 1969. Depth distribution of oceanic shrimps (Decapoda: Natantia) off Oregon. *J. Fish Res. Bd. Canada*, 23: 1135-1143.

Pearcy, W. G. and J. W. Ambler. 1974. Food habits of deep-sea macrourid fishes off the Oregon coast. *Deep Sea Res.*, 21: 745-759.

Ponomareva, L. A. 1963. The euphausiids of the North Pacific, their distribution and ecology. *Dokl. Akad. Nauk. USSR*: 142 pp. Israel program for Sci. Transl., 1966.

Tchindonova, J. G. 1959. Feeding of some groups of macroplankton in the north-western Pacific. *Trud. Inst. Oceanol., 30:* 166-189.

Roger, C. 1974. Les euphausiacés du Pacifique équatorial et sud tropical. *Mémoires ORSTROM., 71:* 1-265.

Vinogradov, M. E. 1968. *Vertical distribution of the oceanic zooplankton.* 399 p. Moscow, Nauka (Israel Program for Scientific Transl., 1970.

Wheeler, J. H., Jr. 1967. Copepod detritus in the deep sea. *Limnol. Oceanogr., 12:* 697-702.

Wickstead, J. H. 1962. Food and feeding in pelagic copepods. *Proc. Zool. Soc. Lond., 139:* 545-555.

Weigman, R. 1970. Zur Ökologie und Erührungsbiologie der Euphausiacea (Crustacea) in Arabiahn Meer. *'Meteor'-Forsh-Ergehnisse Reihe, 5:* 11-52.

III

BIOACOUSTICS

Recommendations of the Working Group on Bioacoustics

G. B. Farquhar and D. V. Holliday, Chairmen

Naval Oceanographic Office and Tracor, Inc.

INTRODUCTION

The goal of this workshop was to establish a clear definition of the problem of ultimately developing a prediction capability for sound scattering in the ocean from information on the physical/chemical/biological environment.

The responsibility of the bioacoustics working group evolved into several distinct activities: 1) to define the important acoustic parameters for which we require prediction; 2) to identify those non-acoustic parameters which are known to affect sound scattering; 3) to identify areas of research into oceanic mechanisms and populations in which acoustic tools can be used to aid in understanding the basic causes of sound scattering; 4) to provide expertise in acoustic techniques and the results of acoustic backscattering measurements to the remaining groups.

Our approach was to: 1) describe the state and content of our present base of knowledge; 2) identify the significant gaps in the data base; 3) provide an outline of necessary research and when possible, alternative or parallel approaches; 4) order the identified research areas in temporal sequence where important and in priority.

In order to better treat the areas of discussion, the deliberations were divided between two subgroups: a) the acoustic tools which could offer aid in understanding the basic process (Backus, Brown, Holliday, Johnson, Scully-Power, Traganza, and Vent). b) the definition of the acoustic parameters to be predicted (Baird, Brooks, Farquhar, Friedl, Love, and Peiper).

ACOUSTIC TOOLS

Part of the subgroup discussions were centered on the hypothesis that an important contribution could be made to a better understanding of the physical/chemical and biological environment in the ocean through the use of acoustics as a tool. Consequently, discussions centered on the identification, development, and use of acoustic methods and measurements to aid in defining the structure of the ocean's biological populations and the other factors which affect the backscattering of sound.

Within the above context, two topics were addressed. The first involved identification of specific problem areas and studies which might benefit from the development and application of an acoustic tool. These applications included, but were not limited to use of acoustic technology which is within the state-of-the-art. In our judgment, they represent realistic goals or objectives. The second topic dealt with the identification of a select number of research areas from which a flow of information to the bioacousticians would be beneficial. In some cases it is possible that major technological developments must be achieved before the required studies can be conducted. In other cases, the technology appears to be adequate.

Potential Future Applications of Acoustics as a Tool to Study Ocean Features

Now it is frequently feasible to design and use an acoustic system for a specific task or target organism. In some cases, a custom acoustic system can provide more information than a standard, general purpose system such as an echo sounder. However, this optimization cannot be achieved without adequate physical or biological information relevant to the task. The acquisition of this information will require an interdisciplinary approach with close cooperation of biologists, chemists and physicists who share a common interest. This cooperative experimental design is critical in understanding regional and temporal patchiness.

Acoustic techniques useful for studying ocean features are important for understanding the basic mechanisms which control sound scattering in the ocean. The most promising topics for such acoustic techniques are:

A. <u>Patchiness and Aggregation in Time and Space</u>. It is generally accepted that the properties and populations of the sea are neither uniformly nor randomly distributed in space, but are in patterns that vary with time. In attempting to study the causes and effects of patchiness, a means of locating and rapidly delineating patch size and distribution could be useful, as could a study of the internal structure of patches. The application of currently employed techniques of sonar mapping of schools of epipelagic fish to zooplankton and phytoplankton are probable future developments.

B. **Net Calibration.** Nets will continue to be used for capturing aquatic organisms in order to meet a variety of requirements including the collection of "ground truth" for acoustic techniques. Understanding avoidance and the spatial distribution of organisms are two problems which may eventually be treated acoustically. In the first case sonar may offer a means of observing avoidance behavior, thereby assisting in a solution of the problem or calibration of the net catch relative to the total organisms in the path. Some questions regarding quantitative assessment of the spatial distribution of organisms with nets may eventually be settled either by direct measurement with narrow beam sonars, or by a forward-looking, narrow beam system mounted at the mouth of a net. Detection of organisms at the net mouth would then be correlated with their arrival at the cod end, thereby assessing the net performance.

C. **Remote Classification and Possible Identification of Organisms.** Under some conditions one type of organism can be distinguished acoustically from other types of fauna and from non-biological targets. This distinction or classification depends upon the evaluation of clues in the scattering response of the target. These clues in turn depend on the physical properties and behavioral traits of the organism. Classification and identification are more likely when non-acoustical information is also used. Success in remote identification of acoustic scatterers will depend on interdisciplinary efforts.

D. **Geographic and Depth Distribution of Organisms.** Acoustic mapping of geographic and vertical distribution of marine organisms is required to gain an estimate of the distribution of populations in the world's oceans on a broad scale. By using acoustic techniques, over large frequency ranges (cd. 100 Hz - 10 mHz) the distribution of nekton, plankton, other invertebrates, and mammals could be obtained; in addition, the distribution of commercial species may be estimated on a temporal basis. The acoustic mapping method would also yield useful information on relative scattering strength for broad geographical areas. We strongly recommend that the acquisition of such acoustic data be made in a high resolution form in both depth and frequency. Different derivative measures can then be obtained for different purposes. Current procedures such as measuring integrated column strength over third octave bands allows no flexibility in the use of the data for purposes other than that for which the original experiment was designed.

E. **Quantitative Biomass Estimation.** For diffuse aggregations, layers or patches, the two most promising techniques are the echo integrator and a procedure based on the statistical character of zero crossings in acoustic echoes. Even in their rather preliminary stages these two acoustic techniques offer implied accuracies in some interesting situations which exceed our ability to validate them by any known direct sampling process. The assessment of biomass in schools and dense aggregations is currently limited to measurement of horizontal dimensions. Several more sophisticated acoustic techniques are recognized as potentially fruitful, but none are currently under active investigation. Remote acoustic identification will be required before quantitative acoustic biomass sampling is fully realized.

F. <u>Behavioral Activity</u>. Acoustic tools could possibly be developed to study faunal behavior such as swimming, reaction to stimuli, shape and size of schools, spacing within a school and migration. Some suggested applications are in studies of locomotion (modes, endurance), dispersal and compaction, tracking of mobile animals with respect to the motion of the water column, reactions to nets, pumps, sound and light, and reactions to other organisms including predation. Perhaps the most important benefit of the acquisition of data on these items is to provide species specific behavioral clues for use in remote identification. Other benefits are definition of the location and mix of organisms, inputs into the biases of patchiness, better estimates of avoidance of direct sampling and for use in energetics studies.

G. <u>Physical Properties and Physiological Characteristics</u>. Acoustic scattering has been used to provide information on volume changes of gas filled swimbladders. Similar techniques can be used to study physical properties such as density, bulk modulus and shear modulus of the body parts of organisms.

H. <u>Direct or Indirect Observation of Physical and Chemical Phenomena</u>. Whenever there is a change in the acoustic impedance of a transmission path, some fraction of the incident energy is backscattered and the associated propagation characteristics are modified. The amount of backscatter is dependent on the amount of change in the acoustic impedance and the frequency of the incident energy. Thus the presence of turbulence, sharp density gradients, fronts, suspended solids and free bubbles should be directly observable in the ocean at particular acoustic frequencies. Indeed, all have been noted, although their documentation is extremely sparse.

Studies dealing with the cause and effect relationships between volume reverberation and chemistry of the oceans should consider ways and means to better define water masses or natural hydrographic regions in terms of their chemistry and natural boundaries, including the chemistry of eddies, upwelling, divergence, convergence, etc. These physical mechanisms of chemical entrainment of life substances may well be identified through acoustic techniques. This capability, if coupled with automated chemical analyzers, could greatly assist in the description of chemical distribution in the ocean.

Information Required to Aid the Development of Acoustic Tools for Studying the Water Column and its Contents

We have sensed an increased need for description of both acoustic and environmental information in a probabilistic rather than in a purely deterministic mode. Just as acoustic classification of fauna will be expressed in terms of probability of correct identification, so we see a need for many biological and chemical parameters and descriptions to be expressed in probability distribution functions. Many of the measurements in priority lists A and B below are of this character. However, recognizing that statistical descriptions of nature can be modified and strengthened by appropriate inclusion of deterministic effects, we feel that continued attention must be given to parameters of this latter type; thus our inclusion of priority lists C and D. Although not quite so critical as measurements in categories A and B, they must not be ignored if we are to develop an ultimately adequate description.

The essential information in order of importance is:

A. <u>Behavior of Organisms</u>. The orientation of animals such as fish and squid affects acoustic signals. How are organisms oriented in the ocean? Does this orientation affect grouping?

At what rates do animals move about? Do they move at a constant rate, in bursts, or in slow movement followed by quiet periods? What are the stimuli which cause them to move?

How do organisms react to physical and chemical parameters in the ocean and what is the distribution of these parameters in time and space?

B. <u>Patchiness</u>. Groupings of organisms are called aggregations, patches, layers, or schools, depending on packing density, orientation, and other factors. The dimensions of these groups vary. What are the dimensions attained and how do they vary in space and time? What is the spacing between individuals in a group or between and among groups? What influences this spacing? Mechanisms for grouping are known, such as feeding and reproductive behavior, reaction to prey and physical oceanographic parameters such as upwelling, currents and mixing. Are there other causes? Do the patches contain more than one type of organism? How are groups of more than one kind of organism arranged in space? Does this arrangement change with time and to what extent? The dimensions of groups and spacing and orientation within groups have direct influence on acoustic scattering, and knowledge of the above will greatly aid modeling efforts.

Patchiness is accepted as the principal problem in the sampling of organisms of all sizes which exhibit that behavior or distribution. In an equally important sense, patchiness in organisms which scatter sound in the ocean significantly affects the statistics of the received echo signals and thereby often determines or limits the performance of sonars in the detection, classification, and localization of discrete targets in the ocean. Any information which can be obtained on the distribution, sizes or internal structure of such patches would be useful in acoustic system design and operation.

From the view of an ocean ecologist, measurements of patchiness can be useful in providing basic information on its causes and effects. These include eddies, areas of strong vertical mixing, upwelling, internal waves, Langmuir circulation, convergences and divergences. Also included are rip tides, currents, physical and chemical boundaries or gradients, reproduction rates, social behavior, chemotactic substances, life histories, locomotion, growth rates, sinking, migration, light and weather.

C. <u>Physical Properties of Organisms</u>. The physical properties of body parts of marine organisms are needed for use in and to refine scattering models. The accuracy desired is about 5%. The most important parameters include the density, bulk and shear modulus of the body parts of the organism, and the elasticity of the surrounding tissue.

D. <u>Fish Gasbladder Dynamics and Morphology</u>. The shape and size of fish gasbladders determine resonant frequency, target strength and mode of vibration. Information about changes in the volume and shape of bladders as a function of pressure and the mechanisms for these changes would assist in the interpretation of acoustic scattering from fish. In addition, the identification of those fish which change bladder gas volume actively in response to environmental changes is essential to a continued refinement of all of the acoustic measurements made on organisms which contain gas bladders.

E. <u>Oceanic Areas Especially Suitable for Sound Scattering Studies</u>. In order to attain their objectives, acoustical backscattering experiments may require certain categories of assemblages of marine organisms such as monospecific schooled populations of fish, crustaceans, etc.; areas of high or low abundance; shallow or deep stratification; mixed populations; populations distributed by size, and so forth. Many of these distributional patterns are caused by life history stages, abundance of nutrients and food, and physical parameters over daily, diel, seasonal and annual cycles. Interested biologists can assist acousticians in defining the best areas and times to conduct experiments.

Major Technological Advances Required

The application of acoustic techniques and the biological measurements necessary to improve and extend these techniques require improvement of the technology presently used. In the case of bioacoustic studies, the use of <u>existing</u> equipment and standard methods from other underwater sound work could produce a <u>significant</u> increase in capability and performance. Use of the best tools currently available will not, however, achieve all the results that are feasible with acoustic techniques. Major technological advances would be useful in at least the following areas involving theoretical studies: modeling for multiple scattering and other effects which depend on target spacing; strategies for measuring the size and spatial distribution of patches; scattering model for fish without gas filled swimbladders; and limits to target classification based on information theory.

In like manner, major technological advances would be useful in the following areas of application technology and equipment development: development of techniques to measure bulk modulus, shear modulus, density, swimbladder size, shape and condition, and sound speed in live organisms; improvements in procedures to conduct physical, acoustical and physiological studies of marine organisms under *in situ* conditions; increased frequency range of bioacoustic studies, at least to 10 mHz; increased spatial resolution through the formation of multiple beams and the use of parametric arrays; techniques for the study of demersal fish; vertically profiling packages; and automatic self-doppler compensation equipment.

ACOUSTIC PARAMETERS

The purpose of this subgroup was to examine the acoustic characteristics of the ocean in relation to physical, chemical and biological processes, assess the interrelationships in these four disciplinary areas, and provide recommendations or define problem areas relative to studies which will lead to better understanding of acoustic sound scattering in the oceans. Much of what we say in this summary statement has been fostered by stimulating discussions with other marine physicists, biologists, and chemists participating in this conference. As scientists concerned primarily with bioacoustic considerations, we do, however, assume full responsibility for our conclusions, recommendations, and statements and trust they will serve as a stimulus to our colleagues in their future research efforts. The hoped-for result will be the identification of environmental (including biological, chemical and physical) oceanic characteristics that are, or may be, related to acoustic scattering properties, and which may be used either to validate models or to provide necessary data to modelers to improve their approach to the development of total "structured ocean" descriptions.

In a far more fundamental sense, we at this conference are trying to determine those ways in which an understanding of acoustic scattering, together with all other oceanic processes, will enhance our collective knowledge of the oceans and our capability for the modeling and prediction of such processes.

Both resonant (fish bladders, etc.) and non-resonant scattering effects are operable. A fish may be an effective resonant scatterer at a particular frequency dependent upon the size of the swimbladder (or gas bubble) it encloses. However, the same fish may also effectively scatter sound at higher frequencies on the basis of its body length with respect to the wavelength of the frequency being used. Zooplankton also scatter sound at the higher frequencies.

The list of problem areas in acoustic scattering includes: frequency; time of day; depth, geographic location; season; resonant and non-resonant effects; size-frequency distributions of all species; spacing of individuals (fish school effects) and patchiness; acoustic cross-sections, and their variations as a function of bubble size, size of scatterer, species, depth in the water column, and tissue effects; contributions by bubble generation through chemical, biochemical, or physical processes, or contributions due to non-living particulate material and physical inhomogeneities; and lack of agreement between measurements and equations that have been developed to predict or model the physics of swimbladder scattering and resonance.

This list is a long and involved one, and from it we can conduct an inferential exercise which will enable us to define somewhat more interdisciplinary problem areas and make some recommendations about attacking those problems. The basic needs or problem areas are as follows:

A. Scattering strength measurements should ideally be of scattering strength per unit volume as a function of depth. This is not to say that column strength data are never useful, but column strength can be derived from profile data, and the converse is not true;

B. Information on the functional and physical morphology of of the sound scattering fishes, including studies which address the problem of dynamic swimbladder changes as a function of depth, the regulatory mechanisms involved, and the role of fat investment of the swimbladder;

C. The roles of the photic, hydrodynamic, thermal and chemical environment in stimulating, regulating, or otherwise controlling behavior patterns of scatterers;

D. More definitive descriptive information on large scale patterns, (e.g., water masses) since the factors of abundance, speciation, evolutionary (gene pool) history, and food web economics, coupled with physical and chemical parameters appear useful as a basis for partitioning the oceans into areas of common acoustic properties;

E. As an aid to more meaningful interpretation of acoustic data, conduct definitive studies of food relationships and feeding behavior of mesopelagic organisms, and other aspects such as breeding cycles and the role of ontogeny in patchiness, size distribution, and abundance of populations;

F. Information on the relationship between behavior of

organisms and acoustic pattern recognition, and how the identification of specific acoustic signatures will be a useful tool in studying such populations;

G. Expansion of acoustic measurement techniques to cover a broad frequency range, at least from 0.1 to 200 kHz, and probably into the megahertz region, so that both resonant and non-resonant scattering effects may be examined, and that the utility of high frequency sources as a tool for examining zooplankton populations may serve, as well, in investigations of this vital part of food chain ecology.

H. From the standpoint of mesopelagic fish distribution, as well as the distributional features of zooplankton populations, investigate the role of all scales of ocean inhomogeneities from microstructure to gyres, their duration statistics, and chemical composition as an aid toward understanding population dynamics and the concomitant relationship that may exist between these phenomena and acoustic scattering variability in the oceans;

I. Mathematical techniques need to be refined to relate results across disciplines. Data, including temporal and spatial variances of measurements, should be gathered and presented in a manner amenable to cross-disciplinary utilization and interpretation and should be routinely available.

In considering the problems listed above, it must be remembered that acoustic techniques do not separate species. Net sampling at all scales must be continued both to determine distributions (total and individual species abundance) and to collect organisms for morphological and physiological studies. Major problem areas still exist in sampling the larger and faster fishes and cephalopods with standard nets, and in sampling over small scales (microstructure and patchiness). Other techniques need to be developed to better sample these organisms for physiological studies, perhaps from submersibles.

It is clear from recent studies that certain specialized environments, which may contain monospecific scattering layers, such as in the Cariaco Trench, may represent good natural laboratories for special tests and studies. The use of such areas should be kept in mind.

The choice of experimental sites must be based on knowledge of the environment and its variability. Only in this way will it be possible to isolate for study the specific factors affecting the overall problem.

Biological Sound Scattering in the Oceans: A Review

G. Brooke Farquhar

U. S. Naval Oceanographic Office

ABSTRACT

This paper reviews the phenomenon of biological sound scattering in the ocean, including investigative procedures and frequency dependence, causative organisms, vertical migration, acoustic patterns of scattering, and environmental considerations. Resonant scattering, such as by bubble-bearing mesopelagic fishes, is shown to be significant over the frequency range of most measurements; non-resonant scattering, and the contributions of planktonic forms, such as crustaceans, at the higher frequencies, also are significant aspects of the scattering problem. Most of the principal scattering groups undergo daily vertical migrations, thus influencing the acoustic patterns of scattering. Regional variations in acoustic scattering are associated with the large scale features of circulation in the ocean, and pronounced changes occur when crossing oceanographic boundaries between water masses.

INTRODUCTION

In April 1970, the Department of the Navy, through the Maury Center for Ocean Sciences, sponsored a three-day international conference on biological sound scattering in the ocean, held at the Airlie House Conference Center in Warrenton, Virginia. That symposium, born out of the Navy's strong interest in sound transmission in the oceans, was organized to review current understanding of the biological, environmental, and acoustic aspects of sound scattering. The proceedings volume from that conference, therefore, represents a good review of the subject through the spring of 1970 (*Farquhar*, 1971).

Interest in sound scattering has continued since that time, of course, and some of the recent published work is particularly germane to the conference which is the subject of this book. This presentation is an attempt to provide background concerning the acoustic scattering patterns that have been observed and to give a feeling for some of the approaches that have been used in relating those patterns to environmental and biological parameters. What this discussion will not be is a lesson in underwater acoustics; it further presupposes the basic facts that when one goes to sea and watches the standard echosounder, operating at standard settings, one sees one or more dark traces at mid-depths (generally, somewhere in the upper 1000 meters) which move toward the surface around sunset and move back to mid-depths around sunrise; that the returning, or backscattered sound energy that causes these traces is produced by layered populations of marine organisms, many or most of which carry some sort of gas bubble; and that the patterns of these traces manifested in the echosounder records may be highly variable, from layers (or portions of layers) which show no diel migration to multiple layers which migrate over a considerable extent (several hundreds of meters). This is, of course, a very simplistic statement of the phenomenon but, since its discovery in the 1940's, this pattern has formed the basis for much of the ensuing work that has been done.

In my view, what we attempted to do at the conference at Asilomar was to examine oceanographic (throughout this paper "oceanographic" is intended to imply physical and chemical aspects), biological, and acoustic patterns, relationships, and variability, and their interactions as we understood them. From a "bioacoustics" viewpoint, my approach is descriptive, and it was my intent to arm the participants of the conference with sufficient information to make the workshop sessions which followed the plenary sessions profitable. These sessions, after all, would be the real heart of the meeting.

This review suffers from one serious deficiency in that it does not deal with aspects of scattering and reflection of sound as they concern commercial fishery interests. I point it out as an acknowledgment of the voluminous literature dealing with fishery-related hydroacoustics. It has been mutually beneficial in the past for fishery scientists and investigators working on volume reverberation to talk with each other, and will be in the future.

METHOD OF INVESTIGATION

It is appropriate to look briefly at those techniques that have been used to learn something about the phenomenon of sound scattering. Over recent years, it has been convenient to say that the dominant scattering organisms are bubble-bearing, and may either be mesopelagic fishes with swimbladders, or siphonophores (Barham, 1963). The main reason for this assumption is centered around the frequency domain of the acoustic measurements, and the knowledge, from the

standpoint of physics, that gaseous structures such as swimbladders in fishes and the pneumatophones of siphonophores are,
indeed, effective scatterers at frequencies close to resonance.

The significance of frequency dependence of sound scattering
cannot be stressed strongly enough. Most acoustic measurements
of scattering have been made in the frequency region where resonant
scattering is the most important contribution. The frequencies
scattered back by such bubble-bearing organisms as lanternfishes
are primarily dependent upon the size of the "bubble", or swimbladder. Bubble size, or bladder size, and its resonant frequency
also vary with depth, and since most of these animals change their
position in the water column twice a day through diel migration,
they may be effective scatterers at more than one frequency.
Figure 1 demonstrates the complexity of the frequency problem with
which we must deal. This is a night dipnet collection from a station in the mid-Atlantic Ocean, of a single species of lanternfish,
Myctophum punctatum. This species undergoes extensive diel migration all the way to the surface at night, and possesses a well-
developed, gas-filled swimbladder. The size distribution in this
single sample, and the fact that each individual encloses a bubble
of different size, implies that a particular species may effectively
scatter sound over a broad range of frequencies.

Clearly, resonant scattering is not the only kind of scattering.
The work of *Bary and Pieper* (1971), *Pieper* (in press), *Hansen and
Dunbar* (1971), and others, has demonstrated that when high frequency echosounders are used (i.e., frequencies of 40 kHz to about
300 kHz) other faunal elements, such as euphausiids and pteropods,
begin to be "seen" in midwater populations. *Castile* (1975) provides some evidence that copepods about 1 mm in length, and even
phytoplankton, may contribute to reverberation at 330 kHz. The
implication of this work is that high frequency acoustic sources may
represent a useful way of examining plankton populations and their
changes with time.

An added complication to the frequency dependence of scattering
is the fact that although a bubble-bearing fish of a given size may
scatter sound most effectively at resonance, it may also contribute
to scattering at higher frequencies, in the geometric region of
scattering (length of fish, L, over acoustic wave length, λ, $\geq$
about 1). Similarly, for any given frequency, a large, non-resonant
fish can have a larger acoustic cross-section than a small, resonant
fish (see, for example, *Love*, 1969, and *Love*, 1971, pp. 351-353).

The approach in most recent studies has been to combine
acoustic measurements with environmental data and net hauls. As
biologists recognize, net hauling can be an inexact way to look at
biological distributions. A net has a built-in filter mechanism
because of mouth and mesh size. A haul of, say, one hour's duration
integrates the distribution of organisms it does catch, and sheds
little light on patchiness or clumped distrbutions without the use
of specialized sampling procedures, such as replicate tows. It has

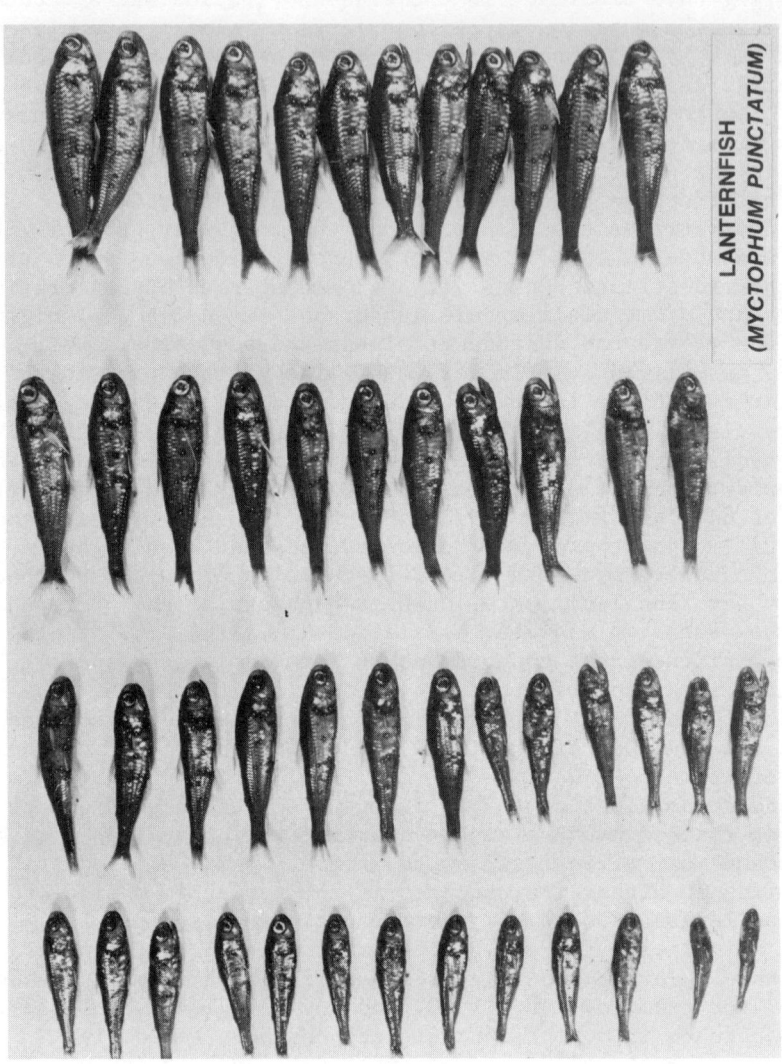

Fig. 1. Single night dipnet collection of the lanternfish, *Myctophum punctatum*, taken in the mid-Atlantic Ocean.

been pointed out by a number of workers that no single net will
effectively sample, representatively, all of the kinds and sizes
of organisms in the water column. The result is that various kinds
of "universes" of oceanic biota are being dealt with, built around
such devices as one-meter ring nets with various mesh sizes, 6-foot
or 10-foot trawls--and even such large nets as the Engel trawl.
Clearly, the development and engineering of instrumented trawls,
the use of which has increased tremendously in the last decade and
a half (e.g., *Aron et al.*, 1964; *Pearcy and Mesecar*, 1971), have
greatly enhanced the value of biological sampling programs in the
deep ocean basins. Isolumes and isotherms can be sampled with tows,
the depth of the trawl in the water can be monitored, and a measure
of filtered water volume obtained. With multi-chambered or multi-
netted cod end units, or even multi-mouthed nets, several samples
can be obtained from different parts of the water column during the
same tow, a tremendous savings in time and money. With this sort
of multi-sampler capability, information has been acquired about
the arrival and departure times of some scatterers into and out of
the surface layers (0-150 m) by holding the trawl at a fixed depth
during a diel migration, and successively opening and closing the
various samplers at certain time intervals. Figure 2 illustrates
the kind of tow path attainable with an instrumented trawl. This
particular tow was made using a 6-foot Isaacs-Kidd Mid-water Trawl
(IKMT) instrumented as described by *Aron et al.* (1964).

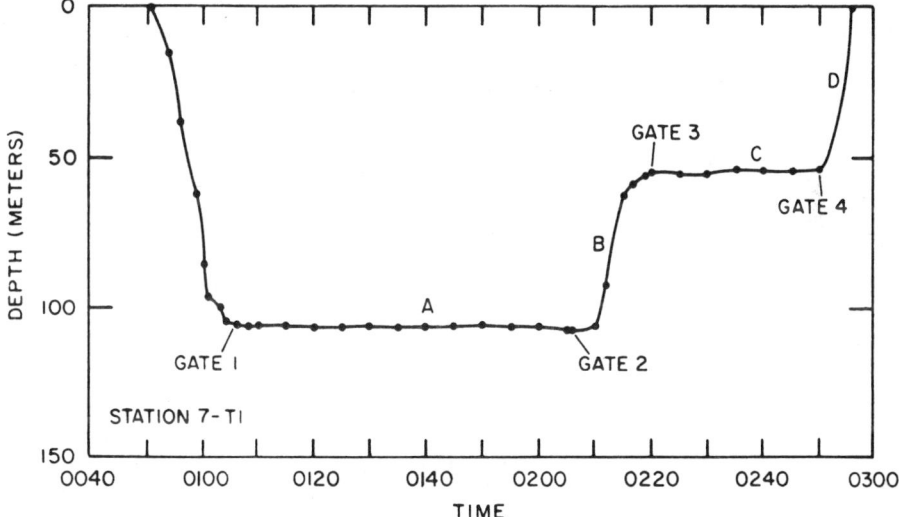

Fig. 2. Towpath using an instrumented 6-ft Isaacs-Kidd Midwater
Trawl, in which two horizontal discrete samples (A and C) and two
oblique samples are obtained.

The use of Deep Submergence Vehicles (DSV) has added significantly to knowledge of scattering layers and to an understanding of the behavioral patterns of mesopelagic organisms. In a biological sense, the latter point is the most significant, because mesopelagic fishes are fragile, and even dipnetted specimens last in shipboard aquaria only a short time. Physiological studies of these animals under controlled conditions, which would provide some insight particularly in the problem area of swimbladder gas mechanisms, thus have never been done.

The knowledge that is possessed of these organisms has been gleaned, first of all, by extrapolation from swimbladder studies of shallow coastal or fresh water species and, more importantly, by direct observation from DSV's. *Barham* (1971, and earlier work) has pointedly shown the value and utility of such observations from a DSV. He, and others, have disclosed a large body of facts about the behavior of these difficult-to-study animals and their probable roles as sound scatterers. For example, *Backus et al.* (1968) in a series of dives in ALVIN, were able to ascribe a peculiar scattering layer, known as Alexander's Acres, to schools of a lanternfish, *Ceratoscopelus madarensis*. *Barham* (1971) described a pattern of at-depth, daytime behavior for several kinds of principal migrators, based on about 50 DSV dives over a period of about 6 years. His observations show that, during the day, many of these organisms remain relatively motionless ("lethargic"), and often are vertically oriented, either head up, or head down. Nighttime observations of the same species, in the upper 100 m, show that most of them are actively swimming and horizontally oriented. Reports by other observers (see, e.g., *Farquhar*, 1971, pp. 116-118), including the data reported by *Backus et al.* (1968), on *Ceratoscopelus madarensis*, substantiate the concept of relative inactivity of many mesopelagic organisms during the day, and high activity at night, perhaps related to feeding behavior.

Acoustic measurement techniques are summarized in Figure 3. In general, the methods employ either a pulsed CW transducer operating at a fixed frequency, or an explosive source and an omnidirectional or directional receiver (*Chapman, et al.* 1971, 1974). Bottom mounted transducers have been employed which look upward (*Swarts*, 1971), and explosive arrays lowered over the side to make *in situ* measurements of volume scattering strengths (*Scrimger and Turner*, 1973; *Love*, 1977, this volume) also have been used successfully.

There are many references which could be cited, but for a concise review of the acoustic aspects of volume scattering, and an example of a joint acoustic and biological experiment, see *Love*, 1975 and 1977.

CAUSATIVE ORGANISMS

The purpose of this conference was to come to agreement on

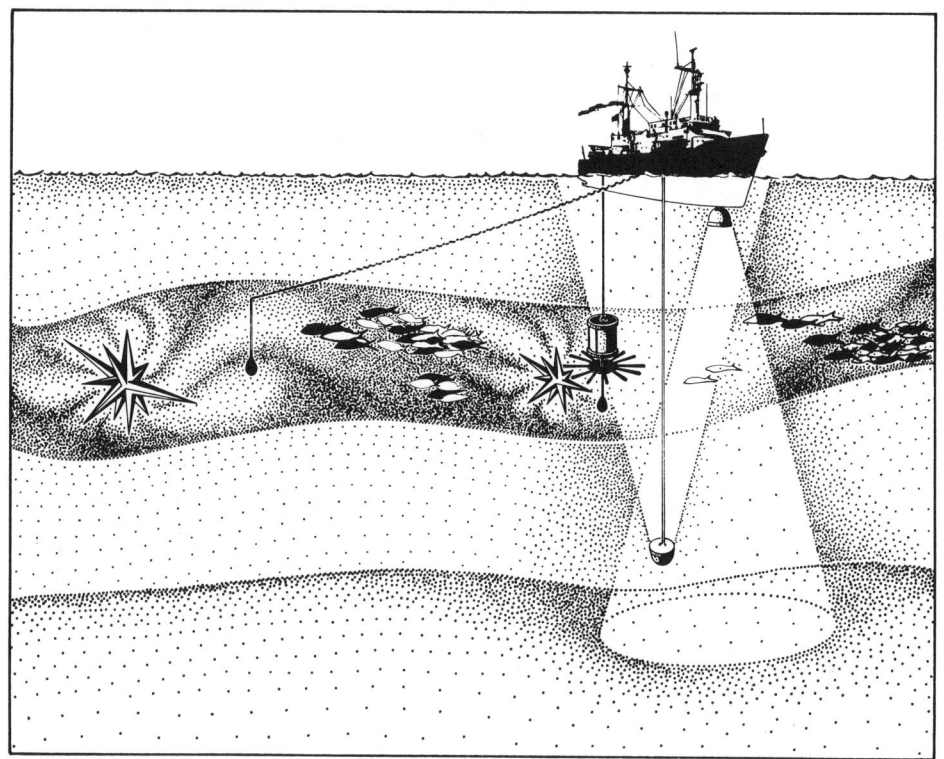

Fig. 3. Schematic of acoustic measurement techniques, including an explosive source and omnidirectional hydrophone (far left), a lowered explosive array, and upward-and downward-looking directional transducers.

those avenues of research, tortuous as they may be at times, which will yield meaningful predictive tools for the applied oceanographer to use in meeting his goals. There were, therefore, a host of "motherhood" terms involved in this complex phenomenon that everyone was involved in examining. All were concerned with "food webs", environmental modeling", "diel vertical migrations", physicochemical relationships", "ontogenetic factors", "swimbladder gas physiology", etc. To examine these problems and interrelations during the week of discussion it seemed important to identify, specifically, those organisms that are known to be sound scatterers. It is a certainty that the energy pathways and relationships which will be most useful in a predictive sense will be critically tied to knowledge about what particular animals (or groups of animals) are the causative sound scatterers.

The ubiquitous character of sound scattering layers led Marshall (1951) to propose four general criteria for assessing the potential of an organism as a contributor to scattering layers:

1. The organism must be widely distributed;
2. It must occur in concentration at depth during the day;
3. It must show pronounced powers of vertical migration (it is now known, of course, that there are non-migratory layers); and
4. It must have the ability to reflect or scatter sound.

Based on these criteria, he proposed mesopelagic fishes as the principal scatterers.

A great many details have been learned about these criteria for scattering potential, particularly in the frequency domain where swimbladder resonance is significant. Of the approximately 850 described mesopelagic fish species (150-1000 m) about one-third have gas-filled swimbladders in the adult stage, including about 180 of the 200 species of lanternfishes (*Marshall*, 1971). Numerically, it is estimated from trawling work that 60 per cent of the fishes caught, and 80 per cent of the migratory species caught will have gas-filled swimbladders (*Marshall*, ibid, based on data from *Backus et al.*, 1969). *Gibbs and Roper* (1971) obtained similar results in the Ocean Acre Program. Their data showed good correlation between certain migratory fish species and 12-kHz scattering layers, and between non-migratory species and non-migratory scattering layers. The picture arising out of all the recent controlled trawling work is that, except for the high arctic and antarctic regions, these fishes are, indeed, common and wide-spread throughout the world oceans.

Open-close discrete depth tows, and numerical solution of the scattering equation by substituting measured scattering strength values have shown that not very many organisms are required to account for the scattering levels observed. For scattering at resonance, typical concentrations are of the order of only a few organisms/1000 m^3. For example, *Chapman et al.* (1974) described a low frequency resonant peak between 1 and 2 kHz off Newfoundland. Calculated concentrations ranged between 0.1 and 1 fish/1000 m^3 (*Farquhar*, 1971, p. 317). *Van Schuyler* (1971), based on acoustic measurements in Hawaiian waters, calculated population densities in the main layers for daytime and nighttime conditions. For fishes with effective swimbladder radii ranging from about 0.1 to 2.5 cm, his calculated concentrations for resonant populations ranged from about 25/1000 m^3 for the small scatterers to about 6/million m^3 for the large ones. To turn the argument around, *Love* (1975), using biological trawl data which had yielded similar population densities (very few /1000 m^3) achieved good agreement in predicting scattering strengths measured over a range from 6.3 to 20 kHz. Agreement between biologically predicted and observed scattering was poor at frequencies below 6.3 kHz, probably due mainly to the sampling gear, which did not effectively catch the larger, low-frequency scatterers

The identification of *Ceratoscopelus maderensis*, by *Backus et al* (1968) as the organism responsible for a special type of scattering layer has already been mentioned. The layer characteristically resides at a depth range of about 330-600 m during the day, and scatters sound at a primary frequency of 12 kHz; at sunset, the layer moves close to the surface, and the resonant frequency shifts to about 3 kHz.

Specialized kinds of layers, or specialized environmental situations appear to be more amenable to definitive scattering studies than the typical deep ocean basins. For example, the Cariaco Trench, north of Venezuela, with its oxic-anoxic interface, shallow sill depth, and resultant depauperate mesopelagic fauna, has provided a special environmental site for significant scattering research. *Wilson* (1972), and *Baird, Wilson, and Milliken* (1973), identified a codlet, *Bregmaceros nectabanus* Whitley, as an organism contributing to a 12-kHz scattering layer in the Cariaco Trench, and the specific causative organism for a 25-kHz layer which migrated through the oxic-anoxic interface, deep into the anoxic, sulfurous waters of the trench, 500-600 meters below the interface. This same 25-kHz layer, which by closing net hauls was shown to be *B. nectabanus* only, migrated upward out of the anoxic layer towards sunset and joined another layer which had migrated downward and remained above the interface. The entire merged layer continued toward the surface, and effectively scattered sound at both 25 kHz and 12 kHz. In some later work (*Baird, Wilson, Beckett, and Hopkins*, 1974) it was established that *Diaphus taaningi* was responsible for a distinct 250-m deep daytime layer in the Cariaco Trench which produced scattering at 12, 25, and 50 kHz. Fish concentrations from net haul data were of the order of $2/1000 \text{ m}^3$. The standard 12-kHz echosounder record showed only a single layer. However, two layers were resolved at 25 and 50 kHz. *Bregmaceros nectabanus*, as well as *Steindachneria argentea* also may have contributed to scattering from these layers, but only *B. nectabanus* migrated through the oxic-anoxic interface to remain in sulfurous waters during the day.

The spate of papers in recent years on scattering in the resonant frequency range (about 1 to 20 kHz) has convincingly shown that swimbladder-bearing mesopelagic fishes, particularly the speciose lanternfishes (Myctophidae), constitute the dominant midwater scattering community. A number of investigators have related particular species to specific layers, some convincingly, some not so convincingly. For example, *Kashkin* (1974), based on assumed relationships between fish length and bladder size, attributes scattering layers over broad reaches of the Atlantic Ocean north of the Azores (40-65°N.) on both eastern and western sides, to the glacier lanternfish, *Benthosema glaciale*. The layers are resonant at 10-15 kHz during the day, at depths of 500-900 m, and at 5-9 kHz at night in the upper 200 m. He states that the swimbladder "is always full of gas". *Zahuranec and Pugh* (1971) however, from data collected during a Naval Oceanographic Office cruise in northern Atlantic Ocean waters, show progressive deposition of adipose tissue with growth, beginning at standard lengths greater than 20 mm until, at about 60-64 mm, the lumen, or gas-filled space in the swimbladder of *B. glaciale* is considerably reduced in size relative to fish length. This is mentioned only to point out that considerable care must be exercised in interpreting catch results for the purpose of identifying individual causative scattering species.

Other fishes have been incriminated as scatterers, particularly in northern latitudes where low frequency scattering peaks have been observed. Chapman et al. (1974) attribute one such prominent peak in the Labrador Sea to the ocean redfish, *Sebastes marinus*, and another slightly higher peak in the Davis Strait to the scaled lancetfish (*Alepisaurus* sp.). Hansen and Dunbar (1971) attributed scattering from a layer they observed in the Beaufort Sea to the polar cod, *Arctogadus glacialis*. This layer scattered sound at both 12 and 100 kHz. Other likely candidates as low-frequency resonant scatterers in northern waters are herring and cod.

Recent use of high frequency echosounders has enabled the identification of certain marine zooplankters as scatterers. Pieper (in press; see also Bary and Pieper, 1971; and Pieper, 1977) has demonstrated that euphausiids caused scattering at 42, 107, and 200 kHz in the Saanich Inlet, but not at 11 kHz. The principal species was *Euphausia pacifica*. Hansen and Dunbar (1971) showed that the thecosomatous pteropod, *Spiratella helicina*, caused scattering at 100 kHz in the Beaufort Sea. It formed a thin layer along the pycnocline. Castile (1975), as mentioned earlier, has attributed scattering he observed at 330 kHz in the North Pacific to copepods, and suggested that even phytoplankton may begin to "show up" at that frequency.

In summary, the principal causative scattering organisms, at least over the range of frequencies used so far, are the mesopelagic fishes and the major zooplankton groups. Most, but not all, undergo diel vertical migrations, they are linked closely together in the food web, and collectively they represent key elements in the energy mechanisms of trophic dynamics in the deep ocean. A careful choice of acoustic sensors, coupled with biological and oceanographic studies, should enhance considerably our understanding of ecological relationships in these two vitally important groups.

VERTICAL MIGRATION

The pronounced diel migratory capabilities of most mesopelagic organisms has been well established. The range of extent of migration varies with species and, for a given species, varies ontogenetically. Among the fishes with swimbladders there are obvious family differences in migration; in general the myctophids and gonostomatids have a considerable migratory range, while the melamphaid and sternoptychid fishes have very limited vertical ranges. In species of the extremely abundant genus *Cyclothone*, the postlarval stage is spent in the mixed layer, and a gas-filled swimbladder is present. As the individuals metamorphose, the swimbladder regresses and becomes invested with fat. They live in deeper waters as adults, and the swimbladder is no longer present (Marshall, 1971).

These migrating, swimbladder-bearing fishes feed principally on zooplankton, mostly copepods, euphausiids, and sergestids, which themselves undergo vertical migrations of varying extent. The established importance of these plankton forms in the feeding ecology of mesopelagic fishes, and the growing awareness that high frequency acoustic sources can provide a useful technique for studying zooplankton populations, have obvious implications in designing studies to explore behavioral and ecological aspects of vertical migration.

From the standpoint of fish migration, vertical changes go far beyond behavioral ecological aspects in assuming significance to sound scattering. The fishes concerned are physoclistous. That is, there is no open communication (pneumatic duct) between the swimbladder and the gut, as in such physostomous fishes as the anchovy. During diel migration, then, these fishes may be considered as vertically moving populations of "gas bubbles", influenced to some degree by the fish tissue surrounding the "bubble". As mentioned earlier, resonant frequency depends primarily upon bubble size and depth (see Love, 1976, this volume, for a detailed discussion). The relationship, from Andreeva (1964), which expresses the resonant frequency of a swimbladder, is:

$$f_r = (2\pi R)^{-1} \left[\frac{3 \gamma P + 4\mu}{\rho} \right]^{1/2} \qquad (1)$$

where f_r is the resonant frequency in Hertz, R is the effective radius of the swimbladder in cm, P is the hydrostatic pressure in dynes/cm^2, γ is the ratio of specific heats of the swimbladder gas at constant pressure and volume, ρ is the density of sea water in grams/cm^3, and μ represents the real part of the complex shear modulus of fish tissue, experimentally determined (Lebedeva, 1965) to be between 10^5 and 10^7 dynes/cm^2.

It can be seen that the resonant frequency varies inversely with bubble size and directly with depth. To phrase the relationship in terms of fishes:

1. At depth, a larger fish (and hence, a larger bubble size) is required for resonant scattering at a particular frequency than in surface waters; and

2. For a fish of a given size, the resonant frequency will increase as the animal descends in the water column, and decrease as it ascends.

Obviously, the fish-swimbladder unit is a living system, and what the fish does or does not do to its swimbladder gas volume during migration will influence its acoustic characteristics. A more detailed consideration of these aspects of fish migration may be found in Alexander (1971), D'Aoust (1971), and Shearer (1971).

Briefly, there would seem to be two alternatives available to the fish:

1. It is neutrally buoyant at its nighttime (nearest-to-the-surface) depth and allows the swimbladder to contract as pressure increases during sunrise descent (the mass of the gas remains constant), maintaining its daytime depth by swimming or opercular movements (see *Barham*, 1971); or

2. It remains neutrally buoyant at all depths during migration by adding or subtracting gas from the swimbladder through secretion or resorption (the volume of the gas remains constant).

Acoustic evidence presented by *Hersey et al.* (1962), in which the frequency shifts of certain resonant layers were monitored, indicated that both constant mass and constant volume mechanisms were operative during migration. In the constant mass case, the resonant frequency shifted as the 5/6 power of the hydrostatic pressure, and as the one-half power of the pressure in the case of constant volume.

It is clear that further study into the physiological and energy mechanisms of the swimbladder and of vertical migrations will contribute significantly towards understanding the complex acoustic features that have been observed.

ACOUSTIC PATTERNS

Volume reverberation, or backscattering varies with frequency, geographic location, depth, time of day, and season. During the acoustic measurement efforts of the last 10-15 years, the use of explosive techniques has resulted in a fairly good idea of the generalized patterns of frequency variation, for both day and night conditions, at least in the frequency range from about 1 to 20 kHz. Geographic coverage is spotty, but fairly broad throughout the world ocean. Because of the use of explosive techniques, our knowledge of depth dependence is relatively poor, although new developments in measurement techniques, and the more frequent use of some type of "scattering profiler", are beginning to build the data base on vertical variability (e.g., *Chapman*, 1974; *Vent*, 1972). With the exception of a very few locations where special studies have been made, our knowledge of seasonal variability is poor at best. Table 1 provides some idea of the geographic coverage, and enables access to the pertinent literature or at least identifies the laboratory making the measurements.

It is convenient to discuss frequency and geographic patterns of variability together. Figure 4 shows nighttime backscattering spectra for the North Atlantic Ocean. The ordinate expresses the integrated or column scattering strength in decibels (dB), and represents the total amount of energy scattered back from a water column 1 m^2 in area and extending from the surface to some depth, d, which is the depth to the bottom of the deepest scattering layer in the water column. The scattering-versus-frequency curves have been separated into four regional groupings (A, B, C, and D) based on the general shape of the curves. The data were collected by the Naval Oceanographic Office.

TABLE 1. Geographic Coverage of Volume Reverberation Measurements, and Source of Measurements

Geographic Area	Reference/Laboratory
Norwegian Sea	Naval Oceanographic Office (NAVOCEANO); Chapman, et al. (1974)
Davis Strait and Baffin Bay	Chapman, et al. (ibid)
Labrador Sea	Chapman, et al. (ibid)
Icelandic waters	NAVOCEANO; Chapman, et al. (ibid)
Northeast Atlantic	Chapman, et al. (ibid); McElroy (1974); NAVOCEANO
Temperate Atlantic	Chapman, et al. (ibid); NAVOCEANO
Northwest Atlantic	Chapman, et al. (ibid); NAVOCEANO (Zahuranec, et al., 1970; Davis, 1971)
Bermuda waters	Fisch and Dullea, 1973, Brown and Brooks, 1974 (Naval Underwater Systems Center); Chapman, et al. (ibid); NAVOCEANO
Sargasso Sea	NAVOCEANO (Gold, 1966, Gold and Van Schuyler, 1966); Chapman, et al. (ibid)
Gulf of Mexico	NAVOCEANO (Van Schuyler and Hunger, 1967); Naval Research Laboratory (Wilson, et al., 1968)
Caribbean Sea	NAVOCEANO; Chapman, et al. (ibid); Naval Research Laboratory (Wilson, 1972); Baird, et al. (1973, 1974)
Mediterranean Sea	NAVOCEANO; Naval Underwater Systems Center; Chapman, et al. (ibid); Jeannin (1971); McElroy and Wing (1971)
Equatorial Atlantic	NAVOCEANO; Chapman, et al. (ibid)
South Atlantic (along 30°W., to Antarctic Waters)	Chapman, et al. (ibid)
Eastern Pacific (Antarctica to Alaska)	Chapman, et al. (ibid)
Northeast Pacific Ocean	Vent, R. J. 1972 (Naval Undersea Center); Chapman, et al. (ibid); NAVOCEANO; Scrimger and Turner (1973)
Hawaiian waters	Van Schuyler, 1971; Scrimger and Turner (ibid)
Gulf of Alaska/Aleutians	Scrimger and Turner (ibid); NAVOCEANO
Northwest Pacific	NAVOCEANO
Tasman Sea, Coral Sea, South China Sea, New Zealand waters, and various parts of the Indian Ocean	Hall (1971, 1973); Batzler, 1975

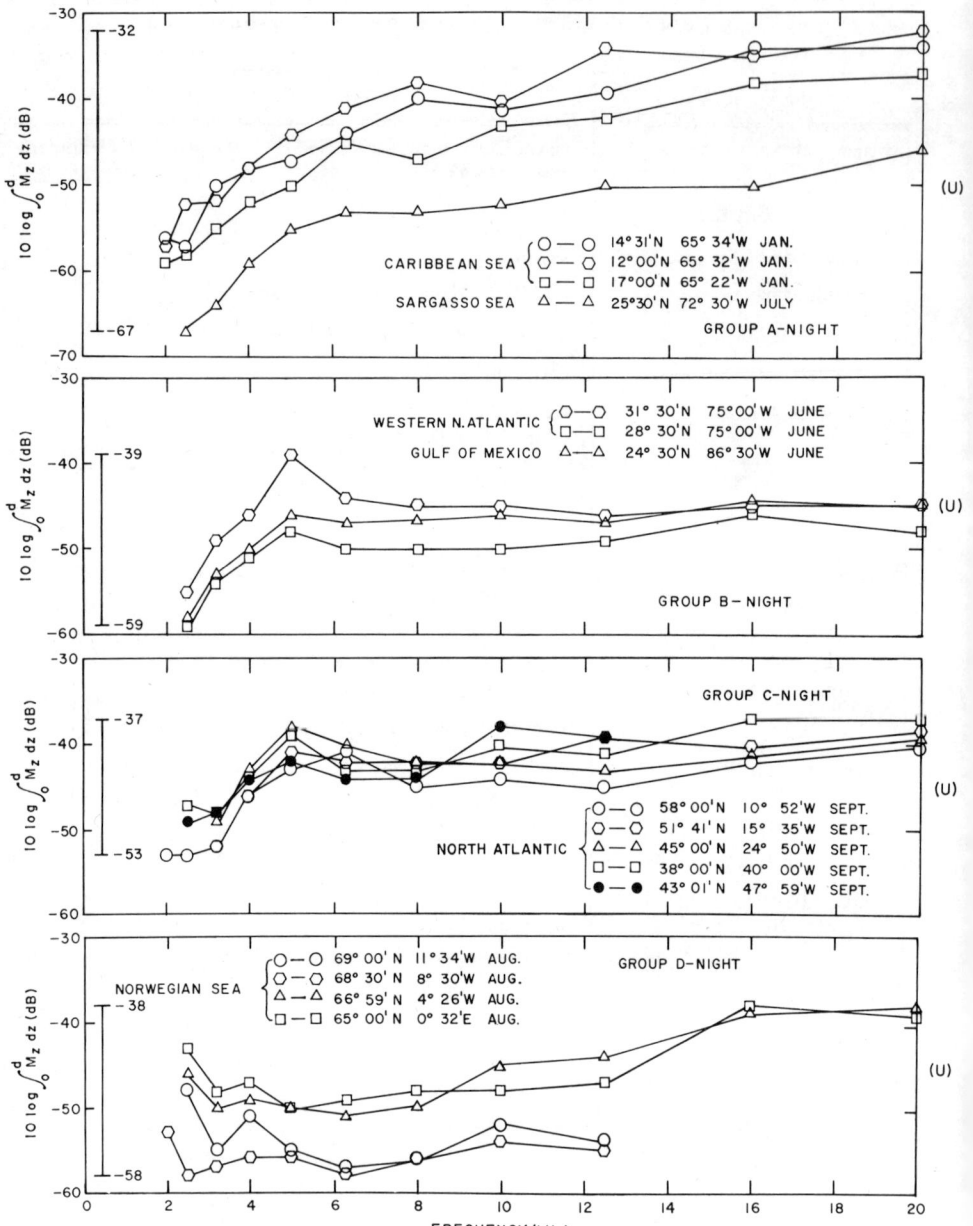

Fig. 4. Nighttime scattering strength spectra from representative areas of the North Atlantic Ocean.

Group A data, representative of the Caribbean and the southern Sargasso Sea, are characterized by a relatively sharp rise in levels with frequency from about 2 to 6 kHz; above 6 kHz the levels continue to increase with frequency, only more gradually. *Chapman et al.* (1974) show similar nighttime curves in these areas. The Sargasso Sea curve, though similar in shape, is displaced by 7-10 dB from the Caribbean Sea levels, and the scattering is the weakest recorded in the Atlantic Ocean. The Group B curves similarly show a sharp increase in scattering with frequency, rising to a peak at 5 kHz, after which the levels remain essentially constant with increasing frequency. Group C data were collected on a transect across the Atlantic, in a northeasterly direction, towards the Norwegian Sea. From about 3.5 to 20 kHz the curves are quite similar to the Group B curves, including a mild peak at about 5 kHz. From 2 to 3.5 kHz, however, there are only minor changes of the order of 1 dB. Group C curves at the lower end of the spectrum seem to represent a transition between Group B and the distinctive curves of Group D data for the Norwegian Sea. In this region, scattering decreases with increasing frequency from 2 to 3.5 kHz. Levels increase again until about 4 kHz, and then drop to a minimum at 6.3 kHz, after which scattering gradually increases with frequency. The obvious separation of the Group D data into two data sets differing in scattering level is relatable to oceanographic conditions in the region, and is discussed in the section on environmental considerations.

The Group D data were collected during August. During a subsequent NAVOCEANO cruise to the Norwegian Sea in July 1970, 17 reverberation stations were occupied in an area extending from about 63°-73°N, and from about 3°W-11°E. The spectral curves for all stations showed the typical Group D shape with the exception that the pattern was shifted to the right along the abscissa. This is, scattering decreased with increasing frequency from 2 to 4 kHz, increased until 6.3 kHz, decreased to a minimum at about 9 kHz, and increased from 9 to 20 kHz, more rapidly than the August data. Processing of the July data was extended to 1 kHz to see if the trend of scattering increase with decreasing frequency at the low end of the spectrum continued. At all but two stations, the scattering levels dropped sharply from 2 to 1 kHz, or at least flattened. The two exceptions were northerly stations located in cold polar water, and the levels continued to increase to a maximum at 1 kHz.

Daytime spectra for the Atlantic are shown in Figure 5, the regional groupings remaining the same. The major difference between night and day spectra in Groups A and B is the overall decrease in scattering levels during the day. Group C daytime scattering differs from the nighttime pattern in that a minimum occurs during the day between 2 and 4 kHz; there is little day-night difference at frequencies above 6.3 kHz. The daytime curves for Group D data show a flattening, but the same characteristic shape persists; there is little day-night difference and, at some frequencies the daytime values exceed the nighttime values.

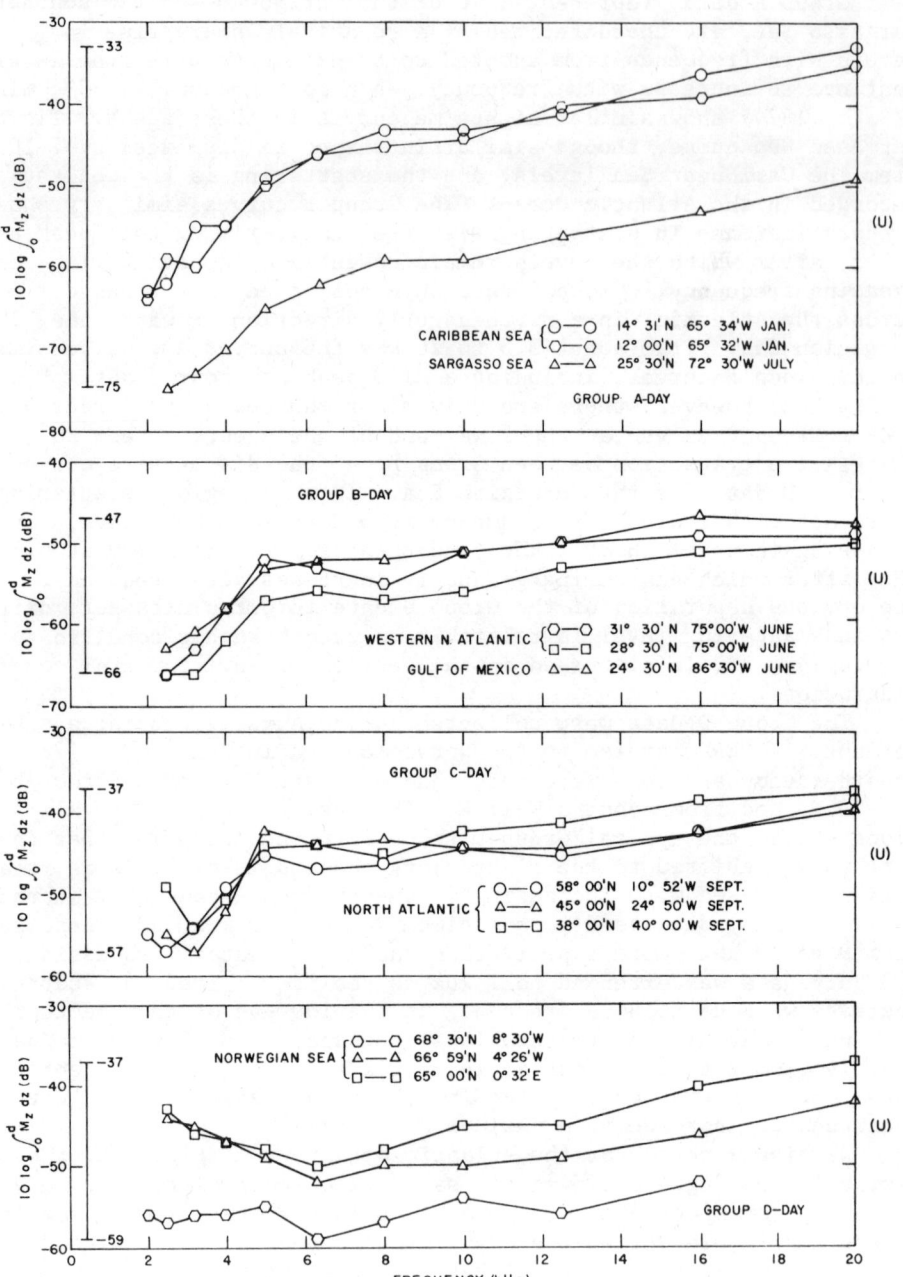

Fig. 5. Daytime scattering strength spectra from representative areas of the North Atlantic Ocean.

As the swimbladder-bearing fishes migrate towards the surface near sunset, their concentrations increase, and they scatter sound at lower and lower frequencies as they move upward in the water column. In general, the scattering strengths increase at night, and the day-night differences are most pronounced at the lower end of the frequency spectrum. Figure 6 shows the results, for two low frequency octave bands, of an explosive sequence made across sunset in the southwestern Sargasso Sea. Migration begins at about 1845-1850 hours, prior to local sunset, indicated by the increase in scattering strengths. Migration continues across sunset and is completed by about 2000 hours, as the curve flattens out at its nighttime level. A reverse of this pattern occurs across sunrise as the animals descend to their daytime depths.

Diurnal variation for the four regional groupings discussed above was determined by subtracting the daytime values from the nighttime values for each frequency examined, and plotting the difference. The results are shown in Figure 7. With the exception of two points, (20 kHz in Groups A and C), night values exceed day values at all frequencies in Groups A, B, and C. Greatest differences occur in the lower portion of the spectrum. Day-night differences gradually decrease with increasing frequency, and tend to become independent of frequency at the higher frequencies. Above 8 kHz, no significant diurnal variation occurs. With a few exceptions, Group D data from the Norwegian Sea show relatively minor day-night differences across the frequency spectrum, and day values exceed night values at several frequencies. At one of the stations (66° 59'N; 4°26'W) the daytime value at 3.2 kHz is 5 dB higher than the nighttime value, and about 7 dB lower at 16 kHz. All of the July stations made in the Norwegian Sea similarly show only minor day-night differences.

To examine the geographic patterns of scattering in more detail, it is useful to plot scattering strength as a function of latitude or longitude. Figure 8 shows scattering strengths vs. latitude for 2.2, 4.5, and 9.1 kHz. The symbols identify the approximate meridian where the measurements were made. For all three frequencies, the general pattern is the same. Scattering decreases with increasing latitude from about 12°-25°N, then increases with latitude until about 40-45°N, and then decreases again. The rate of decrease north of 40-45°N is steeper in the western Atlantic Ocean than on the eastern side, as indicated by the relative slopes of the two lines drawn in the figure, one representing 40°W, the other 25°W. The more gradual rate of decrease on the eastern side reflects the influence of the North Atlantic Drift. At 2.2 kHz, the values indicated by the symbol β show a continued increase well up into the Norwegian Sea, as would be expected from the low frequency pattern of the Group D curves described above. Within the Norwegian Sea, the July 1970 data indicate a decrease from south to north, and from east to west.

McElroy (1974) reported on daytime scattering spectra between 33° and 63°N. Table 2 was prepared from his average spectra for oceanographically and acoustically defined regions on the eastern

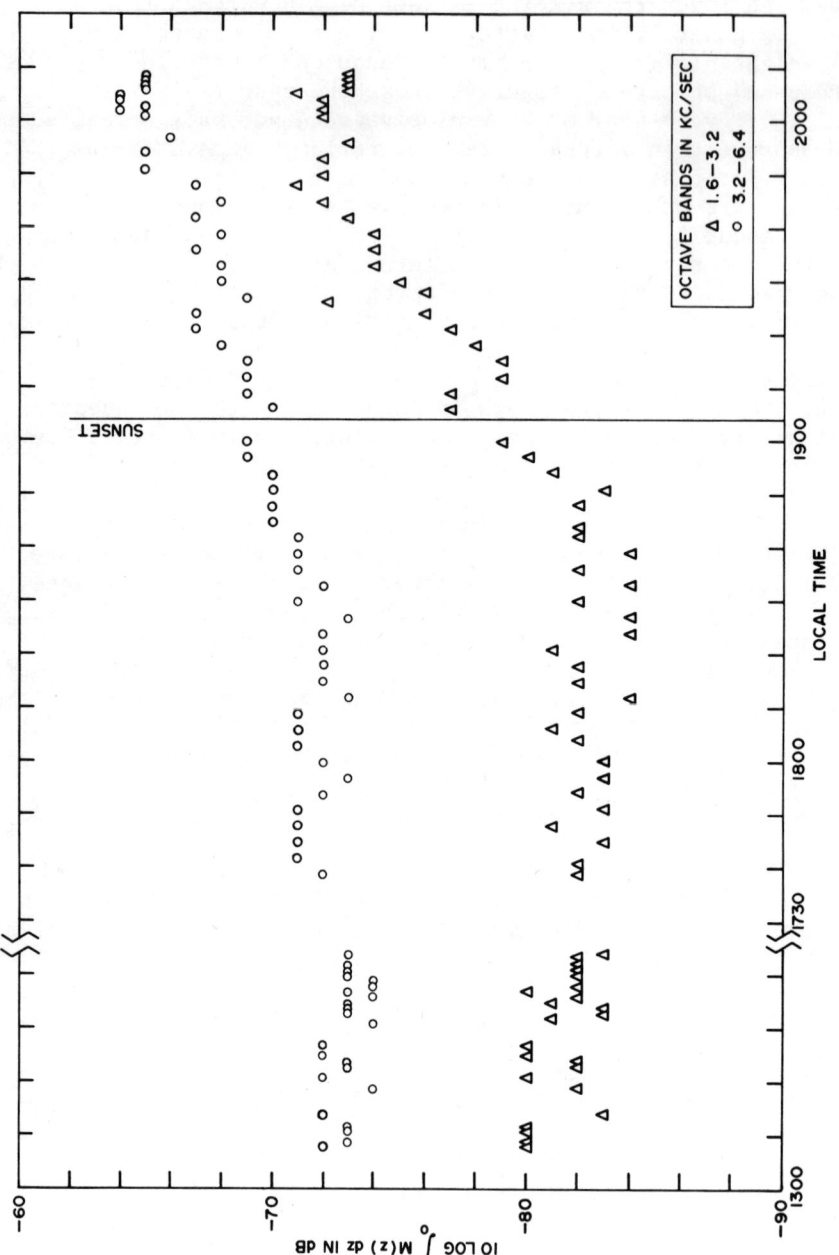

Fig. 6. Scattering strength variations over sunset in the 1.6–3.2 and 3.2–6.4 kHz octave bands, July 1965, Sargasso Sea.

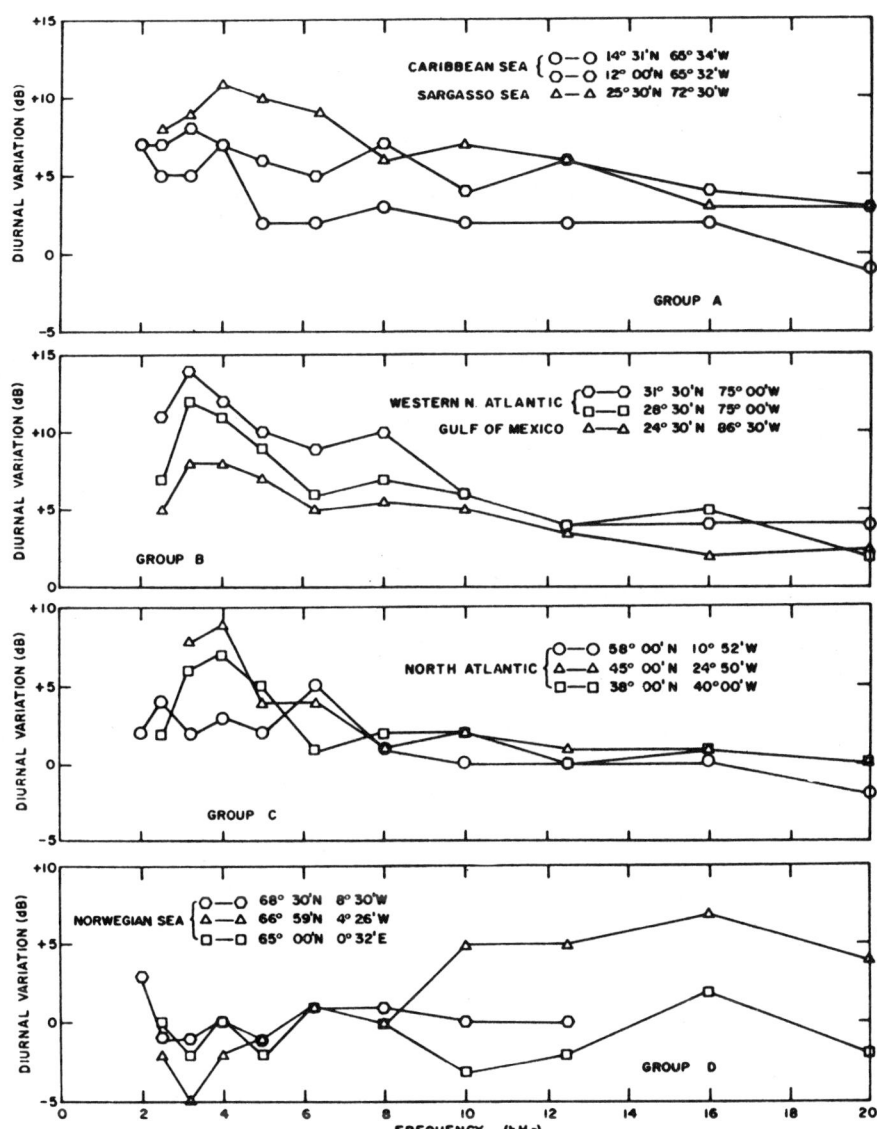

Fig. 7. Diurnal variations (nighttime minus daytime values) of scattering strengths in the North Atlantic Ocean.

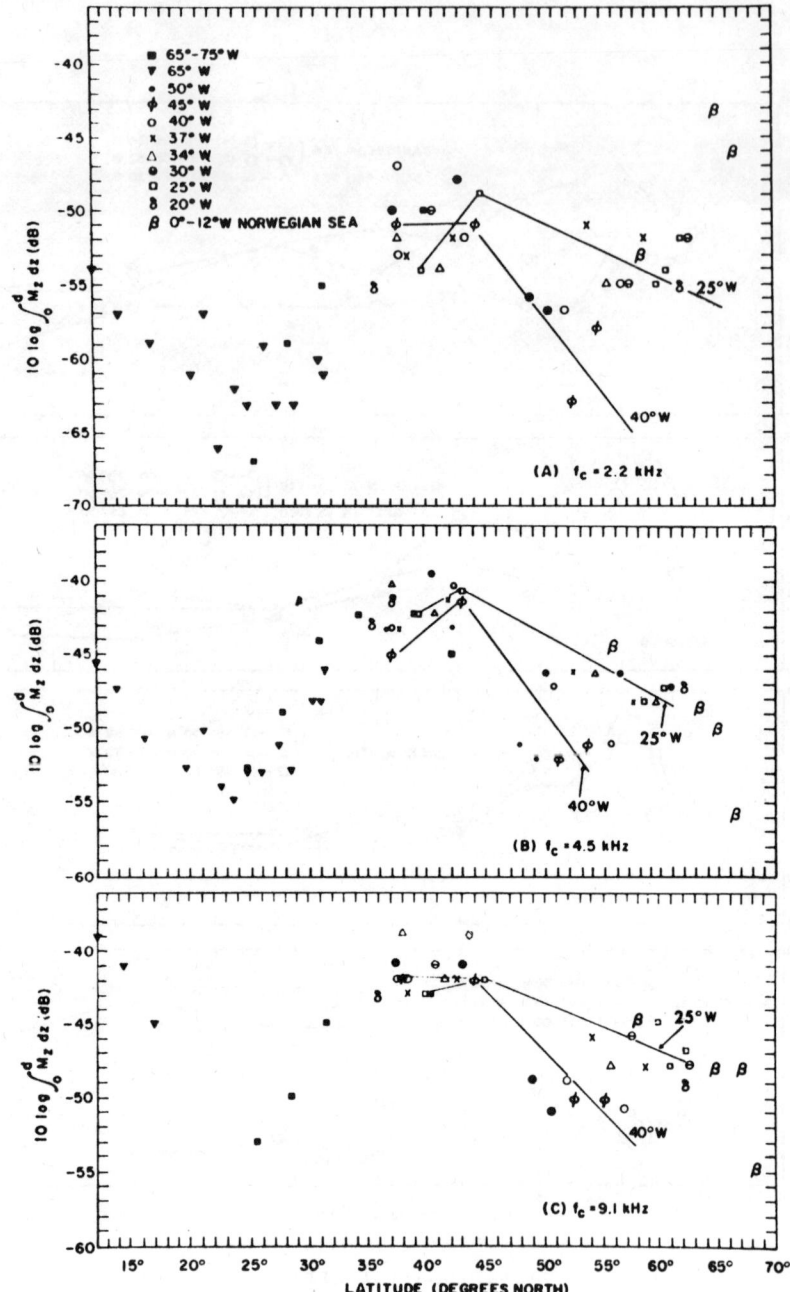

Fig. 8. Nighttime scattering strength variations as a function of latitude in the North Atlantic Ocean, at frequencies of (A) 2.2 kHz, (b) 4.5 kHz, and (C) 9.1 kHz.

TABLE 2. Daytime Scattering Strength (dB) Values at 2.2, 4.5, and 9.0 kHz East of 25 W in the Atlantic (from McElroy, 1974)

Latitude Range of Stations Averaged	Frequency (kHz)		
	2.2	4.5	9.0
$33°-35°N$	-59	-55	-56
$33°-36°30'N$	-59	-61	-61
$38°-15'-40°N$	-64	-62	-61
$43°-51°30'N$	-62	-55	-52
$55°45'-56°30'N$	-65	-57	-53
$61°-63°N$	-52	-61	-65

side from the Azores to the Norwegian Sea. The latitudinal trend of the values suggests greater variability in the mid-latitudes on the eastern side of the Atlantic than in the west.

The comprehensive measurement program of the Canadian Defense Research Establishment Atlantic, under the direction of Dr. R. P. Chapman, has provided a large body of data for comparison with North Atlantic geographic patterns. Figure 9 shows column scattering strengths in the octave band from 3.2-6.4 kHz, extending from the equator to south of the Antarctic convergence along 30°W (Chapman et al. 1971). Keeping in mind that octave band analysis may slur some of the detail available in 1/3-octave analysis, the data show a latitudinal pattern strikingly similar to the North Atlantic. A plot for the eastern North Pacific Ocean, from a line of stations along the west coast of North America (Chapman et al., 1971) shows a similar pattern in the 1.6-3.2 kHz octave band (Figure 10), except that a second peak occurs between equatorial waters and 20-25°N. Unlike the Atlantic Ocean, where the strongest day-night differences were observed in the low-frequency portion of the spectrum, only slight day-night differences occurred along this track.

In some later work, Chapman et al. (1974) provided scattering spectra from stations along 150°W in the Pacific, extending from south of the Antarctic Convergence to the Gulf of Alaska. The spectra reported are for representative stations in major faunal provinces, defined on the basis of fish distribution and physical oceanography. Table 3 was constructed from those spectra (i.e., their Figure 17, p. 1732).

Scattering levels across the measured spectrum (2-20 kHz) appear to be lower in Antarctic waters than anywhere else in the world's oceans where similar measurements have been made, and there is little day-night difference. Unlike the high latitudes of the northern hemisphere, no low frequency peaks or increases are present. Scattering strengths increase sharply north of the Antarctic convergence in the latitudinal belt from 60°-40°S, and the day-night

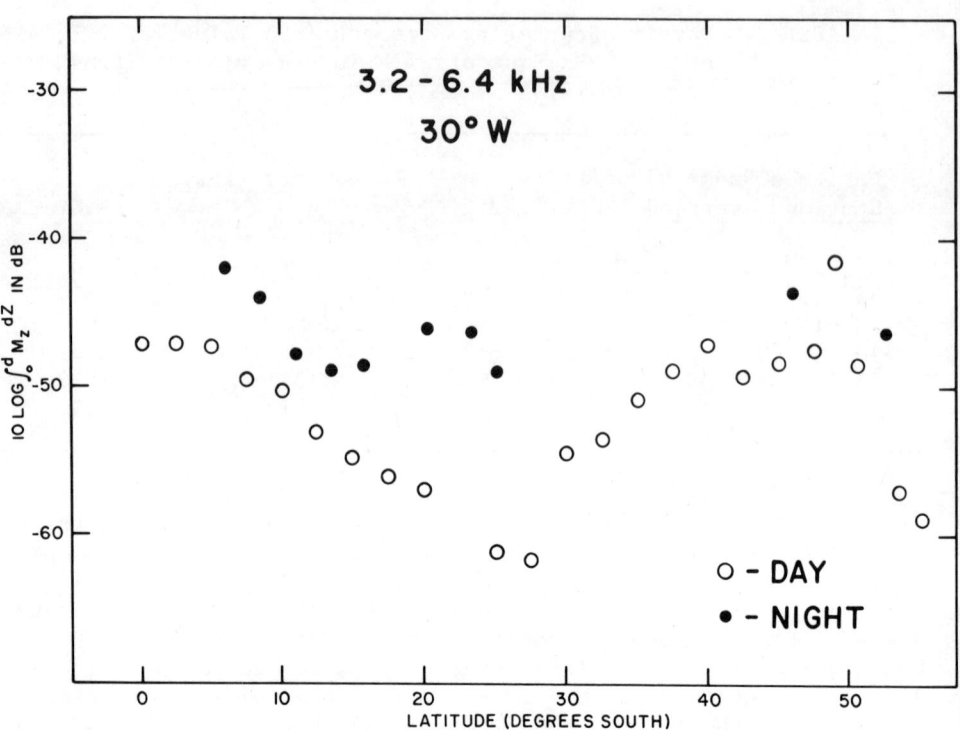

Fig. 9. Column scattering strength vs. latitude in the south Atlantic Ocean (from *Chapman et al.*, 1971).

differences are large below 10 kHz. Levels decrease again in the south subtropical belt bracketing 30°S, and the marked day-night differences continue. Though day-night differences persist all the way to 40°N, scattering levels remain relatively weak until near the equator, where they rise again at frequencies greater than about 2.5 kHz. Between the equator and about 40°N, changes in scattering are less marked than in the North Atlantic Ocean. North of 40°N, day-night differences are negligible, while scattering increases slightly at most frequencies, decreasing only at about 4-5 kHz and 16-20 kHz. North of 53°N, maximum scattering occurs at 2 and 3.5 kHz, but levels are depressed considerably at other frequencies. As in northern Atlantic Ocean and Norwegian Sea waters, day-night differences are minor.

Scrimger and Turner (1973) reported on data collected along transects between Vancouver and Hawaii, and showed decreases with increasing latitude, with distinct changes in the spectra occurring in the transition region between subtropical and subarctic waters, at about 38°-42°N.

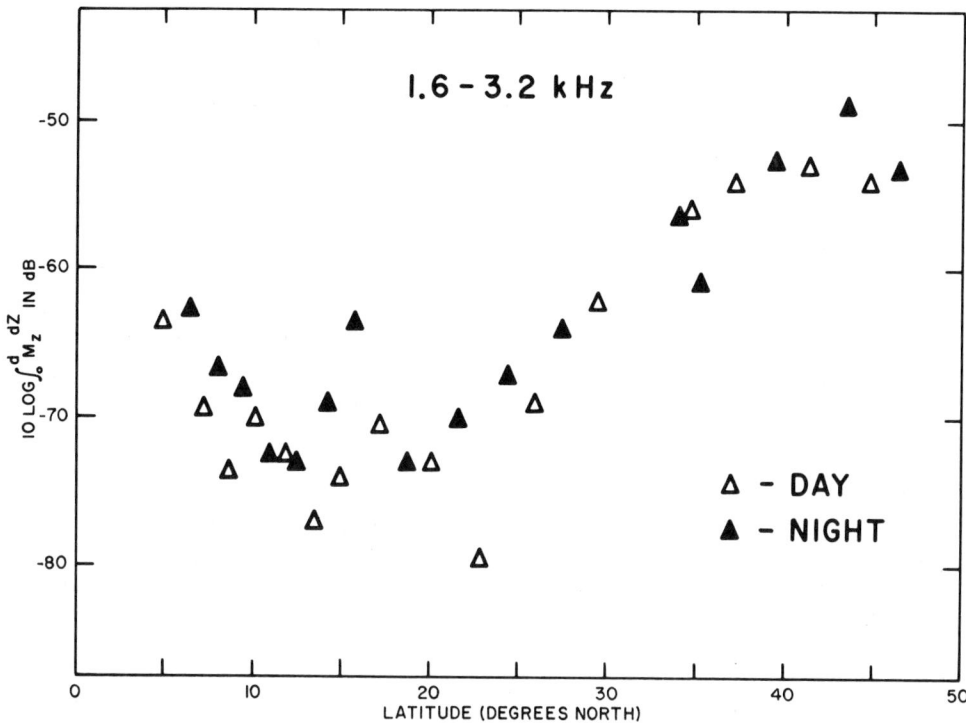

Fig. 10. Column scattering strength vs. latitude in the eastern North Pacific Ocean (from Chapman et al., 1971).

Seasonal data are scanty. A series of monthly measurements taken by the Naval Oceanographic Office over a two-year period in Bermuda waters shows about a thousand fold change in scattering level in the limited frequency band from 0.8 to 3.15 kHz. The levels were highest in July, lowest in January.

Vent (1972) investigated seasonal variations at 3.5, 5 and 12 kHz in the eastern Pacific, at a location 250 miles southwest of San Diego. Scattering at 12 kHz was independent of season; scattering at 5 kHz varied seasonally both in scattering level and layer thickness. His observations show the importance of making acoustic measurements using profiling as well as explosive methods, since he notes that "large (seasonal) changes in scattering strength at a particular depth may not be evident in change in column strength".

TABLE 3. Day and Night Scattering Strength Values at 2.0, 4.5 and 10.0 kHz, from Antarctica to the Gulf of Alaska Along 150°W (from Chapman et al., 1974)

Latitude Range of Pelagic Region (Latitude of Station)*	COLUMN SCATTERING STRENGTH (dB)					
	2.0 kHz		4.5 kHz		10.0 kHz	
	Day	Night	Day	Night	Day	Night
South of 60°S (60°40'S)	-68	-71	-56	-60	-55	-55
60°-40°S (50°S)	-63	-52	-45	-37	-43	-44
40°-25°S (30°S)	-66	-50	-60	-51	-59	-48
25°S-20°N (10°S)	-66	No data	-60	No data	-58	No data
25°S-20°N (Equator)	-66	-57	-55	-42	-45	-42
25°S-20°N (10°N)	-68	-55	-56	-46	-49	-47
20°-40°N (30°N)	-64	-60	-53	-45	-47	-45
40°-53°N (43°N)	-61	-55	-52	-50	-46	-44
Above 53°N (55°N)	-56	-54	-58	-59	-58	-57

*See text.

ENVIRONMENTAL CONSIDERATIONS

Environmental factors obviously are of fundamental importance in determining distribution, abundance, and behavior of organisms in the sea. The delineation of reverberation provinces and the "prediction" of biological scattering thus are tied very closely to our understanding of how these factors operate on and regulate scattering communities. Space does not permit an exhaustive review of environmental effects. As a result only a few points will be touched upon and the discussion will focus, in a general way, on large scale patterns.

It is clear that mesopelagic organisms, especially the vertical migrators, respond in some critical way to light intensity, and light changes (Clarke, 1971; Kampa, 1971). It is equally clear, however, that the responses are complex and that other factors operate during migration. Some scattering layers have been observed to follow an isolume; others do not. Some animals migrate all the way to the surface; others do not. Are those that do not barred from further upward movement by thermal changes, density gradients, or pressure effects, or is it perhaps due to selective feeding behavior? Is the upward migration really initiated by a nutritional cue and reinforced by light stimuli? Sufficient evidence is not yet available to answer these and other questions about the behavioral aspects of vertical migration, and the interplay between the stimulus of light in the sea and other environmental factors, but the importance of light is indisputable.

In the very early, pre-*Challenger* days of marine biology, Edward Forbes hypothesized that life in the ocean was distributed in bathymetric zones; he also suggested that there was probably no life at depths greater than about 300 fathoms because the sea was probably anaerobic below that depth. It was this mistake which served, in part, to stimulate the *Challenger* voyage.

The occurrence of oxygen minimum layers in the sea which may extend over broad areas is now well-known. Though these minima may impose faunal changes (see *Farquhar*, 1971, pp. 316-317), animals do occur there in substantial numbers, and scattering layers have been observed in these regions. *Kinzer* (1967) reported on scattering layers, and catches of zooplankton and large numbers of myctophid and gonostomatid fishes, from the O_2-minimum layer in the Arabian Sea, where oxygen concentrations were less than 0.08 ml O_2/L. *Kanwisher and Ebeling* (1957) found an abundance of hatchetfish (Sternoptychidae) in the O_2-minimum of the eastern tropical Pacific Ocean (less than 0.25 ml O_2/L). The swimbladder gas of mesopelagic fishes is mostly oxygen (*Kanwisher and Ebeling*, 1957), and the swimbladder may act as an auxiliary respiratory organ in fishes living in oxygen-poor waters. As mentioned above, the deepsea codlet, *Bregmaceros nectabanus*, spends each day several hundred meters deep into the anoxic, sulfurous waters of the Cariaco Trench (*Baird et al*, 1973); perhaps an adaptation to avoid predation. Though oxygen minima may act as effective barriers to some organisms, resulting in definite faunal differences in such areas, low oxygen levels nevertheless do not seem to be severely limiting, and scattering layers do occur in such waters.

A rewarding approach in recent years to delineating pelagic regions and faunal provinces is that exemplified by *Backus et al.*, (1971), in which the boundaries of six pelagic regions in the equatorial and western North Atlantic Ocean were defined physically, and shown to have significance as faunal boundaries based upon catches of mesopelagic fishes. Backus and his colleagues have since expanded their efforts to include all of the North Atlantic Ocean north of 30°S. A comparison of acoustic data with their province breakdown is provided in *Backus and Craddock* (1977).

Where the appropriate kinds of data are available, analysis of boundary conditions provides a useful approach in delineating provinces. For example, hydrographic conditions were examined at stations where scattering strength measurements were made along the Polar Front in the northern North Atlantic Ocean. The analysis is based partly on Naval Oceanographic Office data, and on data very kindly furnished by Dr. R. P. Chapman of the Defense Research Establishment in Canada.

Hydrographic data from 21 acoustic stations were used, and the locations were examined for the presence or absence of influence by Atlantic Ocean water. A summary of the stations is shown in Figure 11. The stations fall into two groups and the criterion for separation is based on the temperature-salinity curve for North Atlantic Central Water (labelled NACW in the upper left-hand plot in Fig. 11),

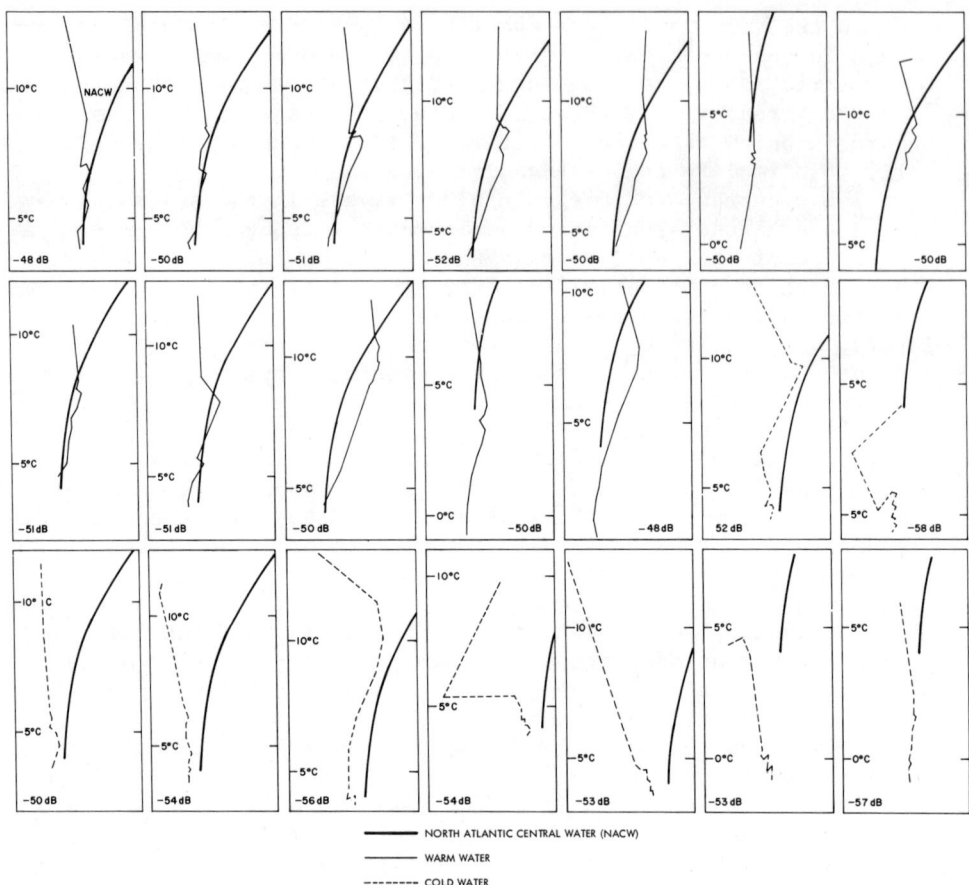

Fig. 11. Temperature-salinity conditions at 21 acoustic stations (nighttime column scattering strength values at 3.5 kHz shown) along the Polar Front in the North Atlantic Ocean. For North Atlantic Central Water (NACW), S = 35.00 °/oo at 4° C and 35.40 °/oo at 10°C.

some portion of which is shown as a heavy solid curve in each station representation. Reference salinity values (abscissa for each station plot) for North Atlantic Central Water are 35.00 °/oo at 4°C and 35.40 °/oo at 10°C. When the T-S plot for a station remains to the left (or colder, less saline) side of the NACW curve, it is a "cold" station (dashed curve), and where the curve is crossed by the station plot, "warm" or Atlantic Ocean water is influencing the location (light solid curve). Figure 12 shows the location of these stations, with a boundary line separating the two classes of water.

Nighttime values of scattering strength at 3.5 kHz, measured at each of these stations, are also shown in the corner of each station

BIOLOGICAL SOUND SCATTERING   519

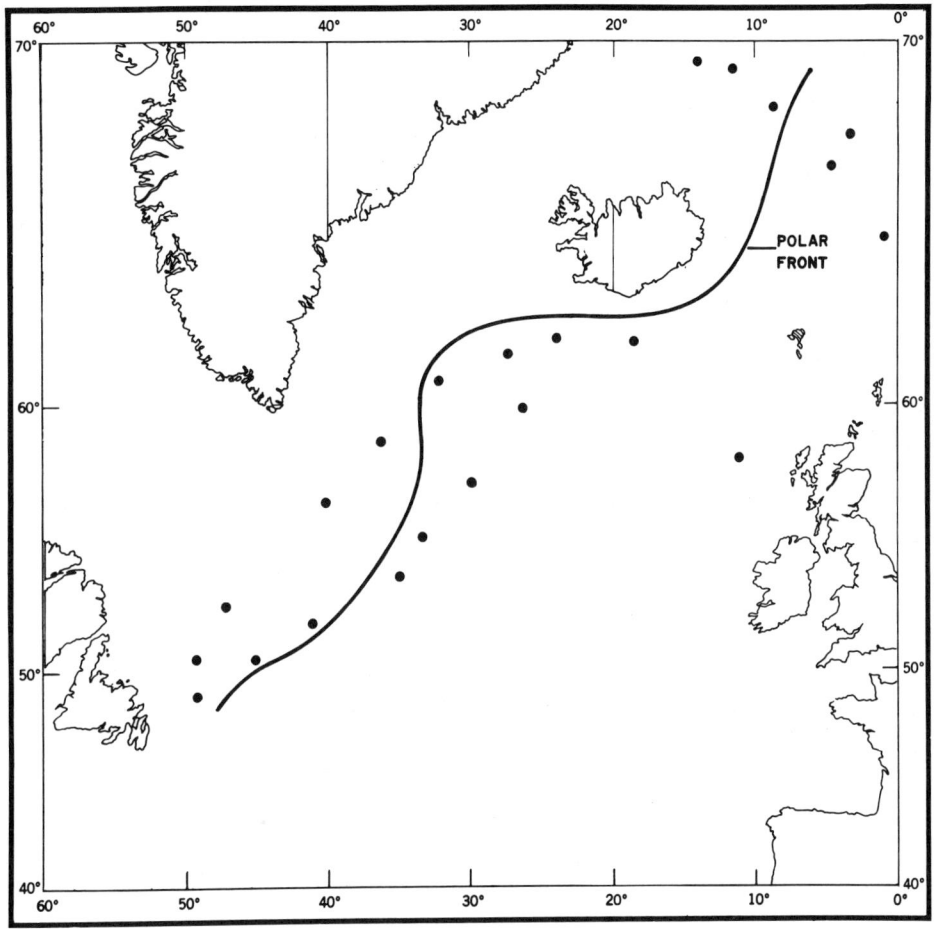

Fig. 12. Locations of "warm" and "cold" water stations along the Polar Front.

plot in Figure 11. The total range for all stations is -48 to -58 dB. The values at the cold water stations are lower, in general, than those for the warm water stations. The respective averages are -54 and -50 dB. To test the significance of the difference between the two groups, the Mann-Whitney "U" test, a rank sum-test, was used. This is a non-parametric test that does not require normally distributed data, and it is a rather conservative test. By applying this test, the null hypothesis is rejected that there is no significant difference in scattering strengths measured in the two classes of water; the difference is significant at $P < 0.01$. In other words, there would appear to be a reliable distinction between reverberation levels along this front, persisting all the way up into the Norwegian Sea.

Two stations from the above groups were examined in more detail. Station 2A, at 68°30'N; 8°30'W, was located in cold water north and east of Iceland, and Station 4, at 65°N; 0°30'E, was located in the warmer water of the North Atlantic Drift. Scattering strength values at night for 3.5 kHz were -57 and -48 dB, respectively, and scattering strength as a function of frequency for both day and night are shown in Figure 13.

There are two striking features of these scattering strength plots. One is the almost negligible diurnal variation of 0, 1, or 2 dB at nearly all frequencies examined. The other is the decrease in scattering strength with increasing frequency until the center frequency of 6.3 kHz, after which the values tend to increase, most markedly at Station 4. The difference in character between these Group-D curves and those shown for other areas of the Atlantic has already been mentioned in the discussion of acoustic patterns.

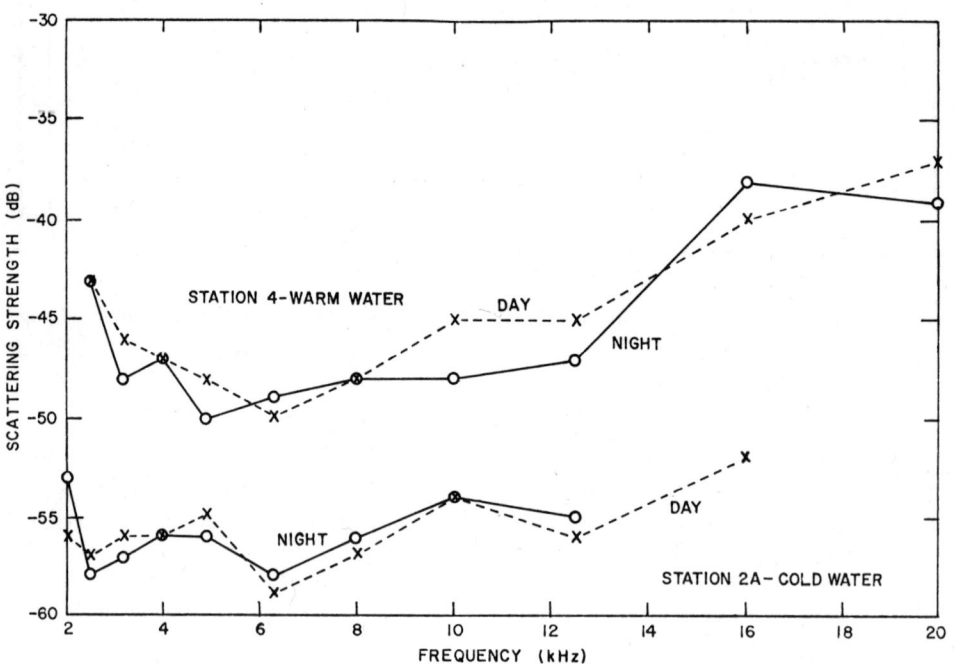

Fig. 13. Day and night scattering strength spectra at warm and cold water locations in the Norwegian Sea.

The greatly reduced diurnal variation in scattering strengths probably is due to a much more limited vertical migration by marine animals in northern waters. Light conditions in these latitudes do not vary much in summer, and the light penetrates to much shallower depths than in lower latitudes. Changing light intensity as a stimulus for migration is thus reduced. The much colder water below the thermocline also may act to limit the extent of vertical movement. Many species that live at rather shallow levels in northerly waters occur at progressively greater depths towards the tropics, where the extent of vertical migration is much greater, and where, too, reverberation levels typically show a large day-night variation.

Part of the measurement program at these two stations included biological sampling with a 6-foot Isaacs-Kidd Midwater Trawl. In 11 hauls at Station 2A, in cold water, only six fish were caught; one 25.4-cm cod and five specimens of a lanternfish, *Benthosema glaciale*. Fishes susceptible to capture by this particular net were extremely sparse at this location. The highest concentration was calculated to be only 0.08 fish/1000 m^3. In contrast to this low catch, nine hauls at Station 4 yielded 392 specimens. One was a small cod, one a small bristlemouth or gonostomatid, and the rest were all *Benthosema glaciale*.

The considerable difference in fish abundance between these two stations may explain the difference in reverberation levels. However, the general similarity in shape of the curves for the two stations, and the same characteristic increase in scattering strength with decreasing frequency in the lower portion of the spectrum (Group D scattering) indicates that some other faunal element is present that is large enough to avoid our net and which accounts for the increase in low frequency scattering. Another explanation would be that so few fishes are present which scatter sound in the mid-frequency region that levels in that portion of the reverberation spectrum are depressed.

Both factors very likely are operative. Herring and cod are ubiquitous in these northern waters during summer, both have well-developed, gas-filled swimbladders and are large enough to avoid a 6-foot net except by chance, which probably explains the capture of only a single 10-inch cod at Station 2A. Echosounder records at 12 kHz for this station show layers made up of group scattering, the traces consisting of discrete scatterers or schools of fish. At this station, lanternfishes and other similar fishes typical of warmer waters were essentially absent. The presence of the larger fishes thus may elevate the reverberation curve at the lower frequencies and the absence of more typical scatterers lowers the curve at the middle frequencies.

It is interesting to examine the relationship between faunal changes and the transition from Group D to Group B scattering. The diversity and numbers of mesopelagic fishes, such as lanternfishes, are greatest in temperate, subtropical, and tropical waters. Aside from *Benthosema glaciale*, and perhaps two or three other species, those mesopelagic fishes which do occur in the Norwegian Sea very likely are expatriates carried outside their normal centers of abundance by the northeast movement of Atlantic Ocean waters. At Station 1, a warm water station about 450 miles southwest of Station 4, scattering strengths were measured. Figure 14 shows the daytime and nighttime values. The curve is typical of Group C scattering. Ten midwater trawl hauls were made at this station, and the total fish catch was 1,496 specimens representing 15 genera and 15 species. *Benthosema glaciale* was still the most abundant, but other lanternfishes, as well as bristlemouths and hatchetfishes, also were well represented in the catches. The mesopelagic fauna here was much more diverse than at Station 4, and the concentrations were considerably higher. Associated with this increased diversity is a reverberation pattern transitory between Group-D scattering in the Norwegian Sea, where relatively few of these fishes occur, and Group-B scattering in the warmer waters of the North Atlantic, where many more species occur.

Subsequent measurements by NAVOCEANO, as well as other studies already discussed above, repeatedly demonstrate the significance of physical boundaries as lines of demarkation between different reverberation provinces. Indeed, temperature alone may be a useful index

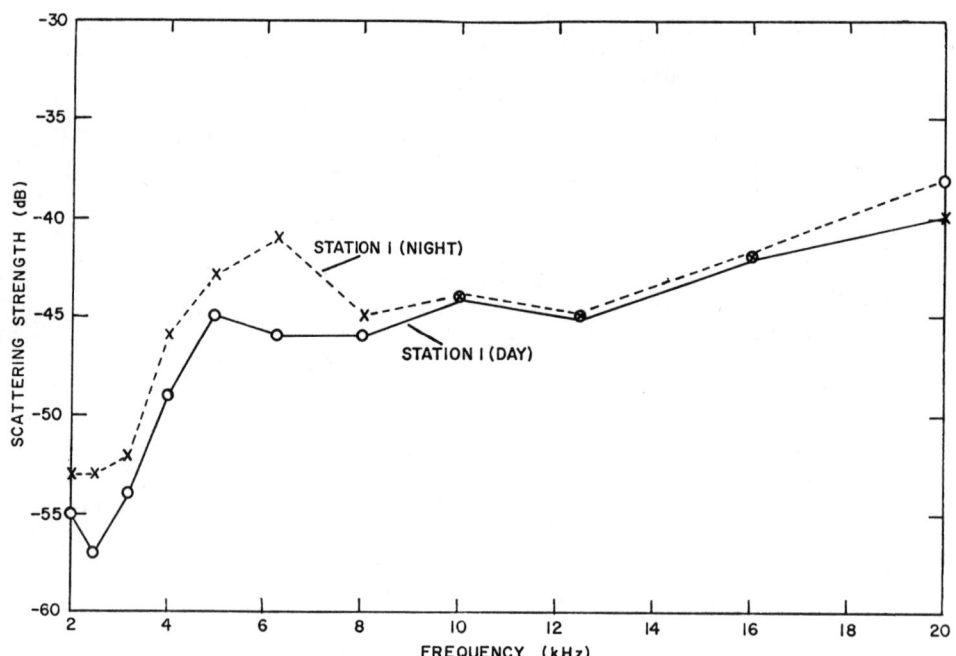

Fig. 14. Day and night scattering strength spectra at a warm water location in the northeastern Atlantic Ocean (58°00'N; 10°52'W).

of reverberation variations. Based upon a limited set of data collected by the Naval Oceanographic Office (*Winokur*, pers. comm.) correlation coefficients between column scattering strength and sea surface temperature, and between column scattering strength and the temperature at 75 meters, were significant at the 0.01 level for 4 out of 5 frequencies examined over the range from 1-16 kHz (i.e., at 2, 4, 8 and 16 kHz, but not at 1 kHz).

SUMMARY

1. Biological sound scattering varies with frequency, depth, time of day, geographic location, and season. Its frequency dependence is of particular importance in studies concerned with identifying causative organisms, since a single species may scatter sound over a relatively broad range of frequencies, and a fish of a given size may scatter sound at more than one frequency.

2. At high frequencies (> 40 kHz) zooplankton groups, such as euphausiids and pteropods, become effective sound scatterers. There is some evidence that copepods about 1 mm long, and perhaps even phytoplankton, may scatter sound at frequencies above 300 kHz.

3. Except in high Arctic and Antarctic Seas, mesopelagic fishes possessing gas-filled swimbladders (about 280 species) are the principal resonant sound scatterers over frequencies from about 1-25 kHz; physonectid siphonophores also may be significant resonant sound scatterers, probably at frequencies above about 10-12 kHz. Concentrations of the order of only a few organisms/1000 m^3 are sufficient to account for observed scattering levels.

4. In high northern latitudes, where mesopelagic fishes are neither speciose nor abundant, larger fishes such as herring, cod, and redfish become significant resonant scatterers at low frequencies (below about 4-6 kHz). Comparable low frequency peaks do not occur in Antarctic Ocean waters because such large fishes do not occur there.

5. Vertical migration is of considerable ecological significance since the major migrating, sound-scattering communities, the fishes and the principal zooplankton groups, occupy adjacent links in the food chain. Their vertical movements assume added importance when trying to understand observed day-night differences in acoustic patterns. This is particularly true of mesopelagic fishes possessing swimbladders, since bubble size and depth determine the resonant frequencies scattered by given fish.

6. Sufficient acoustic backscattering measurements have been made, especially over the frequency range from about 1 to 20 kHz, to delineate generalized patterns, including day-night differences and geographic variation. Data are beginning to accumulate on the depth dependence of scattering but little is known of the seasonal variability.

7. Regional variations in acoustic scattering are associated with the large scale features of circulation in the ocean, and pronounced changes occur when crossing oceanographic boundaries.

REFERENCES

Alexander, R. McN. 1971. Swimbladder gas secretion and energy expenditure in vertically migrating fishes. In: Farquhar, 1971, pp. 74-85.

Andreeva, I. B. 1964. Scattering of sound by air bladders of fish in deep sound-scattering ocean layers. Akust. Zh., 10: 20-24 (English translation in: Soviet Physics-Acoustics, 10: 17-20).

Aron, W., N. Raxter, R. Noel, and W. Andrews. 1964. A description of a discrete depth plankton sampler with some notes on the towing behavior of a 6-foot Isaacs-Kidd midwater trawl and a one-meter ring net. Limnology and Oceanography, 9(3): 324-333.

Backus, R. H., J. E. Craddock, R. L. Haedrich, D. L. Shores, J. M. Teal, A. S. Wing, G. W. Mead, and W. D. Clarke. 1968. Ceratoscopelus madarensis: Peculiar sound-scattering layer identified with this myctophid fish. Science, 160 (3831):991-993.

Backus, R. H., J. E. Craddock, R. L. Haedrich, and D. L. Shores. 1969. Mesopelagic fishes and thermal fronts in the western Sargasso Sea. Mar. Biol. 3(2): 87-106.

Backus, R. H., J. E. Craddock, R. L. Haedrich, and D. L. Shores. 1971. The distribution of mesopelagic fishes in the equatorial and western North Atlantic Ocean. In: Farquhar, 1971, pp. 20-40.

Backus, R. H. and J. E. Craddock. 1977. Pelagic faunal provinces and sound-scattering levels in the North Atlantic Ocean. In: "Oceanic Sound Scattering Prediction", N. R. Andersen and B. J. Zahuranec (Eds.), Plenum, N.Y., (this volume).

Baird, R. C., D. F. Wilson, and D. M. Milliken. 1973. Observations on Bregmaceros nectabanus Whitley in the anoxic sulfurous water of the Cariaco Trench. Deep-Sea Res., 20: 503-504.

Baird, R. C., D. F. Wilson, R. C. Beckett, and T. L. Hopkins. 1974. Diaphus taaningi Norman, the principal component of a shallow-sound-scattering layer in the Cariaco Trench, Venezuela. J. Mar. Res., 32(2): 301-312.

Barham, E. G. 1963. Siphonophores and the deep scattering layer. Science, 140(3568): 826-828.

Barham, E. G. 1971. Deep-sea fishes: Lethargy and vertical orientation. In: Farquhar, 1971, pp. 100-118.

Bary, B. McK. and R. E. Pieper. 1971. Sonic-scattering studies in Saanich Inlet, British Columbia: A preliminary report. In: Farquhar, 1971, pp. 601-611.

Batzler, W. E. 1975. Deep-scattering-layer observations off New Zealand and comparison with other volume scattering measurements. J. Acoustic. Soc. Amer., 58(1): 51-71.

Brown, C. L. and A. L. Brooks. 1974. A summary report of progress in the Ocean Acre Program. NUSC Tech. Report 4643, 44 pp.

Castile, B. D. 1975. Reverberation from plankton at 330 kHz in the western Pacific. J. Acoust. Soc. Amer., 58(5): 972-976.

Chapman, R. P., O. Z. Bluy, and R. H. Adlington. 1971. Geographic variations in the acoustic characteristics of deep scattering layers. In: Farquhar, 1971, pp. 306-317.

Chapman, R. P., O. Z. Bluy, R. H. Adlington, and A. E. Robinson. 1974. Deep scattering layer spectra in the Atlantic and Pacific oceans and adjacent seas. J. Acoust. Soc. Amer., 56(6): 1722-1734.

Clarke, G. L. 1971. Light conditions in the sea in relation to the diurnal vertical migrations of animals. In: Farquhar, 1971, pp. 41-50.

D'Aoust, B. G. 1971. Physiological constraints on vertical migration by mesopelagic fishes. In: Farquhar, 1971, pp. 86-99.

Davis, E. E. 1971. Quasi-synoptic measurements of volume reverberation in the western North Atlantic. In: Farquhar, 1971, pp. 294-305.

Farquhar, G. B. 1971. Editor, Proceedings of an International Symposium on Biological Sound Scattering in the Ocean; March 31 - April 2, 1970, Airlie House Conference Center, Warrenton, Virginia. Dept. of the Navy, Maury Center for Ocean Science, Report No. MC-005; 629 pp.

Fisch, N. P. and R. K. Dullea. 1973. Acoustic volume scattering at the Bermuda Ocean Acre site during spring and summer, 1970, and summer 1971 (Cruises 9, 10, and 12). NUSC Tech. Rept. 4469, 24 pp.

Gibbs, R. H., Jr. and C. F. E. Roper. 1971. Ocean Acre: Preliminary report on vertical distribution of fishes and cephalopods. In: Farquhar, 1971, pp. 119-133.

Gold, B. A. 1966. Measurements of volume scattering from a deep scattering layer. J. Acoust. Soc. Amer. 40(3): 688-696.

Gold, B. A. and P. Van Schuyler. 1966. Time variability of volume scattering in a small oceanic area. J. Acoust. Soc. Amer., 40(6): 1317-1321.

Hall, M. 1971. Volume backscattering in the South China Sea and the Indian Ocean. J. Acoust. Soc. Amer., 50(3, Part 2): 940-945.

Hall, M. 1973. Volume backscattering in the Tasman Sea, the Coral Sea, and the Indian Ocean. J. Acoust. Soc. Amer., 54(2): 473-477.

Hansen, W. J. and M. J. Dunbar. 1971. Biological causes of scattering layers in the Arctic Ocean. In: Farquhar, 1971, pp. 508-526.

Hersey, J. B., R. H. Backus, and J. Hellwig. 1962. Sound scattering spectra of deep scattering layers in the western North Atlantic Ocean. Deep-Sea Res., 8: 196-210.

Kampa, E. M. 1971. Photoenvironment and sonic scattering. In: Farquhar, 1971, pp. 51-59.

Kanwisher, J. and A. Ebeling. 1957. Composition of the swimbladder gas in bathypelagic fishes. Deep-Sea Res., 4: 211-217.

Kashkin, N. I. 1974. Identification of sound-scattering layers from the size structure of the population of mesopelagic sound-scattering fish species. Oceanology, 14(5): 736-739 (Translated from *Okeanology of the Academy of Sciences of the USSR*, pp. 903-907).

Kinzer, J. 1967. Die Verbreitung des Zooplanktons in Echostreuschichten extrem sauerstoffarmen Wassers. *Umschau in Wissenschaft und Technik*, No. 22/67: 733-734.

Lebedeva, L. P. 1965. Measurement of the dynamic shear modulus of animal tissues. *Akust. Zh.*, 11: 197-200 (English translation in: *Soviet Physics-Acoustics*, 11: 163-165).

Love, R. H. 1969. Maximum side aspect target strength of an individual fish. *J. Acoust. Soc. Amer.*, 46(3, Part 2): 746-752.

Love, R. H. 1971. Evening discussion session. In: Farquhar, 1971, pp. 351-353.

Love, R. H. 1975. Predictions of volume scattering strengths from biological trawl data. *J. Acoust. Soc. Amer.*, 57(2): 300-306.

Love, R. H. 1977. Recent developments in volume reverberation research at the Naval Oceanographic Office, this conference. In: *Oceanic Sound Scattering Prediction*, Andersen, N. R. and B. J. Zahuranec (Eds.), Plenum Press, N. Y., (this volume).

Marshall, N. B. 1951. Bathypelagic fishes as sound scatterers in the ocean. *J. Mar. Res.*, 10(1): 1-17.

Marshall, N. B. 1971. Swimbladder development and the life of deep-sea fishes. In: Farquhar, 1971, pp. 69-73.

McElroy, P. and Asa Wing. 1971. Scattering returns in the Mediterranean and Eastern Atlantic--data and instrumentation. In: Farquhar, 1971, pp. 220-231.

McElroy, P. T. 1974. Geographic patterns in volume-reverberation spectra in the North Atlantic between 33°N and 63°N. *J. Acoust. Soc. Amer.*, 36(2): 394-407.

Pearcy, W. G. and R. S. Mesecar. 1971. Scattering layers and vertical distribution of oceanic animals off Oregon. In: Farquhar, 1971, pp. 381-394.

Pieper, R. 1976. A study of the relationship between zooplankton and high-frequency scattering of underwater sound. *Natural History Museum of Los Angelos* (in press).

Pieper, R. 1977. Some comparisons between oceanographic measurements and high-frequency scattering of underwater sound. In: *"Oceanic Sound Scattering Prediction"*, N. R. Andersen and B. J. Zahuranec (Eds.), Plenum, N.Y., (this volume).

Scrimger, J. A. and R. G. Turner. 1973. Backscattering of sound from the ocean volume between Vancouver Island and Hawaii. *J. Acoust. Soc. Amer.*, 54(2): 483-493.

Shearer, L. W. 1971. Comparisons between surface-measured swimbladder volumes, depth of resonance, and 12-kHz echograms at the time of capture of sound-scattering fishes. In: Farquhar, 1971, pp. 453-473.

Swarts, R. L. 1971. Time variations of some acoustic volume reverberation parameters. In: Farquhar, 1971, pp. 241-267.

Van Schuyler, P. and A. A. Hunger. 1967. A volume scattering and oceanographic study of an area in the eastern Gulf of Mexico. *U. S. Naval Oceanographic Office, Informal Report* No. 67-34, 48 pp

Van Schuyler, P. 1971. An acoustically determined distribution of resonant scattering north of Oahu. In: *Farquhar*, 1971, pp. 328-340.

Vent, R. J. 1972. Acoustic volume scattering measurements at 3.5, 5.0, and 12.0 kHz in the eastern Pacific Ocean: Diurnal and seasonal variations. *J. Acoust. Soc. Amer.*, 52 (1, Part 2): 373-382.

Wilson, D. F., L. C. Ricalzone, and R. C. Beckett. 1968. Some investigations of sound-scattering layers near Key West, Florida. *Nav. Res. Lab. Report 6582*: 40 pp.

Wilson, D. F. 1972. Diel migration of sound scatterers into, and out of the Cariaco Trench anoxic water. *J. Mar. Res.*, 30(2): 168-176.

Zahuranec, B. J., W. L. Pugh, and G. B. Farquhar. 1970. Biological sound scattering studies. Part I: Initial investigations in the Gulf of Mexico and western North Atlantic. *Naval Oceanographic Office Technical Report*, TR-224, 35 pp.

Zahuranec, B. J. and W. L. Pugh. 1971. Biological results from scattering layer investigations in the Norwegian Sea. In: *Farquhar*, 1971, pp. 360-380.

Pelagic Faunal Provinces and Sound-Scattering Levels in the Atlantic Ocean

R. H. Backus and J. E. Craddock

Woods Hole Oceanographic Institution

ABSTRACT

An earlier, incomplete scheme of dividing the North Atlantic Ocean into pelagic faunal provinces (Backus et al., 1970) has been shown to have value as a scheme of sound-scattering provinces (Chapman et al., 1974). The present paper extends and refines the earlier scheme of faunal provinces and, based upon the size of midwater fish catches, shows how volume reverberation levels can be expected to change from province to province.

INTRODUCTION

During the last dozen years the authors and their colleagues at the Woods Hole Oceanographic Institution have made midwater collections with a 10-foot Isaacs-Kidd midwater trawl over wide reaches of the deep Atlantic Ocean for the purpose of studying the geographic distribution patterns of the small fishes of the upper 700 or 800 meters, so-called "mesopelagic fishes". One result has been the dividing of the Atlantic north of about 35°S (but not including the Arctic) into a system of pelagic faunal *regions* and *provinces*. An earlier version of this scheme for the tropical and western North Atlantic was given by *Backus et al.* (1970).

Also, it has been argued (*Backus*, 1972) that faunal provinces reasonably can be considered to be sound-scattering provinces. Using certain catch-rate data, expected mean levels of volume reverberation relative to a reference province were calculated for the other provinces. This scheme for dividing the Atlantic Ocean into provinces and the predicted reverberation levels was compared with actual broad-band measurements made over wide reaches of the Atlantic by Chapman, Bluy, and Adlington (1971) and some agreement was found (Backus, 1972).

Later, Chapman et al. (1974) made an elaborate comparison of an even more extensive body of broad-band sound-scattering data with a scheme of Atlantic faunal provinces mainly as set forth by Backus et al. (1970). For an eastern Pacific Ocean comparison they used zoogeographic information from Kort (1967). They concluded in their summary (in part): "Spectra of column strength have been used as a tool for comparing reverberation conditions in a wide range of oceanographic areas. As the spectra often maintained consistent features over distances of hundreds of kilometers but changed dramatically in the neighborhood of known faunal or oceanographic boundaries, the present study supports the concept of dividing the world's oceans into reverberation provinces."

The purpose of the present paper is to set forth our extended scheme of faunal regions and provinces and, by means of catch-rate data for mesopelagic fishes, to suggest how volume reverberation levels will change from province to province. We do not compare our results with actual acoustical measurements (as we did in the earlier paper, Backus, 1972). The reverberation levels derived are considered to be approximations at best, and the point that there is faunal and acoustical correspondence from ocean area to ocean area has been well made by Chapman et al. (1974) as is noted above.

THE DATA AND THEIR COLLECTION

Our collections number 1022 (Figure 1) and total 2510.16 hours of collecting (net in water). Most of the hauls were so-called "horizontal" ones - the net was lowered to some chosen depth, towed more or less at that depth for some time, and then raised. The tows varied by depth and by time of day. Most night-time tows were made at depths less than 200 m. Most day-time tows were made at depths between 240-800 m.

We have written elsewhere (Backus et al., 1965; Backus et al., 1969; and Backus et al., 1970) about the sampling rationale behind these collections. Our aim has been to sample the fish fauna of the mesopelagial as completely as possible over long distances. We have attempted to do this by choosing sampling depths such that a few succesive tows more or less covered the upper few hundred meters of the water column before the ship changed geographic position appreciably. Individual fishing depths were chosen so as to maximize the amount of material caught. We did *not* follow some rigid plan of arbitrarily chosen fishing depths. The latter scheme, commonly used by physical oceanographers, for instance, makes little sense for biological sampling. In choosing fishing depths at which the chances for making large catches were thought

Fig. 1. The geographic distribution of the midwater collections.

good, we were aided most by the echo-sounder and bathythermograph. These let us identify sound-scattering maxima and critical points on the temperature-depth curve that often seemed to coincide with planes of concentration of midwater fishes. Additional details concerning the sampling can be found in Backus et al.

Some midwater fish species come to the very sea surface at night and, thus, are a part of the so-called "neuston". Some of these species, in fact, are very poorly sampled by midwater trawling because of their extreme upward migration. To sample these species better, we towed surface skimming nets (Bartlett and Haedrich, 1968) while midwater trawling; we have data from 531 neuston collections totaling 2209.21 hours of fishing.

Most of our collections have been made on long transects. Temperature sections were made simultaneously, based on hourly bathythermograph observations. Such sections have been invaluable in identifying points of physical change along the collecting transects and in understanding the nature of these changes.

The basic fish datum from these midwater trawls and neuston tows is the number of specimens of a given species in a given collection. From the duration of a collection the catch-per-unit-of-effort for a given lot (all of one species from one collection) can be calculated. We also tabulate for each lot the range of standard lengths and the displacement volume.

The data upon which the present paper is based come from the lanternfishes (Myctophidae), the most speciose family of midwater fishes in the world ocean and also of first or second importance from the standpoint of biomass (Paxton, 1972). In our midwater trawl collections myctophids number 242, 143 specimens in 11,326 lots with 107 species represented. The neuston collections yeilded 41,712 specimens of 35 myctophid species.

THE PELAGIC FAUNAL REGIONS AND PROVINCES OF THE ATLANTIC OCEAN

By a pelagic faunal province we mean a part of the open waters of the deep ocean within which the fauna is thought to be reasonably homogenous. The boundaries between faunal provinces, then, are zones (narrow ones) within which there is a more or less marked faunal change.

Our method for drawing the boundaries of faunal provinces has been: 1) the fish data, most of which have been collected on long transects, are examined for faunal change, 2) locations along the transect of marked faunal change are examined for evidence of physical change, 3) the physical change (which is assumed always to exist and is generally readily detectable in the presence of a marked faunal change) is identified with some widespread circulational feature, and 4) the circulational boundary is plotted, using the data of physical oceanography, and is taken to be the faunal boundary. Of course, it must be understood that while the boundaries between faunal provinces are drawn as lines, they are in reality narrow zones that may shift seasonally or with other rhythms.

For example, if one begins in the Slope Water off southern New England and proceeds south making midwater trawls and bathythermograph observations, the collections will show a faunal change that is simultaneous with a change in the temperature structure of the water column that signals the northern edge of the Gulf Stream. The faunal change noted is assumed to be directly related to the physical change that is the northern edge of the Gulf Stream. It is further assumed that the faunal change noted could be noted at any point of crossing of the northern edge of the Gulf Stream between Cape Hatteras and the Tail of the Grand Banks. Taking some generally accepted criterion for demarking the northern edge of the Gulf Stream (for instance, the 15°C isotherm at 200 m), and resorting to a temperature atlas, the average position of the northern edge of the Gulf Stream is plotted and is taken to be the average position of a faunal boundary.

The Slope Water - Gulf Stream physical and faunal boundary is used as an example because it is conspicuous and familiar. Most of the boundaries of pelagic faunal provinces are less well-marked. Furthermore, in many parts of the Atlantic physical understanding of the ocean is weak and the method described above is difficult to apply.

In order to examine our data for faunal change we have experimented with a method based on the incidence of first-time and last-time captures of species in collections made along transects (Backus et al., 1965). Such captures have a greater incidence near faunal boundaries. The expected incidence of such captures in successive collections from a fauna assumed to be homogeneous can be compared with the observed incidence using a chi-square test. This method fails to take into account changes in abundance within species and also has practical difficulties that limit its usefulness.

Changes in relative abundance have been taken into account by simply tabulating the incidence of "events", collection by collection; "events" include both first and last-time captures and changes in the abundance of species. This way of examining the data is not purely objective. In any case, we have ignored apparent faunal changes that could not be associated with some physical boundary.

For making a bias-free assessment of the fitness of our system of faunal boundaries, 477 shallow night-time tows were studied using cluster techniques, and the grouping of the collections according to the faunal province of their origin was judged good. The faunal provinces are not uniformly different from one another. Rather, they are to be arranged on the basis of faunal similarities into groups called pelagic faunal regions. For a measure of the inter-relatedness of the several pelagic faunal provinces and for help in grouping them into regions, a similarity analysis was made using each province's shallow, night-time collections. The faunal inter-relatedness of provinces agrees well with their physical inter-relatedness, that is, the faunal regions are physically as well as faunally distinct. Our system of Atlantic Ocean pelagic faunal provinces and regions is shown in Table 1 and Figure 2 and is very briefly described below.

TABLE 1. Faunal Regions and Provinces of the Atlantic Ocean (South of the Arctic Sea and North of the South Atlantic Subtropical Convergence).

  I. Atlantic Subarctic Region

    1. Atlantic Subarctic Province

 II. North Atlantic Temperate Region

    2. Northern Gyre (Province)
    3. Slope Water (Province)
    4. Azores-Britain Province
    5. Mediterranean Outflow (Province)
    6. Western Mediterranean Sea (Province)
    7. Eastern Mediterranean Sea (Province)

III. North Atlantic Subtropical Region

    8. Northern Sargasso Sea (Province)
    9. Southern Sargasso Sea (Province)
   10. Northern North African Subtropical Sea (Province)
   11. Southern North African Subtropical Sea (Province)

 IV. Gulf of Mexico (Region)

   13. Gulf of Mexico (Province)

  V. Mauritanian Upwelling (Region)

   18. Northern Mauritanian Upwelling (Province)
   19. Southern Mauritanian Upwelling (Province)

 VI. Atlantic Tropical Region

   12. Lesser Antillean Province
   14. Caribbean Sea (Province)
   15. Amazonian Province
   16. Guinean Province
   23. Straits of Florida (Province)

VII. South Atlantic Subtropical Region

   17. South Atlantic Subtropical Sea (Province)

Briefly, it can be said that in the North Atlantic (exclusive of arctic seas) there are subarctic, temperate, subtropical, and tropical regions plus two special regions - the Gulf of Mexico and the Mauritanian Upwelling. The tropical region extends also, of course, into the South Atlantic Ocean, and south of it there lies a subtropical region analogous to the one of the North Atlantic (and beyond whose southern limit our observations scarcely extend).

PELAGIC FAUNAL PROVINCES 535

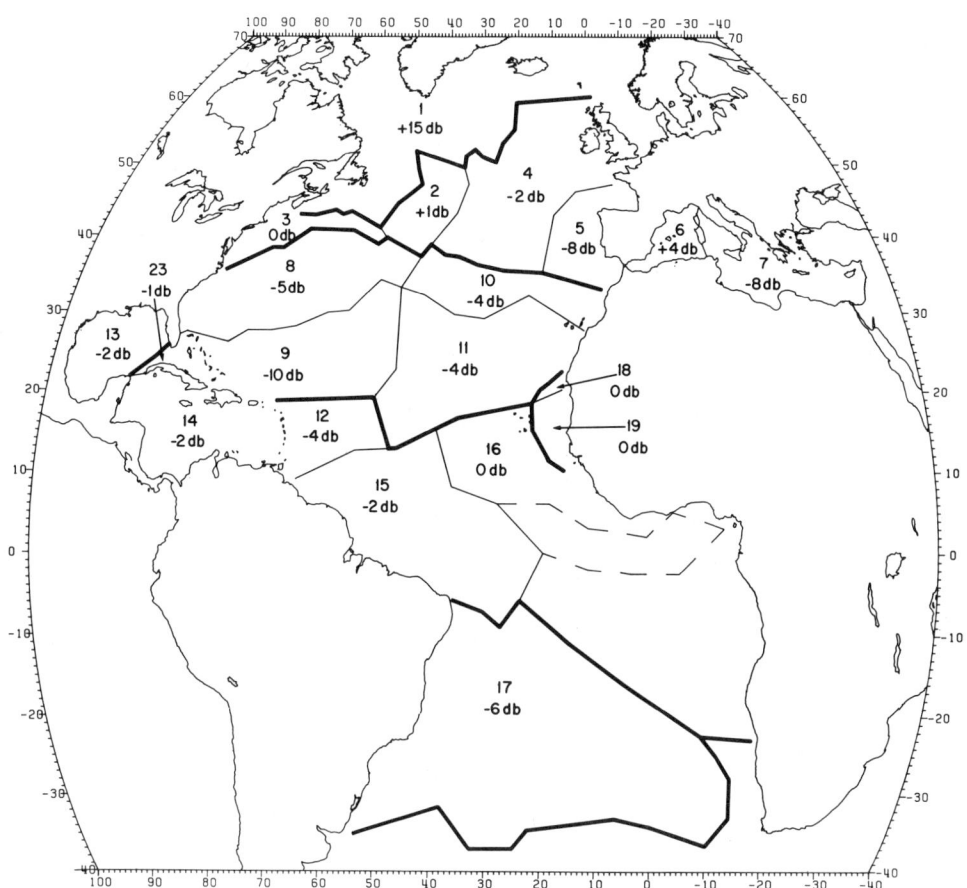

Fig. 2. Chart of Atlantic Ocean pelagic faunal provinces (upper numbers) and provincial volume reverberation levels (lower numbers). The provincial names corresponding to the numbers are: 1) Atlantic Subarctic Province, 2) Northern Gyre, 3) Slope Water, 4) Azores-Britain Province, 5) Mediterranean Outflow, 6) Western Mediterranean Sea, 7) Eastern Mediterranean Sea, 8) Northern Sargasso Sea, 9) Southern Sargasso Sea, 10) Northern North African Subtropical Sea, 11) Southern North African Subtropical Sea, 12) Lesser Antillean Province, 13) Gulf of Mexico, 14) Caribbean Sea, 15) Amazonian Province, 16) Guinean Province, 17) South Atlantic Subtropical Sea, 18) Northern Mauritanian Upwelling, 19) Southern Mauritanian Upwelling, and 23) Straits of Florida. The heavier boundaries are regional ones; the names of the regions so set off can be found in Table 1. See text.

The Atlantic Tropical Region (consisting of Guinean, Amazonian, and Lesser Antillean Provinces and the Caribbean Sea and Straits of Florida) is spanned (except for the last-named province) by the thermal equator and is physically characterized by a very warm and rather constant surface layer which, however, is quite thin in most of the region. Eighteen of 75 mesopelagic Atlantic myctophids have a tropical distribution pattern, that is, they are found principally throughout the tropical region (Figure 3) or rarely, only in parts of it (Figure 4). Some tropical species are swept northward in considerable numbers by the Gulf Stream and so end up in the Northern Sargasso Sea or in the Slope Water (Figure 3).

The North Atlantic Subtropical Region (consisting of the Northern and Southern Sargasso Seas and the Northern and Southern North African Subtropical Seas) is commonly thought of as a single anticyclonic gyre occupying mid-ocean. Compared to the tropics, the subtropics has a cooler surface layer but relatively warm water going to a much greater depth. Consequently, at depths from 200-1000 m or thereabouts the ocean cools both poleward and equatorward from the subtropics. Subtropical waters are the clearest, bluest, and least productive of the North Atlantic. In the South Atlantic there is an analogous gyre - the South Atlantic Subtropical Region.

Thirteen of 75 Atlantic myctophids show the subtropical distribution pattern; that is, they are found principally in the Sargasso and North African Subtropical Seas (Figure 5). Some live also in the Gulf of Mexico and, to a lesser extent, in the Caribbean Sea. At least 11 of the 13 subtropical North Atlantic myctophids are found also in the South Atlantic Subtropical Region (but not in the intervening tropics) (Figure 5) and, thus, are among those species that are called "bipolar" or "biantitropical". For comparison, the distribution of a purely South Atlantic Ocean subtropical species, *Diaphus anderseni*, is shown in Figure 6.

Eighteen of 75 Atlantic myctophids show what can be called the tropical-subtropical distribution pattern (Figure 7); that is, they are found more or less equally in both the tropics and the subtropics. Species of this sort preponderate in the Gulf of Mexico, a region (and province) that is neither clearly tropical nor subtropical, but which must be set apart.

Furthermore, there are five species that appear to live with about equal abundance in the Atlantic Tropical Region and in the equatorward halves of the North and South Atlantic Subtropical Regions. These species, which occupy the area of the northeast and southeast tradewinds and intervening doldrums, are called tropical-semisubtropical species.

The North Atlantic Temperate Region (consisting of Slope Water, Northern Gyre, Azores-Britain Province, Mediterranean Outflow, and Western and Eastern Mediterranean Seas) stretches from the northern edge of the North Atlantic's subtropical sea to the polar front.

PELAGIC FAUNAL PROVINCES 537

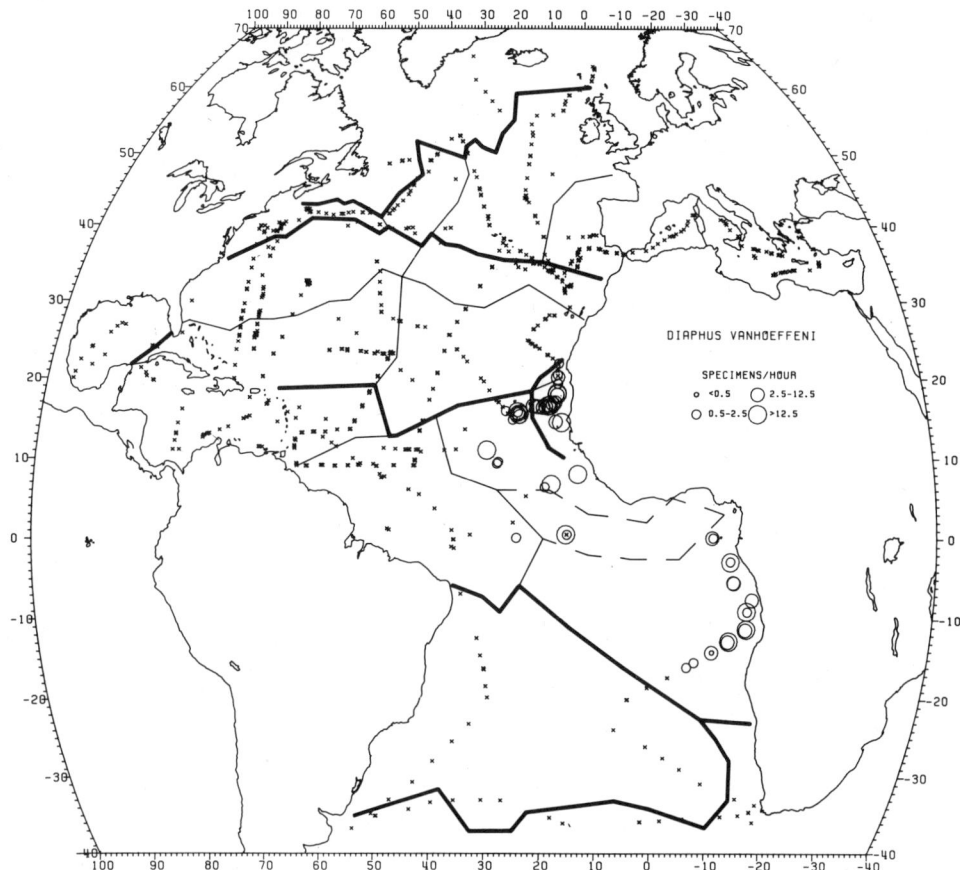

Fig. 3. The distribution of the tropical species *Diaphus dumerili*. It occurs throughout the tropics and is carried in numbers north out of the tropics by the Gulf Stream. The radius of the dot is proportional to catch rate according to the scheme at the top right. The x's are the nagative data; that is, they mark the positions of tows made at proper depths for catching the species in question but that failed in doing so.

The region is bounded approximately by 9° and 15°C isotherms at 200 m. Although there is considerable variation from province to province, the temperate region is generally more productive than the tropics and subtropics (*Ryther*, 1963). Related to this superior productivity is the region's greener color and greater turbidity. Seasonal effects are stronger in the temperate North Atlantic than in the tropics and subtropics. In some respects, the Eastern Mediterranean Sea verges on the subtropical; nevertheless, its fauna is clearly temperate.

Although a few species appear to be restricted to the North Atlantic Temperate Region, most of the species living there live also in the northern provinces of the North Atlantic Subtropical Region (temperate-semisubtropical species, Figure 8) or in the Atlantic Subarctic region (subpolar-temperate species, Figure 9). Altogether, 18 of 75 Atlantic myctophids have these "northern" distribution patterns. Eight of them are bipolar, that is, they are found also at the appropriate latitude in the South Atlantic Ocean.

The Atlantic Subarctic Region (consisting of a single province) lies across the polar front from the North Atlantic Temperate Region.

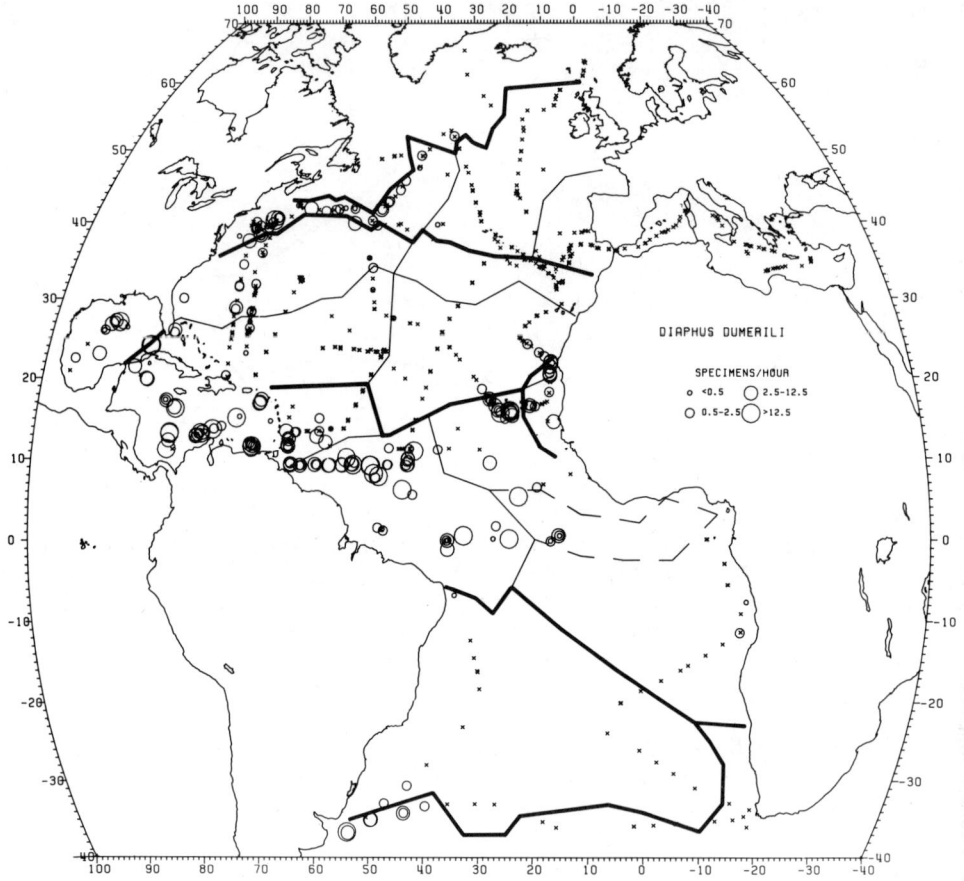

Fig. 4. The distribution of the tropical species *Diaphus vanhoeffeni*. It is absent from the western tropics.

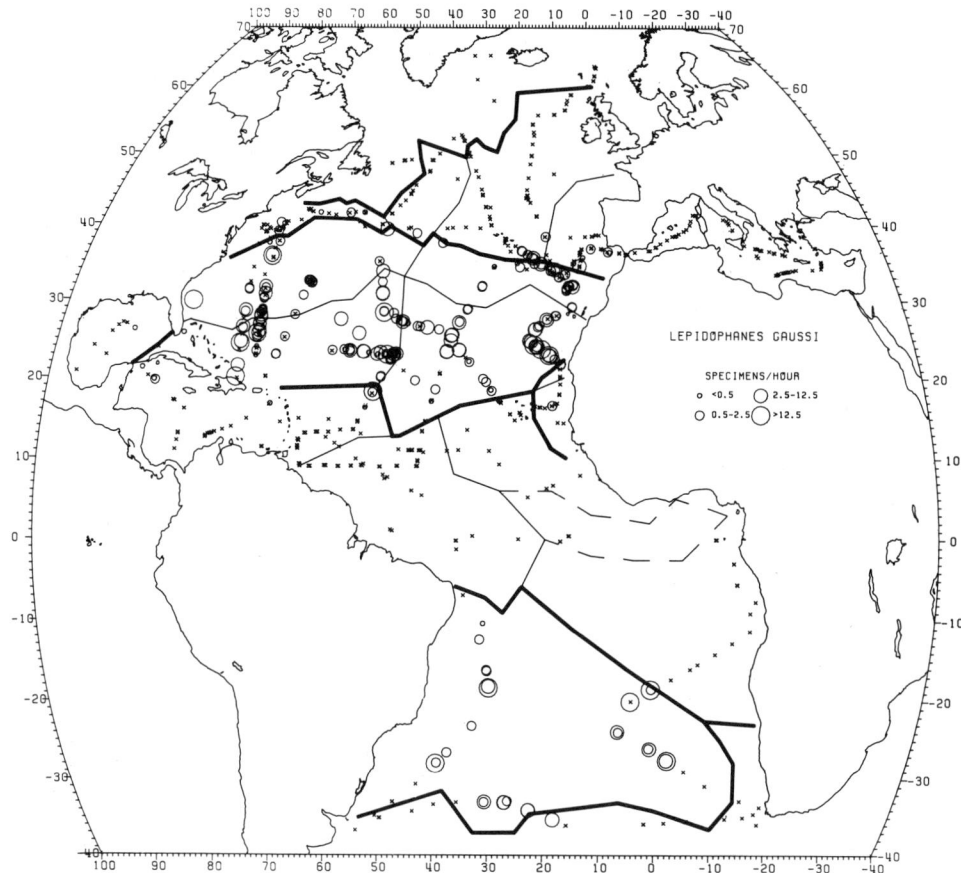

Fig 5. The distribution of *Lepidophanes gaussi*, a bipolar (biantitropical) subtropical species.

The temperature gradient across the front is strong except in the east where the front becomes ill-defined, then disappears. The Subarctic Region is subject to great seasonal change, the most significant part of which, no doubt, is that in solar radiation. The Subarctic Region is poor in species, and there are no myctophids restricted to it.

Finally, the Mauritanian Upwelling Region (consisting of northern and southern provinces), whose name indicates its most significant physical characteristic, proves to be faunally distinct. It is a southern outpost for certain subarctic-temperate and temperate species on the one hand (Figure 9), yet, on the other, has tropical elements in its fauna. One myctophid species (and perhaps a second undescribed one) is endemic to the region.

Although the scheme of faunal regions and provinces just given is based mainly on the distribution of myctophid fishes, our data (Backus et al., 1965; Backus et al., 1969; Backus et al., 1970, and unpublished) indicate that the scheme is applicable to mesopelagic fishes in general. Furthermore, the literature of Atlantic biogeography suggests that the scheme applies to mesopelagic animals as a whole and probably to epipelagic plants and animals as well.

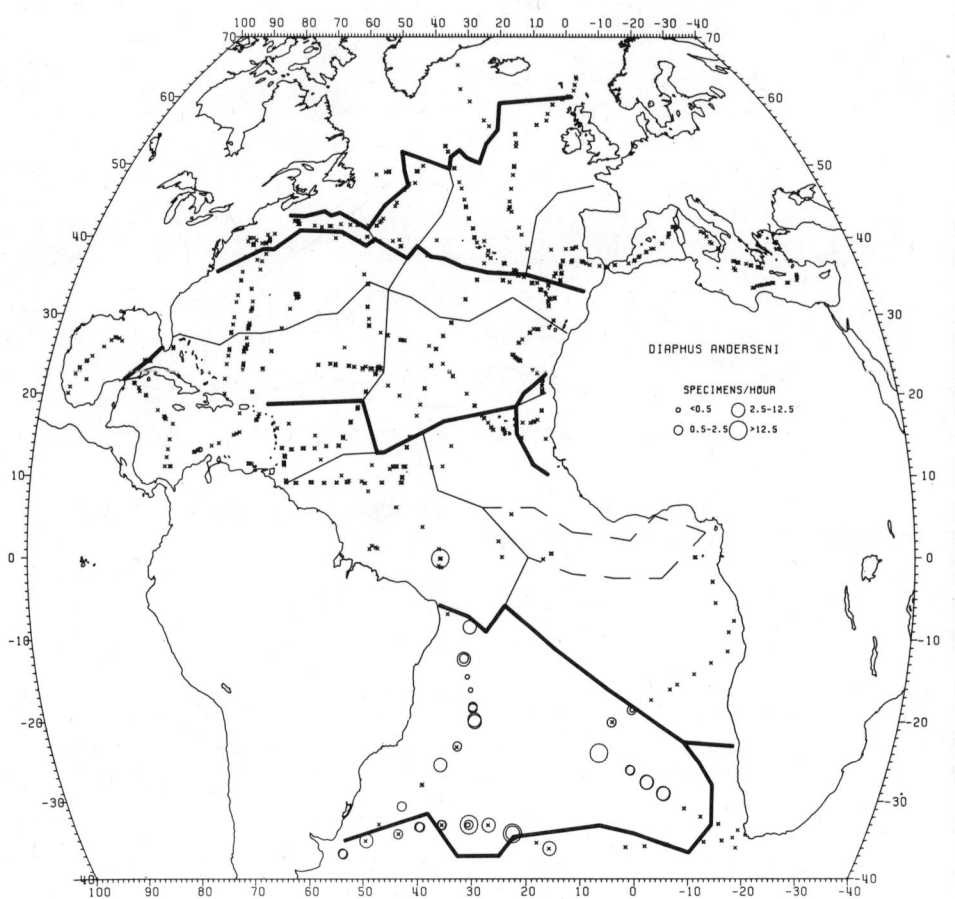

Fig. 6. The distribution of *Diaphus anderseni*, a South Atlantic Ocean subtropical species.

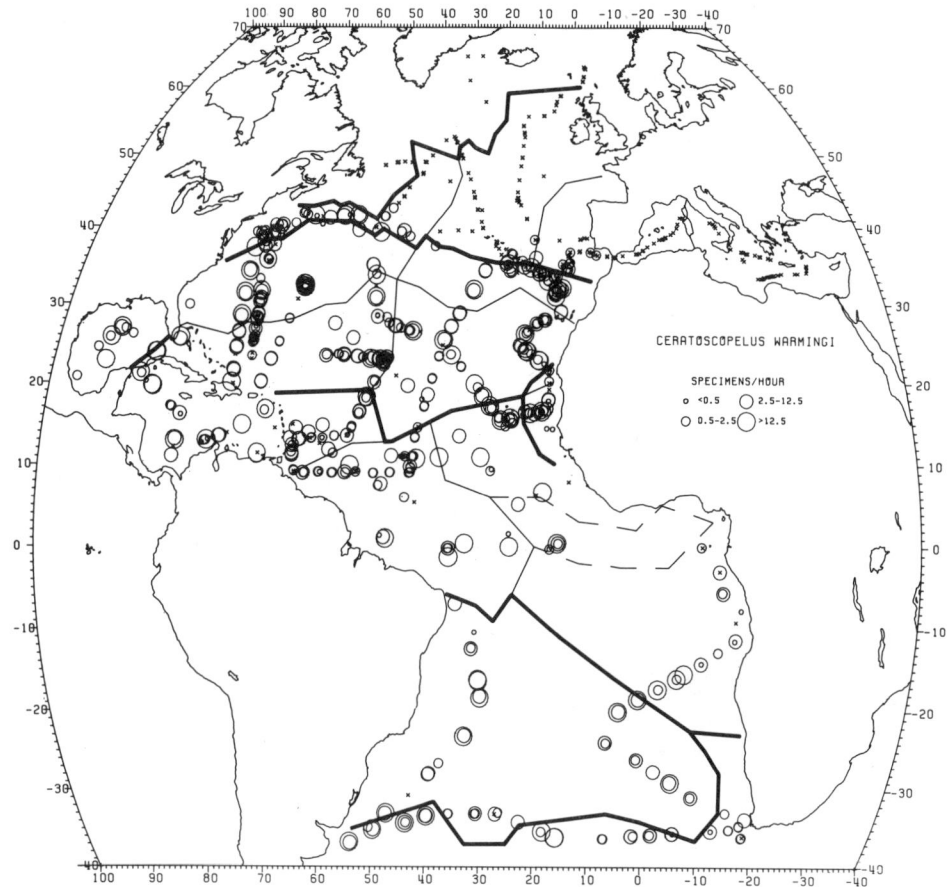

Fig. 7. The distribution of the tropical-subtropical species *Ceratoscopelus warmingi*.

COMPUTATION OF PROVINCIAL SOUND-SCATTERING LEVELS

If it be allowed that myctophids are of preeminent importance as oceanic sound-scatterers or, at least, that they are representative of the distribution of oceanic sound-scatterers in general, then it follows that just as the mass and species composition of the myctophid fauna varies from one part of the ocean to another so will the spectrum of volume reverberation vary. It is suggested, then, that the pattern of volume reverberation in the Atlantic will vary in accordance with the scheme of faunal provinces and regions set out above. Intra-provincial variation will be small compared to inter-provincial variation and provinces belonging to the same region will have certain common properties.

It is also suggested that the overall level of volume reverberation will vary from province to province as the mass of sound-scatterers varies. Figure 2 attempts to show the sign and magnitude of the change in volume reverberation that would be encountered on the average in crossing provincial boundaries. The values shown have been computed by 1) pooling the catches of a province's shallow (<200 m) nighttime collections and dividing by the pooled collection durations to get an average catch rate (ml/hr), 2) normalizing the data by dividing the several catch rates by the catch rate for the Slope Water (Province 3), which thus becomes the reference province, and 3) assuming that doubling the catch doubles the power of the sound scattering and represents a 3 db increase (Table 2). It is assumed that the samples from each province are similarly enough

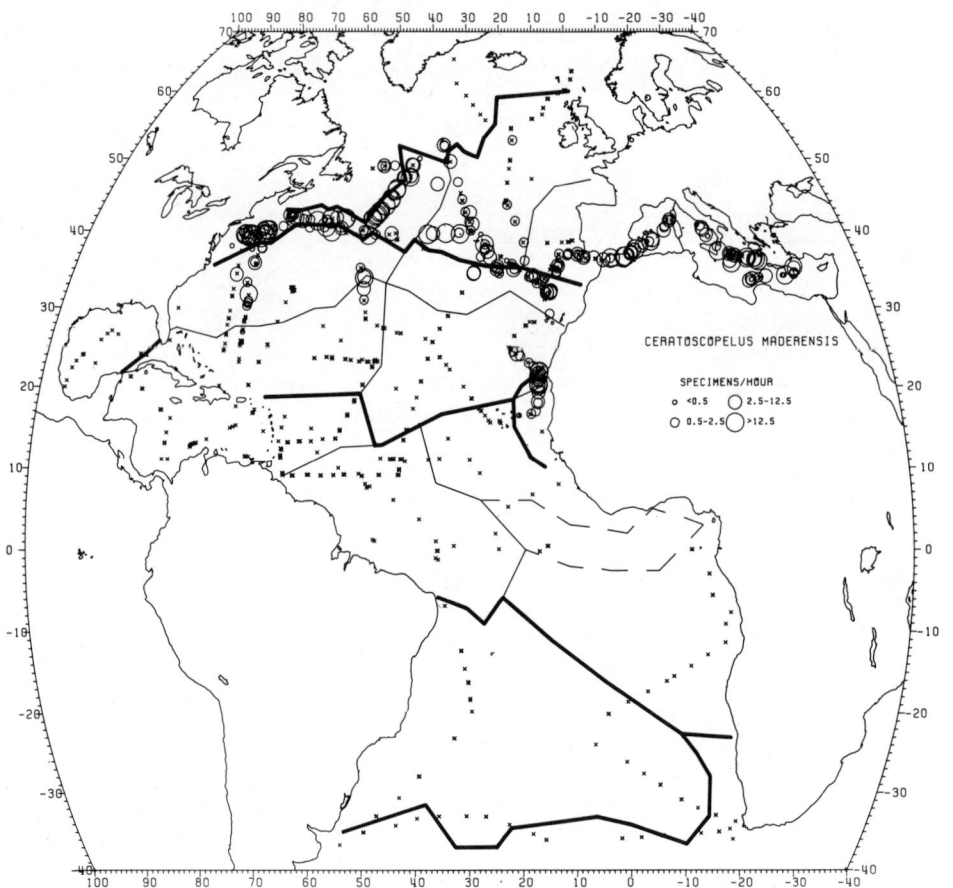

Fig. 8. The distribution of the temperate-semisubtropical species *Ceratoscopelus maderensis*.

PELAGIC FAUNAL PROVINCES 543

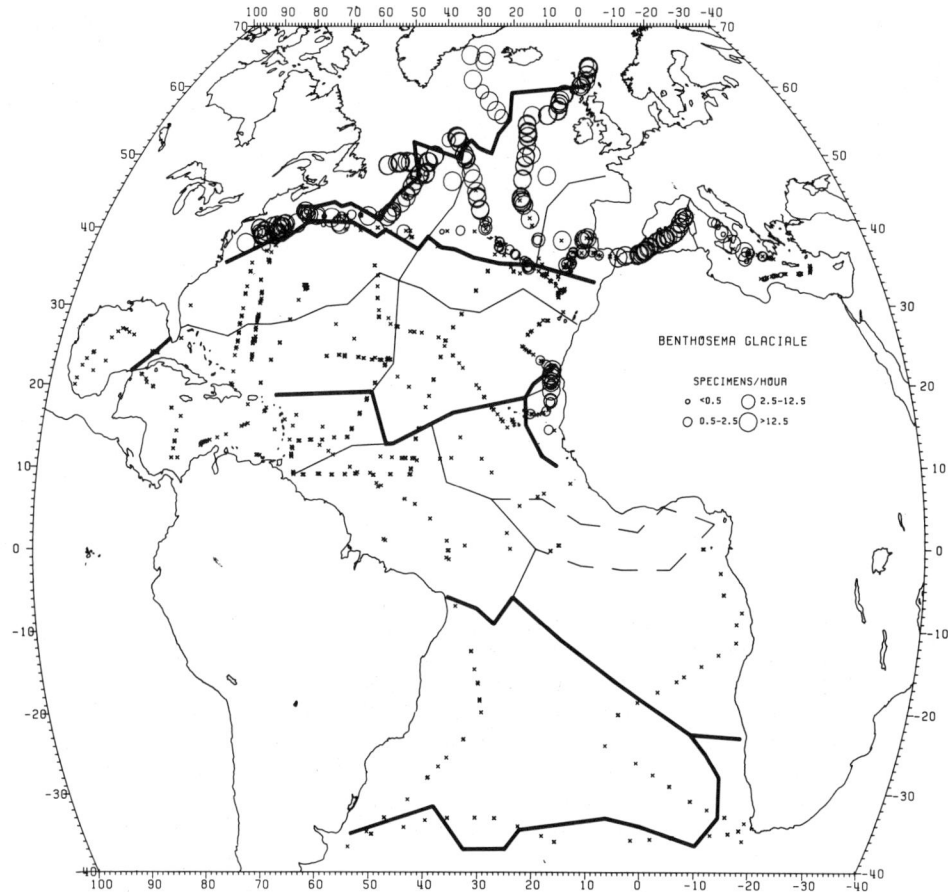

Fig. 9. The distribution of the **subpolar-temperate** species *Benthosema glaciale*.

distributed with respect to depth so that biases due to the depth dependence of catch rate are absent. This assumption is probably unwarranted in the cases of Provinces 2, 12, 13, and 18, where sample size is small.

We have attempted to get some measure of the variability of the data by examining it in the not-necessarily-representative Northern Sargasso Sea (Province 8). Because the catch rate within the shallow nighttime collections varies systematically with depth (Figure 10), an exaggerated notion of the variability of the data is obtained if one proceeds directly to calculate a standard deviation and so on in the usual way. Therefore, we have divided the 36 Northern Sargasso Sea collections into two sets by pairing collec-

TABLE 2. Provincial Volume Reverberation Levels Calculated from Shallow (< 200 m) Nighttime Midwater Trawl Collections.

Province	n	Duration, hrs	Catch, ml/hr	Catch / 83	Level,[**] db
1	20	26.07	485	5.84	+ 15
2	8	23.15	97	1.17	+ 1
3[*]	22	57.95	83	1.00	0
4	59	83.90	53	0.639	− 2
5	31	43.62	13	0.157	− 8
6	20	47.77	193	2.33	+ 4
7	37	92.67	13	0.157	− 8
8	36	74.92	24	0.290	− 5
9	18	38.23	9	0.109	− 10
10	35	57.42	36	0.434	− 4
11	32	59.92	33	0.398	− 4
12	6	11.37	30	0.362	− 4
13	7	24.18	57	0.687	− 2
14	22	71.50	57	0.687	− 2
15	14	37.22	55	0.663	− 2
16	34	66.97	75	0.904	0
17	49	75.92	21	0.253	− 6
18	3	5.15	88	1.06	0
19	12	21.85	76	0.916	0
23	4	13.67	70	0.843	− 1

[*] Reference province

[**] Rounded off to nearest decibel

tions according to depth (Table 3). The differences in catch rate of pairs have been taken as the deviations to be used in calculating a standard deviation and a standard error of the mean. The variance calculated from these deviations (810.46) is halved (because it comes from deviations that are differences between two observations) before extracting the standard deviation ($s = 20.13$) and the standard error ($s_{\bar{x}} = \frac{20.13}{6} = 3.355$). The standard error

PELAGIC FAUNAL PROVINCES 545

has been applied to the mean obtained by dividing the pooled catches by the pooled collection durations. Dividing the mean plus twice the standard error by the mean ([25.85 + 2(3.355)] / 25.85) and converting this ratio (1.2596) to decibels gives approximately the fiducial limits (at the .05 level) for the average reverberation value ($\pm$ 2 db). These limits are probably too narrow in the provinces where sample size is very low (Provinces 2, 12, 13, and 18).

Seasonal variability is expected to be small in the tropical provinces and in the southern provinces of the subtropical region. In the northern provinces of the subtropical region, in the temperate and subarctic regions, in the Mauritanian Upwelling Region, and in the Gulf of Mexico, season should affect reverberation level more. Probably the seasonal effect will be shown to increase with latitude.

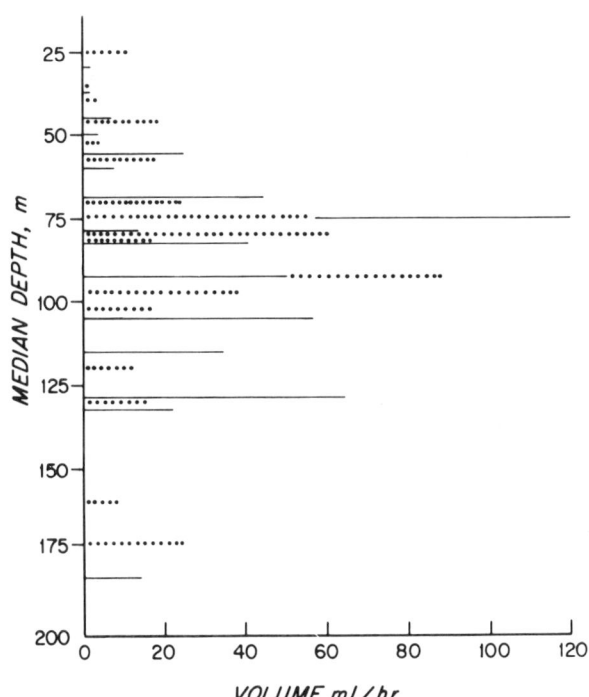

Fig. 10. Catch rate as a function of depth for 36 shallow nighttime tows from the Northern Sargasso Sea. The tows have been divided into two subsets by matching them with respect to depth. See Table 3.

TABLE 3. Variability in 36 Paired Collections from the Northern Sargasso Sea (Province 8).

Collection no.	Depth, m	r_1, ml/hr	Collection no.	Depth, m	r_2, ml/hr	$r_2 - r_1$
1120	50–55	1.47	1119	50–55	2.41	− 0.94
1510	35–35	2.07	1506	36–36	1.13	0.94
1129	70–80	119.94	1127	75–83	59.73	60.21
1740	122–135	64.72	1737	125–135	15.95	48.77
2114	70–80	7.39	1933	70–80	55.20	−47.81
1727	55–65	7.73	1512	55–60	17.20	− 9.47
1736	72–87	12.60	1731	80–85	17.82	− 5.22
1132	52–85	44.21	1052	55–85	22.57	21.64
1513	35–40	1.84	1316	33–46	2.43	− 0.59
1126	45–55	2.80	1123	50–55	3.84	− 1.04
2117	90–95	52.02	2115	90–105	37.44	14.58
1732	125–130	22.12	1728	118–122	12.07	10.05
2118	175–190	14.35	1130	170–175	24.00	− 9.65
2120	100–110	56.22	1505	100–105	16.49	39.73
2119	50–62	25.60	1741	42–50	18.58	7.02
1516	75–90	41.02	1124	85–100	87.62	−46.60
1011	35–55	6.89	1010	22–30	10.80	− 3.91
1935	105–125	34.13	1509	155–165	8.08	26.05

ACKNOWLEDGEMENTS

Thanks are due to many colleagues for their assistance with this project. The work was supported by National Science Foundation Grant No. DES 74-23209 and its predecessors. This is Woods Hole Oceanographic Institution Contribution #3677.

REFERENCES

Backus, R. H. 1972. Midwater fish distribution and sound-scattering levels in the North Atlantic Ocean. *Journal of Underwater Acoustics, 22*(3): 243-255.

Backus, R. H., J. E. Craddock, R. L. Haedrich, and D. L. Shores. 1969. Mesopelagic fishes and thermal fronts in the western Sargasso Sea. *Mar. Biol.*, *3*(2): 87-106.

Backus, R. H., J. E. Craddock, R. L. Haedrich, and D. L. Shores. 1970. The distribution of mesopelagic fishes in the equatorial and western North Atlantic Ocean. *J. Mar. Res.*, *28*(2): 179-201.

Backus, R. H., G. W. Mead, R. L. Haedrich, and A. W. Ebeling. 1965. The mesopelagic fishes collected during Cruise 17 of the R/V Chain, with a method for analyzing faunal transects. *Bull. Mus. Comp. Zool.*, *134* (5): 139-157.

Bartlett, M. R., and R. L. Haedrich. 1968. Neuston nets and South Atlantic larval blue marlin (*Makaira nigricans*). *Copeia, 1968* (3): 469-474.

Chapman, R. P., O. Z. Bluy, and R. H. Adlington. 1971. Geographic variations in the acoustic characteristics of deep scattering layers. In: G. B. Farquhar, Ed., *Proceedings of an International Symposium on Biological Sound Scattering in the Ocean*, pp. 306-317. U. S. Government Printing Office, Washington, D. C.

Chapman, R. P., O. Z. Bluy, R. H. Adlington, and A. E. Robison. 1974. Deep scattering layer spectra in the Atlantic and Pacific Oceans and adjacent seas. *J. Acoust. Soc. Am.*, *56*(6): 1722-1734.

Kort, V. G. 1967. The Pacific Ocean. Biology of the Pacific Ocean, Book III, Fishes of the open waters (Moscow), (Translation 528, USN Oceanographic Office, 1971), 273 pp. (As cited by *Chapman et al.*, 1974).

Paxton, J. R. 1972. Osteology and relationships of the lantern-fishes (family Myctophidae). *Bull. Los Angeles County Mus. Nat. Hist. Sci.*, *No. 13*: 81 pp.

Ryther, J. H. 1963. Geographic variations in productivity. In: M. N. Hill, Ed., *The Sea, 2*: 347-380. Interscience Publishers, New York.

Sound Scattering and Oceanic Midwater Fishes

R. C. Baird and D. F. Wilson

University of South Florida and Don-Wil Electronics

ABSTRACT

A combination of "acoustic profiling" of the water column and quantitative discrete-depth trawling was used to study time-dependent changes in faunal stratification in different midwater environments. Results lead to some descriptions of vertical stratification and migration patterns, probably faunal components involved, and the influence of adaptive strategies, all in relation to observed acoustic phenomena. In the Cariaco Trench, two essentially monospecific vertically migrating, scattering layers were studied. One, consisting of Diaphus taaningi, responded to changes in light intensity while the others, consisting of Bregmaceros nectabanus, did not appear to respond to such changes. Possible factors causing discrepancies between numbers of D. taaningi captured in net hauls and predicted from sound-scattering strength measurements indicate that present equations relating acoustic factors to fish abundance and characteristics and various physical factors are basically valid but require refinement. At stations in the western Caribbean and eastern Gulf of Mexico, both in waters of similar hydrographic properties, acoustic profiles showed the expected heterogeneity in stratification reflecting the greater diversity of species. On the other hand, from stratification patterns revealed by acoustic profiling at a station in the Venezuela Basin portion of the eastern Caribbean, it appears that the midwater scattering community there consists of many more vertical migrators. There appear to be marked differences in faunal composition between the two Caribbean stations, particularly among the lanternfishes, Family Myctophidae. It was concluded that acoustic profiling can be useful to assess the organization of sound scattering, particularly below 100 m, and to *compare between different stations or at different times in patterns of vertical stratification, fish abundance, and vertical migration behavior in relation to environmental and phylogenetic factors. Such information is considered essential in formulating predictive models describing volume reverberation phenomena.*

INTRODUCTION

A substantial literature exists on the theoretical aspects of underwater acoustics and on the actual and potential effects of living organisms on sound propagation in the oceans (e.g., *Weston,* 1967; *Urick,* 1967; *Farquhar,* 1970; *Cushing,* 1973). Measurements have been made of reverberation levels in many oceanic regions (see *Farquhar,* 1970; *Vent,* 1972; *Seligman and Friedl,* 1975). Nevertheless it is apparent that formidable problems remain to be solved before a satisfactory capability is achieved for predicting sound-scattering phenomena in the sea.

Small midwater (mesopelagic) fishes of low individual acoustic target strength contribute significantly to volume-reverberation phenomena and are important in the formation of discrete sound-scattering layers (see, for example, *Farquhar,* 1970). Many of these fishes have gas-filled swimbladders whose resonant frequency lies between 3 and 30 kHz (e.g., *Kleckner and Gibbs,* 1972). Prediction models must, therefore, consider the dependence of scattering patterns on frequency (e.g., *Batzler et al.,* 1975). In this latter connection, it is also likely that some of the problems that will be encountered in obtaining and interpreting the needed data will be different for different frequencies, and also for different geographic areas at a given frequency.

In general there have been difficulties in relating mesopelagic fishes to specific layers and, more important, relating population density from trawl estimates to population density based on acoustic data. Equations have been developed to predict the density of volume scatterers (e.g., *Zahuranec et al.,* 1970; *Van Schuyler,* 1970). Recently, however, more detailed measurements of the sound-scattering properties of mesopelagic fishes related to populations with known swimbladder volumes are being used to formulate predictive models which reconcile acoustic theory with actual observations from various oceanic regions (*Kleckner and Gibbs,* 1972; *Batzler,* 1972; *Batzler et al.,* 1973, 1975). Much of the difficulty in both formulating and testing such models stems from lack of precise collecting techniques, inadequate measurements of swimbladder volume and shape, and lack of knowledge of how these latter may vary with depth.

The use of a variety of adaptive strategies (*sensu Baird,* 1974) in an associated group of species will also complicate their collective contributions to sound-scattering phenomena. Such strategies include species-specific patterns of vertical stratification, diurnal migration, abundance, and seasonal ontogenetic changes (see *Backus* et al., 1969; Pearcy and Mesecar, 1970; Goodyear et al., 1972; Clark, 1973; Seligman and Friedl, 1975). For any given area of the ocean, the acoustic-predictability problem thus involves a spectrum of swimbladder sizes and shapes, fish abundances, and time-dependent patterns of vertical and horizontal distribution.

We have attempted to contribute to the solution of these various problems by using a combination of "acoustic profiling" of the water column (see below) and quantitative discrete-depth trawling to study time-dependent changes in stratification in oceanic environments. Some results are presented that describe the nature of vertical stratification and migration patterns, the possible faunal components involved, and the influence of adaptive strategies on observed acoustic phenomena.

Equipment Used

The acoustic gear consisted of two "acoustic profilers" (Figure 1), which are merely one-kilowatt, downward-looking echosounders equipped with relatively narrow-beam transducers and an A-scan display (Wilson et al., 1968; Wilson, 1972; Baird et al., 1974). They were ordinarily operated at 25 and 50 kHz. The primary readout shows scattering strength in decibels as a function of depth on a cathode-ray oscilloscope; ancillary recording methods (e.g., magnetic tape) are also based on this primary A-scan form of presentation.

Acoustic profilers offer several advantages in studying mesopelagic communities. They essentially "fingerprint" a relatively narrow column of water by indicating, in terms of sound-scattering strength, the depths where major contributions to volume reverberation occur at a given frequency. Within an area, it is possible to compare profiles taken at different seasons or times of day, or made at different frequencies or at slightly different positions. The degree of acoustic stability can be assessed and different areas compared.

For biological sampling in conjunction with the acoustic profiling, modified six- and twelve-foot Tucker trawls with messenger-operated opening and closing mechanisms were used to collect at precisely known discrete depths. The depth at which the trawls were operating was constantly monitored and adjusted from a deck control station where depth data were displayed digitally from a signal originating in a Teledyne-Taber pressure transducer attached to the trawl (Hopkins et al., 1973; Baird et al., 1974; Hopkins and Baird, 1975).

Studies in the Cariaco Trench

Previous evidence from both acoustic and biological studies suggested that the Cariaco Trench offers unique opportunities for

Fig. 1. Photograph of acoustic profilers with A-scan camera attachment.

studies of vertical-migration phenomena and comparisons of population abundance made from acoustic measurements and from trawl catches (Wilson, 1972; Baird et al., 1973, 1974). Anoxic water and the presence of H_2S apparently place severe limitations on faunal diversity; the mesopelagic fauna of the Trench is markedly depauperate, consisting of only three fish species and no large crustaceans (Mead, 1963; Pugh, 1972; Wilson, 1972; Baird et al., 1974). Identification of scattering-layer organisms and factors influencing their vertical migration, and comparison of quantitative trawl catches with scattering-strength values at depth should, therefore, be easier than in more diverse environments.

Two scattering layers revealed by a standard 12-kHz echosounder and precision depth recorder (PDR) were investigated by discrete-depth trawling and acoustic profiling. Figure 2, a PDR record, shows the two layers during morning descent. The shallower layer, which descends to around 270 m, is delimited by arrows in the figure, while a second layer is seen to separate from the simultaneously descending mass at about 0530 local time. At about 0700 this layer merges on the PDR record with a secondary bottom-echo trace occurring at approximately 350 m on the PDR record. Figure 3b shows these layers in a 25-kHz acoustic profile taken at 1030. The shallow layer is represented by a narrow scattering peak at 220-250 m, while the

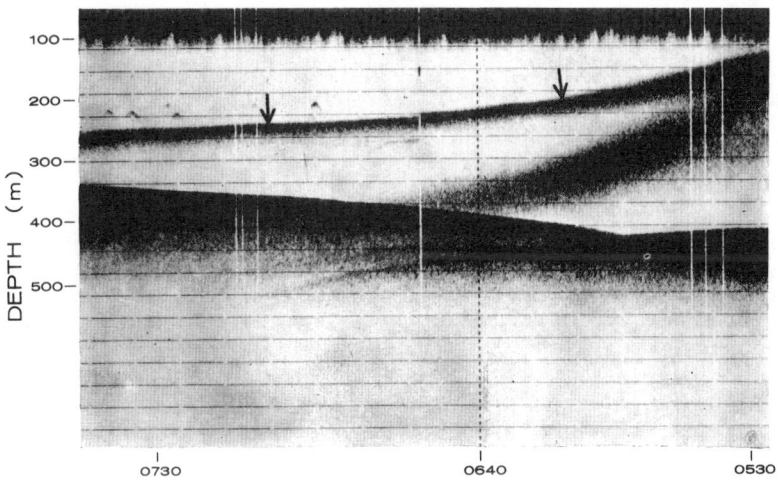

Fig. 2. Photograph of 12 kHz PDR record of scattering layers in Cariaco Trench (arrows delimit D. taaningi layer). Layer recorded in approximately 320 m is an artifact from a second return from the bottom.

deeper layer can be seen as a zone of broad scattering from 600 to 800 m, deep in the anoxic zone. Four discrete-depth trawls were taken in the deep layer in the anoxic zone. The catch consisted of 117 specimens of *Bregmaceros nectabanus* Whitley ranging from 25 to 45 mm (SL), with 70% of the individuals falling between 30 and 40 mm SL. No other organisms were taken. A two-hour trawl in the shallow layer from 220 to 270 m caught 60 individuals of the myctophid *Diaphus taaningi* Norman ranging in size from 35 to 45 mm. Trawls made at other depths (140 - 330 m) during the day did not collect this species (*Baird et al.*, 1973, 1974). On the basis of a few profiles made at 50 kHz with 1-msec pulses and a narrower-beam transducer, the *D. taaningi* layer was estimated to be about 30 m thick with the leading edge varying between 260 and 270 m.

Morphological examination revealed that both species had well-developed swimbladders. Bladder volumes at the surface were calculated for a number of specimens of *D. taaningi*. At a layer depth of 270 m, the calculated mean resonant frequency for the *D. taaningi* population was 14 kHz. The observed sound-scattering strength of the layer at 25 kHz would have required 13 to 20 fish/1000 m^3, whereas estimates from trawl catches were about 2.5 fish/1000 m^3 (see *Van Schuyler*, 1970, for the equations used). Agreement with acoustic estimates of abundance based on our best available 25 kHz data appears good considering the preliminary nature of the studies, possible net avoidance, and some unfortunate limitations in the amount of acoustic data available (see *Baird et al.*, 1974 for details). The results agree within an order of magnitude, however, and suggest that resonant scattering is a major contributor to volume reverberation in this case. The indication also is that the relationships expressed by the equations are basically sound but probably require refinement in certain factors to improve predictability (*Baird et al.*, 1974).

Observations of the timing and rates of vertical migration in the two layers confirm the relative constancy of these parameters from day to day and year to year (at the same season) in the Cariaco Trench (*D. F. Wilson*, unpublished data). The *D. taaningi* layer appears to exhibit small variations in depth correlated with changes in the intensity of the incident sunlight (*D. F. Wilson*, unpublished data). It thus appears to follow a classic pattern in which the population is closely attuned to the photoenvironment, and the variability among individuals in photoresponse (and therefore depth) is low.

The migrating *B. nectabanus* layer cannot be seen in the anoxic zone with the 12 kHz equipment, but it can be followed to near the bottom (around 1000 m) with the 25 kHz profiler. Figure 3a is an acoustic profile taken at night and is remarkable in illustrating the complete absence of scattering at 25 kHz below 400 m in the Trench; the A-scan trace below this depth is similar to a calibration pattern. Figure 3b is a profile taken during the day after the migrating *B. nectabanus* layer has descended well into the anoxic

SOUND SCATTERING AND OCEANIC MIDWATER FISHES 555

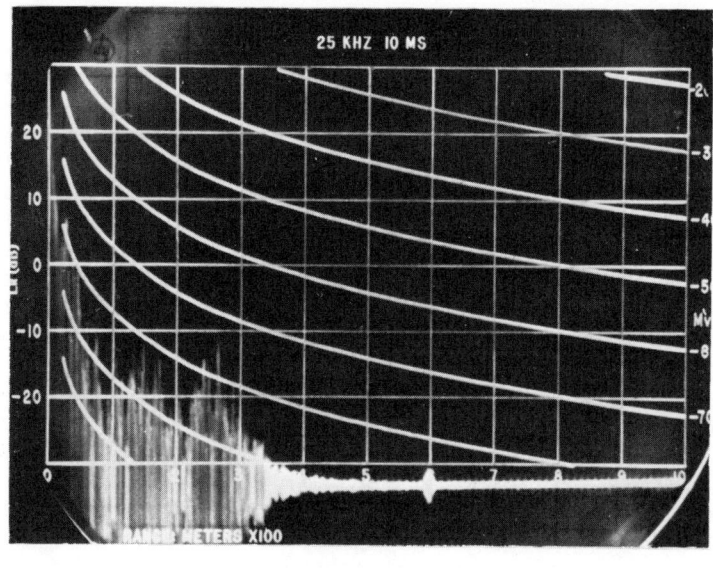

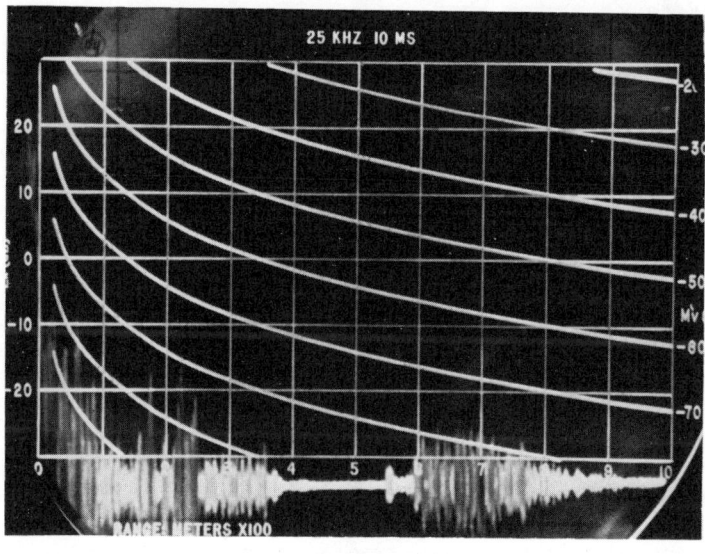

Fig. 3. Photographs of acoustic profiles at 25 kHz in Cariaco Trench (10°38'N; 65°45'W). 3a. Profile taken at 2400 hours local time. 3b. Profile taken at 1030 local time.

zone. Note that instead of a highly pronounced peak (some vertical spreading would be expected even from a narrow layer as differences in slant range increase with signal spread), the scatterers appear to be distributed over several hundred meters (cf Figure 2). Although the B. *nectabanus* layer is more compact as it enters the anoxic zone (see *Wilson*, 1972), it still does not show as a "peak" at any depth. The conclusion is that individuals may not be reacting directly to the photoenvironment (light intensity at the deeper ranges of migration may be below visual thresholds) and, in any case, the variability in individual response to depth is considerably higher. Size stratification may be involved to some extent, but the concentration of most of the population in a narrow size range suggests additional factors are involved. Additional study is clearly needed.

Measurements of scattering strength can be made at short intervals and over relatively short distances to assess the acoustic homogeneity of scattering layers (see *Swarts*, 1970). This has been done in our studies, but the data have not yet been analyzed. However, the variability in mean scattering strength should be a function of the number of scatterers present in the insonified volume and therefore indicative of the extent of patchiness in occurrence of individuals at that depth, since we know the size distribution of the population.

Our data also show that essentially the whole population in both layers migrates to surface waters at night. If moribund or non-migrating individuals are left behind, their contribution to scattering cannot be detected and, given the calculated Acoustic Cross Section and the sensitivity of the equipment, it appears that virtually no individuals remain at depth. It has been suggested that not all of the population of certain mesopelagic species migrate each day (*Farquhar*, 1970; *Clark*, 1973, among others). This does not appear to be the case here in either instance, and further investigation of other species is required to confirm the generality of the patterns observed in this physiologically rigorous environment.

Studies in the Gulf of Mexico and Caribbean Sea

Studies combining acoustic profiling and midwater trawling were also undertaken in the Gulf of Mexico and open Caribbean Sea. Locations were chosen from waters of very low primary production, influenced by the western boundary current system which gives rise to the Gulf Loop Current and Gulf Stream. Two of the locations had hydrographic properties similar to many tropical gyre-or central water-mass environments of broad extent in the world's oceans. Figures 4a-d are acoustic profiles at 25 kHz taken at midday and midnight at a station in the Eastern Gulf of Mexico and one in the Western Caribbean. Note that, unlike in the Cariaco Trench profiles, there is considerable heterogeneity in volume scattering, and that this is spread out over many depths in the water column. Considerable scattering is present below 400 m at night, and a layer of relatively high scattering strength is found in the region from 400 to 500 m at all times. Trawling in this layer confirmed the occurrence of several species known not to undergo extensive vertical migration and which had well developed swimbladders; e.g., *Valenciennellus tripunctulatus* and *Argyropelecus hemigymnus*. Clearly we were dealing with a considerably more complex pattern involving elements of all of the problems of diversity mentioned earlier.

There are, however, a number of similarities in the profiles of the two stations. By day there is a concentrated layer of scattering from 400 to 500 m which remains with somewhat less intensity at night. At night, scattering is fairly uniform and moderately intense from 250 to 500 m, whereas there is a marked reduction in scattering below 600 m compared to the daytime pattern. Volume reverberation is less at the Gulf station than at the Caribbean station. Catch data indicated that while the species composition was very similar between the stations, fish were less abundant at the Gulf station. For example, three trawls were taken in the depth ranges 90-220 m at night at both stations. The Gulf station had a mean catch rate of 33.6 fish per hour while the Caribbean station produced 52.6 fish per hour. The myctophids *Lepidophanes guentheri*, *Diaphus dumerili* and *Notolychnus valdiviae* were among the six most numerous species at each station. The remaining three dominant species were different between the two areas yet were abundant at both stations. The important point here is that two areas of similar hydrographic properties have similar *patterns* of vertical stratification but differences in the fish abundances, and the degree of similarity or difference can be measured quantitatively by acoustic profiling.

Figure 5 presents acoustic profiles taken from a location in the Venezuela Basin of the eastern Caribbean Sea. Noteworthy are the differences in pattern observed (Figure 5) compared to the western Caribbean station (Figures 4a,b). The eastern Caribbean locale is characterized by a region of high backscattering from 300 to 400 m with considerable scattering present rather uniformly to about 700 m at midday. The major differences (cf Figure 4a) are the shallower region of high backscattering and the somewhat higher density of scatterers from 500 to 700 m in the eastern Caribbean. The differences are more pronounced at night. There is considerable reduction in volume reverberation at all depths below 220 m. While some scattering is apparent from 300 to 500 m, it is not as pronounced as that seen in the western Caribbean profiles. It appears that the scattering community in the eastern Caribbean consists of many more vertical migrators and, in general, one could predict that populations of swimbladder-bearing fishes are less dense at night at depths below 220 m. Preliminary evidence from midwater trawl collections made at this station suggests differences in

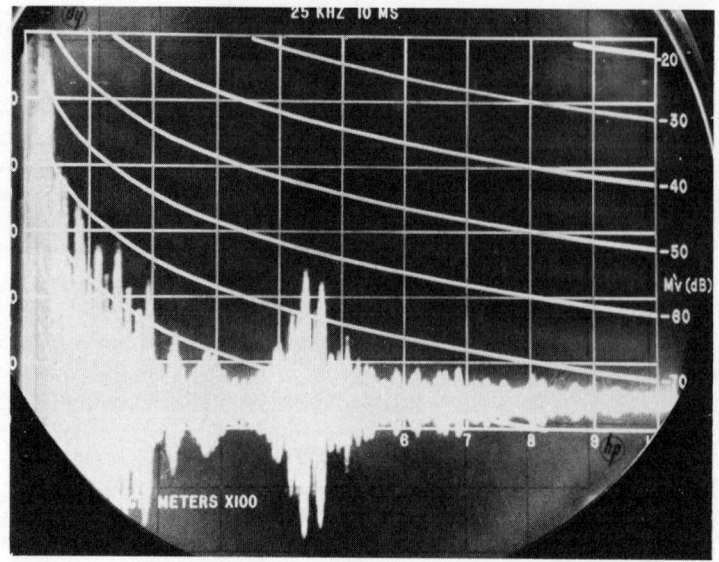

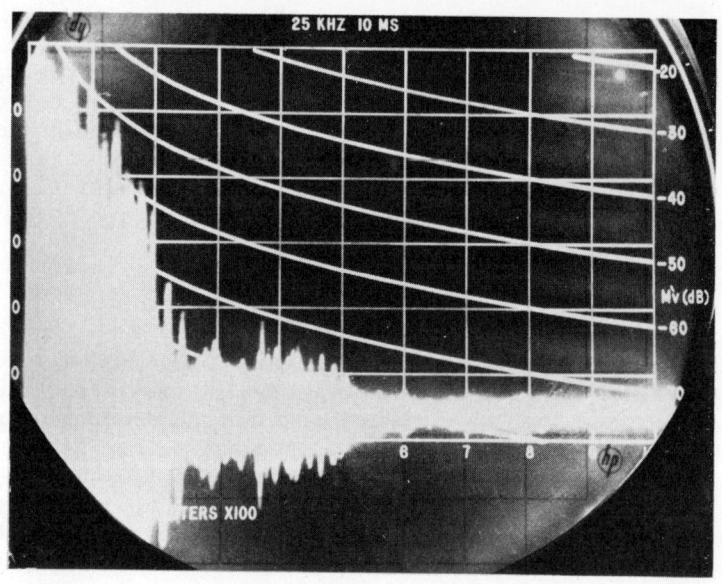

Fig. 4. Photographs of acoustic profiles at 25 kHz from single locations in the Gulf of Mexico and Caribbean Sea. 4a. Western Caribbean (19°59'N; 85°07'W) 1200 local time. 4b. Same station at 2400 local time.

SOUND SCATTERING AND OCEANIC MIDWATER FISHES

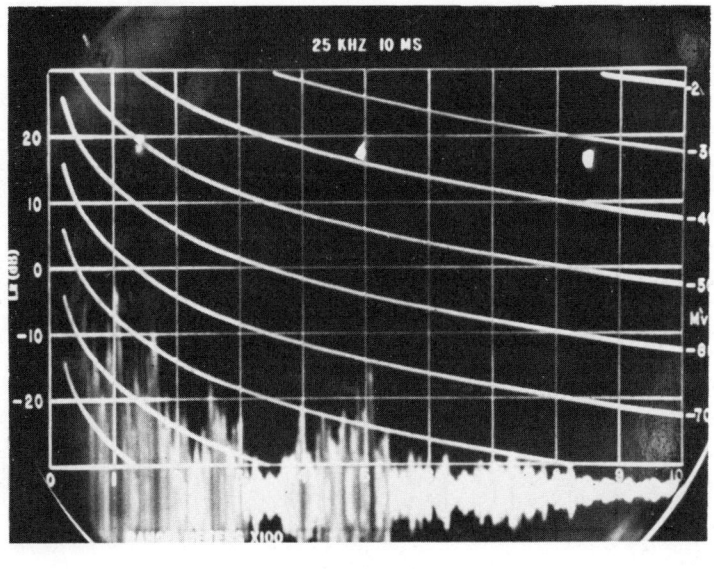

c

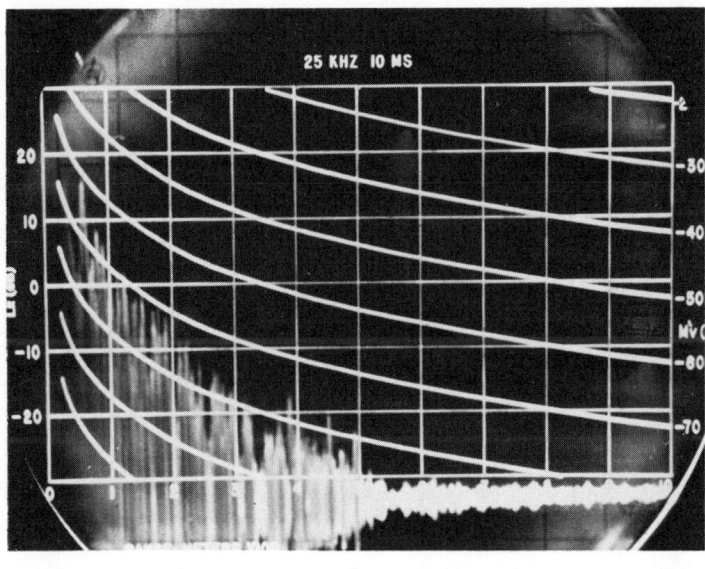

d

4c. Eastern Gulf of Mexico (25°49'N; 85°29'W) 1200 local time.
4d. Same station at 2400 local time.

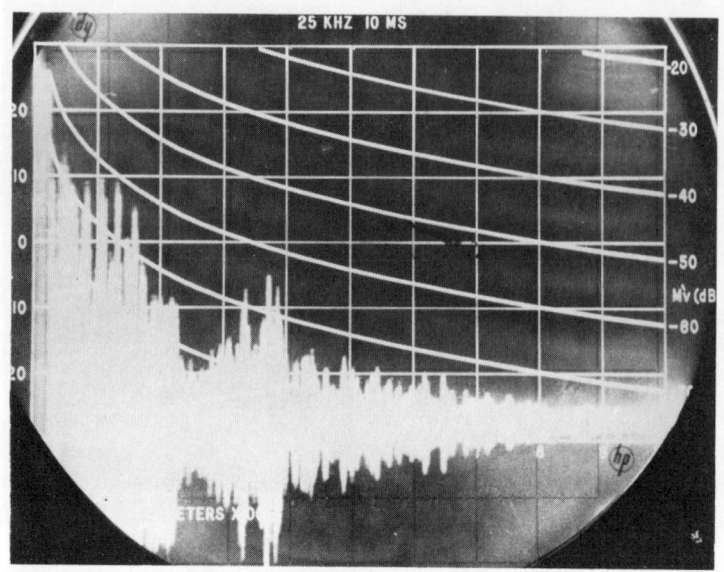

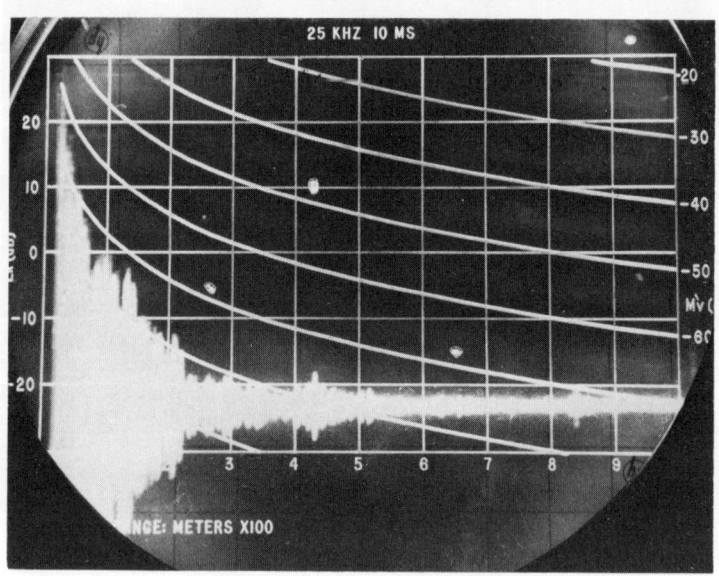

Fig. 5. Photographs of acoustic profiles at 25 kHz from single location in Eastern Caribbean (13°20'N; 66°44'W). 5a. Profile taken at 1200 local time. 5b. Same station at 2400 local time.

faunal composition from that in the western Caribbean, particularly among species and genera of the family Myctophidae.

A great deal more study is, of course, required to confirm the consistency of the patterns observed and the organization and identification of the scatterers responsible for these patterns. However, even at the semi-quantitative level obtained in these preliminary studies, biologically meaningful hypotheses can be set forth which can be tested in future research.

CONCLUSIONS

Our preliminary findings indicate the value of studying one specific area in great detail. The use of acoustic profilers as an initial step in examining the organization of sound scattering, particularly below 100 m, appears especially valuable. One can assess the acoustic homogeneity of an area over a time span and compare profiles with those from other geographic areas. Initially, frequencies in the range 12 to 30 kHz should be used, as most is known about resonant scattering over these frequencies. After a satisfactory predictive capability is developed for that range, then other frequencies could be examined. At higher frequencies attenuation and scattering by small, non-resonant scatterers become increasingly important and more difficult to appraise.

Little has been mentioned about other organisms or physical/chemical parameters. Mesopelagic fishes are components of complex ecosystems with distinct trophic organization and primary production. Predictive models should attempt to incorporate the ecological mechanisms underlying observed patterns (e.g., *Baird et al.*, 1975). It must also be emphasized that many mesopelagic fishes have no gas filled swimbladders or lose them during the course of ontogeny. While these species may not contribute to resonant scattering, they are important components of oceanic ecosystems and can be expected to contribute significantly to gross volume reverberation and to sound scattering at higher frequencies.

ACKNOWLEDGEMENTS

The authors acknowledge with pleasure the significant contributions to this work by T. L. Hopkins, D. M. Milliken, and J. K. Rolfes of the Department of Marine Science, University of South Florida, and by R. C. Beckett of the Ocean Science Division of the Naval Research Laboratory. The data for this paper were obtained during a series of collaborative cruises between the Department of Marine Science, University of South Florida, and the Naval Research Laboratory, utilizing USNS MIZAR (D. F. Wilson, Chief Scientist). Financial support was provided by the State University System Institute of Oceanography, Florida, and by the Oceanography Section,

National Science Foundation, under NSF Grant DES 75-03845 to the senior author, and by the Naval Research Laboratory to the junior author.

REFERENCES

Backus, R. H., J. E. Craddock, R. L. Haedrich, and D. L. Shores. 1969. Mesopelagic fishes and thermal fronts in the western Sargasso Sea. *Mar. Biol.*, 3: 87-106.

Baird, R. C. 1974. Aspects of social behavior in *Poecilia latipinna* (Lessueur). *Rev. Biol. Trop.* 21 (2): 399-416.

Baird, R. C., D. F. Wilson, and D. M. Milliken. 1973. Observations on *Bregmaceros nectabanus* Whitley in the anoxic sulfurous water of the Cariaco Trench. *Deep-Sea Res.*, 20: 503-504.

Baird, R. C., D. F. Wilson, R. C. Beckett, and T. L. Hopkins. 1974. *Diaphus taaningi* Norman, the principal component of a shallow sound-scattering layer in the Cariaco Trench, Venezuela. *J. Mar. Res.*, Vol. 32 (2): 301-312.

Baird, R. C., T. L. Hopkins, D. F. Wilson. 1975. Diet and feeding chronology of *Diaphus taaningi* (Myctophidae) in the Cariaco Trench. *Copeia*, 2: 356-365.

Batzler, W. E. 1972. A model for the prediction of acoustic volume scattering strength. *Naval Undersea Center Tech. Note 859.* 22 pp

Batzler, W. E., W. A. Friedl, and J. W. Reese. 1973. Can acoustic volume scattering be predicted from net haul data? *J. Acoust. Soc. Amer.* 54 (1): 290.

Batzler, W. E., J. W. Reese, and W. A. Friedl. 1975. Acoustic volum scattering: Its dependence on frequency and biological scatterers *Naval Undersea Center TP 442.* 25 pp.

Clark, T. A. 1973. Some aspects of the ecology of lantern fishes (Myctophidae) in the Pacific Ocean near Hawaii. *Fish Bull.* 71 (2) 401-433.

Cushing, D. 1973. Detection of Fishes. Pergamon Press. Oxford.

Farquhar, G. B. (Ed.). 1970. *Proc. Int. Symp. Biol. Sound Scattering in the Ocean.* Maury Center for Ocean Science. MC Rep. 005. U.S. Govt. Printing Office, Washington, D.C. 629 pp.

Goodyear, R. H., B. J. Zahuranec, W. L. Pugh, and R. H. Gibbs, Jr. 1972. Ecology and vertical distribution of Mediterranean midwater fishes. *Mediterranean Biological Studies Final Report,* Smithsonian Institution (ONR Contract N0014-67-A-0399-0007). Vol. I pp. 91-230. Washington, D. C.

Hopkins, T. L., R. C. Baird, and D. M. Milliken. 1973. A messenger-operated closing trawl. *Limnol. Oceanogr.*, 18: 488-490.

Hopkins, T. L. and R. C. Baird. 1975. An analysis of net feeding in mesopelagic fishes. *Fish. Bull.* (in press).

Kleckner, R. C. and R. H. Gibbs, Jr. 1972. Swimbladder structure of Mediterranean midwater fishes and a method of comparing swimbladder data with acoustic profiles. *Mediterranean Biological*

Studies Final Report, Smithsonian Institution (ONR Contract N0014-67A-0399-0007). Vol. 1, pp. 230-282. Washington, D.C.
Mead, G. W. 1963. Observations on fishes caught over the anoxic waters of the Cariaco Trench, Venezuela. *Deep-sea Res., 10:* 733-734.
Pearcy, W. G. and R. S. Mesecar. 1970. Scattering layers and vertical distribution of oceanic animals off Oregon. Maury Center for Ocean Science, MC Rep. 005. U.S. Govt. Printing Office, Washington, D.C. 381-394.
Pugh, W. L. 1972. Collections of midwater organisms in the Cariaco Trench, Venezuela. *Bull. Mar. Sci., 22:* 592-600.
Seligman, P. F., and W. A. Friedl. 1975. Fasor II: Correlative Biological and Acoustical Studies in the North Pacific Ocean. Naval Undersea Center, San Diego, Calif. 1-63, plus figures A-1 to B-4.
Swarts, R. L. 1970. Time variations of some acoustic volume reverberation parameters. Maury Center for Ocean Science. MC Rep. 005. U.S. Govt. Printing Office, Washington, D. C. 245-277.
Urick, R. J. 1967. Principles of underwater sound for engineers. McGraw-Hill, New York. 342 pp.
Van Schuyler, P. 1970. An acoustically determined distribution of resonant scattering north of Oahu. Maury Center for Ocean Science. MC Rep. 005, U.S. Govt. Printing Office, Washington, D.C. 338-359.
Vent, R. J. 1972. Acoustic volume-scattering measurements at 3.5, 5.0 and 12.0 kHz in the eastern Pacific Ocean: Diurnal and seasonal variations. *J. Acoust. Soc. Amer.* 52 (1): 373-382.
Weston, D. E. 1967. Sound propagation in the presence of bladder fish. In: *V. M. Albers Ied. Underwater acoustic II,* pp. 55-88. Plenum Press, New York. 416 pp.
Wilson, D. F. 1972. Diel migration of sound scatterers into, and out of, the Cariaco Trench anoxic water. *J. Mar. Res., 30:* 168-176.
Wilson, D. F., L. C. Ricalzone, and R. C. Beckett. 1968. Some investigations of sound-scattering layers near Key West, Florida. NRL Report 6682.
Zahuranec, B. J., W. L. Pugh, and H. B. Farquhar. 1970. Biological sound scattering studies. Pt. I. Initial investigations in the Gulf of Mexico and western North Atlantic. Tech. Rept. (TR-224), Naval Oceanographic Office. Washington, D.C. 35 pp.

A Study of the Swimbladders of Selected Mesopelagic Fish Species

Albert L. Brooks

Naval Underwater Systems Center

ABSTRACT

The intra- and inter-specific variation in the relationship of fish standard length to swimbladder length, width, and volume was studied in selected mesopelagic fish specimens belonging to 55 species from 9 families. These species are believed to account for most of the volume reverberation occurring throughout a large part of the Sargasso Sea.

Regression equations define the relationship of fish length to swimbladder length, width, and volume. For 40 species, bladder volume increases with increasing standard length. Slopes of the regression lines for 14 species are shown to be insignificant from zero. For one species, bladder volume decreases slightly with increasing standard length. Volume of the swimbladder of a given species of given standard length can vary greatly as can the elevations and slopes of the regression lines for different fish species.

Swimbladder volumes estimated from this study are in agreement with measurements published by N. B. Marshall but are less than volumes estimated by other authors.

INTRODUCTION

Since the discovery by *Eyring, et al.* (1948) of a deep layer in the ocean that scattered sound, a voluminous literature has been written dealing with the acoustics and biology of what is now called the Deep Scattering Layer (DSL). Though organisms such as siphonophores, euphausiids, and cephalopods have, in some cases, been implicated as the cause of sonic scattering layers, overwhelming evidence indicates that volume reverberation in the open ocean is largely due to gas-filled cavities (principally swimbladders) of small mesopelagic fish.

Despite this, only limited quantitative information exists relating swimbladder characteristics to scattering. More fundamentally, few data are available on the relationships between bladder dimensions length, width, and volume; and fish morphometry.

This is a digest of a more comprehensive study (Brooks, 1976) which examines the relationship of fish standard length to swimbladder dimensions. Also studied are the intra- and inter- specific allometric variations encountered in these variables in selected mesopelagic specimens belonging to 55 species from 9 families. These fish represent the predominant bladdered species believed to account for most volume reverberation occurring throughout much of the Sargasso Sea.

METHODS AND MATERIALS

Individuals used in this study were selected from fish specimens collected from an open ocean area off Bermuda known as the Ocean Acre region and from the Mediterranean Sea. Measurements of fish standard length and swimbladder dimensions of Bermuda specimens were furnished by the Division of Fishes at the Smithsonian. Measurements of Mediterranean fish were kindly loaned to me by R. C. Kleckner and formed part of an earlier report on swimbladder structure of Mediterranean mid-water fishes by Kleckner and Gibbs (1972) and an M. S. dissertation by Kleckner (1974).

All fish were preserved in 10-percent sea water formalin after retrieval of each Isaacs-Kidd Midwater Trawl (IKMT) tow. In the laboratory, specimens were washed and transferred through a series of increasingly concentrated solutions of ethanol up to 70 percent. Standard length (snout tip to caudal base) was measured in millimeters with dial calipers. Dissections to reveal the swimbladders were made under a microscope; measurements of bladder length (the anterior-posterior length of the external swimbladder wall) and bladder width (external dimension at the greatest width of bladder) were made with an ocular micrometer.

RESULTS AND DISCUSSION

Table 1 lists the 55 species of gas bladdered mesopelagic fishes included. All of these species will be considered in detail, but representative examples have been chosen to illustrate salient features of swimbladder allometry. Though all of these species occur in the Ocean Acre region, data for the 11 species marked with an asterisk were obtained from Mediterranean fish. For these species Kleckner and Gibbs (1972) showed that the relationship between fish standard lengths and their respective swimbladder lengths and widths was approximately linear.

Table 1. Predominant Bladdered Species Investigated

Myctophidae

Benthosema suborbitale
Bolinichthys indicus
Bolinichthys photothorax
*Ceratoscopelus maderensis
Ceratoscopelus warmingi
Diaphus brachycephalus
Diaphus metapoclampus
Diaphus mollis
Diaphus problematicus
*Diaphus rafinesquei
Diogenichthys atlanticus
*Hygophum benoiti
*Hygophum hygomi
Lampadena chavesi
Lampadena luminosa
Lampadena speculigera
Lampadena urophaos
Lampanyctus alatus
Lampanyctus ater
*Lampanyctus crocodilus
Lampanyctus cuprarius
Lampanyctus festivus
Lampanyctus lineatus
Lampanyctus photonotus
*Lampanyctus pusillus
Lepidophanes gaussi
Lepidophanes guentheri
*Lobianchia dofleini
Lobianchia gemellari
Myctophum nitidulum
Notolychnus valdiviae
Notoscopelus caudispinosus
Notoscopelus resplendens
Symbolophorus rufinus
Taaningichthys bathyphilus
Taaningichthys minimus

Gonostomatidae

Bonapartia pedaliota
*Ichthyococcus ovatus
Pollichthys mauli
Valenciennellus tripunctulatus
*Vinciguerria attenuata
*Vinciguerria poweriae

Sternoptychidae

Argyropelecus aculeatus
*Argyropelecus hemigymnus
Sternoptyx diaphana

Melamphaidae

Melamphaes pumilus
Melamphaes typhlops
Scopeloberyx opisthopterus

Berycidae

Anoplogaster cornuta
Poromitra capito

Bregmacerotidae

Bregmaceros sp.

Melanostomiatidae

Melanostomiatid larvae

Nomeidae

Cubiceps sp.

Opisthoproctidae

Rhynchohyalus natalensis

Table 2. Results of Regression Analysis of Swimbladder Length and Width on Standard Length.

Species	Number of Specimens	Range of STD Lengths	Mean of STD Length L_{STD}	Mean of Swimbladder Length $\bar{L}_{SB}$	Regression Equation L_{SB} on L_{STD}	Correl. Coeff. R	Mean of Swimbladder Width $\bar{W}_{SB}$	Regression Equation W_{SB} on L_{STD}	Correl. Coeff. R
Anoplogaster cornuta	5	11.5-81.5	37.6	2.8	*1.379+0.038L_{STD}	0.85	1.6	0.755+0.022L_{STD}	0.97
Argyropelecus aculeatus	30	9.0-55.9	17.1	2.3	-0.301+0.150L_{STD}	0.98	1.3	0.069+0.073L_{STD}	0.91
Argyropelecus hemigymnus	47	7.2-33.8	20.1	2.7	-0.421+0.156L_{STD}	0.97	1.6	-0.253+0.094L_{STD}	0.96
Benthosema suborbitale	11	11.6-26.0	20.4	4.1	-2.375+0.319L_{STD}	0.84	1.1	-0.607+0.081L_{STD}	0.65
Bolinichthys indicus	31	18.3-39.4	27.6	5.8	1.575+0.153L_{STD}	0.46	1.1	-0.114+0.045L_{STD}	0.50
Bolinichthys photothorax	10	34.6-60.6	49.2	10.6	*0.376+0.208L_{STD}	0.60	2.3	*-0.140+0.050L_{STD}	0.42
Bonapartia pedaliota	26	17.0-59.0	42.9	7.2	-3.147+0.240L_{STD}	0.88	2.0	-0.516+0.059L_{STD}	0.74
Bregmaceros sp.	4	31.2-85.0	57.8	4.1	*2.267+0.032L_{STD}	0.58	1.2	*0.410+0.013L_{STD}	0.51
Ceratoscopelus maderensis	45	7.4-75.3	35.4	3.9	0.879+0.084L_{STD}	0.85	1.3	0.462+0.023L_{STD}	0.53
Ceratoscopelus warmingi	37	18.6-66.0	39.9	4.5	1.532+0.074L_{STD}	0.53	1.2	0.653+0.014L_{STD}	0.43
Cubiceps sp.	6	27.8-81.2	47.5	8.8	3.848+0.104L_{STD}	0.90	1.9	*1.924-0.001L_{STD}	0.03
Diaphus brachycephalus	10	29.9-42.0	35.6	7.6	*2.788+0.134L_{STD}	0.20	1.7	*3.286-0.046L_{STD}	0.26
Diaphus metapoclampus	11	24.5-74.6	61.6	14.6	-1.540+0.262L_{STD}	0.89	4.0	-0.769+0.078L_{STD}	0.74
Diaphus mollis	20	22.1-47.2	36.1	7.5	3.779+0.103L_{STD}	0.45	1.0	*0.893+0.004L_{STD}	0.06
Diaphus problematicus	14	32.0-79.5	61.6	8.3	*7.581+0.012L_{STD}	0.07	1.6	*2.565-0.016L_{STD}	0.33

Table 2. Results of Regression Analysis of Swimbladder Length and Width on Standard Length. (Cont.)

Species	Number of Specimens	Range of STD Lengths	Mean of STD Length L_{STD}	Mean of Swimbladder Length L_{SB}	Regression Equation L_{SB} on L_{STD}	Correl. Coeff. R	Mean of Swimbladder Width W_{SB}	Regression Equation W_{SB} on L_{STD}	Correl. Coeff. R
Diaphus rafinesquei	49	7.4-77.6	50.0	10.5	$-1.486+0.240L_{STD}$	0.95	3.2	$-1.256+0.089L_{STD}$	0.87
Diogenichthys atlanticus	32	11.1-19.4	14.3	2.1	$-1.559+0.259L_{STD}$	0.84	0.6	$-0.439+0.075L_{STD}$	0.73
Hygophum benoiti	70	6.3-44.0	20.8	3.0	$-0.119+0.150L_{STD}$	0.81	1.0	$-0.005+0.050L_{STD}$	0.87
Hygophum hygomi	55	5.7-59.1	38.2	5.0	$0.447+0.119L_{STD}$	0.52	1.4	$0.085+0.035L_{STD}$	0.55
Hygophum taaningi	15	12.4-39.8	22.5	2.8	$0.740+0.091L_{STD}$	0.64	0.8	$-0.147+0.042L_{STD}$	0.71
Ichthyococcus ovatus	26	12.0-35.0	25.1	4.2	$-1.901+0.241L_{STD}$	0.96	1.7	$-0.654+0.095L_{STD}$	0.94
Lampadena chavesi	10	18.5-31.7	22.0	2.3	$-1.371+0.167L_{STD}$	0.85	0.7	$-0.845+0.069L_{STD}$	0.91
Lampadena luminosa	7	18.6-61.5	44.7	4.8	$-0.927+0.129L_{STD}$	0.96	1.0	* $0.638+0.007L_{STD}$	0.29
Lampadena speculigera	6	18.8-21.5	20.6	2.6	$-4.008+0.320L_{STD}$	0.85	0.8	* $1.667-0.041L_{STD}$	0.30
Lampanyctus urophaos	35	22.6-78.3	49.8	6.1	$-0.474+0.132L_{STD}$	0.76	1.7	$-0.204+0.037L_{STD}$	0.60
Lampanyctus alatus	11	30.3-51.2	39.3	5.8	* $0.999+0.123L_{STD}$	0.55	1.3	$-0.908+0.055L_{STD}$	0.61
Lampanyctus ater	35	25.2-97.4	47.2	1.5	$2.364-0.018L_{STD}$	0.34	0.6	$0.854-0.004L_{STD}$	0.34
Lampanyctus crocodilus	65	9.5-171.7	56.1	8.1	$-2.642+0.191L_{STD}$	0.97	2.0	$-0.397+0.042L_{STD}$	0.92
Lampanyctus cuprarius	47	30.3-87.1	53.4	0.9	* $1.026-0.003L_{STD}$	0.12	0.4	$0.604-0.004L_{STD}$	0.41
Lampanyctus festivus	37	25.3-61.3	41.2	4.8	$-0.717+0.135L_{STD}$	0.85	1.3	$-0.163+0.035L_{STD}$	0.69

Table 2. Results of Regression Analysis of Swimbladder Length and Width on Standard Length. (Cont.)

Species	Number of Specimens	Range of STD Lengths	Mean of STD Length L_{STD}	Mean of Swimbladder Length L_{SB}	Regression Equation L_{SB} on L_{STD}	Correl. Coeff. R	Mean of Swimbladder Width W_{SB}	Regression Equation W_{SB} on L_{STD}	Correl. Coeff. R
Lampanyctus lineatus	11	36.3-121.9	79.9	1.3	* 1.991-0.009L_{STD}	0.34	0.6	* 0.652-0.001L_{STD}	0.20
Lampanyctus photonotus	43	21.3-64.3	41.5	6.1	0.036+0.145L_{STD}	0.81	1.3	0.214+0.027L_{STD}	0.52
Lampanyctus pusillus	48	6.4-40.0	27.3	4.0	-0.760+0.172L_{STD}	0.92	1.0	0.061+0.034L_{STD}	0.79
Lepidophanes gaussi	32	15.4-41.9	28.6	4.8	-1.924+0.235L_{STD}	0.87	1.0	-0.169+0.039L_{STD}	0.54
Lepidophanes guentheri	6	16.4-52.8	40.2	7.2	-1.370+0.213L_{STD}	0.95	1.4	-0.033+0.036L_{STD}	0.97
Lobianchia dofleini	58	12.0-42.4	25.9	3.3	0.327+0.116L_{STD}	0.56	1.0	0.146+0.033L_{STD}	0.50
Lobianchia gemellari	42	12.8-99.8	29.7	5.7	-2.462+0.276L_{STD}	0.97	1.2	-0.003+0.040L_{STD}	0.88
Melamphaes pumilus	59	10.2-21.4	17.8	2.0	-1.660+0.208L_{STD}	0.83	1.0	-0.402+0.081L_{STD}	0.77
Melamphaes typhlops	30	9.6-70.8	36.5	5.1	-1.916+0.192L_{STD}	0.89	2.3	-0.435+0.074L_{STD}	0.89
Melanostomiatid larvae	8	11.0-16.0	13.7	1.1	-1.044+0.154L_{STD}	0.89	0.5	*-0.174+0.051L_{STD}	0.51
Myctophum nitidulum	21	16.3-63.6	27.5	3.6	-2.087+0.205L_{STD}	0.97	1.2	-0.413+0.059L_{STD}	0.90
Notolychnus valdiviae	34	14.6-21.7	18.9	2.0	* 2.442-0.022L_{STD}	0.10	0.8	*-0.133+0.047L_{STD}	0.32
Notoscopelus caudispinosus	6	41.5-67.3	55.2	7.1	* 3.403+0.068L_{STD}	0.41	1.7	* 1.638+0.001L_{STD}	0.04
Notoscopelus resplendens	60	21.7-72.6	40.1	4.5	0.897+0.089L_{STD}	0.70	1.0	0.584+0.010L_{STD}	0.37
Pollichthys mauli	35	31.0-48.0	39.8	3.5	* 3.037+0.013L_{STD}	0.08	1.5	* 0.315+0.030L_{STD}	0.32

Table 2. Results of Regression Analysis of Swimbladder Length and Width on Standard Length. (Cont.)

Species	Number of Specimens	Range of STD Lengths	Mean of STD Length L_{STD}	Mean of Swimbladder Length L_{SB}	Regression Equation L_{SB} on L_{STD}	Correl. Coeff. R	Mean of Swimbladder Width W_{SB}	Regression Equation W_{SB} on L_{STD}	Correl. Coeff. R
Poromitra capito	32	12.0-99.1	35.6	3.8	$-1.839+0.158L_{STD}$	0.95	2.0	$-0.605+0.073L_{STD}$	0.92
Rhynchohyalus natalensis	9	16.9-52.5	33.2	6.3	$-3.882+0.308L_{STD}$	0.99	1.2	*$0.429+0.022L_{STD}$	0.63
Scopeloberyx opisthopterus	30	13.0-38.6	27.9	2.8	$0.820+0.069L_{STD}$	0.69	1.6	$-0.392+0.071L_{STD}$	0.83
Sternoptyx diaphana	30	8.9-35.0	20.6	3.4	$-0.853+0.204L_{STD}$	0.90	1.9	$-0.483+0.115L_{STD}$	0.93
Symbolophorus rufinus	26	14.2-85.9	38.5	4.7	$-1.709+0.165L_{STD}$	0.82	1.2	$-0.336+0.040L_{STD}$	0.74
Taaningichthys bathyphilus	25	17.7-65.9	48.6	6.3	$-0.829+0.146L_{STD}$	0.94	2.5	$-0.391+0.060L_{STD}$	0.79
Taaningichthys minimus	31	22.1-54.2	38.7	6.1	$-3.603+0.251L_{STD}$	0.88	1.6	$-1.045+0.068L_{STD}$	0.71
Valenciennellus tripunctulatus	71	11.2-29.5	21.9	3.4	$-0.243+0.166L_{STD}$	0.87	1.4	$-0.266+0.076L_{STD}$	0.70
Vinciguerria attenuata	44	12.4-36.4	20.3	2.6	$-1.057+0.177L_{STD}$	0.97	0.8	$-0.115+0.043L_{STD}$	0.80
Vinciguerria poweriae	34	12.1-33.7	20.6	2.6	$-1.253+0.186L_{STD}$	0.89	0.8	$-0.120+0.044L_{STD}$	0.76

*An F test indicates that the slope of these regression lines is not significant at the 0.05 level.

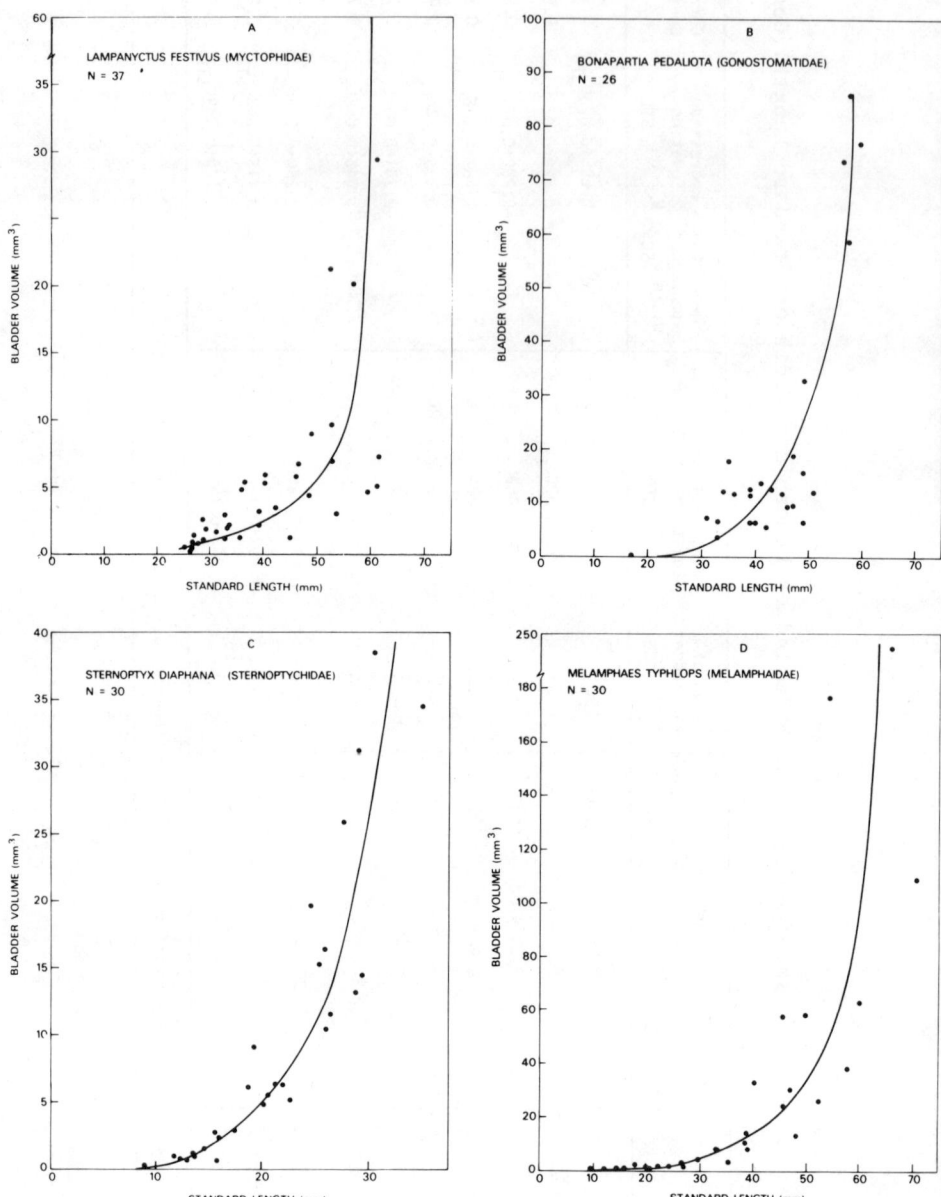

Fig. 1. Arithmetic plots of swimbladder volume against fish standard length for representatives of four important midwater fish families.

REGRESSION ANALYSES
FISH STANDARD LENGTH VERSUS SWIMBLADDER LENGTH/WIDTH

Table 2 presents the results of a linear regression analysis performed on all 55 species. All measurements in the table are in millimeters. The left-hand section of the table lists alphabetically the mesopelagic fish species, total number of specimens measured, range of standard lengths, and the mean standard length $\bar{L}_{STD}$ of individuals included in the analysis. The center section lists the mean of the swimbladder lengths $\bar{L}_{SB}$, the regression equations relating swimbladder length L_{SB} to fish standard length L_{STD} and the correlation coefficients R. An F test was performed to examine the significance of the slope of the regression lines. The right-hand section of the table lists the mean of the swimbladder widths $\bar{W}_{SB}$, the regression equations relating swimbladder width W_{SB} to fish standard length L_{STD}, and the corresponding correlation coefficients. In all cases, an asterisk indicates that the slope of the regression line is not significantly different from zero at the 0.05 level of probability.

Sample correlation coefficients R between swimbladder length and fish standard length equal or exceed 0.70 in 37 of 55 species. The R values between swimbladder width and fish standard length are not as high, with 28 of the 55 species exceeding a value of 0.70. The F test (0.05 level) indicates that there is no significant relationship in 11 species between standard length and bladder length. In addition, for 14 species no significant relationship is found between fish standard length and bladder width in most cases because of the sample size. Inspection of a plot of standard lengths against bladder lengths and widths suggests that, if more data were included in the analysis, a significant slope would result for several species.

For those species where a significant slope is found, it is positive (i.e., bladder length or width or both increase with increasing standard length) for all species except *Lampanyctus ater* and *L. cuprarius*, whose slopes are slightly negative.

REGRESSION ANALYSES
FISH STANDARD LENGTH VERSUS SWIMBLADDER VOLUME

Measurements of the swimbladder lengths and widths of the same specimens used in the previous analyses were converted to volumes using the equation for a prolate spheroid $V = 4/3 \pi ab^2$ where a and b are the semi-major and semi-minor axes, respectively. Arithmetic plots of standard length against swimbladder volumes of four species representing the four most important fish families found in the Bermuda area are shown in Figure 1a-d. A line of best fit drawn by eye through the scatter of points, clearly shows a nonlinear relationship between these variables. Linear regression analysis of these data in raw form would yield unacceptable errors, and curvilinear analysis is burdensome and complicated.

Table 3. Results of Regression Analysis of Swimbladder Volume on Standard Length.

Species	Number of Specimens	Range of STD Lengths	Mean of STD Length L_{STD}	Derived Mean of Swimbladder Volumes $\bar{V}_{SB}$	Regression Equation $Log(V_{SB}+1)=a+bL_{STD}$	Correl. Coeff. R
Anoplogaster cornuta	5	11.5-81.5	37.6	5.6	0.139+0.013L_{STD}	0.97
Argyropelecus aculeatus	30	9.0-55.9	17.1	3.6	-0.231+0.041L_{STD}	0.92
Argyropelecus hemigymnus	47	7.2-33.8	20.1	6.2	-0.436+0.054L_{STD}	0.97
Benthosema suborbitale	11	11.6-26.0	20.4	4.5	-0.654+0.057L_{STD}	0.71
Bolinichthys indicus	31	18.3-39.4	27.6	4.7	-0.286+0.034L_{STD}	0.55
Bolinichthys photothorax	10	34.6-60.6	49.2	39.7	* 0.382+0.021L_{STD}	0.44
Bonapartia pedaliota	26	17.0-59.0	42.9	20.8	-0.371+0.035L_{STD}	0.83
Bregmaceros sp.	4	31.2-85.0	57.8	4.6	* 0.043+0.009L_{STD}	0.54
Ceratoscopelus maderensis	45	7.4-75.3	35.4	6.1	0.041+0.016L_{STD}	0.69
Ceratoscopelus warmingi	37	18.6-66.0	39.9	4.6	0.125+0.012L_{STD}	0.51
Cubiceps sp.	6	27.8-81.2	47.5	19.5	* 1.031+0.003L_{STD}	0.21
Diaphus brachycephalus	10	29.9-42.0	35.6	14.3	* 1.579-0.015L_{STD}	0.14
Diaphus metapoclampus	11	24.5-74.6	61.6	285.3	-0.051-0.033L_{STD}	0.90
Diaphus mollis	20	22.1-47.2	36.1	4.9	* 0.434+0.007L_{STD}	0.16
Diaphus problematicus	14	32.0-79.5	61.6	13.0	* 1.419-0.007L_{STD}	0.24
Diaphus rafinesquei	49	7.4-77.6	50.0	116.9	-0.102+0.035L_{STD}	0.94
Diogenichthys atlanticus	32	11.1-19.4	14.3	0.6	-0.567+0.052L_{STD}	0.82

Table 3. Results of Regression Analysis of Swimbladder Volume on Standard Length. (Cont.)

Species	Number of Specimens	Range of STD Lengths	Mean of STD Length L_{STD}	Derived Mean of Swimbladder Volumes $\bar{V}_{SB}$	Regression Equation $Log(V_{SB}+1)=a+bL_{STD}$	Correl. Coeff. R
Hygophum benoiti	70	6.3-44.0	20.8	3.8	$-0.248+0.033L_{STD}$	0.87
Hygophum hygomi	55	5.7-59.1	38.2	9.2	$-0.042+0.021L_{STD}$	0.56
Hygophum taaningi	15	12.4-39.8	22.5	1.7	$-0.297+0.027L_{STD}$	0.72
Ichthyococcus ovatus	26	12.0-35.0	25.1	12.8	$-0.710+0.062L_{STD}$	0.96
Lampadena chavesi	10	18.5-31.7	22.0	0.8	$-0.857+0.048L_{STD}$	0.92
Lampadena luminosa	7	18.6-61.5	44.7	3.1	$*-0.036+0.012L_{STD}$	0.66
Lampadena speculigera	6	18.8-21.5	20.6	1.0	$* 0.139+0.007L_{STD}$	0.11
Lampadena urophaos	35	22.6-78.3	49.8	16.2	$-0.315+0.024L_{STD}$	0.72
Lampanyctus alatus	11	30.3-51.2	39.3	6.4	$-0.595+0.034L_{STD}$	0.63
Lampanyctus ater	35	25.2-97.4	47.2	0.4	$* 0.252-0.002L_{STD}$	0.29
Lampanyctus crocodilus	65	9.5-171.7	56.1	65.7	$-0.182+0.021L_{STD}$	0.97
Lampanyctus cuprarius	47	30.3-87.1	53.4	0.1	$0.069-0.001L_{STD}$	0.31
Lampanyctus festivus	37	25.3-61.3	41.2	6.0	$-0.388+0.026L_{STD}$	0.79
Lampanyctus lineatus	11	36.3-121.9	79.9	0.3	$* 0.160-0.001L_{STD}$	0.29
Lampanyctus photonotus	43	21.3-64.3	41.5	7.4	$-0.181+0.023L_{STD}$	0.68
Lampanyctus pusillus	48	6.4-40.0	27.3	2.8	$-0.314+0.029L_{STD}$	0.87
Lepidophanes gaussi	32	15.4-41.9	28.6	4.0	$-0.487+0.035L_{STD}$	0.82

Table 3. Results of Regression Analysis of Swimbladder Volume on Standard Length. (Cont.)

Species	Number of Specimens	Range of STD Lengths	Mean of Std Length L_{STD}	Derived Mean of Swimbladder Volumes $\bar{V}_{SB}$	Regression Equation $Log(V_{SB}^+) = a + bL_{STD}$	Correl. Coeff. R
Lepidophanes guentheri	6	16.4-52.8	40.2	15.7	$-0.403 + 0.032 L_{STD}$	0.97
Lobianchia dofleini	58	12.0-42.4	25.9	2.5	$-0.201 + 0.024 L_{STD}$	0.56
Lobianchia gemellari	42	12.8-99.8	29.7	7.0	$-0.134 + 0.025 L_{STD}$	0.93
Melamphaes pumilus	59	10.2-21.4	17.8	1.4	$-0.628 + 0.054 L_{STD}$	0.83
Melamphaes typhlops	30	9.6-70.8	36.5	32.4	$-0.357 + 0.037 L_{STD}$	0.93
Melanostomiatid larvae	8	11.0-16.0	13.7	0.2	*$-0.194 + 0.019 L_{STD}$	0.62
Myctophum nitidulum	21	16.3-63.6	27.5	8.8	$-0.493 + 0.036 L_{STD}$	0.95
Notolychnus valdiviae	34	14.6-21.7	18.9	0.7	*$-0.160 + 0.019 L_{STD}$	0.26
Notoscopelus caudispinosus	6	41.5-67.3	55.2	11.5	* $0.879 + 0.003 L_{STD}$	0.17
Notoscopelus resplendens	60	21.7-72.6	40.1	3.6	$0.056 + 0.011 L_{STD}$	0.54
Pollichthys mauli	35	31.0-48.0	39.8	4.6	* $0.129 + 0.014 L_{STD}$	0.30
Poromitra capito	32	12.0-99.1	35.6	27.3	$-0.398 + 0.032 L_{STD}$	0.97
Rhynchohyalus natalensis	9	16.9-52.5	33.2	6.0	$-0.269 + 0.029 L_{STD}$	0.85
Scopeloberyx opisthopterus	30	13.0-38.6	27.9	4.7	$-0.385 + 0.037 L_{STD}$	0.85
Sternoptyx diaphana	30	8.9-35.0	20.6	10.2	$-0.519 + 0.065 L_{STD}$	0.96
Symbolophorus rufinus	26	14.2-85.9	38.5	8.1	$-0.346 + 0.023 L_{STD}$	0.76

Table 3. Results of Regression Analysis of Swimbladder Volume on Standard Length. (Cont.)

Species	Number of Specimens	Range of STD Lengths	Mean of STD Length L_{STD}	Derived Mean of Swimbladder Volumes $\overline{V}_{SB}$	Regression Equation $Log(V_{SB}{}^+)=a+bL_{STD}$	Correl. Coeff. R
Taaningichthys bathyphilus	25	17.7-65.9	48.6	38.8	$-0.437+0.035L_{STD}$	0.92
Taaningichthys minimus	31	22.1-54.2	38.7	14.3	$-0.886+0.045L_{STD}$	0.85
Valenciennellus tripunctulatus	71	11.2-29.5	21.9	4.0	$-0.453+0.050L_{STD}$	0.78
Vinciguerria attenuata	44	12.4-36.4	20.3	1.3	$-0.390+0.033L_{STD}$	0.90
Vinciguerria poweriae	34	12.1-33.7	20.6	1.3	$-0.358+0.032L_{STD}$	0.86

To overcome these problems, the dependent variable (bladder volume) was transformed logarithmically. Various logarithmic transformations have been applied to fisheries data similar to the present data. Preliminary treatment of measurements provided by Kleckner on 11 species of Mediterranean midwater fishes indicated that a logarithmic transformation satisfied the assumptions underlying the regression technique and the analysis of variance.

Thus, the calculated values for swimbladder volumes V_{SB} were transformed to $log(V_{SB} + 1)$ and regressed against measurements of standard length (Table 3). The first four columns in Table 3 are the same as found in Table 2, because the same individuals were used in each analysis. The fifth column lists the derived mean in cubic millimeters of all the swimbladder volumes V_{SB} used in the analysis of a given species. The regression equation for each species (column 6) relating the transformed swimbladder volumes to standard length is in the form

$$log(V_{SB} + 1) = a + bL_{STD} \qquad (1)$$

where a is the $log(V_{SB} + 1)$ intercept, b is the slope of the line (i.e., the number of units of $log(V_{SB} + 1)$ corresponding to every unit of standard length), and L_{STD} equals the fish standard length. To obtain any swimbladder volume directly in cubic millimeters, the above equation can be converted to

$$V_{SB mm^3} = (10^a \cdot 10^{bL_{STD}}) - 1. \qquad (2)$$

As in Table 2, an asterisk indicates that an F test has shown the slope of that regression line to be not significant from zero at the 0.05 level of probability. The correlation coefficients R, on the whole, are similar in value to those shown in Table 2 and exceed 0.70 in 32 of the 55 species. The slope of the regression line at the 0.05 level was shown to be insignificant from zero for 14 of the 55 species. Eight of these were associated with sample sizes of 10 or less. In those species for which a significant slope is indicated, all were positive, with the exception of Lampanyctus cuprarius, which exhibited a slight decrease in bladder volume with increasing fish standard length.

Evidence for the validity of this transformation and regression analysis can be gained by comparing swimbladder volume regression lines from this analysis with individual volumes calculated from the regressions of bladder length and width against standard length. Six species representing a range of regression line and data characteristics were chosen for this comparison.

In Figure 2, the transformed swimbladder volumes $log(V_{SB} + 1)$ are plotted against fish standard length (shown as open circles) for the 6 selected species. Swimbladder volumes may be read directly from the equivalent scale shown on the axis on the right side of each figure. Superimposed on each graph is the line of best fit specified for each

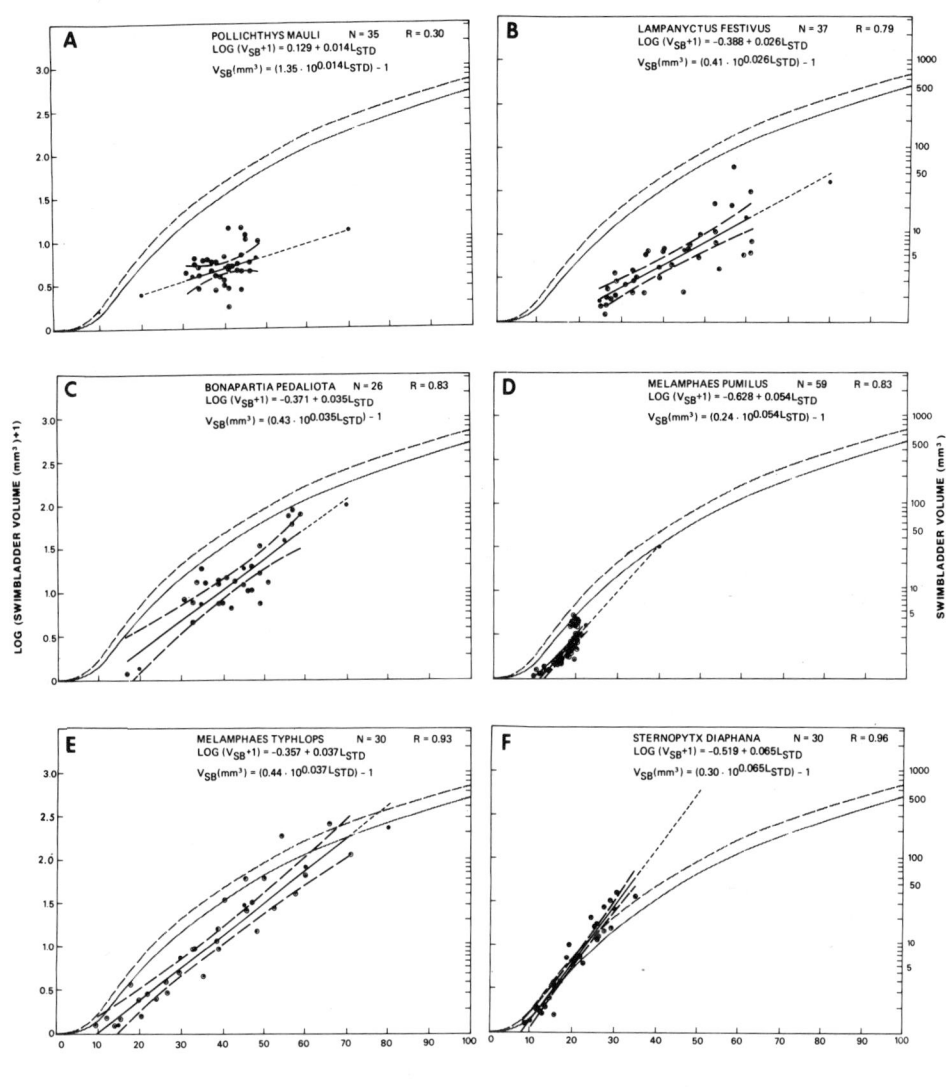

Fig. 2. Swimbladder volume regression lines compared with individual volumes calculated from regressions of bladder length and width against fish standard length and with regression lines from previous work.

species by the regression equations listed in Table 3 and the 95 percent confidence limits for each line. Bladder volumes were also calculated using the equation for a prolate spheroid where the values for bladder length and width for a given species of given standard length were determined from the regression equations given in Table 2. The quantity one was added to each of these calculated bladder volumes; the logarithms of these sums are shown on each graph as closed circles spaced at more or less regular intervals of standard length along each regression line.

With the possible exception of *Melamphaes typhlops*, the bladder volumes estimated by these two separate methods show remarkably close agreement within the limits of the data used to construct the regression line.

In *Lampanyctus festivus*, *Melamphaes typhlops*, and *Sternoptyx diaphana*, swimbladder volumes estimated by the two methods diverge to varying degrees when the uppermost limit of standard length included in the bladder volume regressions is exceeded by 15-20 mm (indicated by dashed line). In other species, namely, *Pollichthys mauli*, *Bonapartia pedaliota*, and *Melamphaes pumilus*, little if any digression is present within these same limits. Because the relationship between fish standard length and bladder length and width is linear, extrapolation of the regression lines relating these variables beyond the actual data will probably furnish a good approximation of what occurs in the population. Bladder volumes can then be calculated by using the equation for a prolate spheroid for specimens outside the limits of the measurement data included in the present analyses.

On the other hand, since the relationship between standard length and bladder volume is curvilinear, estimation of bladder volumes by extrapolation is inappropriate. The two long curved lines drawn on each graph are the solutions to equations derived by *Andreeva and Chindonova* (1964) (upper curve) and by *Haslett* (1962) (lower curve); they will be given and discussed below.

INTRA- AND INTER-SPECIFIC SWIMBLADDER VARIABILITY

As can be seen from the plots in Figure 2, the volume of the swimbladder of a particular fish species of given standard length can vary greatly. In the case of *Bonapartia pedaliota*, three separate fish with a standard length of 49.0 mm had calculated bladder volumes of 6.5, 15.7, and 32.8 mm^3, and even greater potential variability exists for other species. The variability in bladder volume is considerably less, however, in *Sternoptyx diaphana*.

A visual comparison of the regression lines relating bladder volume to fish standard length in different species of the same genus reveals wide differences in the slopes and elevations of these lines.

Figure 3a illustrates these differences for five species belonging to the genus *Diaphus*. The number in parenthesis after each specific name gives the sample size. Next is the correlation coefficient

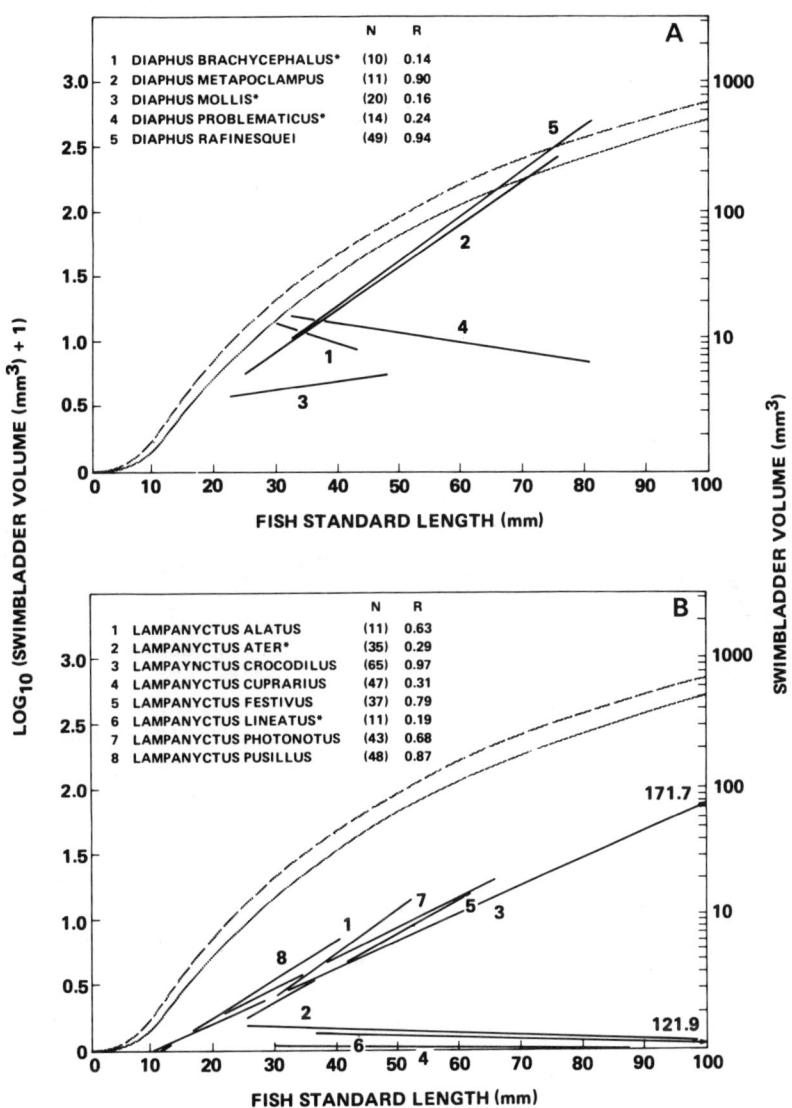

Fig. 3. A comparison of the slopes and elevations of regression lines relating swimbladder volume to fish standard length for 13 species of 2 genera of midwater fishes.

asterisks indicate that the F test showed that the slope was not significantly different from zero. The range in fish standard length on which each analysis is based is indicated by the limits of the regression line for each species. The slope and elevation of the lines for *Diaphus metapoclampus* and *D. rafinesquei* are similar, but they differ markedly from those of the other three species of this genus.

Figure 3b presents a similar comparison for eight species of the genus *Lampanyctus*. As in the former genus, species of *Lampanyctus* appear to separate into two more or less distinct groups: one, in which bladder volumes are small and change little with increasing standard length (*Lampanyctus ater*, *L. cuprarius*, and *L. lineatus*); and the other, where the rate of increase in bladder volume with increasing standard length is similar for the five remaining species.

Clearly, swimbladder volumes relative to fish standard length of these fish are highly variable within a given species, between species of the same genus, and between different genera of the same family. According to *Kleckner and Gibbs* (1972), much of this variability may be due to an ability of some (perhaps all) species of the Myctophidae to contract the bladder within the covering layer of peritoneum.

Some of the regressions shown in Figures 3a and 3b indicate a close similarity in slope and elevation, for example, between *Diaphus metapoclampus* and *D. rafinesquei*, between *Lampanyctus festivus* and *L. photontus*, and between *L. ater*, *L. lineatus*, and *L. cuprarius*. To verify these apparent similarities between regressions, a T test was used to test the hypothesis that samples of R for the respective species were drawn from a common population correlation. The test showed that the regressions for each of the species combinations mentioned above, in fact, did not differ significantly at the 0.05 level of probability. In addition, the test showed that the correlation coefficients for *Diaphus metapoclampus* and *D. rafinesquei* differed significantly from those of *D. brachycephalus*, *D. mollis*, and *D. problematicus*. The same test showed that the respective correlation coefficients for *Lampanyctus festivus* and *L. photonotus* and again for *L. ater.*, *L. lineatus*, and *L. cuprarius* were drawn from a common population correlation, respectively, but that there was a significant difference (0.05 level) between these two species groupings.

Because swimbladders presumably influence the vertical distribution and migration of midwater fish, it would be interesting to learn if similarities or differences in the slopes and elevations of the regressions for the various groups of species could be matched with some aspect of vertical distribution or migration of functional significance. *Bone* (1973) suggested that, in certain myctophids, functional types can be grouped by swimbladder state, lipid content, density, and size of pectoral fins. Unfortunately, none of these seven species was collected in large enough quantities in discrete-depth samples to permit any conclusions about this aspect of swimbladder development.

When adequate additional measurements of standard length and bladder volume become available, a definitive assessment might be provided by using covariance analysis. Application of such a technique would provide a more objective comparison of the regression lines for species included in this study and would, within predetermined probability levels, specify the species whose regressions of bladder volume against standard length exhibited affinities with other species. A more critical evaluation of the functional significance of these relationships could then be undertaken.

SWIMBLADDER VOLUME RELATED TO FISH VOLUME AND LENGTH

Jones (1951) calculated that for a marine teleost to achieve neutral buoyancy, the volume of the swimbladder should be somewhat less than 5 percent of the total fish volume. *Marshall* (1960) and other investigators also feel that the 5-percent ratio is a reasonable figure. On the other hand, more recent measurements of swimbladder dimensions and total fish volume show that the bladder volume of midwater fishes rarely approaches 5 percent of total fish volume (*Capen*, 1967; *Kleckner and Gibbs*, 1972). *Horn* (1975) reports the mean ratio of swimbladder volume to total fish volume for 12 species of stromateoid fish ranged from 0.6 to 3.4 percent. Increasing evidence indicates that lipids play an important role as a buoyancy device in several species of midwater fish (*Butler and Pearcy*, 1972; *Horn*, 1975).

Other authors have related swimbladder volume to the more easily measured fish total length and have derived the following formulas:

$$V_{SB} = 3.4 \times 10^{-4} L_{TL}^3 \quad \text{(Haslett, 1962)} \tag{4}$$

$$V_{SB} = 5 \times 10^{-4} L_{TL}^3 \quad \text{(Andreeva and Chindonova, 1964)} \tag{5}$$

where V_{SB} is the swimbladder volume in cm^3 and L_{TL} is the fish total length in cm. *Haslett's* equation is derived from his studies of six specimens of the whiting, *Gadus merlangus*, and is based on a mean bladder volume equal to 4.1 percent of the total fish volume. According to *Andreeva and Chindonova*, their own equation is "only very approximate" and assumes that fish volume = 0.01 L^3, where L is apparently fish total length in cm and bladder volume equals 5 percent of total fish volume. *Shearer* (1970), who determined swimbladder volumes for 91 fresh specimens belonging to 4 species of mesopelagic physoclistous fishes essentially by the method of *Kanwisher and Ebeling* (1957), reported wide discrepancies and little correlation between estimated bladder volumes for 3 of these species and those calculated from total lengths by either equations 4 or 5.

The present report offers additional comparison with results of *Andreeva and Chindonova*, *Haslett*, and *Shearer*. Equations 4 and 5 are expressed in fish total length. To make them compatible with fish standard length used here, they were converted to equations

$$V_{SB} = 5.2 \times 10^{-4} L_{STD}^{3} \tag{6}$$

and

$$V_{SB} = 7.6 \times 10^{-4} L_{STD}^{3}, \tag{7}$$

respectively, by assuming that fish total length L_{TL} is 15 percent greater than fish standard length L_{STD}. The solutions to these equations are plotted as long curved lines in Figures 2, 3, 4, and 5. These curves in Figures 2 and 3 reveal a poor match (for every species except *Sternoptyx diaphana*) between bladder volumes estimated from the current research and those estimated from either equations 6 or 7.

One may argue, perhaps justifiably, that the correlation coefficients for the regressions of several species (*viz.*, *Diaphus mollis*, *Diaphus brachycephalus*, *D. problematicus*, *Lampanyctus alatus*, *L. ater*, *L. cuprarius*, *L. lineatus*, *L. photonotus*, and *Pollichthys mauli*) are low enough to invalidate such a comparison. The fact remains, however, that a poor match still exists, even for species with high (>0.70) correlation coefficients (*viz.*, *Lampanyctus festivus*, *Bonapartia pedaliota*, *Melamphaes typhlops*, *Melamphaes pumilus*, *Diaphus metapoclampus*, *D. rafinesquei*, *Lampanyctus crocodilus*, and *L. pusillus*). Similar comparisons of *Andreeva and Chindonova's* and *Haslett's* curves with other species included in this study, though not shown here, are equally poor.

Regressions of fish length versus bladder volume for three Ocean Acre species (*Lepidophanes guentheri* (R = 0.97), *Myctophum nitidulum* (R = 0.95), and *Sternoptyx diaphana* (R = 0.96) are compared with regressions presented by *Shearer* (1970) for these same species (Figure 4). No comparison was made with *Shearer's* fourth species, *Diaphus brachycephalus*, because of the low correlation coefficient (R = 0.14) associated with the Ocean Acre data. The original equations given by *Shearer* for the regressions for *Lepidophanes guentheri* (R = 0.46), *Myctophum nitidulum* (R = 0.53), and *Sternoptyx diaphana* (R = 0.88) are

$$V_{SB} = 0.98 L_{TL}^{2.53}, \tag{8}$$

$$V_{SB} = 2.81 L_{TL}^{1.95}, \tag{9}$$

and

$$V_{SB} = 1.52 L_{TL}^{2.63}, \tag{10}$$

respectively, where L is total length in cm and V_{SB} is swimbladder volume in mm^3. These have been converted to make them compatible as follows:

$$V_{SB} = 0.004 L_{STD}^{2.53} \text{ for } Lepidophanes\ guentheri \tag{11}$$

$$V_{SB} = 0.04 L_{STD}^{1.95} \text{ for } Myctophum\ nitidulum \tag{12}$$

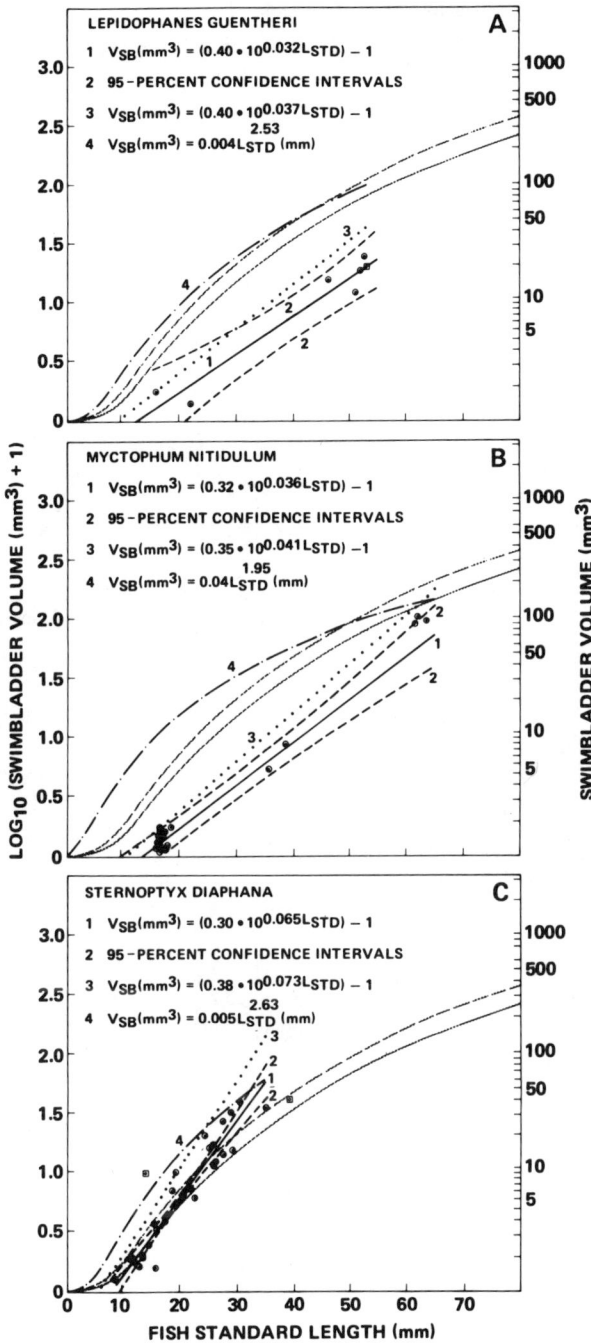

Fig. 4. Bladder volume regressions of Ocean Acre fishes compared with regressions from the equations of Shearer, Andreeva and Chindonova, and Haslett including calculated volumes for Marshall's specimens.

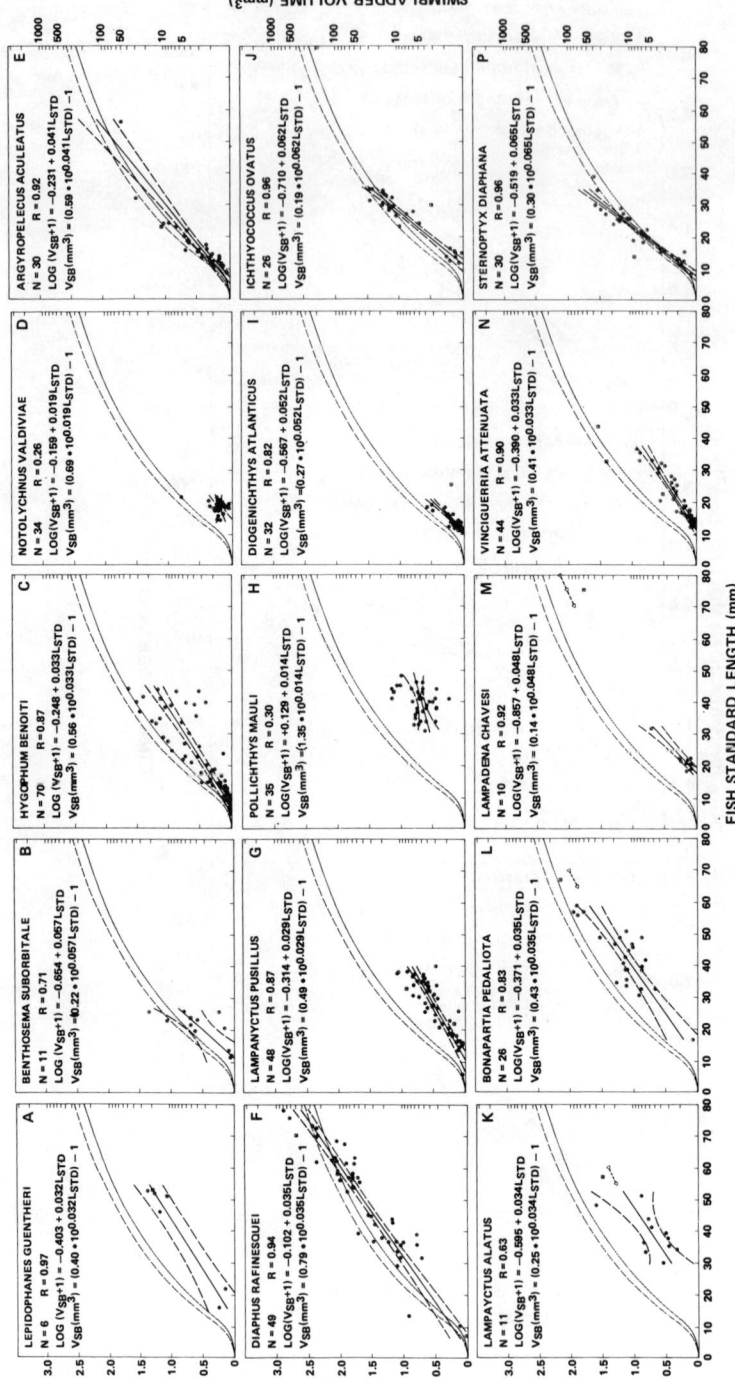

Fig. 5. Bladder volume regressions of Ocean Acre fishes compared with calculated volumes for Marshall's specimens as well as Andreeva and Chindonova's and Haslett's regressions.

and

$$V_{SB} = 0.005 L_{STD}^{2.63} \text{ for } Sternoptyx\ diaphana \tag{13}$$

where L_{STD} is standard length in mm, V_{SB} is volume in mm³, and it is assumed that

$$L_{TL} = 1.15 L_{STD}. \tag{14}$$

Bladder volumes estimated by the two methods differ widely, with *Shearer's* estimates being larger in all three cases. As mentioned above, *Shearer* worked with fresh material; he determined volume by the method of *Kanwisher and Ebeling* (1957). According to *Andreeva* (1964), the linear dimensions of the swimbladders of living fish may be 30 to 40 percent greater than measurements of specimens fixed in formalin. On the other hand, in a recent study of the histology and morphology of stromateoid swimbladders, *Horn* (1975) allowed only 10 percent for shrinkage of preserved swimbladders. It seems likely that 30 percent represents a maximum value for shrinkage allowance. In the following section, this correction is applied to Ocean Acre swimbladder dimensions. The dotted lines on Figure 4 show the effect on the regression line for bladder volume of a 30-percent increase in the linear dimensions of the swimbladder. Although such an increase in linear dimensions yields an increase in bladder volume of almost 120 percent, *Shearer's* volume estimates are mostly still considerably higher than those determined for Ocean Acre specimens. When *Shearer's* results are compared with the long curved lines shown in Figure 4, it can be seen that his estimate of bladder volume for *Lepidophanes guentheri* agrees fairly well with *Andreeva and Chindonova's* and with *Haslett's* estimates but, for the most part, exceeds these estimates for *Myctophum nitidulum* and *Sternoptyx diaphana*.

Marshall (1951, 1960) has reported measurements of standard length and bladder dimensions for several species of midwater fish. Wherever his species are common with the species used here, bladder volumes have been calculated from his bladder dimensions with equation 1. These volumes are compared with *Shearer's* results in Figure 4a and 4c and with bladder volumes resulting from the present study in Figure 5. In these figures, *Marshall's* volume estimates are shown as squares, bladder volumes for Ocean Acre specimens are shown as circles. The least-square regression line (solid straight line) is shown for each Ocean Acre species, as are the 95-percent confidence intervals (curved dashed lines) around the regression line.

In four species (*Lepidophanes guentheri* (Figure 5a), *Benthosema suborbitale* (Figure 5b), *Hygophum benoiti* (Figure 5c), and *Notolychnus valdiviae* (Figure 5d)), bladder volumes calculated from *Marshall's* data fall within the 95-percent confidence intervals around the least-squares line calculated for Ocean Acre specimens. The bladder volume of *Marshall's* specimen of *Lepidophanes guentheri* is considerably less than *Shearer's* estimate (Figure 4a). In four other species (*Argyropelecus aculeatus* (Figure 5e), *Diaphus rafinesquei* (Figure 5f), *Lam-*

panyctus pusillus (Figure 5g), and *Pollichthys mauli* (Figure 5h)), Marshall's swimbladder volumes fall within the range but outside the confidence limits of volumes one may expect to encounter in Ocean Acre specimens.

In two species (*Diogenichthys atlanticus* (Figure 5i) and *Ichthyococcus ovatus* (Figure 5j), Marshall's data show bladder volumes considerably smaller than those calculated from Ocean Acre material. The standard lengths of Marshall's specimens of *Lampanyctus alatus* (Figure 5k), *Bonapartia pedaliota* (Figure 5l), and *Lampadena chavesi* (Figure 5m) exceed those of Ocean Acre specimens included here. They cannot be compared with an extrapolated least-squares line for these species. Instead, the regressions for bladder length and width have been used to estimate the bladder volume for Ocean Acre individuals comparable in size to Marshall's specimens. These volumes are shown in Figures 5k-m as symbols connected with a dashed line. When compared with bladder volumes calculated in this manner, the volumes of Marshall's specimens of *Lampanyctus alatus* and *Bonapartia pedaliota* are somewhat greater. The bladder volume of Marshall's specimen of *Lampadena chavesi* (Figure 5m) is less.

In Figure 5n, one of Marshall's three specimens of *Vinciguerria attenuata* (standard length 22.5 mm) had a calculated bladder volume that fell within the data for Ocean Acre individuals of this species, but two other specimens had bladder volumes that greatly exceeded what might be expected for Ocean Acre specimens.

One final comparison is shown in Figure 5p, where volumes of both of Marshall's specimens of *Sternoptyx diaphana* were outside the range of those calculated for Ocean Acre specimens. As Figure 4c shows, the bladder volume of one of Marshall's specimens of *Sternoptyx diaphana* exceeded Shearer's estimate and the other was less.

Examination of the long curved lines in Figure 5a-p reveals that in every species (except *Sternoptyx diaphana* and possibly *Argyropelecus aculeatus*) swimbladder volumes estimated by equations 6 and 7 exceed volumes estimated for Ocean Acre specimens.

SUMMARY AND CONCLUSIONS

This report examines the relationship of fish standard length to swimbladder dimensions; the intra- and inter-specific variation is also measured in selected midwater fish specimens belonging to 55 species from 9 families. These species are believed to account for most acoustic volume reverberation occurring throughout a large part of the Sargasso Sea.

It is shown that fish standard length is linearly related to swimbladder length and swimbladder width. Linear regression equations are presented to define these relationships.

The relationship of fish standard length to swimbladder volume, however, is found to be curvilinear. To facilitate analyzing this relationship, a logarithmic transformation commonly employed in fisheries work is applied to the bladder volume data, and additional regression equations are presented to relate fish length to bladder volume. The slopes of these regression lines are shown to be insignificant from zero for 14 of the 55 species. For the remaining 41 species, all slopes were positive (with the exception of *Lampanyctus cuprarius*, which showed a slight decrease in bladder volume with increasing fish standard length).

The volume of the swimbladder of a given species of given standard length can vary greatly. Three separate specimens of *Bonapartia pedaliota*, each with a standard length of 49 mm, had calculated bladder volumes of 6.5, 15.7, and 32.8 mm^3. In other species, such as *Sternoptyx diaphana* and *Diogenichthys atlanticus*, variability in bladder volume was considerably less. Comparison of the elevations and slopes of regression lines for different fish species also reveals wide differences.

Swimbladder volumes estimated from this study are in fair agreement with measurements published by *Marshall* (1951, 1960) for like species of fish but, for the most part, are considerably less than volumes estimated by either *Haslett's* (1962), *Andreeva and Chindonova's* (1964), or *Shearer's* (1970) equations.

REFERENCES

Andreeva, I. B. 1964. Scattering of Sound by Air Bladders of Fish in Deep Sound-Scattering Layers. *Akusticheskii Zhurnal*, 10 (1): 17-20.

Andreeva, I. B. and Yu. G. Chindonova. 1964. On the Nature of Sound Scattering Layers. *Okeanologiya*, 4 (1): 112-124.

Bone, Q. 1973. A Note on the Buoyancy of Some Lantern-Fishes (Myctophoidei). *Journal of the Marine Biological Association of the United Kingdom*, 53: 752-761.

Brooks, A. L. 1976. Swimbladder Allometry of Selected Midwater Fish Species. NUSC Technical Report 4983: 44 p.

Butler, J. L. and W. G. Pearcy. 1972. Swimbladder Morphology and Specific Gravity of Myctophids off Oregon. *Journal of Fisheries Research Board of Canada*, 29: 1145-1150.

Capen, R. L. 1967. *Swimbladder Morphology of Some Mesopelagic Fishes in Relation to Sound Scattering*. U.S. Navy Electronics Laboratory Research Report 1447: 1-25, appendices A, B, and C.

Eyring, C. F., R. J. Christensen, and R. W. Raitt. 1948. Reverberation in the Sea. *Journal of the Acoustical Society of America*, 20: 462-475.

Haslett, R. W. G. 1962. Determination of the Acoustic Back-Scattering Patterns and Cross Sections of Fish. *British Journal of Applied Physics*, 13: 349-357.

Horn, M. H. 1975. Swim-Bladder State and Structure in Relation to Behavior and Mode of Life in Stromateoid Fishes. *Fishery Bulletin*, 73 (1): 95-109.

Jones, F. R. H. 1951. The Swimbladder and the Vertical Movements of Teleostean Fishes, I. Physical Factors. *J. Exp. Biol.*, 28 (4): 553-566.

Kanwisher, J., and A. Ebeling. 1957. Composition of the Swimbladder Gas in Bathypelagic Fishes. *Deep-Sea Research*, 4: 211-217.

Kleckner, R. C. 1974. Swimbladder Morphology of Mediterranean Sea Mesopelagic Fishes. M.S. Dissertation, University of Rhode Island, 66 p.

Kleckner, R. C. and R. H. Gibbs, Jr. 1972. Swimbladder Structure of Mediterranean Midwater Fishes and a Method of Comparing Swimbladder Data with Acoustic Profiles. In *Mediterranean Biological Studies, Final Report*, Smithsonian Institution, 346 pp.

Marshall, N. B. 1951. Bathypelagic Fishes as Sound Scatterers in the Ocean. *Journal of Marine Research*, 10 (1): 1-17.

Marshall, N. B. 1960. Swimbladder Structure of Deep-Sea Fishes in Relation to Their Systematics and Biology. *Discovery Reports* 31: 1-222.

Shearer, L. W. 1970. Correlations Between Surface-Measured Swimbladder Volumes, Depth of Resonance, and 12-kHz Echograms at the Time of Capture of Sound Scattering Fishes. In *Proceedings of International Symposium on Biological Sound Scattering in the Ocean*. G. B. Farquhar, Ed. Maury Center for Ocean Science Report 005: 453-471.

THE MINOX PROGRAM: AN EXAMPLE OF A MULTIDISCIPLINARY OCEANIC INVESTIGATION

William A. Friedl, George V. Pickwell, and Robert J. Vent

Naval Undersea Center, San Diego

ABSTRACT

In 1970, personnel at the Naval Undersea Center undertook an investigation of the acoustics, biology, and chemistry of the eastern tropical North Pacific Ocean under a program called MINOX for the MINimal OXygen concentrations typical of the water in that region. The investigation's purpose was to relate the biological and chemical conditions in the region to acoustical measurements in order to provide insight into the environmental conditions producing particular acoustical results. General conclusions and techniques, so developed, could then be used to relate the biological causes and acoustical effects of volume scattering to environmental conditions in other oceanic regions. This report (1) reviews the background, concepts and premises of the MINOX program, (2) summarizes the three cruises completed in the program to date, (3) presents the general findings from those cruises, and (4) evaluates the findings with regard to the hypotheses under which the program was formulated. Conclusions are drawn to relate the MINOX program, in particular, to multidisciplinary oceanic investigations including acoustics, in general.

BACKGROUND

The Naval Undersea Center (NUC) and its predecessor organizations have a history of environmental acoustical studies dating back to the post-World War II period when much of the work of the University of California Division of War Research was continued by the Navy at the Point Loma Laboratory facilities in San Diego. These early efforts were directed toward description of the Deep Scattering Layer (DSL) and results of many of these investigations were summarized by Hersey and Backus (1962) in their review of biologically-caused sound scattering in the ocean volume. Among the early studies of the DSL, those of Boden (1950), Tucker (1951), and Batzler and Westerfield (1953), to name a few, contained elements of the more formalized cross-disciplinary studies of volume reverberation that were to follow. The major aspects of the multidisciplinary basis of the MINOX program are given in the following four sections.

Submersibles

Observations were made from manned submersible vehicles as adjuncts to DSL studies on extensive operations between southern California and Panama during the late 1950s and through the 1960s (Barham and Davies, 1966; Barham et al., 1966 and 1968; Adams, 1967). These observations added considerable insight into the structure and composition of the DSL (Barham, 1963; 1966), but most studies from submersibles lacked concurrent environmental and chemical measurements beyond temperature and salinity observations (see, for example, Pickwell et al., 1970).

Far-Seas Expeditions

The advent of AGOR-class vessels enabled personnel of the Point Loma laboratories to extend their investigations to the central and western Pacific and the eastern Indian Oceans with three FASOR (Forward Area SOnar Research) cruises during the 1960s; an additional FASOR-type cruise was fielded in the eastern Pacific in 1968 (Figure 1). These cruises emphasized acoustical measurements, with some time allotted for biological trawling and hydrographic casts. Environmental information mainly concerned physical parameters that determine sound speed in the sea, although some chlorophyll a measurements were also made. Biological and environmental sampling on these expeditions was largely qualitative and descriptive, limiting the extent to which results could be related to acoustical observations (Batzler and Vent 1967). Thus, FASOR data has yielded a general acoustical/biological characterization for some regions of the western Pacific (Seligman and Friedl, 1975), but multidisciplinary application of the results has not been extensive.

Westfall Seamount Cruises

In 1969, joint efforts of acousticians and biologists at the Point Loma Navy laboratories were consolidated into a coordinated study of the diel and seasonal characteristics of biologically-caused volume scattering at 30°N latitude, 120°W longitude, in the vicinity of Westfall Seamount, some 320 km southwest of San Diego. The study emphasized closely coordinated, quantitative measurements and sought biological explanations for observed acoustical conditions. Some acoustical results from the Westfall study are given in *Vent* (1972) and *Batzler et al.* (1975) used biological results from the study to verify an acoustical model.

Although the Westfall study did not include detailed environmental sampling, it established the pattern of close, joint effort between biologists and acousticians that was essential to the MINOX program.

Chemical Oceanographic Capability

Late in 1969, a chemical oceanography branch was created at NUC to provide more extensive environmental chemical data to augment field efforts in acoustics and biology at the Point Loma laboratories. From within the Chemical Oceanography Branch, the MINOX program was developed and directed.

During the period discussed above, traditional, strictly disciplinary investigations continued at NUC. Sea trials, often conducted in the deep water of the San Diego Trough within twenty miles of Point Loma, produced techniques applicable to multidisciplinary activities. These efforts also provided an understanding of the environmental and acoustical complexities of the waters off southern California. By 1970, it was obvious that an understanding of the diverse environmental and biological factors involved in acoustical scattering in the California Current would only come with considerable effort over a long period of time, if at all.

CONCEPTS

The MINOX program is a multidisciplinary oceanic investigation that seeks to identify environmental conditions that may be related to the magnitude and occurrence of biologically-caused acoustic volume reverberation in the requency range from 1 to 40 kHz. Fundamental to the program is an investigation of the biological causes of volume reverberation and the environmental conditions that may affect how and where the causative organisms live. This investigation involves identification of the relationships, if any, between environmental parameters and biological populations responsible for particular, identifiable acoustical characteristics at specific acoustic frequencies, times of day or year, depths or geographical locations. Because certain organisms, particularly mesopelagic fishes, are known as primary causes of oceanic volume reverberation (see *Farquhar*, 1970), and the distribution of such organisms may be proscribed by environmental factors, acting singly or in combination (*Rass*, 1967; *McGowan*, 1971), quantitative biological samples were required along with acoustical measurements at a variety of frequencies. When integrated with the appropriate environmental measurements, the acoustical and biological data would provide the substance for identification of recurring environmental correlates to help define the broad geographical limits of particular types of volume reverberation for oceanic areas that are inadequately characterized acoustically.

Experience of the acousticians, biologists and chemical oceanographers at NUC indicated that acoustic scattering conditions in tropical regions would be generally less complex than those of the more-convenient San Diego Trough or of the California Current regions near San Diego. Thus the area chosen for the initial MINOX effort was the eastern tropical north Pacific, mainly the region between 10° and 20°N latitude and eastward from 115°W longitude to the Mexican/Central American coast (Figure 1). The extremely low values of dissolved oxygen, characteristic of the proposed area of investigation (Cline and Richards, 1972) had the potential of exerting significant and identifiable influence on the behavior and distribution of organisms that cause volume reverberation in the region.

As the relationships between scattering, biological populations and environmental factors became understood for the tropical region, the insight, principles and techniques developed could be applied to the more complex communities that produce scattering in the temperate waters. With these plans and objectives, the MINOX program commenced in 1970.

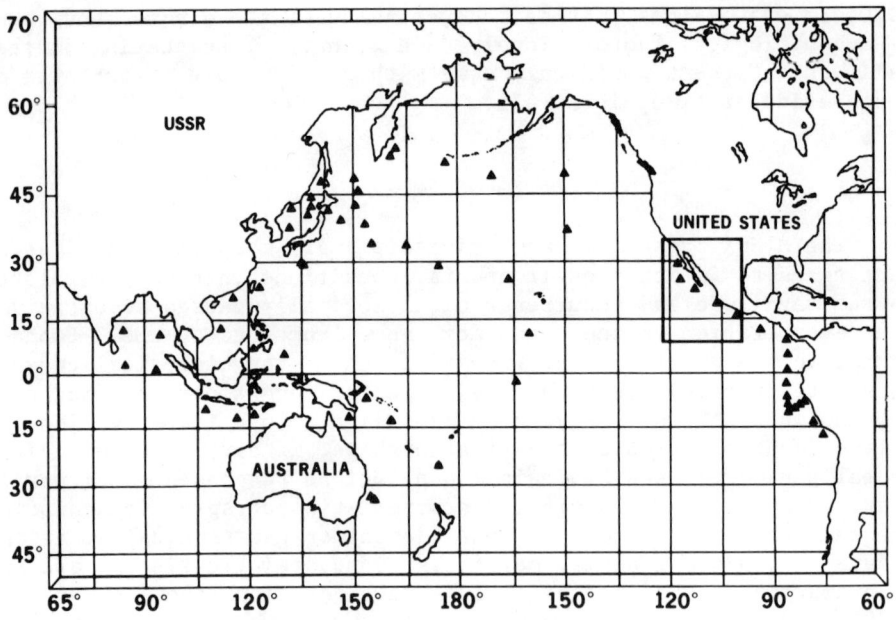

Fig. 1. Location of stations where acoustical measurements were made and biological and chemical samples were taken as part of the Far Seas expeditions fielded by NUC personnel prior to 1970 (triangles). Inset area delineates the region detailed in Figure 2.

METHODS

MINOX cruises were fielded in July, 1970, and March, 1972, on USNS *De Steiguer* (T-AGOR-12) and in October, 1973, aboard USNS *Bartlett* (T-AGOR-13). Stations were occupied between San Diego and the Minimal Oxygen region to obtain comparative measurements of acoustic scattering populations encountered in other oceanographic regions through the comparison of acoustical measurements and biological and environmental samples (Figure 2). Typically, thirty to fifty hours were spent on each station and major events included at least one surface-to-bottom hydrocast series and acoustical measurements and biological sampling during both daytime and nocturnal periods. To

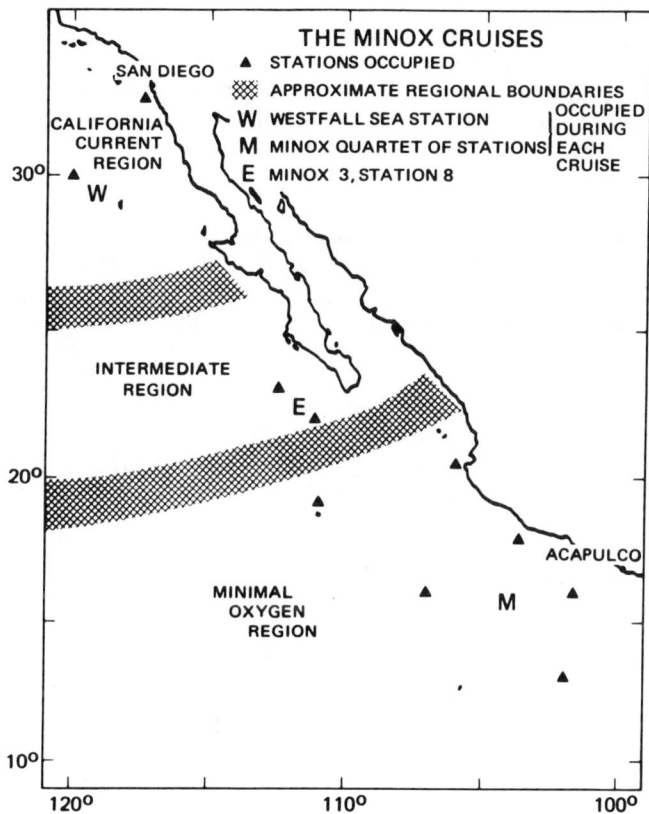

Fig. 2. Location of stations occupied during the MINOX cruises (triangles). Oceanographic regions and their approximate boundaries are generalized from Sverdrup et al. (1942). The Westfall Sea Station (W), in the California Current region, and the quartet of stations south of 18°N latitude, in the Minimal Oxygen region (M), were occupied during each cruise; the station off Cabo Corrientes (20°35'N, 106°W) was not occupied during MINOX-2. All other stations were occupied for one cruise only.

provide a qualitative record of scattering in the upper 800 m of the water column, 12 kHz echosounders were monitored continuously while the ship was underway and whenever echosounder operations would not interfere with acoustical measurements on station. During the third cruise, echosounder records were also obtained at 3.5 kHz while underway and at 24 and 38 kHz on some stations.

Water samples were taken with 8 L PVC bottles equipped with reversing thermometers. Details of the environmental and chemical oceanographic sampling schemes and analytical techniques are given in *Seligman* (1974), *Douglas et al.* (1976) and *Vent and Pickwell* 1977). Dissolved oxygen concentrations were determined by iodometric titration ith a microburette, according to the modified Winkler procedure of *Carpenter* (1965), as adapted to the conditions of the Minimal Oxygen region by *Pickwell* (*Douglas et al.*, 1976). The limits of precise measurement with this technique have been set at ±0.03 ml/L.

Acoustic volume scattering measurements were made at each station using both broad-band explosive and single-frequency continuous wave (CW) tone-pulse methods. Explosive and CW measurements were made around noon and midnight on each station and at other periods on some stations. Acoustical field and analytical procedures are detailed by *Vent and Pickwell* (1977).

Biological samples were taken principally with a Tucker trawl (*Davies and Barham*, 1969) with a mouth 2 m per side, a 1.1 cm mesh body, and a 0.5 mm mesh cod end. The trawl was equipped with a timer actuated opening/closing mechanism, a depth-indicating pinger and digital flowmeters mounted inside and outside the net body proper. Trawling speed was 1 to 2 m/sec and the net was open for 30 to 70 min at depth. Strata to be sampled were identified after discussion with acousticians and examination of 12 kHz echosounder records. Diurnal hauls were made below, through, and above the DSL; nocturnal hauls were made in near-surface scattering layers and any layers present at depth. In the Minimal Oxygen region, at least one nocturnal haul per station sampled the interval between 500 and 100 m, whether scattering was indicated in the interval or not.

Following each haul, fishes were separated from plankton and residue. On MINOX-2, swimbladder volume measurements and gas analyse were conducted on selected fish immediately following completion of a haul (*Valkirs*, 1972, MS). On MINOX-3, swimbladder volumes and certain blood characteristics were determined for a variety of mesopelagic fishes (*Douglas et al.*, 1976; and unpublished). Post-cruise processing of biological samples includes sorting of plankton into major taxonomic groups, identification of fishes to species, measurement of the standard length of fishes, and on selected specimens, examination and measurement of swimbladder characteristics.

RESULTS

This report will not detail all the scientific results of the MINOX program, but will note the findings from the study that particularly relate to the premises underlying the investigation. Certain of the scientific results are available from other sources (For example, Seligman, 1974; Douglas et al., 1976; Vent and Pickwell, 1977) and more will be forthcoming as analyses continue.

The three oceanographic regions identified in Figure 2 have distinctive physical characteristics (Sverdrup et al., 1942; Dobrovol'skiy, 1968) and the oceanographic measurements from the MINOX cruises agreed, on the whole, with historical characterizations of the regions. The relationship of physical conditions to the nutrient and plant pigment distributions observed on MINOX cruises are discussed by Seligman (1974), who found distinctive relationships with depth for each region. Each region also had a distinctive general profile of dissolved oxygen with depth, and although minimum values occurred in the upper 1000 m of the water column in each region, the depth of the minimum, its vertical extent, and its oxygen content were regionally distinctive (Figure 3). At Westfall, values below 1.0 ml/L are found between 500 and 1200 m. In the Intermediate region, a sharply-defined minimum, with values below 0.1 ml/L, occurs near 500 m. In the Minimal Oxygen region, the extent of the extremely-low oxygen water varied from station to station; concentrations become vanishingly small near 100 m and remain below 0.1 ml/L to at least 700 m (Figure 4). Below 1000 m, profiles of dissolved oxygen from each region were less distinctive.

Acoustically, the regions were distinctive as well. Echosounder records showed characteristic patterns of scattering and layer migrations for each region. At Westfall, patterns of migration and nocturnal scattering were particularly complex. A single, diurnal DSL, centered at about 400 m, characteristically split in the late afternoon prior to the evening migration of the upper portion toward the surface; the lower portion of the DSL remained near 400 m as a nocturnal non-migratory layer (NML). On MINOX-3, an intermediate diurnal layer was also recorded between 200 and 250 m. A complex evening migration produced increased scattering above 150 m and left several intermediate scattering layers between 150 m and the NML (Figure 5). In the morning, several distinct layers leave the near surface zone and migrate downward to merge with the NML and form the diurnal DSL. The patterns of scattering at Westfall remained fairly constant from cruise to cruise. The major difference, as noted above, was a diurnal intermediate layer that was recorded only on one cruise.

In the intermediate region, scattering records were quite variable. Typically, the diurnal DSL was not well-developed; it was often intermittently recorded during the day and the record produced was typically blotchy and uneven (Figure 6). Upward migration in the evening lacked distinct layers and had a characteristic saw-tooth pattern (Figure 6). In the morning, during the downward migration from the near-surface zone, the layer typically faded from the 12 kHz record (Figure 7). Nocturnal scattering was concentrated above 100 m and, although not typical, a NML was recorded on several occasions. Intermittent, typically heavy, group scattering—distinct from scattering layers—was recorded in the upper 150 m during the day in this

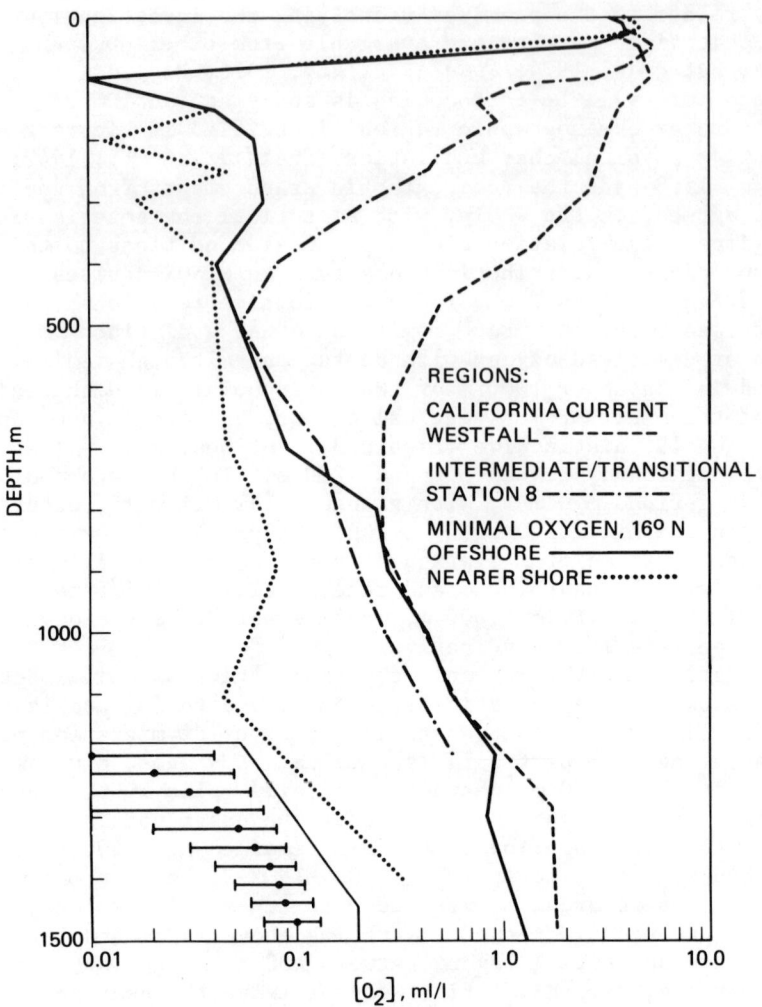

Fig. 3. Composite, semi-logarithmic profiles of dissolved oxygen concentrations with depth for the three oceanographic regions sampled during the MINOX cruises. Data are from MINOX-3; results from the other cruises show similar regional characteristics. Inset presents the graphic range, to scale, of the limits of precise oxygen determinations (±0.03 ml/L) about discrete values, every 0.01 ml/L, for a range of concentrations from 0.01 to 0.10 ml/L.

region (Figure 7). Scattering conditions in the Intermediate region changed significantly during a given day or night, but seldom in a consistent manner.

Echosounder records from the Minimal Oxygen region reveal less-complex scattering patterns that varied, at night, from one cruise to

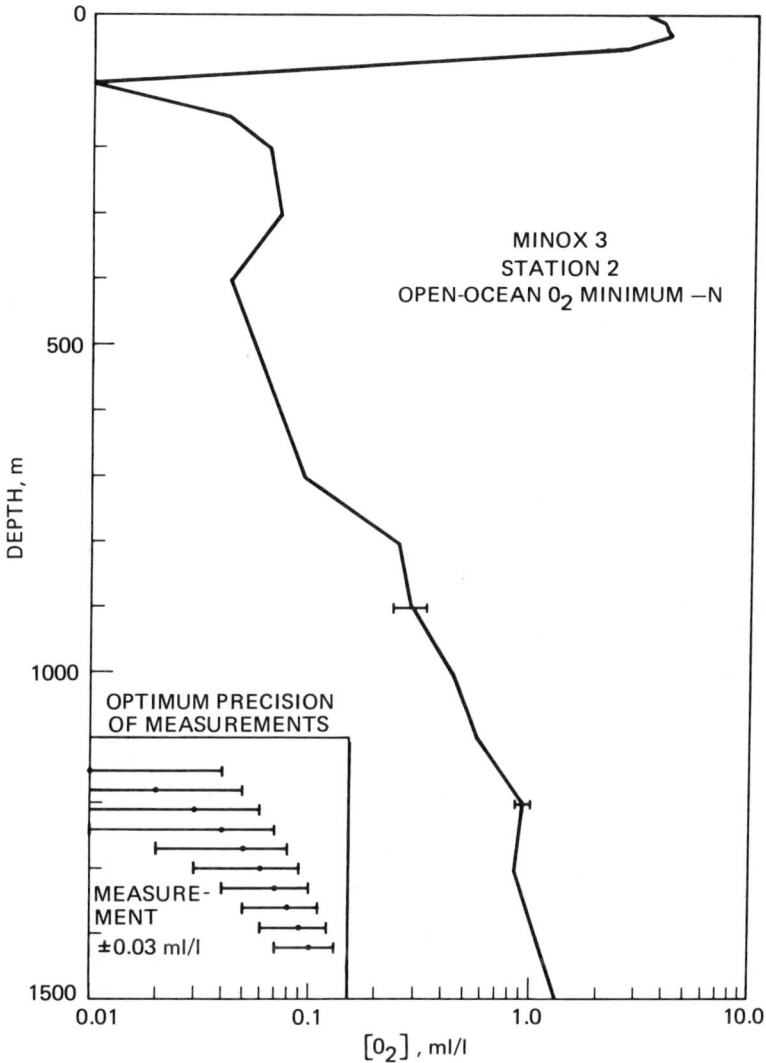

Fig. 4. Semi-logarithmic profile of dissolved oxygen concentration with depth from the Northern O_2 Minimum station (Table 1) in the Minimal Oxygen region; data are the same as used for the "Offshore" station in Figure 3. Horizontal bars connect determinations from the same depths on different casts. Inset presents the graphic range, to scale, of the limits of precise oxygen determinations (±0.03 ml/L) about discrete values, every 0.01 ml/L, for a range of concentrations from 0.01 to 0.10 ml/L.

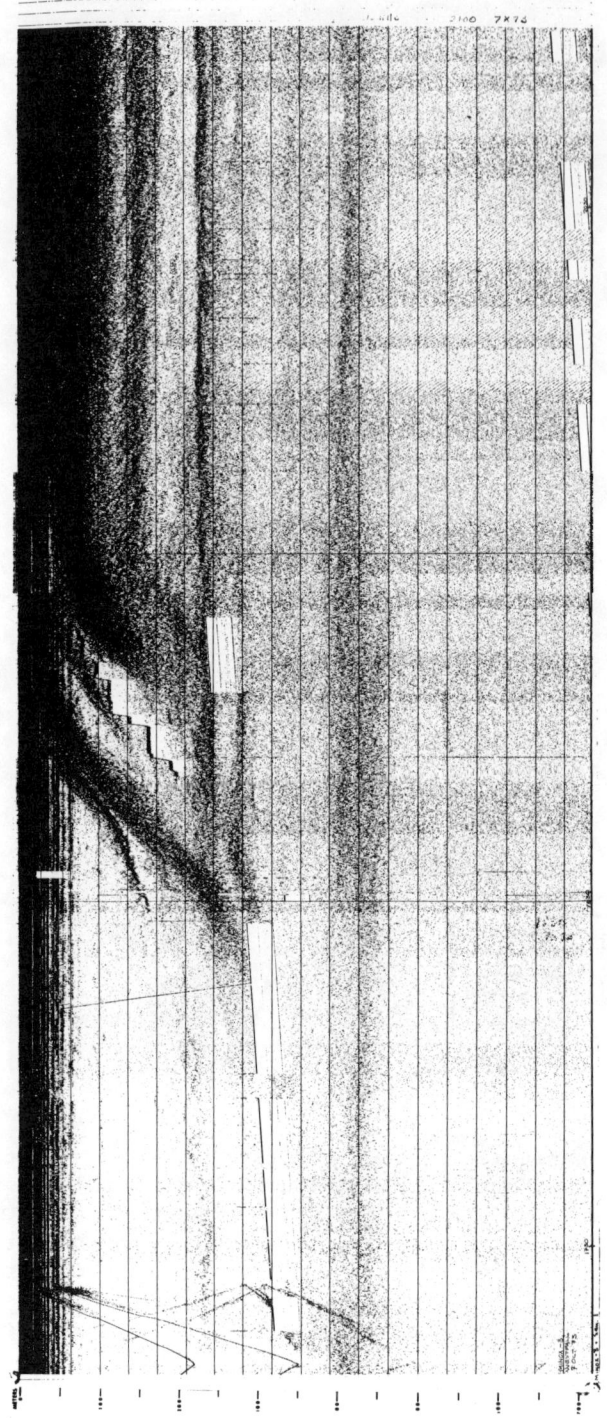

Fig. 5. 12 kHz echosounder record from Westfall Sea Station, 1700 - 2100 (local time), 7 October 1973, showing scattering in the upper 400 fm (741 m). Horizontal lines on the record indicate 20 fm (37 m) intervals. The records were made from left to right, as presented, and earlier times are to the left of the figures and later times to the right. Layers' migration upward and the increase in scattering above 300 m, typical of the region, is evident on the record, as is the non-migratory layer at 400 m. Clear areas near 300 m on the record, before and during the migration, are artifacts of the recording. Step-like trace in the midst of the migration is the return from a SV/STD probe that was in use at the time of the recording.

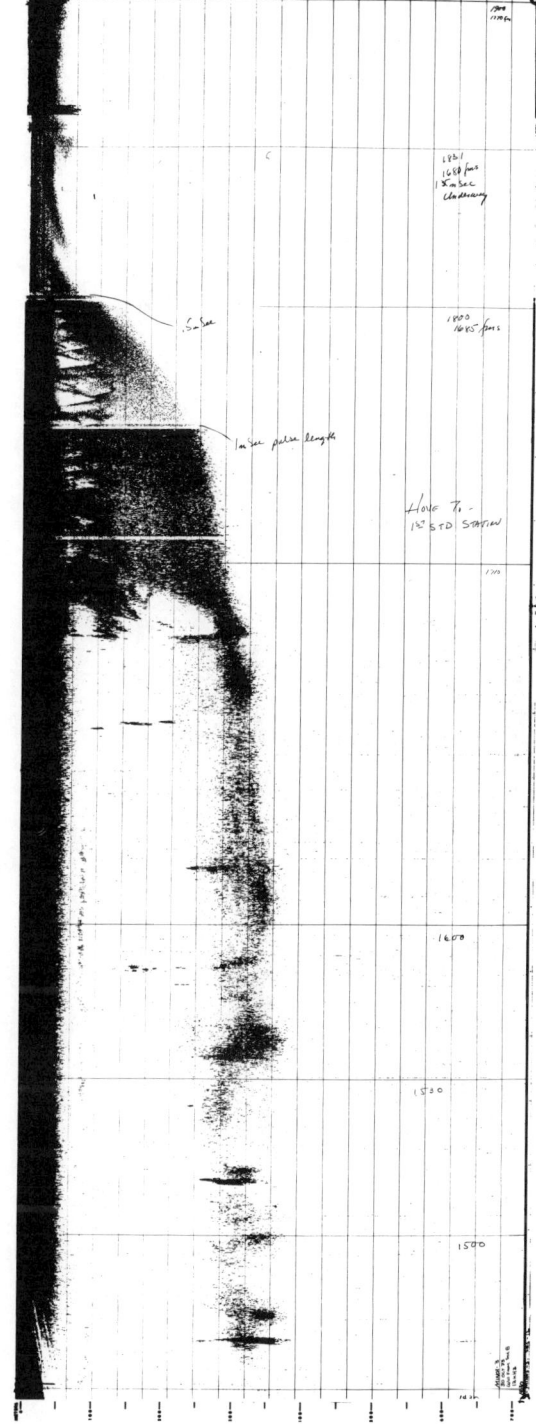

Fig. 6. 12 kHz echosounder record from the Intermediate region, 1430 – 1900, 30 October 1973, same type of record as Figure 5. Irregular diurnal DSL, recorded on the left side of the figure, moves toward the surface with a characteristic "saw tooth" pattern. No scattering was recorded at depth following the ascent of the DSL on this particular evening. The variations in layer intensity on the record during the migration resulted largely from variations in the outgoing pulse length of the signal from the recorder during the migration period.

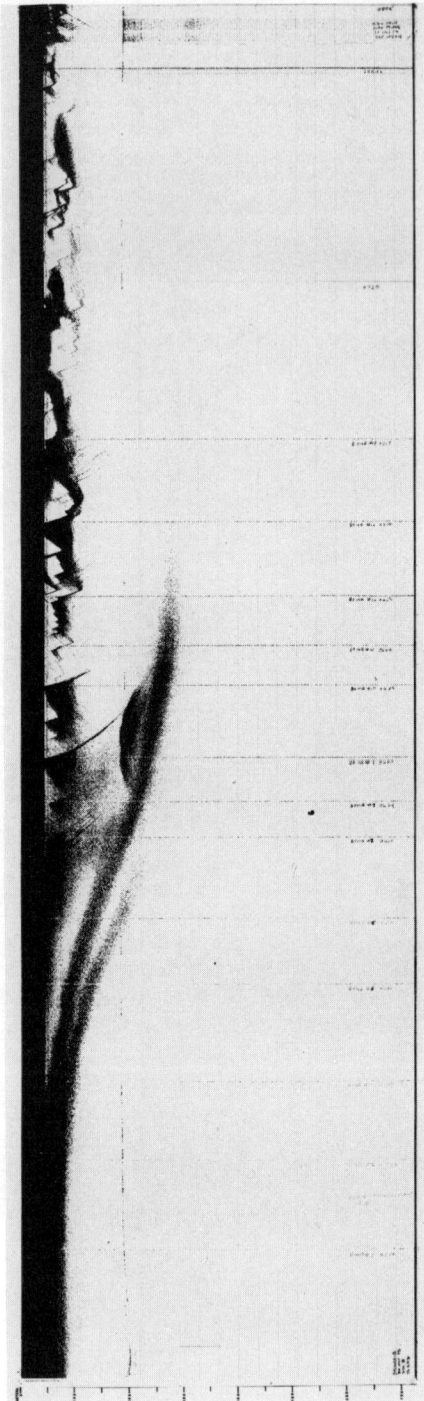

Fig. 7. 12 kHz echosounder record from the Intermediate region, ca. 0500 – 0805, 30 October 1973, same type of record as Figure 5. The SSL moves downward as several distinct elements that fade from the record at depth. Irregular scattering above 150 m is associated with groups of pelagic crabs.

the next. A well-developed diurnal DSL was typical between 250 and 450 m in this region. The layer was characteristically single, and often more distinct in the afternoon than in the morning. Several distinct layers migrate to and from the near-surface zone (above 100 m) in the crepuscular periods (Figures 8 and 9). For at least part of the night, after the evening (upward) migration of the DSL, layers are absent below 100 m, but on two cruises, scattering was recorded between 200 and 300 m some time after the upward migration of the DSL in the evening (Figure 8). This scattering has been dubbed the MINOX migratory layer (MML) because of its unique occurrence in the Minimal Oxygen region. The MML was recorded on approximately 20 nights, total, during the MINOX-1 and -3 cruises. On some occasions, the MML did not exhibit a clear-cut migration from the surface-scattering layer (SSL), as depicted in Figure 8, but rather appeared at depth some time after the evening migration. Typically, however, the MML migrated from the SSL some time during the night and persisted at depth until it was obscured by the layers migrating downward in the morning (Figure 9). The MML is unique, though somewhat ephemeral, characteristic of the Minimal Oxygen region and will be discussed in detail in another paper.

Quantitative acoustical measurements from all three regions had the same general characteristics of scattering strength (S_v) versus depth, namely a high but variable S_v above 600 m, reduced strengths between 600 and 1000 m, and increased values below 100 m. Within this general description, patterns for each region were distinguished on the basis of the 12 kHz CW pulse measurements (Figures 10, 11, and 12). Further illustrations and results must await more complete processing of the acoustical data at other discrete frequencies and from the explosive measurements.

Acoustical measurements made in the California Current during the MINOX cruises reveal a pronounced increase in volume scattering strength between 300 m and the surface at night and slight diel variation in scattering strengths measured between 300 and 500 m on the Westfall Sea Station. Other characteristics of the scattering strength measurements made in the California Current were variable, between acoustic frequencies, from cruise to cruise.

Details of acoustical measurements made in the Intermediate region are given in *Vent and Pickwell* (1977). In general, the pattern of scattering was one of more distinct peaks or absences of scattering than was typical for the California Current (Figure 11). In the Intermediate region, the diurnal scattering maximum was quite pronounced between 250 and 500 m. Nocturnally, scattering increased markedly above 50 m and remained high near 300 m, but the near-surface scattering and mid-depth maximum were separated by an excess of 100 m of water with low S_v; nocturnal scattering remained low between 400 and 1000 m (Figure 11). Column strength (S_c = the volume scattering strength of a column of unit cross section, usually 1 m^2 or 1 yd^2, extending from the surface to a specified depth) values changed but slightly in the Intermediate region, day to night.

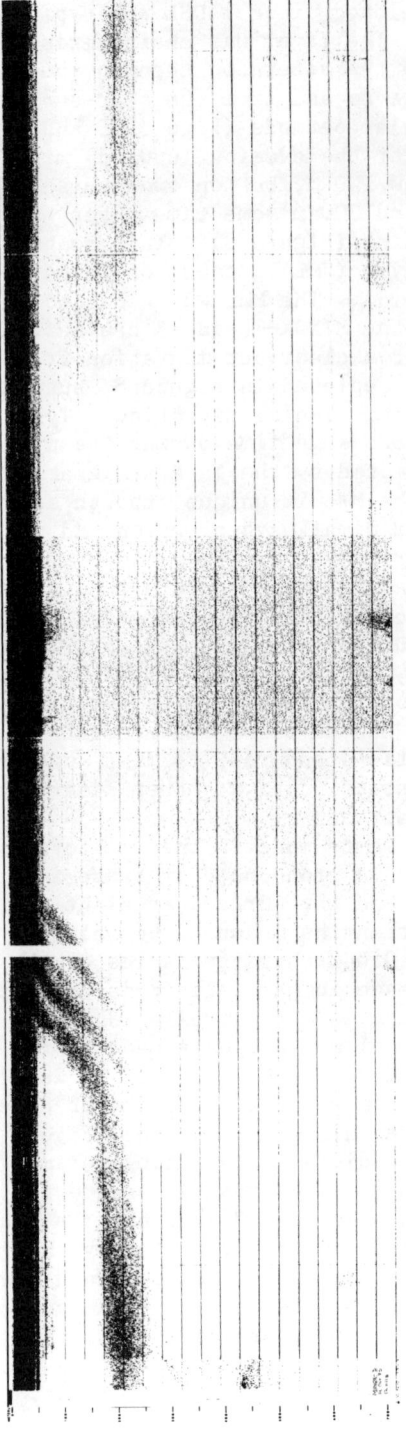

Fig. 8. 12 kHz echosounder record from the Minimal Oxygen region, ca. 1630 - 2212, 14 October 1973, same type of record as Figure 5. Upward migration of the diurnal DSL is as three distinct layers and leaves no NML or partially-migratory layers at depth. About two hours after the upward migration, a MINOX Migratory Layer (MML) leaves the SSL and returns to depth.

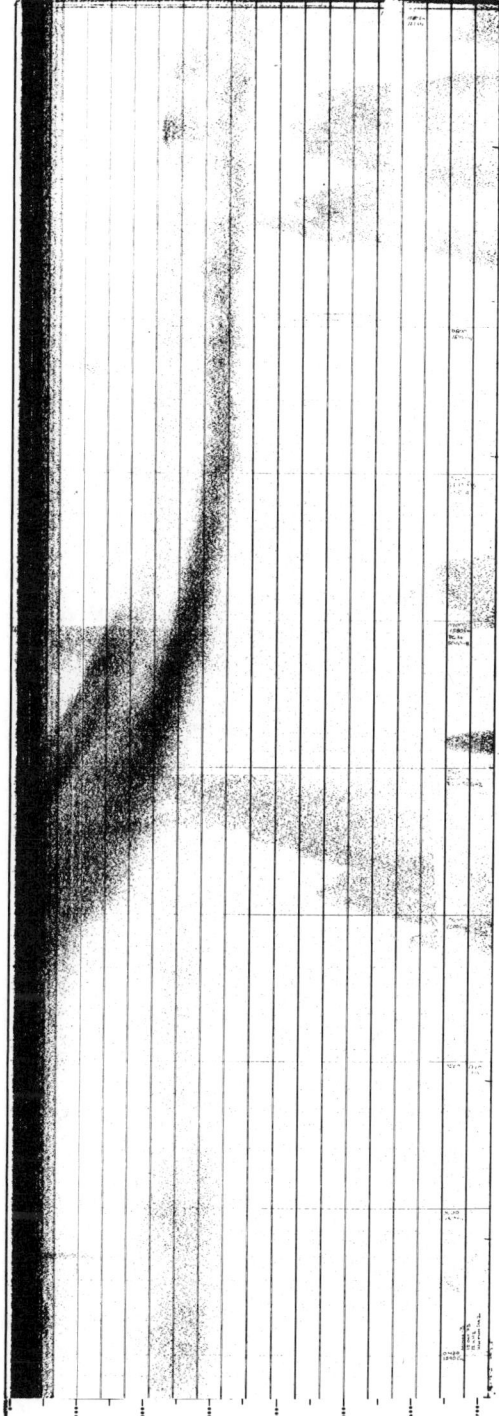

Fig. 9. 12 kHz echosounder record from the Minimal Oxygen region, ca. 0430 - 0910, 15 October 1973, same type of record as Figure 5. Downward migration from the SSL is evident in several distinct layers. The first layer to leave the SSL persists at depth, though the record lightens considerably as the layer attains greater depth; the second layer to leave the SSL fades from the record when it reaches about 200 m. A MML is evident on the record as a faint layer between 200 and 300 m before the downward migration obscures it.

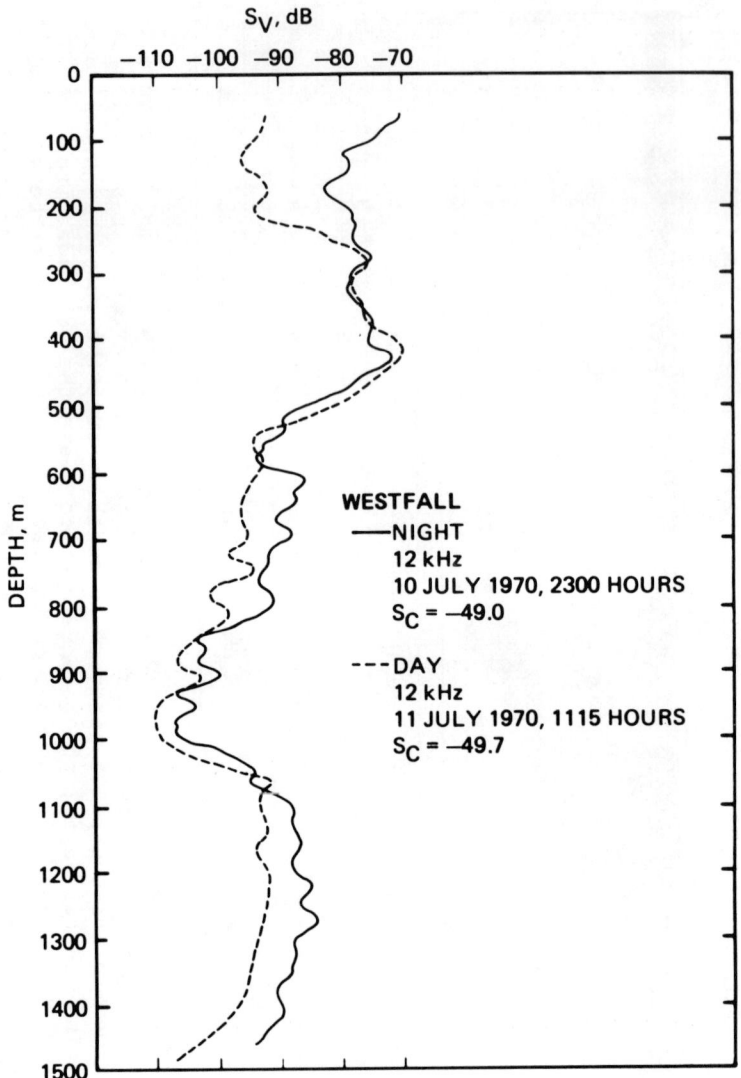

Fig. 10. 12 kHz CW volume scattering strength profiles (S_V in dB re yd^{-1}) from day and night measurements at Westfall Sea Station during MINOX-1.

Qualitative acoustical measurements from the Minimal Oxygen region were characterized by (1) pronounced peaks in the scattering strength profiles day and night; (2) considerable diel change in the depths at which peaks of scattering occurred; (3) widely different values of S_V, night versus day, at depths around 100 and 400 m; and (4) rather high column strength values, with night values in excess

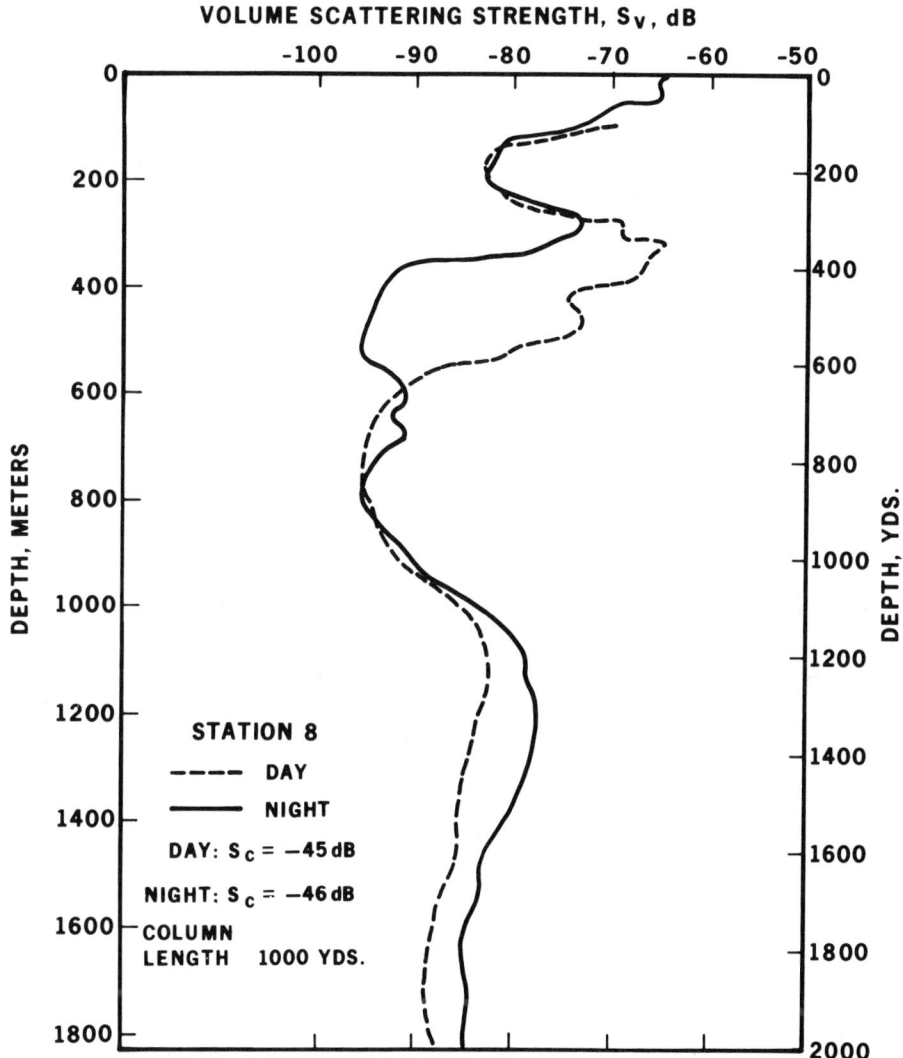

Fig. 11. 12 kHz CW volume scattering strength profiles (S_v in dB re yd^{-1}) from day and night measurements at Station 8 in the Intermediate region during MINOX-3 (from *Vent and Pickwell*, this symposium).

of those measured diurnally (Figure 12). The MML, when present, was identifiable from acoustical records as a scattering peak, distinct from the SSL, and generally in the interval between 200 and 300 m.

Sampling effort and haul distribution, by station and diel period, are summarized for the MINOX cruises in Table 1. The unequal effort, day versus night, reflects a scientific work schedule, on station,

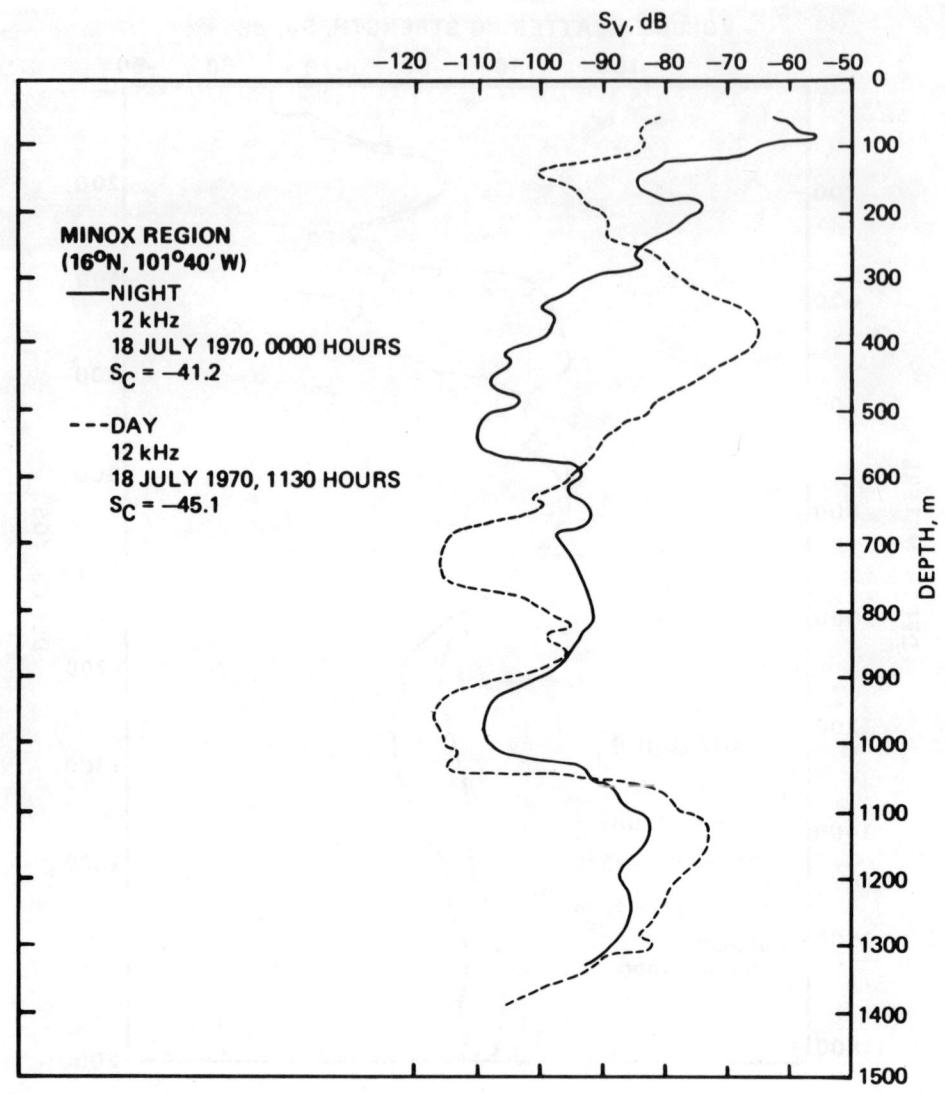

Fig. 12. 12 kHz CW volume scattering strength profiles (S_V in dB re yd^{-1}) from day and night measurements at the Northern O_2 Minimu in the Minimal Oxygen region. Profiles were made from data taken on MINOX-1.

that provided more time at night for biological sampling and a generally reduced total hauling interval, per haul, nocturnally because more hauls were made above 200 m at night and the time required no lower and retrieve the trawl, before and after sampling, was accordingly reduced.

Table 1. Tucker Trawl Summary from the MINOX cruises.

STATION DESIGNATION	LOCATION	CRUISES OCCUPIED	NUMBER OF HAULS		SAMPLING DURATION (Min.)	
			DAY	NIGHT	DAY	NIGHT
[California Current Region]						
San Diego Trough	32°25'N, 117°35'W	1[a]	0	1	—	67
Westfall Sea Station	30° N, 120° W	3	4	7	273	278
[Intermediate Region]						
Cabo San Lucas	23° N, 112°30'W	1[a]	1	2	64	36
Subtropical Front	22° N, 111°06'W	1[b]	3	7	202	408
[Minimal Oxygen Region]						
Cabo Corrientes	20°35'N, 106° W	2[c]	4	6	242	278
Is. Revillagigedo	19°08'N, 111° W	1[a]	1	2	65	97
(MINOX Quartet of Stations)						
Mid-America Trench	17°53'N, 103°54'W	3	6	14	378	758
Northern O$_2$ Minimum	16° N, 107° W	3	3	11	170	442
Trench Edge	16° N, 101°40'W	3	5	7	255	258
Southern O$_2$ Minimum	13° N, 102° W	3	4	13	214	492

[a]Occupied only during MINOX-1
[b]Occupied only during MINOX-3
[c]Not occupied during MINOX-2

Certain results from biological analyses and identifications from MINOX cruise samples have been utilized and are available elsewhere (for example, *Batzler et al.*, 1975; *Valkirs*, 1972, MS; *Douglas et al.*, 1976; and *Vent and Pickwell*, 1977). From these and other analyses and identifications, certain details have emerged about the general biological patterns and characteristics of the biological populations sampled during the MINOX cruises. These characterizations reflect the sampling biases of the Tucker trawl toward capture of small fishes and macroplankton typical of oceanic scattering layers (see *Batzler et al.*, 1975, for discussion). Although the collections are not complete samples of the biological assemblages encountered, they relate well both to the chemical and acoustical measurements made during the cruises and to occurrences and distributions of organisms given in other pertinent works (e.g., *Ebeling*, 1962; *Ebeling and Weed*, 1963; *Berry and Perkins*, 1965; *Rass*, 1967; *Robison*, 1972; and *Brewer*, 1973).

The major biological characteristics of the three oceanographic regions visited during the MINOX cruises are given in Table 2 and were, for the most part, seasonally consistent. The samples from Westfall Sea Station, in the California Current, were diverse and the numbers of species varied markedly, even within a given sampling series during a cruise. Overall, however, the populations seemed to segregate by migratory pattern and depth, so the same species, in varying numbers, appeared in hauls from similar depth strata on different cruises. Hauls in the scattering layers at Westfall typically produced large, diverse catches.

In the Intermediate region, catches had somewhat less variety of types than was typical at Westfall and hauls made above 200 m were marked by the occurrence of the pelagic stages of the Galatheid crab, *Pleuroncodes planipes*. In addition to the species listed in Table 2, other fishes, such as *Triphoturus mexicanus*, *Diaphus pacificus*, *Scopelogadus mizolepis bispinosus* and *Bregmaceros sp.* were taken in this region, but in too few numbers or hauls to establish an associative pattern for them. Generally, this region had a migratory fauna distinct from that of the other regions, while the non- or partially-migratory forms were similar to those at Westfall.

In the Minimal Oxygen region, virtually all the fishes were migratory and the "Other Scattering Layers" of Table 2 are, specifically, the MML discussed earlier. Hauls in the diurnal DSL in this region typically took many *Diaphus pacificus* and *Vinciguerria lucetia*. *Brotuloides emmalas* and Bregmacerotidae were typical of the layer in this region as well. Nocturnal hauls, taken below the SSL and in the absence of a MML, typically caught small squid and a few melamphaids. Against a general background of high numbers of *Vinciguerria lucetia* and *Diaphus pacificus*, hauls took a variety of other species, whose occasional and inconsistent occurrence suggested a patchy, contagious distribution for the forms. Furthermore, members of the gonostomatid genus *Cyclothone* were absent from hauls taken above 800 in the Minimal Oxygen region during the MINOX cruises.

Table 2. Biological Characteristics of the Oceanographic Regions Sampled During the MINOX Cruises. (Regions and Stations are Shown in Figure 2. Most of the Data are from Tucker Trawl Samples.)

REGION	CATCH		
	1. Fishes		2. Non-Fish
	a. From Main Scattering Layers	b. From Other Scattering Layers	
California Current	Fam. MYCTOPHIDAE *Triphoturus mexicanus* *Ceratoscopelus townsendi* *Myctophum nitidulum* *Symbolophorous californiense* Fam. GONOSTOMATIDAE *Vinciguerria lucetia* Fam. STERNOPTYCHIDAE	*Lampanyctus ritteri* *Danaphos oculatus* *Cyclothone spp.* *Argyropelicus pacificus*	Euphausiids abundant in numbers and types; other Crustacea also typical. Considerable amount of mucus-like residue and gelatinous remnants.
Intermediate Region	Fam. MYCTOPHIDEA *Gonichthys tenuiculus* *Diogenichthys laternatus* *Lampanyctus idostigma* Fam. GONOSTOMATIDAE *Vinciguerria lucetia* Fam. STERNOPTYCHIDAE	 *Cyclothone spp.* *Argyropelicus*	Euphausiids and Decapod Crustacea, especially *Pleuroncodes planipes* in hauls above 200 m. Mucus-like residue moderate to light.
Minimal Oxygen	Fam. MYCTOPHIDAE *Diaphus pacificus* (very abundant) *Lampanyctus parvicauda* *Myctophum aurolaternatum* *Symbolophorous evermanni* Fam. GONOSTOMATIDAE *Vinciguerria lucetia* (very abundant) *Diplophos spp.* Fam. OPHIDIIDAE *Brotuloides emmelas* Fam. MELAMPHAIDAE Fam. BREGMACEROTIDAE	 *Scopelogadus mizolepis bispinosus* *Bregmaceros spp. (atlanticus ?)*	Small squids, euphausiids and copepods: numbers quite variable haul-to-haul. *Nautilus* and *Halobates* in nocturnal surface-skimmer* hauls. Hauls largely free of mucus-like residue.

*Night-time samples were taken within 0.5 m of the sea surface with a surface-skimming net with a mouth 100 cm wide by 30 cm deep and a body made of 0.5 mm mesh netting. Surface Skimmer samples were taken for 30 min intervals during Tucker trawl hauls.

Various disbributional gradations are evident from the MINOX samples. For instance, *Vinciguerria lucetia* was taken in each region, but its numbers were consistently high in hauls in the Minimal Oxygen region and more variable at Westfall. Whereas Westfall had a distinctive, depth-related distribution of species, the fishes of the other regions were, more typically, migratory. In addition, the Minimal Oxygen region was distinguished by the presence of *Brotuloides emmelas*, and the absence of *Cyclothone* above 800 m. These characterizations of the MINOX midwater samples are but general examples of the information that is forthcoming as analyses and identifications continue.

DISCUSSION

The MINOX program was conceived and initiated at NUC with broad objectives to relate environmental, biological, and acoustical characteristics of physically-defined oceanic regions. Partial results from three cruises in the eastern north Pacific have revealed a complex system of scattering layers, chemical conditions, biological populations, and acoustical characteristics whose precise causal relationships are, as yet, indistinct.

Central to the MINOX effort are the acoustical measurements. The chemical oceanographic aspects of the program reflect physical/environmental chemistry more than biological/environmental chemistry. Extensive evaluation of biologically-important substances in the water and measurements of primary production were excluded because of limited shipboard space. Similarly, the biological effort was directed toward organisms with a high probability of being causes of oceanic volume reverberation; the smaller or Secondary plankton of *Seligman and Friedl* (1975) were not extensively studied. Thus, before the first MINOX cruise was fielded, the broad-based scheme "investigate the ocean", in which many potential influential factors are evaluated (Figure 13a), became a more practical concept involving three related aspects (Figure 13b). As a result of field experience, the program ultimately adopted a two-level structure, in which the biological causes and acoustical effects of volume scattering are considered together and environmental information is used to explain potential constraints that may act on the organisms to produce the biological conditions manifest in the acoustical observations (Figure 13c). This approach is expanded upon by *Vent and Pickwell* (1977). The emphasis on acoustical results reflects the experiences of the personnel and the likely uses of the study's final results.

The MINOX data are abundant and the available results have revealed a complex and dynamic system of biologically-caused acoustic scattering in each region, but even the processes that have been identified as operative within the systems and regions are, themselves, yet poorly understood.

The complexity of the scattering processes considered here are illustrated by the MINOX migratory layers (MML) encountered in the

THE MINOX PROGRAM

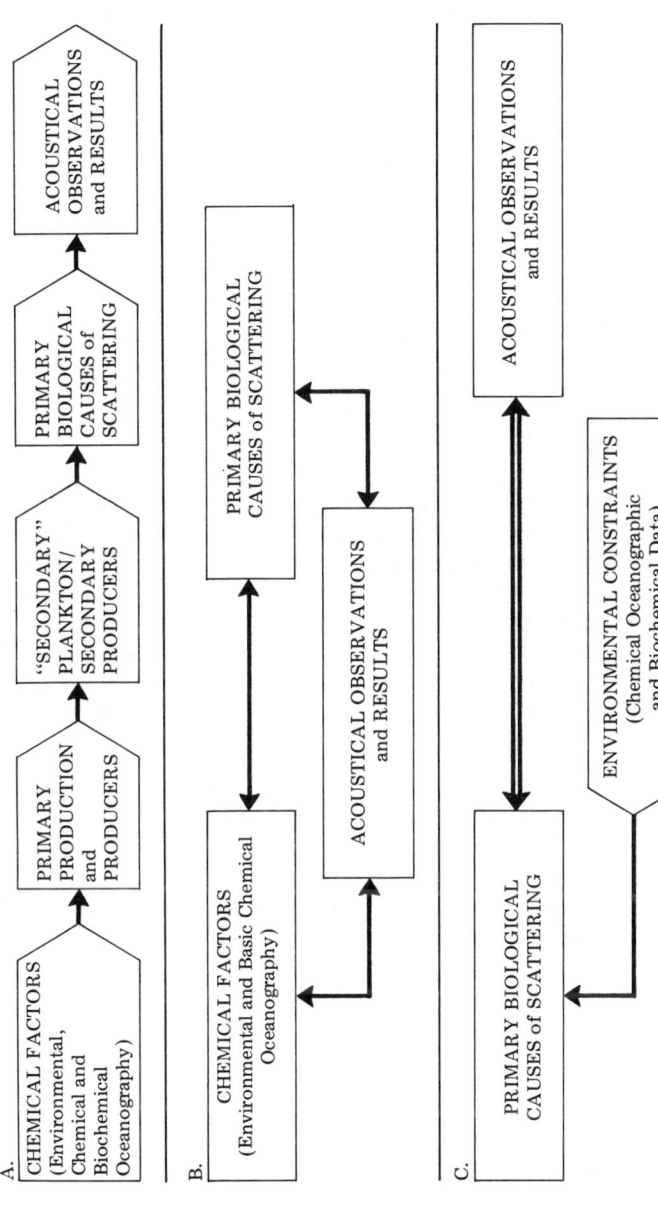

Figure 13. Format of schemes for investigation of the multifaceted problem of biologically-caused volume reverberation, as addressed by the MINOX program. A. A broad-based, linear concept ("investigate the ocean") in which many potential influential factors are considered. B. The initial concept of the MINOX program, with three major, interrelated aspects, each closely related. C. Two-level scheme that has evolved from the three major MINOX cruises. Acoustical conditions are directly related to their probable biological causes and environmental conditions are indirectly related to their probable biological causes and environmental conditions are indirectly related as constraints that act upon the organisms that produce the acoustical effects.

Minimal Oxygen region during MINOX cruises 1 and 3. The layers' occurrence was unexpected and, although cruise plans were modified to include sampling and measurement of the MML, strict ship operating schedules prevented extensive investigation of it during the initial cruise. From limited observations, however, it was evident (1) that the MML did not migrate at the same time each night, (2) that the manifestation of the MML on 12 kHz echosounder records varied greatly, one night to the next, and (3) that the MML could be distinct on acoustical records when it was absent on echosounder records. Paradoxically, the MML was not observed during MINOX-2. During MINOX-3, the MML was again observed and more extensively sampled, thus the causes of the layer are better defined, though still incompletely understood. The foregoing illustrates the complexity of volume scattering in the ocean, even in a geographical area chosen for study because of its supposed simplicity. The mechanisms that produce the MML are suggested from data in hand, but critical experiments and measurements are still required to confirm hypotheses about the MML. As a result, the physical, chemical, biological and acoustical correlations that are the basis of understanding the magnitude and occurrence of volume scattering in the Minimal Oxygen region remain incomplete.

Mechanistic description and prediction underlie all scientific endeavor, but resource limitations prevent state-of-the-art evaluation of all aspects of a phenomenon as multifaceted and complex as biologically-caused volume reverberation. Decisions must be made regarding resource allocation to achieve desired information; these decisions, of necessity, involve the intended use of the information so obtained. Biologically-caused volume reverberation may seriously degrade the performance of acoustic devices in the ocean (Urick, 1967), but the kind of information required to minimize the degradation is often far more specific than can be provided by syntheses or models utilizing chemical or biological oceanographic input, that has too much variability to be of use in applications of specific devices. Multidisciplinary oceanic investigations, such as MINOX, are necessary to contribute to the basic understanding of mechanisms active in the ocean, but complete understanding of volume scattering and its environmental relationships will be gained only slowly and at great effort and expense.

Acoustical theory of volume reverberation is well-advanced (Vent et al., 1974) but critical experiments relating acoustical theory to organisms and organisms to their environment remain to be done. Because the physical properties of organisms determine their acoustical effects, two species, with similar physical/acoustical properties, can affect acoustical measurements similarly and thus remove biological specificity from scattering measurements. Thus, if a project seeks to evaluate the impact of biologically-caused scattering on acoustic devices, effort should be concentrated on making acoustical measurements directly, and biological and chemical information should be used only to indicate the probable causes of the scattering. As long as the relationship between specific organisms and scattering conditions in particular regions remains obscure, this approach will be favored, for it deals directly with the environmental problem posed to acoustical systems and their operators, namely: what conditions are present and how will they affect the systems' operation.

On the other hand, there are the real and important scientific questions relating to the causes of the conditions manifest in the acoustical measurements of volume scattering. Knowledge of the basic structure of the populations of the sea and the role played by various organisms, underlies an understanding of the complex relationships of the environment, plants and animals that produce given acoustical conditions. Such studies require long-term commitments of resources because goals for such an effort are nebulous, relevant data are scarce and the processes involved are obscure; the approach is more general than specific.

CONCLUSIONS

The scientific results of the MINOX project have revealed aspects of volume reverberation, scattering layer populations and chemical oceanography from three physically-defined oceanic regions, but the relationships between the specific acoustical conditions and physical/chemical or biological observations in the regions are yet poorly understood. Without such understanding, many of the fundamental objectives of the MINOX program will remain unfulfilled.

In a general sense, the MINOX results emphasize the tenuous connection between systems-relevant ocean acoustics and environmental chemistry, primary production, zooplankton biomass and similar oceanographic characteristics. Certain biological properties of particular water types are distinctive, readily identified and may serve as qualitative indicators of certain acoustical characteristics. However, detailed description of causal relationships and evaluation of their effect on specific acoustical systems for any oceanic region will require long-term commitments of personnel, equipment and resources.

ACKNOWLEDGMENTS

We thank the scientists and shipboard personnel, who helped us field and execute three MINOX cruises, for their assistance, efforts and enthusiasm. We also gratefully acknowledge the technical review of this paper at NUC by W. E. Batzler and P. F. Seligman. Timely identification of *Brotuloides emmelas* was kindly provided by Daniel M. Cohen, Laboratory Director of the National Marine Fisheries Service Systematics Laboratory, Washington, D. C.

REFERENCES

Adams, R. L. 1967. Bio-acoustic scattering measurements. *Navy Elec. Lab. Deep Submergence Log* (3): 53-56.

Barham, E. G. 1963. Siphonophores and the deep scattering layer. *Science* 140(3568): 826-828.

Barham, E. G. 1966. Deep scattering layer migration and composition: Observations from a diving saucer. *Science* 151 (3716): 1399-1403.

Barham, E. G., and I. E. Davies. 1966. Bio-acoustics. *Navy Elec. Lab. Deep Submergence Log* (1): 31-38:

Barham, E. G., I. E. Davies, and J. W. Wilton. 1966. Bio-acoustics. *Navy Elec. Lab. Deep Submergence Log* (2): 15-26.

Barham, E. C., G. V. Pickwell, and I. E. Davies. 1968. Deepstar bio-scattering program. *Naval Undersea Warfare Cent. Deepstar Log* (4): 13-14.

Batzler, W. E., and E. C. Westerfield. 1953. Sonar studies of the deep scattering layer in the north Pacific. *Navy Elec. Lab. Rep. 334.* 22 pp.

Batzler, W. E., and R. J. Vent. 1967. Volume-scattering measurements at 12 kc/sec in the western Pacific. *J. Acoust. Soc. Amer.* 41(1): 154-157.

Batzler, W. E., J. W. Reese, and W. A. Friedl. 1975. Acoustic volume scattering: Its dependence on frequency and biological scatterers. *Nav. Undersea Cent. Tech. Publ. TP 442.* 25 pp.

Berry, F. H., and H. C. Perkins. 1965. Survey of pelagic fishes of the California Current area. *Fish. Bull.* 65(3): 625-682.

Boden, B. P. 1950. Plankton organisms in the deep scattering layer. *Navy Elec. Lab. Rep. 186, Prob. 2A5.* 29 pp.

Brewer, G. D. 1973. Midwater fishes from the Gulf of California and the adjacent eastern tropical Pacific. *Contr. in Sci., Nat. Hist. Mus., Los Angeles County.* No. 242. 47 pp.

Carpenter, J. H. 1965. The accuracy of the Winkler method for dissolved oxygen analysis. *Limnol. Oceanogr.* 10(1): 135-140.

Cline, J. D., and F. A. Richards. 1972. Oxygen deficient conditions and nitrate reduction in the eastern tropical north Pacific Ocean. *Limnol. Oceanogr.* 17(6): 885-900.

Davies, I. E., and E. G. Barham. 1969. The Tucker opening-closing micronekton net and its performance in a study of the deep scattering layer. *Mar. Biol.* 2(2): 127-131.

Dobrovol'skiy, A. D. (Editor). 1968. Hydrology of the Pacific Ocean, Part One. In V. G. Kort (Editor), *The Pacific Ocean, Vol. 2.* Nauka, Moscow (In Russian). Translation: *Nav. Oceanogr. Off. Contract Transl.* 35. 430 pp. 1969.

Douglas, E. L., W. A. Friedl, and G. V. Pickwell. 1976. Fishes in oxygen minimum zones: Blood oxygenation characteristics. *Science* 191(4230): 957-959.

Ebeling, A. W. 1962. Melamphaidae I. Systematics and zoogeography of the species in the bathypelagic fish genus *Melamphaes* Gunther. *Dana Rep. No. 58, 1962.* 145 pp.

Ebeling, A. W., and Walter H. Weed III. 1963. Melamphaidae III. Systematics and distribution of the species in the bathypelagic fish genus *Scopelogadus* Vaillant. *Dana Rep. No. 60, 1963.* 58 pp.

Farguhar, G. B. (Editor). 1970. Proceedings of an international symposium on biological sound scattering in the ocean. *MC Rep. 005.* U. S. Gov. Printing Off. 629 pp.

Hersey, J. B., and R. H. Backus. 1962. Sound scattering by marine organisms, p. 498-539. In M. N. Hill (Editor), *The Sea, Vol. 1.* Wiley, New York. 864 pp.

McGowan, J. A. 1971. Oceanic biogeography of the Pacific, p. 3-74. In B. M. Funnell and W. R. Riedel (Editors). *The Micropalaeontology of Oceans.* Cambridge Univ. Press. 828 pp.

Pickwell, G. V., R. J. Vent, E. G. Barham, W. E. Batzler, and I. E. Davies. 1970. Biological acoustic scattering off southern California, Baja California and Guadalupe Island, p. 490-507. In G. B. Farquhar (Editor). *Proc. Internat. Symp. Biol. Sound Scattering in the Ocean.* MC Rep. 005. U. S. Gov. Printing Off. 629 pp.

Rass, T. S. (Editor). 1967. The fishes of the open waters of the Pacific Ocean. In V. G. Kort (Editor). *The Pacific Ocean, Vol. 7, Part 3.* Nauka, Moscow. 273 pp. (In Russian). Translation: *Nav. Oceanogr. Off. Transl. 528.* 320 pp. 1971.

Robison, B. H. 1972. Distribution and ecology of midwater fishes of the eastern north Pacific Ocean. *Ph.D. Dissertation,* Stanford Univ. 175 pp.

Seligman, P. F. 1974. The nature of pheo-pigment and chlorophyll distribution and degradation in the California Current and northeastern tropical Pacific. *M. S. Thesis,* San Diego State University. 156 pp.

Seligman, P. F., and W. A. Friedl. 1975. FASOR II: Correlative biological and acoustical studies in the north Pacific Ocean. *Nav. Undersea Cent. Tech. Publ. TP 448.* 63 pp plus appendices.

Sverdrup. H. U., M. W. Johnson, and R. H. Fleming. 1942. The Oceans. Prentice-Hall, New York. 1087 pp.

Tucker, G. H. 1951. Relation of fishes and other organisms to the scattering of underwater sound. *J. Mar. Res. 10(2):* 215-238.

Urick. R. J. 1967. Principles of Underwater Sound for Engineers. pp. 187-233. McGraw-Hill, New York. 342 pp.

Valkirs, A. O. 1972 MS. Swimbladder gas composition of mesopelagic fishes found in low oxygen waters. 15 pp. plus appendices.

Vent, R. J. 1972. Acoustic volume-scattering measurements at 3.5, 5.0 and 12.0 kHz in the eastern Pacific Ocean: Diurnal and seasonal variations. *J. Acoust. Soc. Amer. 52(1) Part 2:* 373-382.

Vent, R. J., and G. V. Pickwell. 1977. Biological and chemical correlatives of acoustic volume scattering observations in the eastern Pacific. In: *"Oceanic Sound Scattering Prediction",*

N. R. Andersen and B. J. Zahuranec (Eds.), Plenum, N.Y., (this volume). (Available in *Nav. Undersea Cent. Tech. Publ. TP 525*, 24 pp., July, 1976.)

Vent, R. J., with W. A. Friedl, W. E. Batzler, and I. E. Davies. 1974. Classification of submerged targets I. Biological sound scattering. *50th Meeting, Undersea Warfare Res. Dev. Council*, Lake Pend Oreille, Bayview, Idaho. 1-3 July 1974. 56 pp.

Extracting Bio-Physical Information from the Acoustic Signatures of Marine Organisms

D. V. Holliday

Tracor, Inc.

ABSTRACT

A formal mathematical procedure is introduced for extracting bio-physical information from the acoustic signatures of layers, aggregations or schools of marine organisms. Several possible generalizations and limitations of the technique are briefly discussed.

INTRODUCTION

An increasing interest is apparent in relating measurements of the acoustic signatures of echoes from a variety of marine organisms to the specific character of the organisms under surveillance. Much of the current interest is in predicting the levels and character of acoustic volume reverberation when the composition and character of the ensonified population is known. The natural result of this interest has been that most published comparisons of biological and acoustic data are based on computation of a predicted acoustic signature, given biological data on species, population mix and the sizes of the organisms under study.

This paper addresses a formal approach to the inverse problem. In this approach, one attempts to develop from the acoustic signature, the character and size distribution of the acoustic scatterers, and then to make a direct comparison of this result with the biological sample or known population in the water column under investigation. By solving this problem, we hope to contribute to the understanding of the basic phenomena. This should aid in the successful achievement of the original goal, *i.e.*, prediction of reverberation levels given a knowledge of local biological populations.

It should be recognized at the beginning, that the end result of the analysis presented here only represents a starting point in the solution of this complex problem. It does, however, provide a formal mathematical basis for future investigations. Perhaps of even more value, it points directly to several key factors which are either missing or which require improvement if a goal of quantitative transfer of the results of acoustic studies into the biological world is to be realized. The desirability of achieving this goal is readily evaluated by comparing the cost and time now necessary to sample a given volume of water acoustically with the time and effort needed to sample it directly with nets.

MATHEMATICAL FORMULATION

We begin by presuming prior acquisition, at M discrete frequencies, of the acoustic cross section for a group of marine organisms near a particular depth in the ocean. This measurement of the cross section (or a proportional quantity) may be made either one frequency at a time with tone bursts, or by narrowband spectrum analysis of broadband echoes.

Let us denote a set of measured cross sections by the column vector S whose elements are S_i. The subscript i indexes the acoustic measurement frequency. Presume that the acoustic cross section for each individual organism in the sample volume near the depth of interest is characterized by a parameter R. In the case of organisms containing swimbladders, the parameter might be a dimension of the swimbladder. In the case of other organisms, the parameter may denote size or some algebraic combination of bio-physical measures which characterize the individual's back scattering cross section. In either case we will quantize the parameter over an arbitrary but physically realistic range by dividing the range into N intervals. We label each interval with the index j, and let the acoustic cross section of one organism with parameter R_j at the frequency ω_i be σ_{ij}. If we let the number of organisms which fall into each interval of the parameter R be denoted by the column vector, n_j, we can write

$$c_i = \sum_{j=1}^{N} \sigma_{ij} \, n_j . \qquad (1)$$

Here we have assumed negligible multiple scattering, no shadowing and simple incoherent summation of the acoustic power in individual echoes. The term c_i represents the theoretical or predicted acoustic cross section for the aggregation or school of organisms under study. The elements of this column vector are indexed by the subscript i, which we associate with each of M measurement frequencies.

BIO-PHYSICAL INFORMATION   621

At this point the usual procedure is to compute C_i (the calculated cross-sections) and compare it with S_i (the measured cross-sections). While this often provides useful information, the results frequently infer an inadequate knowledge of the elements of the column vector n_j. This inference is generally, however, qualitative [Andreeva, 1964; Batzler, et al., 1973; Holliday, 1972; Love, 1975; MacDonald, 1972]. Our aim is to develop an approach to a more quantitative assessment of observed differences between measured and calculated acoustic backscattering. To achieve this end, we form the quantity

$$\sum_{i=1}^{M} (S_i - C_i)^2 \ . \tag{2}$$

Following the standard least squares approach, we choose to minimize the sum of the squares of the difference between the measured and calculated acoustic cross sections at the frequencies for which we have measured values. A condition introduced at this point is that the number of measurement frequencies, M, is equal to or greater than the number of intervals, N, which we have chosen over the range of R.

Our objective is to make a least squares estimate of the column vector n_j. For that reason we wish to vary n_j until the least squares condition shown in Equation (3) is met.

$$\frac{\partial}{\partial n_k} \left[\sum_{i=1}^{M} \left(S_i - C_i \right)^2 \right] = 0 \tag{3}$$

Standard tests can be used to assure a minimum in the sum of the indicated squares.

If we now substitute from Equation (1) into Equation (3) and expand the squared term under the summation on the index i, we can write Equation (4).

$$\frac{\partial}{\partial n_k}\left\{\sum_{i=1}^{M}\left(S_i^2 - S_i \sum_{j=1}^{N}\sigma_{ij}n_j - S_i \sum_{p=1}^{N}\sigma_{ip}n_p\right.\right.$$

$$\left.\left. + \sum_{j=1}^{N}\sigma_{ij}n_j\sum_{p=1}^{N}\sigma_{ip}n_p\right)\right\} = 0 \qquad (4)$$

Taking the indicated derivative and performing a minor algebraic manipulation yields the N equations in N unknowns represented by Equation (5).

$$\sum_{i=1}^{M}\sigma_{ik}S_i = \sum_{i=1}^{M}\sigma_{ik}\sum_{j=1}^{N}\sigma_{ij}n_j \qquad (5)$$

By observing the order of summation on the indices of the elements of each of the individual cross section terms, σ, we can rewrite Equation (5) in matrix notation as follows.

$$\sigma^T S = \sigma^T \sigma n \qquad (6)$$

Here, σ^T is the transpose of the matrix σ. The symbols S and n represent column vectors. The solution to Equation (6) for the vector n can now be written

$$n = (\sigma^T \sigma)^{-1} \sigma^T S \qquad (7)$$

The quantity $(\sigma^T \sigma)^{-1}$ is, of course, the inverse of the matrix $(\sigma^T \sigma)$.

DISCUSSION

This formal solution to the transfer of acoustic information into bio-physical properties is only a first step. As stated in the preceding section, the elements of the matrix σ are restricted to a functional dependence on one parameter, R. Further, we allow only one functional form for the backscattering cross section, e. g., we cannot mix organisms with and without gas-filled swimbladders. These restrictions can be removed by reformulating our original model as in Equation (8).

$$\sum_{i=1}^{M} \left(s_i - \sum_{j=1}^{N} \sigma_{ij}\, n_j - \sum_{p=1}^{N'} \sigma'_{ip}\, n'_p \right) \qquad (8)$$

In this formulation σ and σ' represent two different forms of acoustic backscattering, e.g., swimbladder dominated scattering and scattering from organisms without swimbladders. The vectors n and n' represent the analogous densities within intervals on different biophysical parameters R and R'. Standard techniques for obtaining a least squares fit in a multi-parameter problem will yield a set of simultaneous equations in n and n' which may then be solved to achieve a best fit to the experimental data originally hypothesized. Further similar complications of the model of the biological population may be made and solutions obtained in the same manner at the expense of a rapidly increasing computational problem.

Under most conditions noise will be present in the measurement of the group acoustic cross section as a function of frequency. If the signal-to-noise ratio is not sufficiently large, it may be necessary to constrain the solution for the density vectors such that each $n_k \geq 0$. This constraint arises from obvious physical considerations and the necessity for its use will depend on a number of factors including the signal-to-noise ratio in the measured data, the intervals chosen for quantizing the parameter R, the measurement frequencies used, the dimensions of the matrices inverted, the computational procedure used to obtain the solutions of the simultaneous equations and the number of significant figures carried in the calculations.

The validity and completeness of the mathematical model used to describe the cross section of the individuals in the population under study will also impact the necessity for constraining the solution. It is almost certain that the available models will be inadequate in many cases of interest. It is equally certain that we must consider stochastic models as well as deterministic models in future work. The development of new approaches, both theoretical and experimental, to modeling acoustic backscattering from discrete entities over a wide range of frequencies appears to be a fruitful area for further investigation.

We would end the discussion on a positive note. The form of Equation (7) is such that the matrix coefficient of the measurement vector is independent of the measurement. Thus the quantity

$$(\sigma^T \sigma)^{-1} \sigma^T \qquad (9)$$

can be computed once, stored and applied to each separate measurement of the acoustic signature. For populations whose backscattering varies with some physical parameter such as depth, it may be necessary to calculate a limited table of the matrices indicated in Equation (9). Still, in a potential at-sea application, the major burden of matrix manipulation could be placed on a land-based computer before departure, the coefficient matrices stored on a suitable medium and the relatively simple final multiplication done in a small machine at sea.

SUMMARY

This paper has introduced a formal procedure for extracting bio-physical information from the acoustic signatures of aggregations, layers, schools or other groups of marine organisms. The purpose of the paper, beyond the formal introduction of the mathematical procedure, has been to point out a need for added research in modeling the individual acoustic response of a variety of classes of marine organisms. Without adequate, validated mathematical models, whether empirical, analytical, deterministic or stochastic, both the state-of-the-art in predicting reverberation from the biology and the inverse problem will be severely constrained.

REFERENCES

Andreeva, I. B. July-September 1964. Scattering of sound by air bladders of fish in deep sound-scattering ocean layers. *Soviet Physics-Acoustics, 10,* (1), 17-20.

Batzler, W. E., W. A. Friedl and J. W. Reese. 1973. Can acoustic volume scattering be predicted from net-haul data? *J. Acoust. Soc. Am.* 54, 290 (Abstract).

Holliday, D. V. 1972. Resonance structure in echoes from schooled pelagic fish. *J. Acoust. Soc. Am.* 51 (4), 1322-1332.

Love, R. H. 1975. Predictions of volume scattering strength from biological trawl data. *J. Acoust. Soc. Am.* 57 (2), 300-306.

MacDonald, R. B. September 20, 1972. A comparison of acoustical and biological backscattering strengths at discrete depths for Ocean Acre Cruises 6 and 10. *NUSC Technical Memorandum TA 131-225-72.*

Spectral Models for Biological Sound Scattering

Richard K. Johnson

Oregon State University

ABSTRACT

The characteristics and underlying assumptions of the commonly used biological scattering models including fluid sphere, gas-filled swimbladder, and fish length are compared and discussed. Gaps exist in our knowledge of the form of the models and the magnitudes of their parameters. Frequency responses are predicted for several biological targets.

INTRODUCTION

In discussing the frequency dependence of scattering from biological targets, the scattering parameter used in this paper is the acoustic backscattering cross section, σ, which is related to the target strength, S_t, by

$$S_t = 10 \log (\sigma/4\pi). \tag{1}$$

The value of σ used in calculating a target strength should be in square meters. The acoustic cross section can be compared with the projected area of the target to show what fraction of the intercepted sound energy is scattered. The projected area for a sphere with radius a is πa^2. Thus a sphere with $\sigma/\pi a^2 = 1$ scatters all of the sound energy that it intercepts.

FLUID SPHERE MODEL

Anderson (1950) has presented the analytical solution for sound scattering from a fluid sphere. Measurements by *Beamish* (1971) on

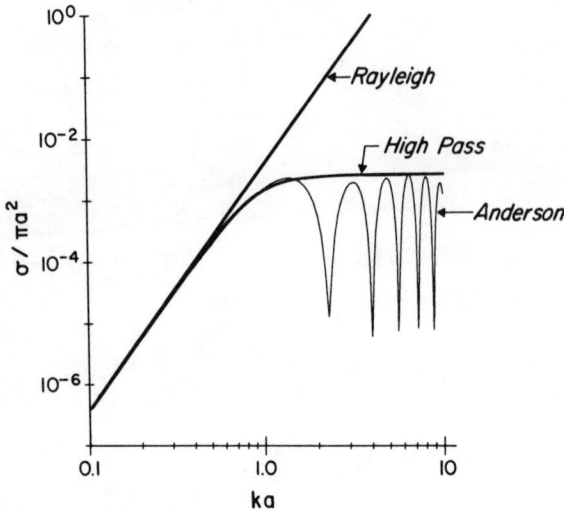

Fig. 1. Calculated response for a fluid sphere with density contrast 1.03 and bulk modulus contrast 1.18 (euphausid values). The *Rayleigh* approximation is within 1dB of the complete solution for $ka \leq 0.6$.

euphausiids and *McNaught* (1969) on cladocerans were reported to be consistent with the back-scattering predictions of the fluid sphere model. The mathematical form of the model and its derivation are available in *Anderson*. Only an example will be given here.

The scattering predicted by the model depends on the density ratio (ρ'/ρ) and the compressibility ratio (κ'/κ) of the target to the medium and on the scale parameter ka (where k is the wave number and a is the target radius). For targets with density and compressibility ratios near 1, the scattering response increases as the fourth power of ka and peaks near $ka = 1.4$, as in Figure 1. In water, $ka =$ when the product of frequency in kilohertz and radius in millimeters is about 240. The maximum level of these curves increases as the density and compressibility contrasts become greater. For biological targets, $\sigma/\pi a^2$ might range from 0.15 to zero. Since the biological targets are not homogeneous nor spherical, it is unlikely that the dips at higher values of ka will occur exactly as predicted for a fluid sphere. Consequently it may be possible to use a simpler model in the form of a high pass filter

$$\sigma/\pi a^2 = 2\alpha(ka)^4 (2 + 3(ka)^4)^{-1} \qquad (2)$$

where

$$\alpha = \left[\frac{4}{9} \frac{\kappa'}{1-\kappa} + \frac{3(\rho'/\rho)}{1 + 2\rho'/\rho} \right] \qquad (3)$$

BUBBLE RESPONSE

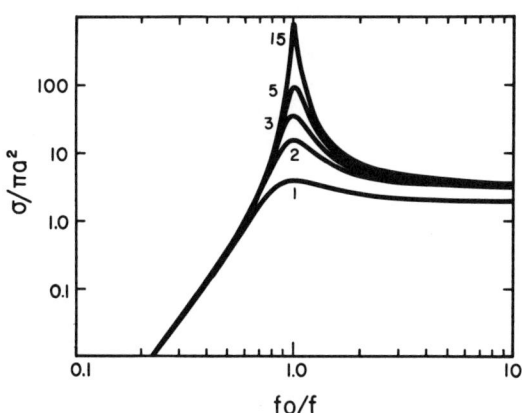

Fig. 2. Frequency response of an air bubble in water with Q as a parameter.

is the *Rayleigh* (1945) coefficient.

Because the scatterers described by the fluid sphere model are small and the frequencies are high, little work has been done to check the predictions. Two things are needed: acoustic measurements over a range of ka values and direct measurements of target compressibility and density.

GAS-FILLED SWIMBLADDER MODEL

The scattering response of a gas bubble in water is described (*Weston*, 1967) by

$$\sigma/\pi a^2 = 4\{[(f_0/f)^2 - 1]^2 + 1/Q\}^{-1}. \qquad (4)$$

This equation is plotted in Figure 2 for several values of the resonance enhancement Q. Predicted values of Q for fish with gas-filled swimbladders range from 3 (*Weston*, 1967) to 10 (*Andreeva*, 1964). There are a few measurements in the range 3 to 5 (*Batzler and Pickwell*, 1970).

According to this expression, $\sigma/\pi a^2$ tends to 4 at high frequencies. A more complete theory shows that $\sigma/\pi a^2$ tends to 1 for frequencies f greater than $240/a$ where f is in kHz and a is in mm. The resonant frequency f_0 is given (*Andreeva*, 1964) by

$$f_0 = (2\pi a)^{-1}\{[3\gamma P + 4\mu_1]/\rho\}^{\frac{1}{2}} \qquad (5)$$

where γ is the ratio of specific heats for the swimbladder gas (1.4), P is the ambient pressure, ($P = (1 + .103D) \cdot 10^6$ dynes/cm^2 with D in

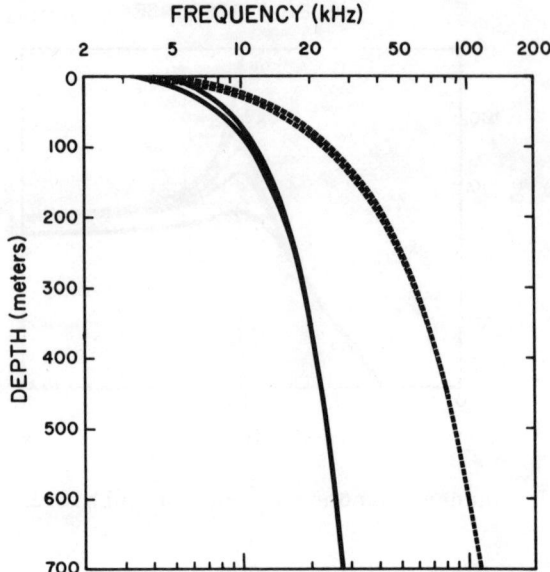

Fig. 3. Depth dependence of the resonant frequency of a 1 mm radius (at the surface) swimbladder for constant volume (-) and constant mass (--). Each curve forks near the surface to illustrate the effect of the real part of the fish tissue shear modulus, μ_1. The left fork is for $\mu_1 = 0$ and the right fork for $\mu_1 = 10^6$.

m), μ_1 is the real part of the shear modulus of fish tissue (10^6 dynes/cm^2), and ρ is the density of sea water (1.026 g/cm^3). Thus the resonant frequency increases with decreasing bladder radius and increasing depth. Figure 3 illustrates the depth dependence of resonant frequency. At the surface the predicted resonant frequency of a 1 mm radius sphere is 4.4 kHz if $\mu_1 = 10^6$ or 3.1 kHz if $\mu_1 = 0$.

The swimbladders of fish are typically not spherical. Weston (1967) has presented a corrected resonant frequency for the case of a prolate spheroid.

$$f_e/f_0 = 2^{\frac{1}{2}} e^{-\frac{1}{3}} \varepsilon^{\frac{1}{2}} [\ln(1 + \varepsilon)/(1 - \varepsilon)]^{\frac{1}{2}} \tag{6}$$

where e is the ratio of the major to the minor axis and ε is the eccentricity $(1 - e^2)^{\frac{1}{2}}$. The frequency f_e is always greater than the resonant frequency for a sphere (f_0) of the same volume, but the effect is not great. For $e = 2$ the increase in frequency is about 2%, for $e = 16$, about 30%.

We do not know whether the fish that migrate allow their swimbladders to expand and contract with pressure (constant mass) or actively regulate for constant volume. It is also not certain that the dependence on μ_1 is correct or even that μ_1 is 10^6 dynes/cm. The

SPECTRAL MODELS

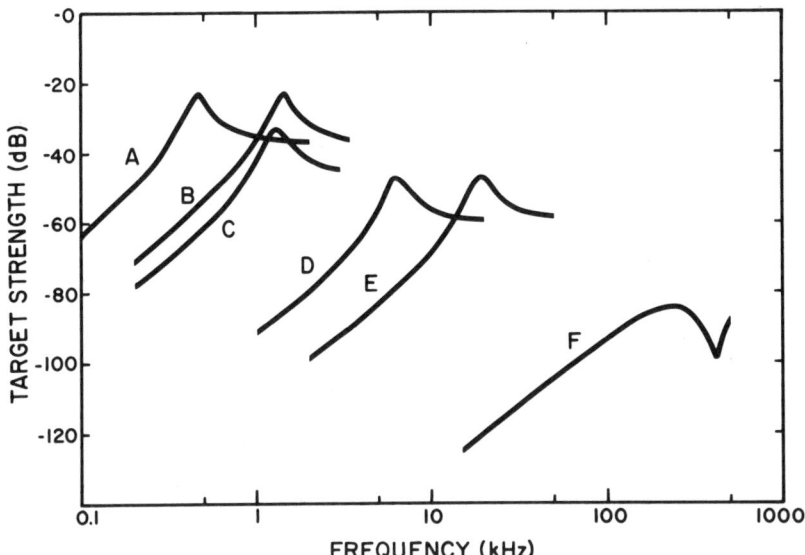

Fig. 4. Predicted target strengths for swimbladders of radius 14 mm (A: hake at 20 m, B: hake at 350 m), 5 mm (C: anchovy at 20 m), 1 mm (D: myctophid at 20 m, E: myctophid at 350 m), and for fluid spheres of radius 1.5 mm (F: euphausiid).

model has, however, been used by *Holliday* (1972) for the identification of fish schools and by *Love* (1975) to predict volume reverberation.

FISH LENGTH MODEL

There is no accepted theoretical model for scattering from the body of a fish. Since fish are typically not spherical and not small compared with the acoustic wavelengths considered here, their scattering is dependent on aspect. *Love* (1969) has analyzed maximum side aspect and dorsal aspect target strength measurements and calculated these regression equations:

Maximum side aspect: $\sigma/\lambda^2 = 0.064 \, (L/\lambda)^{2.28}$ (7)

Dorsal aspect: $\sigma/\lambda^2 = 0.041 \, (L/\lambda)^{1.94}$ (8)

His later work (*Love*, personal communication) indicates that, on the average, target strength for a given fish is independent of frequency in the range $1 \leq L/\lambda \leq 100$ (where L is the fish length). For small L/λ, target strength should increase as $(L/\lambda)^4$. This change in frequency dependence would be useful in distinguishing the scattering of a few large fish from that of many small ones. The total response might be of the form

$$\sigma/L^2 = \beta[\delta + (\lambda/L)^4]^{-1} \tag{9}$$

with $\beta = 0.05$ and $\delta = 1$.

Calculations using this model indicate that the target strength of a fish at large L/λ is about equal to the target strength of its swimbladder above resonance. More measurements are needed at small L/λ for fish both with and without gas-filled swimbladders.

PREDICTED LEVELS

Figure 4 shows predicted scattering for several targets. The scattering strength of a group of targets that are not densely packed should be the sum of the target strengths in the group. Thus, a group of euphausiids would have a volume scattering strength curve with the same shape as the target strength curve F in Figure 4. The difference in level between the two curves would correspond to the population density of the euphausiids. This means that with measurements at appropriate frequencies it is possible to distinguish between a few strong, large scatterers and many weak, small scatterers.

REFERENCES

Anderson, V. C. 1950. Sound scattering from a fluid sphere. *J. Acoust. Soc. Am* 22: 426-431.

Andreeva, I. B. 1964. Scattering of sound by air bladders of fish in deep sound scattering layers. *Akust. Zh.* 10: 20-24.

Batzler, W. E. and G. V. Pickwell. 1970. Resonant acoustic scattering from gas-bladder fishes. In G. B. Farquhar, ed., *Proceedings of an International Symposium on Biological Sound Scattering in the Ocean*: 168-178. U.S. Government Printing Office, Washington, DC.

Beamish, P. 1971. Quantitative measurements of acoustic scattering from zooplanktonic organisms. *Deep-Sea Res.* 18: 811-822.

Holliday, D. V. 1972. Resonance structure in echoes from schooled pelagic fish. *J. Acoust. Soc. Am.* 51: 1322-1332.

Love, R. H. 1969. Maximum side aspect target strength of an individual fish. *J. Acoust. Soc. Am.* 46: 746-752.

Love, R. H. 1975. Predictions of volume scattering strengths from biological trawl data. *J. Acoust. Soc. Am.* 57: 300-306.

McNaught, D. C. 1968. Acoustical determination of zooplankton distributions. *Proc. 11th Conf. Great Lakes Res.*: 76-84.

Rayleigh, Lord. 1945. *Theory of Sound* 2: 277. Dover Publications, New York.

Weston, D. E. 1967. Sound propagation in the presence of bladder fish. In V. E. Albers, ed., *Underwater Acoustics* 2: 55-88. Plenum Press, New York.

Progress in the Correlation of Volume Scattering Strengths to Biological Data

Richard H. Love

Naval Oceanographic Office

ABSTRACT

Progress in correlating volume scattering strengths to biological data is discussed. This includes the development of techniques to obtain scattering strength profiles over broad frequency ranges, a model which utilizes biological data to predict scattering strengths and a theoretical study aimed at improving the predictive model.

INTRODUCTION

The Naval Oceanographic Office (NAVOCEANO) has been investigating the acoustic and biological aspects of volume reverberation in various parts of the world ocean for a number of years. Data collection and analysis techniques have been constantly revised and improved in order to provide improved support to sonar performance prediction studies. Past investigations have shown general agreement between gross biological parameters, such as abundance of swimbladder fishes, and levels of volume reverberation over large ocean areas. Progress in collection and analysis of both acoustic and biological data has led to a much improved ability to correlate volume reverberation to biological data. Correlations of this type can lead to the ability to predict volume reverberation levels from biological data.

This paper presents some of the high-lights of recent progress. These include the development of an improved method for analyzing explosive-source volume reverberation data and a high-frequency volume reverberation measurement system. In addition, a volume reverberation model which utilizes fish distribution data has been developed to permit quantitative predictions of depth dependent scattering strengths. Presently, a theoretical study of resonant scattering from swimbladder fish is being conducted in an attempt to improve this model.

The fundamental quantity of volume reverberation is the backscattering coefficient of a unit volume of ocean, M, which is the ratio of the scattered intensity at a unit distance from the unit volume to the incident intensity (Machlup and Hersey, 1955). M is generally assumed, at least for purposes of calculation, to be only a function of depth, z. The scattering strength of a unit volume of ocean at depth, z, $S_v(z)$, is then defined as

$$S_v(z) = 10 \log M(z). \qquad (1)$$

The integrated scattering strength as a function of depth, $S_c(z)$, is defined as

$$S_c(z) = 10 \log \left[\int_o^z M(z) dz \right]. \qquad (2)$$

The integrated scattering strength of the complete water column, or column scattering strength, S_c, is defined as

$$S_c = 10 \log \left[\int_o^d M(z) dz \right], \qquad (3)$$

where d is the depth of the deepest scatterers.

M can be defined in terms of the scatterers as

$$M = \frac{1}{4\pi V} \sum_{i=1}^{n} \sigma_i \qquad (4)$$

where n is the number of scatterers in volume V and σ_i is the acoustic cross section of an individual scatterer.

TECHNIQUES TO OBTAIN SCATTERING STRENGTH PROFILES OVER BROAD FREQUENCY RANGES

Improved Explosive Source Data Collection and Analysis

Traditionally, scattering strength measurements have most often been made by two techniques. The first technique has essentially utilized a calibrated echo sounder to obtain $S_v(z)$ as a function of depth at a specific frequency. The second technique has utilized an explosive sound source and an omnidirectional hydrophone to obtain S_c over a broad range of frequencies. In order to obtain meaningful correlations between biological and acoustic data, scattering strengths need to be obtained as a function of depth over broad frequency ranges. Only then can acoustic features be positively compared to specific biological features.

Techniques which provide both depth and frequency coverage do exist. One method is to use an array of calibrated echo sounders, each operating at a different frequency. Another method is to develop a broad-band directional receiving array for use with explosive sound sources. For low frequencies, both types of arrays would be large, difficult to handle, and costly to replace. Their use is limited to low sea states unless they are hull-mounted and stabilized.

The technique described herein is an improved technique which processes data collected with an omnidirectional method to provide both depth information and broad frequency coverage at low cost and in moderate sea states, making it competitive under some circumstances with techniques utilizing arrays (*Love*, 1974). The basic data collection technique is not original, a similar technique having been used for several years by the Canadian Defence Research Establishment Atlantic (*Chapman, et al.*, 1970). The method of processing the data to obtain information on depth variations of scattering is new.

Previous measurement techniques employed standard Navy SUS (Signal Underwater Sound), containing an explosive charge of 1.8 pounds of TNT as the sound source. The SUS were detonated at a nominal depth of 18 meters and the acoustic signals received by a broad-band omnidirectional hydrophone which was at the same depth. The detonation produced a shock wave and an oscillating gas globe, or bubble. The pulses produced by the bubble masked any volume reverberation for about 0.4 seconds or about 300 meters range. This 0.4 second signal also caused the depth resolution below 300 meters to be poor. In addition, at low frequencies and high sea states, surface reverberation could predominate over volume reverberation for more than 0.4 seconds. These factors usually made it impossible to infer anything about the depth distribution of the scatterers.

The improved measurement technique employs 0.5 pound blocks of TNT which are electrically detonated at a depth of 0.5 meters. The pulse length is approximately 20 msec (*Gaspin and Shuler*, 1971). The receiver is again a broad-band omnidirectional hydrophone, but at a depth of 9 meters. This experimental configuration eliminates the bubble pulses, since the gas globe vents, and also greatly reduces the effects of surface scattering by reducing the area insonified at high grazing angles.

The received signals are recorded on analog magnetic tape in both techniques. In the laboratory the analog signals are played back through standard 1/3-octave band-pass filters and the filtered signals are displayed as pressure versus time on a logarithmic graphic level recorder. Figure 1 compares typical data collected with the different sources.

Due to its nature, the result produced by the omnidirectional technique is the integrated scattering strength, $S_c(z)$. $S_c(z)$ can be calculated from the following equation (*Marshall and Chapman*, 1964):

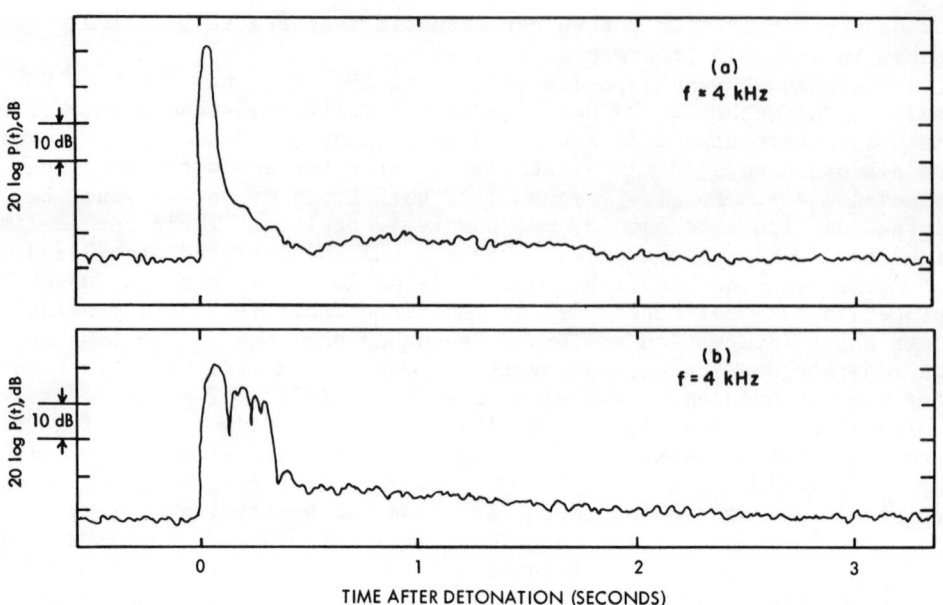

Fig. 1. Typical daytime omnidirectional volume reverberation data obtained employing (a) 0.5 pound blocks of TNT detonated at 0.5 meters and (b) SUS detonated at 18 meters.

$$S_C(z) = 20 \log p(t) + 30 \log t - 10 \log E + \alpha c t - K, \quad (5)$$

where t is the time after detonation, $p(t)$ is the received pressure level in the analysis bandwidth at time t, E is the source energy in the analysis bandwidth, α is an absorption coefficient (*Thorpe*, 1967) c is the speed of sound, and K is a constant which depends on the experimental configuration and the units of the various terms. The limitations in resolution inherent in the technique employing SUS usually make it possible to only obtain the column scattering strength S_C. The technique utilizing surface-venting blocks of TNT eliminates many of the limitations of the older technique and $S_C(z)$ can be determined from the data.

To obtain $S_C(z)$ the analog pressure versus time curves are first digitized utilizing a curve-following digitizer and a digital tape recorder. The digital tape is then computer processed and a plot of $S_C(z)$ versus depth is obtained for each shot and frequency. A second computer program combines and averages the data from all the shots in a data collection sequence and a computer plot of $S_C(z)$ versus depth is obtained for each frequency. Two separate programs are used, so that all shots may be examined and any anomalous data removed before averaging. Presently, the data processing technique is being improved by replacing the pencil-follower digitizer with an analog to digital converter which digitizes the output of the logarithmic graphic level recorder directly.

VOLUME SCATTERING STRENGTHS 635

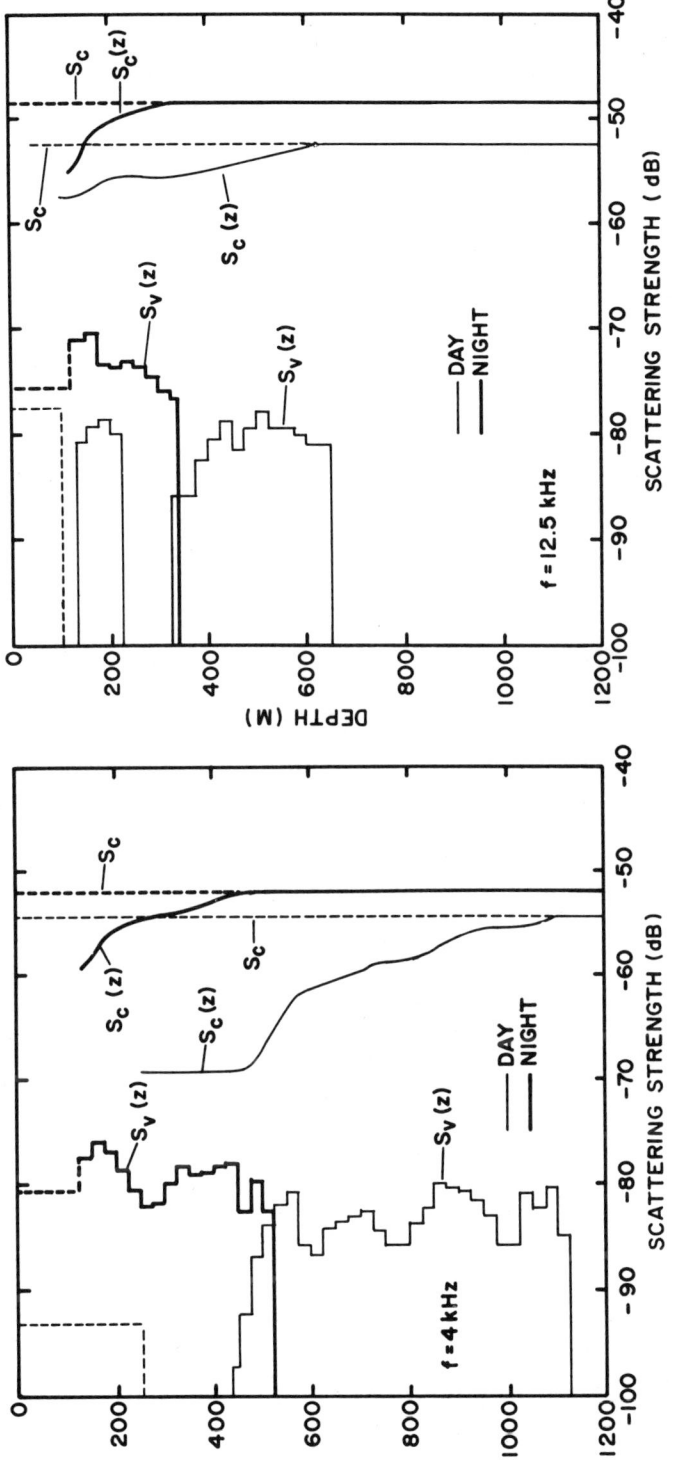

Fig. 2. Comparisons of daytime and nighttime scattering strengths at 4 kHz at a location in the Mediterranean Sea.

Fig. 3. Comparisons of daytime and nighttime scattering strengths at 12.5 kHz at a location in the Mediterranean Sea.

Figures 2 and 3 show profiles of $S_C(z)$ at 4 and 12.5 kHz for daytime and nighttime at a location in the Mediterranean Sea. Column scattering strengths, S_C, the only information available from the SUS-technique, are also indicated. Figures 2 and 3 show that the diurnal variation in S_C is about 3 dB at 4 kHz and 4 dB at 12.5 kHz, which would appear to indicate that there are no substantial day-night changes in scattering at this particular location. However, the $S_C(z)$ profiles indicate that there are large diurnal variations in scattering at certain depths. The scatterers have risen and redistributed themselves at night in such a way as to produce only minor variations in column scattering strength.

Although the $S_C(z)$ profiles contain significantly more information than the single value of S_C obtained with the old measurement technique, the profiles are difficult to correlate to biological parameters and have several acoustically-related deficiencies. One deficiency is that the relative scattering strengths of different profiles at the same depth or of the same profile at different depths are difficult to estimate. Another is that the depth of the deepest scatterers is sometimes difficult to determine. These two deficiencies are related to the logarithmic nature of the profiles. A third deficiency is that no information is provided on scattering shallower than the initial depth of the profile.

In order to permit correlation with biological data and to eliminate or minimize the acoustically-related deficiencies, a technique was developed to numerically differentiate the $S_C(z)$ profiles to obtain $S_V(z)$. $S_V(z)$ can be estimated from $S_C(z)$ by using the following procedure. At any depth $z = z_1$,

$$S_C(z_1) = 10 \log \left[\int_0^{z_1} M(z) dz \right]. \tag{6}$$

If it is assumed that $M(z)$ is constant over $0 \leq z \leq z_1$ and $M(z) = \overline{M(z_1)}$, where the bar denotes the average over depth, then the integration in equation 6 can be performed to give

$$S_C(z_1) = 10 \log \overline{M(z_1)} z_1, \tag{7}$$

from which the value of $\overline{M(z_1)}$ can be determined. Similarly, if over $z_1 \leq z \leq z_2$, it is assumed that $M(z) = \overline{M(z_2)}$, then

$$S_C(z_2) = 10 \log \left[\int_0^{z_2} M(z) dz \right] = 10 \log \left\{ \overline{M(z_2)} [z_2 - z_1] + \overline{M(z_1)} z_1 \right\}, \tag{8}$$

and the value of $\overline{M(z_2)}$ can be determined. In general then,

$$S_c(z_n) = 10 \log \sum_n \overline{M(z_n)} [z_n - z_{n-1}]. \tag{9}$$

Thus $\overline{M(z_n)}$ can be determined and therefore,

$$S_v(z_n) = 10 \log \overline{M(z_n)}. \tag{10}$$

A third computer program was developed to calculate $S_v(z)$ from $S_c(z)$ utilizing equations 9 and 10 and a nominal depth increment of 25 meters.

The $S_v(z)$ profiles calculated from the respective $S_c(z)$ profiles are also shown in Figures 2 and 3. These profiles enable scattering strength values at different depths or of different profiles to be compared to each other or to biological data. In addition, by calculating the average $M(z)$ required to produce the initial $S_c(z)$, an estimate of shallow scattering can be made. However, the $S_v(z)$ profiles have limitations which are due to the logarithmic nature of the $S_c(z)$ profiles from which they are obtained. These limitations stem from the fact that a scattering layer which is much weaker than a shallower layer increases the value of $S_c(z)$ very minimally. Thus, the weakest detectable layer will be no less than 10 dB weaker than the strongest layer above it. Therefore, the minima in the $S_v(z)$ profiles may not be accurate. However, it is the maxima that are important, and they are not seriously affected by these limitations. In addition, the difficulty in determining the depth of the deepest scatterers remains. This can be remedied somewhat by choosing some other depth as a characteristic depth for describing scattering layer depth variations.

The above limitations in the $S_v(z)$ profiles can be eliminated and higher data quality obtained by using a directional receiver. However, this technique has disadvantages related to cost and deployment. Therefore, it must be decided whether or not the increased data quality obtained with the directional technique is worth the increased cost and operational difficulties. In many instances, the new omnidirectional measurement and processing technique may be more suitable.

High Frequency Measurement System

Explosive sources do not provide sufficient acoustic energy for volume reverberation measurements above 20 or 30 kHz. Above this frequency range alternate sources are required. Consequently, a 50-kHz volume reverberation measurement system has been developed to examine scattering in the 0 to 600 meter depth range. Prime criteria in the development of the system were reasonable cost, rapid delivery, ease of handling, and the capability to be readily used on any research vessel. The primary system consisted of two Edo Model 311 circular plane transducers, driven individually by an Edo Model 248C

50-kHz, 2-kw transceiver. The transducers were mounted on the same frame, one looking vertically upward and the other vertically downward. The frame is lowered over the side utilizing 4HO electrical conducting cable, which is standard on all NAVOCEANO research vessels This configuration enables measurements to be made from the surface to about 200 meters below the maximum allowable transducer depth.

Due to the high absorption coefficient at 50 kHz (approximately 15 dB/km at 0°C) and the relatively low power of the system, the transducers are operated at several different depths, nominally 150, 300, and 450 meters, in order to be certain that data is obtained over the complete water column of interest. At each operating depth each transducer is pulsed ten times, with a pulse length of 10 msec. The time spent at any single depth is usually five to ten minutes and a complete sequence takes about one hour.

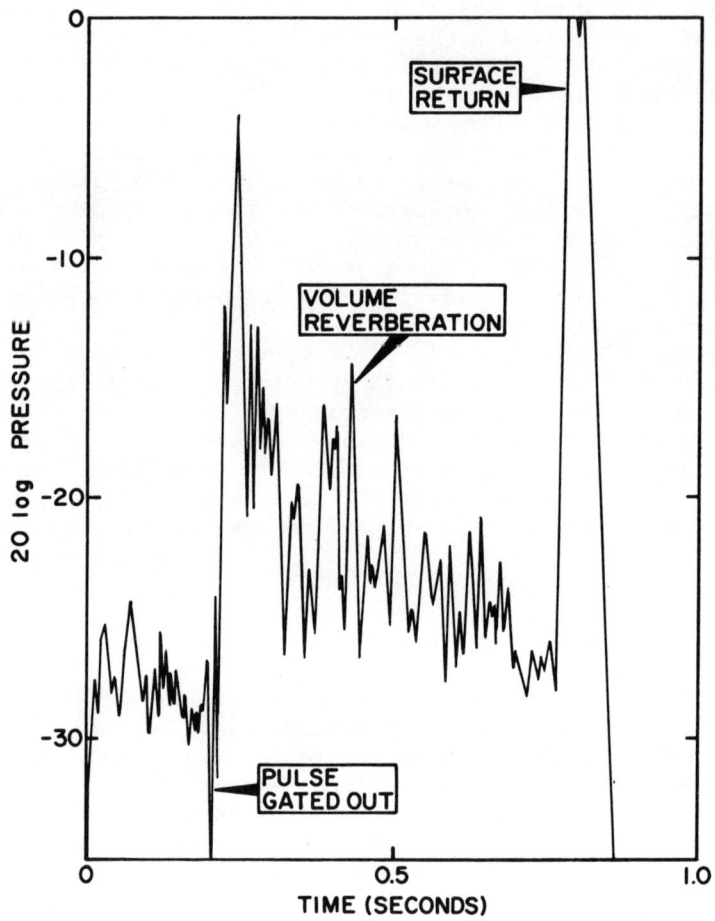

Fig. 4. Typical high-frequency upward-looking digitized signal

The received signals are recorded on analog magnetic tape. In the laboratory the analog signals are converted into a digital format. The analog signals are first fed into a logarithmic graphic level recorder, the output of which is then digitized with an analog-to-digital (A/D) converter. Use of the logarithmic graphic level recorder permits the use of an A/D converter with a fairly slow sampling rate. The present A/D converter samples at a rate of 500 samples/second for 4 seconds. The sampling period is manually initiated about one second before the pulse so that the background noise level is also sampled. The A/D converter also formats the digitized data for recording on digital magnetic tape.

The digital data are then processed in several steps. First, a computer program was developed to transfer the data from magnetic tape to punched cards and to plot the digitized signal. A plot of a typical digitized signal is shown in Figure 4. Comparisons between these plots and the analog records which are obtained directly from the logarithmic graphic level recorder indicate whether or not the signal has been properly digitized.

A second computer program was developed to calculate the scattering strength per unit volume, $S_v(z)$, and to plot the results. $S_v(z)$ is calculated from a modified form of the sonar equation (Urick, 1967):

$$S_v(z) = 20 \log p(t) - SL - 10 \log V + 40 \log \frac{ct}{2} + \alpha ct, \quad (11)$$

where SL is the source level, V is the insonified volume, and the other terms have been previously defined. This program is normally run twice. First, the calculations are done for the complete digitized signal. Then the exact depth is determined from the surface return and the program is run again, eliminating surface and bottom returns and any subsequent data.

Two additional programs were developed to examine the statistics of the $S_v(z)$ profiles. One determined and plotted the mean $S_v(z)$ profile. This program averaged all $M(z)$ values within a sliding 4.5 meter depth range in order to obtain a mean $S_v(z)$ value for every 1.5 meters. Figure 5 shows the daytime and nighttime means for a location in the northern North Atlantic. The other, in order to examine the pulse-to-pulse variability, determined and plotted the minimum, maximum, and quartile values of $S_v(z)$ in each 10 meter depth increment. Figure 6 shows the nighttime quartiles from the same location as Figure 5. It should be noted that, due to the measurement technique, different parts of the curves in Figures 5 and 6 were obtained from unequal numbers of pulses.

This new high frequency measurement system is being employed in conjunction with the omnidirectional explosive-source technique in order to determine relationships between high and low frequency volume scattering. It may then be possible to predict scattering strengths at high frequencies from the vast amount of low frequency data by utilizing these relationships in conjunction with oceanographic and biological information. Presently this system is being expanded to a triple frequency system.

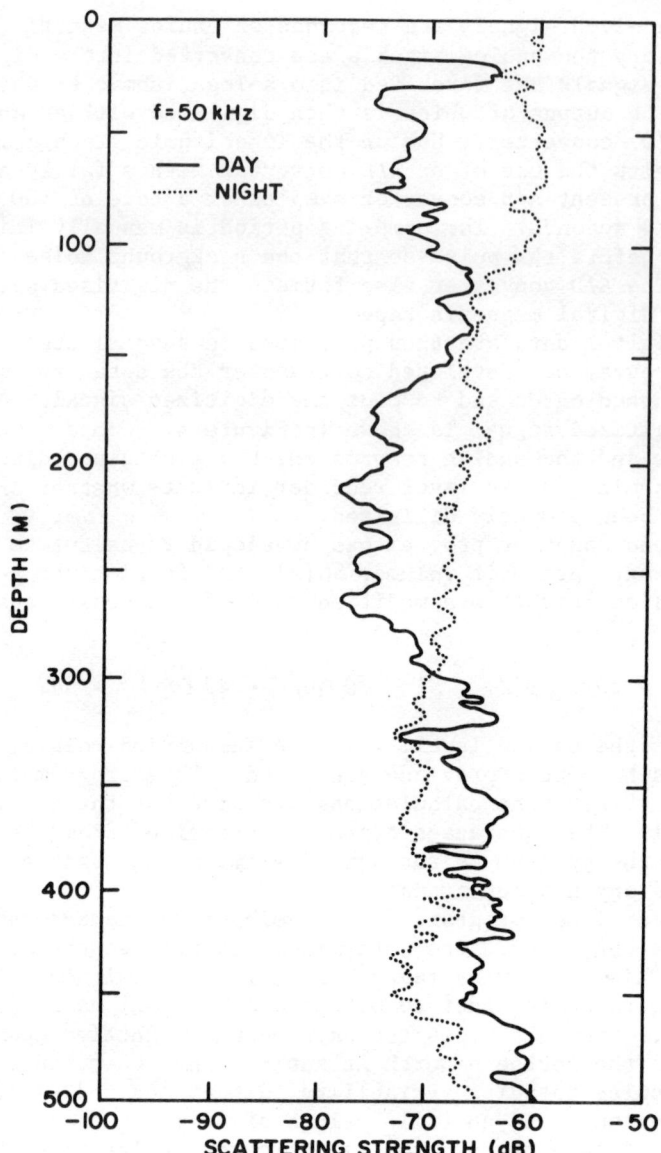

Fig. 5. High-frequency scattering strength profiles at a location in the northern North Atlantic

PREDICTION OF VOLUME SCATTERING STRENGTHS FROM FISH DISTRIBUTION DATA

Although coincident acoustic volume scattering measurements and biological trawl data have been collected for a number of years, few quantitative comparisons have been made between the two types of data

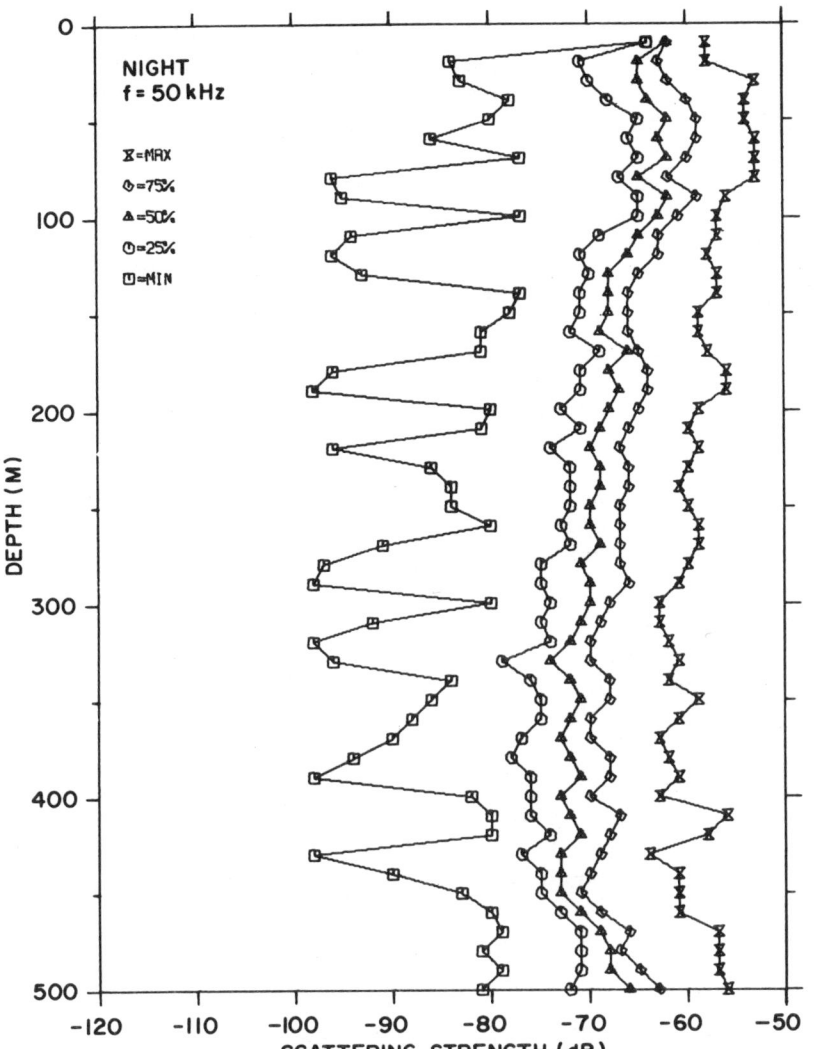

Fig. 6. High-frequency scattering strength quartiles at a location in the northern North Atlantic.

The capability of obtaining $S_V(z)$ profiles over broad frequency ranges coincidentally with biological data now makes quantitative comparisons feasible. Thus a volume reverberation model which utilizes fish distribution data has recently been developed to permit quantitative prediction of $S_V(z)$ (*Love*, 1975).

To validate the model, fish distribution data were obtained at three locations in the Mediterranean Sea with a 10-foot Isaacs-Kidd Midwater Trawl, terminated with a four-chambered discrete-depth cod-end sampler. This sampler provided two horizontal samples and two oblique samples from each tow. The horizontal sampling period was usually 0.5 to 1 hour in duration. Each fish collected was identified and sized and the data entered on computer cards for subsequent processing.

$S_v(z)$ was then predicted utilizing equations 1 and 4, where n is the number of fish collected, σ_i is the acoustic cross section of an individual fish, and V is the volume of water which passed through the trawl. The number of fish and the volume are readily determined, the difficulty in predicting $S_v(z)$ lies in determining the acoustic cross sections.

In order to estimate the acoustic cross section of individual swimbladder-bearing fish near resonance, equations developed by Andreeva (1964) were employed. In an effort to approximate the losses caused by fish tissue surrounding the swimbladder, Andreeva modified the equations for an ideal air bubble in water (Nat. Def. Res. Comm., 1946) by assuming that the bubble was in an infinite elastic medium having the shear properties of fish flesh. For the acoustic cross section, Andreeva obtained

$$\sigma = \frac{4\pi R^2}{\left[\left(\frac{f_R^2}{f^2} - 1\right) + \frac{f_R^2}{Q^2 f^2}\right]}, \qquad (12)$$

where R is the equivalent spherical radius of the swimbladder, f_R is the resonant frequency of the bladder, f is the insonifying frequency and Q is the reciprocal of the damping factor at resonance. The value of Q is given by Andreeva as varying from about 5 to 10 with depth. For the resonant frequency Andreeva obtained,

$$f_R = \frac{1}{2\pi R}\left(\frac{3\gamma P + 4\mu}{\rho}\right)^{\frac{1}{2}}, \qquad (13)$$

where P is the ambient pressure, ρ is the density of sea water, γ is the ratio of specific heats of the gas in the bladder, and μ is the real part of the complex shear modulus of the fish tissue (Lebedeva, 1965).

By utilizing Andreeva's equations and the regression equations determined by Kleckner and Gibbs (1972) which relate fish species, length and swimbladder volume, acoustic cross sections near resonance were estimated for each fish collected. At frequencies far above resonance, the body of a fish becomes as important a contributor to scattering as its swimbladder. In order to take this effect into account, an empirical equation which related acoustic cross sections to fish length and acoustic frequency was utilized (Love, 1971). Then, at frequencies above resonance the acoustic cross sections obtained by the two methods were compared and the larger value chosen.

For each frequency of interest the acoustic cross sections were summed for each species and for all fish collected in a sample. Thus from the trawl data, scattering strengths and dominant species were

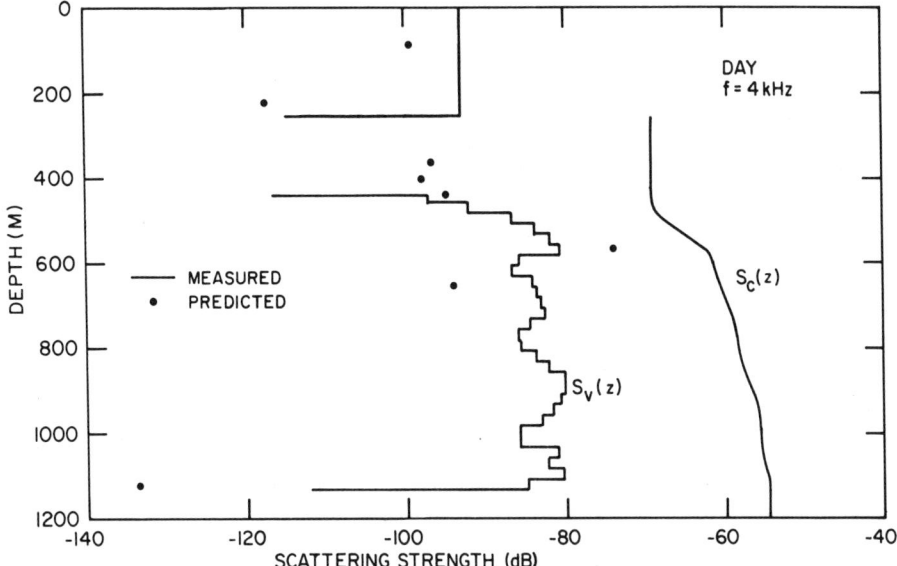

Fig. 7. Comparison of scattering strengths at 4 kHz at a location in the Mediterranean Sea calculated from omnidirectional acoustic measurements and predicted from fish distribution data.

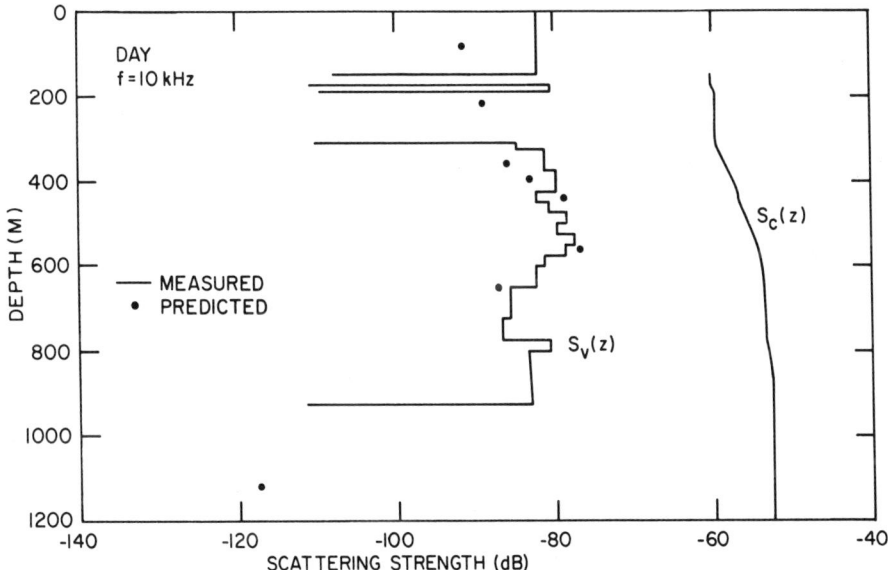

Fig. 8. Comparison of scattering strengths at 10 kHz at a location in the Mediterranean Sea calculated from omnidirectional acoustic measurements and predicted from fish distribution data.

predicted for each trawl depth for each frequency. It should be
noted that the dominant scattering species were not necessarily the
most numerous.

The data on fish distribution obtained in the Mediterranean Sea
were used to predict scattering strengths at frequencies from 2.5 to
20 kHz. These predictions were then compared to $S_V(z)$ profiles obtained coincidentally with the improved omnidirectional explosive-source technique. Figures 7 and 8 give examples of the comparisons.
The scattering strengths predicted from the fish distribution data
averaged 2 dB lower than those obtained from the acoustic measurements and the average absolute difference between the two types of
data was 6 dB. The agreement was best at the higher frequencies,
with the agreement decreasing with frequency below 6.3 kHz. In general, the lower the frequency, the larger the fish responsible for
the scattering. Thus, the increasing disagreement at the low frequencies is primarily caused by the inability of the trawl to catch the
larger fish. Other sources of error include errors in swimbladder
size and in *Andreeva's* equations.

Based on this limited data set, it appears feasible to utilize
biological collections to predict volume scattering strengths. However, it is easier, much faster, and more accurate to determine
scattering strengths directly from acoustic measurements than to predict them from biological collections. Even so, the technique described here can be useful in predicting scattering over very wide
frequency bands, which may be impractical or inconvenient to measure
acoustically. By combining the information on dominant scattering
species with oceanographic information, geographic regions with
similar volume scattering properties can be delineated. In addition,
there are historical biological collections which could be used to
predict scattering strengths in areas or seasons for which no acoustic
data is available. Of course, the accuracy of these predictions
greatly depends on the biological sampling gear used and on the completeness of the sampling program.

IMPROVED MODEL FOR ACOUSTIC CROSS SECTIONS OF RESONANT FISH

In the above study on the prediction of scattering strengths
from fish distribution data, there were several instances where one
species, of limited size range, was the predominant source of reverberation over a wide frequency band. In many of these instances,
near resonance the predicted scattering strengths were higher than
the measured values; whereas away from resonance, the predicted and
measured values were in better agreement, with the predicted values
usually on the low side. This discrepancy would be reduced if the
damping factor were increased; that is, if Q decreased.

Several researchers have conducted experiments to determine the
scattering characteristics near resonance of several species of fish
at relatively shallow depths (*Batzler and Pickwell*, 1970; *McCartney and Stubbs*, 1970; *Sand and Hawkins*, 1973; and *Sundnes and Sand*, 1975)

Batzler and *Pickwell* made measurements on relatively small fish. Their experimental values of resonant frequency agreed well with values predicted from *Andreeva's* equation. However, *Sand and Hawkins*, whose measurements were made on somewhat larger fish, obtained experimental resonant frequencies which were several times higher than those predicted by *Andreeva's* equation. In all cases, the experimental Q*'s* have generally been lower than those predicted by *Andreeva*.

A theoretical study is presently being undertaken in an effort to improve the existing equations for the resonant frequency and acoustic cross section of an individual fish by utilizing an improved model. The model is a viscous, heat-conducting spherical shell in water, containing an air-filled cavity. The shell has the physical properties of fish flesh and the shell-cavity interface supports surface tension.

The study will determine the appropriate wave equations for the model from the linearized equations of motion. The solutions of the wave equations will be determined by standard separation of variables techniques and the appropriate boundary conditions applied. Then, determination of the coefficient of the scattered pressure wave will enable the resonant frequency and acoustic cross section to be determined. It is expected that the equations obtained from this study will improve present volume scattering predictions.

SUMMARY

Progress in the correlation of acoustic and biological volume reverberation data has been discussed. This includes the development of techniques to obtain $S_V(z)$ profiles over broad frequency ranges, a model which utilizes biological data to predict $S_V(z)$, and a theoretical study aimed at improving this predictive model.

REFERENCES

Andreeva, I. B. 1964. Scattering of Sound by Air Bladders of Fish in Deep Sound-Scattering Ocean Layers. *Sov. Phys. Acoust.* 10: 17-20.

Batzler, W. E. and G. V. Pickwell. 1970. Resonant Acoustic Scattering from Gas-Bladder Fishes. *Proceedings of an International Symposium on Biological Sound Scattering in the Ocean*, Rep. 005, Edited by G. B. Farquhar, Maury Center Ocean Sci., Washington, DC: 168-179.

Chapman, R. P., O. Z. Bluy, and R. H. Adlington. 1970. Geographic Variations in the Acoustic Characteristics of Deep Scattering Layers. *Proceedings of an International Symposium on Biological Sound Scattering in the Ocean*, Rep. 005, Edited by G. B. Farquhar, Maury Center Ocean Sci., Washington, DC: 306-317.

Gibbs, R. H., Jr., R. H. Goodyear, R. C. Kleckner, C. F. E. Roper, M. J. Sweeney, B. J. Zahuranec, and W. L. Pugh. July, 1972. Mediterranean Biological Studies, Final Report, Smithsonian Inst., Washington, DC.

Gaspin, J. B. and V. K. Shuler. October, 1971. Source Levels of Shallow Underwater Explosions. Naval Ordnance Lab. Report TR-71-160, White Oak, MD.

Lebedeva, L. P. 1965. Measurement of the Dynamic Complex Shear Modulus of Animal Tissues. *Sov. Phys. Acoust.* 11: 163-165.

Love, R. H. 1971. Dorsal-Aspect Target Strength of an Individual Fish. *J. Acoust. Soc. Am.* 49: 816-823.

Love, R. H. August, 1974. An Improved Method of Explosive-Source Volume Reverberation Data Collection and Analysis. Naval Oceanographic Office Report TN 6130-3-74, Washington, DC.

Love, R. H. 1975. Predictions of Volume Scattering Strengths from Biological Trawl Data. *J. Acoust. Soc. Am.* 57: 300-306.

Machlup, S. and J. B. Hersey. 1955. Analysis of Sound-Scattering Observations from Non-Uniform Distributions of Scatterers in the Ocean. *Deep-Sea Res.* 3: 1-22.

Marshall, J. R. and R. P. Chapman. 1964. Reverberation from a Deep Scattering Layer Measured with Explosive Sound Sources. *J. Acoust. Soc. Am.* 36: 164-167.

McCartney, B. S. and A. R. Stubbs. 1970. Measurements of the Target Strengths of Fish in Dorsal Aspect, Including Swimbladder Resonance *Proceedings of an International Symposium on Biological Sound Scattering in the Ocean*, Rep. 005, Edited by G. B. Farquhar, Maury Center Ocean Sci., Washington, DC: 180-211.

Nat. Defense Res. Comm. 1946. Physics of Sound in the Sea, Part IV, Div. 6 Sum. Tech. Rep. 8, Chap. 28: 460-467.

Sand, O. and A. D. Hawkins. 1973. Acoustic Properties of the Cod Swimbladder. *J. Exp. Biol.* 58: 797-820.

Sundnes, G. and O. Sand. 1975. Studies of a Physostome Swimbladder by Resonance Frequency Analyses. *J. Cons. Int. Explor. Mer.* 36: 176-182.

Thorpe, W. H. 1967. Analytic Description of the Low-Frequency Attenuation Coefficient. *J. Acoust. Soc. Am.* 42, 270(L).

Urick, R. J. 1967. *Principles of Underwater Sound for Engineers.* (McGraw-Hill, NY) p. 193.

Variations in Abundance of Sound Scattering Animals Off Oregon

W. G. Pearcy

Oregon State University

ABSTRACT

Although the same numerically dominant species of micronekton and macroplankton persist in time and space in the oceanic waters off Oregon, abundance of individual species may vary from year-to-year, with seasons, with distance offshore, among depths, and among repeated tows. Some of these changes, such as seasonal and inshore-offshore fluctuations, appear to recur and are hence predictable, others are not. Pronounced changes in oceanographic conditions may or may not be correlated with changes in species composition or abundance. More data are needed on the biology of animals, the influence of advection, and relationships with physical-chemical factors to understand spatio-temporal variations of pelagic animals.

Some major features of biological sound scattering are common to all cruises off Oregon, but the fine structure of volume scattering profiles varies appreciably within and among cruises. Based on our knowledge of abundances and vertical distributions of mesopelagic fishes, we can "predict" what species will be associated with migratory and non-migratory 12 kHz layers. Detailed agreement between volume scattering and midwater trawl catches has been elusive, emphasizing the difficulties of prediction without more data on scattering properties, abundance and distributions of the oceanic community.

INTRODUCTION

The answer to the question, "Can sound scattering be predicted from information on ocean environment?" depends on the degree of resolution or accuracy desired. We can safely predict that much of the low frequency (5-20 kHz) scattering is caused by mesopelagic fishes with gas-filled swimbladders in most oceanic regions of the world (*Batzler, Reese and Friedl*, 1975; *Johnson*, 1976) and that diel vertical migrations result in increased scattering in surface layers at night and mid-depths during the day. Gross prediction of the levels of sound scattering at various frequencies may also be possible from general knowledge of the size structure and standing stocks of pelagic organisms.

To predict detailed features of sound scattering, however, requires detailed knowledge of the sound scattering properties, which in turn requires extensive information on the biological community, including:

- *abundance of individual species*
- *size structure of species*
- *occurrence and sizes of gas inclusions and other effective sound reflecting structures*
- *spatial-temporal variations in the abundance, size, and scattering properties of species, e.g.,*
  - *diel migratory patterns*
  - *seasonal variations*
  - *annual or year-to-year variations*
  - *microscale patchiness*
  - *mesoscale variations*
  - *differences among water masses.*

The structure of extant biological communities is the result of long-term evolutionary processes which we can scarcely predict on the basis of physical-chemical factors. The causal factors of evolutionary change---- mutation, gene flow, natural selection, and genetic drift---- may result in different faunal communities in similar environments. Moreover, environments existing today are certainly not identical to those present during the evolutionary past.

Confounding prediction on smaller time-space scales are processes such as mixing of water types containing different biological communities, succession of the fauna within isolated or expatriated parcels of water, patchiness, and random variability.

Because of these considerations, predicting sound scattering in oceanic communities will depend on:

(1) Identification and characterization of the major biogeographic units which have distinctive biological and acoustical similarities. Water masses may be such units. They are often semi-enclosed systems with unique physical-chemical-biological features resulting from quasi-homogeneous and predictable conditions, and from a common evolutionary history.

(2) Characterization of the degree of heterogeneity within water mass units. Variations exist in time and space in one

area. Knowledge of their extent and cause are obviously necessary for detailed predictions. For example, marked seasonal changes in the physical-chemical environment may not be accompanied by biological changes in sound scattering; or conversely, significant changes in the sound scattering may take place in absence of obvious environmental changes. Scatterers may also have different concentrations or patterns of vertical migrations within one water mass. Patchiness may be large. In addition, characterization of an area requires more than one "representative" midwater trawl sample, one ping from a single-frequency transducer, and one STD cast. Hence variability and averaging of data are always problems.

The purpose of this paper is to describe some of the temporal and spatial variations of the oceanic fauna, their environment, and their sound scattering that occur within one small geographic area in the northeastern Pacific Ocean off Oregon. Special attention is given to the mesopelagic fishes.

METHODS

Pelagic animals were collected with the Isaacs-Kidd Midwater Trawl (IKMT). During the 1960's oblique tows in the upper 200 m or horizontal tows in surface waters were made with a 1.8 m IKMT (see *Pearcy*, 1964). In later years a 2.4 m IKMT was used with a 1 m^2 Multiple Plankton Sampler (MPS) as an opening-closing system. Electrical cable permitted actuation of multiple nets at selected depths and monitoring of depth, temperature and water flow through the net (*Pearcy and Mesecar*, 1971; *Pearcy et al.*, in press). During a cruise in September 1974 this IKMT-MPS system provided samples from horizontal depth strata that usually coincided with areas of high or low acoustical scattering. Most nets during this cruise filtered equal volumes of water so that variability of catches could be more readily analyzed.

Data on acoustical scattering was obtained from 12 kHz and 38.5 kHz echograms and from volume scattering measurements with a nominal 12 kHz transducer.

THE ENVIRONMENT AND FAUNA

The ocean off Oregon is a transitional region near the origin of the California Current, a sluggish eastern boundary current. The waters in the upper 1000 m consist largely of modified Subarctic Water, varying from about 55% to 78% Subarctic at depths of 200-1000 m and from 85-120 km off the coast (*Rosenberg*, 1962; *Pattullo and Lorz*, unpubl.). A permanent halocline between 100 and 150 m and temperature-salinity profiles also indicate subarctic origin of waters off Oregon (*Fleming*, 1958; *Pearcy*, 1972).

The midwater fauna off Oregon has zoogeographic affinities with Subarctic and Transitional Water Masses. Most oceanic species are also found in the Subarctic Pacific, but some are unique to the Transitional region between Subarctic and Central Waters (*Brinton*, 1962; *McGowan*, 1971; *Pearcy*, 1972).

Despite the complex, transitional ocean environment, *Pearcy* (1964) reported that the four common mesopelagic fishes (*Stenobrachius leucopsarus, Diaphus theta, Tarletonbeania crenularis* and *Tactostoma macropus*) were numerically dominant in nearly all midwater trawl collections at night in the upper 200 m regardless of distance from shore, season or latitude off Oregon. The shrimp *Sergestes similis* and the euphausiid *Euphausia pacifica* are also numerically dominant species in most midwater samples (*Hebard*, 1966; *Pearcy and Forss*, 1966; *Pearcy*, 1972).

The relatively stable species structure, and the high dominance, are illustrated in Table 1 which shows the average abundance of common mesopelagic fishes caught during eight seasonal cruises with an opening-closing IKMT during daytime tows in the upper 500 m. Again, the same mesopelagic fishes dominated these collections. The most abundant species was usually *Stenobrachius leucopsarus*. It ranked first in abundance in all but two cruises when it was second in abundance. Though their ranks varied, *Diaphus theta, Tarletonbeania crenularis* and *Tactostoma macropus* were again common species and one of them usually ranked second in abundance. Significant agreement of concordance among cruises was present in the ranks of abundance of the nine most common species in Table 1 ($W = 0.78$, $P < 0.01$), indicating a high correlation among the abundance rankings for the different cruises, and hence a fairly stable species composition of the common mesopelagic fishes.

Year-to-Year Variability

Significant year-to-year fluctuations have occurred in the ocean environment due to general changes in circulation. For example, 1957 and 1958 were exceptionally warm years in the California Current system in general. In 1963 off Oregon, the heat content in the upper 100 m was unusually large (*Pattullo, Burt and Kulm*, 1969) and the per cent Subarctic Water was rather low (Figure 1). Those changes were probably caused by weakened southerly transport from the Subarctic Pacific into the California Current during that year (*Wickett*, 1967). During 1964, on the other hand, meridonal transport to the south was large, heat content off Oregon was low, and the percentage of Subarctic Water was high.

The abundance of several species of pelagic animals in 1963 differed from other years. *Hubbard and Pearcy* (1971) believed that the presence of warm-water salps off Oregon in 1963 was evidence for a weakened California Current and intrusion of water from the

TABLE 1. Averaged Catches of Mesopelagic Fishes (No. Per $10^5 m^3$) in Daytime Opening-Closing IKMT-MPS Net Tows in the Upper 500 m.

	Cruise (Year and Month)							
	6911	7011	7107	7206	7209	7211	7302	7306
Stenobrachius leucopsarus	161.0	158.2	68.7	66.7	9.6	32.9	8.1	18.8
Diaphus theta	22.7	78.1	12.8	23.1	8.2	25.2	11.3	17.6
Tarletonbeania crenularis	3.3	103.7	17.9	7.8	2.3	7.0	3.5	8.6
Tactostoma macropus	3.3	1.3	35.9	13.6	8.8	4.0	2.9	4.5
Protomyctophum thompsoni	9.9	21.5	12.4	7.2	12.1	6.3	1.7	8.3
Lampanyctus ritteri	3.8	12.1	1.7	6.2	2.7	4.3	0.8	0.7
Cyclothone atraria	0	0	14.9	0.3	0.4	2.7	2.0	4.5
Protomyctophum crockeri	1.4	0	0.4	1.2	0.4	0	0	0
Bathylagus pacificus	0	0	4.3	0.3	0	0.3	0	0
Lampanyctus regalis	2.8	0	1.7	0	0	0.3	0	0.7
Bathylagus milleri	0	0	1.7	0.2	0	0	0	0
Aristostomias scintillans	0	0	0	0.2	0	0	0	0
Poromitra crassiceps	0	0	1.7	0	0	0	0	0
Stenobrachius nannochir	0	0	1.3	0	0	0	0	0

TABLE 2. Total Number of Common Mesopelagic Fishes Captured in Standard Oblique 0-200 m IKMT Net Tows at Stations 85 and 120 km off Newport, Oregon, 1963-1964.

	S. leucopsarus	D. theta	T. crenularis	T. macropus	L. ritteri
Jan. 1963	11	24	21	2	1
Feb.	51	22	15	2	2
May	38	9	18	8	7
June	34	9	21	5	3
July	44	14	16	7	7
Aug.	46	28	8	14	3
Oct-Nov.	4	11	11	1	1
Dec.	55	22	10	10	4
Feb. 1964	44	36	6	0	0
Feb.	38	18	11	3	8
Mar.	58	15	15	5	5
Apr.	15	1	3	4	2
June	38	9	9	0	4
July	23	18	12	7	3
Oct.	11	3	6	2	10
Nov.	102	63	27	15	14
Dec.	184	15	13	2	6

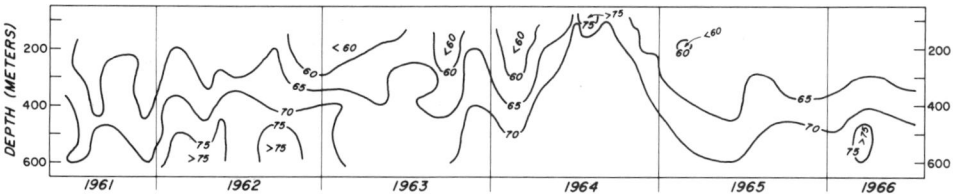

Fig. 1. The percentage of Subarctic Water present at the station 85 km off Newport, Oregon, 1961-1966, calculated by comparing observed temperature-salinity curves with a hypothetical curve for Subarctic Water (from *Pattullo and Lorz*, unpublished).

south. The squid *Abraliopsis felis*, the most abundant species off southern California (*Okutani and McGowan*, 1969), displaced *Gonatus* spp. as the most common squid off Oregon in 1963.

Because 1963 and 1964 were contrasting years, the catches of mesopelagic fishes were examined for changes in species composition or abundance of the common species that could be correlated with the different ocean conditions (Table 2). The five species in Table 2 represented 96% of all the mesopelagic fishes collected. As before, *S. leucopsarus* is the dominant fish; it was most abundant in all but two collections. No change in species composition is evident even between mid-1964, when the per cent Subarctic Water was high, and 1963. However, catches of *S. leucopsarus* were largest during November and December 1964, a trend that is also apparent in Figure 2 which shows the average number of *S. leucopsarus* and *D. theta* collected per 1000 m^3 over the 1961-1966 period. Thus, significant changes of oceanographic factors or abundance of species occurred but they were not always directly correlated.

Seasonal and Inshore-Offshore Variability

The ocean off Oregon is located at fairly high latitudes where seasonal changes in solar insolation and advection produce marked seasonal changes in environmental conditions. The thermal structure in the upper water column changes with seasons, and a well-defined thermocline develops offshore during summer months. *Pattullo, Burt and Kulm* (1969) found that seasonal variations in the heat content of the upper 100 m were also evident, especially in offshore waters beyond 185 km from the coast, where solar heating explains most of the increased heat content. Predictable seasonal cycles of heating were less apparent inshore due to the effects of advection—upwelling of cold water along the coast during the summer and downwelling and inshore transport of relatively warm water during the winter. The Columbia River plume also affects the physical structure and

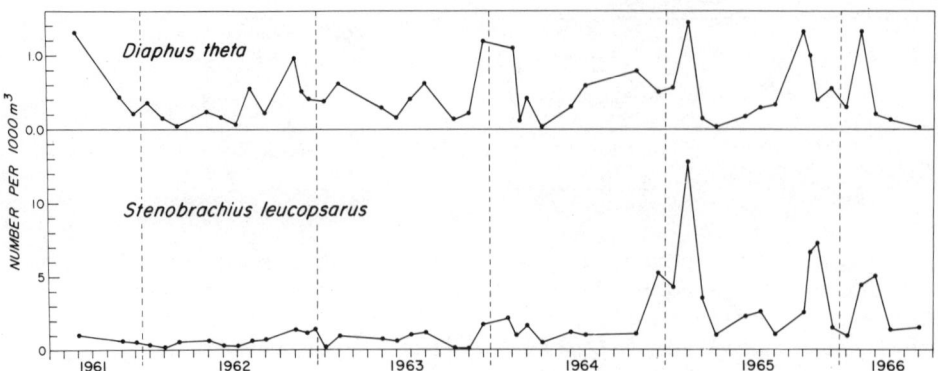

Fig. 2. Average nighttime catches of the two most common mesopelagic fishes in the upper 200 m off Newport, Oregon, 1961-1966.

chemistry of surface waters off Oregon during the summer (*Conomos, Gross, Barnes and Richards*, 1972; *Barnes and Gross*, 1966). As a result of all these processes, there are strong seasonal and inshore-offshore gradients in thermal structure, current velocity fields, physical and chemical properties, and primary productivity (*Stefansson and Richards*, 1963; *Anderson*, 1964; *Wyatt, Burt and Pattullo*, 1972; *Bourke and Pattullo*, 1974; *Huyer, Pillsbury and Smith*, 1975).

Seasonal variations are apparent in the catches of oceanic animals off Oregon. Significant seasonal differences were found in the biomass of micronekton (fishes, squids and shrimps) collected with a 1.8 m IKMT, but not in the biomass of herbivores such as copepods and euphausiids caught in 1 m diameter plankton nets (*Pearcy*, 1976). Distinct inshore-offshore differences were present in the biomass of micronekton. Biomass was largest at 85 km offshore in the summer but at 28-46 km offshore during the winter. Similarly, *Laurs* (1967) found seasonal and inshore-offshore variations in nutrients, chlorophyll concentrations and catches of oceanic animals that he related to upwelling off the southern Oregon coast.

Significantly larger catches of mesopelagic fishes, oceanic squids and shrimp have also been reported during some seasons off Oregon (*Pearcy*, 1965; *Pearcy and Laurs*, 1966; *Pearcy and Forss*, 1969). These seasonal changes may be related to annual reproduction cycles, migrations of nekton, or to advection of water.

Geographic variations in the catches of two common mesopelagic fishes compared with surface temperature and salinity data (Figure 3) show that they were caught in largest numbers over the continental slope or shelf, where both temperature and salinity were modified by upwelling and the Columbia River waters. Both species were absent near the coastline and were found in low numbers in oceanic waters over 100 km offshore. It is not known if the increased abundance of

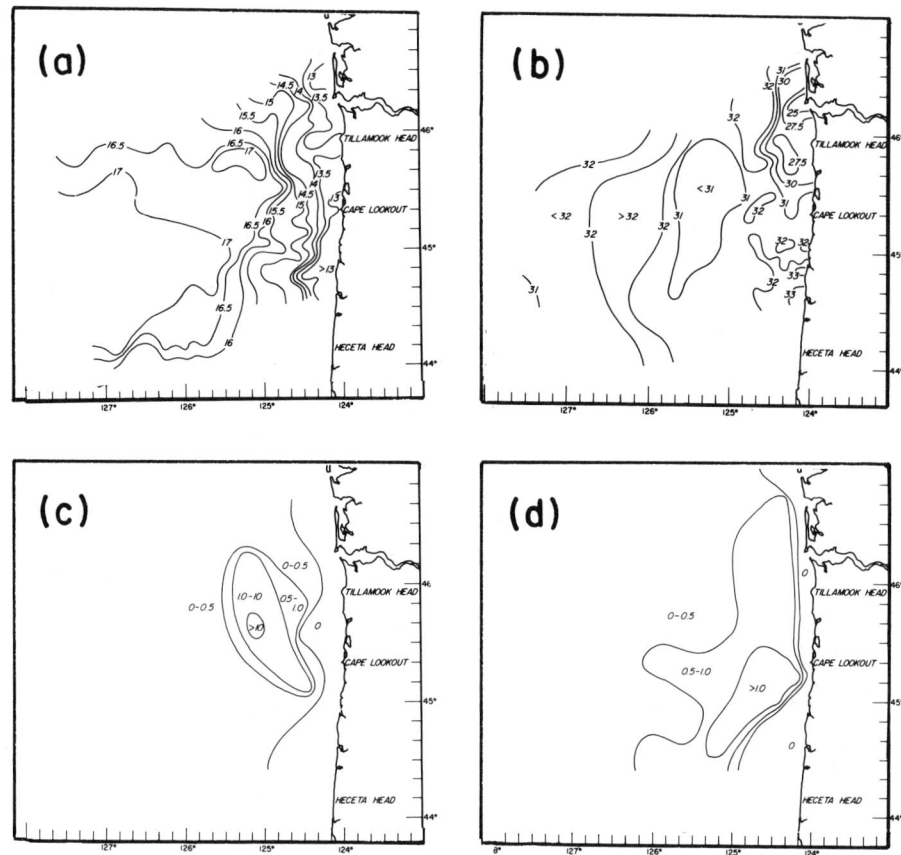

Fig. 3. Results of a cruise during July 31–August 12, 1969. (a) surface temperature, (b) surface salinity, (c) catches of *Tarletonbeania crenularis* per $10^3 m^3$ in 16 nighttime surface IKMT tows, and (d) catches of *Stenobrachius leucopsarus* per $10^3 m^3$ in 20 nighttime 0-200 oblique tows.

mesopelagic animals in nearshore waters shown here or by Pearcy (1976) is a result of passive inshore advection and accumulation of animals in shallow water or an active migration into areas of rich food supplies.

As a result of seasonal patterns of reproduction, the size composition of common species, such as S. *leucopsarus* varies among seasons (Smoker and Pearcy, 1970) and inshore-offshore during the same season. Large individuals of S. *leucopsarus* and *D. theta* comprised a greater proportion of the catches near the coast than in oceanic waters (Pearcy, unpubl.). Because only small S. *leucopsarus* have gas-filled swimbladders (Butler and Pearcy, 1972), differences in size composition of this species undoubtedly could produce seasonal and inshore-offshore differences in acoustical scattering.

Patchiness

Variability in the numbers of organisms captured in repeated collections is often large, resulting in large confidence limits to any measure of central tendency. For example, Pearcy (1964) reported that mesopelagic fishes had clumped or patchy distributions based on repeated oblique IKMT collections.

To assess the degree of variability in a continuous layer at one depth, our IKMT-MPS system was towed horizontally and nets were opened and closed so that equal volumes of water were filtered by each net. The results for three species of fishes show substantial variation both within and among tows (Figure 4). Evidence for vertical migration is present for S. leucopsarus and D. theta but some individuals of both species were captured in deep water during the night indicating that not all individuals migrated each night (see also Figure 6, and Pearcy and Mesecar, 1971). Figure 4 provides no evidence for vertical migration of P. thompsoni.

Coefficients of dispersion ($s^2/\bar{x}$) were plotted against mean abundance for catches of D. theta and small S. leucopsarus in tows for this cruise (Figure 5). Tows were categorized by whether they were made during day or night, during twilight or other periods when light intensity changed by more than an order of magnitude, or if tows were oblique or horizontal. If distributions were random, coefficients of dispersion would be equal to about 1.0. Obviously, spatial distributions are patchy, especially at higher densities. Surprisingly, highest coefficients were not always obtained for periods when light intensity changed, such as the twilight migratory periods.

Diel Variations

Diel vertical migrations of mesopelagic animals are well documented off Oregon (Pearcy and Laurs, 1966; Pearcy and Forss, 1966; Pearcy and Osterberg, 1967; Pearcy and Mesecar, 1971; Pearcy, 1972). Analyses of variance of our data from a series of stratified closing-net tows during various seasons indicate that depth x diel interactions (diel vertical migration) were significant for the five most common mesopelagic fishes (as illustrated in Figure 6), but not for P. thompsoni or Chauliodus macouni. Of the large crustaceans only Sergestes similis, the dominant decapod, and Bentheogennema burkenroadi migrated. Other shrimps and mysids (Hymenodora, Petalidium, Eucopia and Boreomysis) apparently did not undertake appreciable daily vertical migrations (Pearcy et al., in press).

SOUND SCATTERING ANIMALS 657

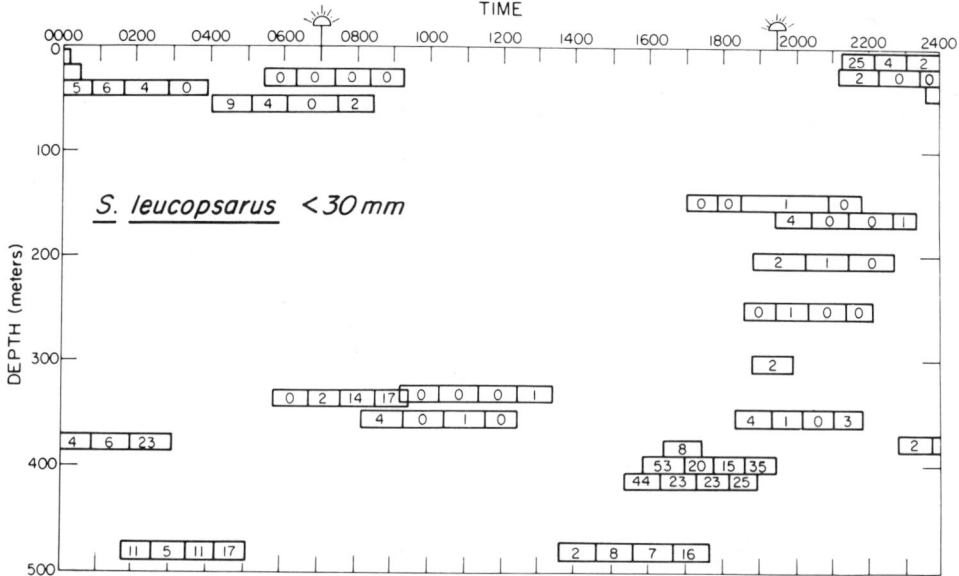

Fig. 4 Actual numbers of three lanternfishes collected in horizontal opening-closing net samples with the IKMT-MPS system, September 1974. Each cell represents one net. All nets filtered the same volume of water, about 32,560 m^3.

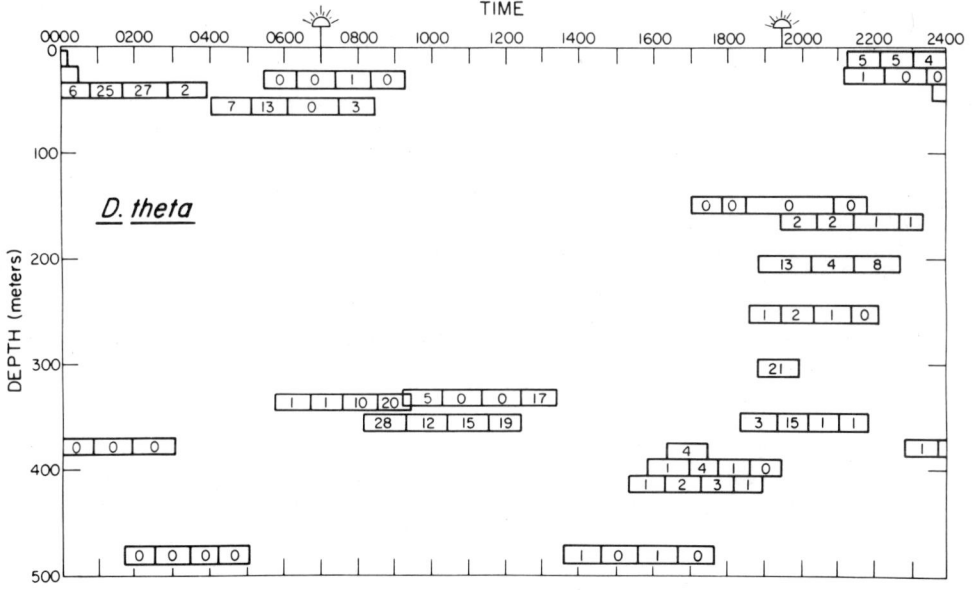

Fig. 4b.

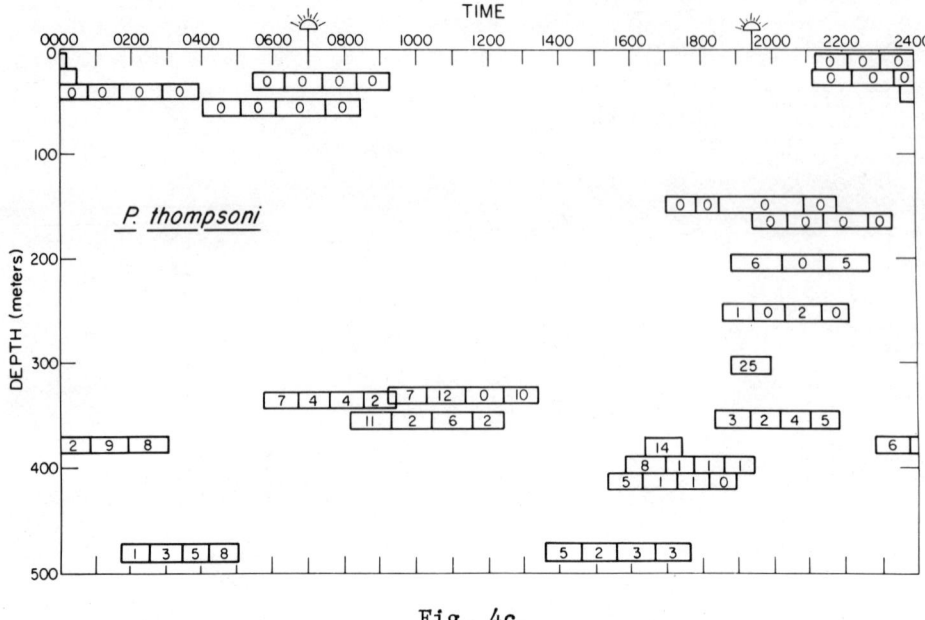

Fig. 4c.

SOUND SCATTERING

Echograms from a 12 kHz echosounder were obtained from eight cruises in 1969-1973. Results are reconstructed in Figure 7 for one typical 24-hr period. Common features of these echograms included both non-migratory and migratory scattering layers. Usually a non-migratory layer was present between 300 and 400 m during both day and night periods. Often a second non-migratory layer was either located above this layer, or two strata of intense scattering were apparent within a broad layer between 250 and 400 m. Migratory layers appeared to merge with or originate from these non-migratory layers. Often two distinct migratory layers were obvious within the upper 200 m during twilight hours, with either 38.5 kHz (Donaldson and Pearcy, 1972) or 12 kHz echosounders. As indicated in Figure 7 fair correlation exists between the rates and times of movement of migratory layers and isolumes. Vertical excursions of the non-migratory layers, however, appear to be independent of irradiance.

Vertical details of 12 kHz volume scattering strengths are illustrated in Figure 8 for two periods of sequential samples. Throughout most of this cruise a deep scattering layer persisted between 300 and 400 m. This layer often appeared to be bimodal with

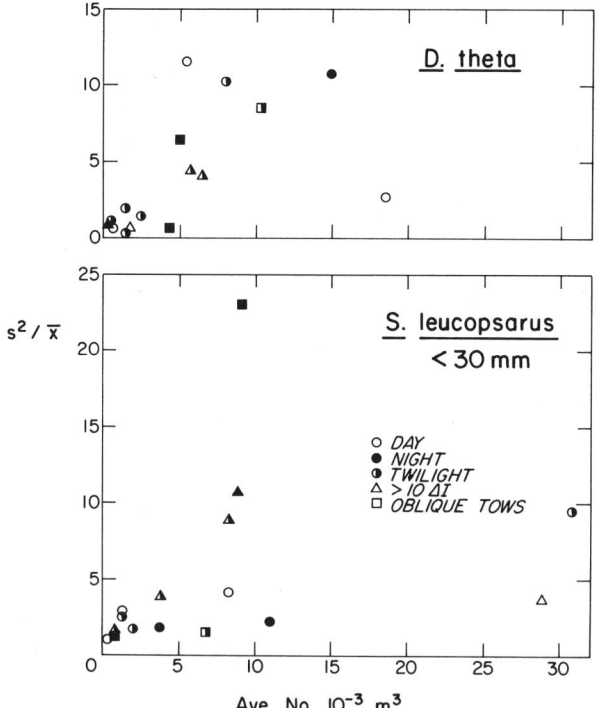

Fig. 5. Coefficients of dispersion ($s^2/\bar{x}$) vs. average numbers of two common mesopelagic fishes captured on the September 1974 cruise. Actual catches for all but the oblique tows are shown in Figure 4.

two peaks in scattering. Considerable variability is evident in the fine structure of scattering throughout these periods; this may be related to spatial differences (since the ship was underway), or temporal differences. Both sequences include the times of a twilight migration. Migration of scatterers from and to the scattering peak at about 300 m can be followed. The isolumes appear to encompass the migratory layers, as well as the peak in scattering at 300 m during the day.

Beklimishev (1964) noted considerable variability in the concentration of sound scattering layers, probably because of advection, during a two-week anchor station on the northern edge of the Kuroshio. He concluded that one must average data over a large area or obtain data from areas where advection is limited in order to describe variations in time. Donaldson and Pearcy (1972) also described some of the variable features of 38.5 kHz sonic scattering off Oregon. The extent of migration of layers, the thickness

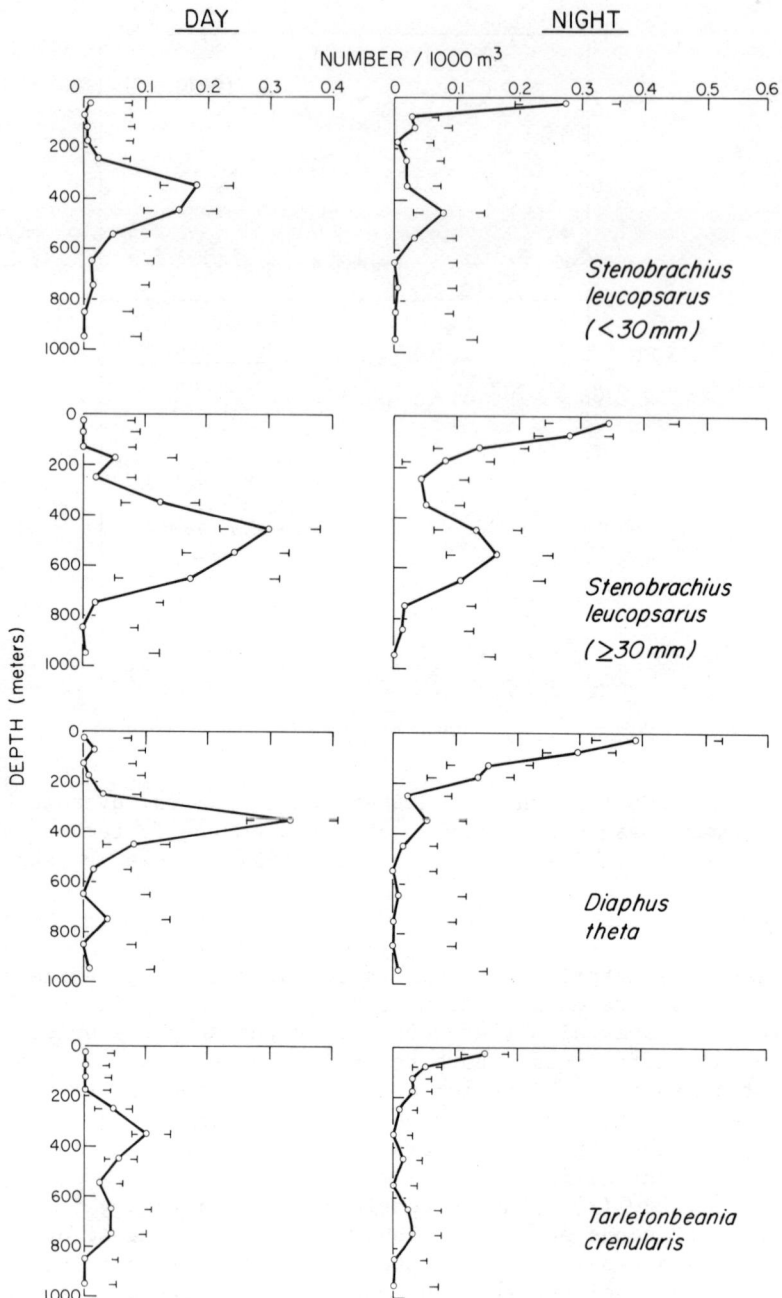

Fig. 6. Average catches (± 2 standard deviations) of common lanternfishes during day and night periods in opening-closing IKMT-MPS tows for 6 cruises, 1970-1973, at stations 120 km off Newport, Oregon. Tows were oblique so that nets fished contiguous depth strata.

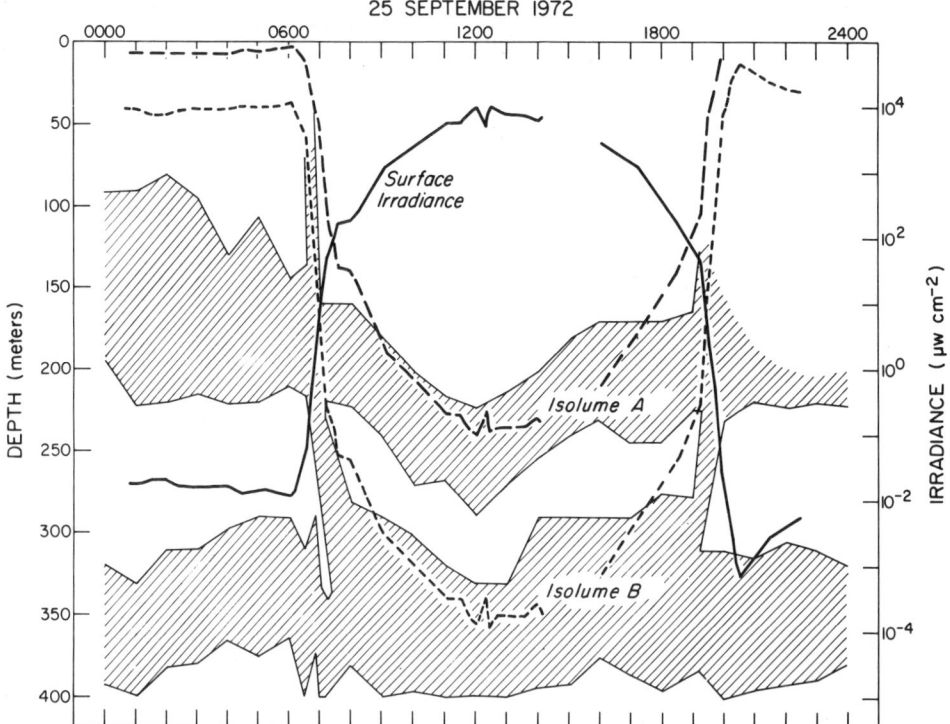

Fig. 7. Reconstruction of a 12 kHz echogram for September 25, 1972. Depths of isolumes were determined from surface irradiance values and attenuation-depth curves for the same station during this cruise.

of the layers present over a diel period, and the thickness of scattering present during different months all fluctuated.

Typical 12 kHz scattering strength profiles on one cruise are shown for day and night periods in Figure 9, along with data on the average catches from horizontal opening-closing net tows of the common myctophids that often contain gas-filled swimbladders, as well as fish and total biomass. Net tows were positioned relative to scattering features. During the day all four myctophids resided at about the same depths as the bimodal deep-scattering layer. *S. leucopsarus* and *T. crenularis* were most abundant at the depth of the deep-mode, and *D. theta* and *P. thompsoni* were most abundant at the depth of the shallow mode. Biomass was large in as well as below the 300-400 m layer. At night, three of the myctophid species migrated into near-surface waters. Only small *S. leucopsarus* and the non-migratory *P. thompsoni* are found in the deep layer at depths

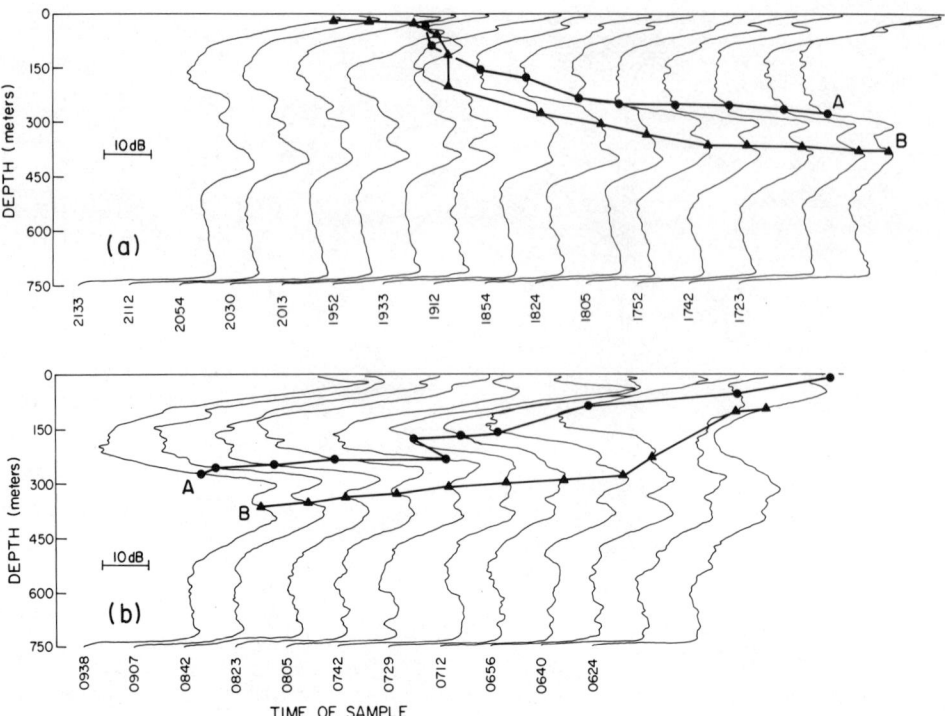

Fig. 8. Variations of profiles of volume scattering and two isolumes during evening and morning migrations, September 1974. Volume scattering is S_v where $S_v = RL - SL + 40 \log r - 10 \log V$ (Urick, 1975). Since the abscissa is scattering strength, profiles are not separated by a constant time interval, hence isolumes may appear to cross each other.

of 300-400 m. Maximal biomass, however, was found at about 200 m at night, between the intense surface and deep scattering layers.

General features of sound scattering and catches of mesopelagic fishes are often correlated. Peaks in the abundance of fishes with gas-filled swimbladders correlate with peaks in scattering at various depths. As indicated in Figure 9, small *S. leucopsarus* and *P. thompsoni* are often found at depths of the non-migratory layer at night, *D. theta* and *T. crenularis* may also be located at these depths by day. The three migratory myctophids have peak numbers in the vicinity of the surface scattering layer at night.

Despite these correlations, quantitative agreement among scattering and fish abundance or distribution is often lacking. For example, scattering levels were not always largest during the times or at the depths of largest catches of mesopelagic fishes

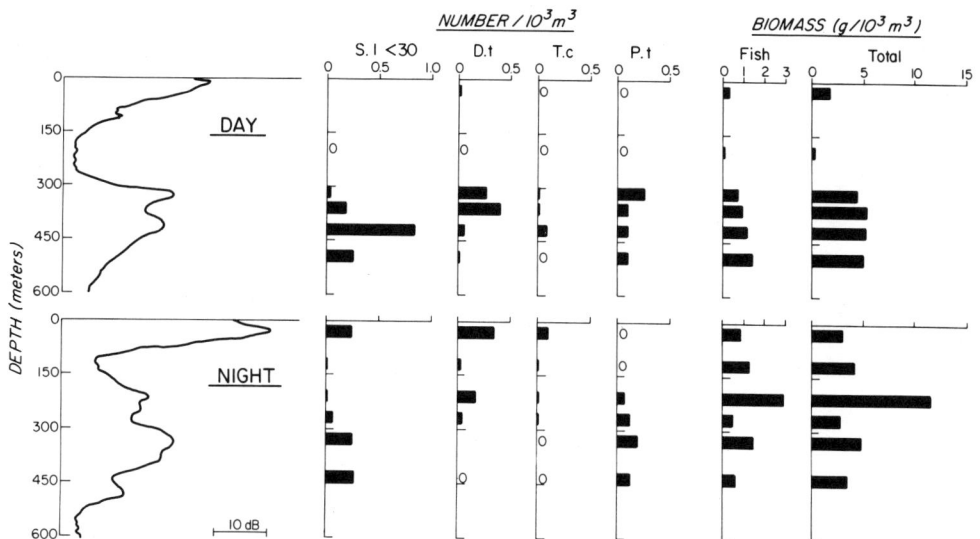

Fig. 9. Typical volume scattering profiles for the September 1974 cruise, along with average number of four lanternfishes containing gasfilled swimbladders, and fish and total biomass per $10^3 m^3$ caught in our IKMT-MPS nets (S.1.<30 = *Stenobrachius leucopsarus* <30 mm *standard length*, D.t. = *Diaphus theta*, T.c. = *Tarletonbeania crenularis*, P.t. = *Protomyctophum thompsoni*.

with gas-filled swimbladders (see Figure 4 for examples of the variation in fishes among tows). Differences in scattering levels among cruises or day-night periods also may not be correlated with abundance of captured fishes. Moreover, few animals are collected in midwater trawls that can account for scattering at low (10 kHz and below) frequencies (Batzler, Reese and Friedl, 1975; Johnson, pers. comm.). Obviously, more detailed data are needed on acoustical properties and actual abundances of individual species and sizes of midwater animals in order to explain details of scattering, or to predict the quantity and quality of biological sound scattering.

ACKNOWLEDGEMENTS

This research was supported by the Office of Naval Research, contract N00014-67-A-0369-0007 under project NR083-102.

REFERENCES

Anderson, G. C. 1964. The seasonal and geographic distribution of primary productivity off the Washington and Oregon coasts. *Limnol. Oceanogr.* 9: 284-302.

Barnes, C. A. and M. G. Gross. 1966. Distribution at sea of Columbia River water and its load of radionuclides. In: *Disposal of Radioactive Wastes into Seas, Oceans and Surface Waters*. IAEA. 291-302.

Batzler, W. E., J. W. Reese, and W. A. Friedl. 1975. Acoustic volume scattering: Its dependence on frequency and biological scatterers. *Naval Undersea Center Report, TP 442*: 22 pp.

Beklimishev, K. V. 1964. Echosounding records of macroplankton concentrations and their distribution in the Pacific Ocean. *Trudy Inst. Okeanol. SSSR 65*: 197-229 (U. S. Naval Oceanographic Office Transl. 343).

Bourke, R. H. and J. C. Pattullo. 1974. Seasonal variation of the water mass along the Oregon-northern California coast. *Limnol. Oceanogr.* 19: 190-198.

Brinton, E. 1962. The distribution of Pacific euphausiids. *Bull. Scripps Inst. Oceanogr. 8 (2)*: 51-270.

Butler, J. and W. G. Pearcy. 1972. Swimbladder morphology and buoyancy of northeastern Pacific myctophids. *J. Fish. Res. Bd. Canada 29*: 1145-1150.

Conomos, T. J., M. G. Gross, C. A. Barnes, and F. A. Richards. 1972. River-ocean nutrient relations in summer. In: A. T. Pruter and D. L. Alverson (Eds.), "*The Columbia River estuary and adjacent ocean waters*", Univ. Wash. 151-175

Donaldson, H. A. and W. G. Pearcy. 1972. Sound-scattering layers in the North-eastern Pacific. *J. Fish. Res. Bd. Canada 29*: 1419-1423.

Fleming, R. H. 1958. Notes concerning the halocline in the northeastern Pacific Ocean. *J. Mar. Res. 17*: 158-173.

Hebard, J. F. 1966. Distribution of euphausiacea and copepoda off Oregon in relation to oceanographic conditions. Ph.D. thesis, Oregon State University, Corvallis, Oregon.

Hubbard, L. T. and W. G. Pearcy. 1971. Geographic distribution and relative abundance of Salpidae off the Oregon coast. *J. Fish. Res. Bd. Canada 28*: 1831-1836.

Huyer, A., R. D. Pillsbury, and R. L. Smith. 1975. Seasonal variation of the alongshore velocity field over the continental shelf off Oregon. *Limnol. Oceanogr. 20*: 90-95.

Johnson, R. K. 1976. Deep scattering layer spectra in the eastern North Pacific. *J. Acoust. Soc. Am.* 59: 465-466.

Laurs, R. M. 1967. Coastal upwelling and the ecology of lower trophic levels. Ph.D. thesis, Oregon State University, Corvallis, Oregon.

McGowan, J. A. 1971. Oceanic biogeography of the Pacific. In: B. M. Funnell and W. R. Riedl (Eds.), "*The Micropaleontology of Oceans*", Cambridge Univ. Press. 3-74.

Okutani, T. and J. A. McGowan. 1969. Systematics, distribution and abundance of epiplanktonic squid decapod (Cephalopoda) larvae from the California Current April, 1954-March, 1957. *Bull. Scripps Inst. Oceanog.* 14: 1-90.

Pattullo, J. G., W. V. Burt, and S. A. Kulm. 1969. Oceanic Heat Content off Oregon: Its variations and their causes. *Limnol. Oceanogr.* 14: 279-287.

Pearcy, W. G. 1964. Some distributional features of mesopelagic fishes off Oregon. *J. Mar. Res.* 22: 83-102.

Pearcy, W. G. 1965. Species composition and distribution of pelagic cephalopods from the Pacific Ocean off Oregon. *Pac. Sci.* 19: 261-266.

Pearcy, W. G. 1972. Distribution and ecology of oceanic animals off Oregon. In: A. T. Pruter and D. L. Alverson (Eds.), "*The Columbia River Estuary and Adjacent Ocean Waters*". Univ. Wash. Press. 351-371.

Pearcy, W. G. 1976. Seasonal and inshore-offshore variations in the standing stocks of micronekton and macroplankton off Oregon. *NOAA Fish. Bull.* 74: 70-80.

Pearcy, W. G. and C. A. Forss. 1966. Depth distribution of oceanic shrimps (Decapoda; Natantia) off Oregon. *J. Fish. Res. Bd. Canada* 23: 1135-1143.

Pearcy, W. G. and C. A. Forss. 1969. The oceanic shrimp *Sergestes similis* off the Oregon coast. *Limnol. Oceanogr.* 14: 755-765.

Pearcy, W. G., E. Krygier, R. Mesecar, and F. Ramsey. In Press. Vertical distribution and migration of oceanic micronekton off Oregon. *Deep-Sea Res.*

Pearcy, W. G. and R. M. Laurs. 1966. Vertical migration and distribution of mesopelagic fishes off Oregon. *Deep-Sea Res.* 13: 153-165.

Pearcy, W. G. and R. S. Mesecar. 1971. Scattering layers and vertical distribution of oceanic animals off Oregon. In: "*Proceedings of the International Symposium on Biological Sound Scattering in the Ocean*", G. B. Farquhar (Ed.), Maury Center for Ocean Science Report 005. 381-394.

Pearcy, W. G. and C. L. Osterberg. 1967. Depth, diel, seasonal, and geographic variations in zinc-65 of midwater animals off Oregon. *Int. J. Oceanol. Limnol.* 1: 103-116.

Rosenberg, D. H. 1962. Characteristics and distribution of water masses off the Oregon coast. M.S. thesis, Oregon State University, Corvallis, Ore.

Smoker, W. and W. G. Pearcy. 1970. Growth and reproduction of the lanternfish *Stenobrachius leucopsarus*. *J. Fish. Res. Bd. Canada* 27: 1265-1275.

Stefansson, U. and F. A. Richards. 1963. Processes contributing to the nutrient distributions off the Columbia River and Strait of Juan de Fuca. *Limnol. Oceanogr.* 8: 394-410.

Urick, R. J. 1975. Principles of Underwater Sound. McGraw-Hill, New York, 384 pp.

Wickett, P. W. 1967. Ekman transport and zooplankton concentration in the North Pacific. J. *Fish. Res. Bd. Canada* 24: 581-594.

Wyatt, C., W. V. Burt, and J. G. Pattullo. 1972. Surface currents off Oregon as determined from drift bottle returns. J. *Phys. Oceanogr.* 2: 286-293.

Some Comparisons Between Oceanographic Measurements and High-Frequency Scattering of Underwater Sound

Richard E. Pieper

University of Southern California

ABSTRACT

High-frequency scattering of underwater sound at frequencies of 100 kHz and 200 kHz is discussed in relation to hydrographic macrostructure and microstructure (vertical) in the ocean. Data are presented which show scattering layers to be located at or near depths of density interfaces and chlorophyll maxima. Patches of scattering are also discussed and are related to the non-homogeneous distribution (patchiness) of organisms.

INTRODUCTION

The relationships between oceanographic parameters (both biotic and abiotic) and sound scattering should be compared on a three dimensional basis for all oceans, as well as over all times of the day and year. In practice, most observations and comparisons are made (a) over a vertical section (vertical profile) at one or more points in space and time or (b) horizontally at different depths to look at the distributions on the X-Y plane (non-random distributions or patchiness). Both types of information for many times and locations are necessary to piece together a three dimensional pattern.

Comparisons between oceanographic measurements and scattering in the past have been concentrated around the vertical-profile type of measurement and the acoustic frequencies used have usually been below 30 kHz. A review of these studies is given by *Hersey and Backus* (1962) and more recent papers are presented by *Farquhar* (1970). The primary scattering organisms at or below 30 kHz are probably fishes, especially those with swim bladders (e.g., *Farquhar*, 1970; *Valdez and Cushing*, 1966) or physonectid siphonophores which have gas-filled floats (*Barham*, 1963). Although zooplankton are often collected at depths which correspond to these low-frequency scattering layers, they are probably not the scatterer responsible for the observed layer (see for example *Hersey and Backus*, 1962). Where physical or chemical stratification occurs in the water column zooplankton may be separated vertically from the fishes, and scattering (12 kHz) will correspond to the fish distribution, but not zooplankton (*Bary*, 1966 a,b).

Scattering layers recorded at frequencies below 30 kHz have often been reported at or near discontinuities (thermoclines, haloclines, pycnoclines, oxyclines) in the water column (e.g., *Cushing et al.*, 1956; *Frassetto et al.*, 1962; *Herdman*, 1953). Similarly, some scattering layers which migrate diurnally in response to changing light levels (*Kampa and Boden*, 1954) appear to be stopped by such interfaces (e.g., *Wilson*, 1972).

High frequency echo sounders (frequencies greater than 30 kHz) have been used more recently to look at small-scale phenomenon and to record scattering from smaller organisms. *Northcote* (1964) correlated *Chaoborus* movements in a lake with a 200 kHz scattering layer and noted that downward movement of this layer stopped at the thermocline. Zooplankton distributions in Gull Lake, Michigan, were also correlated with a 200 kHz scattering layer and both were found to be concentrated immediately above and below the thermocline (*McNaught*, 1968, 1969).

Zooplanktonic scattering at high frequencies has also been reported from the oceanic environment and has often been correlated with oceanographic discontinuities. A 200 kHz scattering layer in the North Pacific was found to correspond to large numbers of the copepod *Calanus cristatus* (*Barraclough et al.*, 1969). This layer was arrayed at depths between 20 and 40 m which would be in the mixed layer just above the usual spring thermocline (reported to be around 30 m by *Dodimead et al.*, 1963). The importance of the thermocline in delimiting 100 kHz scattering was also shown by *Hansen and Dunbar* (1970). They correlated this 100 kHz layer with the presence of the pteropod *Spiratella helicina* which was at the interface between Arctic Surface Water and Arctic Intermediate Water. The abundance, distribution, and scattering-cross section of euphausiids were discussed in relation to 42, 100 and 200 kHz scattering in Saanich Inlet, British Columbia, by *Bary and Pieper* (1970), *Beamish* (1971), and *Pieper* (in press). When surface water (0 to about 100 m) became stratified with respect to the deeper waters in the inlet, as indicated by the presence of an oxycline around 100 m, scattering layers associated with euphausiids stopped and compressed against this interface.

The distributions of scattering and organisms may not, however, be correlated with specific oceanographic measurements. Fish school or shoals need not be limited by any specific interface or current system. Similarly, the larger zooplankton may be distributed vertically in relation to light or some other factor rather than a density discontinuity. The distributions of these larger zooplankters, as with most of the other plankton, are also often clumped or non-random. Such clumped or patchy distributions are similarly seen in sound scattering records.

Plankton, by definition, are drifters and their distribution would, therefore, be expected to be controlled to a large extent by oceanographic parameters. Recent work on patchiness in phytoplankton has shown that their patchy distribution in the mixed layer (patch size less than 100 m) may be correlated with water turbulence generated by the wind (*Platt*, 1972). Patch sizes larger than 100 m could not be correlated with this wind driven turbulence (*Powell et al.*, 1975).

Patchiness in zooplankton has been recognized by a number of authors (e.g., *Cassie*, 1963; *O'Connell*, 1971). *Wiebe* (1970) noted that zooplankton patches off Baja California had a mean radius of 13.6-15.6 meters during the day and 38.4-73.1 meters during the night. He also found that the average density per patch varied from 2.6-5.1 times the background density. *Stavn* (1971) showed experimentally that *Daphnia* accumulated at certain specific locations relative to induced currents similar to Langmuir spirals which were described in nature by *Langmuir* (1938). The existence of Langmuir circulation may explain some of the larger scale patchiness found above the thermocline. At present, however, the reason for this clumped or patchy distribution is not well understood, nor well described. Similarly, the relationship between the vertical distribution of organisms, hydrography, and sonic scattering has not been well defined.

Present research involves quantification of 100 kHz scattering to determine scattering energies. These energies will then be compared to the distribution of potential scatterers (primarily euphausiids, larger crustaceans, and fishes) to determine the extent to which high-frequency scattering can be used to determine standing stock estimates of zooplankton. The non-random distribution of zooplankton (patchiness) will also be studied.

This paper is a preliminary report on the study. The potential use of high-frequency scattering studies is discussed and some data are presented (qualitative) on comparisons with oceanographic parameters and scattering patchiness.

MATERIALS AND METHODS

The data presented were taken on three different oceanographic cruises, two off southern California and one in Saanich Inlet, British Columbia. The echograms were recorded on two different Ross echo sounders (Ross Laboratories, Seattle, Washington) operated at frequencies of 100 kHz (California) and 200 kHz (B.C.). Both sounders had a narrow beam angle (full-angle, half-power point of 3.5°, 100 kHz; 5° x 7°, 200 kHz). The pulse lengths used were 1.0 msec (100 kHz) and 0.5 msec (200 kHz). Calibration of the 200 kHz sounder has been completed (see Pieper, in press), as well as a preliminary calibration of the 100 kHz sounder. A complete calibration of the 100 kHz system will be completed shortly.

Water salinity was measured with a Beckman model RS-7B Induction Salinometer. Chlorophyll was measured by continuous flow fluorometry using a model 111 Turner Fluorometer. The temperature information with the Saanich Inlet data was measured by a thermister attached to the instrumented, plankton-sampler, Catcher II (Bary and Frazer, 1970). Biological samples from the Saanich study were taken with the Catcher II (CAT) and are reported by Pieper (in press) Biological collections off southern California were taken with a 10' Isaacs-Kidd Midwater Trawl or a 6', opening-closing, modified Tucker Trawl (Davies and Barham, 1969). Data analysis from these trawls are not yet completed and will be published, along with the quantitative acoustic measurements, in a later report.

RESULTS AND DISCUSSION

Echograms at 100 kHz from 0-125 m are compared to salinity (Figure 1) and chlorophyll a (Figure 2). The scattering profile in Figure 1, recorded from the Catalina Basin off southern California, shows surface scattering and major layer from 38-78 meters. This layer appears to be just above the halocline. Figure 2 shows the echogram and a chlorophyll a profile taken aboard the David Starr Jordan (National Marine Fisheries Service) approximately 37 miles West of Pt. Loma, southern California. Scattering is recorded from 0-8 meters, 20-40 meters, and other layers centered around 55, 70 and 95 meters. The chlorophyll a profile indicates a maximum near the surface, minimum around 12 meters (where no scattering is recorded) and an increase at the depth where the scattering is strong.

These echograms and others from the high-frequency sounders show a tremendous increase in the amount of sonic scattering recorded when compared to low frequency measurements (see Bary and Pieper, 1970; Pieper, in press). A greater number of specific layers is also evident. The fact that much of the scattering is associated with discontinuity layers is not too surprising, nor is the correlation between chlorophyll a and scattering. The accumulation of phytoplankton, detritus and micro-organisms at density interfaces is well known (e.g., Cushing et al., 1956; Kriss, 1960). The association of zooplankton with discontinuities also is common (e.g., McNaught, 1968). This correlation might result from passive sinking or upward movement which is stopped by the interface. It is also likely, however, that the local food web interacts to the extent to lead zooplankton to the phytoplankton (high-frequency

OCEANOGRAPHIC MEASUREMENTS SOUND SCATTERING 671

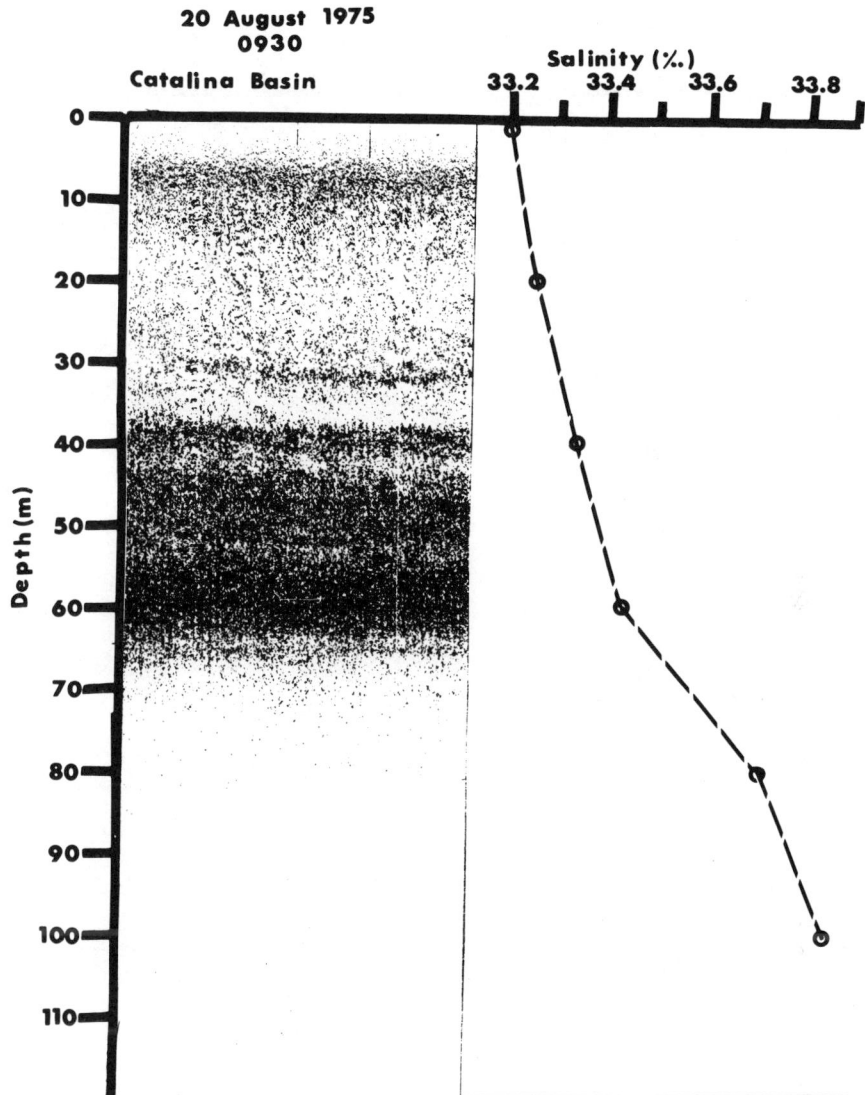

Fig. 1. Echogram at 100 kHz and a salinity profile; Santa Catalina Basin, 20 August 1975, 0930 P.D.T.

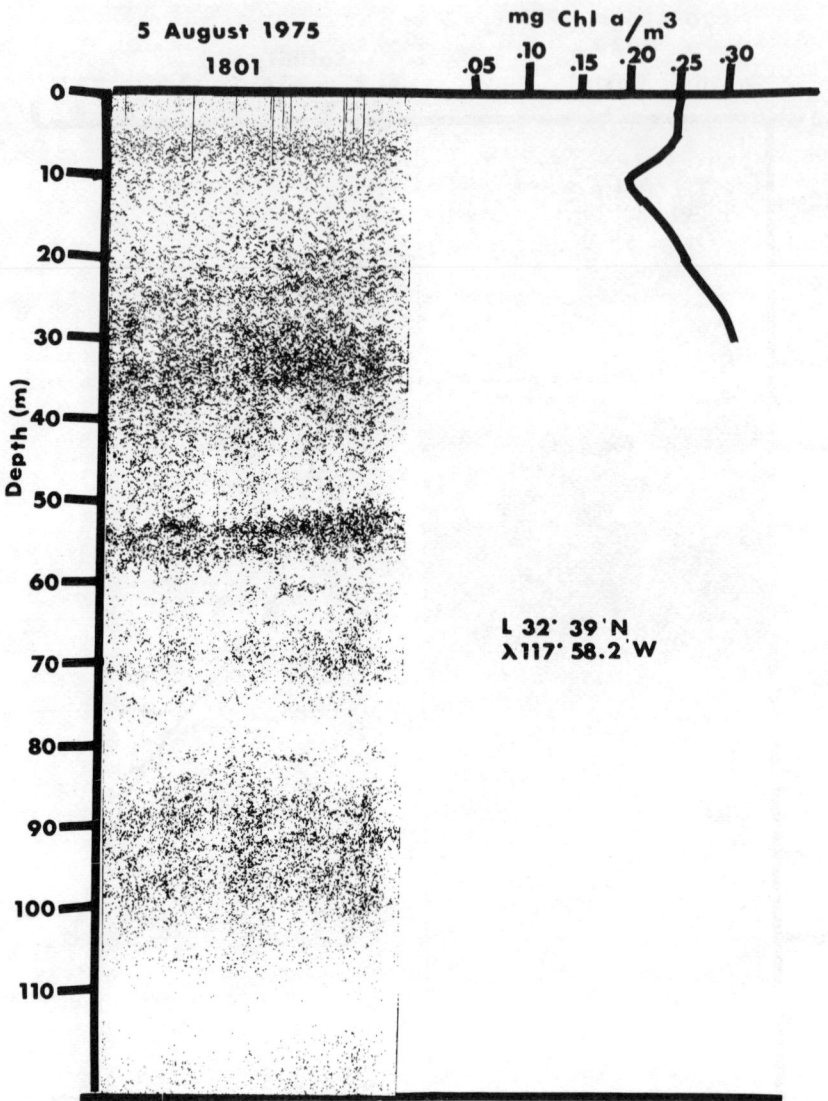

Fig. 2. Echogram at 100 kHz plus a chlorophyll *a* profile; 37 miles West of Pt. Loma, 5 August 1975, 1801 P.D.T.

scattering) and fishes to the zooplankton (low-frequency scattering). Thus, correlations between scattering profiles and hydrographic stratification may be extremely common in the euphotic zone and/or mixed layer. Further work, expecially with the high-frequency echo sounders, should better define this interaction.

Stratification in the deeper depths below the pycnocline also occurs. Much of this in the past has been correlated with light or isolumes (e.g., *Kampa and Boden*, 1954) or oxygen minimum zones (*Childress*, 1968). The relationships between scattering and the distribution of biological organisms with hydrographic structure (micro-structure) at these deeper depths are poorly defined, but they might exist.

Patchiness or non-random distributions of plankton and fishes have plagued oceanic studies for many years. Such patchiness is also recorded on echograms. Acoustically-dense patches from the high-frequency echo-sounders are shown in Figues 3 and 4. The "patch" (Figure 3) corresponds to a mean depth of 180 meters in the Santa Catalina Basin off southern California. This patch is similar to many others recorded from the local area; it has an approximate size of 830 meters in the X or Y direction and is 25 meters thick. Biological collections at this depth show high concentrations of euphausiids which vary in number from 1.0 to 10.8 per lineal meter of water traversed (6' Tucker Trawl; if the angle of the net is assumed to be 45°, this euphausiid concentration would equal 0.4-4.6/m^3). Quantitative acoustic data are presently being collected which will be compared with the biological collections to determine whether or not euphausiids can account for the observed scattering. A 28 kHz echo sounder is being used to determine the presence of absence of fish scattering at the depth of these high-frequency scattering layers.

A similar type of patchiness is shown at 200 kHz in Saanich Inlet (Figure 4). This layer has been definitely associated with the presence of the euphausiid, *Euphausia pacifica* (*Bary and Pieper*, 1970; *Pieper*, in press). The patch in the inlet is located at a mean depth of 85 meters and its size is approximately 15 meters deep by about 500 m in horizontal extent. Euphausiid concentrations in this layer varied from 40-380/m^3 (CAT).

A trace of the temperature variation at the depth of the biological sampler (CAT) is also shown in Figure 4. It is interesting to note that the temperature decreases from 9.2° C to 8.7° C as the gear moves from an area where no "patch" occurs to an area where a "patch" exists (even though the depth is decreasing and a temperature increase would be expected). From this data it is tempting to suggest some sort of localized upwelling to explain the existence of the "patch" shown on the echogram and the corresponding high euphausiid concentrations. This suggestion must, however, remain very tentative due to the lack of data.

There is a surprising similarity between patch structure as reflected in the echograms off Southern California (Figure 3) and

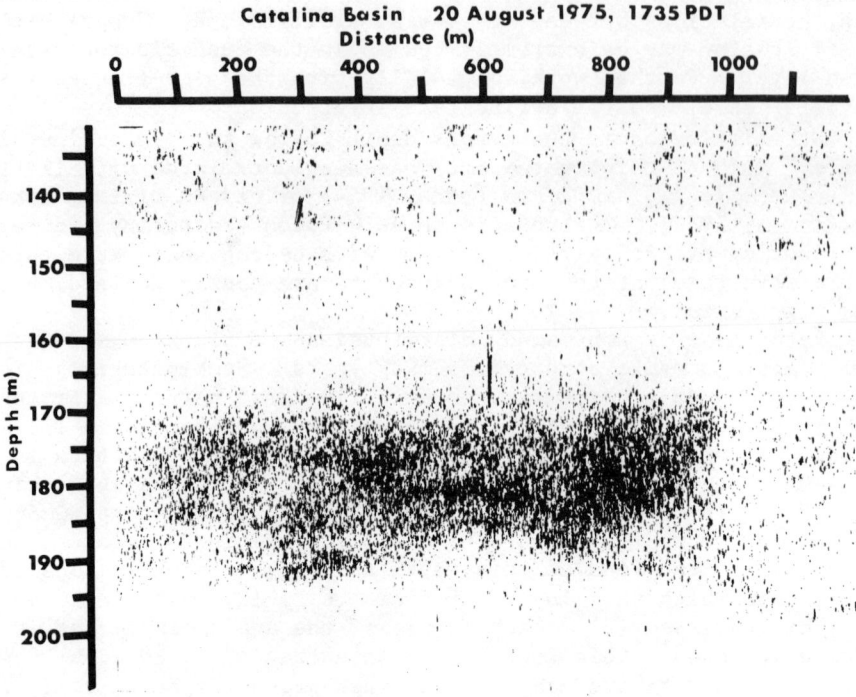

Fig. 3. Echogram at 100 kHz; Santa Catalina Basin, 20 August 1975, 1735 P.D.T.

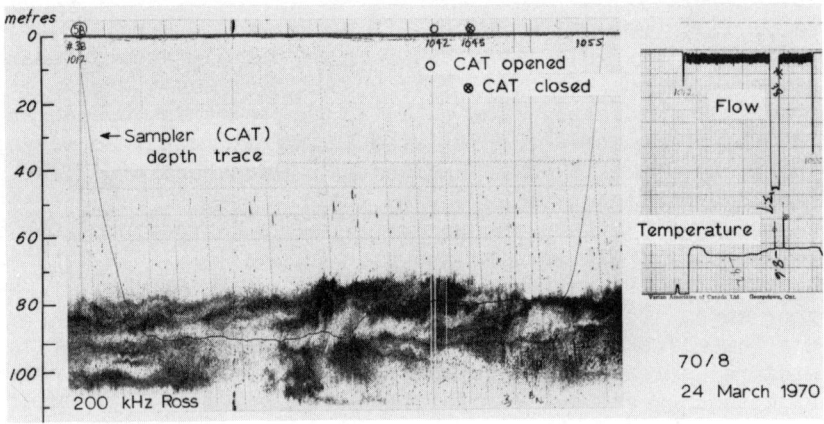

Fig. 4. Echogram at 200 kHz during biological sampling operations with a depth trace of the sampler, and a chart record of the temperature variations detected by a thermister on the sampler and recorded at the time of sampling; Saanich Inlet, B.C., 24 March 1970, 1040 P.S.T.

in Saanich Inlet (Figure 4) even though the sizes are slightly different. Due to the beam angle of the echo sounders, a symmetrical 15 meter "patch" or aggregation as described by *Wiebe* (1970) might not be discernable. Such a "patch" might be evident if the echo sounder transducer were lowered to depths near the source of scattering. Similarly, variations in numbers (approximately 10 fold off southern California and in Saanich; about 5 fold by Wiebe) may be a function of the times of collection represented in the present studies. Euphausiid concentrations are known to be low early in the day (more diffuse aggregations) and then to increase during the day as the shoal compresses before migrating to the surface (*Pieper*, in press).

The existence of patchiness in the ocean has been shown by many studies; the explanations as to how and why are, however, still in question. Such aggregations may be largely controlled by abiotic processes such as upwelling, or may be biologically controlled by, for example, group behavior similar to that found in schooling fishes.

SUMMARY AND CONCLUSIONS

The use of high-frequency echo sounders should produce more detailed scattering profiles, and these vertical profiles should often correspond to measurable hydrographic parameters. Thus, the locations of this type of scattering could possibly be predicted based on such hydrographic measurement, especially in the surface (or near surface) areas.

Patchiness, on the other hand, will still present problems to the biologist and acoustician. Such aggregations are still poorly understood and their relative locations are unmapped. Continued use of echo sounders at the higher frequencies may help to better define and understand these distributions.

The data presented in this paper are very preliminary and, due to their qualitative nature, only appropriate for producing generalizations. Quantitative scattering data are needed at a wide range of frequencies (especially from the higher frequencies) to produce a better understanding of structure in the oceanic system.

ACKNOWLEDGEMENTS

This work is supported in part by the Office of Naval Research, contracts number N00014-67-A-0269 and N00014-75-C-0683.

REFERENCES

Barham, E. E. 1963. Siphonophores and the deep scattering layer. *Science, 40*: 826-828

Barraclough, W. E., R. J. Le Brasseur, and O. D. Kennedy. 1969. Shallow scattering layer in the subarctic Pacific Ocean: detection by high-frequency echo sounder. *Science, 166*: 611-613.

Bary, B. McK. 1966a. Back scattering at 12 kc/s in relation to biomass and number of zooplankton organisms in Saanich Inlet, British Columbia. *Deep-Sea Res., 13*: 655-666.

Bary, B. McK. 1966b. Qualitative observations of scattering of 12 kc/s sound in Saanich Inlet, British Columbia. *Deep-Sea Res., 13*: 667-677.

Bary, B. McK. and E. J. Frazer. 1970. A high-speed, opening-closing plankton sampler (Catcher II) and its electrical accessories. *Deep-Sea Res., 17*: 825-835.

Bary, B. McK. and R. E. Pieper. 1970. Sonic-scattering studies in Saanich Inlet, British Columbia: A preliminary report, pp. 601-611. In: G.B. Farquhar (ed.), *Proceedings of an International Symposium on Biological Sound Scattering in the Oceans*, Maury Center for Ocean Science, Washington, D. C.

Beamish, P. 1971. Quantitative measurements of acoustic scattering from zooplanktonic organisms. *Deep-Sea Res., 18*: 811-822.

Cassie, R. M. 1963. Microdistribution of plankton, In: H. Barnes (ed.) *Oceanogr. Marine Biol., 1*: 223-252.

Childress, J. J. 1968. Oxygen minimum layer: vertical distribution and respiration of the mysid *Gnathophausia ingens*. *Science, 160*: 1242-1243.

Cushing, D. H., A. J. Lee, and I. D. Richardson. 1956. Echo traces associated with thermoclines. *J. Mar. Res., 15*: 1-13.

Davies, I. E. and E. G. Barham. 1969. The Tucker opening-closing micronekton net and its performance in a study of the deep scattering layer. *Mar. Biol., 2*: 127-131.

Dodimead, A. J., F. Favorite, and T. Hirano. 1963. Salmon of the North Pacific Ocean. Part II. Review of oceanography of the subarctic Pacific region. *International North Pacific Fisheries Commission, Bulletin No. 13.* 195 p.

Farquhar, G. B. (ed.). 1970. *Proceedings of an International Symposium on Biological Sound Scattering in the Ocean*, Maury Center for Ocean Science, Washington, D. C.

Frassetto, R., R. H. Backus, and E. Hays. 1962. Sound-scattering layers and their relation to thermal structure in the Strait of Gibraltar. *Deep-Sea Res., 9*: 69-72.

Hansen, W. J. and M. J. Dunbar. 1970. Biological causes of scattering layers in the Arctic Ocean. In: G.B. Farquhar (ed.), *Proceedings of an International Symposium on Biological Sound Scattering in the Ocean*, Maury Center for Ocean Science, Washington, D. C. 508-526.

Herdman, H. F. P. 1953. The deep scattering layer in the sea: association with density layering. *Nature, 172*: 275-276.

Hersey, J. B. and R. H. Backus. 1962. Sound scattering by marine organisms. No. 13, pp. 498-539. In: M. Hill (ed.), *The Sea: Ideas and Observations on Progress in the Study of the Seas*. Volume 1. Interscience, N.Y.

Kampa, E. M. and B. P. Boden. 1954. Submarine illumination and the twilight movement of a sonic scattering layer. *Nature, 174*: 869-874.

Kriss, A. E. 1960. Micro-organisms as indicators of hydrological phenomenon in seas and oceans I. *Deep-Sea Res., 6*: 88-94.

Langmuir, I. 1938. Surface motion of water induced by wind. *Science, 87*: 119-123.

McNaught, D. C. 1968. Acoustical determination of zooplankton distribution. *Proc. 11th Conf. Great Lakes Res., 1968*: 76-84.

McNaught, D. C. 1969. Developments in acoustic plankton sampling. *Proc. 12th Conf. Great Lakes Res., 1969*: 61-68.

Northcote, T. G. 1964. Use of a high-frequency echo sounder to record distribution and migration of *Chaoborus* larvae. *Limnol. Oceanogr., 9*: 87-91.

O'Connell, C. P. 1971. Variability of near-surface zooplankton off Southern California, as shown by towed pump sampling. *Fish. Bull., 69*: 681-697.

Pieper, R. E. In press. A study of the relationship between zooplankton and high-frequency scattering of underwater sound. *Museum of Natural History of Los Angeles County Science Series*.

Platt, T. 1972. Local phytoplankton abundance and turbulence. *Deep-Sea Res., 19*: 183-187.

Powell, T. M., P. J. Richerson, T. M. Dillon, B. A. Agee, B. J. Dozier, D. A. Godden, and L. O. Myrup. 1975. Spacial scales of current speed and phytoplankton biomass fluctuations in Lake Tahoe. *Science, 189*: 1088-1090.

Stavn, R. H. 1971. The horizontal-vertical distribution hypothesis: Langmuir circulations and *Daphnia* distributions. *Limnol. Oceanogr., 16*: 453-466.

Valdez, V. and D. H. Cushing. 1966. The diurnal variations in depth and quantity of echo traces and their distribution in area in the southern bight of the North Sea. *J. Cons. Perm. Int. Explor. Mer., 30*: 237-254.

Wiebe, P. H. 1970. Small-scale spacial distribution in oceanic zooplankton. *Limnol. Oceanogr., 15*: 205-217.

Wilson, D. F. 1972. Diel migration of sound scatterers into, and out of the Cariaco Trench anoxic water. *J. Mar. Res., 30*: 168-176.

On the Prediction of Sound Scattering in the Oceans from Fish Capture Data

Paul Scully-Power

Royal Australian Naval Research Laboratory

ABSTRACT

Based on the results of Project Ocean Acre, avoidance of towed-net samplers is shown to be the major factor contributing to the discrepancy between acoustically measured and biologically derived sound scattering strength profiles in the ocean. The theory is developed of the maximum likelihood of capture which, together with Barkley's (1972) theory, provides both an upper and lower bound to the probability of capture by towed-net samplers.

INTRODUCTION

In assessing any model of the complex ecological chain, stretching from physico-chemical properties of the ocean to biological community structure and its resultant sound scattering properties at various frequencies, one of the greatest areas of inexactitude lies in adequately determining the transfer function between nekton population and depth dependent volume scattering strength. Since the principal features of the observed acoustic responses can be explained by the swimbladder resonance (Holliday, 1972), the volume scattering strength can be calculated from the product of the acoustic cross-sections and the fish densities.

The major uncertainties in estimating the acoustic cross-sections are in assigning Q-values to the sharpness of resonance and in determining the swimbladder systematics, i.e., whether swimbladders operate under constant mass or constant volume conditions. Since these parameters are functions of pressure, the inaccuracies in determining acoustic cross-sections can be categorized as depth dependent factors. On the other hand, a lack of knowledge of the effectiveness of towed-net samplers is the greatest impediment to accurately assessing nekton populations. Since net effectiveness is a combined measure of filtering efficiency and selectivity which, to a first order, are not functions of pressure, the inaccuracies in determining fish densities can be categorized, conversely, as depth independent factors. Thus there are two distinct categories of major factors affecting the accurate prediction of volume reverberation in the ocean from estimates of nekton population. It remains to rank order these factors as to their importance in the evaluation of the transfer function and to ascertain if one category dominates the other.

Although very few programs have compared fish capture data to simultaneous *in situ* measurements of volume scattering strength profiles with depth, one of the most intensive efforts to date, the Ocean Acre Project (*Brown and Brooks*, 1974), produced a most remarkable result. It was observed that there was generally an excellent correlation between the shapes of the acoustically measured and biologically derived scattering strength profiles (*MacDonald*, 1972) although the biologically derived values consistently underestimated the acoustically measured values. A typical example is given in Figure 1. The fact that the shapes of these profiles were coherent with depth and that the biologically derived profile was always an underestimate indicated a probable marked improvement in the model prediction if only the depth independent factors could be better determined.

As a first step, the net filtering efficiency of the 3-meter Isaacs-Kidd midwater trawl used on Ocean Acre was carefully determined (*Brooks, Brown and Scully-Power*, 1974). The measured efficiency of 92%, however, could not account for the discrepancy in magnitude between the biological and acoustical profiles, so attention was then turned to consideration of better estimates of net selectivity.

Net catches are usually biased and hence selective because some fish avoid the net and others are lost through the meshes. This bias is in general strongly bimodal; avoidance being paramount for larger more motile fish while mesh escapement is important for very small fish. Escapement can be modified by suitable selection of mesh size for the population being sampled and some estimate can be made of this factor (*Vannucci*, 1968). Avoidance, however, is much more difficult to parameterise, and recent predictive models of volume scattering strength (*Love*, 1975) do not include this factor. Hence an adequate avoidance model is required before sound scattering in the oceans can be fully predicted.

AVOIDANCE

An excellent summary of the avoidance of samplers is given by *Clutter and Anraku* (1968) who point out the high variability in test

THE PREDICTION OF SOUND SCATTERING

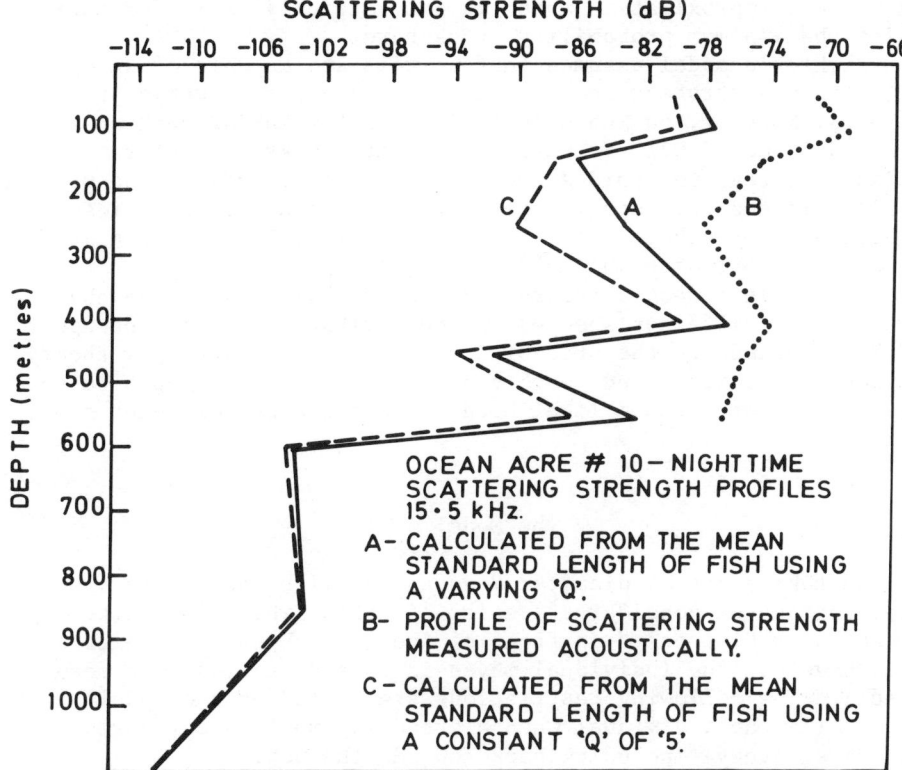

Fig. 1. Comparison of profiles generated from biological data with acoustically measured profile. (After *Brown and Brooks*, 1974).

results on the calibration of catching efficiency of nets and the difficulty in formulating deterministic models because of the varying behavioral responses of different animals.

Barkley (1964) propounded the first theoretical model of avoidance in which he calculated the minimal escape velocity relationships for an organism in terms of the pertinent kinematic parameters. In an extension of that work, *Barkley* (1972) derived a relationship for the theoretical minimum probability of certain capture by constructing a model in which "animals which could possibly escape do so". This resulted in a reduction of the geometrical considerations to two dimensions and a reasonably simple analytical solution.

A similar derivation to that of Barkley was given by *Laval* (1974), although in trying to generalize the results to escape directions other than the optimum direction he retains the two dimensional approximation which is only strictly valid for estimating the minimum probability of capture.

Barkley's model assumes that animals are capable of an optimum avoidance strategy and that in fact they adopt such a strategy. There is, however, no known justification for making such an assumption and it could well be that, in reality, nekton adopt some suboptimum strategy to avoid a towed net or that they adopt no stragegy at all. Indeed some species may have developed better avoidance strategies than others (*Fleminger and Clutter*, 1965; *McGown and Fraundorf*, 1966; *Aron and Collard*, 1969).

Hence, in order to relate net captures with avoidance characteristics of different species, we must first know both the upper and lower bounds to the probability of capture. Barkley's theory provides the lower bound to this probability; we develop here the upper bound, or maximum likelihood, of capture by towed-net samplers.

The Model

We take a net of diameter D advancing with speed U in one direction and we consider a species of nekton which is alerted to the net at a distance d in front of the net. Upon being alerted we assume that the individual moves in some direction at a mean speed u. The following restrictions are placed on the model:

(a) the individual when alerted may move in any direction (random strategy) including back towards the net,

(b) once it enters the net the individual is trapped, i.e., it does not swim out nor are there any net losses,

(c) each individual reacts separately to the presence of the net, i.e., there are no schooling effects,

(d) in reacting to the net, each individual moves in a straight line,

(e) only those individuals in the swept out volume of the net react to the net,

(f) the net has a circular mouth opening.

Evaluation of the Capture Probability

At a distance d in front of the net there is a circle of possible locations from each of which an individual may move in any direction at a speed u. In plan view the geometry of the situation is:

THE PREDICTION OF SOUND SCATTERING

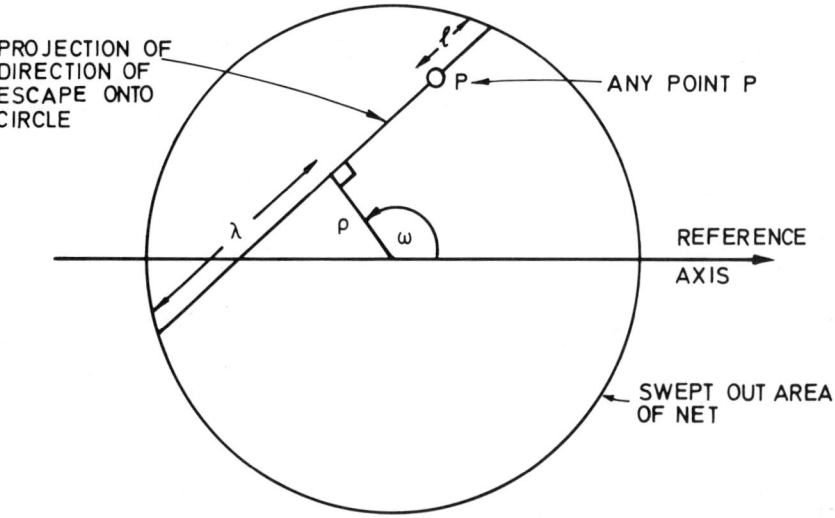

Fig. 2. Geometry at distance d from the net.

It is shown in Annex A that the probability ν_1 of an individual being at position P and moving in a direction such as that indicated above is given by

$$\nu_1 = (\pi^2 R^2)^{-1} \, d\omega \, d\ell \, d\rho \qquad (1)$$

The probability of escape must be calculated for all directions which have the same projection onto the above circle, i.e., the probability of escape in the plane represented by:

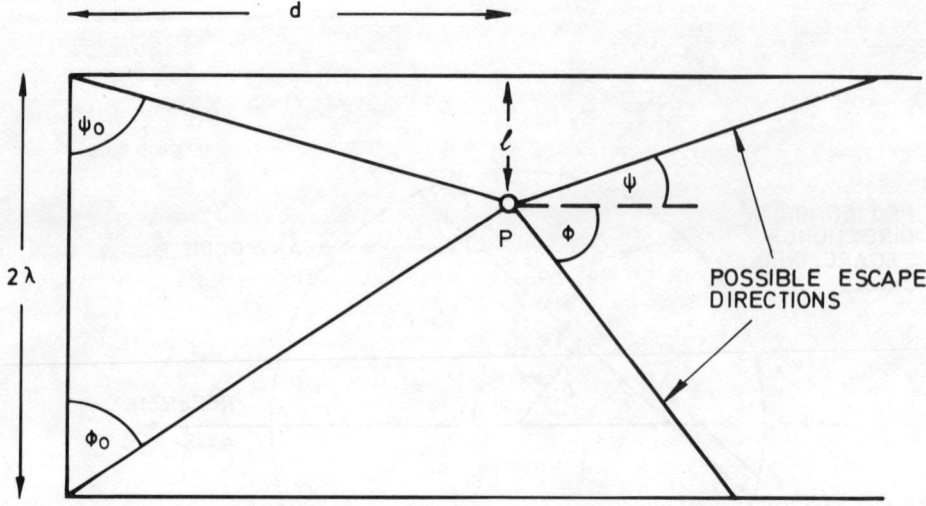

Fig. 3. Geometry in the plane of motion of the individual.

Using this geometry, it is shown in Annex B that the probability of escape ν_2 in this plane is given by

$$\nu_2 = \pi^{-1} \left[\text{arc cos} \left(\frac{\cos \psi_0}{a} \right) + \text{arc cos} \left(\frac{\cos \phi_0}{a} \right) \right] \quad (2)$$

Hence, by integrating the total elemental probability $\nu_1 \nu_2$ the total probability of escape is obtained;

$$P_E = (\pi^3 R^2)^{-1} \int_0^R \int_0^{2(R^2-\rho^2)^{1/2}} \int_0^{2\pi} \left[\text{arc cos} \left(\frac{\cos \psi_0}{a} \right) \right.$$

$$\left. + \text{arc cos} \left(\frac{\cos \phi_0}{a} \right) \right] dwd\ell d\rho \quad (3)$$

where R is the radius of the net.

The evaluation of equation (3) is by no means straight forward. Indeed it does not have an analytical solution. However, it is shown in Annex C that equation (3) can be transformed into a single integral and, when converted into a probability of capture P_c, it takes the form:

THE PREDICTION OF SOUND SCATTERING

For $a < 1/p$

$$P_c = 1 - (8/\pi^2)s \int_0^a \frac{(1-p^2x^2)^{1/2}}{(1-x^2)^2} \arccos\left(\frac{x}{a}\right) dx$$

For $1/p \leq a \leq 1$

$$P_c = (8/\pi^2)s \int_0^{1/p} \frac{(1-p^2x^2)^{1/2}}{(1-x^2)^2} \arcsin\left(\frac{x}{a}\right) dx$$

(4)

For $a > 1$

$$P_c = (4/\pi^2)s \int_0^{1/p} \frac{(1-p^2x^2)^{1/2}}{(1-x^2)^2} \arcsin\left(\frac{x}{a}\right) dx$$

$$+ (4/\pi^2)s \int_0^{1/p} \frac{(1-p^2x^2)^{1/2}}{(1-x^2)^2} \arcsin x \, dx$$

where:

$a = u/U$

$s = d/D$

$p = (1+s^2)^{1/2}$

At first sight, the splitting of the solution into three ranges appears unusual. However, there is a physical basis (as well as a mathematical basis) for this partition. It only needs to be noted that for any point P, i.e., for any distance ℓ on the "λ plane" of Figure 2, there must exist an optimum strategy. Hence, as d diminishes (or equally u diminishes) there must arise a situation in which only the optimum strategy will suffice for escape. The case where $\ell=0$, yet the value of u for a given d is just great enough for the individual to traverse the λ plane and escape (i.e., it does have an escape angle ϕ in Figure 3) is given by $a = 1/p$ in the limiting case when $\lambda = R$. Hence, the first partition.

As u increases (and hence a increases) the individual does not have to adopt the optimum strategy in order to escape. In this case, therefore, there is a finite angle in which the individual may escape both in the ψ-direction and the ϕ-direction. As u is increased further, there comes a point in which these escape angles (ψ and ϕ) begin to overlap. This point is reached when $a = 1$, i.e., the escape speed of the individual equals the net speed. Hence, the second partition.

DISCUSSION

The solution given in (4) has been evaluated numerically and the results plotted in Figure 4 as a function of capture probability against a for various values of s. As has been indicated earlier, this is the maximum probability of capture for any given species. It may well be argued that any given species would not adopt a random stragegy as it would soon learn to adopt a strategy which would give it a better chance of survival.

However, as soon as a species adopted a non-random strategy, its probability of capture would diminish until such time that it always adopted the optimum strategy in which case the capture probability would be the minimum as given by *Barkley* (1972). A quantification of the avoidance strategy adopted by any given species, however, is far from being completed. Nevertheless, in the interim, a knowledge of both the minimum and maximum probabilities of capture will enable data on nekton capture by towed-net samplers to be analyzed at least in a semi-quantitative manner.

Annex A:

CALCULATION OF PROBABILITY THAT INDIVIDUAL IS AT POSITION P AND MOVES ON THE "λ-PLANE"

The area of segment of the circle given in Figure 2 is

$$A = 1/2\pi R^2 - \left[\rho(R^2 - \rho^2)^{1/2} + R^2 \arcsin(\rho/R)\right] \tag{A1}$$

So,

$$dA/d\rho = -(R^2 - \rho^2)^{1/2} + \rho^2/(R^2 - \rho^2)^{1/2} - R^2/(R^2 - \rho^2)^{1/2}$$
$$= -2(R^2 - \rho^2)^{1/2} \tag{A2}$$

Hence, the probability that P lies on a chord distant ρ from the center of the circle is, for any given w,

$$\frac{2(R^2 - \rho^2)^{1/2}}{1/2\pi R^2} d\rho$$

$$= \frac{4}{\pi R^2} (R^2 - \rho^2)^{1/2} d\rho$$

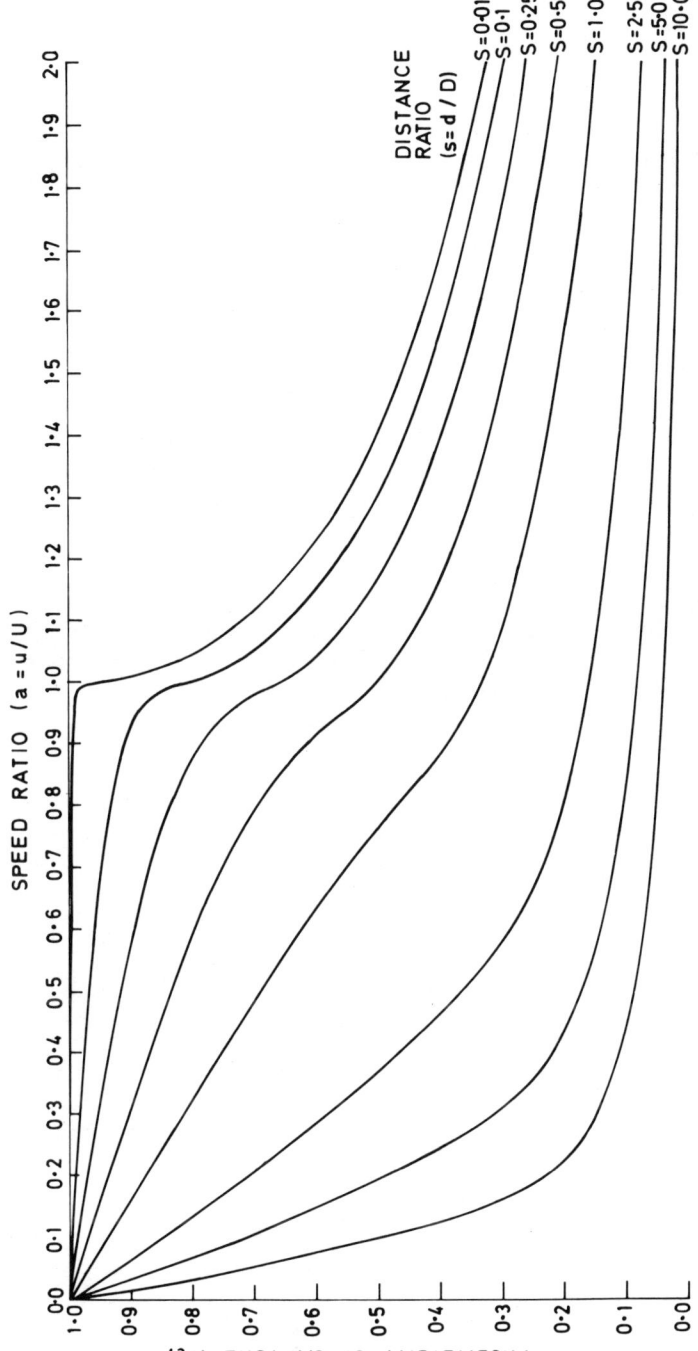

Fig. 4. Probability of capture versus speed ratio for various values of distance ratio.

Also, the probability that P lies at length ℓ from a given end of this chord is

$$1/2 (R^2 - \rho^2)^{-1/2} d\ell$$

And, the probability that the chord is at an angle given by ω is

$$\frac{\omega}{2\pi}$$

Hence, the total elemental probability ν_1 is given by

$$\nu_1 = (\pi^2 R^2)^{-1} d\omega d\ell d\rho \qquad (A3)$$

Annex B:

CALCULATION OF PROBABILITY OF ESCAPE ON THE "λ-PLANE"

To escape, an individual must travel to the outside of the "swept out" volume of the net in the time the net travels to that point. Hence, from Figure 3, an individual on a λ-trajectory must travel a distance of $\ell \operatorname{cosec} \psi$ while the net travels a distance $d + \ell \cot \psi$. For escape, then, the escape velocity u_e is given by

$$u_e = \frac{U \ell \operatorname{cosec} \psi}{d + \ell \cot \psi}$$

$$= \frac{U \operatorname{cosec} \psi}{\tan \psi_0 + \cot \psi}$$

$$= \frac{U \cos \psi_0}{\cos (\psi - \psi_0)} \qquad (B1)$$

Hence, the minimum escape velocity is $U \cos \psi_0$ and is associated with the ψ-trajectory given by $\psi = \psi_0$. Thus, for an individual having an alerted speed $u(>u_e)$, it will escape on all ψ-trajectories in the range

$$\psi = \psi_0 \pm \operatorname{arc} \cos \left(\frac{\cos \psi_0}{a}\right) \qquad (B2)$$

So, the total escape angle for a ψ-trajectory is

$$2 \operatorname{arc} \cos \left(\frac{\cos \psi_0}{a}\right)$$

THE PREDICTION OF SOUND SCATTERING

Similarly, the total escape angle for a ϕ-trajectory is

$$2 \text{ arc cos}\left(\frac{\cos \phi_0}{a}\right)$$

Since the total possible angle in which an individual may move is 2π, the probability ν_2 of escape on the λ-plane is

$$\nu_2 = \pi^{-1}\left[\text{arc cos}\left(\frac{\cos \psi_0}{a}\right) + \text{arc cos}\left(\frac{\cos \phi_0}{a}\right)\right] \quad (B3)$$

Annex C:

EVALUATION OF THE TOTAL CAPTURE PROBABILITY INTEGRAL

From (3), the probability of escape is given by:

$$P_E = (\pi^3 R^2)^{-1} \int_0^R \int_0^{2(R^2-\rho^2)^{1/2}} \int_0^{2\pi} \left[\text{arc cos}\left(\frac{\cos \psi_0}{a}\right) + \text{arc cos}\left(\frac{\cos \phi_0}{a}\right)\right] dw\, d\ell\, d\rho \quad (C1)$$

Integrating (straight forwardly) with respect to w yields

$$P_E = 2(\pi^2 R^2)^{-1} \int_0^R \int_0^{2(R^2-\rho^2)^{1/2}} \left[\text{arc cos}\left(\frac{\cos \psi_0}{a}\right) + \text{arc cos}\left(\frac{\cos \phi_0}{a}\right)\right] d\ell\, d\rho \quad (C2)$$

Examining the inner integral

$$\int_0^{2(R^2-\rho^2)^{1/2}} \text{arc cos}\left(\frac{\cos \psi_0}{a}\right) d\ell$$

and making the transformation;

$$x = \cos \psi_0 = (1 + d^2/\ell^2)^{-1/2}$$

the following is obtained

$$\ell = xd(1-x^2)^{-1/2}$$

$$d\ell/dx = d(1-x^2)^{-3/2}$$

Hence

$$\int_0^{2(R^2-\rho^2)^{1/2}} \arccos\left(\frac{\cos\psi_0}{a}\right) d\ell = d\int_0^{x_\rho} (1-x^2)^{-3/2}$$

$$\arccos(x/a)\,dx \qquad (C3)$$

where

$$x_\rho = \left[1 + d^2/4(R^2-\rho^2)\right]^{-1/2}$$

Similarly, by making the transformation

$$y = \cos\phi_0 = \left[1 + d^2/\left(2\left[R^2-\rho^2\right]^{1/2} - \ell\right)^2\right]^{-1/2}$$

It can be shown that

$$\int_0^{2(R^2-\rho^2)^{1/2}} \arccos\left(\frac{\cos\phi_0}{a}\right) d\ell = d\int_0^{x_\rho} (1-x^2)^{-3/2}$$

$$\arccos(x/a)\,dx \qquad (C4)$$

Thus, by combining the results of (C3) and (C4) in equation (C2)

$$P_E = 4d(\pi^2 R^2)^{-1} \int_0^R \int_0^{x_\rho} \left[(1-x^2)^{-3/2} \arccos(x/a)\right] dx\,d\rho \qquad (C5)$$

is obtained.

Integration of the inner integral

$$F(x_\rho) = \int_0^{x_\rho} (1-x^2)^{-3/2} \arccos(x/a)\,dx$$

THE PREDICTION OF SOUND SCATTERING

gives

$$F(x_\rho) = \left[2(R^2 - \rho^2)^{1/2}/d\right] \text{arc cos}(x_\rho/a) + \text{arc sinh}$$

$$\left(a\left[1 - a^2\right]^{-1/2}\right) - \text{arc sinh}\left(\left[a^2 - x_\rho^2\right]\left[1 - a^2\right]^{-1}\right)^{1/2}$$

and it is noted that

$$F(x_\rho) = 0 \qquad \text{for } \rho = R;$$

and

$$\left[d/dx_\rho\right] F(x_\rho) = (1 - x_\rho^2)^{-3/2} \text{arc cos}(x_\rho/a)$$

and since

$$\left[d/d\rho\right] F(x_\rho) = \left[dx_\rho/d\rho\right]\left[d/dx_\rho\right] F(x_\rho)$$

the following is obtained

$$\left[d/d\rho\right] F(x_\rho) = \left[-2\rho(1 - x_\rho^2)^{3/2}/d(R^2 - \rho^2)^{1/2}\right]$$

$$\times \left[(1 - x_\rho^2)^{-3/2} \text{arc cos}(x_\rho/a)\right]$$

$$= -2\rho/d(R^2 - \rho^2)^{1/2} \text{arc cos}(x_\rho/a)$$

(C5) can now be written in the form:

$$P_E = 4d(\pi^2 R^2)^{-1} \int_0^R F(x_\rho) \, d\rho$$

$$= 4d(\pi^2 R^2)^{-1} \left\{\left[\rho F(x_\rho)\right]_0^R + 2/d \int_0^R \rho^2 (R^2 - \rho^2)^{-1/2} \text{arc cos}\right.$$

$$(x_\rho/a) \, d\rho\right\}$$

$$= 8(\pi^2 R^2)^{-1} \int_0^R \rho^2 (R^2 - \rho^2)^{-1/2} \text{arc cos}(x_\rho/a) \, d\rho$$

$$= 8(\pi^2 R^2)^{-1} \int_{1/p}^0 \rho^2 (R^2 - \rho^2)^{-1/2} \text{arc cos}(x_\rho/a) (d\rho/dx_\rho) \, dx_\rho$$

$$= 4d(\pi^2 R^2)^{-1} \int_0^{1/p} \rho \left(1 - x_\rho^2\right)^{-3/2} \text{arc cos}(x_\rho/a) \, dx_\rho$$

$$= 4d(\pi^2 R)^{-1} \int_0^{1/p} \left(1 - p^2 x_\rho^2\right)^{1/2} / (1 - x_\rho^2)^2 \text{arc cos}(x_\rho/a) \, dx_\rho$$

Or, more simply

$$P_E = (8/\pi^2) \, s \int_0^{1/p} (1 - p^2 x^2)^{1/2} / (1 - x^2)^2 \text{arc cos}(x/a) \, dx \qquad (C6)$$

where

$$s = d/D$$

$$p = (1 + s^2)^{1/2}$$

It is noted in (C6) that the integrand is defined only for $x \leq a$. Hence, if $a < 1/p$ we can only integrate up to $x = a$.

Thus, for $a < 1/p$,

$$P_E = (8/\pi^2) \, s \int_0^a (1 - p^2 x^2)^{1/2} / (1 - x^2)^2 \text{arc cos}(x/a) \, dx$$

and so the probability of capture is given by

$$P_c = 1 - (8/\pi^2) \, s \int_0^a (1 - p^2 x^2)^{1/2} / (1 - x^2)^2 \text{arc cos}(x/a) \, dx \qquad (C7)$$

Now, for $1/p \leq a \leq 1$,

$$P_E = (8/\pi^2) \, s \int_0^{1/p} (1 - p^2 x^2)^{1/2} / (1 - x^2)^2 \text{arc cos}(x/a) \, dx$$

THE PREDICTION OF SOUND SCATTERING

$$= 1 - (8/\pi^2) \, \& \int_0^{1/p} (1 - p^2 x^2)^{1/2} / (1 - x^2)^2 \, arc \, sin(x/a) \, dx$$

and so, the probability of capture is given by

$$P_c = (8/\pi^2) \, \& \int_0^{1/p} (1 - p^2 x^2)^{1/2} / (1 - x^2)^2 \, arc \, sin(x/a) \, dx \quad (C8)$$

As mentioned previously, when $a>1$ the possible angles of escape (ψ and ϕ trajectories) begin to overlap. At this stage, the total angle for a ψ-trajectory escape is

$$\psi_0 + arc \, cos \left(\left[cos \, \psi_0 \right] \right) / a$$

$$= arc \, cos \, (cos \, \psi_0) + arc \, cos \left(\left[cos \, \psi_0 \right] / a \right)$$

Similarly, the total angle for a ϕ-trajectory escape is

$$arc \, cos \, (cos \, \phi_0) + arc \, cos \left(\left[cos \, \phi_0 \right] / a \right)$$

Hence, the probability ν_2 of escape on the λ-plane becomes

$$\nu_2 = (2\pi)^{-1} \left[arc \, cos \, (cos \, \psi_0) + arc \, cos \, (cos \, \phi_0) \right.$$
$$\left. + arc \, cos \left(\left[cos \, \psi_0 \right] / a \right) + arc \, cos \left(\left[cos \, \phi_0 \right] / a \right) \right]$$

A similar derivation for P_c yields, for $a>1$,

$$P_c = (4/\pi^2) \, \& \int_0^{1/p} (1 - p^2 x^2)^{1/2} / (1 - x^2)^2 \, arc \, sin \, (x/a) \, dx$$

$$+ (4/\pi^2) \, \& \int_0^{1/p} (1 - p^2 x^2)^{1/2} / (1 - x^2)^2 \, arc \, sin \, x \, dx. \quad (C9)$$

ACKNOWLEDGEMENTS

This work was commenced while I was on exchange to the Naval Underwater Systems Center (NUSC), New London, Connecticut. I would like to thank Charles Brown, Jr. and Albert Brooks of NUSC who not only introduced me to Project Ocean Acre but provided many stimulating discussions over a long period, and to William Von Winkle and Raymond Hasse, Jr. of the same institution who supported and encouraged my involvement in the project. I would also like to thank Bernard J. Zahuranec of the Office of Naval Research, Neil R. Andersen of the National Science Foundation and Mary-Frances Thompson of the American Institute of Biological Sciences for their assistance in many ways. This paper is a contribution to Project Ocean Acre.

REFERENCES

Aron, W. and S. Collard. 1969. A study of the influence of net speed on catch. *Limnol. Oceanogr.*, 14: 242-249.

Barkley, R. A. 1964. The theoretical effectiveness of towed-net samplers as related to sampler size and to swimming speed of organisms. *J. Cons.*, 29: 146-157.

Barkley, R. A. 1972. Selectivity of towed-net samplers. *Fish. Bull.*, U.S. 70(3): 799-820.

Brooks, A. L., C. L. Brown, Jr., and P. D. Scully-Power. 1974. Net filtering efficiency of a 3-meter Isaacs-Kidd midwater trawl. *Fish. Bull.*, U.S. 72(2): 618-621.

Brown, C. L. and A. L. Brooks. 1974. A summary report of progress in the Ocean Acre program. *Tech. Rpt. 4643*, Naval Underwater Systems Center, New London, Connecticut.

Clutter, R. I. and M. Anraku. 1968. Avoidance of samplers. In: D. J. Tranter (ed.), Part I, Reviews on zooplankton sampling methods, UNESCO. *Monogr. Oceanogr. Methodol.*, 2: 57-76.

Fleminger, A. and R. I. Clutter. 1965. Avoidance of towed nets by zooplankton. *Limnol. Oceanogr.*, 10(1): 96-104.

Holliday, D. V. 1972. Resonance structure in echoes from schooled pelagic fish. *J. Acoust. Soc. Am.*, 51(4): 1322-1332.

Leval, Ph. 1974. Un modèle mathématique de l'évitement d'un filet à plancton, son application pratique, et sa verification indirecte en recourant au parasitisme de l'amphipode hypéride *vibilia armata* Bovallius. *J. exp. mar. Biol. Ecol.*, 14: 57-87.

Love, R. H. 1975. Predictions of volume scattering strengths from biological trawl data. *J. Acoust. Soc. Am.*, 57(2): 300-306.

MacDonald, R. B. 1972. A comparison of acoustical and biological backscattering strengths at discrete depths for Ocean Acre cruises 6 and 10. *Tech. Memo. TA131-225-72*, Naval Underwater Systems Center, New London, Connecticut.

McGowan, J. A. and V. J. Fraundorf. 1966. The relation between size of net used and estimates of zooplankton diversity. *Limnol. Oceanogr.*, *11*(4): 456-469.

Vannucci, M. 1968. Loss of organisms through the meshes. D. J. Tranter (ed.), Part I, Reviews on zooplankton sampling methods, UNESCO. *Monogr. Oceanogr. Methodol.*, *2*: 77-86.

Acoustic Volume Scattering Measurements with Related Biological and Chemical Observations in the Northeastern Tropical Pacific

R. J. Vent and G. V. Pickwell

Naval Undersea Center

ABSTRACT

Measurements of acoustic volume scattering from 1 to 40 kilohertz were made in the Eastern Pacific Ocean with accompanying biological trawls and chemical determinations of several oceanographic parameters of the water column. This report summarizes results from one of the several stations occupied. Acoustic continuous-wave pulse measurements at 2.0, 3.5, 4.0, 4.3, and 12.0 kilohertz indicate that the daytime scattering is frequency dependent, originating from a layer whose average depth is 320 meters, and is caused primarily by midwater fishes containing swimbladders. Explosive source measurements of scattering from 1 to 20 kilohertz made during two migration periods (descent followed by ascent) indicate that the largest swimbladders, which contribute significantly to the low frequency scattering, migrate at constant mass in both directions. In addition, comparison of day and night acoustic observations with those obtained during migratory periods indicates that these same fishes secrete or resorb amounts of gas into or out of the swimbladder in an attempt to control buoyancy.

INTRODUCTION

During the early 1970s, a cooperative effort between biologists, chemists, and acousticians was initiated at NUC to study the effects of scattering on acoustic propagation in the ocean. Since scattering can affect acoustic propagation by producing background levels and possible phase alterations which corrupt signal processing techniques, scattering from the volume of deep ocean areas at midband frequencies (1 to 40 kilohertz) was of prime interest. Because principal scattering agents in this frequency band are mainly biological entities, it was believed that simultaneous measurements of acoustic volume scattering, biology, and chemistry could reveal interrelated factors having predictive possibilities.

Since it is not possible to survey completely the world's oceans in all areas, depths, and seasons of interest, the approach in this program was to identify where possible those aspects of the environment that affect acoustic reverberation. Once identified, data relating to these aspects can be obtained by survey ships on routine missions and then rapidly analyzed while underway. Such parameters, either singly or in combination, may control the occurrence and behavior of biological targets; some of the physical and chemical features may be the fundamental parameters of a recognized water mass. For example, it was of interest to learn if the ultralow dissolved oxygen, characteristic of certain regions, exerted a dominant and predictable effect on the occurrence and behavior of biological targets, whether aggregated into components of the deep scattering layer (diffuse scattering) or as individual discrete false targets (large near-surface fish or schools).

The intent of this report is to discuss results from one of several stations* (Figure 1). Nearly simultaneous measurements of acoustic reverberation and biological samples at discrete depths were taken, as well as hydrographic casts for water samples, to provide chemical, nutrient, and temperature data. The station is located within the subtropical front commonly known as the Cape San Lucas Front. It was chosen because the data gathered from this area presently provide the most complete information for time spent on station, in particular, for acoustic measurements made during migrations of the biological scatters.

DATA GATHERING PROCESSES

Biology

Biological samples were taken with a modified Tucker trawl with a 2-meter-per-side square mouth (*Davies and Barham*, 1969). It was constructed of two types of netting: 1.1-centimeter mesh Marlon (stretched), 5 meters long, was used in the forward position and 0.05-millimeter mesh Nytex plankton netting, a cone 2 meters long with a mouth diameter of 0.5 meter, was used in the after position. The trawl terminated in a stainless-steel cod end bucket. Two flowmeters on the axis of the trawl just inside and outside the mouth permitted determination of the volume of water filtered. The

---

* The suite of stations comprised part of the MINOX (minimum oxygen) program.

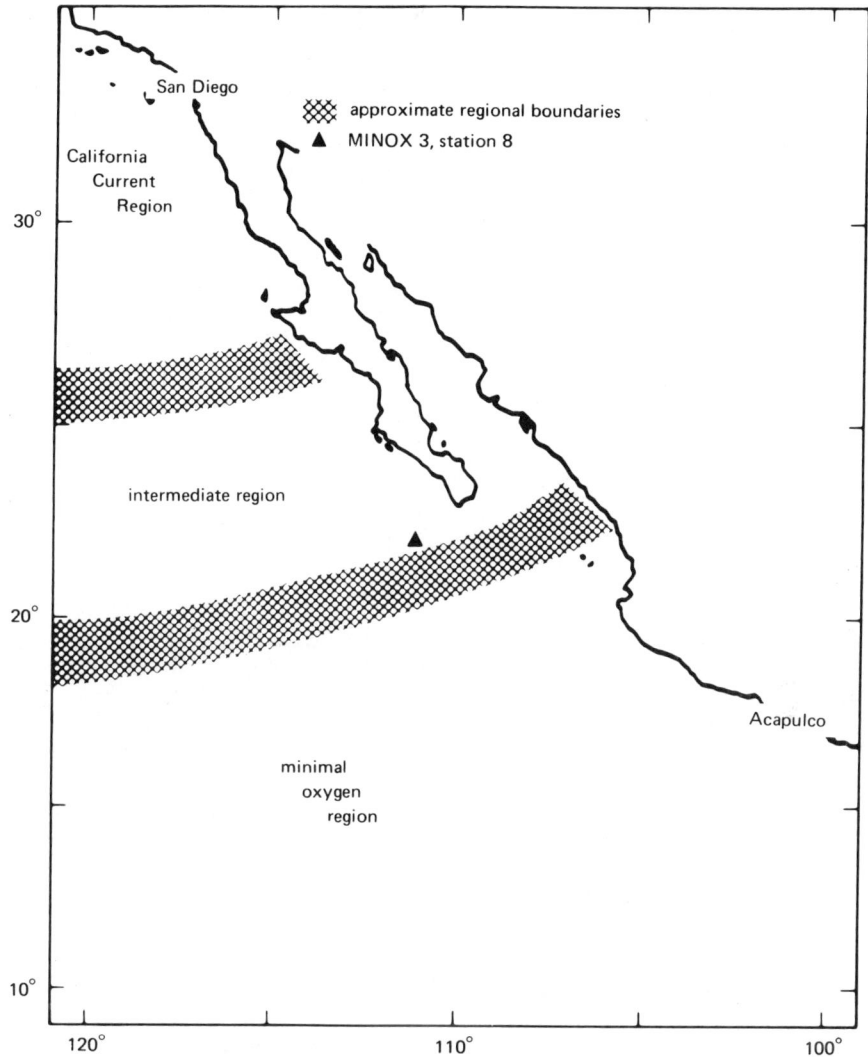

Fig. 1. Station location and water mass boundaries. Data are from station 8 which is located in the San Lucas Front west of the tip of Baja California.

trawls were of the opening-closing type. They sampled a predetermined depth interval of the water column, normally accomplished during midday and midnight hours. All biological data from station 8 are summarized in Figure 2.

The results from a typical catch are in Table 1. This haul (number 36) is a day haul that covers the depth interval of 270 to 350 meters, which corresponds to the approximate depth of the scattering layers as observed on the acoustic recordings.

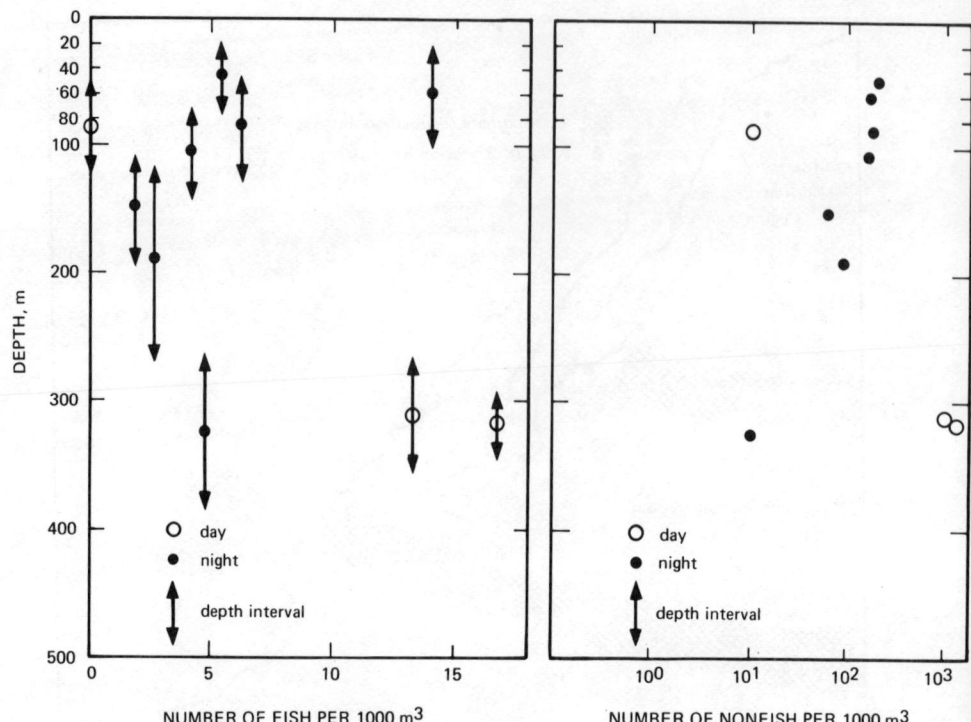

Part A. Concentration of fish netted in 10 trawls at various depths.

Part B. Concentration of non-fish organisms netted at various depths for the same 10 trawls.

Fig. 2. Biological trawls from station 8.

Chemistry and Supporting Oceanography

A hydrocast and two STD lowerings provided water samples and salinity and temperature data from various depths. Particular care was taken to define a temperature inversion occurring persistently at or near 100 meters. Other measurements included density, nutrients, oxygen, chlorophyll, and pheopigment.

Presented in Figure 3 are profiles of temperature, salinity, and density. Macrostructure in the water column is evident at 100 meters and suggested again at 130 meters.

All nutrients were analyzed according to the methods of *Strickland and Parsons* (1968). Quantities of reactive silicate, nitrate, nitrite, and dissolved phosphorus were determined. The observed nutrients (Figure 4) are within the range of expected values

TABLE 1. Results of Typical Catch.

Cruise: MINOX 3   Date: 28 October 1973
Station: 8–transition frontal   Haul: 36
Gear: Tucker o/c trawl
Haul duration: start 1509 finish 1657 total 108 min
Sampling time: open 1528 close 1643 total 75 min
Maximum wire out: —— m   Sampling interval: 270 to 350 m
Sampling speed: 2.25 knots; 70.1 m/min   Volume filtered 21,035 m^3

Organisms	Count
Fishes	
Myctophids	29
Cyclothone	8
Vinciguerria	247
Fragments	2
Others	
Copepoda	20,246
Amphipoda	4
Penaeidea	37
Euphausiacea, adults	1393
Ostracoda	1
Stomatopoda	3
Cephalopoda	1

when compared with those from previous studies of the eastern Pacific (Sverdrup et al., 1942).

A modified microWinkler technique was used to determine the dissolved oxygen (Carritt and Carpenter, 1966). Accuracy was increased by the use of 100-milliliter gastight syringes and vaccine stoppers for collecting the seawater without air contamination; reagents were then injected directly into the syringe.

Pigment analysis was accomplished by utilizing fluorometric techniques developed by Yentsch and Menzel (1963), modified by Holm-Hansen et al., 1965, and further modified by Seligman (1974). Chlorophyll included both chlorophyll a and chlorophyllide a and the pheopigments included pheophytin a and pheophorbide a (Figure 4).

Acoustics

Continuous-wave (CW) pulses and explosive sources were used to obtain the measurements of acoustic volume scattering. Measurements were made as near to midday and midnight as possible and during two migration periods (upward and downward).

Single frequency pulse data at 2.0, 3.5, 4.0, 4.3, 12.0, and

38.0 kilohertz were obtained using directional transducers. All transducers had conical beams and were suspended from the ship at a depth of approximately 7 meters with the sound beam directed vertically downward. A more complete description of experimental procedure and data analysis techniques can be found in *Batzler*, 1975; *Vent*, 1972; and *Batzler and Vent*, 1967. Data obtained using directional transducers and CW pulses are presented as scattering strength profiles, that is, volume scattering strength as a function of depth (Figure 5).

Each profile was based on the average backscattering from 10 pulses, transmitted over a period of several minutes. After correcting for noise, the data were treated as follows:

$$S_v(z) = 10 \log \left(\frac{\sum_{i=1}^{10} p_i(z)}{10 p_0} \right)^2 + 2H - 10 \log V(z) \qquad (1)$$

where $S_v(z)$ is the volume scattering strength in decibels referenced to meter^{-1}; p_0 is the root-mean-square pressure emitted by the

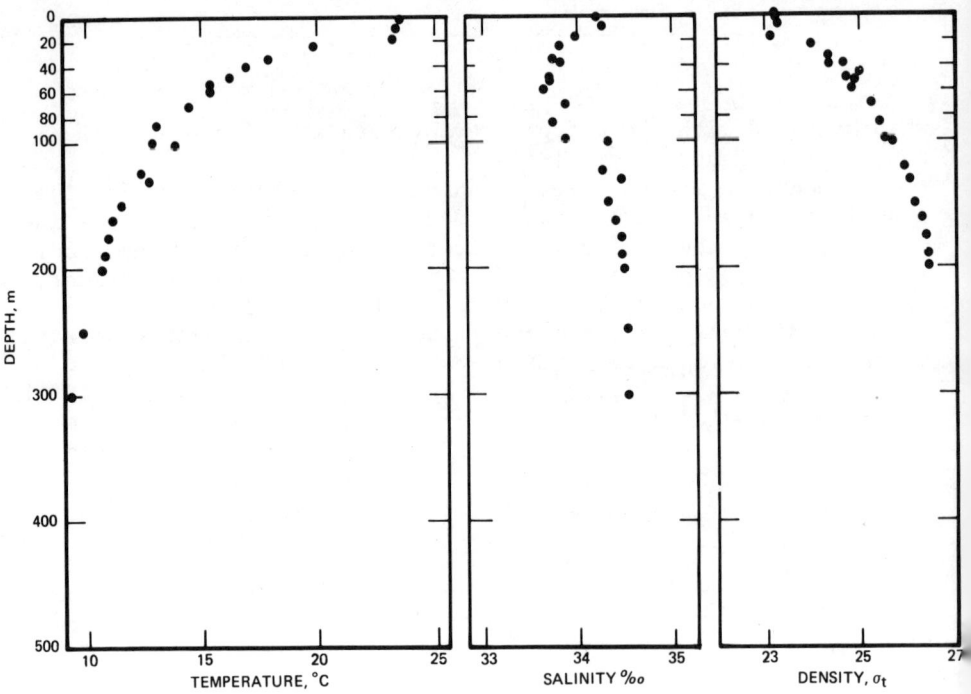

Fig. 3. Temperature, salinity, and density profiles from station 8; determined from hydrocast data.

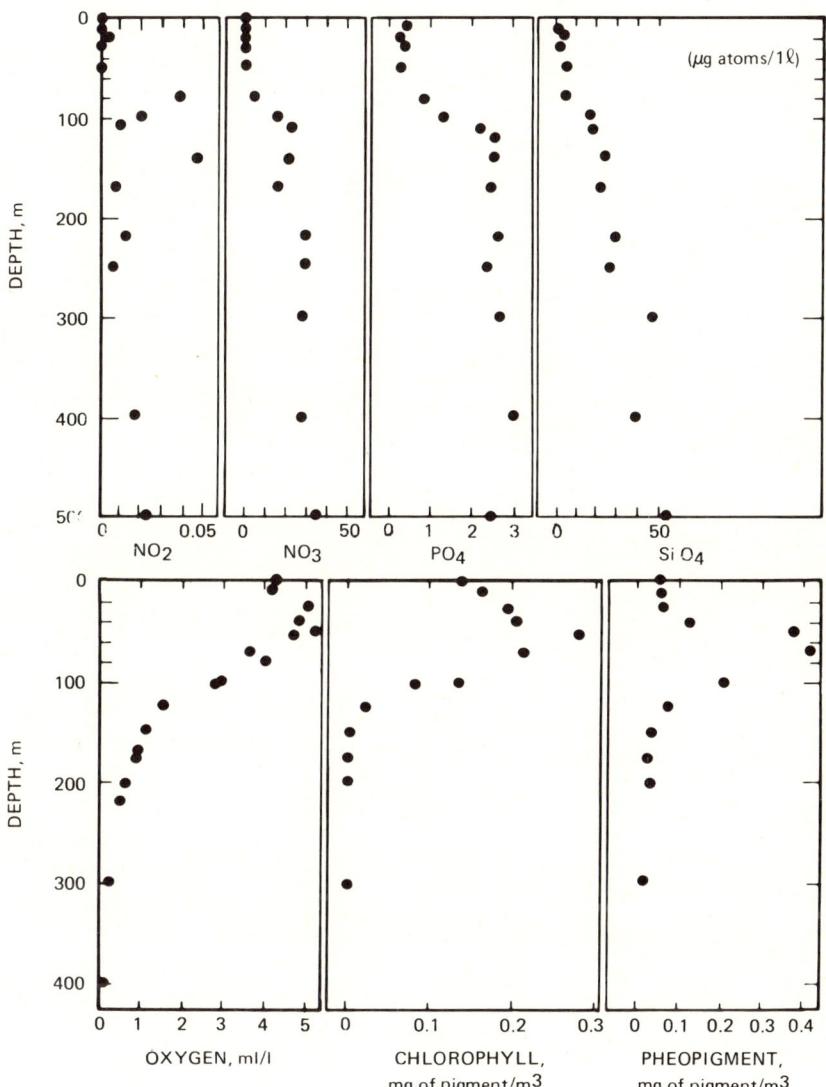

Fig. 4. Chemical features of the water column from station 8.

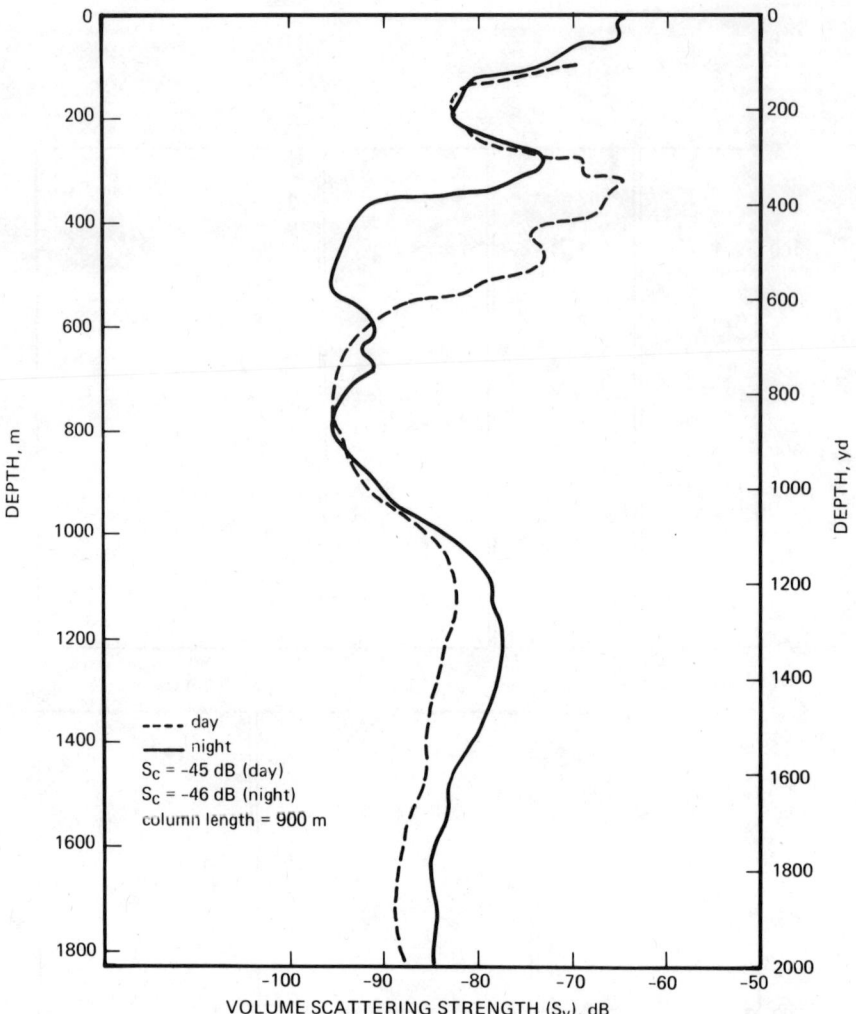

Fig. 5. Acoustic volume scattering profiles at 12 kilohertz for station 8. The night profile was extended to the sea surface by lowering the transducer to 300 meters and looking vertically upward.

source in microPascals at 1 meter; p_i is the root-mean-square pressure backscattered from the water volume in microPascals; V is the ensonified volume at range z from the source; and $2H$ is the two-way transmission loss in decibels (spherical spreading plus attenuation). Examples of profiles for day and night backscattering at 12 kilohertz are in Figure 5, where scattering strength is

ACOUSTIC VOLUME SCATTERING MEASUREMENTS 705

considered as the average target strength of an ensemble of scatterers in a cubic meter of water.

An integration of S_v in antilog form over a column of water produces the column scattering strength S_c.

$$S_c = 10 \, \log \int_{Z_1}^{Z_2} 10^{(S_v/10)} \, dZ. \qquad (2)$$

Values for S_c are in Figure 5 for day and night profiles. Several factors should be noted regarding these profiles.

1. The night profile extends to the sea surface, i.e., on this station the transducer was turned to face up and lowered to 300 meters to permit additional scattering data to be obtained from a depth of 200 meters to the sea surface. Consequently the profile is a combination of the upward- and downward-looking situations.

2. The column strengths result from an integration of the profiles. For better comparison with the explosive data, they are for column lengths not exceeding a depth of 900 meters.

3. A difference in day and night backscattering is evident for the layer near 320 meters. During the day, the main layer was at this level, but during the night it decreased 10 decibels in magnitude and was displaced upward to 275 meters. Total migration may not have occurred. If it is considered that the scatterers in the layer had approximately the same target strength, the reduction of 10 decibels implies a reduction in the number of scatterers by 10 $\log$ 10 or by a factor of 10.

4. The existence of a deep and rather thick layer beginning at 1100 meters is indicated. This layer has also been observed on all other stations.

Profiles for pulse data taken with 50-millisecond pulse duration on station 8 are in Figure 6. The level of backscattering observed from the layer at 320 meters is obviously frequency dependent. Also included are the column strengths for each profile.

Data using broadband explosives and omnidirectional hydrophones were obtained as follows.* A hydrophone was placed at a horizontal distance of 100 meters from the ship and at a depth of 16 meters. Mk 61 SUS bombs (0.82 kilogram of TNT) were detonated at 18 meters in close proximity to the hydrophone. The resulting spherical shells (one direct and one surface reflected) provided the ensonified volume, the geometric center being approximately at the receiving hydrophone. Figure 7 is a stylized representation of the experimental

---

* See *Batzler* (1975) for a more complete description of the data gathering and analysis procedure.

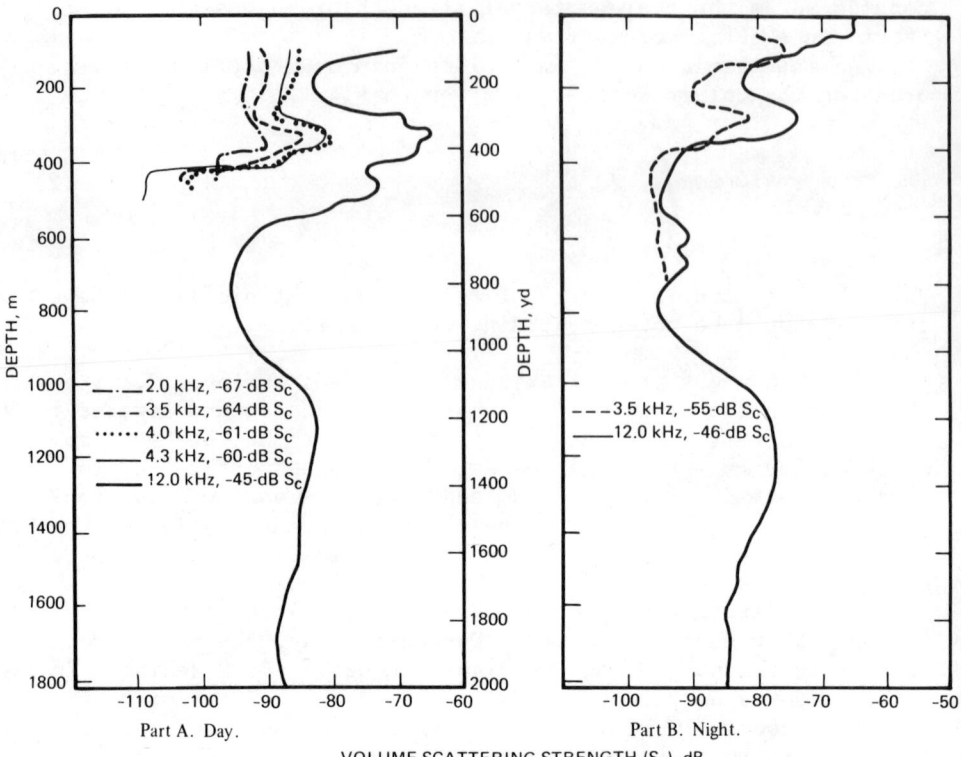

Fig. 6. Summary of scattering profiles obtained by using directional CW pulsed transducers.

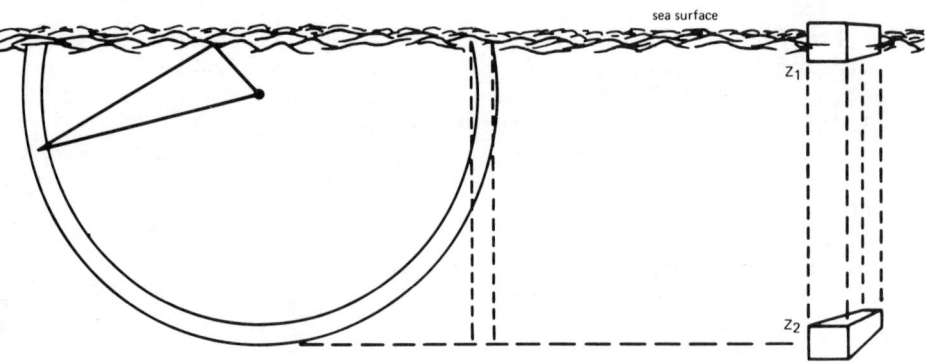

Fig. 7. Explosive source procedure. The column scattering strength or the scattering strength of a water column 1 square meter in cross section extends over a defined depth interval (Z_2-Z_1). The depth interval extends from the sea surface to 900 meters.

procedure. (Explosive scattering data can also be processed to obtain the column strength.)

Between 1.10 and 1.30 seconds after detonation, four 10-millisecond time samples were taken by a Spectral Dynamics 301 Real Time Analyzer. An individual sample was processed by a bank of filters, each 150 Hertz wide. The output from each filter was then averaged over the four time samples. The resulting sound pressures were treated in the following manner:

$$S_c = 10 \log \left(\frac{\sum_{i=1}^{4} p_i(f_j)}{4} \right)^2 - 10 \log E(f_j) - K(f_j) + 30 \log t \quad (3)$$

where $S_c(f_j)$ is the column scattering strength at frequency f_j in decibels; p is the received sound pressure in a 150-Hertz band centered at f_j; E is the source energy in ergs per square centimeter; K is the proportionality factor (including receiver sensitivity, attentuation, and conversion factors); and t is time in seconds. The frequency range of data was from 1 to 20 kilohertz. The energy levels for the 0.82-kilogram charges were taken from *Stockhausen* (1964). During the time period considered (1.10 through 1.30 seconds) the signal resulting from the explosive was considered as stationary; an average range of 900 meters was used to determine the transmission loss (spherical spreading plus attenuation). Sound ray diagrams and propagation loss calculations based on sound velocity profiles revealed no transmission anomalies for this station.

The scattering results using explosives are presented as column scattering strength as a function of frequency in Figure 8 (column lengths of 900 meters). The data points are 500 Hertz apart; the original points were treated with a weighted running average.

RESULTS AND DISCUSSION

The scattering observed in the frequency band of 1 to 20 kilohertz is primarily caused by biological entities, particularly fishes with swimbladders. Although earlier attempts to relate geographically the occurrence of nektonic fishes to the standard chemical nutrients or chlorophyll *a* in seawater were not always successful (*Blackburn* 1966) it was decided that it would be wise to again consider these features in the area of this study. Predictably, no simple and obviously direct correlation was apparent between the occurrence of the target fishes and the profiles for nutrients, chlorophyll *a*, or oxygen (Figures 2, 4, and 5). This basically results because such correlations tend to appear in the horizontal dimension while the possible effects of these basic parameters on the vertical distribution of target organisms appear less likely (oxygen being a possible exception) and have been

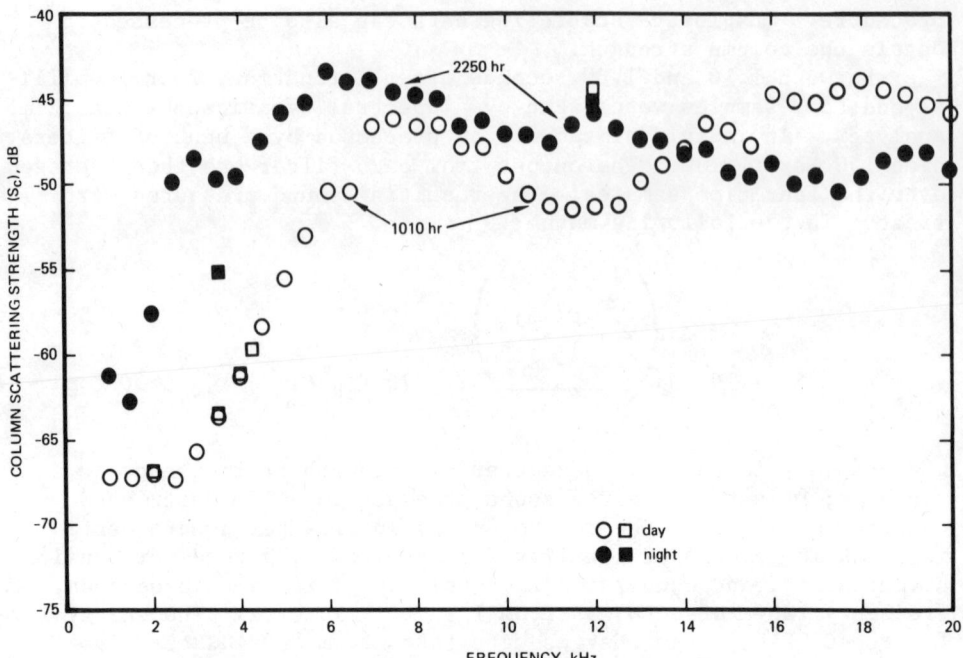

Fig. 8. Column scattering strength using explosive sources for station 8 (each data set is an average of four shots). Standard deviation about each data point was less than 2 decibels in all cases. The open and closed circles refer to midday and midnight values; the open and closed squares refer to day and night values of column strength determined from the CW pulse data of Figure 6.

little studied. However, it is clear that at least some of the target fishes were performing daily vertical migrations into and out of a highly developed oxygen minimum and traversing in the process a dissolved oxygen gradient amounting to two orders of magnitude (Figure 9). It was tempting to speculate that this zone of ultralow dissolved oxygen in some way controlled the behavior of the fishes being monitored. However, even in areas where the vertical extent of the oxygen minimum was still greater, it has not yet been possible to identify the governing factor or factors (Douglas et al 1976). Whi dissolved oxygen may well be a controlling feature, the process involved remains unclear.

An important feature of the mesopelagic fishes was described by Backus et al., 1970).

"Sampling the fishes of the mesopelagial, the dimly lit upper midwaters of the open ocean between about 100 and 1000 m is difficult Like all fishing gear, midwater nets are selective. Moreover, most mesopelagic species are daily vertical migrators, and there is evidence that they are continually altering their depth, although

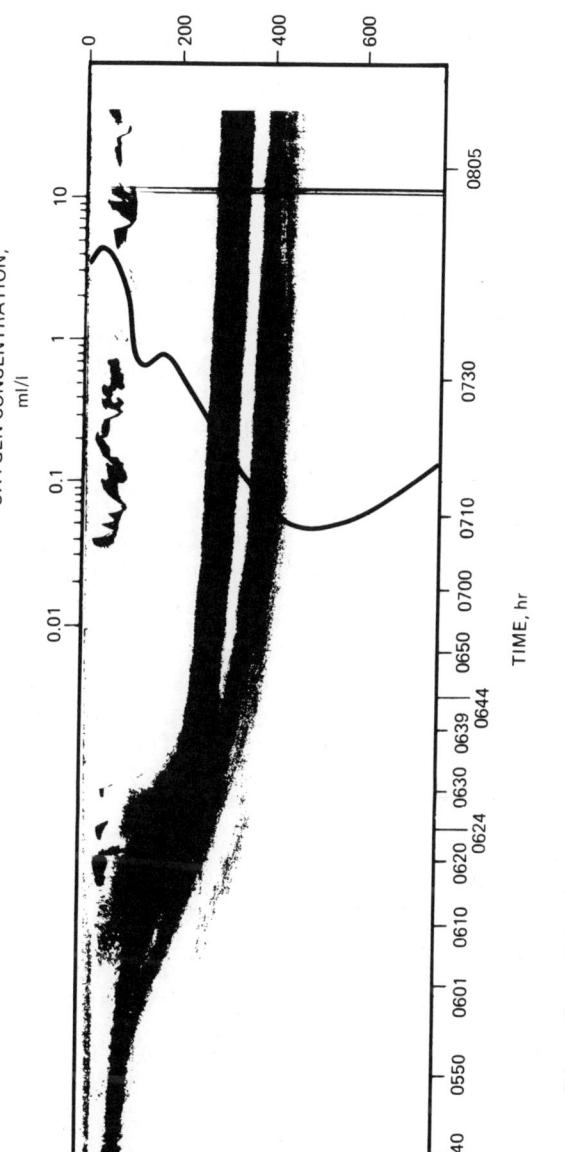

Fig. 9. Echo sounder recording of a downward migration at 38 kilohertz.

most of the changes come at dusk and at dawn. Furthermore, the physical factors that appear to control the depths at which these fishes lie (transparency, temperature, and the like) vary horizontally over distances of a few miles. In short, the arrangement of fishes in the water column is continually changing from moment to moment and from place to place."

These remarks should be considered when reading the remainder of this section.

Migration Periods

Based on echo sounder recordings similar to those in Figure 9, most if not all scattering during the day originates from biological layers centered at approximately 320 meters. These layers appear to be fairly stable in position during the day. Some scatterers migrate upward at night towards the sea surface, a certain number possibly stopping at different depths for unknown periods of time while others gather at shallower depths.

From Figure 8 it can be seen that this change in position affects the magnitude of the column scattering strength. In the 1- to 12-kilohertz band, the largest swimbladders are the prime scatters, causing the f^4 dependence. Beyond 12 kilohertz, the scattering pattern is a composite of lower frequency swimbladders, swimbladders resonant above 12 kilohertz, and entities other than swimbladders.

The dyanamics of the scattering patterns are quite evident during migration periods. Both upward and downward migrations were monitored using explosives. Figure 9 shows the downward migration as recorded at 38 kilohertz on a depth recorder. The bomb detonation times can be identified at different positions along the recording. The heavy scattering patches near the surface at times greater than 0620 hours were suspected to be caused by pelagic red crabs (*Pleuroncodes*) which were numerous in all near-surface net hauls at station 8.

The data gathered using explosives during both migration periods were treated as previously described. As the layers migrated the lowest frequency peak could be seen to shift and change magnitude. In Figure 10, the lowest peak frequency is plotted as a function of layer depth for 11 of the bombs deotnated during this downward migration period. The straight line which the plotted points tend to follow is of the form $f = K(d + 10)^{5/6}$ indicating that the resonance frequency of the largest swimbladders is proportional to $P^{5/6}$, where P is the absolute static pressure. This suggests that scattering at these low frequencies was caused by gas cavities which changed with depth, but maintained a constant mass of gas. An identical trend was evident for the upward migration.

Measurements of column strength for the two migration periods are in Figure 11. Part A summarizes the night-to-day migration,

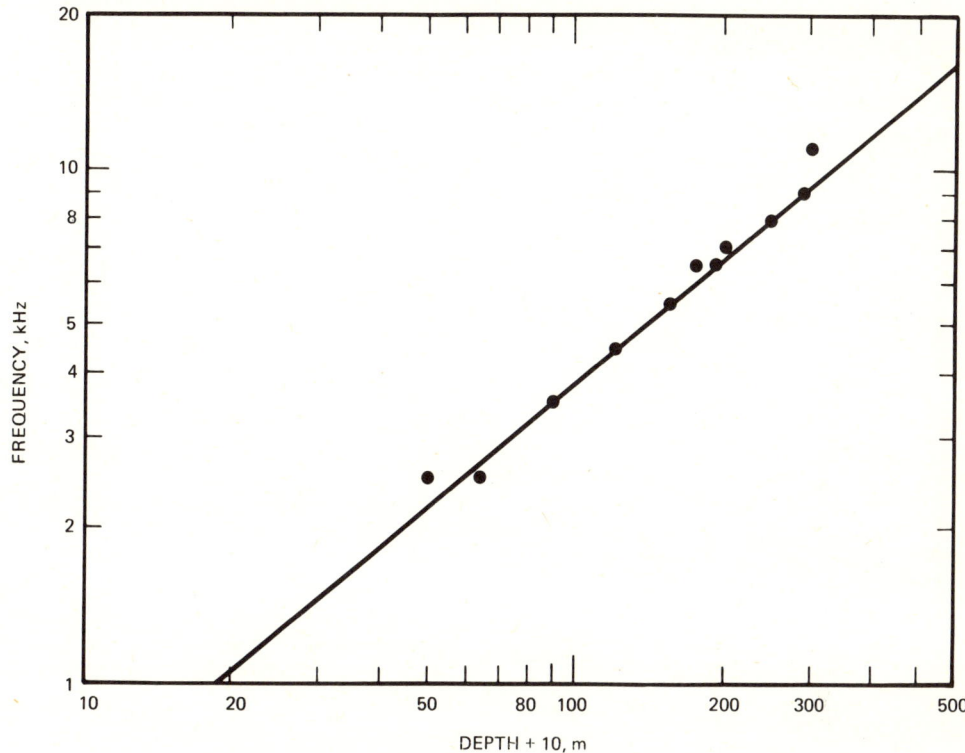

Fig 10. Lowest peak frequency as a function of depth for a downward migration observed on station 8. The straight line is proportional to $P^{5/6}$, suggesting that the migration was accomplished by gas cavities with constant masses of gas.

and part B the day-to-night migration. Most change in the scattering pattern is seen within the band of 1 to 12 kilohertz. The ensemble of largest swimbladders near the sea surface at night migrated downward at nearly constant mass (a bladder resonant at 2.5 kilohertz at 30 meters will be resonant at 13 kilohertz at 320 meters for a constant mass migration).* The reverse process of constant mass migration is displayed for upward migration in part B. These data support the idea that neutral buoyancy for a fish containing a gas bladder may occur at or near the sea surface. However, the data in parts C and D reveal changes in the scattering pattern that suggest a "pumping up" effect by the fishes once they have

---

* Alexander (1972) presents detailed discussions of the resonant frequency expression for gas bladder fishes including migratory effects.

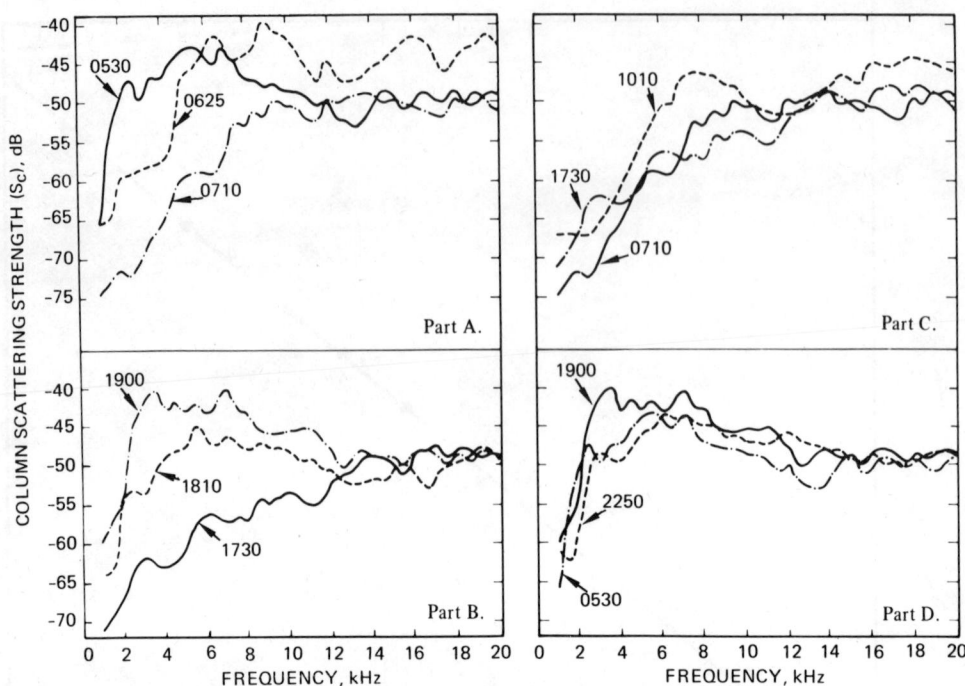

Fig. 11. Measurements of column strength for migration periods. Column lengths extend from the sea surface to a depth of 900 meters. Part A: Column strengths measured before (0530 hours), during (0625 hours), and after (0710 hours) the downward migration. Part B: Column strengths measured before (1730 hours), during (1810 hours), and after (1900 hours) upward migration. Part C: Column strengths measured after downward migration (0710 hours), during the day (1010 hours), and just prior to upward migration (1730 hours). Part D: Column strengths measured after upward migration (1900 hours), during the night well beyond the unstable migration period (2250 hours), and just prior to downward migration (0530 hours).

reached the day depth. For example, again consider the gas bladder resonant at 2.5 kilohertz at 30 meters: An equal volume of gas at the daytime depth of 320 meters will be resonant at 6.5 kilohertz. In part C the bulge in the 1010 hours scattering pattern at 7 kilohertz suggests a pumping up effect; the other curves also show somewhat similar changes. These results may indicate that the fishes with the largest swimbladders migrate downward under constant mass and that upon arriving at the daytime depth secrete gas into the bladder. However, as time progresses the gas is evidently removed from the bladder in preparation for the upward ascent (1730 hours). Upward migration seems to occur under constant mass. The amount of gas removed prior to upward migration may not be enough to permit neutral buoyancy at or near the surface. After the fish have migrated upward, the net effect is a slight overpressurization of the gas within the bladder, compared with the static pressure external to the fish. Data in part D support this hypothesis. The major difference between the three curves in part D appears as a peak at 3.5 kilohertz for the 1900 hours recording. At 2250 and 0530 hours this peak is absent. It is believed that this peak results from the largest bladder fishes which have migrated upward under nearly constant mass. After the fishes ascend past that point in depth where they are neutrally buoyant the bladder is subject to an overpressure which tends to increase the resonant frequency, compared to an equal volume of gas whose pressure is equal to the static pressure at that depth. By 2250 hours, the excess gas appears to have been removed and the ensembles of scatterers remain constant until morning migration.

The changes in magnitude for the column strength at one frequency for different times on this same day are in Figure 12.

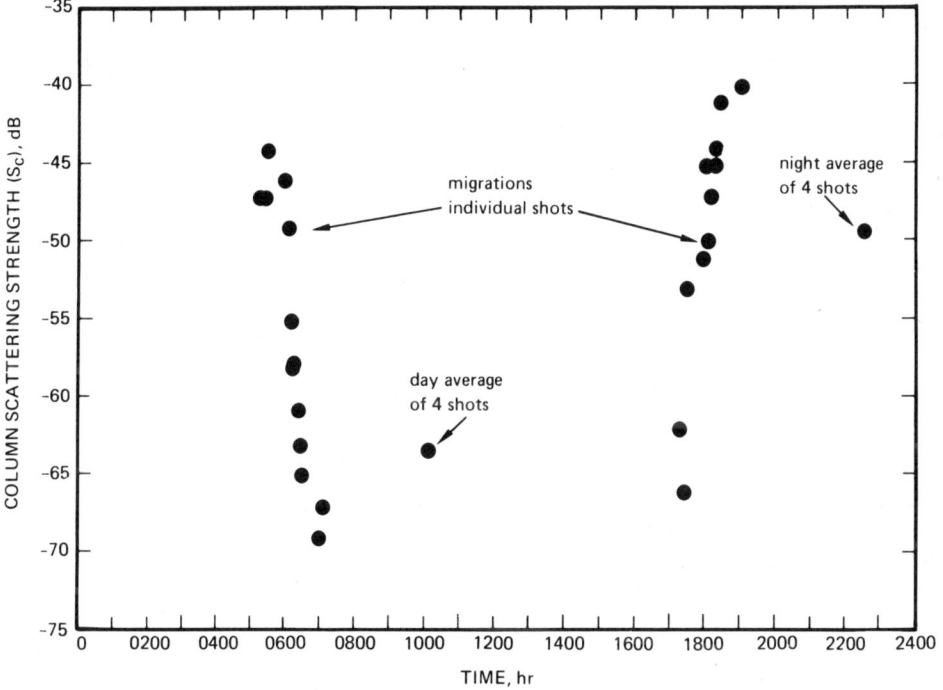

Fig. 12. Column scattering strength as a function of time of day. Strengths at 3.5 kilohertz are presented as determined at different times over a 15-hour period for all data obtained using explosives. All column lengths are 900 meters.

The day and night data points are based on an average of four shots of explosives whereas each of the migration data points represents one explosive shot. Although only a small portion of the 24 hour period was sampled, the changes in column strength reflect the dynamic nature of volume scattering. The difference between night and day values of S_c is 14 decibels; however, the maximum difference observed for the migration times is approximately 26 decibels. The maximum value is -40 decibels, which is 9 decibels greater than the night value.

Energetics

Additional comments concerning the energetics of vertical migration by small midwater fishes with gas-filled swimbladders are needed. *Alexander* (1972) has provided cogent arguments in favor of the possibility that vertically migrating myctophids with swimbladders face a difficult energetics requirement at daytime depths. Regardless of whether they maintain position by swimming with a deflated swimbladder (the constant mass condition, assuming neutral buoyancy at the upper limit of their vertical migration) or by achieving neutral buoyancy through gas secretion (the constant volume condition), the fishes must meet what seem to be rather severe energetics demands. In fact, Alexander suggests that the constant volume condition (inflating the swimbladder at depth) may be the most costly to the fish since it requires possibly more than 25 cubic centimeters of oxygen per kilogram per fish per hour (or about 50 microliters of oxygen per hour for a 2-gram myctophid). This report presents the strongest evidence available that at least some degree of reinflation of the swimbladder is accomplished by these fishes at daytime depths, apparently despite the low levels of ambient dissolved oxygen. Thus, there appears to be a paradox wherein the best available data suggest that small migratory midwater fishes with oxygen-filled swimbladders are following the most costly course available in terms of metabolism energetics. This is further complicated when it is noted that the measured levels of dissolved oxygen at the fishes' daytime depths were only in the range of 50 to 200 microliters of oxygen per liter of seawater (Figures 4 and 9). Possibly the fishes are required to maintain some degree of activity to simply obtain the required oxygen. Yet it was recently shown that some of these same migratory fishes have blood oxygen affinities that fall so low that at daytime depths the hemoglobin may be only marginally functional (*Douglas et al* 1976). Thus there is a series of paradoxes that arise from an intriguing situation where data seem to indicate that these midwater fishes cannot accomplish what in fact they do accomplish.

It seems likely that future high resolution acoustic analyses from additional areas and times of day, when coupled with sufficiently precise chemical and biological determinations, will help in further understanding this complex problem.

ACKNOWLEDGEMENTS

The authors would like to thank I. E. Davies who assisted in analyzing the CW pulse data and supplied the recording in Figure 9; W. Friedl who supplied the biological information; and W. Batzler who critically reviewed and made helpful comments on the data.

REFERENCES

Alexander, R. 1972. The energetics of vertical migration by fishes. In: The effects of pressure in organisms, M. A. Sleigh and A. G. MacDonald, ed., *Academic Press*: 273-294.

Backus, R. H., J. E. Craddock, R. H. Haedrich, and D. L. Shores. 1970. The distribution of mesopelagic fishes in the equatorial and western North Atlantic Ocean. In: *Proc. Int. Symp. Biol. Sound Scattering Ocean*, Maury Center Rep. 005, G. Brooke Farquhar, ed.: 20-40.

Batzler, W. E. and R. J. Vent. 1967. Volume scattering measurements at 12 kc/sec in the Western Pacific. *J. Acoust. Soc. Am.*, 41: 154-157.

Batzler, W. E. 1975. Deep-scattering layer observations off New Zealand and comparison with other volume scattering measurements. *J. Acoust. Soc. Am.*, 58:1: 51-71.

Blackburn, M. 1966. Relationships between standing crops at three successive tropic levels in the eastern tropical Pacific. *Pacific Science*, 20:1: 36-59.

Carritt, D. E. and J. H. Carpenter. 1966. Modifications of the Winkler method for dissolved oxygen in sea water: a comparison and evaluation. *National Res. Council. Comm. on Oceanography*, Draft Report. 1-37.

Davies, I. E. and E. G. Barham. 1969. The tucker opening-closing micronekton net and its performance in a study of the deep scattering layer. *Marine Biol.*, 2:2: 127-131.

Douglas, E. L, W. A. Friedl, and G. V. Pickwell. 1976. Fishes in oxygen minimum zones: blood oxygenation characteristics. *Science* 191: 957-959.

Holm-Hansen, O., C. J. Lorenzen, R. W. Holmes, and J. D. Strickland. 1965. Fluoromatic determination of chlorophyll. *J. Cons. Perm. Int. Explor. Mer.*, 30:1: 3-15.

Seligman, P. F. 1974. The nature of pheo-pigment and chlorophyll distributions and degradation in the California current and northeastern tropical Pacific. *Masters of Science Thesis*, San Diego State University, pp i-ix: 1-160.

Stockhausen, J. H. 1964. Energy per unit area spectrum of the shock wave from 1-1b TNT charges exploded underwater. *J. Acous. Soc. Am.*, 36: 1220-1221.

Strickland, J. D. H. and T. R. Parsons. 1968. A practical handbook of seawater analysis. *Bull. Fish. Res. Bd. Can.*, 167: 1-311.

Sverdrup, H. U., M. W. Johnson, and R. H. Fleming. 1942. The Oceans, Prentice-Hall, Inc.: 1-1087.

Vent, R. J. 1972. Acoustic volume-scattering measurements at 3.5, 5.0, and 12.0 kHz in the eastern Pacific Ocean: diurnal and seasonal variations. *J. Acoust. Soc. Am.*, 52: 373-382.

Yentsch, C. S. and D. W. Menzel. 1963. A method for the determination of phytoplankton chlorophyll and phaeophytin in fluorescence. *Deep-Sea Res.*, 10: 221-231.

IV

MODELING

Recommendations of the Working Group on Modeling

James J. O'Brien, Chairman

Florida State University

RECOMMENDATIONS OF OCEANOGRAPHERS

Whereas hydrodynamical models are currently available with such precision as to forecast qualitative phenomena on "a priori" grounds, the state-of-the-art in biological-hydrodynamic modeling lags far behind in this regard. Hence, the assistance that modelers can render the acousticians in forecasting volume reverberation in the sea is minimal at this time.

It is more likely that the modeler will be the major beneficiary of any exchange, using the acoustic data to construct elements of the evolving models of oceanic ecosystems. In particular, it seems that data from the newer high-frequency technologies will be of special interest to modelers because of their abilities to elucidate temporal-spatial patterns of zooplankton and, possibly, phytoplankton. Eventually, of course, biological models will evolve to give some degree of prediction of nektonic distributions, at least in a probabilistic sense. Volume reverberation can prove a great help in elucidating qualitative mechanisms to be included in these models.

At a minimum, the models should include the relevant nutrient, phytoplankton, and zooplankton compartments. To be useful to the acousticians it is imperative that diurnal, seasonal, and spatial variation be included in the model. Spatial resolution is imperative in the vertical direction and should be deterministic in nature. Resolution in the horizontal plane most likely will be possible only in a stochastic sense. Because of the problem of time scaling, seasonal variation will probably be accomplished by discrete models for each system during each discernible season. Hopefully, during the development of these models the connection between the lower trophic phenomena and the nektonic species will become apparent, at least in a correlative sense, thus effecting some degree of prediction for volume reverberation in the 1-20 kH range.

The physical terms in oceanic models are quite complicated while the biological terms are extremely elementary. That is because biology does not have its Newton as yet! There are no basic formulative laws. Modelers need to develop and determine general principles. Efforts do not need to be spent trying to "predict" what is going to happen at a given point at a given time until governing factors of the ecosystem are known. Once governing relations are known, they can be employed to do simulations.

Modeling efforts should be focused on elucidating classes of processes which apply to the various domains in the sea rather than on attempts to simulate observed distributions of organisms in particular geographic locations.

Improvements should be made in formulations for time variable physiological processes such as nutrient uptake and incorporation into phytoplankton biomass, relation between light and photosynthetic carbon fixation, and growth of phytoplankton. Improved formulations for feeding relations and for population dynamics of zooplankton and fishes should be developed which incorporate time delays where appropriate. The quality of biological formulation used in state of the art models for biological processes is not as high as the quality of physical formulations used in the models. Some variability in biological processes has been included in several models and has resulted in improved model response fit to field data. The usefulness of non-spatial biological numerical simulation models is now questionable, since the interaction of advection and diffusion with biological processes has been demonstrated to be highly significant in controlling patterns in model response.

Mechanisms which can lead to patchiness in the distribution of marine organisms should be investigated using models. Attention should be focused on the relation between advection, diffusion, and biological processes for different observed combinations of magnitudes and scales of the factors appropriate to different hydrographic regimens, with emphasis on the large ocean gyres. Grid spacing of th models must be chosen so that patchiness scales of interest are properly resolved and are not embedded within the grid spacing. Events such as the injection of nutrients through the seasonal thermocline accompanying oceanic frontal dynamics, lateral translation of eddies, and intense local vertical wind mixing of the upper layers of the sea should be investigated. This should be done through the use of numer ical simulation methods to ascertain their significance in driving patchiness in chemical and biological factors relative to the signifi cance of biological processes, such as feeding and vertical migration of organisms which are known to contribute to patchiness.

Simulation of dynamics of several species at a trophic level should also be a goal of the next generation of models. This feature is significant both for models of phytoplankton communities where seasonal changes in species composition can be extreme, and for model of domains such as upwelling regions where pronounced species succession occurs. Advection and intermittent vertical mixing of the kind envisioned in the vicinity of oceanic fronts could modify phytoplankton species distributions on short time and space scales.

Models which incorporate stochastic elements or which are wholly stochastic should be devised and used to investigate the problem of patchiness dynamics.

RECOMMENDATIONS OF BIOACOUSTICIANS

The modeling requirements of bioacousticians differ somewhat from the approaches of those modelers of ecological systems employing sets of coupled differential equations and various rate parameters supplied from experiment. The primary bioacoustical requirements fall into two separate categories: (1) Probabilistic descriptions of natural processes, and (2) deterministic descriptions of the acoustical scattering properties of individual targets. However, some slight shifts in emphasis on the part of the eco-modelers might provide some needed information to the acoustician.

"Patchiness" is perhaps a good example illuminating both the differences in approach and the common ground of the ecosystem modelers and the bioacousticians. The spatial distribution, size, and amplitude of patches are major determinants of the character of the scattered sound. Therefore, they are of great interest to the acoustician. While one particular configuration of patches might be used as the scattering object in a calculation or experiment, such a choice would be too specialized to be natural. In most situations, the acoustician will find it preferable to marry theory and experiment, if he employs a statistical description of the patch parameters.

Alternatively, the ecosystem modeler might offer a prediction of patch shape and separation as a function of certain environmental parameters. Then, a knowledge of the distribution function of those environmental parameters would imply the distribution of patch parameters.

Another application of the more deterministic ecosystem modeler's approach to patchiness might be in helping to describe zooplankton's location of food based on gradients in trace quantities of chemical compounds emitted by the phytoplankton (i.e., a chemotaxic response). Probably some minimum level of trace meterial is necessary for response by the zooplankton, and some minimum gradient necessary for direction finding to the food resource. These minimum levels and gradients determine the volume from which the zooplankton are drawn and hence their clustering. A mathematical model might elucidate this process and thus provide by perspective on zooplankton clustering (i.e., patch creation). This line of argument can be conceptually (if not in practice) carried up the food chain to include carnivores.

The deterministic models in which the acoustician has interest are those of individual targets. They have interest both in their own right and as a means of refining statistical models in such areas as classification. The inputs to such models are acoustical and biological in character.

Another theoretical problem requiring adequate modeling is the scattering from schools of high fish density. Since the target distributions are still somewhat random in such schools, a probabilistic approach is dictated. Account must be taken of the shadowing of fish further from the sound source by those nearer, and of multiple scattering effects whereby and individual fish is insonified by both the source and other fish reradiating.

Empirical schemes for relating acoustical, biological, chemical, and physical oceanographic data appear as a useful way of determining significant interrelationships, even if the reasons for the relationships are not understood. Once a relationship is noted, it can be examined in a more analytical way. Such empirical schemes include regression (single, multiple, and non-linear) and various schemes from pattern recogition such as principal components analysis.

The tasks outlined above could easily require decades for completion. However, the naval engineer does not require definitive mechanistic explanations to achieve his purposes. Meanwhile, it is quite possible, that the empirical (a posteriore) modeling methodologies would prove useful in achieving and engineering progonstication of volume reverberation.

On the Large Scale Circulation of the Ocean: A Discussion for the Unfamiliar

James C. McWilliams

National Center for Atmospheric Research

    I am pleased to be able to speak to you about a current view of physical processes in the ocean. The present years are exciting ones for the study of currents in the deep ocean: a great deal of the traditional wisdom must now be revised to account for a much more active and energetic ocean than was previously imagined. Since I am not familiar with your knowledge of ocean currents, I have tried to imagine that you are some ten years out of touch with research results and would perhaps enjoy a description of what has been learned, or at least imagined, about large-scale, middle ocean, low frequency currents. I shall place an emphasis on how these currents might transport chemical and biological properties.
    We begin with a very broad point of view. If the planet earth did not have an active atmosphere and oceans, then very great variations in temperature would occur over the globe, due simply to the equatorial concentration of the sun's radiative energy input. Our climate would be more like that of the moon, with its great extremes of temperature, except for differences due to different rates of planetary rotation and thermal inertia. Life, if possible at all, would be extraordinarily harsh.
    It has been known for a long time that the atmospheric motions on earth cause an equalizing of temperature over the globe, transferring the radiative excess of heat from the equatorial regions to the poles, where there are radiative deficits. For several centuries, it was believed that some time mean, poleward air flow was the principal carrier of this heat. In the first half of this century, however, it was proposed and finally verified that it is the eddies (i.e., fluctuations about the winds averaged in time and longitude) which accomplish most of this transport in the middle latitudes. In other words, it is the net effect of all the high and low pressure centers, such as are seen on daily weather maps, which keeps us from

being solidly frozen here in Monterey.

Unfortunately, we now know that even this description is incomplete. It is unfortunate, that is, from the point of view of simplicity, but fortunate from oceanographers' pride in the power of the sea. Satellite measurements of radiation and improved statistics of the eddy transport of heat in the atmosphere clearly implicate the ocean as an equal partner in this poleward transport of heat. This is shown in Figure 1 where the northward energy fluxes for both the ocean and the atmosphere are plotted as a function of Northern Hemisphere latitude. Near the equator neither fluid transfers much heat. Between 10°N and 30°N, though, the ocean contribution is dominant, by as much as a factor of 2½, and it remains an important contributor as far north as 60°N, which is also where severe constrictions of both the North Atlantic and Pacific Basins occur. We have thus come to realize that the oceans play a central role in the equable temperatures of our environment, not just locally as along the California coast, but globally as well. Furthermore, since the atmospheric contribution is mainly through its eddies, we are led by analogy to ask whether the same is true in the ocean.

But does the ocean even have any eddies? Not just surface transience due to local winds--this effect is confined within too thin a layer to have any global heat transporting consequences--but something involving the full water column, penetrating even through the main thermocline. I know of no historical references pronouncing on this, but I can tell you a story. Two English oceanographers named John Swallow and James Crease planned an experiment to directly measure deep currents in the middle of the ocean during the late 1950s. They placed neutrally buoyant floats at depths of 2000 and 4000 m in an Atlantic Ocean region several hundred kilometers west of

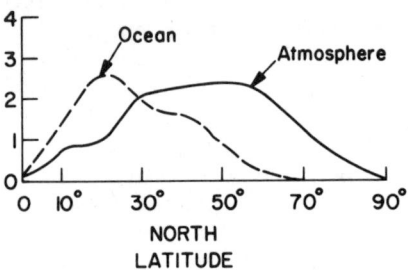

Fig. 1. Variation of net energy transport with latitude for the Northern Hemisphere. The atmospheric transport is directly measured; the oceanic transport is inferred as a residual between the total radiational heating (from satellite measurements) and the atmospheric transport. This figure is adapted from Von der Haar and Oort (1973).

Bermuda. These floats were supposed to move with the local horizontal currents. Their expectation of these deep currents was speeds of one mm/sec or so, moving mainly in a meridional direction, which would be consistent with traditional interpretations of the large-scale patterns of water density and chemical elements. They planned a many month experiment, with several transits between Bermuda and the predictably and slowly translating floats. From my build-up, perhaps you can guess the result: their floats scattered as if in a maelstrom. With the resources available, the floats were trackable only for a brief time. Current speeds were many cm/sec; kinetic energy levels were three or four orders of magnitude higher than expected. Swallow and Crease were so surprised that the full results were not published for twelve years (*Swallow*, 1971), until the idea of energetic ocean eddies had gained a larger following.[1] This, then, is one of the ways in which it was learned that the mid-ocean does indeed have quite energetic eddy motions.

I cannot tell you whether these eddy motions account for the greater part of the northward heat transport by the ocean that is shown in Figure 1. However, because of their great energy level relative to that of mean motions--proportionally this disparity is much greater in the mid-ocean than in the atmosphere--it seems reasonable to expect that the eddy contribution is considerable. Furthermore, dynamically passive chemical and biological properties may often be transported much in the same way as heat--especially below the thermocline. For those who are interested in the distributions of these properties, turbulent eddy transport may be more important in many cases than advection by mean currents.

To illustrate in a simplified manner the revolution which has been caused by our recent ability to directly measure ocean currents, I have prepared two figures. The first, Figure 2, depicts the traditional conception of the large scale circulation, as inferred indirectly from the geographical distributions of water mass characteristics. It was generally agreed that the atmosphere was violently active, but in some manner all but the surface layers of the ocean were buffered from much of this violence--please do not accept too literally the depicted mechanism. While the atmosphere was recognized as the driving force for the ocean circulation, it was thought that most of the water column only felt this driving as a slow and regular grinding like that of a mill stone. The consequent water motion was slow and mostly confined to near the surface; for particular regions this was thought of as a sticky, highly frictional phenomenon. In the deep, the water was at rest. For the spreading and mixing of contaminants, both biological and chemical, the process was viewed as uniform and stately diffusion. However, the extent of the spreading was known not to be small.

---

[1] They did not, however, avoid communication. A brief summary appeared in *Crease* (1962).

Fig. 2. The traditional view of the oceans as inferred from the geography of water masses.

Fig. 3. The current view of the oceans as inferred from moored current meter time series.

Measurements of the water motion in the mid-ocean, both by drifting floats and moored current meters, have forced us to a different view of large-scale, low frequency currents, as shown in Figure 3. We do not view the atmosphere as any more energetic as before, but the ocean is surely not so well buffered from its effects. As viewed from below, it is perhaps as if in the surface layers dancers were tramping on a tin roof above the rest of the ocean. Occasionally the dancers may even fall through. The currents both above and below the thermocline are strong and strongly interacting. We imagine this interaction occurs both laterally as well as vertically. Because of significant motions in the deeper layers, in contrast to the earlier conception of quiescence, the existence of sharp sea floor relief must also be important. The spreading of contaminants is perhaps no greater in extent than was previously thought, but the manner of this spreading cannot be so stately and uniform; it probably is much more irregularly turbulent.

Perhaps I should present these ideas less figuratively. The topics which follow include a description of the traditional view of the mean circulation, a recent revision of this view, illustrations of the nature of ocean eddies, and speculations about their role in influencing both the mean circulation and the spreading of passive properties.

Firstly, the traditional view of the mean currents. Figure 4 shows the Atlantic circulation from *Stommel* (1966). In the North

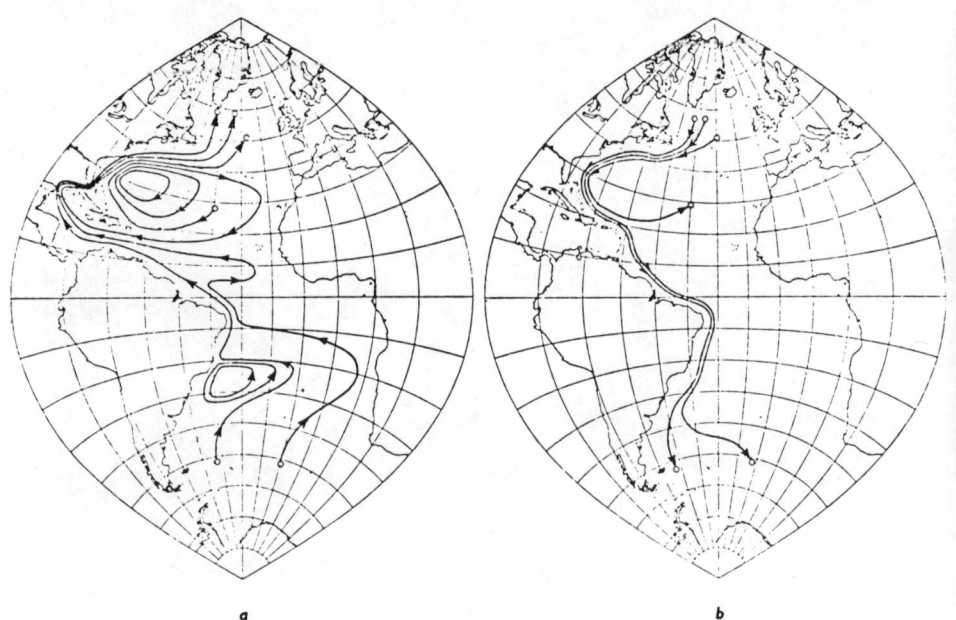

Fig. 4. Schematic charts of the transport (a) above the thermocline and (b) below it (from *Stommel*, 1966, Figure 82). Transport contours are in increments of 10^{13} cm^3/sec.

Fig. 5. The schematic pattern of horizontal transport beneath the thermocline for the world's oceans (from Stommel, 1958).

Atlantic Subtropical Gyre the currents above the thermocline, shown on the left, have a broad, weak, southward drift in the mid-ocean with a rapid return flow as the Florida current, the Gulf Stream, and the eastward Gulf Stream Extension. At greater depths there is principally a weak, southward, western boundary current which carries cold water from sinking regions near the pole to where it very slowly rises in a broad region in the mid-ocean. This happens in order to dilute the warmer surface waters and preserve the thermal equilibrium of the deep layers against diffusion from above. Greater detail in this deep circulation is shown in Figure 5, taken from Stommel (1958), where the pattern of motion is rather uniformly spread over the basin except again for strong boundary currents. The deep circulation is thus not wholly quiescent, but the interior horizontal currents are quite weak (~1 mm/sec).

The basis for these circulation diagrams is widely spaced sections of vertical profiles of temperature and salinity, such as is shown in Figure 6. This is a meridional section of temperature along 50°W longitude, from Newfoundland on the left to Brazil on the right. If one assumes that (1) the deep layers are at rest, (2) nothing important is occurring on spatial scales smaller than are resolved by the sampling here (i.e., 100 km), and (3) the hydrostatic and geostrophic relations are correct, then one can infer patterns such as Stommel's--in this particular section, there is a strong eastward Gulf stream in the North, associated with southward dips in the isotherm depths, and a weak, broad westward flow confined to near the surface over most of the rest of the section. The tilting of isotherms is generally small at depth (geostrophically implying small

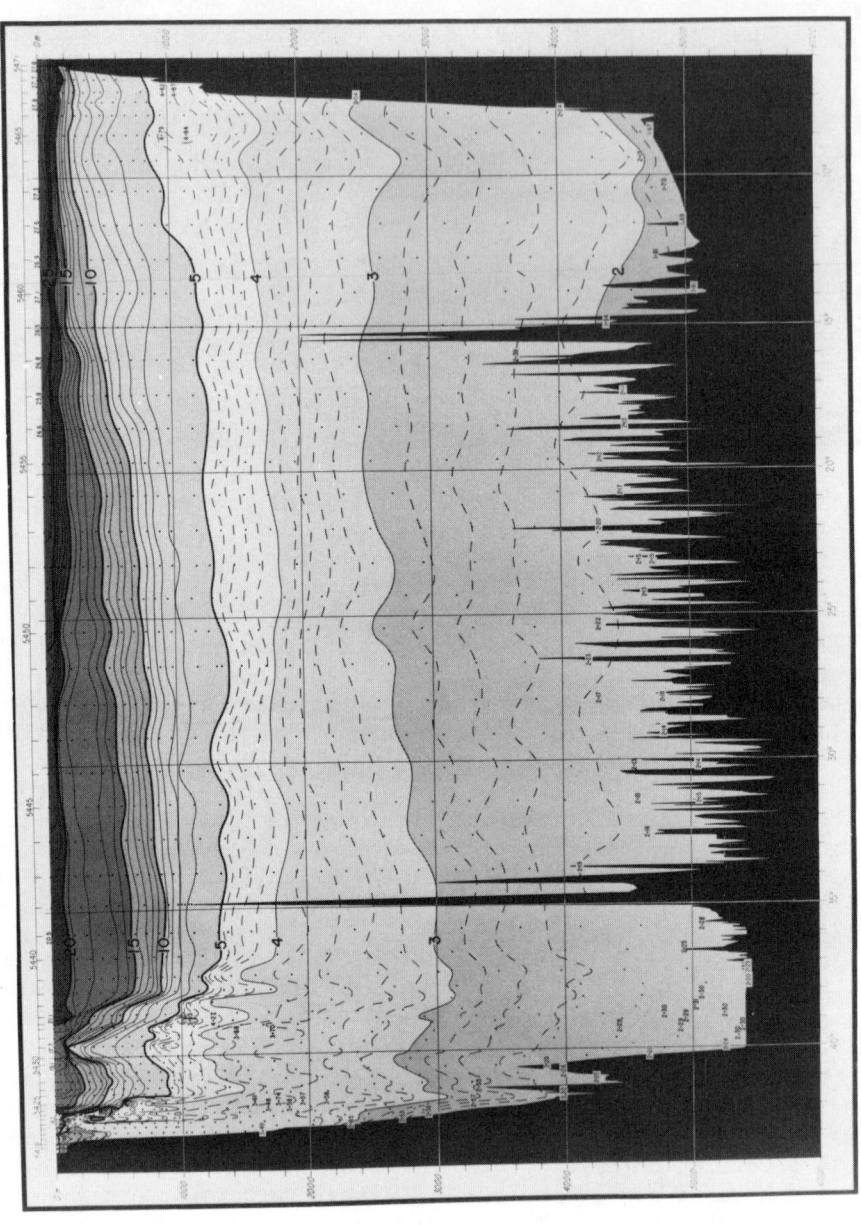

Fig. 6. A section of vertical profiles of temperature along 50°W (from *Fuglister*, 1960).

velocity shears, but not necessarily small velocities). I will not try to dissuade you of the correctness of the hydrostatic and geostrophic relations, but the other two assumptions listed above certainly do need reconsideration. I would finally remark that numerical simulations, using only time independent wind and thermal atmospheric forcing and very large, spatially uniform frictional coefficients, were able to reproduce qualitatively patterns such as are shown in Figures 4 and 5 (e.g., *Bryan and Cox*, 1967).

There has been a recent reconsideration of the North Atlantic mean circulation. It has come, not from any explicit inclusion of the eddy processes, but rather from a very careful consideration of the observed distribution of chemical properties in the gyre, plus an abandoning of the assumption of no motion in the deep oceans, where necessary, for consistency with these chemical distributions. The mean flow inferred on this basis is by *Worthington* (1976). Figure 7 shows his pattern for the circulation below 4°C, or below approximately 1500 m depth. In three very important ways it differs from its predecessors: (1) this deep layer is not nearly quiescent but quite vigorous, with a maximum transport of some 62 Sverdrups in the Gulf Stream and speeds, in the mean, of up to ten cm/sec;[2] (2) the region of recirculation is spatially restricted, occupying only a small fraction of the basin; and (3) there are two anti-cyclonic gyres --the Gulf Stream gyre and what *Worthington* calls the Northern gyre --not just the traditional single one. There is nothing in the mean atmospheric heating or wind stress, nor in the ocean basin shape, which would easily explain either the spatial concentration of the mean circulation or its two gyres if we model the response using constant frictional coefficients. An explanation I believe must come from the internal dynamics of the ocean, involving eddy motions as well as time mean processes.

Worthington's departures from tradition are not confined to the deep layer. Figure 8 shows his pattern for the main thermocline circulation. It has geographically expanded somewhat, but is basically similar to the deeper pattern. An example of the type of evidence upon which these patterns are based is shown in Figure 9, which is relevant to the question of the existence of two separate gyres. Shown in the lower panel is the distribution of hydrographic stations in the regions of the two gyres. Above this panel are the volumetric censuses of the amount of water found with particular combinations of salinity and potential temperature in each of the gyres. What is most striking here, particularly at the upper and lower ends of the temperature range, is the dissimilarity of the temperature and salinity characteristics between the two gyres. It is argued by *Worthington* that no plausible process would permit water from the Gulf Stream to flow into the Northern gyre region while changing its characteristics rapidly enough to match those present in the new region.

---

[2] In fact, the southward western boundary current in Figure 4 is very much weaker than the recirculating gyres in Figure 7. Much more is occurring in the deep ocean than the minimum transport required to maintain the mid-latitude thermocline.

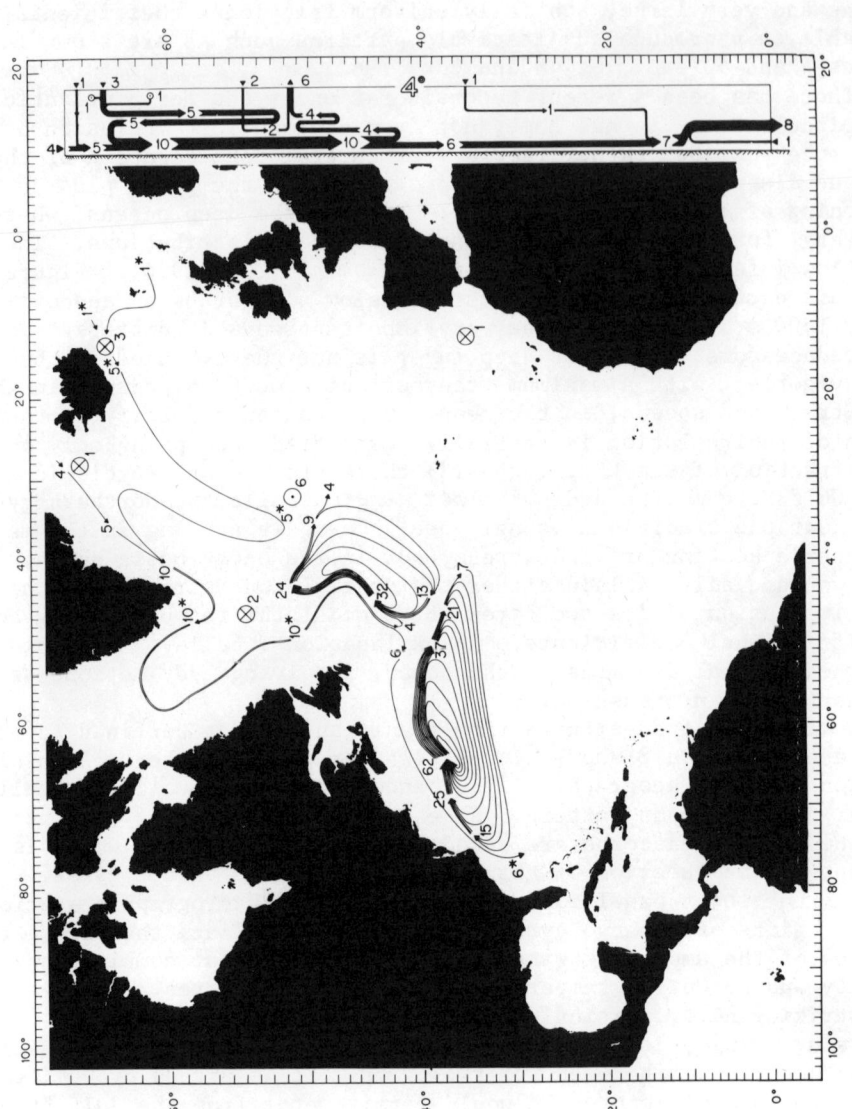

Fig. 7. Circulation diagram for the deep ($T \leq 4°C$) circulation in the North Atlantic (from Worthington, 1976, Figure 11).

LARGE SCALE CIRCULATION

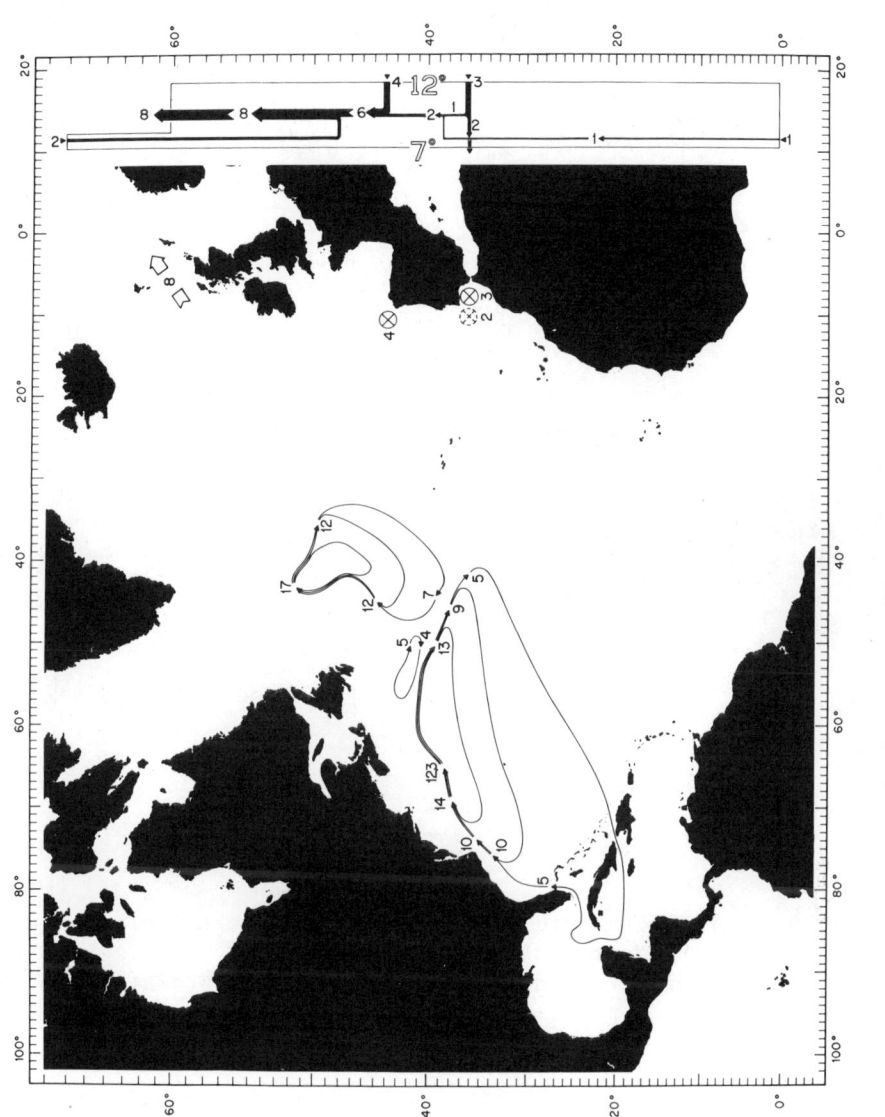

Fig. 8. Circulation diagram for the thermocline ($7° \leq T \leq 12°C$) circulation in the North Atlantic (from *Worthington*, 1976, Figure 26).

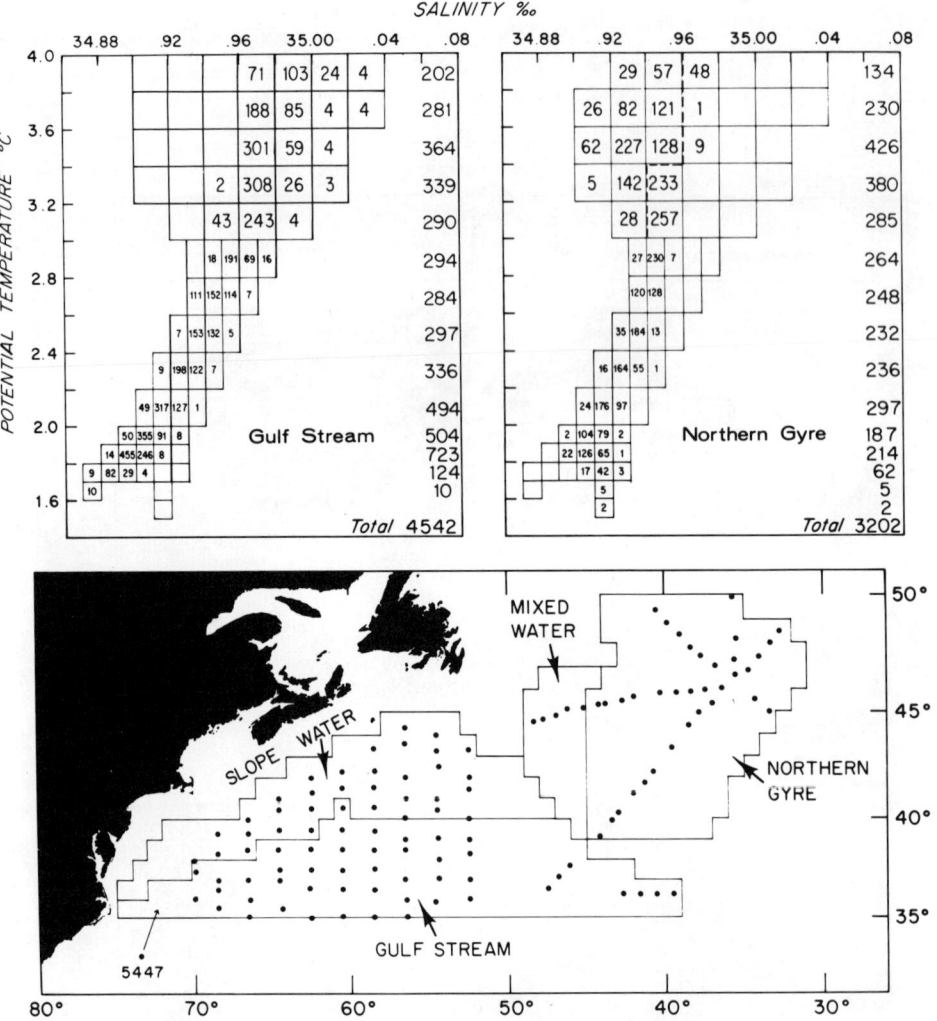

Fig. 9. Top: Volumetric potential temperature/salinity diagram for two areas in the North Atlantic (in units of 10^6 km^3). Bottom: Station positions and the limits of the areas (from Worthington, 1976, Figure 14).

Hence, he concludes that the recirculations must be largely independent.

Views about the mean circulation in the North Atlantic are rapidly changing. I suspect that revisions and alternatives to Worthington's scheme will appear in the next several years. Likely none will be so slow, so spatially uniform, and so broad scale as earlier ones. In particular, the next major achievement must be the devising of circulation patterns which are consistent with the dis-

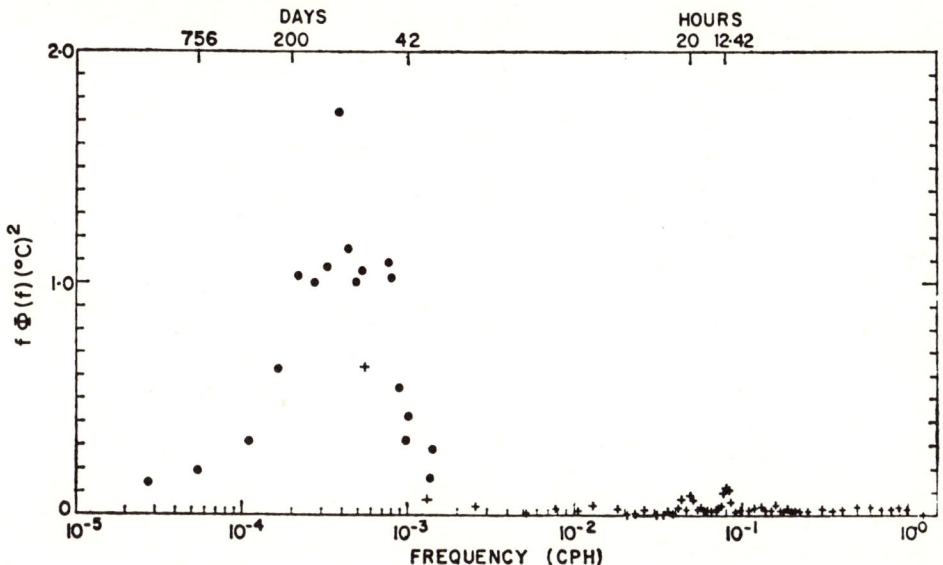

Fig. 10. The power spectral density times frequency (such that area under the curve represents contribution to the variance for upper ocean temperature measurements taken near Bermuda (from *Wunsch*, 1972). The symbols + and · indicate different time series used in the spectral estimation.

tribution and influence of the energetic eddies, the large-scale turbulence in the ocean.

Let us now consider a few examples of what is known about the mid-ocean eddy field. Figure 10 shows a frequency spectrum of upper ocean temperature changes in a region near Bermuda. The axes have been chosen so that the area under a curve connecting the calculated point values would represent the contribution to the total thermal variance for a particular frequency band. The dominant energy peak is the broad one associated with periods of 40 to 250 days, a regime of motions which is often referred to as mesoscale eddies. The mesoscale eddies in this region are very much more energetic than either the very low frequency motions (the "mean") or the relatively much faster ones--inertial and tidal and internal gravity and surface gravity currents. These latter can be seen as having energies hardly distinguishable from zero on the right side of this figure.

There are similar defining scales for mesoscale eddies in the vertical and horizontal directions. A vertical sequence of temperature time series is shown in Figure 11; each time series extends over a three month period. They were measured from fixed moorings in the Western North Atlantic Ocean. A considerable variation occurs with depth in the amplitudes of the temperature time changes, and over

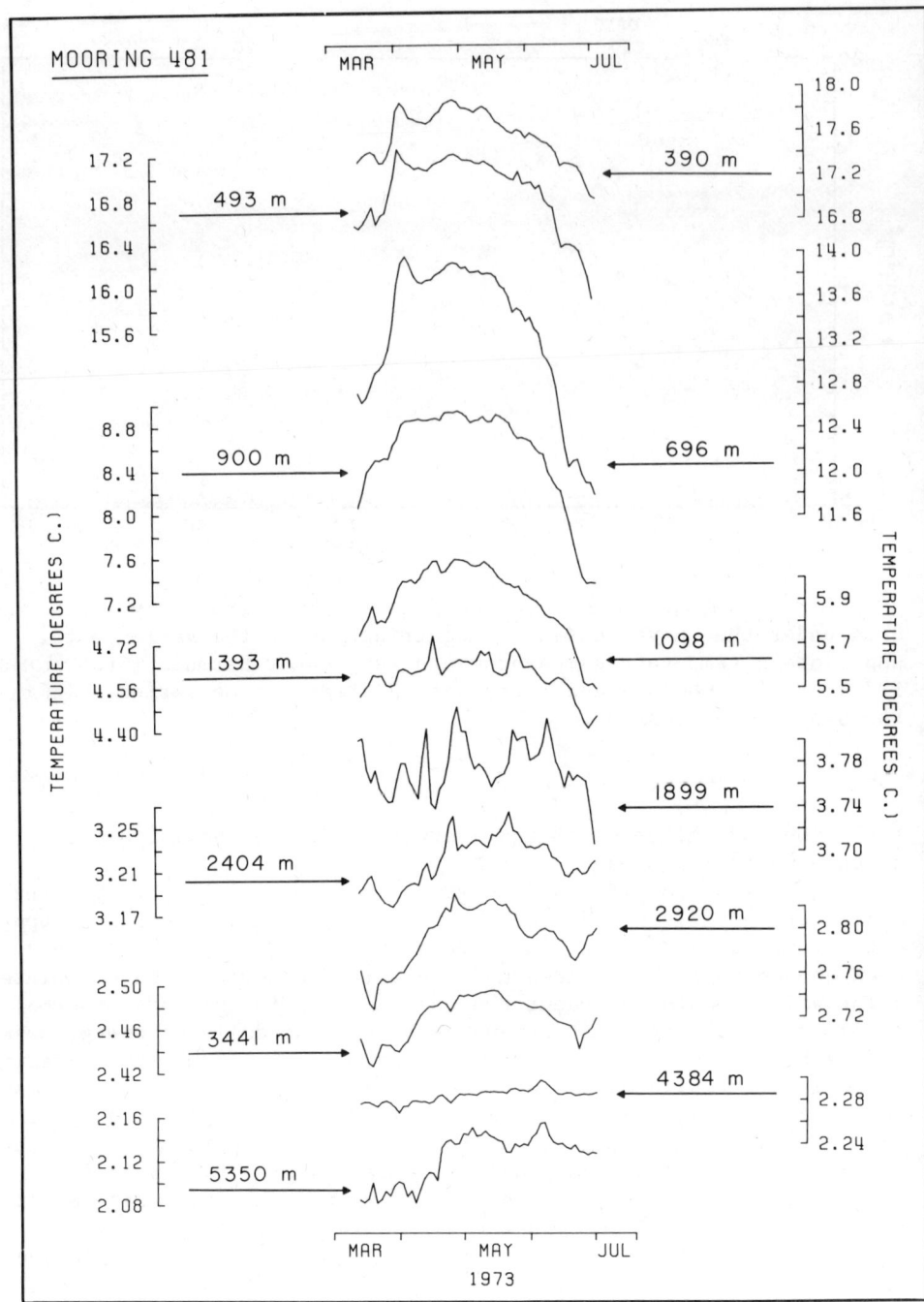

Fig. 11. Temperature time series at several depths at the central mooring during MODE-I (i.e., 28°N, 69°40'W) (from Drs. William Schmitz and Carl Wunsch, personal communication).

LARGE SCALE CIRCULATION

restricted depth ranges there are features on time scales shorter than the dominant one. On the whole, however, the signal associated with a generally warm middle period is coherent and self similar throughout the water column. Finally, in Figure 12, the horizontal patterns of pressure at 150 m depth are drawn for three different days (each separated by a month). These maps are drawn in regions nearly 500 km on a side. The eddies that can be seen are crudely circular and closely packed together. Their radii are about 100 km. The most apparent motion of these eddies is towards the west at a rate of a few km/day (not in particular the best resolved feature, the central high, which also caused the major temperature changes in Figure 11). The intensity of the circular currents for these figures is on the order of 10-20 cm/sec, and they imply vertical displacements of the main thermocline of 100 m. Considering that the main thermocline level only changes a few hundred meters across the whole of the North Atlantic, these local eddy variations are hard to ignore. I would also imagine that they have an important influence locally on the level at which a mesopelagic biological population would reside.

For the sake of further argument, let us assume that the important turbulence for the large scale general circulation is that due to mesoscale eddies. This certainly seems plausible on the basis of their dominant fraction of the total turbulent energy. We have just described their defining scales, but what about their role in large scale diffusion?

There is a traditional physical formalism for describing the effect of turbulent eddies upon large scale flow and property distributions. I am not convinced that this formalism is adequate for the deep ocean circulation, but it at least provides a language for some simple comments. Let us assume that, as classical physicists, we wish to describe the horizontal spreading of the concentration $c(\underline{x},t)$ of some property—biological, chemical, or perhaps even dynamical (such as temperature)—that is carried by currents. Then a likely physical law for this process is

$$\frac{\partial c}{\partial t} + \underline{v} \cdot \nabla c - \nabla \cdot (K \nabla c) = sources \ and \ sinks \qquad (1)$$

in some region with a suitable prescription of what happens on the boundaries. The time change of the concentration is caused either by what is physically carried by the mean velocity $\underline{v}$, by sources or sinks of the property, or by the diffusion of the property.

For this latter process the crucial quantity is the diffusivity $K(\underline{x})$, which we will consider as a function of spatial position in general. If there were no turbulence, then such a term could represent molecular Brownian motion, if we chose a small, positive magnitude and no spatial dependence for K. For spatially homogeneous turbulence—this is the usual hypothesis when one is ignorant of the turbulence—the diffusivity is often considered the same as molecular except for a greatly increased magnitude (e.g., for the ocean the

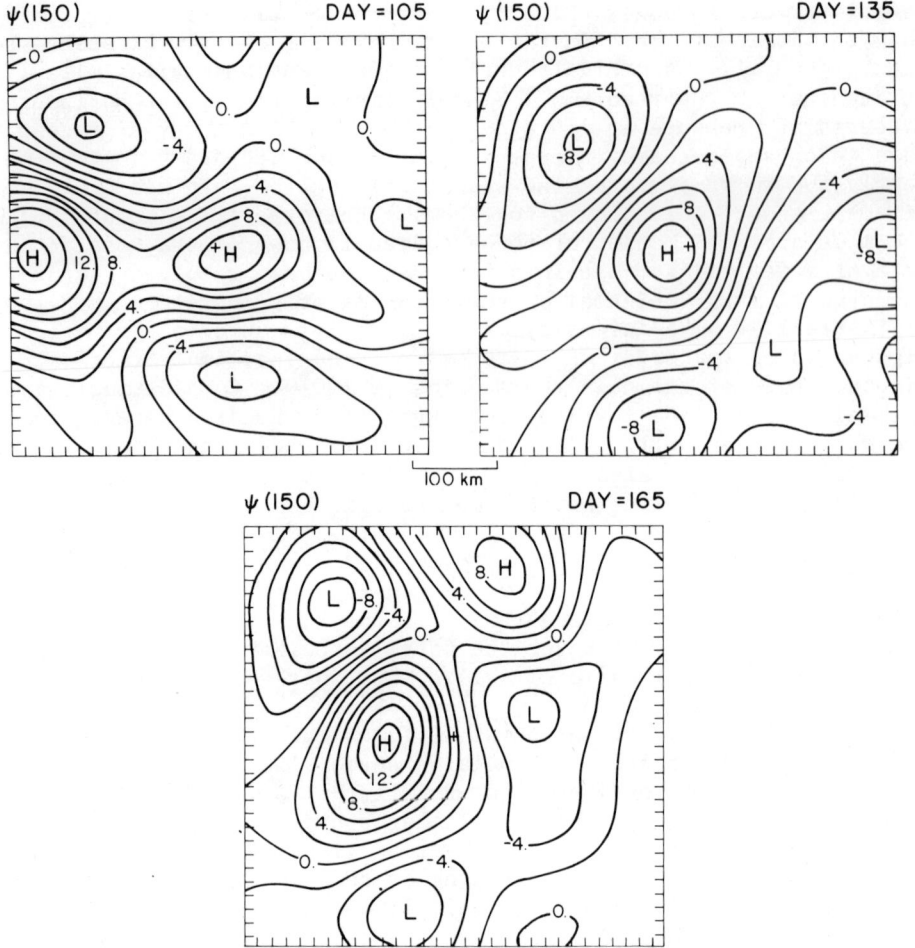

Fig. 12. Maps of geostrophic streamfunction (i.e., pressure) at 150 m depth for three year days in 1973. Units are cm of pressure head. The box centers (marked with a cross) are located at 28°N, 69°40'W, and the dimensions are 465 km. Observations are from density profiles (Crease et al., 1976), and maps are from McWilliams (1976).

increase assumed by mathematical modelers has usually been from a molecular value of 10^{-3} cm^2/sec to a turbulence value of $10^6 - 10^8$ cm^2/sec). A heuristic characterization for this constant eddy diffusivity is that the effect of turbulent eddies is to transport a water parcel (with the local value for concentration, c) a finite distance and then mix it up with its new neighboring parcels. The mathematical consequence of such a law is to diminish irregularities in concentration distributions. For example, let us consider a region with a steady flux through it of the property whose concentration we are calling c. Initially the distribution within the region might have some irregularities such as are shown in Figure 13a. If the diffusivity were a positive constant--the usual representation--then the irregularities would diminish with time, and the distribution for c would tend towards a uniform gradient, with decreasing values in the direction of the flux. At some later time the concentration would appear as depicted in Figure 13b.

So much for traditional simplicity. In terms of the turbulent velocity v',[3] which must cause the spreading of the property represented by c, *Taylor* (1921) defined the diffusivity of water parcels as the time integral of the velocity covariance function,

$$K = \int_0^\infty d(t_1 - t_2) \overline{v'(t_1)v'(t_2)}, \qquad (2)$$

where the overbar indicates an ensemble average over all acceptable manners of occurrence of the turbulence. The covariance is usually presumed to depend not explicitly on the times t_1 and t_2, but only on the magnitudes of the interval between them, $t_1 - t_2$. Thus K depends upon the intensity of the eddies and the manner in which the turbulent velocity at one time is related to itself at another.

As a matter of practice it can be quite difficult to determine K from measurements of a velocity time series. Since the covariance function is the Fourier transform of the variance spectrum $E(\omega)$,

$$\overline{v'(t_1)v'(t_2)} = \frac{1}{2\pi} \int_{-\infty}^{\infty} d\omega e^{i(t_1 - t_2)} E(\omega), \qquad (3)$$

then the diffusivity can also be written as

$$K = \tfrac{1}{2} E(0). \qquad (4)$$

Thus, to know K, one must know the zero frequency variance, and an infinitely long time series is required in principle. Under circumstances where the covariance function vanishes uniformly after not too long an interval $(t_1 - t_2)$, then the estimation of K from a finite time series may not be too inaccurate.[4] In any event we shall

---

[3] v' is defined as the difference between the instantaneous total velocity and the large-scale, "mean" v in the preceding equation.

[4] There is a further difficulty: the velocity covariance function required is a Lagrangian one. The flow must be such that $K(\underline{x})$ varies over broader scales than a typical Lagrangian trajectory in order for the preceding definition to make sense.

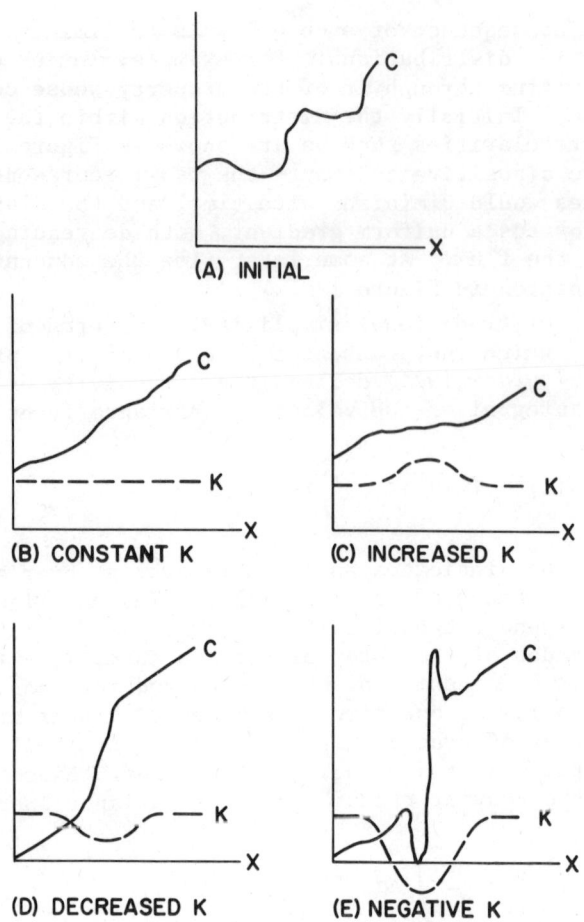

Fig. 13. Schematic solutions for the one-dimensional diffusion equation $(c_t - \partial_x(KC_x) = 0)$ in a region with uniform concentration flux at the boundaries. From an initial distribution (a), then the concentration c at a later time has one of the following forms: that due to (b) a constant diffusivity K; (c) a locally enhanced K; (d) a locally diminished K; (e) a locally negative K.

neglect practical considerations here and make use of the diffusivity to illustrate possible turbulent effects.

Now imagine a situation where the eddies were structurally similar everywhere, but in one part of our region they were more intense. With no change in the shape of the covariance function but with an increase in eddy velocities, K would locally be enhanced. In such a case the local irregularities in the initial concentration distribution would be even more rapidly damped out. However, the longer time tendency would not be towards a uniform concentration gradient across the whole region as before. Instead, a weakening of that gradient would occur where K was largest; that is, where the eddies were stronger (see Figure 13c). Conversely, if there was a local region where K was smaller due to weaker eddies, then initial irregularities would disappear more slowly and there would also be an enhancement of the concentration gradient in the anomalous region (Figure 13d).

In its effects, the local diffusivity gradient acts more like an additional advecting velocity than a mixing and smoothing agent. This can be seen by rewriting the advective and diffusive terms in Equation 1 in the following manner:

$$\underset{\sim}{v}\cdot\nabla c - \nabla\cdot(K\nabla c) = (\underset{\sim}{v} - \nabla\underset{\sim}{K})\cdot\nabla c - K\nabla^2 c. \qquad (5)$$

For the case of a locally increased K, this effective velocity is divergent and causes a flattening of the gradient; when K is locally decreased, the velocity is convergent and increases the gradient.

Unfortunately, in thinking about the oceanic relevance of this, it is pure imagination to say that the diffusivity must scale with the eddy energy level. In the previous definition for K, details of the eddy structure--the nature of the shape of the time covariance function--also mattered. If this quantity is different when the eddy energy is different, then we cannot predict *a priori* the consequence for K. So let us for the moment simply complete a survey of the mathematical possibilities associated with non-uniform diffusion. Imagine that in part of our region the eddies change so drastically that K is not only decreased, but becomes negative as well (Figure 13e). Then there would be a tendency towards an increased concentration gradient in this region as before (Figure 13d), but the initial irregularities would be locally enhanced rather than damped with time. If we were to combine this effect with mean currents carrying concentration irregularities in and out of such a region as well, then the evolution of the property distribution could be quite complicated indeed, and not at all consistent with most people's intuition about diffusion.

If the ocean eddy field is statistically stationary--that is, even though the eddies do move around and change with time, in any two periods sufficiently separated in time their structural characteristics are similar--then probably the preceding equation for the large scale concentrations changes is a useful model. However, if the eddy diffusivity either has important spatial gradients or regions with negative values, then very important changes from the traditional constant diffusion can occur. Additionally, the mean velocity $\underset{\sim}{v}$ may itself be strongly influenced by eddy transport processes. This further consequence of the eddies would influence property distribution as well.

We can make a cruide estimate of whether mesoscale eddies have the strength to act as the diffusion of large-scale properties. Consistent with the heuristic description given above of turbulent mixing, K can be thought of as the product of the mesoscale velocity which varies a property and the length it is carried before being mixed. For a 10 cm/sec velocity and a 100 km horizontal length, the diffusivity would be 10^9 cm^2/sec. This is a safe upper bound for values previously used in general circulation models of the ocean as well as studies of property distributions (e.g., Veronis, 1975). Mesoscale eddies seem at least potentially able to provide the large-scale diffusion.

Given this assumption, what can be said from present observations about the way in which they might provide the diffusion? A usual indication of diffusion is the dispersion of particles initially close together. One can imagine the dispersion of a drop of ink in a pot of boiling water. Alternatively, think of freely drifting floats at a given depth in the ocean. The results of such an experiment in the Western Atlantic are shown in Figure 14 (Freeland et al., 1975). This is a superposition of all the trajectories for floats initially placed in a region some 200 km in diameter (i.e., ~2° of latitude and longitude) at a depth of 1500 m. The total time covered is greater than two years, though no single float track lasts that long. The process certainly has the appearance of dispersion, although the displacements from the central region are by no means always rapid or systematic if one follows individual trajectories.

Now consider the structure of the mesoscale eddy kinetic energy. An example of the spatial variability of the time-averaged energy is shown in Figure 15. The data source and depth level is the same here as in the preceding figure. The energy level varies considerably in the North-South direction along a meridian near 70°W. It changes by a factor of five over a distance on the order of 500 km, rising from a central minimum both towards the north slightly and the south considerably. On a broader scale the mesoscale eddy energy level changes even more. I have tried to indicate this on a map of the North Atlantic Ocean (Figure 16) by indicating the eddy kinetic energy levels at the locations where they have been measured by moored current meters set out by the Woods Hole Oceanographic Institution. These data are for a depth of 4000 m. The range of energy variation here is two orders of magnitude; it is conceivable that if a larger region had been covered the range would have been even greater. The largest value (100 cm^2/sec) is from beneath the Gulf Stream and the smallest values (≤ 1 cm^2/sec^2) come from a region roughly equidistant between North America on the west, the mid-Atlantic ridge on the east, the Gulf Stream extension on the north, and the North Equatorial Current on the south. I suspect that this relative eddy quiescence is not accidentally associated with being well away from these bounding features. Notice that the region near 70°W, 30°N is one of intermediate energy; it is from this general area that the data for Figures 10-14 were obtained. The energy levels in Figure 15 are somewhat weaker than those in Figure 16 in this region; this is consistent with 1500 m being generally less energetic than 4000 m. The energy increase to

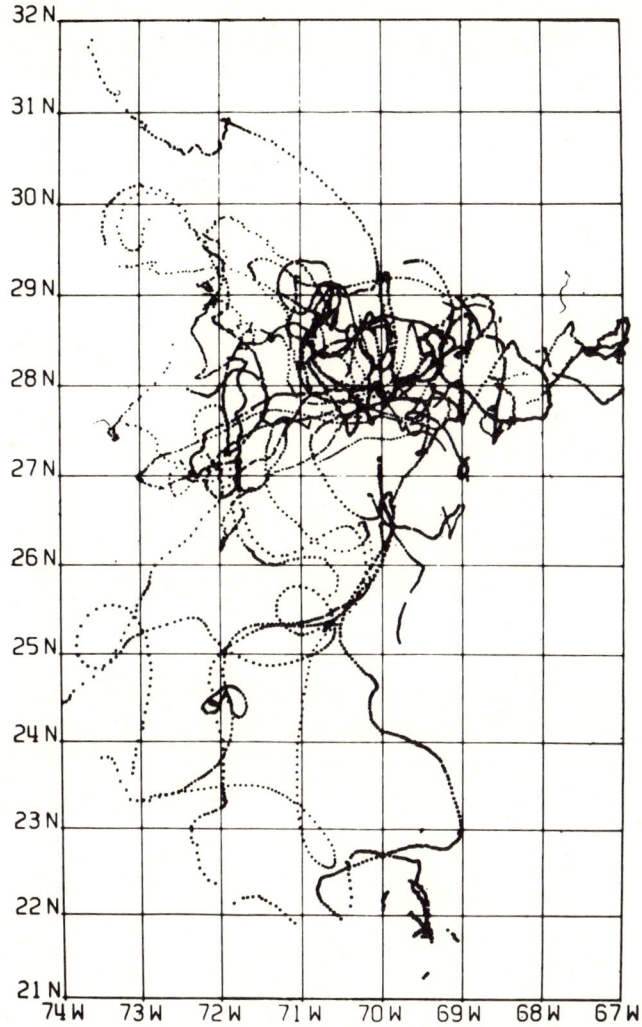

Fig. 14. A superposition of 1500 m float trajectories from the MODE-I experiment and an extended period following it (*Freeland et al.*, 1975).

the south along 70°W, as measured at 1500 m, has not yet been confirmed by deeper measurements.

If we make the approximation described earlier--namely, that the greatest spatial changes in the eddy diffusivity are due to spatial changes in the eddy energy level--then these final two figures are

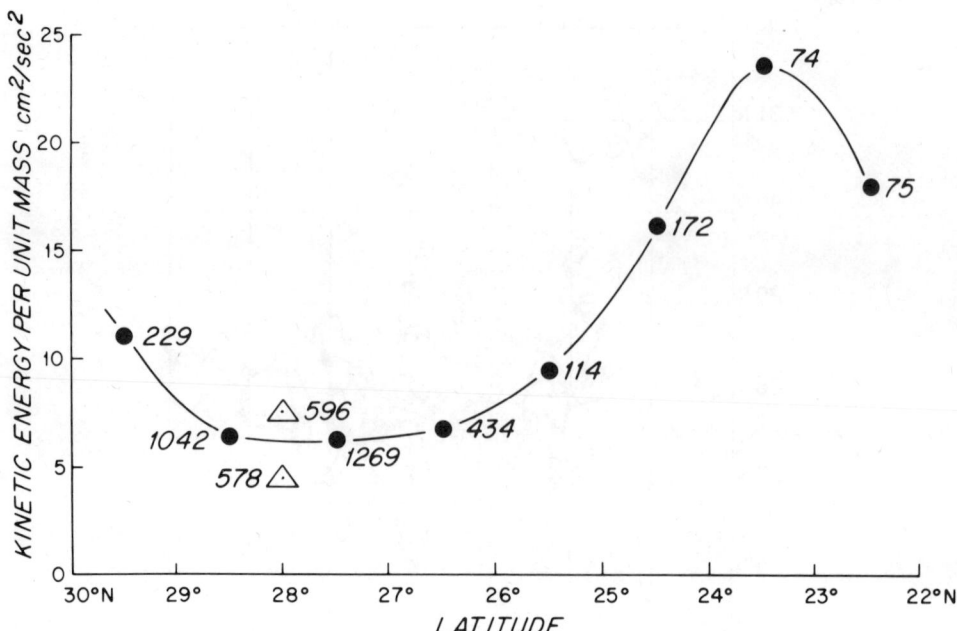

Fig. 15. Latitudinal variation of kinetic energy from 22-30°N along roughly 70°W at a depth of 1500 m (*Freeland et al.*, 1975). The circles are calculated from float velocities; the triangles are from moored current meters (at longitudes a degree apart). The corresponding integers are data days (i.e., the sum of all operating instruments on each of the days there were any).

strong evidence for strong diffusivity gradients. The type of behavior shown in Figure 13c is thus reasonably likely: there will be a tendency for rapid changes in the concentration of chemical and biological properties away from the regions of larger diffusivity due to more intense eddies. This would be more evident the greater the horizontal flux of the property through the ocean gyre. The patterns of deep mean circulation proposed by *Worthington* (Figures 7 and 8) are also confined to the region of stronger eddies. This is surely not fortuitous, but what exactly the causal relationship is between strong eddies and strong mean flows in a relatively small region is a very difficult question, though obviously an important one. I propose no present answer, except to comment that theoretical models do exist both in which strong mean currents can be unstable, and thus, spontaneously generate eddies and in which turbulent eddies can force mean currents (e.g., *Pedlosky*, 1964; and *Rhines*, 1974).

The other example of aberrant diffusivity effects which was discussed above is the increase in concentration irregularities and gradients due to negative diffusivities. Several years ago *Webster*

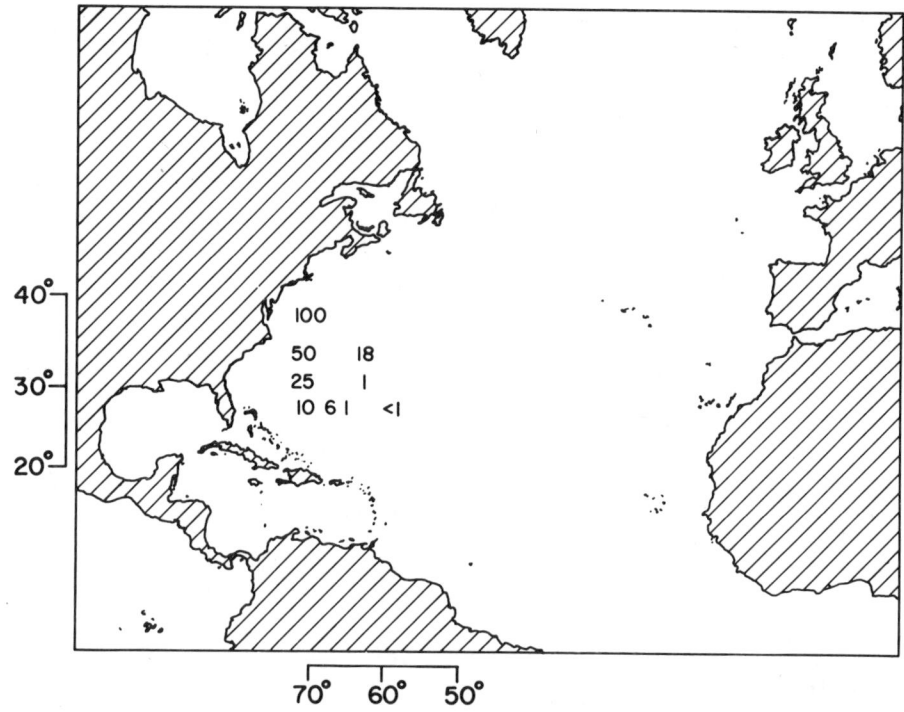

Fig. 16. The time-averaged eddy kinetic energy at 4000 m at selected locations where it has been measured (*Schmitz*, 1976).

(1961) estimated a negative diffusivity for momentum in the Gulf Stream. Unfortunately, the uncertainties in this result are too large for it to be conclusive. It is, however, a persistent characteristic of those oceanic numerical models which explicitly calculate the mesoscale eddies, as well as the general circulation, that small regions of negative diffusivity are predicted. In particular, the mean flow in deep levels may well be mostly driven by the resulting negative eddy diffusivities in these models (e.g., *Holland and Lin*, 1975).

I have tried to describe a point of view about the large scale circulation and property distributions within enclosed ocean basins at middle latitudes. This viewpoint emphasizes the influence that mesoscale eddies are likely to have, acting as frictional and dispersive mechanisms for the larger scales, but not necessarily as simple, uniform friction. The crucial issue to be considered may well be the prevalence and importance of spatial inhomogeneities--in the eddies, in the resulting diffusivities, in the general circulation, and in property distributions. I have no doubt that this aspect of ocean currents deserves more consideration than it has had previously. I also hope I have made apparent our present level of ignorance about satisfactory answers to these questions. We do not even know whether the relatively more intense eddies near the Gulf Stream increase or decrease the local diffusivity!

ACKNOWLEDGEMENTS

My participation in the MODE and POLYMODE programs have provided the circumstances for developing the opinions expressed here. In particular, Francis Bretherton--partly through a recent review article of his which had some similar aims (Bretherton, 1975)--has been very helpful. Howard Crosslen did the art work for Figures 2 and 3. Financial support has been provided by the National Science Foundation through its grants to the National Center for Atmospheric Research. This manuscript is MODE Contribution Number 47.

REFERENCES

Bretherton, F. 1975. Recent developments in dynamical oceanography. *Quarterly Journal of the Royal Meteorological Society* 101: 705-722.

Bryan, K. and M. Cox. 1967. A numerical investigation of the oceanic general circulation. *Tellus* 19: 54-80.

Crease, J. 1962. Velocity measurements in the deep water of the Western North Atlantic. *J. Geophysical Research* 67: 3173-3176.

Crease, J., A. Leetma, R. Scarlet, and W. Sturges. 1976. Observations of the density field during MODE. In preparation.

Freeland, H., T. Rossby, and P. Rhines. 1975. Statistical observations of the trajectories of neutrally buoyant floats in the North Atlantic. *J. Marine Research*, 33: 383-404.

Fuglister, F. 1960. *Atlantic Ocean Atlas*. Woods Hole Oceanographic Institution, Woods Hole. 209 p.

Holland, W. and L. Lin. 1975. On the generation of mesoscale eddies and their contribution to oceanic general circulation. *J. Physical Oceanography* 5: 642-657.

McWilliams, J. 1976. Maps from the MODE Experiment. I. Geostrophic streamfunction. *J. Physical Oceanography*, 6 (6).

Pedlosky, J. 1964. The stability of currents in the atmosphere and oceans. *J. Atmospheric Sciences* 21: 201-219.

Rhines, P. 1974. β-plane mean flows: A simple theory. *MODE Hot Line News* 56: unpublished document.

Schmitz, W. 1976. Eddy kinetic energy in the deep western North Atlantic. *J. Geophys. Res.*, in press.

Stommel, H. 1958. The abyssal circulation. *Deep-Sea Research* 5: 80-82.

Stommel, H. 1966. *The Gulf Stream*. U. of California Press, Berkeley. 248 p.

Swallow, J. 1971. The Aries current measurements in the western North Atlantic. *Phil. Trans. of the Royal Society of London A-270*: 451-460.

Taylor, F. 1921. Diffusion by continuous movements. *Proceedings of the London Mathematical Society, Series 2, 20*: 196-211.

Veronis, G. 1975. The role of models in tracer studies. *Numerical Models of Ocean Circulation*. National Academy of Sciences, Washington, DC. 364 p.

Von der Haar, T. and A. Oort. 1973. New estimate of annual poleward energy transport by Northern Hemisphere Oceans. *J. Physical Oceanography 3*: 169-172.

Webster, F. 1961. The effect of meanders on the kinetic energy balance of the Gulf Stream. *Tellus 13*: 392-401.

Worthington, V. 1976. *On the North Atlantic Circulation*. Johns Hopkins Press, Baltimore: in press.

Wunsch, C. 1972. The temperature spectrum near Bermuda from two minutes to two years. *Deep-Sea Research 19*: 577-593.

On Persistence of Aquatic Ecosystems

T. G. Hallam

Florida State University

ABSTRACT

Fundamental properties of a three level differential equation model of an aquatic ecosystem are investigated. The three tier ecomodel includes an essential conservative nutrient; a plant population, phytoplankton; and a herbivore population, zooplankton and other herbivores.

The behavior of the model is completely determined analytically; tools employed include differential inequalities, linear phase plane analysis and perturbation theory. Emphasis is upon well-posed aspects of the model which indicate consistency with ecological foundations.

The main result states a necessary and sufficient condition for the persistence of the aquatic ecosystem and demonstrates that the ecomodel is nutrient controlled.

INTRODUCTION

The goal and approach used in this article evolve from a mathematical modeling perspective. The ultimate goal of modeling an ecosystem is the determination of the systems' governing characterizations, an approach which requires that essential features of the ecosystem be included in the model but that the basic framework be as uncomplicated as possible. With this format as a premise, a marine or limnotic ecomodel, extracted from more complex formulations in the literature, is explored for ecological consistency.

Many biological, chemical, and physical factors influence the dynamics of aquatic ecosystems. An excellent introduction, from a mathematical modeling viewpoint, to the governing features of marine ecosystems has been given by *Steele* (1974). A survey of the plankton model literature up to 1968 has been done by *Patten* (1968).

A basic assumption imposed here requires that the ecomodel have as an essential component a nutrient that is conserved in dynamic interactions. *Dugdale* (1967) employed a conserved essential nutrient concept in estimating biomass in terms of a biologically limiting nutrient. *Verhoff and Smith* (1971) explore a four tier conservative nutrient system. The specific form of our model was motivated by work of *O'Brien and Wroblewski* (1973) and *Wroblewski and O'Brien* (1975) where a conservative nutrient is considered in a spatial environment.

The above authors have studied models through numerical simulations of several biologically relevant situations. The methodology of this paper is completely analytical. Elementary, but theoretical, aspects of differential equations are gainfully employed to completely determine the behavior of the model and its governing characteristics.

A detailed description of the model is given in the following section. Differential inequalities are used to obtain relationships between parameters that guarantee the extinction of a component of the system and, thus, demonstrate model consistency; stability analysis of the equilibrium points is performed; synthesis of the complete phase triangle associated with the model is presented; and ecological implications about the persistence of the ecosystem as deduced from the solution of the model are discussed in the succeeding sections.

DESCRIPTION OF THE MODEL

The ecosystem model studied in this paper is a two trophic level system consisting of a resource component, nutrient; a plant component, phytoplankton; and a herbivore component, zooplankton.

As mentioned above, the nutrient is required to be conservative and fundamental in the sense that a modification of the total amount of nutrient can have significant effects upon the balance of the system. In the sea, nitrate is apparently the best example of such a nutrient. Silicate could also be used as it is necessary for diatom growth. In fresh water systems, phosphate is known to be a conservative, essential nutrient. Trace elements such as potassium, manganese and iron may effect growth of phytoplankton but their concentrations are usually not sufficiently reduced by uptake so as to become limiting.

Let $N = N(t)$ denote the concentration of a conservative, biologically limiting nutrient in the euphotic zone at time t (seconds). Let $P = P(t)$ denote the concentration of the nutrient in the phytoplankton component at time t. Let $Z = Z(t)$ denote the concentration of the nutrient in the herbivorous grazer population. The total amount of available nutrient, N_T, is conserved; hence, we have the closure property

$$N_T = N(t) + P(t) + Z(t), \quad t \in R_+ = (0, \infty). \tag{1}$$

The dynamics of the phytoplankton population are governed by

$$\frac{dP}{dt} = \frac{V_m NP}{K+N} - BP - EPZ, \quad t \in R_+. \tag{2}$$

The first term on the right side of (2) represents uptake of nutrient by phytoplankton analogous to Michaelis-Menten enzyme kinetics. In particular, V_m denotes the maximum uptake rate (sec^{-1}) of nutrient by phytoplankton and K is the concentration of nutrient that supports one-half the maximum uptake rate. This hyperbolic uptake rate was proposed by Dugdale (1967) and has been widely employed; O'Brien and Wroblewski (1973), Wroblewski and O'Brien (1975), Steele (1974), and others. Various laboratory and field experiments indicate this hyperbolic uptake response to nitrate, ammonium, and phosphate (Eppley and Coatsworth, 1968; Eppley, Rogers and McCarthy, 1969; Eppley and Thomas, 1969; MacIsaac and Dugdale, 1969; Dugdale and MacIsaac, 1971) is reasonably appropriate.

In the second term on the right side of (2), B denotes the rate at which nutrient is lost from the phytoplankton component. The third term in (3) represents loss of nutrient from the phytoplankton component through herbivore grazing.

The dynamics of the herbivores are determined by the equation

$$\frac{dZ}{dt} = EPZ - DZ, \quad t \in R_+. \tag{3}$$

where D is the collective excretion and death rate of the herbivores.

The dynamics of the nutrient are given by

$$\frac{dN}{dt} = -\frac{V_m NP}{K+N} + BP + DZ, \quad t \in R_+. \tag{4}$$

The system of differential equations (2), (3) and (4) contains five parameters. This system shall be transformed to a nondimensional one by scaling t by V_m^{-1} and the components N, P, Z by N_T. The substitutions

$$\tau = t V_m; \quad n = N/N_T; \quad p = P/N_T; \quad z = Z/N_T;$$
$$\beta = B/V_m; \quad \varepsilon = EN_T/V_m; \quad \delta = D/V_m; \quad \alpha = K/N_T; \tag{5}$$

transform the system (2), (3) and (4) into

$$\frac{dp}{d\tau} = \frac{np}{\alpha+n} - \beta p - \varepsilon z p \qquad (6a)$$

$$\frac{dz}{d\tau} = \varepsilon z p - \delta z \qquad (6b)$$

$$\frac{dn}{d\tau} = -\frac{np}{\alpha+n} + \beta p + \delta z \qquad (6c)$$

The nondimensional system (6) contains only four parameters explicitly. The closure relationship (1) is transformed into

$$n(t) + p(t) + z(t) = 1, \quad t \in R_+. \qquad (7)$$

Although the setting of system (6) is in the three dimensional space (n,p,z), the closure relationship (7) implies the problem is a two dimensional one. The theory of two dimensional autonomous systems of differential equations is well developed and will be exploited in the analysis of the model. May (1974) and *Hirsch and Smale* (1974) have used this theory in an ecological setting.

LIMITATIONS TO THE PARAMETER SPACE

The theory of differential inequalities has produced relevant and elegant results in differential equations (Lakshmikantham and Leela, 1969). These techniques do not appear to have been employed significantly in mathematical modeling of ecosystems. In this section by using elementary differential inequalities, conditions are found which guarantee the extinction of a population of the ecosystem. The results where interpreted ecologically demonstrate necessary consistency aspects for applicability of the model.

The main limitations of the parameters will be stated and discussed. The proofs of the subsequent results are religated to Appendix A.

Theorem 1. If the parameters α and β satisfy the inequality

$$(\alpha + 1)^{-1} < \beta \qquad (8)$$

then corresponding to the p-component of each solution of (6) there exists a constant p_0 such that

$$p(\tau) \leq p_0 \exp\left[-\beta + (\alpha + 1)^{-1}\right] \tau, \quad \tau \in R_+. \qquad (9)$$

In particular, there exists a τ_p, $0 \leq \tau_p \leq \infty$, such that $p(\tau_p) = 0$.

To interpret this result, the nondimensional inequality (8) is transformed by using the substitutions (5) to obtain

$$\frac{V_m N_T}{K+N_T} < B. \qquad (10)$$

This inequality, (10), coupled with the conclusion of Theorem I demonstrates that if the (Michaelis-Menten) uptake rate of the phytoplankton component evaluated at the maximum total available nutrient, N_T, is less than the loss rate of nutrient, B, of the phytoplankton component, then the phytoplankton population is eliminated. If the phytoplankton population vanishes, so does the herbivore population. From the parameter relationship (8), it then follows that the ecosystem does not persist. For this parameter range, the terminal equilibrium state of the model is $N = N_T$, $P = 0$, $Z = 0$.

Theorem II. If the parameters ε and δ satisfy $\varepsilon \le \delta$, then corresponding to each solution of (6) there exists a τ_z, $0 \le \tau_z \le \infty$, such that the z-component of the solution satisfies $z(\tau_z) = 0$. If $\varepsilon > \delta$ then corresponding to the z-component of each solution of (6) there exists a T_z, $0 \le T_z \le \infty$, such that $z(T_z) = 1 - \delta/\varepsilon$ and if $T_z < \infty$ then $z(\tau) \le 1 - \delta/\varepsilon$ for $\tau \ge T_z$.

The parameters ε and δ are given in (5) as $\varepsilon = EN_T/V_m$ and $\delta = D/V_m$. The inequality $\varepsilon \le \delta$ is equivalent to $EN_T \le D$. The term EN_T represents the maximum rate at which the nutrient can be obtained by the grazing of the herbivores. If this is less than the rate at which zooplankton lose nutrient then, as the theorem concludes, the zooplankton component is eliminated. Thus, the parameter range $\varepsilon \le \delta$ leads to nonpersistence of the ecosystem.

The final conclusion of the theorem is useful in the following circumstances. For any initial population of zooplankton, $z(0)$, there is an eventual upper bound namely, $1 - \delta/\varepsilon$, on the nutrient present in the zooplankton component.

Theorem III. Let $\gamma \equiv \min(\beta, \delta)$ satisfy $1 \le \gamma$; then corresponding to any solution of (6), there exists a $\tau_0 = \tau_0(p,z)$, $0 < \tau_0 \le \infty$, such that the p and z components of the solution satisfy $p(\tau_0) = 0$ and $z(\tau_0) = 0$. If $(\alpha + 1)^{-1} > \gamma$ then corresponding to any solution of (6) there exists a τ_n, $0 < \tau_n \le \infty$, such that the nutrient component of the solution satisfies $n(\tau_n) = \alpha\gamma/(1 - \gamma)$ and if $\tau_n < \infty$, $n(\tau_n) \le n(\tau)$ for all $\tau \ge \tau_n$.

Implications and interpretations of Theorem III are now commented upon.

If $\gamma \ge 1$ then $\beta \ge 1$ and $\delta \ge 1$; that is, $B \ge V_m$ and $D \ge V_m$. This situation is not feasible for persistence of the ecomodel.

The last sentence of the theorem yields an eventual lower bound for the nutrient pool under the assumption that the nutrient loss rate of the herbivores and the nutrient loss rate of the phytoplankton component is less than the maximum uptake rate of the phytoplankton. In other words, a certain quantity of nutrient, namely $\alpha\gamma/(1 - \gamma)$, is always present in the n-component of the model.

The conclusions of the theorems are for the most part, biologically obvious. It is important from a well-posed modeling approach that such conclusions can be rigorously demonstrated.

The above results are summarized in the following theorem.

Theorem IV. If the parameters α, β, δ, ε satisfy any of the inequalities: $\varepsilon < \delta$, $(\alpha + 1)^{-1} < \beta$, *or* $1 \leq \gamma$; *then the ecosystem as described by system (6) cannot persist.*

It remains to consider parameters in the complement ranges. This set is investigated in the remainder of the paper and leads to the biologically interesting conclusions about persistence of the ecosystem.

STABILITY ANALYSIS OF THE EQUILIBRIUM POINTS

The model of the marine ecosystem (6), coupled with the closure condition, (7), is a two dimensional dynamical system. The relationship (7) with $n \geq 0$, $p \geq 0$, $z \geq 0$ is a triangular sector, Δ, of the plane, where all dynamical motions of (6) can be viewed.

The theory of two dimensional systems of ordinary differential equations is very well developed (*Coddington and Levinson*, 1955; *Lefschetz*, 1963). This theory will be utilized to discuss the persistence of the ecomodel (6).

Stability analyses and especially numerical computations are more readily performed on a two dimensional system in a geometrically pleasing and analytically recognizable form. The **transformations**

$$n = \frac{1}{\sqrt{3}}\eta - \frac{1}{\sqrt{2}}\rho - \frac{1}{\sqrt{6}}\zeta + \frac{1}{2}$$

$$p = \frac{1}{\sqrt{3}}\eta + \frac{1}{\sqrt{2}}\rho - \frac{1}{\sqrt{6}}\zeta + \frac{1}{2} \quad (11)$$

$$z = \frac{1}{\sqrt{3}}\eta \qquad + \sqrt{\frac{2}{3}}\zeta$$

take the coordinate system (n,p,z) into the system (η,ρ,ζ) where

$$\eta = 0,$$

$$\rho = \frac{1}{\sqrt{2}}(p - n),$$

$$\zeta = \sqrt{\frac{3}{2}}\,z.$$

PERSISTENCE OF AQUATIC ECOSYSTEMS

In this setting, the basis for the plane (7) is determined solely by the coordinates ρ and ζ.

The Equilibrium Points

There are three equilibrium points of the system (6) that lie on the triangle Δ. They are

I. $n_0 = 1$, $p_0 = 0$, $z_0 = 0$;

II. $n_0 = \dfrac{\alpha\beta}{1-\beta}$, $p_0 = 1 - \dfrac{\alpha\beta}{1-\beta}$, $z_0 = 0$;

III. $n_0 = -\dfrac{(\alpha\varepsilon - \beta - \varepsilon + \delta + 1)}{2\varepsilon} + \left[\dfrac{(\alpha\varepsilon - \beta - \varepsilon + \delta + 1)^2 + 4\alpha\varepsilon(\beta + \varepsilon - \delta)}{2\varepsilon}\right]^{1/2}$,

$p_0 = \delta/\varepsilon$, $z_0 = 1 - p_0 - n_0$.

The points I and II are located at the base of the triangle Δ and are readily seen to be biologically meaningful. The critical point III is, in general, located in the interior of Δ.

The linear analysis used to determine the nature of an equilibrium point consists of translating the origin of the coordinate system to the equilibrium point, determining the behavior of the linear part of the system, and then applying standard perturbation theorems to obtain the behavior of the nonlinear system. The perturbation results which are needed may be found in *Coddington and Levinson* (1955) or *Ross* (1974).

The Behavior of Equilibrium Point I

The system (6) when linearized about the equilibrium point I has linear part

$$\frac{d\rho_1}{d\tau} = (-\beta + \frac{1}{\alpha+1})\rho_1 + \frac{1}{\sqrt{3}}(\beta - \delta - \frac{1}{\alpha+1})\zeta_1;$$

$$\frac{d\zeta_1}{d\tau} = -\delta\zeta_1. \qquad (12)$$

The characteristic equation associated with the linear system (12) is $(-\delta - \lambda)(-\beta + (\alpha+1)^{-1} - \lambda) = 0$, and the characteristic roots of (12) are $\lambda_1 = -\delta$ and $\lambda_2 = -\beta + (\alpha+1)^{-1}$. The root λ_1 is always negative while the characteristic root λ_2 is positive provided

$\beta < (\alpha + 1)^{-1}$. The origin of (12) is therefore a saddle point assuming $\beta < (\alpha + 1)^{-1}$. The perturbation theorem (Ross, 1974; p. 570) implies that the origin is also a saddle point for the non-linear system; that is, equilibrium point I is a saddle point for system (6) whenever $\beta < (\alpha + 1)^{-1}$.

The Behavior of Equilibrium Point II

The topological nature of the critical point II: $n_0 = \alpha\beta/(1 - \beta)$, $p_0 = 1 - \alpha\beta/(1 - \beta)$, $z_0 = 0$, will now be determined. Throughout this discussion, it shall be tacitly assumed that the parameters α and β satisfy $\beta < (\alpha + 1)^{-1}$. This parametric relationship implies $\beta < 1$ and that $0 < n_0 < 1$, $0 < p_0 < 1$.

This equilibrium point is on the base line of Δ. Translation of the coordinate system to the point II and linearization leads to the approximation

$$\frac{d\rho_2}{d\tau} = -[\beta(\beta - 1) + \frac{(1 - \beta)^2}{\alpha}]\rho_2$$

$$+ \frac{1}{\sqrt{3}}[\beta - \delta - (1 - \frac{\alpha\beta}{1 - \beta})\varepsilon - (\frac{(1 - \beta)^2}{\alpha} + \beta^2)]\zeta_2, \quad (13)$$

$$\frac{d\zeta_2}{d\tau} = [-\delta + \varepsilon(1 - \frac{\alpha\beta}{1 - \beta})]\zeta_2.$$

The linear equation (13) has characteristic equation

$$[-\delta + \varepsilon(1 - \frac{\alpha\beta}{1 - \beta}) - \lambda](-\beta(\beta - 1) - \frac{(1 - \beta)^2}{\alpha} - \lambda] = 0, \quad (14)$$

with characteristic values given by

$$\lambda_1 = -\delta + \varepsilon(1 - \frac{\alpha\beta}{1 - \beta}), \quad (15)$$

$$\lambda_2 = \frac{(1 - \beta)(\alpha + 1)}{\alpha}\left(\beta - \frac{1}{\alpha + 1}\right). \quad (16)$$

The root λ_2 is always negative under our assumptions while the sign of λ_1 depends upon the relative size of $\alpha, \beta, \delta, \varepsilon$. If

$$1 - \frac{\alpha\beta}{1 - \beta} > \frac{\delta}{\varepsilon}, \quad (17)$$

then λ_1 is positive and the equilibrium point $(0,0)$ of (13) is geometrically classified as a saddle point. Thus, if (17) holds, point II is also a saddle point for the system (6).

Whenever $1 - \alpha\beta/(1 - \beta) < \delta/\varepsilon$ then λ_1 is negative and the origin is an asymptotically stable node of the linear system (13). Perturbation theory implies that the equilibrium point II is also an asymptotically stable node for (6).

It is interesting to observe that the quantities composing inequality (17) are the p_0-coordinates of equilibrium point II and III. Stability implications obtained from the relative location of the equilibrium points will be explored in the discussion on biological interpretation and summary.

The Behavior of Equilibrium Point III

A linear analysis of the system about the point III:

$$n_0 = -\left(\frac{\alpha\varepsilon-\beta-\varepsilon+\delta+1}{2\varepsilon}\right) + \left[(\alpha\varepsilon-\beta-\varepsilon+\delta+1)^2 + 4\alpha\varepsilon(\beta+\varepsilon-\delta)\right]^{1/2}/2\varepsilon, \quad (18)$$

$$p_0 = \delta/\varepsilon, \quad (19)$$

$$z_0 = 1 - \delta/\varepsilon - n_0;$$

leads to

$$\frac{d\rho_3}{d\tau} = -\frac{1}{2}\left[\frac{2p_0}{\alpha + n_0} - \frac{2n_0 p_0}{(\alpha + n_0)^2} - \varepsilon z_0\right]\rho_3$$

$$-\frac{1}{2\sqrt{3}}\left[\frac{2p_0}{\alpha + n_0} + 2\varepsilon p_0 + \varepsilon z_0 + 2\delta - \frac{2n_0 p_0}{(\alpha + n_0)^2}\right]\zeta_3, \quad (21)$$

$$\frac{d\zeta_3}{d\tau} = \frac{\sqrt{3}}{2}\varepsilon z_0 \rho_3 - \frac{1}{2}\varepsilon z_0 \zeta_3.$$

The characteristic equation associated with this linear system can be written as

$$\lambda^2 + \left[\frac{p_0}{\alpha + n_0} - \frac{n_0 p_0}{(\alpha + n_0)^2}\right]\lambda + \frac{\varepsilon z_0}{4}\left[\frac{4p_0}{\alpha + n_0} - \frac{4n_0 p_0}{(\alpha + n_0)^2}\right.$$

$$\left. + 2\varepsilon p_0 + 2\delta\right] = 0. \quad (22)$$

The quadratic formula and some simplification give the roots of (22) as

$$2\lambda = -\frac{\alpha p_0}{(\alpha+n_0)^2} \pm \left\{\frac{\alpha^2 p_0^2}{(\alpha+n_0)^4} - \varepsilon z_0 \left[\frac{4\alpha p_0}{(\alpha+n_0)^2} + 2\varepsilon p_0 + 2\delta\right]\right\}^{1/2}. \quad (23)$$

These characteristic roots can theoretically be either real numbers or complex conjugates.

In the complex conjugate case, the discriminant is negative and the real part of λ, $-\alpha p_0/2(\alpha+n_0)^2$, is negative. The topological structure of the origin of the linear system (21) is an asymptotically stable spiral.

If the discriminant is nonnegative then both characteristic roots are negative. This follows immediately since the form of the roots is $\lambda = -Q/2 \pm (1/2)(Q^2 - R)^{1/2}$ where Q and R are positive. In this case, the origin of the linear system (21) is an asymptotically stable node. The asymptotically stable spiral and asymptotically stable node for the linearized system are, respectively, an asymptotically stable spiral and an asymptotically stable node for the nonlinear system. In summary, equilibrium point III can be either a spiral or a node; however, independent of the topological structure, the point III is asymptotically stable.

STABILITY ANALYSIS OF THE MODEL

Under the assumption that the parameters α, β, δ, ε satisfied the inequalities $\delta < \varepsilon$, $\beta < (\alpha+1)^{-1}$, the following conclusions were obtained in the previous discussion

The equilibrium point I: $n_0 = 1$, $p_0 = 0$, $z_0 = 0$ of the nonlinear system (6) is always a saddle point.

The equilibrium point II: $n_0 = \alpha\beta/(1-\beta)$, $p_0 = 1 - \alpha\beta/(1-\beta)$, $z_0 = 0$ of the nonlinear system (6) is either
(i) an asymptotically stable node (whenever $1 - \alpha\beta/(1-\beta) \le \delta/\varepsilon$); or
(ii) a saddle point (whenever $1 - \alpha\beta/(1-\beta) > \delta/\varepsilon$)

The equilibrium point III: $n_0 = -(\alpha\varepsilon-\beta-\varepsilon+\delta+1)/2\varepsilon + [(\alpha\varepsilon-\beta-\varepsilon+\delta+1)^2 + 4\alpha\varepsilon(\beta+\varepsilon-\delta)]^{1/2}/2\varepsilon$, $p_0 = \delta/\varepsilon$, $z_0 = 1 - p_0 - n_0$, of the nonlinear system (6) is either

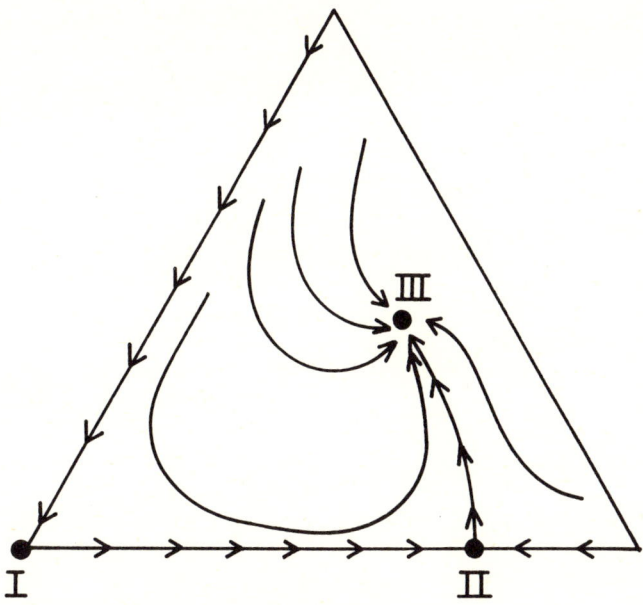

Fig. 1. Phase triangle diagram of (6), (7) when equilibrium points I & II are saddle points and III is an asymptotically stable node.

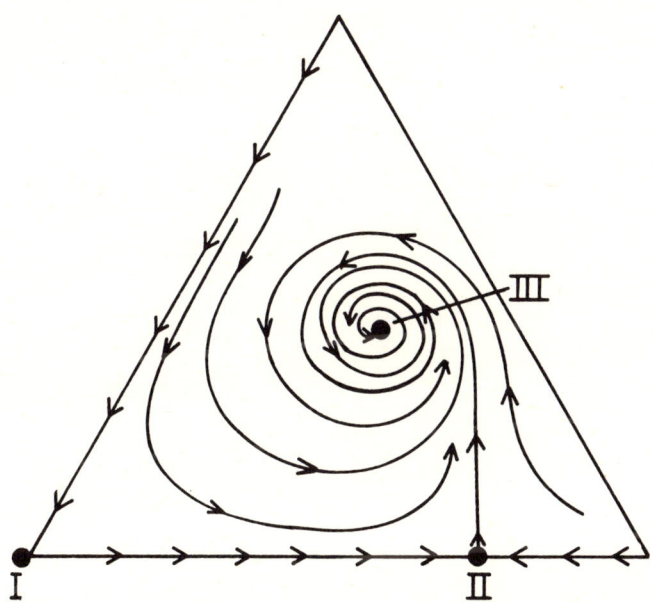

Fig. 2. Phase triangle diagram of (6), (7) when equilibrium points I and II are saddle points and III is an asymptotically stable spiral.

(i) an asymptotically stable spiral; or
(ii) an asymptotically stable node.

In this section the dynamic geometrical configuration of the phase triangle Δ will be described. The objective is to fit the topological structures of the equilibrium points together in an appropriate manner.

The equilibrium point I is located on the n-axis on the base of the triangle Δ. The point II is situated on the base of Δ also. The equilibrium point III is, in general, located in the interior of Δ. It is of fundamental importance to determine when III lies in the interior of Δ for this is the critical factor governing persistence of the ecosystem.

The sign of the z_0 coordinate of III is determined by the relative size of the values of $1 - \alpha\beta/(1 - \beta)$ and δ/ϵ. This coordinate can be written as $z_0 = 1 - p_0 - n_0 = \{(\epsilon-\delta+\alpha\epsilon-\beta+1) - [(\epsilon-\delta+\alpha\epsilon-\beta+1)^2 + 4(\alpha\epsilon\beta+\beta\epsilon-\epsilon+\delta-\beta\delta)]^{1/2}/2\epsilon$. The quantity $\alpha\epsilon\beta+\beta\epsilon-\epsilon+\delta-\beta\delta$ is positive if $1 - \alpha\beta/(1 - \beta) < \delta/\epsilon$, zero provided $1 - \alpha\beta/(1 - \beta) = \delta/\epsilon$, and negative when $1 - \alpha\beta/(1 - \beta) > \delta/\epsilon$. Thus, the quantity in the brackets exceeds the positive initial term in the expression for z_0 provided $1 - \alpha\beta/(1 - \beta) < \delta/\epsilon$ and is less than the initial term provided $1 - \alpha\beta/(1 - \beta) > \delta/\epsilon$. The parameter range where the ecosystem can persist is determined by the inequality

$$1 - \frac{\alpha\beta}{1 - \beta} > \frac{\delta}{\epsilon} , \qquad (24)$$

since only in this range is z_0 positive, and hence, in the interior of the phase triangle Δ.

Inequality (24) has a simple geometrical interpretation. The left side of the inequality is the p_0 coordinate of equilibrium point II and the right side is the p_0 coordinate of equilibrium point III. Inequality (24) locates the relative position of these p_0 coordinates and determines persistence. As indicated in the conclusions of equilibrium point II above, the geometrical position of these components also determines the topological structure of the equilibrium point II.

The solution space configurations as determined by the possibilities delineated in the conclusions for equilibrium point I, II and III above are now presented. Figure 1 illustrates the phase triangle portrait of nonlinear system (6), (7) when equilibrium point II is a saddle and III is an asymptotically stable node located in the interior of Δ.

Another persistent situation occurs when I and II are saddle points and III is an asymptotically stable spiral. This is illustrated in Figure 2.

The next and final possibility for a phase plane portrait of system (6) occurs when the equilibrium point III is not relevant to Δ; that is, either III is the same as II or III is not on the phase triangle. In this case, II is an asymptotically stable node.

This configuration is valid over the parameter range $1 - \alpha\beta/(1 - \beta) \leq \delta/\varepsilon$ or, in other equivalent terms, as long as the z_0-coordinates of equilibrium points satisfy the inequality $z_0(\text{III}) \leq z_0(\text{II}) = 0$.

The behavior of the system (6), (7) when the parameters satisfy $\beta = (\alpha + 1)^{-1}$ is now indicated with little detail. A calculation shows that equilibrium points I and II are identical whenever $\beta = (\alpha + 1)^{-1}$. The argument given previously in this section concerning the sign of z_0 shows that in this setting $z_0 < 0$ and, hence, III is not on the phase triangle Δ. Consideration of the phase diagram reveals that this is a stable but not persistent situation.

BIOLOGICAL INTERPRETATION AND SUMMARY

In previous sections, it was established that parameters in the ranges $\delta \geq \varepsilon$, $\beta \geq (\alpha + 1)^{-1}$, or $\gamma \geq 1$ lead to extinction of a species component of the system (6), (7), and, hence, to the nonpersistence of the ecomodel.

The restrictions $\delta < \varepsilon$ and $\beta < (\alpha + 1)^{-1}$ were imposed in a previous discussion. The essential factor that determines persistence of the ecomodel in this setting was indicated to be the relationship between the values of $1 - \alpha\beta/(1 - \beta)$ and δ/ε. The situation covered topologically by the configurations of Figures 1 and 2 is one of ecomodel persistence and, additionally, of the model tending towards an equilibrium state. The alternative illustrated in Figure 3 is one of nonpersistence in as much as the herbivore component was eliminated. However, the model was still stable in the sense that an equilibrium state was approached.

Remarks are summarized by the following result:

Theorem V. The aquatic ecosystem as modeled by (6), (7) can persist if and only if the parameters α, β, δ, ε satisfy the relationship

$$1 - \alpha\beta/(1 - \beta) > \delta/\varepsilon \qquad (25)$$

In view of the strength of this theorem it seems appropriate to discuss the persistence inequality (25) in additional, biologically relevant, detail. Transforming (25) to original dimensional parameters by employing (5), we find the equivalent expression

$$N_T > \frac{KB}{V_m - B} + \frac{D}{E}. \qquad (26)$$

The terms on the right side of (26) can be roughly interpreted as net proportion loss by gain of nutrient by the phytoplankton and herbivore components. The inequality implies that there will always be available resource as represented by the nutrient component. From the opposite viewpoint, the reverse inequality $N_T \leq KB/(V_m - B) + D/E$

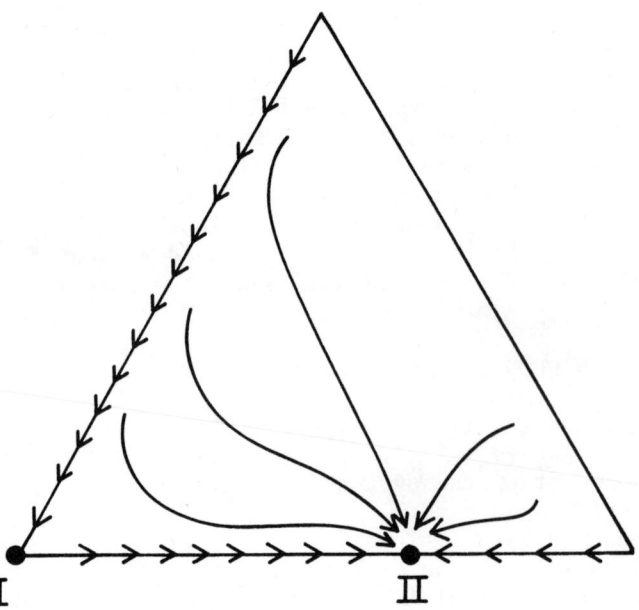

Fig. 3. Phase triangle diagram of (6), (7) when equilibrium points I is a saddle point and II is an asymptotically stable node.

indicates that the total amount of nutrient present is not sufficient for the requirements of the system as it is dynamically represented and hence, the ecosystem cannot be maintained.

APPENDIX A

In this appendix, Theorems I, II, III and IV will be proved by using differential inequalities. The differential inequality method starts with the given system and finds bounds for the derivatives. The solutions of the inequalities are compared with solutions of a comparison differential equation. The comparison differential equation is, hopefully, more amiable to interpretation than the original system. From the comparison, conclusions about the original differential system are obtained.

Theorem I. If the parameters α and β satisfy the inequality $(\alpha + 1)^{-1} < \beta$ then corresponding to the p-component of each solution of (6) there exists a constant p_0 such that

$$p(\tau) \leq p_0 \exp\left[-\beta + (\alpha + 1)^{-1}\right]\tau, \quad \tau \in R_+ .$$

In particular, there exists a τ_p, $0 \leq \tau_p \leq \infty$, such that $p(\tau_p) = 0$.

Proof. Using $z \geq 0$, $p \geq 0$ in the differential equation (6) we find

$$dp/d\tau \leq np/(\alpha + n) - \beta p.$$

As long as $p \neq 0$

$$p^{-1} dp/d\tau \leq 1 - \beta - \alpha/(\alpha + n). \qquad (1A)$$

The nutrient component $n(\tau)$ satisfies $0 \leq n(\tau) \leq 1$ so that $-\alpha/[\alpha + n(\tau)] \leq -\alpha/(\alpha + 1)$. Inserting this into (1A) yields

$$p^{-1} dp/d\tau \leq 1 - \beta - \alpha/(\alpha + 1).$$

An integration from 0 to τ leads to

$$\ln\left[p(\tau)/p(0)\right] \leq \tau\left[1 - \beta - \alpha(\alpha + 1)^{-1}\right]$$
$$= \tau\left[-\beta + (\alpha + 1)^{-1}\right].$$

Solving for p gives the conclusion of the theorem.

Theorem II. If the parameters ε and δ satisfy $\varepsilon \leq \delta$ then corresponding to each solution of (6) there exists a T_z, $0 \leq T_z \leq \infty$, such that the z-component satisfies $z(T_z) = 0$. If $\varepsilon > \delta$ then corresponding to the z-component of each solution of (6) there exists a T_z, $0 \leq T_z \leq \infty$ such that $z(T_z) = 1 - \delta/\varepsilon$ and if $T_z < \infty$ then $z(\tau) \leq 1 - \delta/\varepsilon$ for $\tau \geq T_z$.

Proof. The inequality $p \leq 1 - n - z \leq 1 - z$ implies that $\varepsilon z p \leq \varepsilon z(1 - z)$. From equation (6b) it follows that

$$dz/d\tau \leq (\varepsilon - \delta)z - \varepsilon z^2.$$

Comparing this differential inequality with the equation

$$du/d\tau = (\varepsilon - \delta)u - \varepsilon u^2, \qquad (2A)$$

we find that whenever $z(0) \leq u(0)$ then $z(\tau) \leq u(\tau)$ for $\tau \geq 0$. The solution space of (2A) has the structure indicated in Figure 4.

The z-component of any solution of the system (6) satisfies $z(\tau) \leq u(\tau)$ where u is any solution of (2A) with $z(0) \leq u(0)$. This gives the desired conclusions of Theorem II since in the range $\varepsilon \leq \delta$, all solutions u of (2A) with $u(0) \geq 0$ approach zero as τ approaches infinity. This forces z to vanish at some, possibly infinite, point. The conclusion for $\varepsilon > \delta$ follows in an analogous manner.

Theorem III. Let $\gamma \equiv \min(\beta, \delta)$ satisfy $\gamma \geq 1$; then, corresponding to any solution of (6), there exists a $\tau_0 = \tau_0(p, z)$, $0 \leq \tau_0 \leq \infty$, such that the p and z components of the solution satisfy $p(\tau_0) = 0$ and $z(\tau_0) = 0$. If $(\alpha + 1)^{-1} > \gamma$ then corresponding to any solution

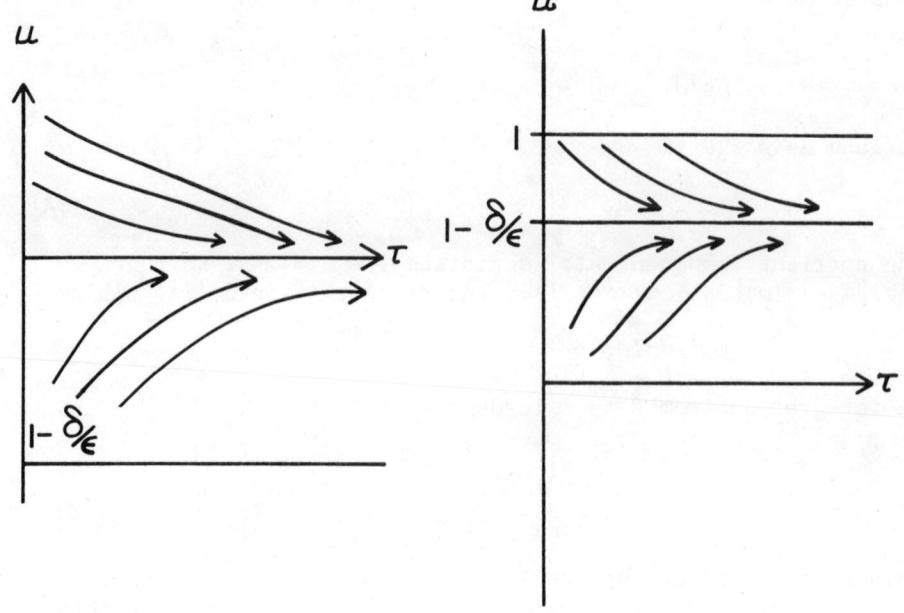

Fig. 4. a. Solution space structure of $du/d\tau = (\varepsilon - \delta)u - \varepsilon u^2$ for $\varepsilon \leq \delta$.
b. Solution space structure of $du/d\tau = (\varepsilon - \delta)u - \varepsilon u^2$ for $\varepsilon > \delta$.

of (6) there exists a τ_n, $0 \leq \tau_n \leq \infty$, such that the nutrient component of the solution satisfies $\overline{n}(\tau_n) = \alpha\gamma/(1 - \gamma)$ and if $\tau_n < \infty$, $n(\tau_n) \leq n(\tau)$ for all $\tau \geq \tau_n$.

Proof. From the differential equation (6) we have

$$\begin{aligned} dn/d\tau &= - np/(\alpha + n) + \beta p + \delta z \\ &\geq - np/(\alpha + n) + \gamma(p + z) \\ &\geq - np/(\alpha + n) + \gamma(1 - n) \quad (A3) \\ &\geq - (1 - n)n/(\alpha + n) + \gamma(1 - n) \end{aligned}$$

When $\gamma \geq 1$ we obtain

$$\begin{aligned} \frac{dn}{d\tau} &\geq \frac{\alpha(1 - n)}{\alpha + n} \\ &\geq \frac{\alpha}{\alpha + 1}(1 - n). \end{aligned} \quad (A4)$$

This differential inequality leads to $\lim_{\tau \to \infty} n(\tau) = 1$ which in turn implies extinction of both p and z.

The condition $(\alpha + 1)^{-1} > \gamma$ is equivalent to $\alpha\gamma/(1 - \gamma) < 1$. With this hypothesis, we find that

$$\frac{dn}{d\tau} \geq \frac{-(1 - n)(1 - \gamma)}{\alpha + n}\left[n - \frac{\alpha\gamma}{1 - \gamma}\right]. \quad (A5)$$

Employing the comparison equation

$$\frac{du}{d\tau} = \frac{-(1 - \gamma)(1 - u)}{\alpha + u}\left[u - \frac{\alpha\gamma}{1 - \gamma}\right], \quad (A6)$$

we obtain a slope field indicating $\lim_{\tau \to \infty} u(\tau) = \alpha\gamma/(1 - \gamma)$, which leads to the last conclusion of the theorem.

ACKNOWLEDGEMENTS

My original interest in this problem was nurtured by Professor James J. O'Brien who provided both encouragement and support for this project. Some of the initial research was done while I was a visiting member of the Instituto de Ciências Matemáticas de São Carlos, Universidade de São Paulo, São Carlos, S. P., Brasil. I am grateful to the Instituto for support and to Professors Nelson Onuchic, Antonio F. Izé and the other members of the Instituto for their kind hospitality. Mr. Joseph S. Wroblewski kindly read the manuscript and made many helpful suggestions.

Finally, I am pleased to acknowledge that this research was completed with support provided by the Office of Naval Research under Contract N00014-76-C-0006.

REFERENCES

Coddington, E. and N. Levinson. 1955. *Theory of Ordinary Differential equations*. New York: McGraw-Hill.
Dugdale, R. C. 1967. *Limnol. Oceanogr.* 12: 685.
Dugdale, R. C. and J. J. MacIsaac. 1971. *Invest. Pesq.* 31(5): 299.
Eppley, R. W. and J. L. Coastworth. 1968. *J. Physiol.* 4: 151.
Eppley, R. W., J. N. Rogers, and J. J. McCarthy. 1969. *Limnol. Oceanogr.* 14: 912.
Eppley, R. W. and W. H. Thomas. 1969. *J. Physiol.* 5: 375.
Hirsch, M. W. and S. Smale. 1974. *Differential Equations, Dynamical Systems, and Linear Algebra*. New York: Academic Press.
Lakshmikantham, V. and S. Leela. 1969. *Differential and Integral Inequalities*, Vol. 1. New York: Academic Press.
Lefschtz, S. 1963. *Differential Equations, Geometric Theory*. New York: Wiley.

MacIsaac, J. J. and R. C. Dugdale. 1969. *Deep-Sea Res.* 16: 45.
May, R. M. 1974. *Stability and Complexity in Model Ecosystems. 2nd Edition.* Princeton: University Press.
O'Brien, J. J. and J. S. Wroblewski. 1973. *Invest. Pesq.* 37(2): 193.
Patten, B. C. 1968. *Int. Revue ges. Hydrobiol.* 53(3): 357.
Ross, S. L. 1974. *Differential Equations (2nd Ed.).* Lexington: Xerox.
Steele, J. H. 1974. *The Structure of Marine Ecosystems.* Harvard: University Press.
Verhoff, F. H. and F. E. Smith. 1971. *J. Theor. Biol.* 33: 131.
Wroblewski, J. S. and J. J. O'Brien. 1975. *J. Mar. Biol.* 35: 161.

Mesoscale Oceanic Phytoplankton Patchiness Caused by Hurricane Effects on Nutrient Distribution in the Gulf of Mexico

Richard L. Iverson

Florida State University

ABSTRACT

Phytoplankton dynamics function at low, steady-state levels over most of the area of the oceanic province during most of the year. Upwelling or vertical mixing of the water column caused by short term, high wind stress events such as hurricanes are processes which transport new nitrogen to the photic zone in tropical and in subtropical latitudes.

Variation in hurricane latitudinal position, ground speed, radius of hurricane winds, and hurricane eye diameter interact to cause differences in the depth of vertical mixing thus causing variation in nitrate input to the photic zone along the hurricane track. Variation in nitrate input can cause spatial differences in phytoplankton growth rate leading to mesoscale oceanic phytoplankton patchiness. Calculations based on the Pollard, Rhine, and Thompson model for deepening of the wind-mixed layer suggest that during hurricane Hilda (1964), up to twice the summer photic zone nitrate-nitrogen mass, up to twice the total particulate nitrogen mass, and up to three times the steady-state water column phytoplankton particulate nitrogen mass could have been transported into the photic zone along the track of Hilda in the Gulf of Mexico.

INTRODUCTION

Spatially heterogeneous distributions of plankton result from the interaction of physical and biological processes along different length and time scales in the sea. It is difficult to quantitatively generalize the interplay of physical and biological processes (*Denman and Platt*, 1975), since in addition to the coexistance of physical processes of different scales and periods in the ocean as discussed by *Stommel* (1963), the magnitudes of biological rates are time variable. Strong turbulence and advection are always capable of dominating biological rates in controlling the distribution of the plankton.

Physical processes which dissipate or which advect plankton populations range in length scale from turbulence on the order of centimeters (*Cassie*, 1962) to the large scale advection processes in the ocean gyres (*McGowan*, 1972).

The interplay between diffusive processes and biological growth rates appears to be particularly important in controlling local plankton blooms in the sea. The size and shape of chlorophyll a patches off Peru changed with time with a patch increasing in largest length from about 15 kilometers to about 60 kilometers over four days before losing its identity (*Beers et al.*, 1971). Theoretical investigations have clarified relations between horizontal mixing rates and biological rates which control local plankton blooms (*Kierstead and Slobodki*, 1953; *Platt and Denman*, 1975; *Wroblewski, O'Brien, and Platt*, 1975). *Steele and Yentsch* (1960) considered the interrection of vertical diflusion and biological processes which affect the vertical distribution of chlorophyll. In contrast to physical processes which act to decrease local plankton populations, physical processes which carry nutrient-rich water into the photic zone facilitate plankton growth. A ten year average of maxima in plankton distribution off the California coast suggested distances from 200 to 500 kilometers off shore were affected by consequences of upwelling processes which occurred over distances about 0.4 times the length of the plankton maxima zones (*Cushing*, 1971). *Walsh et al.* (1971) mapped chlorophyll a distributions which appeared to be in quasi steady-state over a distance of about 40 kilometers extending outward from upwelling centers near the Peruvian coast.

Wind stress on the sea surface causes upwelling, horizontal advection, and vertical mixing, all of which affect the distribution of the plankton. *Sverdrup* (1953) quantized the general effect of vertical turbulence in controlling onset of spring phytoplankton blooms in the sea. *Steele* (1966) and *Parsons et al.* (1966) showed how plankton blooms began in the spring as the mixed layer progressively became shallower than the critical depth in the North Atlantic and Northeastern Pacific Oceans, respectively. During summer in an Alaskan estuary, wind mixing of a stratified water column produced conditions which caused plankton blooms (*Iverson et al.*, 1974). High sustained winds reduced local phytoplankton populations in Puget Sound by rapid horizontal advection (*Winter, Banse, and Anderson*, 1975).

Hurricanes occur in tropical and in subtropical latitudes during months when phytoplankton growth rates are at a minimum due to nutrient depletion in the photic zone of highly stratified oceanic water columns. A 50 year record of hurricane data revealed an average of

4 hurricanes per year in the Gulf of Mexico through 1950 (*Dunn and Miller*, 1964). Since 1954 there have been an average of 2 hurricanes per year in the Gulf of Mexico (*U.S. Department of Interior*, 1976). The radius of hurricane force winds (wind speeds greater than 33 m/sec) is the order of 100 kilometers (including a quiscent eye of the order of 30 kilometers in diameter for large hurricanes). Since hurricanes commonly travel several degrees of latitude during their lives, it might be expected that hurricanes could cause significant vertical transport of nutrient-rich water to the photic zone along their tracks. Hurricanes have been shown to cause conditions which resulted in increased phytoplankton photosynthetic rates (*Franceschini and El-Sayed*, 1968). In this paper, the effects of vertical mixing processes are calculated for hurricane Hilda (1964) in an attempt to assess the significance of a hurricane in driving mesoscale heterogeneity in phytoplankton distribution in the Gulf of Mexico.

HURRICANE EFFECTS ON THE MIXED LAYER DEPTH IN THE GULF OF MEXICO

Decreases in sea surface temperature along hurricane tracks were observed by *Fisher* (1968), and by *Jordan and Frank* (1964). The temperature decreases were attributed primarily to effects of upwelling by *Fisher* (1958), and primarily to effects of vertical mixing by *Jordan and Frank* (1964). *O'Brien* (1967) and *O'Brien and Reid* (1967) developed and numerically solved theoretical models for upwelling driven by hurricane winds. *O'Brien* (1969) suggested that upwelling strength should vary as a function of the speed with which a storm moves over the sea surface. *Geisler* (1970) showed that for a linear, 2-layer model, speed of movement of a hurricane over the sea surface governs the response of the water column to wind stress. Upwelling would only occur for hurricane ground speeds less than the internal long wave speed, $c = (gh\Delta\rho/\rho)^{1/2}$; where g is gravity and h is the thickness of the surface mixed layer. *Leipper* (1967) obtained temperature data at various locations before and after hurricane Hida crossed the Gulf of Mexico. The data indicated that warm surface layers were transported outward from the hurricane center, cooling and mixing as they moved to a convergence region outside the central storm area. The data suggested that cold water upwelled along the hurricane path from depths of about 60 meters. Using data from *Leipper's* (1967) observations on the undisturbed Gulf of Mexico, $c = 2.3$ km/h. Under the *Geisler* criterion and allowing for errors in data used to calculate the internal long wave speed, upwelling could only be possible for the last portion of hurricane Hilda in the Gulf of Mexico (segment 7, Table 1).

There appears to be a conflict between *Geisler's* (1970) upwelling criterion and the interpretation of the data obtained by *Leipper* (1967). This paper is concerned with the estimation of nutrient input into the Gulf of Mexico photic zone, which may be considered

Table 1. Hurricane Hilda (1964) data together with ground speed and π/δ values.

Data Segment	Ground Speed km/hr	Radius to 33 m/sec wind speed km	Eye Diameter km	Max Wind m/sec	π/δ hr
1	12.8	93	18.5	41	14.7
2	10.0	130	37.0	50	14.4
3	9.3	130	28.0	64	14.1
4	9.4	130	33.0	59	13.8
5	8.9	130	18.5	56	13.4
6	10.6	130	46.0	46	12.9
7	3.7	130	74.0	54	12.6

the 75 meter depth at which light intensity is one percent of the surface value (El-Sayed et al., 1972). For that reason, attention will be confined to those regions where vertical mixing can be expected to occur outside the central storm area.

Pollard, Rhines, and Thompson (1973) proposed a model for deepening of the wind mixed layer. The model was intended to hold for at most a few days and in an area without strong horizontal contrasts. Both of these conditions are met for hurricane passage over the open Gulf of Mexico. The maximum mixed layer depth, h_{max}, would be reached after application of wind stress for a time at least equal to one half the local inertial period.

$$h_{max} = 1.7 \left(\frac{\tau}{\rho f N}\right)^{\frac{1}{2}} \qquad (1)$$

where ρ is density, $f = 2\omega \sin\phi$, and $N^2 = g\, \partial\rho/\bar{\rho}\, \partial z$. Wind stress, was computed for use in this paper using the relation

$$\tau = \rho_a C W^2 \qquad (2)$$

where ρ_a = air density, C is the drag coefficient (1.3×10^{-3}), and W is the wind speed.

A wind speed distribution curve taken from O'Brien and Reid (1967) was used together with different wind speed maxima (Table 1) to compute wind speed distribution from the storm center at various locations along the hurricane path (Figure 1). Since the Gulf of Mexico photic zone depth was taken as 75 meters, and since we were concerned with estimating nutrient transport into the photic zone, radial extent of 80 meter, 90 meter, and 100 meter mixed layer depths were computed for each storm segment data set using wind speeds obtained from Figure 1, together with the formulation of Pollard, Rhines, and Thompson (1973) (Figure 2). The π/f criterion was calculated for different latitudinal hurricane data segment locations and was multiplied by the hurricane ground speed for corresponding storm segments to give the minimum ground track length necessary to achieve h_{max} for each storm segment. This length was used to determine the lateral distance (relative to the storm track) from the hurricane center at which the 80 meter, 90 meter, and 100 meter mixed layer depths would be located as the hurricane moved over the sea surface (Figure 3). Using an assumption of symmetry for the Hilda wind field in the absence of data, contours of the lateral extent of the three mixed layer depths as the storm moved over the Gulf of Mexico were prepared (Figure 4).

NITRATE INPUT TO THE PHOTIC ZONE BY VERTICAL MIXING

Nitrate data for summer months were obtained for the open Gulf of Mexico from Collier (1968) (Table 2). Nitrate-nitrogen mass in the upper 75 meters was calculated by integrating the profiles for mean

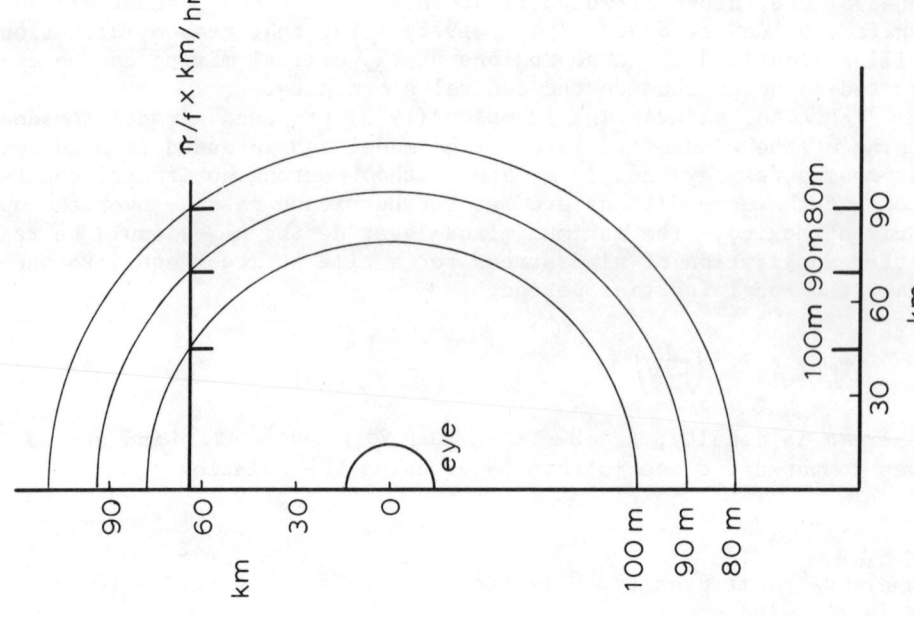

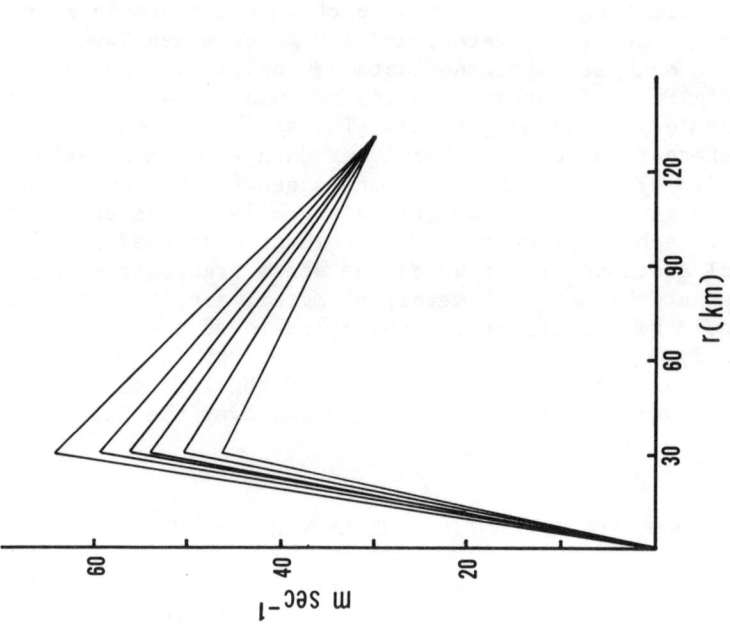

Fig. 1 (above). Wind speed structure for hurricane Hilda (1964).

Fig. 2 (right). Radius of 80 m, 90 m, and 100 m mixed layers for data segment 3 (Table 1). Criterion $\pi/f \times$ km/hr was used to locate distance from the storm track where each mixed depth would be expected.

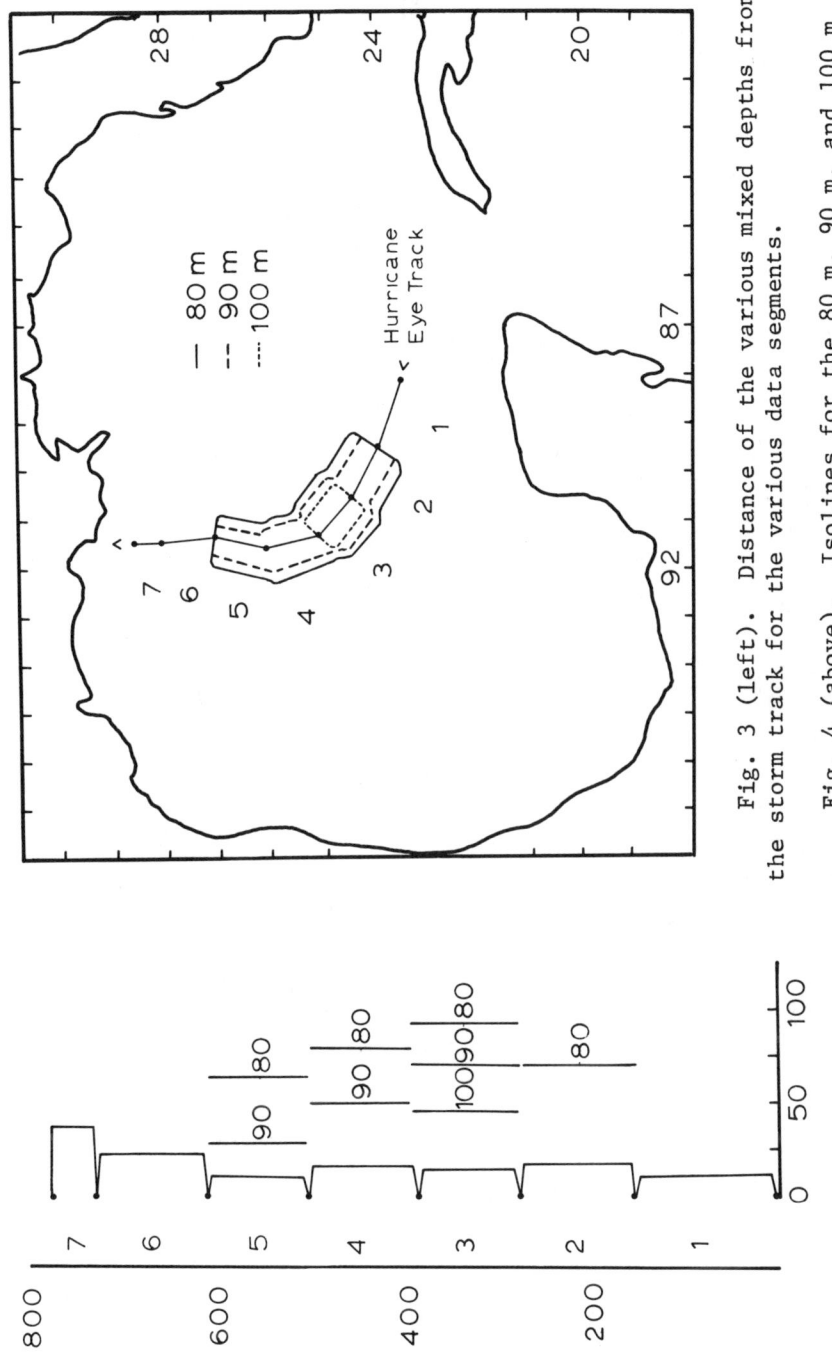

Fig. 3 (left). Distance of the various mixed depths from the storm track for the various data segments.

Fig. 4 (above). Isolines for the 80 m, 90 m, and 100 m mixed layers in the Gulf of Mexico for hurricane Hilda (1964).

Table 2. Nitrate data from the Gulf of Mexico (Collier, 1958) μ

Date	9/1/52	10/1/52	5/25/52	5/26/52	5/26/52	5/26/52	Mean	S
Latitude	27 18	26 42	26 39	26 00	25 19	24 53		
Longitude	92 35	91 59	87 30	87 30	87 45	87 00		
0 m	0	0	1	0	2	0	.50	.84
10 m	0	0	2	0	2	0	.67	1.03
20 m	0	0	1	1	1	0	.50	.55
30 m	0	1	0	1	1	0	.50	.55
50 m	0	0	1	0	2	0	.50	.84
75 m	3	5	2	3	3	5	3.50	1.22
100 m	12	15	5	7	4	10	8.83	4.26
150 m	22	17	9	12	12	10	13.67	4.93

nitrate-nitrogen concentration. The open Gulf of Mexico photic zone contained 1.2 gN/m^2. In comparison, *Menzel and Ryther* (1966) reported nitrate concentrations which yield an estimate of 1.9 gN/m^2 in the upper 75 m of the oligotrophic Sargasso Sea during the summer. Using linear interpolation of nitrate concentration between depths and integrating to each mixed layer depth, 0.3 gN/m^2 would be carried into the photic zone by mixing to 80 m, 1.2 gN/m^2 by mixing to 90 m, and 2.3 gN/m^2 by mixing to 100 m. Photic zone nitrate-nitrogen would be approximately doubled by vertical mixing to 90 m and would be approximately tripled by vertical mixing to 100 m.

An upper limit for mean chlorophyll a values in the region of the Gulf of Mexico transversed by Hilda is 21 mg chlorophyll a/m^2 (*El-Sayed et al.*, 1972). An average carbon to chlorophyll a ratio of 180 was calculated from data collected in the mixed layer of the oligotrophic North Pacific gyre (*Eppley et al.*, 1973). Using this ratio for the oligotrophic open Gulf of Mexico together with the Redfield carbon to nitrogen ratio of 5.6 on a mass basis, 0.67 g phytoplankton N/m^2 is estimated for the open Gulf of Mexico. *Fredericks and Sackett* (1970) reported mean particulate organic carbon values of 0.050 gC/ℓ for the upper 90 m and 0.028 g C/ℓ for depths greater than 90 m in the open Gulf of Mexico. Since the depth to which chlorophyll a values were integrated was not clear, and since chlorophyll a was observed at 200 m in the open Gulf of Mexico (*El-Sayed et al.*, 1972), the particulate organic carbon values of *Fredericks and Sackett* (1970) were integrated to 200 m. Using the Redfield ratio, 1.3 g particulate N/m^2 is estimated for the upper 200 m of the open Gulf of Mexico. Vertical mixing to 100 m would transport about twice the particulate nitrogen mass and about three times the phytoplankton nitrogen mass into the photic zone.

DISCUSSION

The radius of hurricane strength winds was estimated in the data obtained from the National Hurricane Center and is a very important factor in determining whether or not the $\pi/6$ criterion for wind stress duration is satisfied. *Leipper's* (1967) bathythermograph data provides a cross check on that estimate which appears to be accurate to first order.

For the first segment after the tropical storm became known as hurricane Hilda, windspeed and radial extent of winds were not sufficient to satisfy the $\pi/6$ criterion for any of the chosen mixed layer depths (Figure 4). As the hurricane developed to maximum intensity, the various mixed layer depths were located at greater distances from the storm center. As the hurricane lost intensity, the mixed layer depths were located closer to the hurricane center until, in segment 6, the $\pi/6$ criterion was again not satisfied. For a 100 meter surface mixed layer, using data from the undisturbed open Gulf of Mexico (*Leipper*, 1967), the internal long wave speed, c, was 2.3 km/hr which is greater than the *Geisler* criterion. Allowing for error in calculation of the long wave speed, segment 7 is the only hurricane segment where upwelling could possibly have occurred under the *Geisler* criterion since other hurricane segment ground speeds are at least four times as great as the *Geisler* criterion (Table 1).

Since the *Geisler* upwelling criterion is exceeded by at least four times for hurricane data segments one through six, we are confident that if the criterion is correct, we are justified in applying a mixing model to calculate vertical nitrate transport. Uncertainty about the nature of physical dynamics near the hurricane center could be clarified through examination of temperature and nutrient distribution in profiles collected along hurricane tracks. Vertical mixing would affect temperature and nutrient distribution in the same manner while upwelling or a combination of upwelling and mixing would be shown by differences in temperature and nutrient gradients.

Hurricane wind fields are commonly asymmetrical in the Pacific Ocean with the coldest water and maximum wind strength observed to the right of hurricane tracks (*Jordan and Frank*, 1964). *Leipper* (1967) observed cold water on both sides of the track of Hilda with the coldest water observed on the left side of the track. Symmetry was assumed for the wind structure of Hilda in this paper since wind distribution data were not available.

Most of the Gulf of Mexico's photic zone nitrate in the region crossed by Hilda is located in the lower 25 meters. The effects of vertical mixing of the surface layers would be to distribute this nitrate into shallower depths as well as to mix new nitrogen (in the sence of *Dugdale and Goering*, 1967) into the photic zone. The significance of vertical mixing of nitrate into the photic zone at any given point in the open Gulf of Mexico would depend on local conditions. For example, one percent light levels were observed at 75 ± 12 meters, $n = 11$ (*El-Sayed et al.*, 1972). Variations evident in the nutrient data (Table 2) may reflect both variation in the photic zone depth and variation in the growth rate of phytoplankton due to differences in standing crop magnitude or species composition. Unfortunately, zooplankton data from the region of the Gulf of Mexico transversed by Hilda does not exist in quality sufficient to calculate the effect of the input of nitrate by hurricane mixing on the plankton food chain. It is probable that the zooplankton do respond with increased growth to an increased phytoplankton crop. It is possible that mesopelagic fishes could aggregate in larger than usual numbers in locations where plankton maxima occur as a consequence of hurricane effects. A mechanism for this aggregation has been proposed by *Isaacs, Tont, and Wick* (1974).

ACKNOWLEDGEMENTS

I thank N. Frank of the National Hurricane Center for providing data on the hurricane Hilda (1964). Conversations with J. J. O'Brien and W. Sturges during preparation of the manuscript were most helpful

REFERENCES

Beers, J. R., Stevenson, M. R., Eppley, R. W., and E. R. Brooks. 1971. Plankton populations and upwelling off the coast of Peru, June 1969. *Fish. Bull.* 69: 859-876.

Cassie, M. R. 1962. Microdistribution and other error components of C^{14} primary production estimates. *Limnol. Oceanogr.* 7: 121-130.

Collier, A. 1958. *Gulf of Mexico physical and chemical data from Alaska cruises.* Special scientific report: Fisheries 249, Washington, DC. 417 p.

Cushing, D. H. 1971. Upwelling and the production of fish. In Sir Fredrick S. Russell and Sir Maurice Yonge (eds.) *Advances in Marine Biology.* Academic Press, London. 567 p.

Denman, K. L. and T. Platt. 1975. Biological prediction in the sea. Contribution to NATO advanced study institute on "Modeling and prediction of the upper layers of the ocean." *Urbino.* Italy, Sept., 1975.

Dugdale, R. C. and J. J. Goering. 1967. Uptake of new and regenerated forms of nitrogen in primary productivity. *Limnol. Oceanog.,* 12: 196-206.

Dunn, G. E. and B. I. Miller. 1964. *Atlantic Hurricanes.* Louisiana State University Press. 377 p.

El-Sayed, S. Z., Sackett, W. M., Jeffrey, L. M., Fredericks, A. D., Saunders, R. P., Conger, P. S., Fryxell, G. A., Steidinger, K. A., and S. A. Earle. 1972. Chemistry, primary productivity, and benthic algae of the Gulf of Mexico. Serial atlas of the marine environment. Folio 22, American Geographical Society. New York. 29 p.

Eppley, R. W., E. H. Renger, E. L. Venrick, and M. M. Mullin. 1973. A study of plankton dynamics and nutrient cycling in the central of the North Pacific Ocean. *Limnol. Oceanogr.* 18: 534-551.

Fisher, E. L. 1958. Hurricanes and the sea-surface temperature field. *J. Meteor.* 15: 328-333.

Franceschini, G. A., and S. Z. El-Sayed. 1968. Effect of hurricane Inez (1966) on the hydrography and productivity of the Western Gulf of Mexico. *Deut. Hydrog. Zeit.* 21: 193-202.

Fredericks, A. D. and W. M. Sackett. 1970. Organic carbon in the Gulf of Mexico. *J. Geophys. Res.* 75: 2199-2206.

Geisler, J. E. 1970. Linear theory of the response of a two layer ocean to a moving hurricane. *Geophys. Fluid Dynamics* 1: 249-272.

Isaacs, J. D., S. A. Tont, and G. L. Wick. 1974. Deep scattering layers: vertical migration as a tactic for finding food. *Deep Sea Res.* 21: 651-656.

Iverson, R. L., H. D. Curl, Jr., and J. L. Saugen. 1974. Simulation model for wind-driven summer phytoplankton dynamics in Auke Bay, Alaska. *Mar. Biol.* 28: 169-177.

Jordan, C. L. and N. Frank. 1964. On the influence of tropical cyclones on the sea surface temperature field. Final report, National Science Foundation. 31 p.

Kierstead, H. and L. B. Slobodkin. 1953. The size of water masses containing plankton blooms. *J. Mar. Res.* 12: 141.

Leipper, D. F. 1967. Observed ocean conditions and hurricane Hilda, 1964. *J. Atm. Sci.* 24: 182-196.

McGowan, J. A. 1972. The nature of oceanic ecosystems. In *Ecosystems*, C. B. Miller (ed.) *The Biology of the Oceanic Pacific*. Oregon State University Press. pp 119-28.

Menzel, D. W. and J. H. Ryther. 1960. The annual cycle of primary production in the Sargasso Sea off Bermuda. *Deep-Sea Res.* 6: 351-367.

O'Brien, J. J. 1967. The non-linear response of a two-layer, baroclinic ocean to a stationary, axially-symmetric hurricane: Part II. Upwelling and mixing induced by momentum transfer. *J. Atm. Sci.* 24: 208-215.

O'Brien, J. J. 1969. The response of the ocean to a slowly moving cyclone. *Annal. Meteor.* 4: 60-65.

O'Brien, J. J. and R. Reid. 1967. The non-linear response of a two-layer, baroclinic ocean to a stationary, axially-symmetric hurricane: Part I. Upwelling induced by momentum transfer. *J. Atm. Sci.* 24: 197-207.

Parsons, T. R., L. F. Giovando, and R. J. LeBrasseur. 1966. The advent of the spring bloom in the eastern subarctic Pacific ocean. *J. Fish. Res. Bd. Canada* 23: 539-546.

Platt, T. 1972. Local phytoplankton abundance and turbulence. *Deep-Sea Res.* 19: 183-187.

Platt, T. and K. L. Denman. 1975. A general equation for the mesoscale distribution of phytoplankton in the sea. *Proc. Sixth Liege Coll. on Ocean Hydrodynamics, Memoires de la Societe Royale des Sciences de Liege*: 31-42.

Pollard, R. T., P. B. Rhines, and R. O. R. Y. Thompson. 1973. The deepening of the wind-mixed layer. *Geophys. Fluid Dynamics* 3: 381-404.

Steele, J. H. and C. S. Yentsch. 1960. The vertical distribution of chlorophyll. *Mar. Biol. U.K.* 39: 217-226.

Stommel, H. 1963. Varieties of oceanogrpahic experience. *Sci.* 139: 572-576.

Sverdrup, H. U. 1953. On conditions for the vernal blooming of phytoplankton. *J. du Conseil* 18: 287-29

U.S. Department of the Interior. 1976. Final environmental statement, Outer continental shelf. Oil and general lease sale, Gulf of Mexico. *OCS Sale #41*. Washington, DC.

Walsh, J. J., J. C. Kelley, R. C. Dugdale, and B. W. Frost. 1971. Gross features of the Peruvian upwelling system with special reference to possible diel variation. *Invest. Pesq.* 35: 25-42.

Winter, D. F., K. Banse, and G. C. Anderson. 1975. The dynamics of phytoplankton blooms in Puget Sound, a fiord in the northwestern United States. *Mar. Bio.* 29: 139-176.

Wroblewski, J. S., J. J. O'Brien, and T. Platt. 1975. On the physical and biological scales of phytoplankton patchiness in the ocean. *Proc. Sixth Liege Coll. on Ocean Hydrodynamics, Memoires de la Societe Royale des Sciences de Liege*: 43-57.

How Can the Methodologies of Pattern Recognition Aid in Modeling Volume Reverberation Processes?

P. T. McElroy

Bolt Beranek and Newman Inc.

ABSTRACT

Principal components analysis, factor analysis, discriminant analysis, and correspondence analysis are powerful tools for reducing large bodies of data to understandable proportions. Subsuming these statistical techniques under the general title of pattern recognition, their application to the problem of making quantitative comparisons between volume scattering measurements in the ocean and the large variety of environmental measurements which can be made of biological, chemical, or physical-oceanographic features is discussed. A general groundwork is laid, using the ideas of distance measures, projection measures, and the variance-covariance matrix. Vector representations are used to enhance understanding and are coupled with matrix representations. The characteristic features of the data are described in terms of eigenvectors.

Specific applications by the author of distance measures and correspondence analysis are discussed. Possible extensions beyond that work are also discussed; they include measurements at a single site, use of the symmetry properties of correspondence analysis to relate complex faunal and acoustic measurements, and the extension of these techniques to multidimensional matrices.

INTRODUCTION

The common goal of the contributors to this volume is to explore various models that could improve our understanding of the relationship of physical, chemical, and biological oceanographic features of the ocean to acoustic energy scattered by the process of volume reverberation. While in a preliminary stage such understanding can be qualitative, ultimately it must be quantitative to permit the prediction of acoustic scattering from the environmental variables.

I propose to explore the gains in quantitative understanding that might be expected if the techniques falling under the general rubric of pattern recognition are employed. Pattern recognition is a statistical methodology; the individual techniques are known under a variety of names such as principal components analysis (Seal, 1964), correspondence analysis (Benzecri, 1973), factor analysis (Harman, 1967), discriminant analysis (Sebestyen, 1962; Cooley and Lohnes, 1971), and multidimensional scaling (Shepard, 1972). Some might choose to group these methods under a general title other than pattern recognition. So be it; the concern here is not with semantics but with those techniques in which each measurement is described as a vector in a multidimensional space, the full set of measurement vectors is grouped into a data matrix, and the characteristic features of the matrix are extracted by matrix diagonalization techniques, finally yielding a comparison of the original data vectors and the characteristic features of the matrix.

As statistical techniques, these methods require large bodies of data so that the "averages" and "variances" extracted by the process can have statistical significance. In fact, it is in the area of large data arrays that these methods are invaluable, where data which appear indigestible may be reduced to a size that can be understood and where the results can be visualized in two- or three-dimensional graphs. Such "visualization" is central to the success of these techniques (Shepard, 1972). A challenging and interesting part of the process is determining the meaning of the two or three dimensions to which the data have been reduced.

The techniques can be used to group measurements or measurement sites, or, alternatively, this clustering may be given a priori by the experimental design. In either case, once the clusters have been defined, these same general methods can be used to minimize the size of individual clusters and to maximize the distance between different clusters using discriminant analysis (Sebestyen, 1962). Then, new measurements can be assigned to an appropriate cluster based on probabilities.

These methods are described as statistical. They provide a viable alternative to deterministic models but need not be totally separated from them. The more an understanding is gained from a deterministic viewpoint, the more the statistical model can be refined, bringing it closer to reality. For instance, an initial statistical comparison of biological catches and acoustic data might use only the numbers of fish in each species caught, while a more refined model would use a mathematical transformation of those numbers, coupled with size information, into a total scattering cross section. This involves only a change in the variable which is subject to the statistical process.

The methodologies of pattern recognition go back many years. Factor analysis was an active field in the 1930's (*Harman*, 1967). Correspondence analysis (*Benzecri*, 1973), seemingly a new variant, actually was first conceived under a different name in the early 1950's (*Williams*, 1952). Now, decades later, the documentation for these techniques is extensive, and it would not be appropriate to repeat it here, except in its most general terms, so that a common terminology can be employed. I have found a geometrical interpretation the best for quick understanding and useful, in addition, for suggesting new experiments and procedures. It will be used here and will be coupled with the matrix formalism necessary for extraction of the characteristic features.

THE METHODOLOGIES

Vector and Matrix Representations

The first step is the interpretation of a set of measurements as a corresponding set of vectors in a multidimensional space. Since a data set of volume reverberation spectra have been used in previous work (*McElroy*, 1974; *McElroy and Smith*, 1975), spectra will be used as a specific example. Such spectra are, of course, central to the concerns of this symposium as well.

Two hypothetical spectra are shown in Figure 1(a). The number of frequencies is limited to three, so that the data can be plotted as a vector, as shown in Figure 1(b). The value along each of the frequency axes is the spectral value shown in Figure 1(a). These vectors are known as station vectors. Alternatively, the spectral data can be represented as a matrix of general element, k_{ij}, where i is the frequency index ranging from 1 to 3 and j the station index taking on the values 1 and 2 in the example. These matrices are shown in Figures 1(c) and (d) for the general case and with the specific values of the example.

Station-space is complementary to the frequency-space shown in Figure 1(b); it is shown in Figure 1(e) for the same matrix elements listed in Figure 1(d). In this example, station-space is two-dimensional and there are three frequency vectors. As in Figure 1(b), the value measured along each of the axes is the spectral value. Vector representations in both spaces are essential to a full understanding of the pattern recognition techniques.

Distance Measures

Distance measures can be quite simple. The so-called *Euclidean* distance measures the separation of station 1 and 2 in frequency-space as the simple distance separating the tips of the arrows in Figure 1(b). This is

$$D_{1,2} = \left[(k_{11}-k_{12})^2 + (k_{21}-k_{22})^2 + (k_{31}-k_{32})^2 \right]^{1/2}. \quad (1)$$

If station 1 and 2 had identical spectra, $D_{1,2}$ would equal zero. The more dissimilar they are, the greater $D_{1,2}$. A technique very similar to this is the one employed earlier (McElroy, 1974) in analyzing spectra gathered at 34 stations in the North Atlantic so as to cluster the stations. The clustered stations were then assigned to geographical regions based on physical oceanographic considerations.

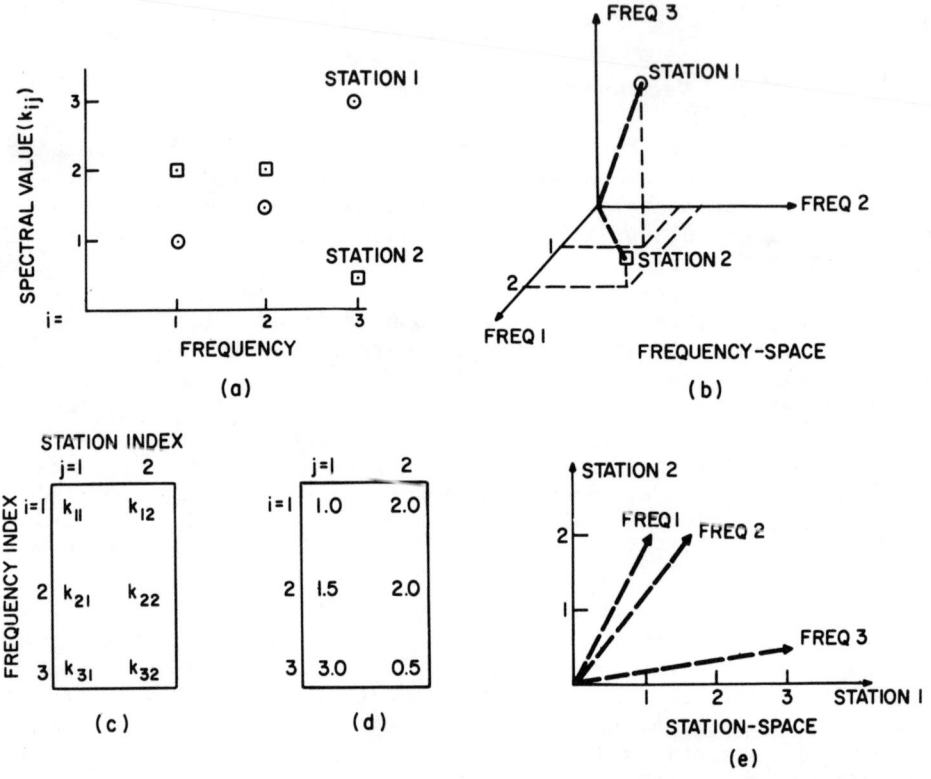

Fig. 1. The vector and matrix representations of spectral data: (a) graph of hypothetical spectra at three frequencies and two stations; (b) representation of these two spectra for stations 1 and 2 as vectors in a three-dimensional frequency-space; (c) 3 x 2 matrix of the spectral values in terms of the spectral value k_{ij}; (d) the particular values of the matrix shown in (a); and (e) plot of three frequency vectors in the complementary station-space for the same data.

PATTERN RECOGNITION

The *Euclidean* distance is convenient since it is independent of any translation or rotation of the coordinate system. However, there are instances when other distance measures are more appropriate. One example is the city-block metric, so called since in a city the shortest distance between two points is not a straight line, but is the path defined by the blocks (*Sebestyen*, 1962). The distance is

$$D_{1,2} = \left|k_{11} - k_{12}\right| + \left|k_{21} - k_{22}\right| + \left|k_{31} - k_{32}\right|. \tag{2}$$

The relevance of the city-block or other distance metrics to the volume reverberation problem has not yet been determined. As a result, this paper will consider only the *Euclidean* distance.

Projection Measures

An alternative way of expressing the relationship of stations 1 and 2 is by projecting one vector on the other. The more correlated they are, the higher the value of this vector dot product. But, while the *Euclidean* distance measure may be independent of the position of the center of the coordinate system, the projection measure is not. Thus, the center needs to be defined in a unique and unambiguous way; this is usually done by 1) selecting one of the two spaces (frequency-space will be chosen) and 2) placing the center of coordinates in that space at the mean position of the vectors defined as

$$\bar{k}_i \equiv \frac{1}{Jmax} \sum_{j=1}^{Jmax} k_{ij}, \tag{3}$$

where i, the frequency index, ranges over its set of values, namely 1 to 3 ($Jmax = 3$) in our example. $Jmax$ is the number of stations. When $Imax = 2$, the center of coordinates is particularly simple, lying half-way between the two station vectors.

It is normal to scale the data as well as to shift the center of coordinates. As seen below, this scaling permits interpreting the data matrix as the variance-covariance matrix so extensively used in statistical analyses. Scaling by the standard deviation, s_i, the transformed matrix element ρ_{ij}^{PC} is defined, where PC implies principal components,

$$\rho_{ij}^{PC} = \frac{k_{ij} - \bar{k}_i}{(Jmax)^{1/2} s_i}, \tag{4}$$

and where

$$s_i = \left[\sum_{j=1}^{Jmax} (k_{ij}-k_i)^2/Jmax \right]^{1/2} \quad (5)$$

This particular form of ρ_{ij} is the one employed in most of the pattern recognition techniques—namely, those that assume a Gaussian distribution of the data and seek to partition or divide the variance into its important constituents. Principal components is a specific example.

ρ_{ij} is thus the component along the ith axis of the transformed station vector for station j. An alternative representation for ρ_{ij} which displays this vector character is $\hat{i}\cdot\vec{\rho}(j)$ where $\hat{i}$ is the unit vector along the ith axis. Explicitly,

$$\rho_{ij} \equiv \hat{i}\cdot\vec{\rho}(j) \quad (6)$$

ρ_{ij} can be used for computing distance measures of the transformed data or in computing projection measures. The projection of $\vec{\rho}(j)$ on $\vec{\rho}(j')$ is

$$\vec{\rho}(j)\cdot\vec{\rho}(j') = \sum_{i=1}^{Imax} \rho_{ij}^{PC}\rho_{ij'}^{PC} \equiv g_{jj'}^{PC} \quad (7)$$

Once the center of coordinates is defined, the projection and distance measures (in the *Euclidean* metric) have a specific relationship to one another (*Torgerson*, 1958; *Carroll and Wish*, 1974). Namely,

$$g_{jj'} = -\frac{1}{2}[d_{jj'}^2 - d_{\cdot j'}^2 - d_{j\cdot}^2 - d_{\cdot\cdot}^2] \quad , \quad (8)$$

where $d^2_{jj'}$ is the distance measure and

$$d^2_{jj'} = \sum_{i=1}^{Imax} \frac{(k_{ij}-k_{ij'})^2}{(Jmax)s_i^2} = \sum_{i=1}^{Imax} \left(\rho_{ij}^{PC}-\rho_{ij'}^{PC}\right)^2 \quad (9a)$$

and is like the earlier distance measure $D^2(j,j')$, except for the scaling by the standard deviation. In addition,

$$d^2_{j'} = (1/Jmax) \sum_{j=1}^{Jmax} d^2_{jj'} \qquad (9b)$$

$$d^2_{j} = (1/Jmax) \sum_{j'=1}^{Jmax} d^2_{jj'} \qquad (9c)$$

and

$$d^2_{..} = (1/Jmax)^2 \sum_{j=1}^{Jmax} \sum_{j'=1}^{Jmax} d^2_{jj'} \quad .* \qquad (9d)$$

The investigator may choose to examine distances or projections dependent on the experimental design. While the spectral data used here as an example are in a form such that either distances or projections could be computed directly, there are some experiments where $d_{jj'}$ is the quantity actually measured (Carroll and Wish, 1974).

Up to this point the projection of one station vector on another in frequency-space has been discussed. However, the projection of one frequency vector on another in the complementary space, station-space (Figure 1(e)), can just as well be considered. That is, consider the quantity $g^{PC}_{ii'}$.

$$g^{PC}_{ii'} = \sum_{j=1}^{Jmax} \rho^{PC}_{ij} \rho^{PC}_{i'j} = \sum_{j=1}^{Jmax} \frac{(k_{ij}-k_i)(k_{i'j}-k_i)}{(Jmax) s_i s_{i'}} \quad . \quad (10)$$

This is just the variance-covariance matrix showing the correlations of the original data k_{ij}, and it is this matrix which is partitioned, disclosing its principal axes, in the application of principal components methods.

---

*The definition of $d^2_{..}$ in Eq. 13 of Carroll and Wish (1974) is incorrect as a comparison with the original formalism in Torgerson (1958) shows.

In summary, distance and projection measures in one space (frequency-space) have been considered as measures of similarity and their equivalence has been pointed out. Projection measures of the same data in the <u>complementary</u> space (station-space) yield the common variance-covariance data matrix.

Before completing this subsection, a contrast should be made with the ρ_{ij} used in correspondence analysis, (C.A.). It is

$$\hat{i} \cdot \vec{\rho}(j) = \rho_{ij}^{C.A.} = \frac{\frac{b_{ij}}{b_j} - b_i}{b_i^{1/2}}, \qquad (11)$$

where

$$b_{ij} = \frac{k_{ij}}{k}, \qquad (12a)$$

$$k = \sum_i^{Imax} \sum_j^{Jmax} k_{ij}, \qquad (12b)$$

$$b_i = \sum_{j=1}^{Jmax} b_{ij}, \qquad (12c)$$

and

$$b_j = \sum_{i=1}^{Imax} b_{ij}. \qquad (12d)$$

Here no assumptions are made concerning the probability distribution. Rather, the data matrix is viewed as a contingency table and it is a chi-square measure of the data which is divided up (*Williams*, 1952; *McElroy and Smith*, 1975). While this may appear arbitrary to those accustomed to partitioning variances, it has important consequences in suggesting ways of comparing biological and acoustic data, which are explored in a later section of this paper (An Additional Application and Two New Approaches).

In correspondence analysis, a weighted projection of station vectors $\vec{\rho}(j)$ and $\vec{\rho}(j')$ in frequency-space is used rather than the unweighted form in station-space given in Eq. 10 for use in principal components studies. The weighting arises from the chi-square formulation of the data (McElroy and Smith, 1975, especially Eqs. 4 and 5).*

$$g_{jj'}^{C.A.} \equiv \sum_{i=1}^{Imax} \rho_{ij}^{C.A.} \rho_{ij'}^{C.A.} (\delta_j \delta_{j'})^{1/2} \quad . \tag{13}$$

Matrix Diagonalization

A data matrix with entry ρ_{ij} has been defined for both correspondence analysis and principal components analysis. The ρ_{ij} are derived from the original data matrix, whose entry was defined as k_{ij}. The characteristic features of the matrix ρ_{ij} are now extracted. These characteristic features are known as eigenvectors and are determined by the process of matrix diagonalization.

While a matrix need not be square to be diagonalized (Eckart and Young, 1936), the usual procedure is to create from ρ_{ij} the square projection matrix of element $g_{ii'}^{PC}$, discussed above.

If its elements are real (Eckart and Young, 1936), $g_{ii'}$ can be diagonalized. This condition is met if the elements k_{ij} are real in the case of principal components and if the k_{ij} are, in addition, positive definite in the case of correspondence analysis (necessitated by the square roots found in Eq. 11). This diagonalization

---

*Certain readers may be concerned about the definition of the matrix with element $g_{jj'}^{CA}$, which is to be diagonalized. It differs from that found in Eq. 9 of McElroy and Smith (1975), where the matrix element is

$$\sum_i \frac{\delta_{ij'}}{\delta_i^{1/2} \delta_{j'}^{1/2}} \cdot \frac{\delta_{ij}}{\delta_i^{1/2} \delta_j^{1/2}} \quad .$$

The two formulations are equivalent. The formulation shown in this paper has the center of coordinates, defined by the set $\delta_i^{1/2}$, subtracted from the data and will yield a set of eigenvectors one less than the rank of the matrix. Conversely, the formulation in Eq. 9 of McElroy and Smith (1975) contains all eigenvectors, including $v_{\alpha i} = \delta_i^{1/2}$, which is the zeroth eigenvector, defining the center of coordinates.

is expressed as

$$g_{ii'} = \sum_{\alpha} v_{\alpha i} \lambda_{\alpha} v_{\alpha i'} \qquad (14)$$

where λ_{α} are the eigenvalues and are the components of a diagonal matrix, and the v's are eigenvectors defined in frequency space. $v_{\alpha i}$ is the ith component of the eigenvector corresponding to the α eigenvalue, λ_{α}. Eigenvectors are perpendicular to one another and are arranged in a hierarchy by the corresponding eigenvalues. Thus, the first eigenvalue, λ_1, has the greatest value, and the associated eigenvector $\hat{v}_{\alpha=1}$ points in a direction such that as much of the variability of the data is accounted for (i.e., the variability lies roughly along the direction of that vector) as is possible. Figure 2 is a geometric representation of this idea in frequency-space. The dots are the tips of station vectors. The eigenvector $\hat{v}_1$ points at the densest cluster of dots. The next eigenvector, $\hat{v}_2$, is orthogonal to $\hat{v}_1$ and points in the direction best suited to account for the remaining variability. There will be as many eigenvectors as the rank

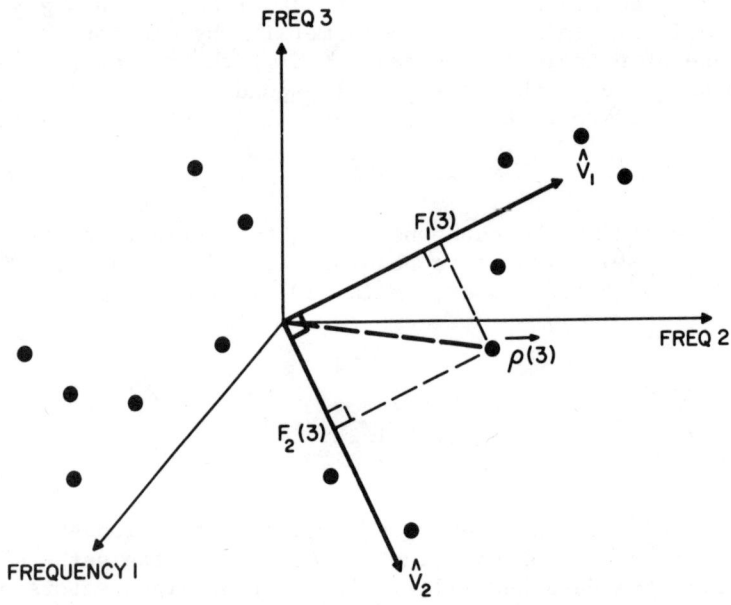

Fig. 2. A common plot of a station vector $\vec{\rho}(3)$, eigenvectors $\hat{v}_1$ and $\hat{v}_2$, and the factor scores $F_1(3)$ and $F_2(3)$ which result from the projection of the station vector on the eigenvectors. The plot is shown in frequency-space. The dots represent other stations.

of the matrix where rank is the lesser of two numbers, the number of frequencies and number of stations.

It can be stated precisely in what sense the eigenvectors "account for the variability of the data". Generally the data will not be distributed through frequency-space so randomly that all of the frequency dimensions (three in the example in Figure 1) would be needed to describe them. For instance, if all the station vectors lay on a straight line, only one dimension would be required, or, if in a plane, only two. If the data are described by a number of dimensions less than the matrix rank (and this is usually done so that the data can be understood and visualized), then the eigenvectors should be used to define those directions. This will be the best possible choice in a least-squares sense (Eckart and Young, 1936) to account for the data variability. And the fraction of the variability that has been accounted for with n dimensions, where $n \leq$ rank of the matrix, is

$$\frac{\sum_{\alpha=1}^{n} \lambda_\alpha}{\sum_{\alpha=1}^{Rank} \lambda_\alpha} \quad (15)$$

The eigenvectors have a meaning of their own. In the example of spectra at many stations, each eigenvector is a spectrum. Each one might be attributable to something simple in the environment such as the spectrum resulting from a predominant species of fish. Or perhaps the eigenvector is a linear combination of such spectra. Alternatively, it may be better described by environmental features of a chemical or physical-oceanographic nature, which in turn determine the fish distributions. Whichever of these situations applies, the eigenvector defines a significant <u>dimension</u> of the problem under study. The task of identifying the variables which determine that dimension, be they biological, chemical, or physical-oceanographic, is the central one in meaningful use of these techniques; the factor scores defined below aid in that identification.

Factor Scores

Factor scores are easily defined in terms of the geometry of Figure 2 which shows both eigenvectors (e.g., $\vec{v}_1$) and station vectors ($\vec{\rho}(3)$). The factor score $F_\alpha(j)$ is the projection of station vector j on the eigenvector v_α. In Figure 2 $F_1(3)$ and $F_2(3)$ are shown. The eigenvectors are unit vectors, so that the factor scores can be viewed as that portion of a station vector (e.g., the jth) that is accounted for by an eigenvector (e.g., the αth). If the projection is high and positive, the station is very much like the eigenvector; if high and negative, the station has characteristics opposite to the eigenvector (the inverse spectrum is then like the station). And if the projection is zero, then that eigenvector is not relevent to describing the station.

Regression and Clustering

Regression techniques are useful in identifying dimensions. But to be successful, the investigator must have measured the significant variables in addition to the acoustic data. Thus, these methods cannot replace an understanding of the basic physical processes which might affect the measured spectral values.

Factor scores of a given eigenvalue (i.e., one dimension) are compared with other measured variables using linear or multiple regression. A high correlation with a variable provides the means of identifying the dimension. For example, factor scores for the first eigenvector in an analysis of benthic fauna were plotted against depth with very low scatter (*Grassle and Smith*, 1975); the first dimension was then identified as depth. It is difficult to identify more than two dimensions in many experimental situations. The reason can be either the randomness inherent in the data-taking process or failure to identify and measure during the original data taking what might be important variables.

In comparing biological populations with acoustic spectra, we might regress the first factor score against the numbers of fish captured in some predominant species, the series of points in the regression corresponding to stations at which both biological and acoustic data taking took place. In situations where few species are present, this might succeed. In more complex situations, the new approach suggested later (i.e., in "An Additional Application and Two New Approaches") may be more productive.

Even if one is unable to identify dimensions, the factor scores are still of great help in clustering the stations. One typically examines the factor scores for each station for the two most important eigenvectors, $\hat{v}_1$ and $\hat{v}_2$, by plotting them on a graph such as shown in Figure 3. Recall that these two eigenvectors (by definition) account for the largest possible fraction of the data variability. Note that some of the stations cluster closely (clusters A and B), while others appear more randomly distributed (stations 3 and 16). Thus, like groups of stations are identified and contrasted with others; one hopes that the clusters have some sense to them on a geographic or other basis. Such meaningful clustering did occur with volume reverberation data (*McElroy and Smith*, 1975, Figure 4) where Gulf Stream stations were concentrated in a small wedge of the factor score plot.

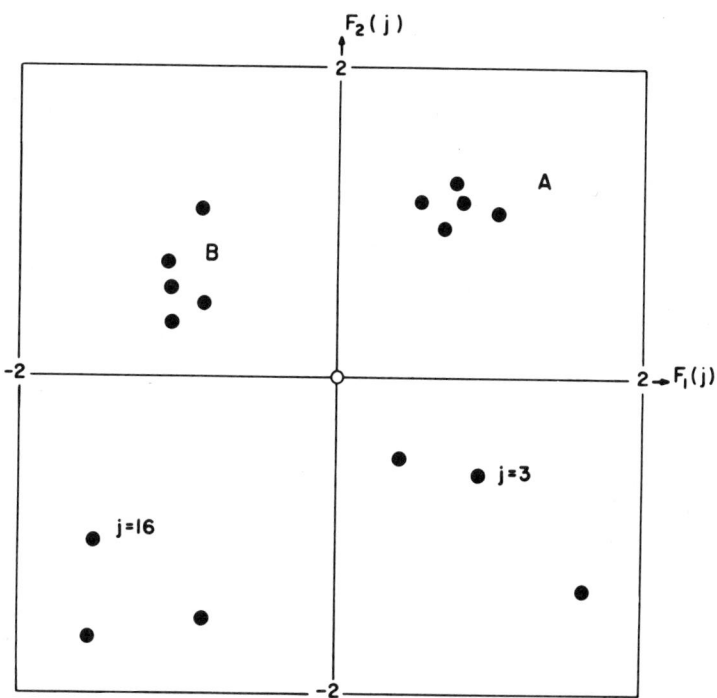

Fig. 3. Plot of hypothetical factor scores for a series of stations of index j. The two axes are the factor scores for the first two (most important) eigenvectors. A and B designate stations clustered by this process, while $j = 3$ and 16 are two (of a number) of stations which are not clustered. Even though 3 and 16 are not clustered, they are well separated from the other stations.

Discriminant Analysis

Once an investigator has decided which stations he wishes to assign to a common cluster, he can apply the techniques of discriminant analysis (Sebestyen, 1962). This assignment to clusters might be on the basis of the groupings suggested by the factor score plot or determined on some *a priori* basis. Examples of the latter are groups of spectra taken at two different depth intervals or at different times of day, or groups of spectra taken from different geographic regions which vary significantly in their environmental characteristics.

Discriminant analysis is like principle components analysis in that it examines variance in the data rather than the chi-square measure of correspondence analysis. (It seems, however, that discriminant analysis could be generalized to treat the data using the correspondence analysis formalism.) Gaussian probability densities are assumed, with the result that curves of equal probability for a given cluster (in, say, frequency-space) are multidimensional ellipses. Discriminant analysis distorts the ellipsoids into spheres, maximizes the separation between spheres, and gives a new set of eigenvectors. These eigenvectors are those directions, again arranged in order of decreasing importance, which provide the best basis for deciding to which cluster a new measurement might belong. These ideas are illustrated in Figures 4a and 4b. Figure 4a shows the unmodified probability curves, while Figure 4b shows the resultant spheres and the first two eigenvectors. Projection of the new measurement on an eigenvector is compared with the projections of points in first one and then another cluster. Classification to one of these three clusters is on the basis of the minimum mean square difference in these projections. Projection on $\hat{v}_1$ in Figure 4b would eliminate the new measurement from cluster A, while projection on $\hat{v}_2$ would assign it to cluster B.

The importance of discriminant analysis lies in the assignment of new spectra to clusters. One has all the apparatus of decision theory in determining to which group a new measurement most probably belongs. Refinements of discriminant analysis permit some variety in the decision criteria actually used.

This ability to classify according to preexistent clusters presupposes that a sufficiently large body of data has been collected to define the clusters well. The author's data set of volume reverberation spectra is probably not that large.

Nonlinear Methods and Questions

The translations and rotations of the data matrix discussed so far are linear operations. These are unduly restrictive. The nonlinear operations have not been examined in detail here. However, such transformations exist, resulting in generalized discriminant functions (*Sebestyen*, 1962, Chapter 3). It might be found in time that such operations should be considered.

The city-block distance metric cited earlier is an example of a nonlinear operation.

Of equal or greater importance is the question of nonlinear relationships among experimental variables. Statistical ecologists have recently been alerted to the hazard of ignoring this in their work (*Noy-Meir and Austin*, 1970; *Austin and Noy-Meir*, 1971), which could profitably be reviewed should it be found that similar problems are encountered in studying acoustic spectra and their relationship to environmental variables. Insofar as they have responded to this problem, they have restricted variables showing nonlinear relationships to small ranges where the linearity assumptions hold.

PATTERN RECOGNITION

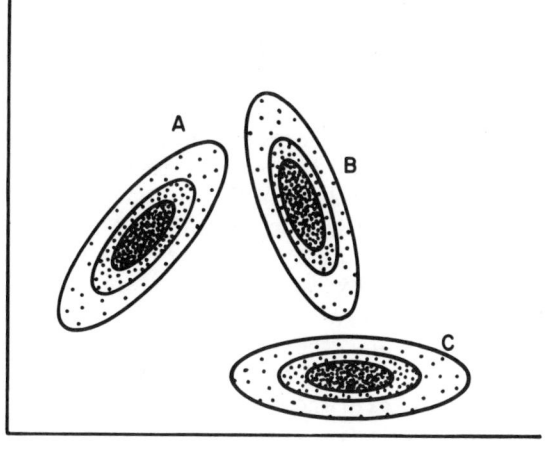

(a)

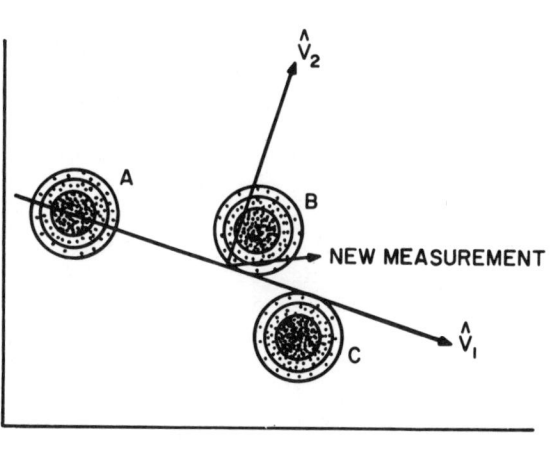

(b)

Fig. 4. Two-part sketch indicating the results of discriminant analysis. (a) Three clusters of data points, designated A, B, and C. The ellipses are curves of equal probability, and require assuming the data are normally distributed. (b) Following discriminant analysis, the clusters are separated and compressed into spheres. Two eigenvectors $\hat{v}_1$ and $\hat{v}_2$ are shown. Projections of a new measurement onto these eigenvectors, compared with the projections of the points within the clusters, provide the basis for assignment of the new measurement to one of the clusters.

APPLICATIONS TO VOLUME REVERBERATION

The author has applied some of these techniques. An initial concern was characterizing the similarity of column strength spectra so that pelagic regions of like spectra might be defined. Qualitaive statements such as that spectrum A was like C and F but unlike B and D were not satisfying. A distance measure was evolved to quantify the similarity (McElroy, 1974), independent of the wealth of work which had been expended in formalizing such measures. The measure proved useful in defining regions whose boundaries were delimited on the basis of physical-oceanographic considerations.

For some time, I had wanted to apply factor analysis to the same data. When the chance arose, there was an alternate technique available--correspondence analysis. After some thought, correspondence analysis was selected in preference to principal components (which is a subset of factor analysis), due to feeling that it was more "natural" for the treatment of spectra.

The results (McElroy and Smith, 1975) were satisfying. Clustering was similar to that noted earlier (McElroy, 1974); Gulf Stream system stations were neatly grouped. An analysis of the Gulf Stream stations alone showed some interesting correlations between the first two factor scores and latitude and longitude. The regression was statistically significant, although too poor to be very useful for prediction. A *symmetric* property of correspondence analysis (Benzecri, 1973; Hill, 1973; McElroy and Smith, 1975), permitted an identification of those clusters of stations with the spectral features of specific frequency ranges which most uniquely characterized them.

The identification of eigenvectors was not possible since the biological information was not fully processed. It is probable that they have no simple meaning, since over 200 species of mesopelagic fish have been identified in the hauls taken over the same sections of ocean where the 34 acoustic stations were located. Another fact complicating interpretation is that the spectra are column strengths which average the entire water column, rather than being scattering strengths taken at one single depth.

With these statistical techniques in mind, it is possible to define new experiments where the meaning of eigenvectors would be more transparent. Namely, measure acoustic scattering strength spectra at one depth (or really one small range of depths) and compare with biological sampling made in the same depth range.

AN ADDITIONAL APPLICATION AND TWO NEW APPROACHES

The basic problem of this symposium is one of relating acoustic information with biological, chemical, or physical-oceanographic parameters. How might these statistical methods of pattern recognition be applied to establish this relationship? A particular experiment is initially discussed which would rely on regression to establish the relevant dimensions (i.e., parameters). Next, the symmetry properties of correspondence analysis suggest a new approach for relating biological and acoustic information. And, finally, what might be accomplished by examining matrices of more than two dimensions is explored.

Regression

The use of linear and multiple regression as a generalized method was previously discussed (The Methodologies). Factor scores for various eigenvectors of the acoustic spectra are compared with parameters the investigator has measured, feeling that they may be physically important in the process.

Regression of factor scores and the numbers, cross-sectional area, or total scattering cross section of predominant species may work if the number of species is low. Where the number is high, the consideration of a single species or even two or three species may be unproductive. Then, the symmetry methods to be discussed below are appropriate.

Regression against physical-oceanographic variables **must** be approached with caution. If the acoustic data are taken over a wide geographic expanse, there may be more than one physical-oceanographic province with "discontinuous" changes in the variables at the boundaries. Attempts at a regression may then fail because of this discontinuity. If the acoustic data are all taken within a single province, then the environmental changes from station to station may be more smooth and the regression feasible. Even under these conditions we must be alert to nonlinear response to the environmental variables.

Even more satisfactory is a series of acoustic measurements done over time at a single site. The Ocean Acre (*Gibbs and Roper*, 1971) studies are a natural body of existent data for this type of study, provided sufficient environmental data were taken.

Symmetry of Correspondence Analysis

The definition of the matrix element ρ_{ij}^{CA}, which was previously given, (Eq. 11), may seem very arbitrary to some readers. However, out of this seeming arbitrariness flows a symmetry between the determination of eigenvectors in frequency-space and a set of eigenvectors in station-space, where the axes are the several stations and the data points (measurement vectors) are frequencies. The first eigenvector in frequency-space (with components $v_{\alpha i}$) has the same eigenvalue, λ_1, as the first eigenvector in station-space (components $v_{\alpha j}$). Determination of the eigenvectors in either one of these spaces is given by matrix diagonalization; the eigenvectors in the other are then immediately given by the so-called transition formulas (McElroy and Smith, 1975; Benzecri, 1973).

$$v_{\alpha j} = \sum_i \frac{b_{ij}}{b_i^{1/2} b_j^{1/2}} \frac{v_{\alpha i}}{\lambda_\alpha^{1/2}} \qquad (16a)$$

$$v_{\alpha i} = \sum_j \frac{b_{ij}}{b_i^{1/2} b_j^{1/2}} \frac{v_{\alpha j}}{\lambda_\alpha^{1/2}} \qquad (16b)$$

These two spaces are shown in Figure 5a and b, with the first two eigenvectors indicated in a hypothetical example.

For a given eigenvalue, there is thus a correspondence between the particular set of stations contributing to the eigenvector in station-space and the particular set of frequencies contributing to the eigenvector in frequency-space. For instance, v_1 is made up of equal parts of stations 1 and 2 in station-space, and none of station 3, while v_2 is made up entirely of station 3 (Figure 5b). These correspond to $\hat{v}_1$ in frequency space, which is made up of equal parts of all three frequencies, and v_2 which is a combination of equal parts of frequencies 1 and 2.

The importance of this symmetry becomes apparent when we consider both sets of acoustic spectral measurements and sets of biological sampling done at the same set of stations. It is most natural to compute eigenvectors for the spectral measurements in frequency space where the eigenvectors are spectra (Figure 5a). Suppose now we apply correspondence analysis to our biological data as well, where, for simplicity, let each of the axes be labeled by a species name and let the value for that axis be the number of that species which is taken at the station. Figure 5c shows the station vector for station 2, where there are 3 *Ceratoscopelus maderensis*, 5 *Lampedena luminosa*, and 3 *Gonostoma atlanticum*. This is the station vector of the raw data, as yet unscaled in accord with the formalism of correspondence analysis. After this scaling, and after the determination of eigenvectors, the vectors can be displayed in species-space and correspond to the particular linear combination of species which best describe the data.

We now have two sets of eigenvectors, one for acoustic data in frequency-space and one for fish hauls in species-space. The spaces are noncommensurate. However, they have a third space in common--station-space. And the eigenvectors in station-space for each type of measurement will correspond to the vectors in frequency- (or species-) space in accord with the symmetry we have noted above. Figure 5d shows hypothetical eigenvectors for each set of

PATTERN RECOGNITION

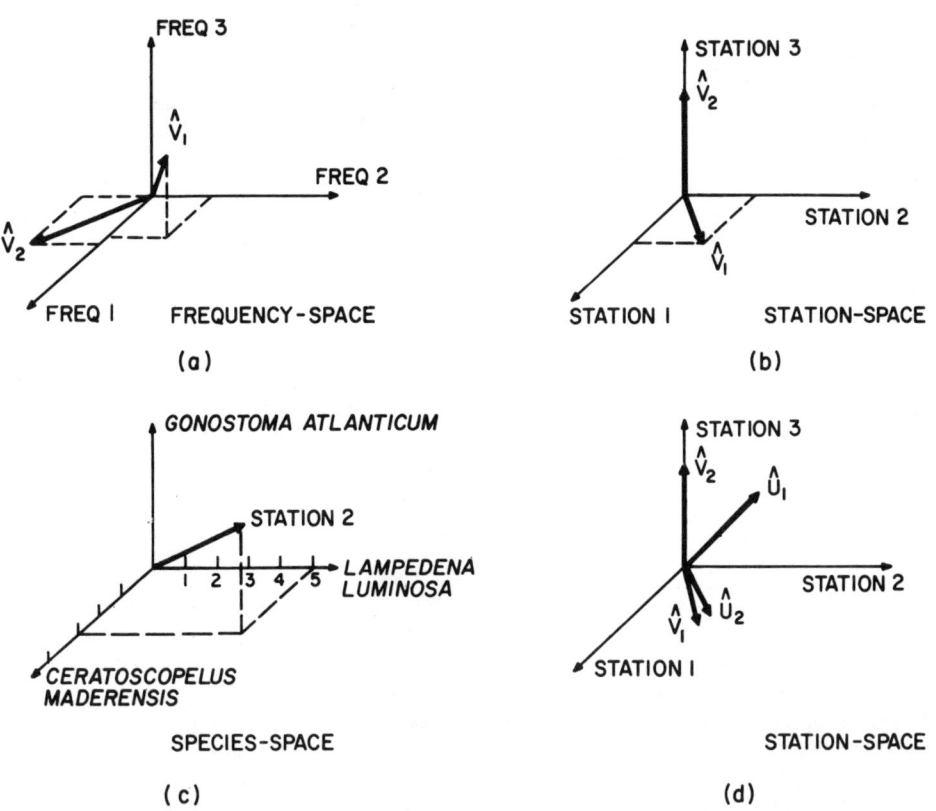

Fig. 5. Symmetry properties of correspondence analysis:
(a) Plot of the first two eigenvectors in frequency-space of a
spectral analysis of sound in a hypothetical example; (b) Plot of
the corresponding two eigenvectors in the complementary station-
space; (c) Raw-data vector of species collected at sites coincident
with the sites of the sound scattering measurement; and (d) Eigenvectors of the spectral analysis ($\hat{v}$'s) and of the faunal analysis
($\hat{u}$'s) plotted in their common space, station-space. Projections of
the $\hat{u}$-vectors on the $\hat{v}$-vectors can be computed to determine correspondence of spectral and faunal features.

measurements in this common space. $\hat{v}$'s have been used for acoustic
data vectors and $\hat{u}$'s for species data vectors. Both sets are unit
vectors.

We are now in position to define the correspondence of acoustic
and biological data by projecting the $\hat{v}$ vectors on the $\hat{u}$ vectors.
For instance, in Figure 5d, $\hat{v}_1$ is nearly coincident with $\hat{u}_2$. Then,
by transforming back to the symmetric spaces, we determine what
spectral curve ($\hat{v}_1$ in frequency-space) corresponds with what relative

proportions of species ($\hat{u}_2$ in species-space). In general, the projections will not be exactly 1, and appropriate linear combinations will need to be taken.

In review, this procedure would extract the most important or characteristic features of both acoustic and biological data (namely, the eigenvectors of their respective data matrices) and provide a unique way of tying these two sets of characteristic features together. Thus we have the means of identifying those species which are important acoustically.

In the example above, too simple a description of the fish has been used. What is of importance is probably not the numbers of fish of a given species, but their total scattering cross section. This is the type of refinement to the statistical model which a more deterministic understanding of the scattering process can give. As this deterministic understanding increases, the correspondence between the biological and acoustic measurements, as determined by the statistical procedures, can be expected to improve.

Three-Way Matrices

The regression techniques described above are one way of taking into account variables in addition to frequency and station number. In that case, a two-dimensional or two-way matrix was considered. Another method is to consider three-way or higher order matrices. As an example, in Figure 6a a three-dimensional matrix is shown. The raw data entries corresponding to the k_{ij}, considered earlier, are $k_{ij}^{(\ell)}$, which are again spectral values. Note that i is the frequency index, j the station index, and ℓ an index describing a particular depth interval. Thus, our matrix entries would be volume scattering strengths rather than column strengths. Can these matrices be manipulated in a way analogous to the two-dimensional matrices? Are eigenvalues and eigenvectors meaningful?

Carroll and Wish (1974) have considered this question in their studies of psychological variables. They have developed a technique which they call three-way multidimensional scaling. Rather than the matrix shown in Figure 6a, they considered that shown in Figure 6b, which is symmetric in the i and i' directions. If attention is restricted for the moment to just one depth interval, the typical data entry could be just the quantity $g_{ii'}$ which was defined earlier, based either on a principle components approach or a correspondence analysis method. Recall that $g_{ii'}$ was diagonalized by the equation

$$g_{ii'} = \sum_\alpha v_{\alpha i} \lambda_\alpha v_{\alpha i'} \quad . \tag{17}$$

PATTERN RECOGNITION

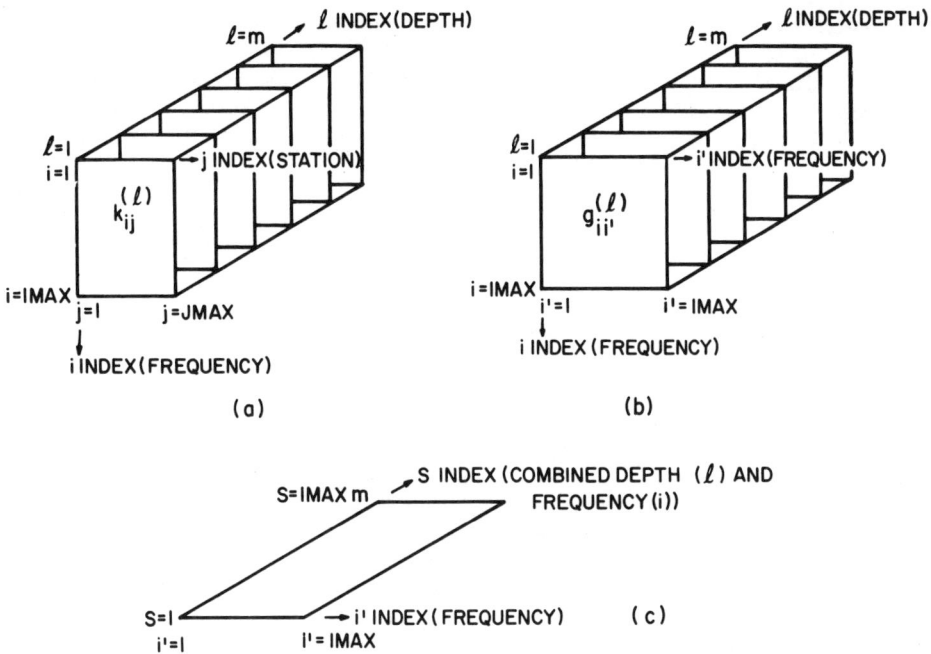

Fig. 6. Three-way matrices in an example taken from volume reverberation studies: (a) raw data matrix of spectral values of volume scattering strength determined at a variety of frequencies, stations, and depths; (b) transformed three-way matrix which is to be "diagonalized" by a least-squares procedure (for a given depth, this is just like the variance-covariance matrix discussed in two-way matrix diagonalization); and (c) one of three iterative steps in determining the least-squares solutions to the three-way matrix; the matrix is collapsed into a two-dimensional matrix by combining depth and frequency indices. The other combinations of indices are used to collapse each pair in turn.

Let this be generalized by adding the ℓ superscript for depth interval

$$g_{ii'}^{(\ell)} = \sum_\alpha v_{\alpha i} \lambda_{\alpha \ell} v_{\alpha i'} \qquad (18)$$

Recall that in the two-dimensional formalism, λ_α was a kind of weighting function showing the relative importance of the different eigenvectors. By analogy, $\lambda_{\alpha\ell}$ is a weighting function for both the eigenvectors and the depth intervals. *Carroll and Wish* (1974) actually call it a weight, $w_{\alpha\ell}$.

Within this formalism, the $\hat{\nu}$'s are like the former eigenvectors. Note that they apply to the entire data matrix, not just to one depth interval, ℓ. However, vectors can be defined for a given depth interval as follows.

$$y_{\alpha i}^{(\ell)} \equiv (\lambda_{\alpha\ell})^{1/2} \nu_{\alpha i} \quad . \tag{19}$$

There is really no rigorous justification for all this; the ν's and y's are not true eigenvectors, nor are the $\lambda_{\alpha\ell}$ true eigenvalues. Nonetheless, *Carroll and Wish* (197) treat them as if they were. They note that determination of eigenvectors provides a best fit to the data set $g_{ii'}$ in a least squares sense, so they proceed to make a least squares fit to $g_{ii'}^{(\ell)}$ by an iterative procedure. This procedure requires that one collapse the three-dimensional matrix of Figure 6b, defined by axes i, i', and ℓ into two-dimensional matrices (Figure 6c) by combining two indices, such as i and ℓ into a new index, s, which is (*Carroll and Chang*, 1970)

$$s = Imax(\ell-1) + i \quad 1 \leq s \leq Imax \cdot m \tag{20}$$

where

$$1 \leq i \leq Imax, \ 1 \leq \ell \leq m \tag{21}$$

In the next iteration, two other indices are combined, and so on until a least-squares fit is attained. One word of warning: there are local minimum problems so that the true minimum may not be attained unless one starts the iteration at a number of different points.

One product of this least-squares process is the pseudoeigenvectors $\nu_{\alpha i}$ and the pseudoeigenvalues or weights $\lambda_{\alpha\ell}$. The ν's are representative of the entire data set, namely all depth intervals and all stations, while the y's individualize the results to discrete depth intervals.

It is natural to view this whole procedure with suspicion. In defense, the truly remarkable results of *Carroll and Wish* (1974) is cited. They have reanalyzed some color perception data which had first been analyzed by other methods. The original author had difficulty in describing the phenomena of color perception in terms of a small number of dimensions; in fact he was required to employ 10 "points-of-view" (i.e., 10 dimensions). Conversely, *Carroll and Wish* (1974) found only two significant dimensions (blue-vs-yellow perception and red-vs-green perception). Individual differences were then accounted for by the different weights $(\lambda_{\alpha\ell})$ assigned to each subject (which would correspond to depth interval in our example).

Higher way matrices can be dealt with using the same approach discussed above, namely by noting the analogy to the two-dimensional least-squares fit and collapsing the data matrix in stages. However, only the three-way procedure has been programmed and extensively tested (*Carroll and Wish*, 1974).

The three-way approach is not recommended in a first examination of data. But should the normal two-way matrix yield useful results, then extension to higher ways could follow, if the requisite data are available.

CONCLUSIONS

A number of the techniques employed in pattern recognition have been reviewed and their applicability suggested to 1) analyzing patterns in acoustic spectra and 2) noting relationships between those spectra and other variables of importance in the ocean which may determine the spectral characteristics. These techniques have great strength in dealing with vast bodies of data which otherwise seem intractable. They reduce the important features of such data to a small number of dimensions which can often be identified. The characteristic features of the spectra, described by eigenvectors, are extracted.

These powerful tools have been of immense help in many disciplines. It is suggested that they might aid in the task posed by this symposium as well.

REFERENCES

Austin, M. P. and I. Noy-Meir. 1971. The problem of non-linearity in ordination experiments with two-gradient models. *J. Ecology* 59: 763-773.

Benzecri, J. P. (editor). 1973. *L'Analyse Des Donnees, tome 2: L'Analyse Des Correspondances* (Dunod, Paris), especially pp. 18-51.

Carroll, J. D. and J. Chang. 1970. Analysis of individual differences in multidimensional scaling via an N-way generalization of "Eckart-Young" decomposition. *Psychometrika* 35: 283-319, especially pp. 283-288.

Carroll, J. D. and M. Wish. 1974. Models and methods for three-way multidimensional scaling. In: *Contemporary Developments in Mathematical Psychology, Vol. II*, D. H. Krantz et al., Eds. (W. H. Freeman, San Francisco). 57-105.

Cooley, W. W. and P. R. Lohnes. 1971. *Multivariate Data Analysis* (Wiley, New York), 364 pp.

Eckart, C. and G. Young. 1936. The approximation of one matrix by another of lower rank. *Psychometrika* 1: 211-218.

Gibbs, R. H. and C. F. E. Roper. 1971. Ocean Acre: Preliminary report on vertical distribution of fishes and cephalopods. In: *Proceedings of an International Symposium on Biological Sound Scattering in the Ocean*, G. Brooke Farquhar, Ed. (U. S. Government Printing Office, Washington, D. C.), 119-133.

Grassle, F. J. and W. Smith. 1975. Woods Hole Oceanographic Institution. Personal communication.

Harman, H. H. 1967. *Modern Factor Analysis, 2nd ed.* (University of Chicago Press, Chicago), 474 pp.

Hill, M. O. 1973. Reciprocal averaging: an eigenvector method of ordination. *J. Ecology* 61: 237-249.

McElroy, P. 1974. Geographic patterns in volume reverberation spectra in the North Atlantic between 33°N and 63°N. *J. Acoust. Soc. Amer.* 56: 394-407.

McElroy, P. and W. Smith. 1975. Clustering of volume reverberation spectra - an application of correspondence analysis. *J. Acoust. Soc. Amer.* 58: 1243-1256.

Noy-Meir, I. and M. P. Austin. 1970. Principal component ordination and simulated vegetational data. *Ecology* 51: 551-552.

Seal, H. 1964. *Multivariate Statistical Analysis for Biologists*, (Wiley, New York), 207 pp.

Sebestyen, G. S. 1962. *Decision-Making Processes in Pattern Recognition*, (MacMillan, New York), 162 pp.

Shepard, R. 1972. Introduction to Volume 1, and A taxonomy of some principal types of data and of multidimensional methods for their analysis. In: *Multidimensional Scaling: Theory and Applications in the Behavioral Sciences, Vol. 1, Theory*, R. Shepard, A. K. Romney, and S. Nerlove, Eds. (Seminar Press, New York), pp. 1-20 and 21-47.

Torgerson, W. S. 1958. *Theory and Methods of Scaling* (Wiley, New York), 254-259.

Williams, E. J. 1952. Use of scores for the analysis of association in contingency tables. *Biometrika* 39: 274-289.

BIOLOGICAL PREDICTION IN PELAGIC MARINE ECOSYSTEMS

Trevor Platt and Kenneth L. Denman

Bedford Institute of Oceanography

ABSTRACT

Several recent mathematical models representing the structure of the pelagic ecosystems are discussed. It is concluded that to offer a general synthesis of the structure and function of the pelagic ecosystems based on the work accomplished in numerical modeling is still not possible. Moreover, while the models that are available are of considerable ecological interest, their relevance to the problem of the prediction of sound scattering in the ocean is rather nebulous.

INTRODUCTION

The available mathematical models of the pelagic marine ecosystem constitute an assemblage of low diversity. There is a certain sameness about them, and it is hard to tell the individuals apart. The temptation is strong to draw parallels between the models themselves and the organisms they purport to represent. As in other low diversity systems, the selective pressure on these models seems to be of the r- rather than the K-type: growth rate rather than efficiency is favored. With the invasion of a relatively unexploited domain of scientific investigation, we have seen an explosive proliferation in the numbers of models, without regard for the metabolic cost. But unfortunately (and here the analogy breaks down), a general tendency to increased size of individual models has not always proved to be symptomatic of increased maturity.

Why should this be so? The answer is, of course, that it is an extremely complex problem. Anyone who is brash enough to consider a general pelagic model has first to contend with the difficulties of constructing a physical model of the advective-diffusive environment, a task that keeps many applied mathematicians busy full-time. Then he has to graft on to it a system of coupled differential equations describing the biological interactions (of which the number tends to infinity) for which neither the functional form nor the rate-parameters are well-known. Finally, he makes a numerical integration of his model knowing that, in addition to systematic errors arising from its analytic shortcomings, propagation of random errors arising from uncertainty in the initial conditions and from aliasing of effects on scales smaller than the grid size will limit the validity of the solution. He will then be obliged to ask himself how applicable is his model to other situations. Reductionists rush in where holists fear to tread.

It is not necessary here to give a comprehensive account of the mathematical representation of the structure of the pelagic ecosystem: *Steele* (1974) has already done this. Instead, a few quite recent models are discussed to illustrate the ways in which the problem is being attacked, and some general conclusions that are emerging from these studies are pointed out.

Walsh (1975)

In recent years, a great deal of effort has been devoted to the study of upwelling ecosystems. The methodology used on these cruises represents the state-of-the-art in continuous sample processing at sea and real-time analysis of the digitized data, including contouring and simulation modeling. The most recent example of this work is *Walsh* (1975), in which is presented a simulation model (two dimensional) of the Peru upwelling ecosystem with a spatial resolution of 10 km in the horizontal and 10 m in the vertical. The domain of interest extends from the surface to a depth of 50 m along a 100 km transect directed offshore.

The basic model comprises eight coupled partial differential equations solved by the finite-difference method for the steady-state. For description of the biological relationships, the common unit is nitrogen. It was considered that an adequate description of the upwelling ecosystem could be given by writing equations for: the four phytoplankton nutrients, nitrate, ammonia, silicate, and phosphate; for the flux of nitrogen through three trophic levels, phytoplankton, zooplankton, and anchoveta (an omnivorous fish); and for detrital nitrogen. Both phytoplankton growth and zooplankton grazing were allowed to vary periodically over 24 hours. Turbulent diffusion across the offshore axis is included. A rough count indicates that some 30 parameters need to be specified to operate the model.

Some degree of correspondence was found between the solutions to the model and the 14 surface observations made along the transect, but no objective criterion was used to test the goodness-of-fit for the 8 dependent variables. Inclusion of the diurnal variation of photosynthesis and grazing, and specification of time of day at which samples were taken improved the apparent fit. Partial agreement was claimed between the model constructed for one set of data and a set of data collected in the same area 3 years earlier. Unfortunately, some apparent errors of consistency in the equations affect the clarity of the presentation. In addition, it is not made absolutely clear how much freedom has been exercised in the selection of arbitrary parameters.

A major weakness of grid models is that they are insensitive to the (nonlinear) effects of microstructure. In the model under discussion, microstructure would be defined as structure on scales smaller than 10 km in the horizontal and smaller than 10 m in the vertical. Unless microstructure is included explicitly in the model, its implications will be undetermined. It is not sufficient to ascribe differences between observed and predicted variables to patchiness, a *post facto* argument of convenience: to understand patchiness we have to model patchiness directly.

Three potential uses are proposed for this model and others of the same ilk: to organize and interrelate the data, to give new insights into old problems, and to define new lines of investigation. It cannot be denied that simulation modeling is a way to organize large data sets. But it is not the only way. Again, *Walsh* (1975) does not make a convincing case that it is the *simulation model* which leads to his new comments on the apparently contradictory calculations of *Ryther* (1969) and *Cushing* (1969). Finally, in view of the admitted weakness of this approach, it would not, perhaps, be wise to rely on it too heavily as a way to suggest new research problems. In particular, it is not immediately obvious how such grid models could help in the design of research programs aimed at understanding sound scattering in the ocean.

Steele (1974)

The model of *Walsh* (1975) discussed above is very complex even though it presents a highly simplified picture of the upwelling ecosystem. One wonders whether it is already too complex to be useful; the line between sophistication and sophistry is not well-drawn. *Steele* (1974) has presented the simplest possible simulation model of the marine ecosystem using the North Sea as a reference: it has 18 degrees of freedom.

Steele (1974) is interested primarily in stability of the ecosystem against perturbations. His criterion of success is whether, after perturbation, the simulated values of the dependent variables are of the right order of magnitude when compared with field data. Because the strongest naturally-occurring perturbation of the system is spring bloom of phytoplankton, the simulation of the post-bloom picture is considered to be a suitable test of the assumptions on which the model is based. It is not worthwhile to attempt to simulate small changes in response since observational data are almost never comprehensive enough to test them. *Steele* (1974) also stresses the lack of generality of numerical simulation, pointing out that interpretation of the results of the models will depend on general conclusions attained by analytical studies of simple interactions.

Recognizing that, typically, validation data are collected non-synoptically over large areas, *Steele* (1974) suggests that a phase-plane representation (joint phytoplankton-zooplankton probability distribution) might be a more testable product than separate time-series for phytoplankton and zooplankton. Thinking only in terms of time could lead to difficulties if the phase of development differed from station to station over the sampling area.

Steele (1974) considered that omission of stochastic effects, both in time (the physical variables) and space (plankton patchiness), placed severe limitations on the validity of comparisons between model and observations, but at the same time it has to be admitted that the extreme simplicity of the model and the lack of resolution in space and time of the observations make it unlikely that, for models of large areas such as the North Sea, the response of the model to these random effects will be verified in the foreseeable future.

Wroblewski and O'Brien (1976)

An alternative approach to numerical simulation is to attempt qualitative analysis of the differential equations describing the ecosystem. *O'Brien and Wroblewski* (1973) examined the properties of a simple three-dimensional equation describing phytoplankton growth to determine the conditions under which the advective term is important. The method chosen was to reform the original equation in non-dimensional terms. A similar approach was used in *Platt and Denman* (1975) where the relative importance of the following terms was computed: upwelling, sinking, horizontal and vertical diffusion, vertical light gradient, grazing rate, and grazing threshold.

More recently, *Wroblewski and O'Brien* (1976) have made a mathematical sensitivity analysis of a mesoscale phytoplankton model. This work is an attempt to generalize the *Kierstead and Slobodkin* (1953) method for the persistence of a phytoplankton patch. While vertical migration of zooplankton is included, other important vertical effects (vertical mixing, cell sinking, attenuation of available light with depth) are omitted. The general sensitivity analysis seeks to calculate the relative rate of change of the (scaled) dependent variable with respect to small changes in the system parameters. In this way, *Wroblewski and O'Brien* (1976) were able to show, for example, that of the two grazing parameters the rate constant was considerably more important than the grazing threshold.

The paper by *Wroblewski and O'Brien* (1976) also includes numerical simulation of their equation for a variety of initial conditions. Like *Walsh* (1975), the authors were impressed with the importance of diurnal periodicity in growth and grazing rates. A particularly interesting case treated by *Wroblewski and O'Brien* (1976) is the effect of voluntary horizontal migration of the herbivores. The assumption made is that the migration velocity would be a function of the horizontal gradient of available food (phytoplankton). If swimming were directed towards higher concentrations of food, zooplankton would tend to accumulate in the phytoplankton patches. Thus, the simulation shows a marked reduction in phytoplankton biomass within the patches. But the interesting aspect is that this reduction is attributable less to the increase in the number of grazers than to the increase in the grazing rate of herbivores in those areas with more favorable concentrations of food. Zooplankton swim too slowly for this mechanism to have a profound influence on the mesoscale distribution of phytoplankton, but at smaller scales it could be quite significant.

Winter, Banse, and Anderson (1975)

Of the attempts to represent in temporal detail the phytoplankton production in a specific location, the best is probably that of Winter et al. (1975) which treats the dynamics of blooms in a stratified fjord (Puget Sound). This study is noteworthy for the completeness of the observational data and for the choice of the functional forms specifying the individual terms in the state equation.

Algal growth within the fjord was limited by a combination of hydrodynamic factors and modulation of available light through self-shading. (It is worth mentioning that algal self-shading and the diurnal periodicity of photosynthesis were the two factors identified as most significant in an x-z model of phytoplankton growth developed by Radach and Maier-Raimer (1975).) The ecosystem of Puget Sound, and the model of Winter et al. (1975), are particularly sensitive to meteorological transients. Several consecutive days of bright sunshine are sufficient for the development of substantial phytoplankton blooms. Horizontal advection following sustained winds will remove the products of growth from the fjord, and a succession of cloudy days leads to the decline of blooms. Biological effects such as grazing and cell sinking are thought to be of relatively minor importance.

Winter et al. (1975) were able to conclude that the best of the available functional forms and parameter values describing phytoplankton metabolism were marginally adequate to use in a short time-scale model. But at the same time they doubted whether we were yet in a position to construct a model giving reliable results for any given fjord. In longer time-scale models provision would have to be made for the parameter values to change in adaptive response to changing environmental conditions.

GENERAL CONCLUSIONS

We are far from being able to offer a general systhesis of the structure and function of the pelagic ecosystem based on the work accomplished in numerical modeling. We can, however, point to certain affinities in the conclusions reached by different approaches. One recurring feature is the general improvement in the matching of theory and observations which is noticed if the parameter values for phytoplankton growth and zooplankton grazing are allowed to vary harmonically over 24 hours. This has been emphasized by *Walsh* (1975), *Wroblewski and O'Brien* (1976), and *Radach and Maier-Raimer* (1975). *Steele* (1974) also stressed the confounding of spatial variability with temporal processes in pseudo-synoptic studies. Inhibition of light-penetration by the products of phytoplankton growth (self-shading) has been shown to be of crucial significance in those models which included it (*Radach and Maier-Raimer*, 1975; *Winter et al.*, 1975). There remain areas of contention, but perhaps these may be resolvable in terms of the nature of the system being modeled. Thus, for the offshore ecosystem, *Walsh* (1975) and *Wroblewski and O'Brien* (1976) considered grazing by herbivores to be a highly important variable. In a fjord, however, *Winter et al.* (1975) considered grazing to be a secondary factor compared to the more-or-less random effect of meteorological forcing. In the North Sea, *Steele* (1974) thought that the omission of these stochastic influences was a grave weakness of his model. In the inshore, at least, a totally stochastic approach to ecosystem modeling might be rewarding.

RELEVANCE TO SOUND SCATTERING IN THE OCEAN

While the models described above are of considerable ecological interest, their relevance to the problem of the prediction of sound scattering in the ocean is rather nebulous. It seems that what is required is explicit modeling of the formation and dissipation of aggregations of organisms of all types. Modeling of the mean values will not be too informative.

REFERENCES

Cushing, D. H. 1969. Upwelling and fish production. *F.A.O. Fisheries Biology Technical Papers*, 84: 40 pp., Rome.

Kierstead, H. and L. B. Slobodkin. 1953. The size of water masses containing plankton blooms. *Journal of Marine Research*, 12: 141 pp.

O'Brien, J. J. and J. S. Wroblewski. 1973. On advection in phytoplankton models. *Journal of Theoretical Biology*, 38: 197-202.

Platt, T. and K. L. Denman. 1975. A general equation for the mesoscale distribution of phytoplankton in the sea, J. Nihoul (Ed.). *Mémoires Société Royale des Sciences de Liège*, 6e série, 7: 31-42.

Radach, G. and E. Maier-Raimer. 1975. The vertical structure of phytoplankton growth dynamics. A mathematical model. *Mémoires Société Royale des Sciences de Liège*, 6e série, 7: 113-146.

Ryther, J. H. 1969. Photosynthesis and food production in the sea. *Science, 166:* 72-76.
Steele, J. H. 1974. *The Structure of Marine Ecosystems*. Harvard University Press, Cambridge, 128 pp.
Walsh, J. J. 1975. A spatial simulation model of the Peru upwelling ecosystem. *Deep-Sea Research, 22:* 201-236.
Winter, D. F., K. Banse, and G. C. Anderson. 1975. The dynamics of phytoplankton blooms in Puget Sound, a fjord in the North Western United States. *Marine Biology, 29:* 139-176.
Wroblewski, J. S. and J. J. O'Brien. 1976. A spatial model of phytoplankton patchiness. *Marine Biology 35:* 161-175.

Daytime Depths of Sound Scattering Layers in the Major Biogeographic Regions of the Pacific Ocean

Sargun A. Tont

Scripps Institution of Oceanography

ABSTRACT

The daytime depth of sound scattering layers monitored with 12 and 18 kHz echo sounders in the major biogeographic regions of the Pacific Ocean vary between 80 and 540 m. Layers shallower than 170 m are found only in the eastern tropical Pacific. Subarctic region layers are confined to a relatively narrow range of 170 to 380 m. Deepest layers are most frequent in the North Central Pacific.

INTRODUCTION

Ever since sound scattering layers (SSL) were reported in 1942, they have been investigated extensively. The majority of these studies may be divided into two main categories: (1) the composition of the layers, and (2) the response of the layers to various stimuli such as light and temperature.

Sound scattering layers have been observed in almost every part of the world oceans except the Antarctic. In general, they have been shown to be caused by the presence of organisms that can scatter sound. Since no single fauna of vertically migrating organisms inhabits the entire world ocean, the composition of the layers must differ from location to location. Various scattering layers have been shown to respond to variations in light intensity by migrating vertically. However, there is not enough information about the biological composition of SSLs to ascertain whether any single stimulus, such as solar irradiance, accounts for their diel migrations everywhere or even over large parts of a single ocean.

Few attempts have been made to examine sound scattering characteristics on an ocean-wide basis. *Beklemishev* (1964) divided the

Pacific Ocean into 17 major biotic regions, the boundaries of which were determined on the basis of the fauna caught in net tows, biological scattering, as well as water masses and currents. McGowan (1974) proposed other biogeographic regions, based on the distribution of a variety of planktonic and nektonic species. His regions differed significantly from those proposed by Beklemishev.

The aim of this paper is to present the distribution of SSL daytime depths in the major biogeographic regions of the Pacific as defined by McGowan (Figure 1). The physical and chemical environments occupied by the SSLs during both day and night as well as their migratory characteristics are discussed in more detail by Tont (in press).

MATERIALS AND METHODS

Data collected on a number of cruises of the Scripps Institution of Oceanography and the National Oceanic and Atmospheric Administration (NOAA) during the period of 1958-1972 have been used in this study. Echograms were obtained during a number of segments of cruise tracks (Figure 2).

Records were obtained using a Precision Depth Recorder (PDR) at a dominant frequency of 12 kHz during Scripps cruises. A Simrad Recorder at a dominant frequency of 18 kHz was used during NOAA cruises.

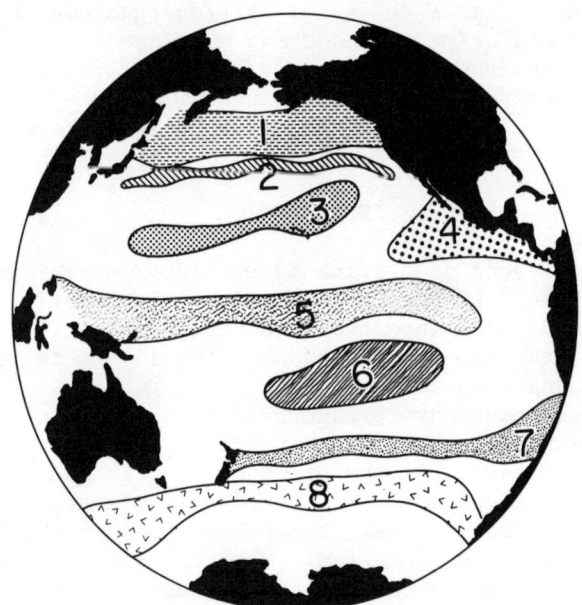

Fig. 1. Patterns of the basic (100% "core" regions) biotic provinces of the oceanic Pacific. (1) Subarctic; (2) Transition Zone, North; (3) North Central; (4) Eastern Tropical Pacific; (5) Equatorial; (6) South Central; (7) Transition Zone, South; (8) Subantarctic. Redrawn from McGowan (1974).

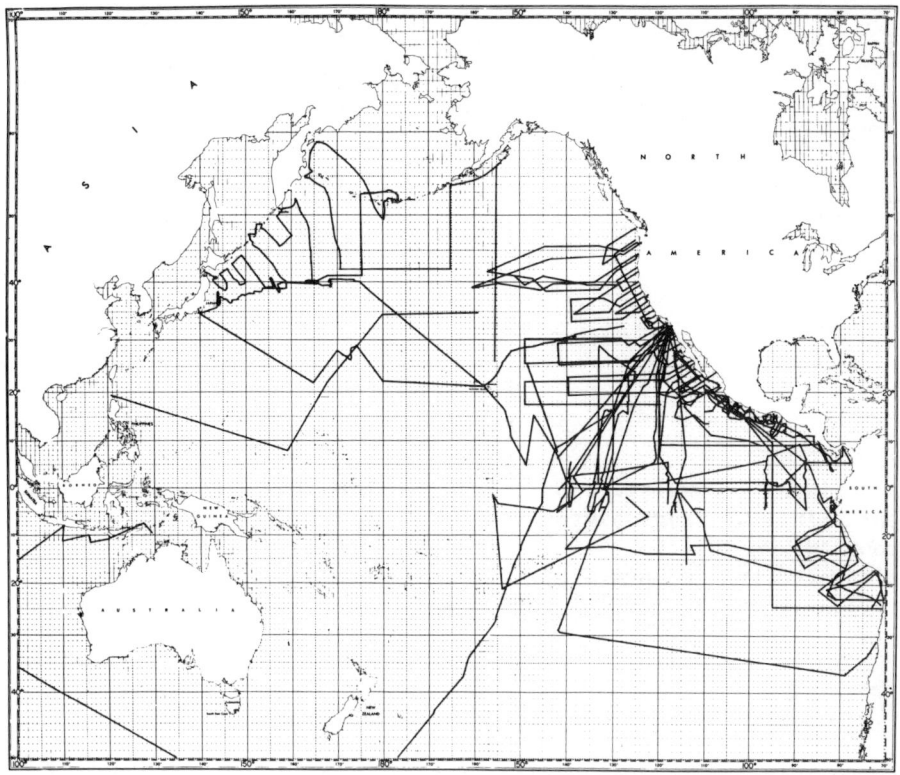

Fig. 2. Locations of cruise tracks where SSL depths have been obtained.

No significant differences were observed between the two frequencies in ascertaining daytime SSL depths within a region.

In this paper, sound scattering layers (SSL) are defined as acoustical scattering, presumably from organisms, where the sound scattering appears as a continuous layer on an echo sounder and where individual organisms cannot be resolved. SSL midday depth is defined as the distance between the top of the SSL to the ocean surface, taken within a few minutes of local apparent noon. If data around local apparent noon were not available, any reading between 1000 and 1400 hours local time was substituted. Only the layers which occupy a midday depth greater than 40 m are considered in this study. Forty meters is roughly the distance below the surface where the outgoing acoustic signal cannot be distinguished from the incoming signal, preventing an accurate SSL depth determination above this depth. Only layers found within the "core" biogeographic regions (Figure 1) are included.

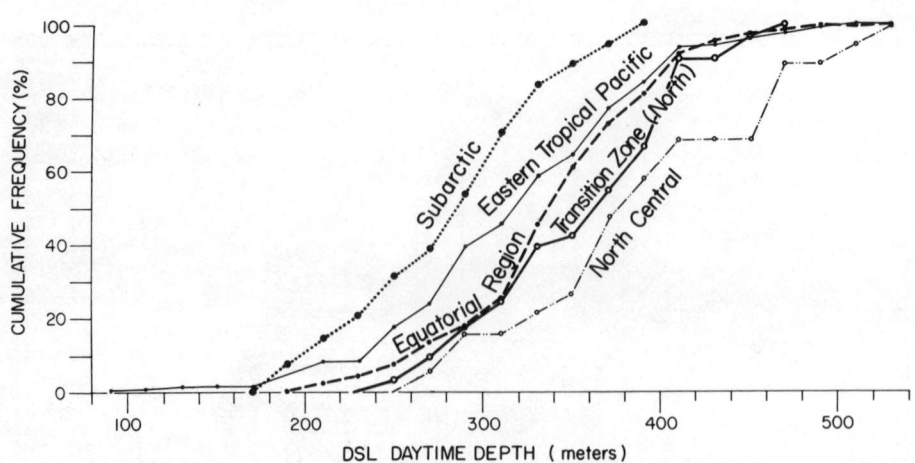

Fig. 3. Cumulative frequency spectra of all the SSL daytime depths in some major biotic regions of the Pacific Ocean. Only the echograms where SSLs have been recorded are included. Spectra is calculated on one record-per-day basis. Subarctic, 54 days; Eastern Tropical, 130; Equatorial, 76; Transition (North), 31; North Central, 14.

RESULTS AND DISCUSSION

Cumulative frequency distributions are included for all SSL daytime depths for all regions except those in the South Pacific where data were scarce (Figure 3). The median depth (50% cumulative frequency) was 285 m in the Subarctic, 320 m in the Eastern Tropical Pacific, 335 m in the Equatorial Pacific, 360 m in the Transitional region, and 375 m in the North Central region.

An important and obvious feature of Figure 3 is the large variability of the daytime SSL depths in all regions. Layers in the Eastern Tropical Pacific region show the largest variability, 90 to 540 m, whereas layers in the Subarctic region show the smallest depth range, 70 to 380 m. Layers shallower than 170 m were only found in the Eastern Tropical Pacific. Thus a distinct latitudinal increase in daytime SSL depths was not obvious. Beklemishev (1967) and Dunlap (1970), on the other hand, reported that SSLs are found at greater depths toward the Equatorial region.

These differences can be explained when the variability of SSL daytime depths within a single region is considered. Although certain transects, such as the ones taken near 155°W show deepening of the layers near the equator, others taken near 90°W do not. Tont (in press) reported that the offshore-onshore gradient is more pronounced along the western coasts of North and South American than the north-south gradient in the mid-ocean.

Little quantitative information can be obtained about the number and distribution of the organisms from echograms, either in the biogeographic regions or the adjoining areas. Nevertheless, some qualitative statements can be made. Monitored under similar gain settings of the acoustic recorders, layers off the coast of Peru were the thickest with clear sharp vertical boundaries. Acoustic layers off California, in the Eastern Tropical, North Transition, and South Transition Zones, as well as in areas near the Kuroshio Current also were well formed, but were not as thick as the ones found off the Peruvian coast. Although two major crossings have been made in the South Central region, SSLs have been recorded only twice during noontime. In both instances the layers were extremely diffuse as compared to layers in other regions at comparable depths (410 m).

Small scale variations, both temporal and spatial are undoubtedly important and the magnitudes of these variations can very well be greater than the variations among the regions. Therefore, an effort should be made to monitor the DSL depths at fixed locations for long periods of time. This, perhaps, can be accomplished by a network of sonic buoys anchored at specific locations.

ACKNOWLEDGEMENTS

I thank the following individuals for their help with this study: John Isaacs, William Pearcy, Richard Schwartzlose, Eric Barham, Paul Smith, Thomas Chase, Gerald Wick, Stephen Tucker, John McGowan, Walter Bryan, Norio Shiora, Anthony Tubbs, Norman Anderson, Donna DeCamp, Meredith Sessions, and Frances Wilkes.

REFERENCES

Beklemishev, K. V. 1967. Echo-sounding records of macroplankton concentrations and their distribution in the Pacific Ocean. *N.O.O. Trans. 343,* Naval Oceanogr. Office, Washington, DC: 49 p.

Dunlap, C. R. 1971. A reconnaissance of the deep scattering layers in the eastern tropical Pacific and Gulf of California. In Proc. An International Symposium on Biological Sound Scattering in the Ocean. M.C. *Report 005,* Dept of the Navy, Washington, DC.

McGowan, J. A. 1974. The nature of oceanic ecosystems. In C. B. Miller (ed.) *The Biology of the Oceanic Pacific*: 9-28. Oregon State Univ. Press, Corvallis.

Tont, S. A. (In press). Deep scattering layers: Patterns in the Pacific. *Calif. Coop. Oceanic Fish. Invest. Reports, 18.*

VERTICALLY MIGRATING HERBIVOROUS PLANKTON – THEIR POSSIBLE ROLE IN THE CREATION OF SMALL SCALE PHYTOPLANKTON PATCHINESS IN THE OCEAN

J. S. Wroblewski[1]

Florida State University

ABSTRACT

The ability of vertically migrating copepods and euphausiids to create small scale, spatial inhomogeneities in the phytoplankton standing stock of the open ocean is explored numerically. Model results suggest herbivore patches of size and concentration often observed in nature can reduce the plant crop to a grazing threshold-level during one night's grazing activity. In nutrient-limited oceanic areas excretion and egestion by nocturnally grazing zooplankton may liberate enough nutrient that phytoplankton growth is enhanced the following day.

A new nonlinear formulation of herbivore egestion rate as a function of food abundance is proposed. Mathematical expressions which simulate continuous and/or nocturnal grazing stresses upon the phytoplankton are presented, along with a formulation for chemical gradient-orientated, zooplankton migration.

INTRODUCTION

Little is known about the spatial and temporal scales of plankton patchiness in the ocean. Only recent advances have made possible quasisynoptic mapping of sea surface chlorophyll fields (Clarke et al., 1970; Platt, 1972a; Walsh, 1972; Denman and Platt, 1975). New sound scattering technology now enables detection of zooplankton spatial inhomogeneities below the sea surface (Beamish, 1971). Acoustic frequencies between 120 and 330 kHz can now

---

[1]Present address: Dalhousie University, Canada

delineate the presence and relative abundance of planktonic animals such as copepods and euphausiids (Barraclough et al., 1969; Bary and Pieper, 1970; Castile, 1975). An improving data base has stimulated interest in the theory of patchiness.

In biologically productive areas of the ocean, abundant plant life supports large standing stocks of herbivores and carnivores. As the ocean's productivity varies geographically, so does the sound scattering strength of the water column. McGowan (1974) has outlined five major biotic provinces in the North Pacific Ocean: the subarctic region, the transition zone centered about 40° N, the central gyre extending from 40° N to 15° N, the eastern tropical Pacific Ocean, and the equatorial Pacific Ocean. The standing stocks of phytoplankton and zooplankton, the dissolved nutrient concentration, and the acoustic properties of each region differ. Only the subarctic and central gyre regions will be discussed here.

The sound scattering strength in the central gyre is considerably lower than in the subarctic region (Beers, personal communication). The observed low concentrations of phytoplankton, nutrient, and zooplankton in the central gyre are typical of an open ocean little influenced by boundary currents, coastal runoff, or coastal upwelling. Phytoplankton are limited by nitrogen and possibly phosphorus (Eppley et al., 1973). The subarctic province is more seasonally variable than the central gyre in phytoplankton, zooplankton, and nutrient characteristics. Plant levels in the subarctic region are always higher with light, rather than nutrients, limiting phytoplankton growth, especially during the winter. Acoustically important macrozooplankton (such as euphausiids) are also more abundant with a peak in biomass in late spring (Ponomareva, 1966; McGowan and Williams, 1973).

Below this geographical variability of acoustic scattering strength is a spectrum of smaller scale spatial variation. Length scales of 10 to 1000 meters tend to reoccur in observations of small scale patchiness. Barraclough et al. (1969) used a high frequency (200 kHz) echo sounder in conjunction with a Longhurst-Hardy sampler to measure zooplankton patches of the order 100 meters in the subarctic region of the North Pacific Ocean. These patches were composed mostly of vertically migrating Calanus cristatus in concentrations up to twenty-five times the background concentration. Copepods which continuously reside in the euphotic zone often form patches with length scales of 10 to 100 meters (Wiebe, 1970). Calanus sp. have been reported to maintain patches approximately 100 meters in size even during strong wind mixing (Kawamura, 1974). Several subarctic Pacific Ocean species of euphausiids (e.g., Thysanoessa raschii, T. inermis, and Euphausia pacifica) are known to form discrete swarms as a feature of their behavior (Mauchline and Fisher, 1969). Densities of several thousand animals per cubic meter are maintained even during vertical migration. Swarming behavior is uncommon for euphausiid species in the central gyre. These animals may prefer to maintain low densities where food is limited.

While zooplankton can maintain a patch to a degree relative to their swimming capabilities, phytoplankton are truly planktonic. Platt (1972b) has shown small scale phytoplankton spatial variability is due in part to turbulent fluctuations in the physical environment. Variation in plant growth rates and *advective and diffusive fluxes* become important in determining the spatial structure of the phytoplankton above the microscale (~ 1 meter).

The spatial and temporal scales on which phytoplankton and zooplankton interact is of special interest here. The purpose of this paper is to demonstrate how numerical modeling can clarify and help evaluate the complex biological and physical relationships involved in oceanic sound scattering. Specifically, the hypothesis that a sufficiently concentrated patch of vertically migrating herbivores can impart a patchy character to the phytoplankton distribution through grazing will be explored.

The Dynamics of Phytoplankton Patchiness

One model cannot be equally successful in resolving the dynamics of patchiness on all temporal and spatial scales. The modeler must determine *a priori* what resolution will be necessary to answer the questions posed. In exploring the role of vertically migrating herbivores in creating small scale phytoplankton patchiness, the temporal and spatial scales of interest are hours and meters, respectively.

The dynamics which determine the concentration of phytoplankton $P(x,y,z,t)$ at any arbitrary point in the ocean may be expressed quite generally as

$$\frac{\partial P}{\partial t} + u \frac{\partial P}{\partial x} + v \frac{\partial P}{\partial y} + w \frac{\partial P}{\partial z} = \frac{\partial}{\partial x}(K_h \frac{\partial P}{\partial x}) + \frac{\partial}{\partial y}(K_h \frac{\partial P}{\partial y}) + \frac{\partial}{\partial z}(K_v \frac{\partial P}{\partial z})$$

$$+ \textit{biological terms} \quad (1)$$

The velocity components $u, v,$ and w are the x-directed, y-directed (horizontal) and z-directed (vertical) vectors of organized oceanic motion. The four terms on the left hand side of equation (1) collectively represent the change in P following a moving water parcel. The first term is the local change in P and the next three terms are the advective changes. The first three terms on the right hand side of equation (1) are the turbulent dispersion terms approximated by a Fickian diffusion law (Fofonoff, 1962). Parameters K_h and K_v are the horizontal and vertical eddy diffusivities (cm^2 sec^{-1}), respectively. The biological terms include the processes such as growth, predation, etc.

Whether advection can be neglected in equation (1) depends on the magnitude of the advective fluxes in relation to the turnover time of the phytoplankton. From equation (1), *O'Brien and Wroblewski* (1973) derived the nondimensional number

$$S = U(V_m L)^{-1} \tag{2}$$

where U is the characteristic speed ($cm\ sec^{-1}$) of the organized fluid motion, V_m is the maximum growth rate (sec^{-1}) of the phytoplankton, and L is the length scale of interest. If $S \ll 1$, the advective terms in equation (1) may be neglected. Mean surface currents in the central gyre of the North Pacific Ocean and in the middle of the subarctic water mass are of the order of $1\ cm\ sec^{-1}$ (*Reid*, 1961). If V_m equals 1.7 doublings day^{-1} or $2 \times 10^{-5}\ sec^{-1}$ (*Eppley*, 1972), then $S \leq 1$ for length scales ≥ 500 meters. For $L < 500$ meters, advection becomes increasingly important relative to biological processes and turbulence in determining the spatial configuration of the phytoplankton.

Equation (1) may be simplified even more upon the assumption of radial symmetry in the horizontal and a well mixed euphotic zone. One then has a one-dimensional equation for phytoplankton patchiness,

$$\frac{\partial P}{\partial t} = \frac{\partial}{\partial x}\left(K_h \frac{\partial P}{\partial x}\right) + biological\ terms. \tag{3}$$

As K_h is a function of length scale (*Okubo*, 1971), the value chosen for K_h must be consistent with the patch size. The assumption of a constant K_h becomes invalid as the patch size changes by several orders of magnitude. It has long been known that Fickian diffusion fails to accurately describe turbulence in the sea (*Stommel*, 1949). The development of a simple yet sophisticated prescription based on physical understanding of microscale turbulence is eagerly awaited by ecosystem modelers.

Kierstead and Slobodkin (1953) began with an equation similar to equation (3) in deriving a fundamental relationship between the temporal and spatial scales of phytoplankton patchiness. For the patch to be stable in time, the patch size must be

$$L_c = \pi \left[\frac{K_h}{b}\right]^{1/2}. \tag{4}$$

For a patch of this length scale, the losses due to turbulent dispersion just balance the phytoplankton growth rate, b. A patch smaller than this critical length, L_c, would be dispersed by turbulent diffusion, and a larger patch would increase in size with time.

Consideration of Ivlev-type herbivore grazing in the *Kierstead-Slobodkin* dynamics modifies this critical length scale to

$$L_c = \pi\left[\frac{K_h}{b-R\Lambda}\right]^{1/2}, \qquad (5)$$

where R is the maximum herbivore grazing ration ($conc.\ sec^{-1}$) and Λ is the Ivlev constant ($conc.^{-1}$) for zooplankton grazing (*Platt and Denman*, 1975; *Wroblewski et al.*, 1975). The effect of grazing is to increase the patch size necessary for stability. Where phytoplankton growth and herbivore grazing are temporally and spatially dependent, the critical length scale for the patch depends on the initial distribution and concentration of the limiting nutrient and herbivore biomass, as well as the daily averaged magnitude of the growth and loss terms (*Wroblewski and O'Brien*, 1976). With insight from this latter work, the theoretical influence of vertically migrating herbivores on small scale phytoplankton patchiness is investigated.

FORMULATION OF THE MODEL

The Phytoplankton Equation

Dugdale (1967) has shown that the growth rate of phytoplankton, V, expressed in terms of uptake of the biologically limiting nutrient, N, can be described as a hyperbolic function of nutrient concentration,

$$V = V_m \frac{NP}{k+N}, \qquad (6)$$

where V_m is the maximum uptake rate of the nutrient by phytoplankton, P, and k is the Michaelis half-saturation constant (i.e., the concentration of N supporting half the maximum uptake rate). In adopting this formulation for the phytoplankton growth dynamics, one assumes the plant community is dominated by one or more species whose V_m and k are known.

Uptake of nutrients by phytoplankton exhibits diel periodicity, as uptake is indirectly coupled to light-dependent photosynthesis (*Goering et al.*, 1964; *Dugdale and Goering*, 1967). To simulate this behavior, the nutrient uptake term is multiplied by the function

$$f(t) = \frac{\pi}{2\theta}\sin\left[\frac{\pi\ mod(t,24)}{\theta}\right], \qquad (7)$$

where θ is the daylight fraction of a day and t is time in hours. When $f(t) < 0$, one sets $f(t) = 0$, to simulate no growth at night. Note that if N is constant, the total uptake over one day is the same with or without multiplication by this function, since

$$\int_0^\theta \delta(t) = 1. \tag{8}$$

The function acts to condense the daily phytoplankton growth into the daylight portion of a day, assuming dawn begins at $t = 0$.

The grazing of herbivores, Z, upon the phytoplankton is assumed to follow a modified Ivlev function (*Parsons et al.*, 1967)

$$g(P,Z) = R_m Z \left(1 - \exp\left[-\Lambda(P - P_t)\right]\right); \quad P > P_t \tag{9}$$

$$g(P,Z) = 0 \qquad\qquad\qquad ; \quad P \leq P_t \tag{10}$$

where R_m is the maximum rate of herbivore grazing (sec^{-1}), Λ is the Ivlev constant ($conc^{-1}$), and P_t is the threshold concentration of P below which the grazing behavior of the herbivores ceases (Figure 1). Both herbivore and phytoplankton biomass will henceforth be discussed in units of concentration of the biologically limiting nutrient (e.g., nitrogen).

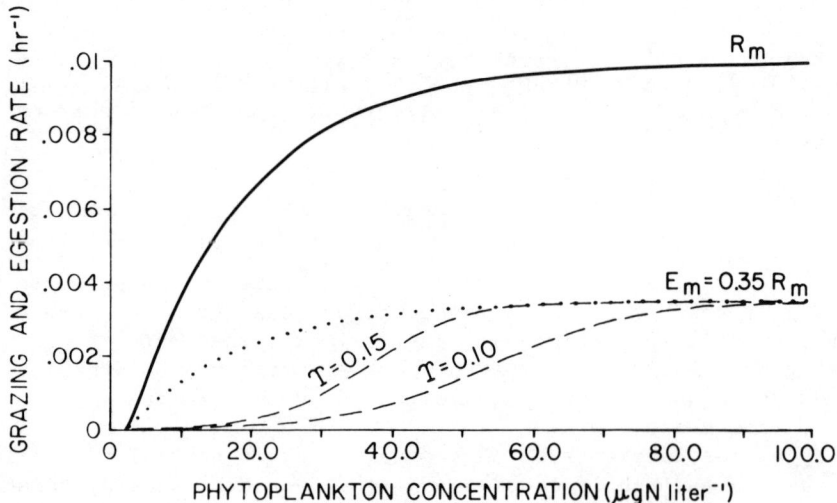

Fig. 1. Ingestion and egestion rates as functions of phytoplankton concentration. The ingestion rate curve (solid line) for nocturnally grazing *Calanus pacificus* is given by an Ivlev function where $R_m = 0.01\ hr^{-1}$, $\Lambda = 0.06\ (\mu gat\ N/\ell)^{-1}$ and $P_t = 2.5\ \mu gat\ N\ \ell^{-1}$ (from *Parsons et al.*, 1967). Egestion as a constant fraction (0.35) of ingestion is represented by the dotted hyperbolic curve. The sigmoid shaped (dashed) curves for egestion rate result from the function proposed in the text with $E_m = 3.5 \times 10^{-3}\ hr^{-1}$, $\delta = 2 \times 10^{-5}\ hr^{-1}$, and $T = 0.15$ and $0.10\ (\mu gat\ N/\ell)^{-1}$.

A second phytoplankton loss term, $-\eta P$, represents loss of nutrient from the cells by extracellular release. Since it has been assumed that $\partial/\partial z = 0$, phytoplankton sinking out of the euphotic zone is expressed as a linear loss, $-w_s P$, where w_s is the loss rate (sec^{-1}) due to sinking.

Upon substitution of these functions for the biological terms, equation (3) becomes

$$\frac{\partial P}{\partial t} = K_h \frac{\partial^2 P}{\partial x^2} + \delta(t) V_m \frac{NP}{k+N} - R_m Z (1 - \exp[-\Lambda(P - P_t)]) - (\eta + w_s) P \quad (11)$$

The Herbivore Equation

In many oceanic areas the herbivorous trophic level is composed of vertically migrating copepods and euphausiids. Large concentrations of these nocturnally feeding animals may impose a highly time-dependent grazing stress on the phytoplankton. Not every individual makes the diel vertical migration each night, and many of the smaller species continuously reside in the euphotic zone (Banse, 1964). The herbivore community is therefore divided into two fractions for modeling purposes. The herbivores present in the euphotic zone at time, t, are labeled Z_1, and those residing at lower depths make up the fraction called Z_2.

The equations describing the change of herbivore biomass within these fractions with time are:

$$\frac{\partial Z_1}{\partial t} = R_m Z_1 (1 - \exp[-\Lambda(P - P_t)]) (1 - \gamma) - \Psi_1 Z_1$$
$$+ \Phi Z_2 \exp[-(t - t_2)^2/\sigma^2] - \Phi Z_1 \exp[-(t - t_1)^2/\sigma^2] \quad (12)$$

and

$$\frac{\partial Z_2}{\partial t} = \Phi Z_1 \exp[-(t - t_1)^2/\sigma^2] - \Phi Z_2 \exp[-(t - t_2)^2/\sigma^2] - \Psi_2 Z_2, \quad (13)$$

where γ is the fraction of the grazing ration ($conc\ sec^{-1}$) which is egested as fecal pellets, Ψ_1 is the metabolic excretion rate (sec^{-1}) of the animals while feeding at the surface, and Ψ_2 is the excretion rate (sec^{-1}) while at depth. Time t_2 is the point within the day about which ascent into the surface waters is centered (usually sunset), and t_1 is the point (after dawn) about which descent from the surface is centered. Parameter σ is the fraction of the day over which the migrational transition takes place. The average vertical distance through which many species of euphausiid migrate is 100-300 meters, usually taking 2 to 4 hours to travel this distance (*Mauchline and Fisher*, 1969).

In surveys with horizontally towed nets, the ratio of zooplankton volumes captured at night as compared to daytime hauls was found to vary greatly with geographic location and season (*Banse*, 1964). At Ocean Station P (50° N; 145° W) in the subarctic region of the North Pacific Ocean, *McAllister* (1961) found an average night/day ratio of 8 (range: 0.2 to 37) for tows with a 0.33 millimeter net aperature. *Eppley et al.* (1973) found a night/day ratio of approximately 2 in the central gyre, where non-migrating protozoa make up a considerable portion of the herbivore biomass. Parameter Φ in equations (12) and (13) allows one to adjust the fraction of the herbivore community which undergoes migration to fit an oceanic area's night/day ratio.

Conover (1966) suggests the fraction of food assimilated $(1-\gamma)$ is independent of food concentration for *Calanus hyperboreus* feeding on diatoms in concentrations of cells normally found in the sea. At low phytoplankton concentrations, however, egestion may be a non-linear function of food ingested (Figure 1). The sigmoid shaped egestion rate curves shown in Figure 1 were computed from the proposed expression

$$E = \frac{E_m \delta \, e^{T(P-P_t)}}{E_m + \delta \left[e^{T(P-P_t)} - 1 \right]} , \qquad (14)$$

where P is again phytoplankton concentration; T ($conc^{-1}$) governs the rate of increase in egestion with P; δ is the egestion rate (sec^{-1}) at the grazing threshold, P_t; and E_m is the maximum egestion rate (sec^{-1}). Egestion rate as a hyperbolic curve following Conover's hypothesis is shown in Figure 1 for comparison.

Steele (1974) and *Frost* (1974) recognized that efficiency of assimilation, defined as (ingestion-egestion)/ingestion, may be high when food is scarce, with efficiency possibly decreasing as food concentration increases. In Figure 2 the assimilation efficiency is calculated for a range of phytoplankton concentrations, where ingestion rate is given by the Ivlev grazing curve and egestion rate is expressed by the sigmoid curve. Above the grazing threshold where the assimilation efficiency is zero by definition, the efficiency rapidly increases, then decreases to a minimum value.

Obtaining data on nitrogen egestion by herbivores in the presence of phytoplankton is made difficult by the rapid uptake of nitrogen by the cells (*Hirota*, personal communication). For lack of laboratory measurements of T and δ, the simpler egestion relationship graphically displayed as the hyperbolic curve in Figure 1 will be used in this paper.

Finally, note that there is no term for diffusion of zooplankton

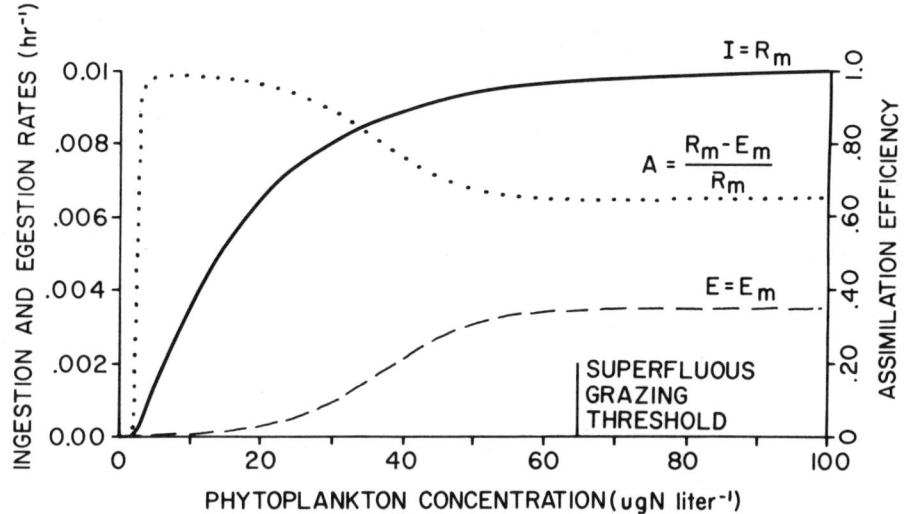

Fig. 2. Assimilation efficiency as a function of food availability. Ingestion rate (I) is defined by the Ivlev grazing curve (solid line) with the same parameter values as in Figure 1. Egestion rate (E) is determined by the sigmoid curve (dashed line) where $T = 0.15$ $(\mu gat\ N/\ell)^{-1}$. Assimilation efficiency (dotted line) is calculated as $(I-E)/I$. The threshold for superfluous grazing is defined where the assimilation efficiency reaches $(R_m-E_m)/R_m$.

in equations (12) and (13). The ability of adult copepods and euphausiids to maintain patches has been well documented (Cushing, 1963; Komaki, 1967; Kawamura, 1974), yet little is known about the swarming mechanism. As zooplankton can sense a pheromone gradient (Kittredge et al., 1974), a swimming behavior to maintain a certain patch density is plausible.

The Nutrient Equation

The model equation for the horizontal diffusion, uptake, resupply, and recycling of limiting nutrient, N, in the euphotic zone is

$$\frac{\partial N}{\partial t} = K_h \frac{\partial^2 N}{\partial x^2} - \delta(t)\, V_m \frac{NP}{k+N} + \psi_1 Z_1 + rP - m(N-N_2) \qquad (15)$$

where N_2 is the aphotic zone nutrient concentration, and m is the coefficient of vertical mixing between the euphotic and aphotic zones. Steele (1958) used a similar formulation for vertical mixing in a two-layered phytoplankton model. Note in equations (11), (12), (13), and (15) there is a net loss of nutrient from the euphotic zone in the form of sinking fecal pellets and sinking phytoplankton cells. There is also a continuous transfer of nutrient from the surface layers to depth by vertically migrating, metabolite excreting animals. The surface layers become depleted of nutrient if there is no resupply by vertical diffusion or upwelling. One can calculate the rate of resupply of nutrients necessary to balance the vertical loss terms in the model. The vertical mixing term $m(N-N_2)$ must equal $w_s P + \Psi_2 Z_2 + \gamma R_m Z_1 (1-\exp[-\Lambda(P-P_t)])$. Substituting these latter terms for the former, equation (15) becomes

$$\frac{\partial N}{\partial t} = K_h \frac{\partial^2 N}{\partial x^2} - \delta(t) V_m \frac{NP}{k+N} + \Psi_1 Z_1 + \Psi_2 Z_2 + (w_s + r)P + \gamma R_m Z_1 (1-\exp[-\Lambda(P-P_t)]). \quad (16)$$

Let $P + Z_1 + Z_2 + N = N_t$, the total amount of limiting nutrient in the euphotic zone.

SCALING OF THE EQUATIONS

Equations (11)-(13) and (16) describe the general interaction of phytoplankton and zooplankton in the euphotic zone of an open ocean. As discussed above, geographic differences in primary production and the standing stock of higher trophic levels are related to the total concentration of limiting nutrient cycling between the biotic components, N_t. If one scales the concentrations of P, Z_1, Z_2, and N in equations (11)-(13) and (16) by N_t, the numerical solutions of these equations can be applied to any oceanic region. For the same percentage of N_t in the phytoplankton component, the standing crop of plants may be relatively high where N_t is large (as in the subarctic North Pacific Ocean), or low where the total amount of limiting nutrient in the euphotic zone is small (as in the central gyre region).

The maximum phytoplankton growth rate, V_m, under optimal light and nutrient conditions is about 1.7 doublings day^{-1} in both the central gyre and subarctic region during spring (Eppley, 1972). If one scales time, t, by V_m, all other biological processes can be examined relative to the turnover time of the phytoplankton.

Using the scaling relationships listed in the Appendix, equations (11)-(13) and (16) in nondimensional space become,

$$\frac{\partial P'}{\partial \tau} = \frac{\partial^2 P'}{\partial x'^2} + \delta(\tau)\frac{N'P'}{\alpha+N'} - \beta Z'_1 \left[1-\exp(-\lambda P'+\lambda P^*)\right] - (\varepsilon + \omega)P' \quad (17)$$

$$\frac{\partial Z'_1}{\partial \tau} = \beta Z'_1 \left[1-\exp(-\lambda P'+\lambda P^*)\right](1-\gamma) - \psi_1 Z'_1 \quad (18)$$
$$+ \Phi Z'_2 \exp\left[-(\tau-\tau_2)^2/\sigma^2\right] - \Phi Z'_1 \exp\left[-(\tau-\tau_1)^2/\sigma^2\right]$$

$$\frac{\partial Z'_2}{\partial \tau} = \Phi Z'_1 \exp\left[-(\tau-\tau_1)^2/\sigma^2\right] - \Phi Z'_2 \exp\left[-(\tau-\tau_2)^2/\sigma^2\right] - \psi_2 Z'_2 \quad (19)$$

$$\frac{\partial N'}{\partial \tau} = \frac{\partial^2 N'}{\partial x'^2} - \delta(\tau)\frac{N'P'}{\alpha+N'} + \psi_1 Z'_1 + \psi_2 Z'_2 + (\varepsilon + \omega)P'$$
$$+ \gamma\beta Z'_1 \left(1-\exp\left[-\lambda(P'-P^*)\right]\right) \quad (20)$$

Hereupon, the primes are dropped for convenience. Note also that $P + N + Z_1 + Z_2 = 1$.

The scaling process results in several important nondimensional ratios. The ratio of the herbivore grazing rate, R_m, to the maximum phytoplankton growth rate, V_m, is a fundamental parameter in any oceanic region. If β is greater than one, the herbivores consume more phytoplankton than are produced in one day. As a result, the phytoplankton standing stock must decline.

Another ratio which arises from the scaling of all concentrations by N_t is $\alpha = k/N_t$. MacIsaac and Dugdale (1969) suggest that the Michaelis constant, k, appears to be related to the nutrient availability of a region. If k increases proportionately with N_t, the ratio α varies little between major biotic provinces.

Note by scaling, the eddy diffusivity coefficient, K_h, is no longer explicit in the model equations. Yet, interpretation of the numerical solutions still depends on the value of K_h. The ratio of the turbulent dispersion parameter, K_h, to the maximum biological growth rate, V_m, defines the length scale on which our solutions apply.

INVESTIGATION OF THE MODEL

The solutions of equations (11)-(13) and (16) depend on the initial conditions of P, Z_1, Z_2 and N, and on the parameter values used. Most parameters (e.g., r, P_t, R_m, k, Λ, Ψ_1, and Ψ_2) are limited to a small range after scaling by N_t and V_m. Parameters such as t_1 and t_2, the times of descent and ascent of migrating zooplankton, and σ, the migrational period, show little geographic variation. Values for the daylight length, θ, and the fraction, Φ, of the total herbivore community which undergoes migration, change with season and latitude.

Subarctic North Pacific Case

To determine how well the model simulates the observed standing stocks of phytoplankton, zooplankton, and dissolved nutrient, equations (17)-(20) are solved using parameter values characteristic of the subarctic region. Central gyre parameter values will be used later to seek geographical differences in the solutions. Both regions are modeled for the late spring, when grazing is important (*Eppley et al.*, 1973; *Parsons and de Lange Boom*, 1974) and resupply of nutrients to the euphotic zone is nearest to quasi-balance with vertical loss terms (*Tully and Barber*, 1960; *McGowan and Williams*, 1973).

The concentration of nitrate below the euphotic zone at Ocean Station P is 15 to 20 $\mu gat\ NO_3^-\ \ell^{-1}$ (*Anderson et al.*, 1969). It is assumed N_t = 18 $\mu gat\ N\ \ell^{-1}$ for the North Pacific Ocean subarctic region. MacIsaac and Dugdale (1969) suggest the Michaelis half-saturation constant for the region is 1.0 $\mu gat\ NO_3^-\ \ell^{-1}$. Nondimensional α or k/N_t therefore equals 0.05.

Parsons and Seki (1970) suggest the release of extracellular products by healthy cells amounts to 15% or less of the carbon fixed, with higher loss rates up to 30% for stressed cells. As the phytoplankton in the subarctic gyre are stressed by light limitation, ε or r/V_m equals 0.3.

May and June mark a peak in zooplankton growth rate and biomass at Ocean Station P (*McAllister*, 1969), and thus, β or $R_m/V_m \geq 1$. The Ivlev constant is species-specific. Assuming *Calanus pacificus* is the dominant herbivorous grazer in a vertically migrating patch, the value chosen for Λ is 0.01 $(\mu g\ C/\ell)^{-1}$. Assuming a C/N ratio of 6/1 for phytoplankton (*Strickland*, 1960), Λ = 0.06 $(\mu gat\ N/\ell)^{-1}$.

Upon scaling by N_t, $\lambda = 1.08$. *C. pacificus* is reported by McAllister (1970) to have grazing thresholds > 10 µgat N ℓ^{-1}. This suggests the herbivore grazes primarily in dense phytoplankton patches. Values of $P_t \geq 0$ will be used in testing the response of the model to this controversial parameter.

Animals migrating through a 200 meter water column in the subarctic region of the North Pacific Ocean are subjected to temperatures between 7 and 20°C in summer. In the central gyre during summer, sea surface temperatures may reach 25°C, with temperatures of 10 to 15°C at 200 meters depth (Sverdrup et al., 1942). Grazing activity and higher temperatures at the surface imply a nitrogen excretion rate, Ψ_1, of near 10% of the animal's body weight per day (Corner et al., 1965; Small and Hebard, 1967). For animals at depth the rate is nearer to 2% per day (Cowey and Corner, 1966; Corner and Davies, 1971). Thus, nondimensional $\psi_1 = 0.05$ and $\psi_2 = 0.01$.

McAllister (1969) suggested an assimilation efficiency of 65% for the zooplankton population at Ocean Station P. As the remaining proportion is egested as fecal pellets, parameter $\gamma = 0.35$.

The one parameter for which there is little observational data is the loss rate of phytoplankton biomass from the euphotic zone by sinking. The depth of the thermocline (75 meters) in late spring is well below the euphotic zone depth (Anderson et al., 1969) and phytoplankton are continuously mixed into and out of the lighted zone. A best estimate is $w = 0.1$, or 18% of the plant biomass is lost through sinking daily.

For initial conditions one may use the observed springtime, standing stocks of phytoplankton (5 µgat N ℓ^{-1}), zooplankton (1 µgat N ℓ^{-1}), and dissolved nutrient (12 µgat N ℓ^{-1}) as reported by McAllister (1969) and Anderson et al., (1969) for Ocean Station P. Figure 3a shows the approach to steady state of the model where all spatial variations have been ignored. The steady state solution is $P = 0.087$ or 1.57 µgat N ℓ^{-1}, total $Z = 0.447$ or 8.04 µgat N ℓ^{-1} and $N = 0.466$ or 8.39 µgat N ℓ^{-1}.

Nutrient recycling by the grazing and excreting herbivores leads to the high concentrations of dissolved nutrient at steady state. The zooplankton biomass rapidly increases during the short period of phytoplankton availability. This behavior of the model tends to confirm the theory that zooplankton grazing controls the phytoplankton standing stock during the summer months in the subarctic region (McAllister, 1969; Parsons and de Lange Boom, 1974).

Figure 3b displays the model behavior where diel periodicity in phytoplankton growth, a time-dependent grazing stress, and vertical herbivore migration are included in the solutions. Since McAllister (1961) reports a night/day ratio of zooplankton biomass at the surface equal to 8, it is initially assumed $Z_1 = 11\%$ of 0.477, Z_2 is 8.9% of 0.447, and $\phi = 1.1$. Note that since $V_m \approx 2$ day^{-1}, $\tau = 2$ is approximately one day in Figure 3. The daily averaged, steady state standing stocks are now $P = 0.116$ or 2.09 µgat N ℓ^{-1}, $Z = 0.667$ or 12.01 µgat N ℓ^{-1} and $N = 0.217$ or 3.91

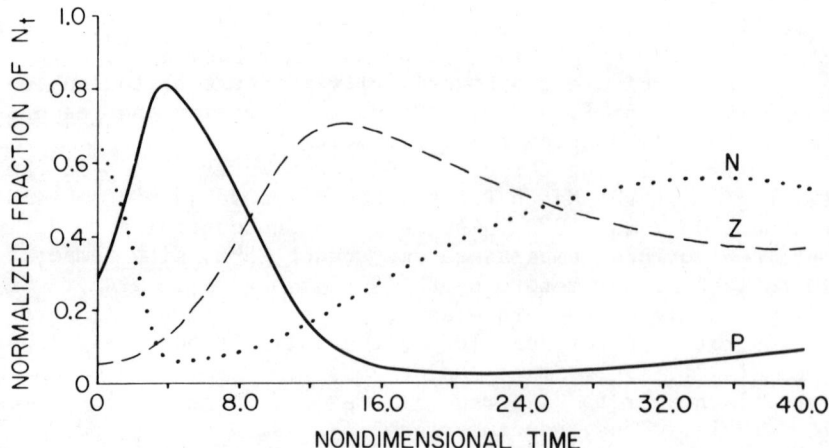

Fig. 3a. Time dependent, standing stock concentrations of phytoplankton, P (solid line), zooplankton, Z (dashed line), and nutrient, N (dotted line). The abscissa is nondimensional time ($\tau = tV_m$). Since $V_m \approx 2$ day^{-1}, one day is approximately two time units. The ordinate is the concentration of the biotic component as a fraction of the total amount of limiting nutrient in the euphotic zone, N_t. The nondimensional biological parameter values characteristic of the North Pacific subarctic gyre are $\alpha = 0.05$, $\beta = 1.0$, $\gamma = 0.35$, $\varepsilon = 0.3$, $\lambda = 1.08$, $\psi_1 = 0.05$, $\psi_2 = 0.01$, $\omega = 0.1$, and $P^* = 0$.

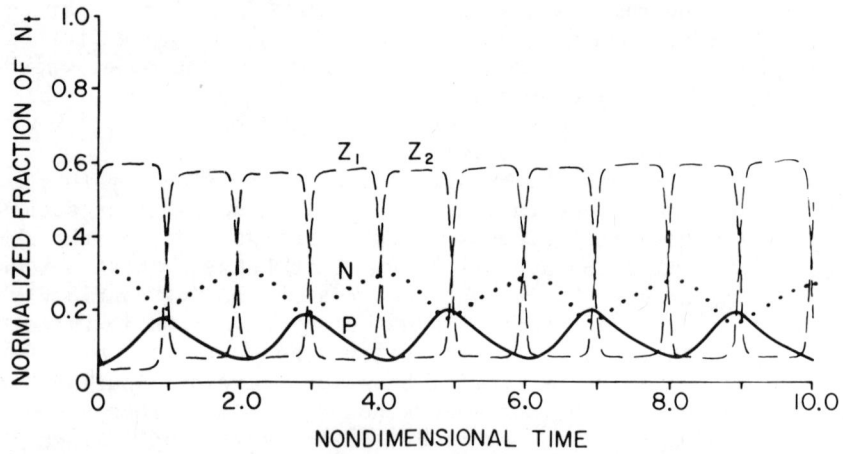

Fig. 3b. Same as Figure 3a except nutrient uptake by phytoplankton and herbivore grazing are functions of time. The zooplankton component, Z, is partitioned into grazing euphotic zone herbivores, Z_1, and nongrazing aphotic zone herbivores, Z_2. Additional nondimensional parameters are τ_1 = dawn, τ_2 = sunset, $\theta = 0.5$, $\sigma = 1.0$, and $\Phi = 1.1$.

$\mu gat\ N\ \ell^{-1}$. It is interesting to note the phytoplankton standing stock is near the reported grazing threshold for *Calanus pacificus*, although $P^* = 0$ in the model.

Nocturnal grazing has increased the steady state herbivore production by 50%. McAllister (1969) demonstrated that different grazing schemes result in large differences in secondary production. The model results support this theory.

Phytoplankton Patchiness in the Subarctic Gyre

Returning to the hypothesis that a vertically migrating patch of herbivores can impart a patchy character to the phytoplankton field, the full spatial and temporal model, equations (17)-(20), are investigated. One begins with the spatially homogeneous phytoplankton, zooplankton, and limiting nutrient background concentrations fluctuating about the steady state shown in Figure 3b. Superimposed on this background is a phytoplankton patch of 500 meters width (Figure 4a). This patch size is above the critical length providing the maximum herbivore grazing, $R_m Z_1 < 3.25 \times 10^{-4}\ \mu gat\ N\ \ell^{-1}\ sec^{-1}$, since

$$L_c = \pi \left(\frac{K_h}{b - R\Lambda}\right)^{\frac{1}{2}}$$

$$= \pi \left[\frac{125\ cm^2\ sec^{-1}}{(2 \times 10^{-5} sec^{-1}) - (3.25 \times 10^{-4} \mu gat\ N\ \ell^{-1} sec^{-1})(0.06\ \ell/\mu gat\ N)}\right]^{\frac{1}{2}}$$

$$= 500\ m.$$

A patch of herbivores 250 meters width and 72 $\mu gat\ N\ \ell^{-1}$ in biomass concentration (15 times the observed average density of zooplankton) migrates into the euphotic zone at sunset (Figure 4b). *Isaacs et al.*, (1974) present evidence that the response to light intensity of vertically migrating herbivores in the subarctic region inadvertently leads the animals to rise into areas of high plant concentration. Thus, the herbivores are allowed to rise directly into the phytoplankton patch. In one night the animals graze the phytoplankton to near extinction (Figure 4a). As they graze, the herbivores excrete metabolites and egest fecal pellets (Figure 4c). Diffusion smooths the **phytoplankton and nutrient** gradients and prevents the local extinction of phytoplankton in the heavily grazed area (Figures 4a and c).

At dawn, the herbivore patch migrates out of the euphotic zone (Figure 4b) and the phytoplankton begin to take up nutrients. As phytoplankton diffuse into the void created by grazing (Figure 4a), they rapidly utilize the local pool of excreted and egested nutrient. There is little uptake in the center of the pool (Figure 4c) as the phytoplankton population there remains small.

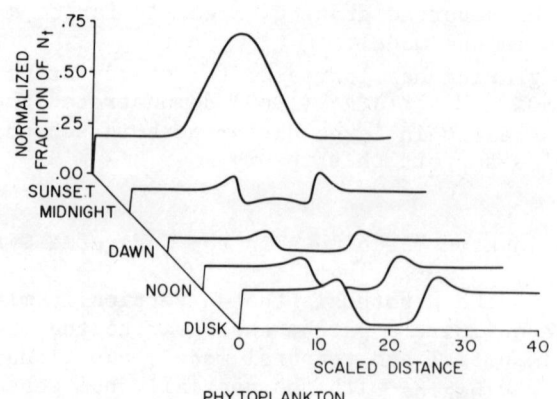

Fig. 4. Time dependent, spatial solution of equations (17)-(20) using subarctic gyre parameters values. Each nondimensional spatial unit $x = x' (K_h/V_m)^{1/2}$ or 25 meters.
 a. Spatial distribution and concentration of phytoplankton with time. The stresses upon the phytoplankton patch are nocturnal grazing by vertically migrating zooplankton, diel periodicity in phytoplankton growth, and turbulent diffusion.

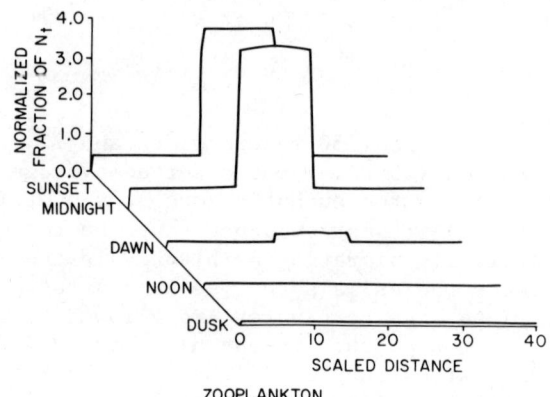

Fig. 4b. Spatial distribution and concentration of herbivores in the euphotic zone with time. A vertically migrating zooplankton patch 200 meters in diameter with a maximum concentration of $4N_t$ or 72 μgat N ℓ$^{-1}$ rises at sunset and descends from the euphotic zone at dawn. A low background concentration of continuously grazing herbivores is always present.

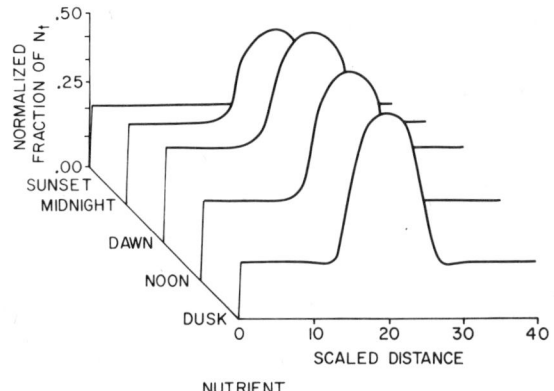

Fig. 4c. Spatial distribution and concentration of dissolved limiting nutrient (nitrogen) excreted as metabolic products and egested as fecal pellets by herbivores grazing in the euphotic zone. The concentration builds with time in the locality of the zooplankton patch until dawn. After the zooplankton descend, uptake by phytoplankton decreases the concentration.

This case suggests phytoplankton can be grazed to near extinction in one night if the herbivore biomass and grazing rate is sufficiently high. Had grazing been discontinued at a threshold level of phytoplankton, that threshold would be the phytoplankton concentration surviving. Steele (1974) and Walsh (1976) discuss the importance of P_t in preventing overgrazing of the primary producers.

The author has used no grazing threshold to demonstrate the role of diffusion in keeping a seed population of plants continually present. The complete local disappearance of phytoplankton is possible only if $P = 0$ everywhere.

North Pacific Central Gyre Case

The same dynamics are used to describe the phytoplankton-herbivore-nutrient relationships in the central gyre. The geographical difference in the value of N_t alters several parameter values, while others scaled by V_m remain the same.

Due to slow vertical mixing (Anderson et al., 1969), and the low concentrations of limiting nutrient below the euphotic zone (Eppley et al., 1973), N_t for the central gyre is only 3.25 µgat N ℓ$^{-1}$. McIsaac and Dugdale (1969) report a Michaelis constant of 0.2 µgat NO_3^- ℓ$^{-1}$ in this oligotrophic area. Alpha or k/N_t then equals 0.06, little changed from its subarctic region value of 0.05.

Phytoplankton cells in the central gyre are stressed by nutrient, rather than light limitation. Extracellular release remains high. The author assumes ε equals 0.15 (i.e., 14% of the nutrient taken up in one doubling time is lost through the cell walls). Because of slow vertical mixing, phytoplankton losses due to cell sinking might be thought considerable. However, Smayda (1970) suggests sinking by nutrient stressed phytoplankton puts the cells in contact with high nutrient concentrations near the bottom of the euphotic zone. Anderson (1969) reports a subsurface chlorophyll maximum in the North Pacific central gyre. Nutrients which diffuse above the permanent halocline are rapidly utilized at the chlorophyll maximum which occurs between 120 and 150 m during the summer (Eppley et al., 1973). In light of this benefit, phytoplankton losses from cell sinking are regarded as nil (i.e., ω is assumed zero).

Eppley et al. (1973) suggest grazing limits phytoplankton production below that possible with the nutrient available. This implies β or $R_m/V_m > 1$. The herbivores of this oligotrophic region are most likely adapted to grazing on low concentrations of phytoplankton. Phytoplankton patch concentrations in the central gyre are seldom more than several times the background concentration of < 1 µgat N ℓ^{-1} (Venrick, 1972; Eppley et al., 1973). If the Ivlev constant, Λ, for the herbivore population were 0.2 (µgat N/ℓ)$^{-1}$, the plant concentration at which grazing were 2/3 the maximum would be 5 µgat N ℓ^{-1}. Scaling by N_t, nondimensional λ or Λ/N_t becomes 0.65. The grazing threshold for this region would expectedly be near the lowest observed concentration of phytoplankton, 0.2 µgat N ℓ^{-1} (Eppley et al., 1973). Nondimensional P^* or P_t/N_t is then ≤ 0.06. Metabolic excretion parameters, ψ_1 and ψ_2, and the assimilation efficiency, γ, are unchanged for lack of data to the contrary.

The spatially averaged, steady state solution of equations (17)-(20) using these parameter values is $P = 0.123$ or 0.40 µgat N ℓ^{-1}, $Z = 0.773$ or 2.5 µgat N ℓ^{-1}, and $N = 0.104$ or 0.34 µgat N ℓ^{-1} (Figure 5a). These numbers compare quite well with observational data.

Eppley et al. (1973) report a night/day ratio of surface herbivore biomass equal to 2 for the central gyre. Figure 5b displays the steady state solution where half the zooplankton migrate vertically and phytoplankton exhibit a diel periodicity in growth rate. The new solution shows a 4% increase in herbivore production, with little change in the daily averaged phytoplankton and dissolved nutrient concentrations.

Phytoplankton Patchiness in the Central Gyre

The central gyre differs from the subarctic region in the magnitude of phytoplankton patchiness (Venrick, 1972). A homogeneous phytoplankton field is used as an initial condition in Figure 6a. The initial phytoplankton, nutrient, and herbivore concentrations

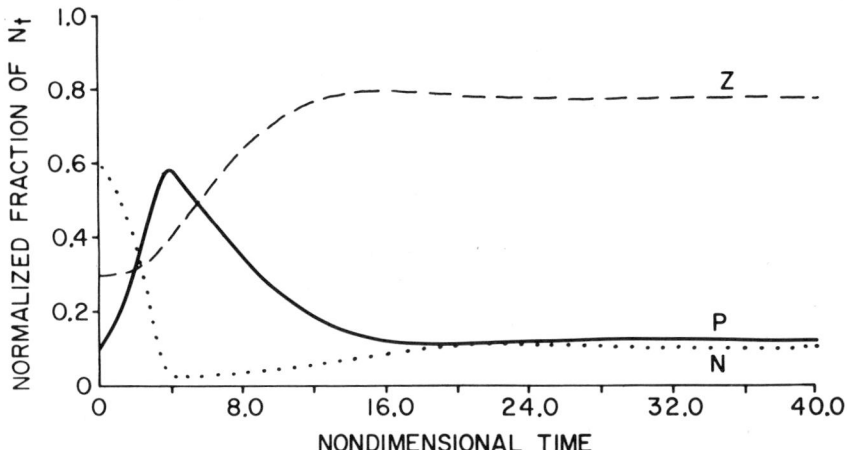

Fig. 5a. Same as Figure 3a except the nondimensional biological parameters are characteristic of the North Pacific Ocean central gyre: $\alpha = 0.06$, $\beta = 1.0$, $\gamma = 0.35$, $\varepsilon = 0.15$, $\lambda = 0.65$, $\psi_1 = 0.05$, $\psi_2 = 0.01$, $\omega = 0$, and $P^* = 0$. One day is approximately two nondimensional time units.

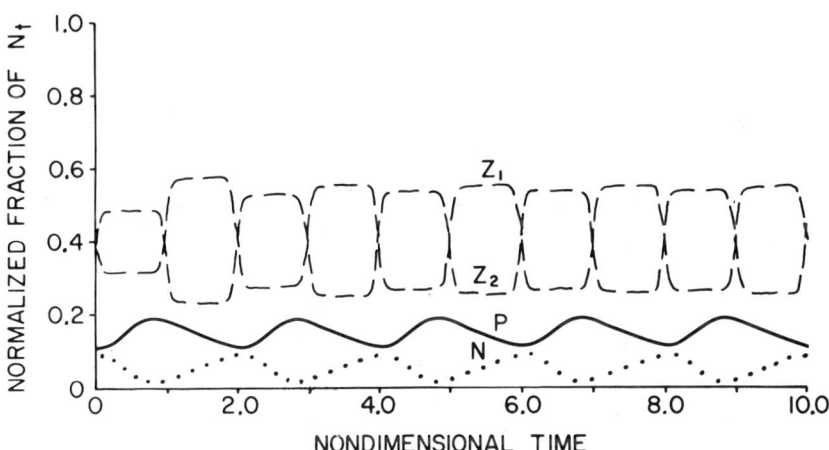

Fig. 5b. Same as Figure 3b except the biological parameters are characteristic of the central gyre and $\Phi = 0.4$, whereby the night/day ratio of euphotic zone herbivore biomass equals two.

fluctuate about the steady state shown in Figure 5b. At sunset a patch of herbivores 250 meters in length and 8.1 µgat N ℓ^{-1} in biomass concentration (fifteen times the observed, average density of zooplankton) rises to the surface (Figure 6b).

The herbivore density and the grazing stress ($\beta = 0.5$) are such that the phytoplankton are not extensively grazed as in the subarctic region case. Phytoplankton surviving in the grazed area utilize the herbivore excreted and egested nutrient (Figure 6c) the following daylight period. Growth is so rapid that the plants attain a concentration at day's end higher than the background concentration (Figure 6a). This is an important result. The recycling of nutrients from decomposing fecal pellets and the metabolic excretion of zooplankton can significantly alter the nutrient chemistry of oligotrophic waters within hours. Nutrient-limited phytoplankton can respond to this chemical anomaly with a potentially measurable increase in production the following day. Dissipation of the resulting bloom by turbulent diffusion will not be completed for several days if the anomaly is of large enough scale. *Beers and Kelley* (1965) attributed short term variation in ammonia concentration in the euphotic zone of the oligotrophic Sargasso Sea to metabolic excretion by vertically migrating animals at night and ammonia uptake by plants during the day. The model simulates this phenomenon nicely.

Patchiness Dynamics with Horizontally Migrating Herbivores

Finally, the theoretical consequences of horizontal migration by grazers on phytoplankton patchiness is explored. *Bainbridge* (1949) and more recently *Kittredge et al.* (1974) demonstrated that marine copepods employing a cycloid swimming pattern can follow a chemical gradient. Marine herbivores may migrate toward a maximum in phytoplankton concentration by following the gradient of plant cell products. Migration toward a food source would be of particular advantage to continuous grazers which lack the mechanism by which vertically migrating herbivores position themselves in phytoplankton rich areas.

To formulate gradient-orientated horizontal migration, one must first define u_s, the swimming speed of the animals. If one scales u_s by $(K_h V_m)^{-\frac{1}{2}}$, the quantity becomes dimensionless. One can write the horizontal migration term as

$$\frac{\partial Z'_1}{\partial \tau} = -u'_s \frac{\partial Z'_1}{\partial x'} \qquad (21)$$

where primes again denote dimensionless variables.

As swimming activity may be a function of food availability, one formulates the instantaneous velocity u'_s as a function of the

VERTICALLY MIGRATING HERBIVOROUS PLANKTON

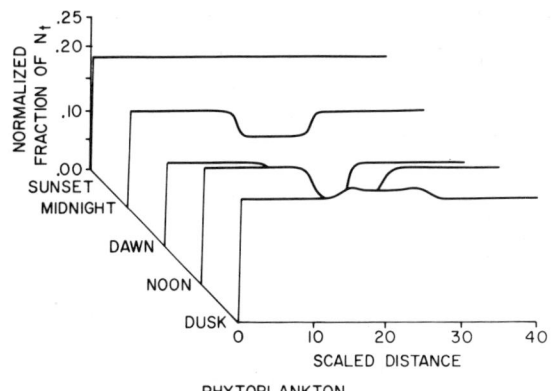

Fig. 6. Time dependent, spatial solution of equations (17)-(20) using central gyre parameter values. Each spatial unit is 25 meters.
   a. Spatial distribution and concentration of phytoplankton with time. The initial homogeneous phytoplankton concentration is locally grazed down by herbivores. However, the herbivore excreted nutrients leads to a phytoplankton patch in the same area the following day.

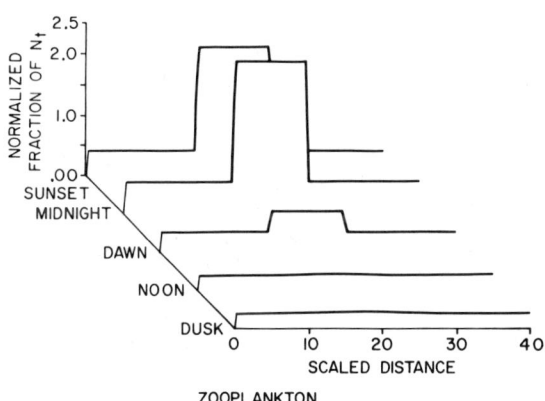

Fig. 6b. Spatial distribution and concentration of herbivores in the euphotic zone with time. The vertically migrating zooplankton patch is 250 meters in width and has a maximum concentration of 8.1 µgat N ℓ^{-1}. The ratio R_m/V_m equals 0.5.

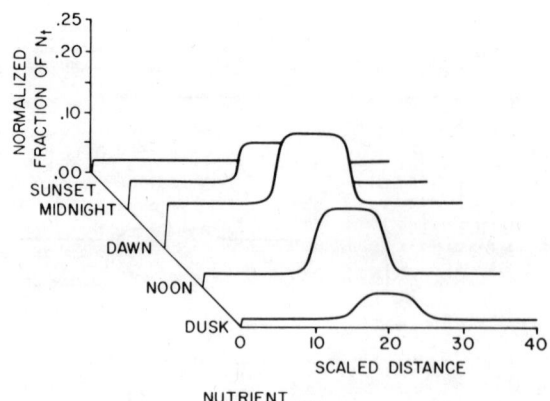

Fig. 6c. Spatial distribution and concentration of herbivore excreted and egested nutrient in the euphotic zone with time. The local concentration peaks at dawn and decreases during the day as phytoplankton utilize the nutrient.

maximum swimming speed, u'_m, and the concentration of phytoplankton, P'. Let

$$u'_s = u'_m \exp(-\zeta/\rho), \qquad (22)$$

where ζ is the previously defined Ivlev grazing rate and ρ is the ingestion rate at which ζ falls to one third the maximum grazing rate, β. Thus,

$$u'_s = u'_m \exp(-3[1-\exp(-\lambda P' + \lambda P^*)]) \qquad (23)$$

such that u'_s equals u'_m when $P' = P^*$ and u'_s is near zero when $P' \gg P^*$.

While several vertically migrating zooplankton species attain swimming speeds of the order 1 cm sec^{-1} (Banse, 1964), progress may be considerably slower when following a chemical gradient (Kittredge et al., 1974). Velocities at which herbivores migrate toward a food source are more likely of the order 0.1 cm sec^{-1}. As an upper bound, let $u_m = 1.25$ cm sec^{-1}. Choosing $K_h = 125$ cm^2 sec^{-1} and $V_m = 2 \times 10^{-5}$ sec^{-1},

$$u'_m = \frac{1.25\ cm\ sec^{-1}}{[(125\ cm^2\ sec^{-1})(2 \times 10^{-5}\ sec^{-1})]^{\frac{1}{2}}}$$

$$= 25.$$

The temporal and spatial scales relevant here are again hours and meters. Figure 7a depicts a phytoplankton patch of 200 meters width located in an area of high nutrient concentration. On either side of the patch are areas high in herbivore concentration. Assuming the animals no longer maintain a specific patch concentration in their search for food, the zooplankton are diffused in the same manner as phytoplankton and nutrients. The animals excrete metabolites and continually graze on available phytoplankton but do not vertically migrate. Diel periodicity in phytoplankton growth is omitted for clarity of solutions.

With initially ample nutrient for growth the phytoplankton patch grows despite losses due to grazing and turbulent diffusion (Figures 7b and c). Thus, $L > L_c$. Then herbivores migrate into the locality of the phytoplankton patch (Figures 7b and d). The zooplankton concentration increases quickly near the patch boundaries where the gradient in P' is greatest and the swimming speed is near u'_m.

After 1.5 doubling times, or 21 hours, the grazing stress and the depletion of nutrients has altered the critical length such that now $L < L_c$, and the phytoplankton patch begins to decay (Figure 7d).

These results suggest that horizontal herbivore migration can significantly affect the dynamics of small scale phytoplankton patchiness. The conclusion is the same using either North Pacific Ocean central gyre or subarctic region parameter values. Bainbridge (1953) recognized these results qualitatively.

SUMMARY

The author has attempted to model the dynamics of one dimensional, small scale (<1 km) phytoplankton patchiness in the open ocean. The ability of vertically migrating herbivores to create spatial inhomogeneities in the phytoplankton standing stock has been explored numerically. The simulations suggest the following:

1) Herbivore patches of size and concentration often observed in nature can rapidly reduce phytoplankton standing stocks to the grazing threshold. Diffusion will supply a seed phytoplankton population to the grazed area for eventual recovery, even if no grazing threshold exists.

2) In oligotrophic areas herbivores can significantly alter the dissolved nutrient chemistry in one night's activity through excretion of metabolites and egestion of fecal pellets. The plants may so flourish the next day on these regenerated nutrients that a phytoplankton patch may form in the grazed area.

3) Migration by herbivores in the direction of a positive food gradient can significantly increase the grazing stress on a phytoplankton patch < 200 meters in size within a short time. Horizontal migration has a negligible effect on

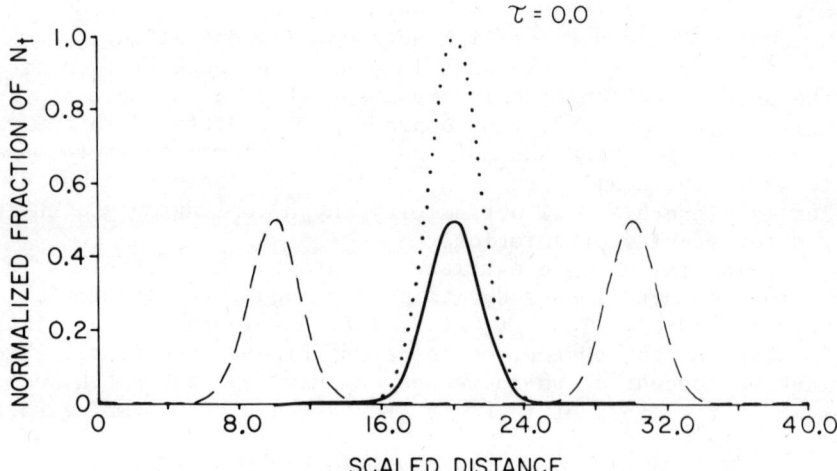

Fig. 7. Time dependent, spatial solution of equations (17)-(20) incorporating the formulation of horizontal migration of herbivores in the direction of a positive phytoplankton gradient. The swimming speed of the zooplankton is a function of food availability (see text for details).

a. The solid line delineates the initial phytoplankton patch. The dotted line represents the concentration and distribution of the limiting nutrient. The dashed line describes the herbivore concentration and distribution. Each nondimensional spatial unit is 25 meters.

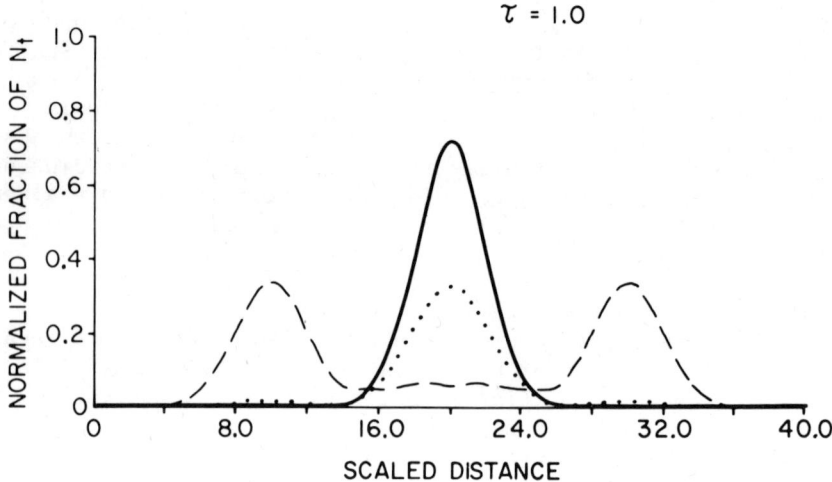

Fig 7b. The solution at $\tau = 1.0$ or 14 hours elapsed time.

VERTICALLY MIGRATING HERBIVOROUS PLANKTON

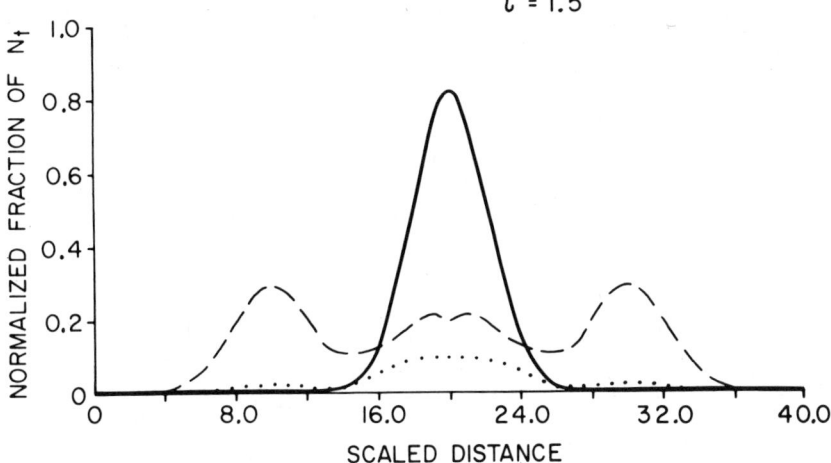

Fig. 7c. The solution at $\tau = 1.5$ or 21 hours elapsed time.

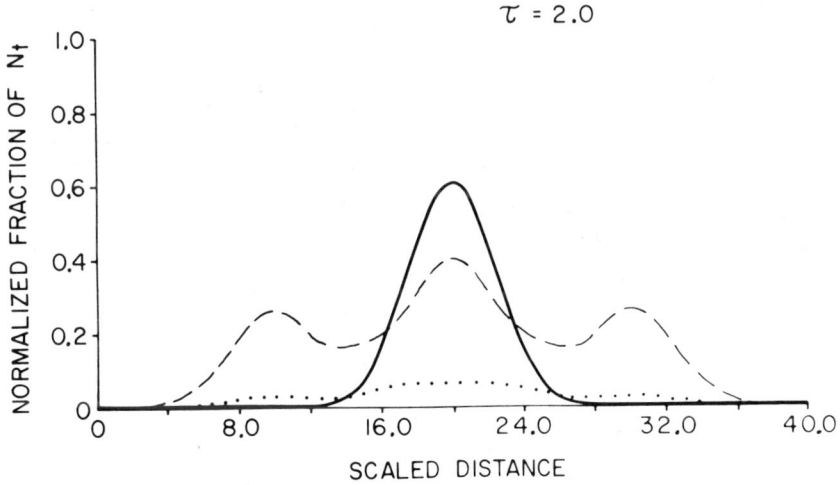

Fig. 7d. The solution at $\tau = 2.0$ or 28 hours elapsed time.

patches of longer length scale due to the slow progression speed of most zooplankton when following a chemical gradient.

However, one should be cautious in applying conclusions based on one-dimensional patchiness theory to actual oceanic phenomena. In the real ocean the dynamics of plankton patchiness are much more complex than considered here.

ACKNOWLEDGEMENTS

This is a contribution to the Coastal Upwelling Ecosystems Analysis Program, sponsored by the National Science Foundation. Support for this research has been provided by the Office of Naval Research contract N00014-75-C-0201 and NSF Grant GA-43265. Computing time was provided by the Florida State University Computing Center.

I am indebted to Drs. J. O'Brien, G. Knauer, J. Beers, P. Wiebe, J. Hirota, and R. Iverson for stimulating discussions.

This is contribution No. 118 of the Geophysical Fluid Dynamics Institute, Tallahassee.

REFERENCES

Anderson, G. C. 1969. Subsurface chlorophyll maximum in the northeast Pacific Ocean. *Limnol. Oceanogr.*, 14: 386-391.

Anderson, G. C., T. R. Parsons, and K. Stephens. 1969. Nitrate distribution in the subarctic Northeast Pacific Ocean. *Deep-Sea Res.*, 16: 329-334.

Bainbridge, R. 1949. Movement of zooplankton in diatom gradients. *Nature*, 163: 910-911.

Bainbridge, R. 1953. Studies on the interrelationships of zooplankton and phytoplankton. *J. Mar. Biol. Ass. U.K.*, 32: 385-447.

Banse, K. 1964. On the vertical distribution of zooplankton in the sea, In: M. Sears (Ed.) *Progress in Oceanography, Vol. 2*, New York, MacMillan Co. 55-125.

Bary, B. McK. and R. E. Pieper. 1970. Sonic-scattering studies in Saanich Inlet, British Columbia: a preliminary report, In: G. G. Farquhar (Ed.), *Proceedings of an International Symposium on Biological Sound Scattering in the Ocean.* Washington, U. S. Gov. Printing Office. 601-609.

Barraclough, W. E., R. J. LeBrasseur, and O. D. Kennedy. 1969. Shallow scattering in the subarctic Pacific Ocean: detection by high-frequency echo sounder. *Science*, 166: 611-613.

Beamish, P. 1971. Quantitative measurements of acoustic scattering from zooplankton organisms. *Deep-Sea Res.*, 18: 811-822.

Beers, J. R. and A. C. Kelley. 1965. Short term variation of ammonia in the Sargasso Sea off Bermuda. *Deep-Sea Res.*, 12: 21-25.

Castile, B. D. 1975. Reverberation from plankton at 330 kHz in the Western Pacific. *J. Acoust. Soc. Am.*, 58: 972-976.

Clarke, G. L., G. C. Ewing, and C. J. Lorenzen. 1970. Spectra of backscattered light from the sea obtained from aircraft as a measure of chlorophyll concentration. *Science, 167*: 1119-1121.

Conover, R. 1966. Factors affecting the assimilation of organic matter by zooplankton and the question of superfluous feeding. *Limnol. Oceanogr.*, 11: 346-354.

Corner, E. D. S. and A. G. Davies. 1971. Plankton as a factor in the nitrogen and phosphorus cycles in the sea, In: F. S. Russell and M. Yonge (Eds.), *Advances in Marine Biology*, 9: 101-204.

Corner, E. D. S., C. B. Cowey, and S. M. Marshall. 1965. On the nutrition and metabolism of zooplankton, III, Nitrogen excretion by *Calanus*. *J. Mar. Biol. Ass. U. K.*, 47: 259-270.

Cowey, C. B. and E. D. S. Corner. 1966. Amino acids and some other nitrogenous compounds in *Calanus finmarchicus*. *J. Mar. Biol. Ass. U. K.*, 43: 485-493.

Cushing, D. H. 1963. Studies on a *Calanus* patch, I, The identification of a *Calanus* patch. *J. Mar. Biol. Ass. U. K.*, 43: 327-337.

Denman, K. L. and T. Platt. 1975. Coherences in the horizontal distributions of phytoplankton and temperature in the upper ocean. *Mem. Soc. Roy. Sci. de Liège*, 7: 19-30.

Dugdale, R. C. 1967. Nutrient limitations in the sea: Dynamics, identification, and significance. *Limnol. Oceanogr.*, 12: 685-695.

Dugdale, R. C. and J. J. Goering. 1967. Uptake of new and regenerated forms of nitrogen in primary productivity. *Limnol. Oceanogr.*, 12: 196-206.

Eppley, R. W. 1972. Temperature and phytoplankton growth in the sea. *Fish. Bull.*, 70: 1063-1085.

Eppley, R. W., E. H. Renger, E. L. Venrick, and M. M. Mullin. 1973. A study of plankton dynamics and nutrient cycling in the central gyre of the North Pacific Ocean. *Limnol. Oceanogr.*, 18: 534-551.

Fofonoff, N. P. 1962. Dynamics of ocean currents, In: M. N. Hill (Ed.), *The Sea, Vol. 1*, New York, Interscience. 323-395.

Frost, B. W. 1974. Feeding processes at lower trophic levels in pelagic communities, In: C. B. Miller (Ed.), *The Biology of the Oceanic Pacific*, Corvallis, Oregon State Univ. Press. 59-77.

Goering, J. J., R. C. Dugdale, and D. W. Menzel. 1964. Cyclic diurnal variations in the uptake of ammonia and nitrate by photosynthetic organisms in the Sargasso Sea. *Limnol. Oceanogr.*, 9: 448-451.

Isaacs, J. D., S. A. Tont, and G. L. Wick. 1974. Deep scattering layers: Vertical migration as a tactic for finding food. *Deep-Sea Res.*, 21: 651-656.

Kawamura, A. 1974. Food and feeding ecology in the southern Sei whale. *Sci. Rep. Whales Res. Inst.*, 26: 25-144.

Kierstead, H. and L. B. Slobodkin. 1953. The size of water masses containing plankton blooms. *J. Mar. Res.*, *12*: 141-147.

Kittredge, J. S., F. T. Takahashi, J. Lindsey, and R. Lasker. 1974. Chemical signals in the sea. *Fish. Bull.*, *72*: 1-11.

Komaki, Y. 1967. On the surface swarming of euphausiid crustaceans. *Pacif. Sci.*, *21*: 433-448.

McAllister, C. D. 1961. Zooplankton studies at ocean weather station "P" in the northeast Pacific Ocean. *J. Fish. Res. Bd. Canada*, *18*: 1-29.

McAllister, C. D. 1969. Aspects of estimating zooplankton production from phytoplankton production. *J. Fish. Res. Bd. Canada*, *26*: 199-220.

McAllister, C. D. 1970. Zooplankton rations, phytoplankton mortality, and the estimation of marine production, In: J. H. Steele (Ed.), *Marine Food Chains*. Edinburgh, Oliver and Boyd. 419-457.

McGowan, J. 1974. The nature of oceanic ecosystems, In: C. B. Miller (Ed.), *The Biology of the Oceanic Pacific*. Corvallis, Oregon State Univ. Press. 9-28.

McGowan, J. A. and P. M. Williams. 1973. Oceanic habitat differences in the north Pacific. *J. Exp. Mar. Biol. Ecol.*, *12*: 187-217.

MacIsaac, J. J. and R. C. Dugdale. 1969. The kinetics of nitrate and ammonia uptake by natural populations of marine phytoplankton. *Deep-Sea Res.*, *16*: 45-57.

Mauchline, J. and L. R. Fisher. 1969. The biology of euphausiids, In: F. S. Russell and M. Yonge (Eds.), *Advances in Marine Biology*. Academic Press, London. 454 pp.

O'Brien, J. J. and J. S. Wroblewski. 1973. On advection in phytoplankton models. *J. Theor. Biol.*, *38*: 197-202.

Okubo, A. 1971. Oceanic diffusion diagrams. *Deep-Sea Res.*, *18*: 789-802.

Parsons, T. R. and B. R. de Lange Boom. 1974. The control of ecosystem processes in the sea, In: C. B. Miller (Ed.), *The Biology of the Oceanic Pacific*. Corvallis, Oregon State Univ. Press. 29-57.

Parsons, T. R. and H. Seki. 1970. Importance and general implications of organic matter in aquatic environments, In: D. W. Hood (Ed.), *Organic Matter in Natural Waters*. Institute of Marine Science, Occ. Publ. No. 1. 1-27.

Parsons, T. R., R. J. LeBrasseur, and J. D. Fulton. 1967. Some observations on the dependence of zooplankton grazing on cell size and concentration of phytoplankton blooms. *J. Oceanogr. Soc. Japan*, *23*: 10-17.

Platt, T. 1972a. The feasibility of mapping the chlorophyll distribution in the Gulf of St. Lawrence. *Fish. Res. Bd. Canada Tech. Rep. #332*. 8 pp.

Platt, T. 1972b. Local phytoplankton abundance and turbulence. *Deep-Sea Res.*, *19*: 18-187.

Platt, T. and K. L. Denman. 1975. A general equation for the mesoscale distribution of plankton in the sea. Mem. Soc. Roy. Sci. de Liège, 7: 31-42.

Ponomareva, L. A. 1966. Quantitative distribution of euphausiids in the Pacific Ocean. Dokl. Akad. Navk. SSSR, 6: 690-692.

Reid, J. L., Jr. 1961. On the geostrophic flow at the surface of the Pacific Ocean with respect to the 1,000-decibar surface. Tellus, 13: 489-502.

Small, L. F. and J. F. Hebard. 1967. Respiration of a vertically migrating marine crustacean Euphausia pacifica Hansen. Limnol. Oceanogr., 12: 272-280.

Smayda, T. J. 1970. The suspension and sinking of phytoplankton in the sea. Oceanogr. Mar. Biol. Ann. Rev., 8: 353-414.

Steele, J. H. 1958. Plant production in the Northern North Sea. Mar. Res. Scot., No. 7.

Steele, J. H. 1974. The structure of marine ecosystems. Harvard Univ. Press. 128 pp.

Stommel, H. 1949. Horizontal diffusion due to oceanic turbulence. J. Mar. Res., 8: 199.

Strickland, J. D. H. 1960. Measuring the production of marine phytoplankton. J. Fish. Res. Bd. Canada Bull., 122: 172 pp.

Sverdrup, H. V., M. W. Johnson, and R. H. Fleming. 1942. The oceans, their physics, chemistry, and general biology. Prentice Hall, Englewood Cliffs.

Tully, J. P. and F. G. Barber. 1960. An estuarine analogy in the subarctic Pacific Ocean. J. Fish. Res. Bd. Canada, 17: 91-112.

Venrick, E. L. 1972. Small-scale distributions of oceanic diatoms. Fish. Bull., 70: 363-372.

Walsh, J. J. 1972. Implications of a systems approach to oceanography. Science, 176: 969-975.

Walsh, J. J. 1976. A biological sketchbook for an eastern boundary current, In: J. H. Steele, J. J. O'Brien, E. D. Goldberg, and I. N. McCave (Eds.), The Sea, Vol. 6. Wiley Interscience, New York (in press).

Wiebe, P. H. 1970. Small-scale spatial distribution in oceanic zooplankton. Limnol. Oceanogr., 15: 205-217.

Wroblewski, J. S. and J. J. O'Brien. 1976. A spatial model of phytoplankton patchiness. Mar. Biol., 35: 161-175.

Wroblewski, J. S., J. J. O'Brien, and T. Platt. 1975. On the physical and biological scales of phytoplankton patchiness in the ocean. Mem. Soc. Roy. Sci. de Liège, 7: 31-42.

APPENDIX

Definition of Symbols and Scaling Relationships

Dimensional Quantity	Definition	Scaling Factor	Nondimensional Quantity
k	Half-saturation constant	k/N_t	α
K_h	Horizontal eddy diffusivity coefficient	-	-
N	Limiting nutrient concentration in the euphotic zone	N/N_t	N'
N_t	Total amount of biologically limiting nutrient in the system	-	-
P	Phytoplankton biomass	P/N_t	P'
P_t	Herbivore grazing threshold	P_t/N_t	P^*
r	Phytoplankton extracellular release coefficient	r/V_m	ε
R_m	Herbivore maximum grazing ration	R_m/V_m	β
t	Time	tV_m	τ
t_1	Dawn	-	τ_1
t_2	Sunset	-	τ_2
u_s	Herbivore horizontal swimming velocity	$u_s/(K_h V_m)^{1/2}$	u_s'
V_m	Phytoplankton maximum specific growth rate	-	-
w_s	Phytoplankton sinking coefficient	w_s/V_m	ω
x	Distance	$x/(K_h/V_m)^{1/2}$	x'
Z_1	Herbivore biomass in the euphotic zone	Z_1/N_t	Z_1'

Dimensional Quantity	Definition	Scaling Factor	Nondimensional Quantity
Z_2	Herbivore biomass in the aphotic zone	Z_2/N_t	Z'_2
γ	Fraction of grazing ration egested	-	-
θ	Daytime fraction of a nondimensional day	-	-
Λ	Ivlev constant	ΛN_t	λ
σ	Fraction of a nondimensional day over which herbivore migration takes place	-	-
Φ	The fraction of the total herbivore biomass which undergoes migration	-	-
Ψ_1	Metabolic excretion coefficient of herbivores in euphotic zone	Ψ_1/V_m	ψ_1
Ψ_2	Metabolic excretion coefficient of herbivores in aphotic zone	Ψ_2/V_m	ψ_2

LIST OF CONTRIBUTORS AND PARTICIPANTS

NEIL R. ANDERSEN, Marine Chemistry, National Science Foundation, Washington, D. C. 20550
MARTIN V. ANGEL, Institute of Oceanographic Sciences, Brook Road, Wormley, Godalming, Surrey, GU8 5UB, England
RICHARD BACKUS, Woods Hole Oceanographic Institution, Woods Hole, Massachusetts 02543
JULIAN BADCOCK, Institute of Oceanographic Sciences, Brook Road, Wormley, Godalming, Surrey, GU8 5UB, England
RONALD C. BAIRD, Geo-Marine, 777 South Central Expressway, Suite 2G, Richardson, Texas 75080
RICHARD T. BARBER, Duke University Marine Laboratory, Beaufort, North Carolina 28516
HUGO BEZDEK, Naval Ocean Research and Development Activity, Code 460, Bay St. Louis, Mississippi 39520
MAURICE BLACKBURN, Scripps Institution of Oceanography, La Jolla, California 92037
ALBERT L. BROOKS, New London Laboratory - Ft. Trumbull, Naval Underwater Systems Center, New London, Connecticut 06320
CHARLES F. L. BROWN, Oceanography, Naval Underwater Systems Center, New London Laboratory, New London, Connecticut 06320
JAMES CHILDRESS, Biology Department, University of California, Santa Barbara, California 93106
THOMAS A. CLARKE, Hawaii Institute of Marine Biology, University of Hawaii, Kaneohe, Hawaii 96744
JAMES E. CRADDOCK, Woods Hole Oceanographic Institution, Woods Hole, Massachusetts 02543
KENNETH L. DENMAN, Biological Oceanography, Marine Ecology Laboratory, Bedford Institute of Oceanography, Dartmouth, Nova Scotia, Canada B24 4A2
A. DEVOL, Department of Oceanography, WB-10, University of Washington, Seattle, Washington 98195
RICHARD C. DUGDALE, Bigelow Laboratory of Ocean Sciences, McKnown Point, W. Boothbay Harbor, Maine 04575
DAVID DYRSSEN, University of Gothenburg, FACK CTH/GU, S-402 20 Goteborg 5, Sweden
T. SAUNDERS ENGLISH, Department of Oceanography, University of Washington, Seattle, Washington 98195
G. BROOKE FARQUHAR, U. S. Naval Oceanographic Office, Washington, D. C. 20373

WILLIAM A. FRIEDL, Undersea Sciences Department, Naval Undersea Center, San Diego, California 92132

ROBERT B. GAGOSIAN, Woods Hole Oceanographic Institution, Woods Hole, Massachusetts 02543

ROBERT H. GIBBS, JR., Department of Vertebrate Zoology, Smithsonian Institution, Washington, D. C. 20560

THOMAS G. HALLAM, Department of Mathematics, Florida State University, Tallahassee, Florida 32306

ERICH HOCHBERG, Santa Barbara Museum of Natural History, 2559 Puesta Del Sol Road, Santa Barbara, California

VANCE HOLLIDAY, TRACOR San Diego Laboratory, TRACOR, Inc., 2923 Canon Street, San Diego, California 92016

OSMUND HOLM-HANSEN, Institute of Marine Resources, University of California, San Diego, La Jolla, California 92037

THOMAS L. HOPKINS, Department of Marine Science, University of South Florida, St. Petersburg, Florida 33701

RICHARD IVERSON, Department of Oceanography, Florida State University, Tallahassee, Florida 32306

RICHARD K. JOHNSON, School of Oceanography, Oregon State University, Corvallis, Oregon 97331

CHARLES KARNELLA, Division of Fishes, U. S. National Museum of Natural History, Washington, D. C. 20560

JOHANNES KINZER, Institut fur Meereskunde an der, Universitat Kiel, D-23 Kiel, Dusternbrooker Weg 20, Federal Republic of Germany

R. R. LEVIN, Office of Naval Research, 1030 East Green Street, Pasadena, California 91106

ALLAN G. LEWIS, Institute of Oceanography, University of British Columbia, Vancouver, B.C., Canada V6T 1W5

JOHN LOEFER, Office of Naval Research, 1030 East Green Street, Pasadena, California 91106

RICHARD LOVE, U. S. Naval Oceanographic Office, Washington, D. C. 20373

JOHN MAUCHLINE, Dunstaffnage Marine Research Laboratory, P.O. Box 3, Oban, Argyll, Scotland PA34 4AD

LAWRENCE E. McCRONE, Department of Oceanography, University of Washington, Seattle, Washington 98195

PAUL McELROY, Bolt, Beranek and Newman, 50 Moulton Street, Cambridge, Massachusetts 02138

JOHN A. McGOWAN, Scripps Institution of Oceanography, La Jolla, California 92037

JAMES C. McWILLIAMS, National Center for Atmospheric Research, P.O. Box 3000, Boulder, Colorado 80303

N. R. MERRETT, Institute of Oceanographic Sciences, Brook Road, Wormley, Godalming, Surrey, GU8 5UB, England

H. J. MINAS, Centre Universitaire de Luminy, Marseille, France

RICHARD Y. MORITA, Department of Microbiology, Oregon State University, Corvallis, Oregon 97331

TAKAHISA NEMOTO, Ocean Research Institute, University of Tokyo, 1-15 Nakano, Tokyo, Japan

LIST OF CONTRIBUTORS

JAMES J. O'BRIEN, Florida State University, Tallahassee, Florida 32306

T. OWENS, Department of Oceanography, WB-10, University of Washington, Seattle, Washington 98195

THEODORE T. PACKARD, Department of Oceanography, University of Washington, Seattle, Washington 98195

WILLIAM G. PEARCY, School of Oceanography, Oregon State University, Corvallis, Oregon 97331

GEORGE V. PICKWELL, Naval Undersea Center, San Diego, California 92132

RICHARD E. PIEPER, Allan Hancock Foundation and Department of Biological Sciences, University of Southern California, Los Angeles, California 90007

TREVOR PLATT, Biological Oceanography, Marine Ecology Laboratory, Bedford Institute of Oceanography, Dartmouth, Nova Scotia, Canada B24 4A2

THOMAS POWELL, Division of Environmental Studies, University of California, Davis, Davis, California 95616

JOSEPH L. REID, Scripps Institution of Oceanography, La Jolla, California 92093

FRANCIS A. RICHARDS, Department of Oceanography, University of Washington, Seattle, Washington 98195

PETER RICHERSON, Division of Environmental Studies, University of California, Davis, Davis, California 95616

GUNNAR I. RODEN, Department of Oceanography, University of Washington, Seattle, Washington 98195

PAUL SCULLY-POWER, RAN Research Laboratory, Garden Island, New South Wales, 2000, Australia

ROBERT E. STEVENSON, Office of Naval Research, Scripps Institution of Oceanography, La Jolla, California 92038

RONALD C. TIPPER, Naval Ocean Research and Development Activity, Bay St. Louis, Mississippi 39520

SARGUN TONT, Scripps Institution of Oceanography, La Jolla, California 92032

EUGENE TRAGANZA, Department of Oceanography, U.S. Naval Postgraduate School, Monterey, California

ROBERT E. ULANOWICZ, University of Maryland, Chesapeake Biological Laboratory, Solomons, Maryland 20688

ROBERT VENT, Naval Undersea, Center, San Diego, California 92132

DONALD F. WILSON, Don-Wil Electronics, P. O. Box 2055, Holmes Beach, Florida 33509

J. D. WOODS, Oceanography Department, University of Southampton, Southampton SO9 5NH, England

J. S. WROBLEWSKI, Department of Oceanography, Florida State University, Tallahassee, Florida 32306

BERNARD J. ZAHURANEC, Naval Ocean Research and Development Activity, Bay St. Louis, Mississippi 39520

RONALD J. ZANEVELD, School of Oceanography, Oregon State University, Corvallis, Oregon 97331

INDEX

abundance estimates, 277
Acartia clasuii, 200, 202
acceleration potential, 51
acoustic cross sections, 644, 679
   measurement techniques, 498
   patterns, 504
   profiling, 551, 557
   scattering strength, 818
   spectra, 790
active metabolism, 350
activity level, 313
alkalinity, 67, 71, 75, 79, 80, 119
amino acids, 75
animal distributions, 250
anodic stripping voltammetry, 121
anomalous areas, 4
anoxic environments, 118, 451
anticyclonic gyres, 19, 22, 57
apparent oxygen utilization, 70
Arctogadus glacialis, 502
Argyropelecus aculeatus, 226, 227, 268, 335, 343, 236
A. hemigymnus, 262, 268, 338, 222, 227, 557
A. olfersi, 268
A. sladeni, 268
Artemia salena, 187
assimilation efficiency, 829
Atlanta peroni, 343
atmospheric motions, 723
avoidance, 221, 680

bacteria, 448
bacterial biomass, 446

baleen whales, 233
Bathylagus stilbius, 221, 338
bathypelagic associations, 182
Bentheuphausia amblyops, 190, 256, 268, 272, 332, 382, 386, 501, 521, 522
B. suborbitale, 268, 272, 343
billow turbulence, 135, 141
bioassay, 395
biocoenosis, 179
bio-dynamics, 102
bioenergetics, 347
biogenic opal, 65
biological membranes, 67
   populations, 790
biomass, 339
   distributions, 284
   levels, 179
bio-physical information, 619
bladdered species, 567
body lengths, 190
Bonapartia pedaliota, 580
brassicasterol, 94
breeding strategies, 241
Bregmaceros nectabanus, 327, 340, 501, 554, 556
broad size, 196
buffer solutions, 81
buoyancy ranges, 135

Calanoides acutus, 190
Calanus cristatus, 818
C. finmarchicus, 190, 200, 341
C. helgolandicus, 183, 200, 341
C. hyperboreus, 824, 190, 200
C. pacificus, 828, 829
caloric equivalency, 341

campesterol, 89
capture probability, 682
carbon: nitrogen: phosphorous ratios, 118
carbonate concentrations, 66, 67, 68
  equilibria, 79
  speciation, 66
carbonic anhydrase, 67
catch rates, 363
causative organisms, 498
CEPEX, 66, 76
Ceratoscopelus maderensis, 58, 498, 500, 786
Chaetoceros, 183
Chauliodus sloani, 339
chelator, 395
chemical composition of
  midwater animals, 304
  control of biological process, 6
  gradient, 836
chlorophyll, 183, 240, 241, 775, 817
Chlorophyta, 88, 94
cholesterol, 88, 89, 96
circulation patterns, 734
Clione limacena, 187, 197
colloidal material, 76
column strength, 530
community, 218
community ecosystems, 424
compensation depth, 68
competitive interactions, 344
composition, 291, 305, 385
Conchoecia spinerostris, 222
copper, 66, 67
coprophagy, 449
correspondence analysis, 780, 795
Coulter Counter, 234
current shears, 240
cyclonic gyres, 21, 22
Cyclotella mana, 89
Cyclothone, 189, 251, 256, 263, 264, 268
C. *acclinidens*, 256, 264, 268, 338
C. *alba*, 256

C. *braueri*, 180, 222, 256, 262, 264
C. *livida*, 256
C. *microdon*, 222, 256, 264, 268
C. *pallida*, 256, 264, 268
C. *pseudopallida*, 256, 264, 268

day: night ratios, 222
decapod crustaceans, 179
deep circulation, 729
  layers, 205
Deep Scattering Layer, 565, 591, 603
deep-sea metabolism, 113
defining food chains, 182
22-dehydrocholesterol, 94
denitrification, 118, 122
depth ranges, 289
desmosterol, 88
detritus, 76
dialysis, 67
Diaphus brachycephalus, 190, 582, 584
D. *doffleini*, 190
D. *dumerili*, 336, 338, 557
D. *metoplampus*, 582
D. *mollis*, 582, 584
D. *problematicus*, 582, 584
D. *taaningi*, 85, 327, 338, 338, 340, 341, 347, 501, 554
D. *theta*, 332
diel variations, 656
diets, 186, 293, 326, 332, 389
diffusion, 139, 796
diffusion induced instability, 240
digestion rates, 338
digestive tracts, 461
Diplospinus multistriatus, 185
discriminant analysis, 780, 791
dissolved organic carbon, 86
dithigone extraction, 76
diurnal depths, 272
  periodicity, 806
  variations, 509, 521
dynamics of phytoplankton populations, 725

ecological grouping, 230
  stability, 233
ecosystem modeling, 750

ectocrine compounds, 449
eddies, 135, 142, 237, 778, 791
eddy motion, 725
edge index, 185
Ekman convergence, 39
Electron Transport System
   Activity, 103, 122
energetics, 714
energy, 188
energy flux, 148
environmental factors, 272
ergosterol, 94
essential nutrients, 750
Euchaeta japonica, 781
Euclidean distance, 781
Eukrohnia bathypelagica, 470
E. fowleri, 470
E. hamata, 470
Euphausia brevis, 39, 42, 43
E. crystallorophias, 408
E. diomedeae, 39, 44
E. distinguenda, 40, 45
E. exima, 407
E. frigida, 408
E. krohnii, 407
E. pacifica, 39, 41, 193, 194,
   197, 202, 206, 407, 408,
   410, 418, 502, 818
E. superba, 183, 408, 416
E. tenera, 40, 46
E. tricantha, 408
E. vallentini, 408
excretion, 200
expatriation, 377

factor analysis, 780
faunal provinces, 424
feeding chronology, 336, 388
   periodicity, 340
   structures, 461
fish length model, 629
fluid sphere model, 625
fluorometers, 235
food conversion, 188
   particle size, 184
Fourier space, 133
frequency dependence, 495
   echosounders, 502
   space, 781, 796

frontal upwelling, 7
fronts, 138, 142
fucosterol, 88
functional morphology, 184

Gadus merlangus, 583
Gausian probability, 792
general metabolism, 197
Gennadas valens, 227
geographic patterns of scattering, 509
geophysical influences, 134
geopotential anomaly, 16, 62
Gonostoma atlanticum, 786
G. elongatum, 343
grazing, 241
grid models, 805
growth, 189, 191
growth curves, 420
   efficiencies, 200
   factors, 408, 415, 420
   rates, 193, 401
   sequences, 408
gyres, 22, 428, 731, 818

Halocypria globosa, 222
heat flux, 143
herbivore equation, 823
Heterokrohnia bathybia, 470
high frequency measurement
   system, 637
horizon sampling approach, 275
horizontal distributions, 268
   flux, 744
   pressure gradients, 18
hurricanes, 768
hurricane effects, 769
hydrogen sulfide, 120, 553

inertial ranges, 135
intermoult period, 403
Isaacs-Kidd Midwater Trawl, 221,
   229, 362, 498, 521, 566, 680
isotopic turbulence, 133
Ivlev-type herbivore grazing, 820

Katsuwonus pelamis, 221

labile organic compounds, 86

Lampanyctus alatus, 342, 343, 584
L. ater, 582, 584
L. cuprarius, 227, 582, 584
L. festivus, 580, 582
L. lineatus, 582, 584
L. maderensis, 190
L. photontus, 582, 584
Laseognathus saccostoma, 326
L. waltoni, 326
Lepidophanes guentheri, 336, 338, 343, 557
Lepomis macrochirus, 338
Leuroglossus stilbius, 57, 60, 326, 327, 332
lifetime and length scales, 136
light, 67
light intensity, 516
lightfield, 153
linear analysis, 755
Lobianchia dofleini, 227, 343, 362
Longhurst Hardy Plankton Recorder, 219, 234
luminous migrating organisms, 273

Macrorhamphosus, 222
M. scolopax (Linnaeus), 259, 262
main thermocline circulation, 731
mapping, 11
mathematical sensitivity analysis, 806
matrix diagonalization, 787
measurement of nutrient distributions, 8
Meganyctiphanes norvegica, 195, 202, 406, 408, 414, 419
Melamphaes typhlops, 580
M. pumilus, 580
Melamphaidae, 251
metabolic rate, 313
metabolism, 101
metal hydroxides, 65
22-methylenecholesterol, 94
24-methylenecholesterol, 89

microbes as food, 448
microbial activity, 449
micronekton, 179, 181
midwater fishes, 249
migration periods, 710
migrations, 373
mixing, 139, 749
Monochrysis lutheri, 206
moulting sequences, 408
multidimensional scaling, 780
Myctophidae, 57
Myctophum punctatum, 495

Nematoscelis difficilis, 403, 406
N. gracilis, 39, 45
N. mecrops, 40, 47
Neocalanus cristatus, 200
N. plumchrus, 200
N. tonsus, 190
Neomysis interger, 200
N. rayii, 200
nepheloid layer, 94
net catches, 680
  feeding, 326
nitrate, 65, 66, 71, 75, 94, 120, 450, 771, 776, 828, 839
nitrate anomalies, 119
  deficits, 119
  reduction, 118, 122
nitrogen: phosphorous ratios, 124
Nitzschia closterium, 89
non-linear methods, 792
norcholestadienol, 94
Notolychnus valdiviae, 343, 557
numerical modeling recommendations, 719
nutrient equation, 825
  limitation, 235
  regeneration, 399
  supply, 7
nutrients, 6, 69, 140, 450, 804, 821, 825
nutritional values, 186
Nyctiphanes couchii, 403, 408
N. simplex, 407

Ocean Acre, 368
ocean circulation, 729, 731, 745
  currents, 725

oceanic habitats, 38
Oithona similus, 200
ontogenetic stages, 365
optics, 5
organic matter, 86
oxygen consumption, 110, 310
  degradation, 69, 79
  demand, 118
  microstructure, 237
  minimum zone, 119, 517
  utilization rates, 108

Pacific Oceanic Fishery
  Investigations, 285
Parathemisto gaudichaudi, 185
Pareuchaeta japonica, 190
particle counters, 234
particulate matter, 69, 76
  organic carbon, 86
patchiness, 233, 236, 365,
  373, 656, 669, 673
patchiness dynamics, 836
pattern recognition, 780
pelagic ecosystem models, 803
  faunal regions, 496
  nekton, 296
pH, 79
Phaephyta, 88
phosphate, 36, 37, 38, 54, 65,
  68, 94, 119, 450
photooxidation, 396
photosynthesis, 67
physiological studies, 302
phytoplankton equation, 821
plankton patchiness, 297, 767,
  817, 819, 831, 834
Pleurobrachia bachei, 184, 187
Pleuragramma, 343
P. antarcticum, 326
P. gracilis, 343
P. piseki, 343
polarization, 154
Pollichthys mauli, 580,584
population structures, 271
potential energy, 348
prey populations, 343
prey-preditor relationships, 476
primary production, 71, 79, 285

principal components analysis,
  780
probability of escape, 688
productions of populations, 201
productivity variations, 113
projection measures, 783
Pseudocalanus menutus, 200

quasi-geostrophic flow, 18

radiance, 147
ration, 339
Rectangular Midwater Trawl, 382
redox potential, 117
regeneration zone, 68
regression, 795
regression and clustering, 790
remineralization, 94
resonant frequency, 503
resource partitioning, 344
respiration, 197
respiratory control ratio, 108
resting periods, 190
Rhincalanus gigas, 190
rings, 236
rotation and buoyancy ranges, 135,
  137

Sagitta elegans, 470
S. hispida, 200
S. macrocephala, 470
S. zetesios, 470
Salpa gerlachei, 193
S. thompsoni, 193
sampling, 343, 362
sampling size, 218
scaling of equations, 826
scattering layer, 55, 56, 57, 178,
  181, 250, 385, 499, 550, 553,
  565, 603 668
  levels, 513, 541
  potential, 500
  strength, 520, 556, 631, 540,
  661
seasonal changes, 332
  cycles, 190
  differences, 289
  variations, 23, 515, 545, 654

selective feeding, 341, 388
Sergestes similus, 184, 326
sex ratios, 276, 370
shadowing effects, 68
silicate, 65, 69, 70, 120, 450
β-sitosterol, 89, 91, 94
size frequency, 222
SOFAR channel, 23, 31
sound scatterers, 55, 129, 133, 283, 325, 494, 668
  scattering, 138, 143, 658
  speed, 22
species abundance, 257
  diversity, 257, 262
spectral analysis, 236
  distribution, 134
  energy flux, 151
  irradiance, 151
  windows, 131
Spiratella helicina, 187, 502
S. retroversa, 187
standard metabolism, 349
station-space, 781, 796
Steindachneria argentea, 501
Steinobrachius leucopsarus, 190
Sternoptyx diaphana, 185, 332, 336, 341, 343, 580, 584
sterols, 86
sterol distribution, 91, 93
  esters, 97
  sampling and analysis, 87
  sources, 88
  structure, 91
stigmasterol, 89
stochastic effects, 806
Stokes parameters, 147
Stolephorus purpureus, 219
stomach contents, 183, 185, 293, 340, 385, 460
Styiola subula, 343
Stylocheiron elongatum, 40, 49
S. maximum, 40, 48
submersibles, 278
succession, 394
sulfate reduction, 118
swimbladder, 256, 495, 503, 521, 573
swimbladder model, 627
  variability, 580

  volume, 578, 580, 582, 583, 587
Systellaspis debilis, 222

Tarlentonbeania crenularis, 332
Temora longicornis, 200
temperature maximum, 24, 33
three-way matrices, 798
Thysanoëssa gregaria, 40, 50
T. inermis, 818, 408
T. longipes, 200
T. rashii, 818
T. spinifera, 407, 408
Thysanopoda, 185
T. cornuta, 457, 460
T. cristata, 227
T. egregia, 457, 460
T. monacantha, 227
T. spinicaudata, 457, 460
Tigriopus, 187
T. brevicornis, 200
Trace Element Intercalibration Study, 76
trace metals, 65, 75, 76, 77, 123, 393
transport of nutrients, 8
Triphoturus mexicanus, 338, 339
trophic relations, 291
Tucker Trawl, 551, 596, 698
turbulence, 9
turbulent diffusion, 236
  motions, 141
  velocity fluctuations, 129

upwelling, 79, 81, 113, 114, 141, 804, 293
upwelling ecosystems, 804

Valenciennellus tripunctulatus, 222, 227, 262, 336, 340, 342, 343, 344
variability of nutrient and biological factors, 9
vector and matrix representations, 781
velocity field, 130
vertical and horizontal mixing, 236
vertical distribution, 370, 385
  migration, 225, 502, 554

zonation, 230
Vincigeurria lucetia, 57, 59
volcanic debris, 77
volume reverberation, 790, 794
  scattering, 596
voluntary horizontal migration, 806

wind mixed layer, 238, 771

yeasts, 89

zinc, 76
zoogeographic distribution, 229
zooplankton biomass, 294
  volume, 39, 52, 53